Remote Sensing and Atmospheric Ozone

Human Activities versus Natural Variability

Arthur P. Cracknell and Costas A. Varotsos

Remote Sensing and Atmospheric Ozone

Human Activities versus Natural Variability

Springer

Published in association with
Praxis Publishing
Chichester, UK

PRAXIS

Professor Arthur P. Cracknell
Department of Applied Physics and Electronic Engineering
University of Dundee
Dundee
U.K.

Professor Costas A. Varotsos
Faculty of Physics
Department of Environmental Physics and Meteorology
University of Athens
Athens
Greece

SPRINGER–PRAXIS BOOKS IN ENVIRONMENTAL SCIENCES
SUBJECT *ADVISORY EDITOR*: John Mason, M.B.E., B.Sc., M.Sc., Ph.D.

ISBN 978-3-642-10333-9 ISBN 978-3-642-10334-6 (eBook)
DOI 10.1007/978-3-642-10334-6
Springer Heidelberg New York Dordrecht London

Library of Congress Control Number: 2011937034

Cover design: Jim Wilkie
Project management: OPS Ltd., Gt. Yarmouth, Norfolk, U.K.

Printed on acid-free paper

Springer is part of Springer Science+Business Media (www.springer.com)

Contents

Preface xi

List of figures xv

List of tables xxiii

List of abbreviations and acronyms xxvii

1 The traditional measurement of ozone concentration in the atmosphere . . 1

1.1 Introduction 1

1.1.1 Observations of the total ozone column 4

1.2 Ground-based instrumentation for TOC observations 5

1.2.1 The Dobson ozone spectrophotometer. 7

1.2.2 Intercomparison of Dobson spectrophotometers 10

1.2.3 Interference of SO_2 and NO_2 in Dobson TOC measurements 19

1.2.4 Influence of stray light on Dobson TOC measurements . 24

1.3 Brewer Spectrophotometer. 34

1.4 Filter ozonometers M-83/124/134 37

1.5 Secondary ground-based instrumentation for TOC observations . 39

1.5.1 System for Analysis of Observation at Zenith (SAOZ) . . 42

1.5.2 MICROTOPS II (Total Ozone Portable Spectrometer) . 42

1.5.3 High-Resolution Visible/Ultraviolet Absorption Spectroscopy. 43

1.5.4 Fourier transform spectrometer (FTS) 43

1.5.5 System for the Monitoring of Stratospheric Compounds (SYMOCS) 46

1.5.6 Star Pointing Spectrometer (SPS) 46

1.5.7 MDR-23 (a Russian commercial device) 46

1.5.8 Scanning spectrometer (EVA) 47
1.5.9 Solar IR spectroradiometer 47
1.5.10 Ground-based UV radiometer (GUV) 47
1.5.11 Spectrometer–Ozonometer PION 48
1.5.12 SPectrometer for Atmospheric TRAcers Monitoring (SPATRAM) 48
1.6 Observations of ozone vertical profile (OVP) 50
1.6.1 Primary ground-based instrumentation for OVP observations. 50
1.6.2 Dobson Umkehr measurements and inversion 51
1.6.3 Brewer Umkehr measurements 54
1.6.4 Secondary ground-based instrumentation for OVP observations 55
1.6.5 Lidar 55
1.6.6 Microwave radiometry 56
1.6.7 Ground-based Millimeter wave Ozone Spectrometer (GROMOS). 57
1.6.8 Stratospheric Sounding by Infrared Heterodyne Spectroscopy (SIRHS) 57
1.6.9 Ground-based microwave radiometers 58
1.6.10 Ground-based infrared solar spectroscopy 59
1.6.11 Stratospheric Ozone Monitoring Radiometer (SOMORA) 59
1.7 Airborne instrumentation for OVP observations 61
1.7.1 Electrochemical ozonesondes 61
1.7.2 Optical ozonesondes 64
1.7.3 Other balloon instrumentation 67
1.7.4 Aircraft instrumentation 71
1.8 Surface ozone measurements 77
1.8.1 Chemiluminescence method 77
1.8.2 Electrochemical potassium iodide method 77
1.8.3 UV absorption method 77

2 Satellite systems for studies of atmospheric ozone 79
2.1 Satellite remote sounding of TOC 82
2.2 Direct absorption measuring instruments 83
2.2.1 TIROS Operational Vertical Sounder (TOVS); GOES . . 83
2.2.2 Laser Heterodyne Spectrometer (LHS)/Tunable Diode LHS (TDLHS) 85
2.2.3 OZON-MIR 86
2.3 Indirect absorption measuring instruments 86
2.3.1 Total Ozone Mapping Spectrometer (TOMS) 86
2.3.2 Ozone Monitoring Instrument (OMI) 90
2.3.3 Advanced Earth Observing Satellite (ADEOS I, II) . . . 93
2.3.4 Solar Backscattered Ultraviolet Radiometer (SBUV) . . . 93
2.3.5 Global Ozone Monitoring Experiment (GOME) 96

2.3.6 ESA ENVISAT, GOMOS 98
2.3.7 The Ozone Mapping and Profiler Suite (OMPS) and the NPOESS 99
2.3.8 Ozone Dynamics Ultraviolet Spectrometer (ODUS) . . . 101
2.3.9 Ozone Layer Monitoring Experiment (OLME) 101
2.3.10 Interferometric Monitor for Greenhouse Gases (IMG) . 101
2.3.11 Infrared Atmospheric Sounding Interferometer 102
2.4 Observed variability in total ozone column 102
2.4.1 Latitudinal variation of TOC 102
2.4.2 Longitudinal variation of TOC 106
2.5 Satellite instrumentation for OVP observations 106
2.5.1 Direct-absorption measuring instruments 107
2.5.2 Scattering-measuring instruments 116
2.5.3 Emission-measuring instruments 118
2.5.4 Summary of ozone-monitoring satellites 130
2.6 Observed variability in vertical ozone distribution 130
2.6.1 EASOE 141
2.6.2 SESAME 142
2.6.3 THESEO 142
2.6.4 SOLVE 142
2.6.5 ORACLE-O_3 143
2.6.6 SCOUT-O3 144
2.6.7 Match 144
2.6.8 ARC_IONS 145

3 Intercomparisons between various atmospheric ozone datasets 149
3.1 Introduction 149
3.2 Total ozone measurements over Athens: intercomparison between Dobson, TOMS (version 7), SBUV, and other satellite measurements 152
3.3 Geophysical validation of MIPAS-ENVISAT operational ozone data 161
3.3.1 Introduction to MIPAS 161
3.3.2 MIPAS ozone data 163
3.3.3 Comparison of MIPAS data with WMO/GAW ground-based measurements 164
3.4 Comparison of MIPAS data with stratospheric balloon and aircraft measurements 190
3.4.1 MIPAS-B2 190
3.4.2 FIRS-2 and IBEX 193
3.4.3 SPIRALE 198
3.4.4 MIPAS-STR, SAFIRE-A, and FOZAN on board the M-55 *Geophysica* aircraft 200
3.4.5 ASUR 207

3.5 Comparison with satellite measurements . . . 208
3.5.1 Comparison of MIPAS data with SAGE II O_3 profiles . 211
3.5.2 Comparison with POAM III O_3 profiles . . . 213
3.5.3 Comparison with Odin-SMR O_3 profiles . . . 216
3.5.4 Comparison with ACE-FTS O_3 profiles . . . 220
3.5.5 Comparison with HALOE O_3 profiles . . . 223
3.5.6 Comparison with GOME O_3 profiles . . . 229
3.5.7 Comparison with SCIAMACHY and GOMOS . . . 231
3.6 Comparison of MIPAS data with ECMWF assimilated fields . . . 234
3.7 Summary of MIPAS comparisons . . . 236
3.8 Other intercomparisons between various ozone-monitoring systems . . . 244
3.8.1 TOMS, GOME, GOMOS, and SCIAMACHY data . . . 245
3.8.2 MLS data . . . 248
3.8.3 SAGE data . . . 249
3.8.4 TES data . . . 250
3.8.5 ACE and IASI data . . . 250
3.8.6 Ozonesonde intercomparisons . . . 254

4 The dynamics of atmospheric ozone . . . 255
4.1 Total ozone trends . . . 257
4.2 Ozone vertical profile variability . . . 273
4.3 General features of ozone global distribution . . . 287
4.3.1 Stratosphere–troposphere exchange . . . 293
4.3.2 Low-ozone pockets . . . 298
4.4 The non-linear nature of ozone variability; detrended fluctuation analysis (DFA) . . . 306
4.4.1 Long-memory processes in global ozone and temperature variations . . . 307
4.4.2 Long-term memory dynamics of total ozone content . . . 314
4.4.3 Scaling behavior of the global tropopause . . . 317
4.4.4 Scaling properties of air pollution at the surface; surface ozone (SOZ) . . . 321
4.5 Impacts of the solar eclipse of March 29, 2006 on surface ozone and related air pollutants . . . 322
4.6 Long-term coupling between TOC and tropopause properties . . 327
4.6.1 Occurrence frequency of tropopause height . . . 329
4.6.2 Association between tropopause properties and TOC . . 333
4.6.3 The tropopause; summary . . . 336

5 The Montreal Protocol . . . 339
5.1 Introduction . . . 339
5.2 The proposition by Molina and Rowland of human releases of CFCs being responsible for ozone depletion . . . 340
5.3 The science from 1974 to 1985 . . 344

5.4 The Ozone Hole 351
5.5 The role of remote sensing in the lead-up to the Montreal Protocol 358
5.6 The NOZE and AAOE expeditions 358
5.7 Theories of the Ozone Hole 363
5.8 Diplomacy, 1974–1989; formulation and ratification of the Montreal Protocol 366
5.9 Reasons for the success in reaching international agreement in Montreal 368
5.10 Ratification of the Montreal Protocol 372

6 The study of atmospheric ozone since 1987 379
6.1 Introduction 379
6.2 The reduction of ozone-destroying chemicals in the atmosphere . 379
6.2.1 Ozone depletion potential (ODP) 382
6.2.2 Equivalent Effective Stratospheric Chlorine (EESC) 385
6.3 Ground-based and ozonesonde data on ozone depletion 388
6.4 Piecewise linear trends in ozone depletion 390
6.5 Recovery of the ozone layer; the polar regions 406
6.5.1 Sudden stratospheric warmings 415
6.5.2 Observation of sudden stratospheric warmings detected in deep underground muon data 417
6.5.3 The role of the diffusion of gases in ice or an amorphous binary mixture in the polar stratosphere and the upper troposphere 420
6.5.4 Experimental studies of the Antarctic ozone hole and ozone loss in the Arctic 425
6.5.5 Antarctic ozone hole predictability; use of natural time series 441
6.6 Long-term monitoring of the ozone layer 448
6.6.1 Measurement of TOC and the OVP 452
6.6.2 The use of models to predict ozone concentration 454
6.6.3 Ozonesonde networks 469
6.6.4 Trends in TOC and tropopause properties 474
6.7 Scientific assessment of ozone depletion 2010 478

7 Atmospheric ozone and climate 485
7.1 Introduction 485
7.2 Radiative-forcing calculations 492
7.2.1 Estimates of changes in RF from pre-industrial times to the present 492
7.2.2 Detailed studies of changes in RF in recent decades . . . 497
7.2.3 Contribution of the transport sector 515
7.3 Ozone-induced climatic impacts 517
7.3.1 The health impacts of changes in ozone concentration . . 543

7.4 Conclusions on tropo-stratospheric variability 544
7.4.1 Stratospheric ozone dynamics and its determining factors 545
7.4.2 Tropospheric processes 546
7.5 New climate research aspects deduced from global ozone dynamics research and remote sensing 547
7.5.1 Climate modeling and atmospheric ozone 547
7.5.2 Role of phase transitions in climate system dynamics. . . 549
7.5.3 Nambu dynamics and ozone–climate modeling 550
7.5.4 Dissipation-induced instabilities in ozone and climate fields 551
7.5.5 Deterministic, chaotic, or stochastic ozone–climate time series 553
7.6 WMO/UNEP Scientific Assessment 2010 554

References and bibliography 559

Index 653

Preface

Ozone is so rare in the atmosphere (there are only about three ozone molecules per 10 million air molecules) that if it was brought down to the Earth's surface and compressed to standard temperature and pressure, it would form a layer about as thick as a small coin. In spite of this, the importance of ozone, particularly in protecting the biosphere from the harmful effects of solar ultraviolet radiation, vastly exceeds what one might expect from this minor trace gas in the atmosphere. In the mid 1970s some scientists alerted the world to the fact that atmospheric ozone (which is mostly in the stratosphere) was being destroyed by manufactured chemicals released into the atmosphere. This led in 1987 to the Montreal Protocol to protect the ozone layer, which has been ratified by virtually every country in the world.

Remote sensing, which is mostly concerned with using Earth-orbiting satellites to study the Earth, played a vital role in the events leading up to the Montreal Protocol and since then has continued to play an important role in monitoring the implementation of the Montreal Protocol. The era of satellite remote sensing began on October 4, 1957, when the former Soviet Union launched Sputnik 1, the world's first artificial satellite (a 55 cm diameter sphere that weighed 83 kg with four antennae attached to it). It circled the Earth once every 96 minutes and transmitted radio signals that could be received on Earth. On November 3, 1957, a second satellite, Sputnik 2, was launched carrying a dog named Laika. The United States launched its first satellite, Explorer 1, on January, 31 1958, and its second, Vanguard 1, on March 17, 1958.

The launch of Sputniks 1 and 2 over 50 years ago marked the beginning of a long period of successful launches of a large number of Earth-orbiting unmanned spacecraft for environmental observation. Earth-orbiting satellites now play a crucial role in studying resources and environmental conditions in the atmosphere, biosphere, cryosphere, geosphere, and the oceans; in particular, in the present context they are involved in studying atmospheric ozone. Traditional measurements of atmospheric ozone from the ground, using ultraviolet spectrophotometers, and from ozonesonde balloons were being made for many years before remote-sensing satellites came to be developed and operated. However, these traditional methods only measured ozone concentrations at a few points on the surface of the Earth. The advantage of remote sensing is that it provides frequent coverage of the whole Earth.

This book is concerned with showing how remote sensing contributes to the study of atmospheric ozone and the consequences of human activities on atmospheric ozone. We cover the traditional measurement of ozone concentration in the atmosphere and give a comprehensive description of the large number of different satellite systems for the study of atmospheric ozone. This is followed by a consideration of the consistency of the ozone observations obtained with different methodologies and techniques and the importance of intercomparisons between various atmospheric ozone datasets and between various ozone-monitoring systems. We give a summary of the state of knowledge of the dynamics of atmospheric ozone and then consider the Montreal Protocol, how it came about and how its implementation is being monitored.

Remote sensing played a key role in the scientific research that led to the Montreal Protocol on ozone-depleting substances and it continues to play an important role in the monitoring of the success of the Protocol (we edited a special issue of the *International Journal of Remote Sensing* on Remote Sensing and the Implementation of the Montreal Protocol, Volume 30, Numbers 15–16, August 2009). Our introductory paper in that special issue (pp. 3853–3873) forms the basis of Chapter 5 of this book. In this context, the topic on the evolution of the Antarctic ozone hole is discussed based on the proposition by Mario Molina and Sherwood Rowland that human manufacture and release of CFCs (chlorofluorocarbons) into the atmosphere causes ozone depletion.

In view of various criticisms of the Intergovernmental Panel on Climate Change (IPCC) and a general feeling of depression induced by the apparent slowness of the international community in addressing the problems of global warming and climate change, it is perhaps instructive to consider the relation between the Montreal Protocol and the Kyoto Protocol and we do discuss this question.

We end with a summary of the results of research work since 1987 (i.e., since the Montreal Protocol was formulated). This involves the study of ozone depletion, both the steady decline in concentration amounting to a few percent per decade, depending on location, and the catastrophic temporary development of the ozone hole for a few months each year in the Antarctic and, more recently, in the Arctic. It also involves the relation between changes in ozone and climatic conditions.

In some ways this book is a sequel to *Atmospheric Ozone Variability: Implications for Climate Change, Human Health and Ecosystems*, by K.Ya. Kondratyev and C.A. Varotsos (Springer/Praxis, 2000). However, a great deal has happened since that book was written, both in terms of new remote-sensing systems and results and in connection with long-term datasets and ozone depletion. We now concentrate here on the science of studying atmospheric ozone itself, principally using remote-sensing techniques. In writing the present book, we have relied on many sources of information, including of course papers from research journals, the proceedings of the various Quadrennial Ozone Symposia and the Reports on the Scientific

Assessment of Ozone Depletion by WMO/UNEP (the World Meteorological Organization and the United Nations Environment Program), which we have listed in our Table 6.1. However, when it came to the 2008 Quadrennial Ozone Symposium held in Tromsø, Norway, no proceedings volume was published; only a set of abstracts was produced and it was left to participants to publish their work in refereed journals. We have made extensive use of these abstracts and they enabled us to find many recently published papers that we have cited, but we have not cited any of the Tromsø abstracts themselves.

We are grateful to many present and former students and colleagues who have contributed to our understanding of remote sensing and atmospheric ozone.

Inevitably, we have reproduced a large amount of copyright material and we are grateful to the various copyright holders who have given their permission and who are acknowledged *in situ* in the text, figures, or tables. Every effort has been made to trace original copyright holders of previously published material but with mergers in the publishing industry it is not always possible to track them down, so we offer our apologies where any have been overlooked. We are very grateful to Clive Horwood, of Praxis Publishing, for his help and encouragement during the writing of this book and to the staff of OPS for their production work.

Arthur P. Cracknell
Costas A. Varotsos
September 2011

Figures

1.1	Sputnik 1 (a basketball in diameter with mass 83 kg), and ENVISAT with various instruments on board	2
1.2	The vertical distribution of atmospheric ozone: partial pressure of ozone and height.	2
1.3	TOC at Table Mountain, California, as a function of season from three different TOC datasets from the 1920s to 1940s	6
1.4	The O_3 spectrum at room temperature	6
1.5	Prof. G.M.B. Dobson with his co-worker Dr. D. Walshaw around 1970, and Dobson spectrophotometer optical diagram.	8
1.6	Photo from the Dobson Intercomparison held in Arosa (July–August 1990)	16
1.7	$\Delta N_i = (N_i$ for Dobson D065 $- N_i$ for Dobson D118), $i =$ A, C, D, on August 5, 1990, versus $\mu = \sec\theta$; and Dr. D. Walshaw with Dr. C. Varotsos around 1985	17
1.8	N_C/μ_C versus μ_C for both D065 and D118, and the same but for the pairs D and A, respectively.	18
1.9	$\Delta N_i = (N_i$ for Dobson D064 $- N_i$ for Dobson D118), where $i =$ A, C, D on July 11, 2006, versus μ	19
1.10	N_C/μ_C versus μ_C for both D064 and D118, and the same but for the pairs D and A, respectively.	20
1.11	Mean daily percentage error in TOC caused by SO_2 for December to March and May to August of the 2-year period 1989–1990	23
1.12	Mean daily percentage error in TOC caused by NO_2 for December to March of the 2-year period 1989–1990.	24
1.13	Illustration of the effect of stray light on log intensity ratios and on the determination of extraterrestrial constants	26
1.14	Variation of ozone error X_{AD} as a function of airmass, μ, for various values of model parameters, as calculated by the stray light model	29
1.15	Comparison of experimental data with stray light models	30
1.16	Variation of ozone error X_{AD} as a function of airmass μ for $R_0 = 10^{-4.0}$ as calculated by Basher's stray light model	31
1.17	As Figure 1.16, but for $R_0 = 10^{-3.0}$.	31

1.18 Comparison between experimental data and stray light model values for November 27, 1992 and December 7, 1992 33
1.19 Brewer Mark IV mechanical assembly (top view) 34
1.20 Map showing the location of total ozone stations where data were submitted to WMO within the period of 1979–2006 38
1.21 Schematic diagram of M-124 Ozonometer 39
1.22 Differential absorption cross-sections of some trace gases absorbing in the UV/visible wavelength region 41
1.23 Optical diagram of a Michelson interferometer; in one of the arms the mirror can be moved to lengthen or shorten the optical path 44
1.24 SPATRAM (SPectrometer for Atmospheric TRAcers Monitoring) installed since April 2004 at the University of Évora, Portugal 49
1.25 The SOMORA instrument 60
1.26 A schematic diagram of the Japanese ozonesonde and of an electrochemical concentration cell ozonesonde 65
1.27 A schematic diagram of the Indian ozonesonde 66
1.28 The M55 *Geophysica* high-altitude aircraft 72
1.29 Photograph of the GAMS spectrometer 76
1.30 Advanced Pollution Instruments (API) Model 400E Ultraviolet Absorption Ozone analyzer 78
2.1 A NASA Global Hawk lands at Dryden, and the A-Train 81
2.2 TOMS optical diagram and the Nimbus-7 spacecraft which carried the first TOMS instrument, before its launch in 1978 88
2.3 Schematics of GOME Optics 97
2.4 Monthly trends in total ozone in the Mediterranean region and at Athens compared with the trends derived from zonally averaged data 105
2.5 A cross-section of monthly trends in column ozone over the latitude belt 35°N–45°N derived from the 13 years of Nimbus-7 TOMS measurements 106
2.6 Trends of differences between ozone measurements made by various ozone-profiling instruments and SAGE-II 109
2.7 HALOE instrument 111
2.8 The ACE-FTS instrument 113
2.9 Photograph of the MAESTRO instrument, showing occultation (solar) and backscatter viewing ports 114
2.10 HIRDLS components 123
2.11 The MetOp satellite carrying the IASI instrument 126
2.12 The scanning mirror of IASI directing emitted infrared radiation from one swath into the uncovered interferometer 127
2.13 Cloud detail view from an image taken in Aberdeen (Scotland) with a digital camera 140
2.14 Locations of IONS sites 145
2.15 Comparisons of several Canadian air quality models with IONS-04 sondes at the surface to 4 km 146
3.1 Scatter diagram for TOMS (version 7) and Dobson observations at Athens, Greece (March 1, 1991–December 31, 1991). All days 154
3.2 Scatter diagram for SBUV (version 7) and Dobson observations at Athens, Greece (March 1, 1991–December 31, 1991). All days 154
3.3 Scatter diagram for TOMS (version 7) and Dobson observations at Athens, Greece (March 1, 1991–December 31, 1991). Cloud-free days 155

3.4 Scatter diagram for SBUV (version 7) and Dobson observations at Athens, Greece (March 1, 1991–December 31, 1991). Cloud-free days 155
3.5 Intercomparison of column ozone observations performed by satelliteborne and ground-based instrumentation in Athens, Greece (2002–2006). 158
3.6 Comparison of column ozone observations performed by satelliteborne and ground-based instrumentation in Athens, Greece during 2002–2008 159
3.7 Comparison of monthly mean column ozone observations performed by satelliteborne and ground-based instrumentation in Athens, Greece during 1991–2008. 159
3.8 Time series of the percentage relative difference in ozone partial column (75–35 hPa) between MIPAS and correlative ozonesonde data at five western and central European stations for 2003 171
3.9 Vertically resolved statistics of the absolute differences between MIPAS O_3 data and NDACC and WOUDC measurements in the Arctic. 173
3.10 Vertically resolved statistics of the relative differences between MIPAS O_3 data and NDACC and WOUDC measurements in the Arctic. 174
3.11 Results of the comparison between MIPAS O_3 profiles and ground-based lidar measurements matching the coincidence criteria of 400 km and 10 h. 177
3.12 Results of the comparison between MIPAS O_3 profiles and ground-based lidar measurements: zonal averages. 181
3.13 Time series of ozone partial columns. 184
3.14 Statistical means and standard deviations of the relative differences between MIPAS and FTIR O_3 profiles 186
3.15 Results of statistical analysis for MIPAS O_3 bias and precision determination by comparison with matching measurements from mid-latitude ozone soundings 191
3.16 Results of the comparison between coincident MIPAS-ENVISAT and MIPAS-B2 ozone measurements at mid-latitudes and in the Arctic region 195
3.17 Absolute difference between MIPAS-ENVISAT and MIPAS-B2 ozone volume mixing ratios averaged over all the available collocations 196
3.18 Comparison between MIPAS v4.61 and FIRS-2 (October 30, 2002 and September 19/20, 2003) ozone measurements 197
3.19 Comparison between MIPAS v4.61 and IBEX (July 29–30, 2002) ozone measurements 198
3.20 Comparison of MIPAS O_3 profiles from orbit 4,678, scan 6 with the *in situ* profiles acquired during the SPIRALE flight 199
3.21 Comparison of MIPAS O_3 profiles for orbit 4,677, scan 20 with the *in situ* profiles acquired during the SPIRALE flight 201
3.22 Results of the comparison between MIPAS-ENVISAT v4.61 ozone data and correlative measurements performed by the remote-sensing and *in situ* payload of the M-55 *Geophysica* aircraft during mid-latitude flights on July 22, 2002 and on October 24, 2002 from Forlì, Italy 205
3.23 Results of the comparison between MIPAS-ENVISAT v4.61 ozone data and correlative measurements performed by the remote-sensing and *in situ* payload of the M-55 *Geophysica* aircraft during high-latitude flights on March 2, 2003 and March 12, 2003 from Kiruna, Sweden 206
3.24 False-color map of potential vorticity on the isentropic surface $\theta = 420$ K and the M-55 *Geophysica* route during the ENVISAT validation flight from Kiruna on March 12, 2003 207

3.25 The absolute and percentage difference between the MIPAS and ASUR ozone profiles in the tropics, mid-latitudes, the Arctic 209
3.26 Comparison between MIPAS and SAGE II: statistics over all the collocated O_3 profiles 213
3.27 Comparison between MIPAS and SAGE II: zonal averages 214
3.28 Comparison between MIPAS and SAGE II: seasonal averages 215
3.29 Comparison between MIPAS and POAM III: statistics over all the collocated O_3 profiles 217
3.30 Comparison between MIPAS and POAM III O_3 profiles: zonal averages . . . 219
3.31 Comparison between MIPAS and POAM III: seasonal averages for the Northern Hemisphere (60°N–90°N) and Southern Hemisphere high latitudes 220
3.32 Comparison between MIPAS and Odin-SMR: statistics over all the collocated O_3 profiles 222
3.33 Comparison between MIPAS and ACE-FTS: statistics over all the collocated O_3 profiles 224
3.34 Comparison between MIPAS and HALOE: statistics over all the collocated O_3 profiles 225
3.35 Comparison between MIPAS and HALOE O_3 profiles: zonal averages 226
3.36 Comparison between MIPAS and HALOE O_3 profiles: seasonal averages . . 227
3.37 Comparison between MIPAS and GOME O_3 profiles. 231
3.38 Comparison between MIPAS and GOME O_3 profiles: zonal averages 232
3.39 MIPAS v4.61 and ECMWF O_3 VMR mean profiles: global averages and corresponding mean relative difference, standard deviation, and MIPAS errors 236
3.40 Comparison between MIPAS v4.61 and ECMWF O_3 VMR mean profiles: zonal and seasonal averages 239
3.41 Summary plot of global mean relative differences between MIPAS O_3 VMR profiles and coincidence measurements by concurrent satellite sensors 240
3.42 Bruker 125HR installed at Polar Environment Atmospheric Research Laboratory (PEARL), located on Ellesmere Island, Canada 251
3.43 Monthly average of the relative differences of collocated ozone column measurements of SBUV(/2) with lidar, SAGE II, and Umkehr at 15.8–10 hPa 253
3.44 Comparison of ozone measurement, relative to a standard ozone photometer instrument, as measured by different combinations of ECC instrument type (ENSCI or SPC) and sensing solution types in intercomparisons 253
4.1 SBUV(/2), Dobson, and TOMS seasonal total ozone trends in percent per decade by latitude to May 1991 266
4.2a Time series of decadal mean total column ozone (1850–1859 to 2000–2009) . 269
4.2b Decadal annual means for tropospheric column ozone 270
4.3 Average total ozone in polar regions. 271
4.4 Monthly distribution of the occurrence of each group of ozone soundings for the period from November 1996 to April 1997 over Athens, Greece. 276
4.5 Ozone partial pressure, relative humidity, and temperature over Athens on January 24, 1997 and November 23, 2004 277
4.6 Ozone partial pressure, relative humidity, and temperature over Athens on April 6, 1997 and June 1, 2005 278
4.7 Ozone partial pressure, relative humidity, and temperature over Athens on January 10, 1997 and April 18, 2005. 279
4.8 Ozone partial pressure, relative humidity, and temperature over Athens on April 1, 1997 and February 18, 2005 280

4.9 Ozone partial pressure, relative humidity, and temperature over Athens on February 6, 1997 and March 15, 2005. 281

4.10 A series of BUV instruments have been measuring ozone vertical distribution since 1970; vertical distribution of Arctic and Antarctic ozone. 286

4.11 Vertical–latitudinal cross-sections of past values and long-term changes for annual mean vertical residual velocity and ozone using an ensemble mean of three simulations from the CMAM 287

4.12 Latitude–time cross-sections of total ozone deviations from 1964 to 1980 "norms" in percent 289

4.13 Variation in total ozone with latitude and season for two time intervals: 1964–1980 and 1984–1993 290

4.14 Difference between total ozone values for the two periods (1964–1980 and 1984–1993) reported in Figure 4.12. 291

4.15 Barnes map of gridded output ozone from a Lagrangian chemical model. . . 300

4.16 Equivalent solar latitude along 14-day diabatic back trajectories for air parcels used in the chemistry–trajectory model on March 9, 1993 at the 850 K surface 301

4.17 TOC mean monthly values in Dobson units during 1964–2004, over the belt 25°S–25°N derived from the WMO Dobson Network. 310

4.18 Log-log plot of the TOC root-mean-square fluctuation function versus temporal interval for deseasonalized TOC values 311

4.19 Log-log plot of the TRT root-mean-square fluctuation function versus time interval for deseasonalized TRT values 313

4.20 Log-log plot of the RMS fluctuation function for the smoothed and detrended reconstructed TOC time series, versus time interval and the respective best fit equation and correlation coefficient at Arosa and Poluy River. 317

4.21 Log-log plot of tropopause height root-mean-square fluctuation function versus time interval for the detrended and deseasonalized monthly *Z* values observed by the radiosonde network over the equator and the globe during 1980–2004. 319

4.22 Map with the locations of the monitoring stations 324

4.23 Surface ozone measurements and the expected surface ozone values as derived from the fitted curve of measurements before, during, and after the eclipse event of March 29, 2006 at four stations 325

4.24 SUVR-B measurements derived from the MICROTOPS II sun-photometer before, during, and after the solar eclipse of March 29, 2006, at Athens. . . . 326

4.25 SUVR measurements derived from the VLX-3W radiometer before, during, and after the solar eclipse of March 29, 2006, at Athens 327

4.26 Nitrogen dioxide measurements before, during, and after the solar eclipse of March 29, 2006 at four stations 328

4.27 Occurrence frequency for the first tropopause height classes at Athens during 1984–2002. 330

4.28 Seasonal variability of the first and second thermal tropopause over Athens, Greece 331

4.29 Relative frequency of occurrence of the second tropopause over Athens, Greece 332

4.30 Temperature trends of the first and second tropopause over Athens, Greece . 333

4.31 Relation between TOC and tropopause height for average TOC values of different tropopause height classes, as deduced from Athens observations. . . 334

4.32 Seasonal variability in ozone partial pressure in the troposphere and the lower stratosphere at Athens as deduced from ozonesonde ascents throughout the period 1991–2001 . . . 334
5.1 Schematic diagram of the CFC impact on the ozone layer . . . 348
5.2 Differences in average ozone concentrations 1970–1988 versus 1931–1969. . . 350
5.3 Percentage change in ozone 1976–1986 versus 1965–1975, Northern Hemisphere Dobson stations . . . 350
5.4 South Pole ozone profiles in 2006 before the ozone hole developed and for the sounding that displayed the minimum ozone . . . 353
5.5 Total ozone over Halley Bay, Antarctica, in October (southern spring) . . . 355
5.6 Schematic diagram of the role of NO_2 in ozone depletion . . . 360
5.7 Schematic diagram of polar ozone destruction . . . 364
5.8 Production and consumption trends for CFCs . . . 376
5.9 Production and consumption trends for halons . . . 377
6.1 Antarctic ozone in October (averaged poleward of 70°) for four different simulations from the NASA/GSFC coupled chemistry–climate model . . . 386
6.2 Hockey stick statistical trend model in the style of Reinsel *et al.* (2002, 2005) 391
6.3 Monthly TOC over Athens, Greece, from 1979 to 1999 . . . 392
6.4 Trends calculated for SAGE I/II for 1979–1991 . . . 393
6.5 Latitude–altitude cross-section of stratospheric ozone trends from various data sources . . . 394
6.6 Vertical profile of ozone trends over northern and southern mid-latitudes estimated from ozonesondes, Umkehr, SAGE I, SAGE II, and SBUV(/2) for the period 1979–2004 . . . 395
6.7 Ozone anomalies after subtraction of annual cycle, QBO, and solar cycle effects at altitude range from 30 to 40 km . . . 397
6.8 Ozone variations for 60°S–60°N estimated from ground-based data and individual components that comprise ozone variations; and deseasonalized area-weighted total ozone deviations estimated from ground-based data adjusted for solar, volcanic, and QBO effects, for 60°S–60°N . . . 402
6.9 Mean seasonal variations of monthly TOC averaged over the NWR, SWR, WS, ES, FE, and Transcaucasus and Central Asia regions over the period 1973–2002 404
6.10 Propagation paths of ozone anomalies to Europe . . . 409
6.11 Vertical profile of ozone trends over northern mid-latitudes estimated from ozonesonde, Umkehr, and SBUV(/2) measurements for the period 1979–2008 412
6.12 Monthly ozone anomalies for Europe as measured by ozonesondes, SBUV(/2), and Umkehr at three pressure layers . . . 414
6.13 Differences in the ozone and temperature vertical profiles from 2001 (cold stratosphere, low ozone) to 2002 (warm stratosphere, high ozone) . . . 416
6.14 Southern hemisphere total ozone content during September 21–27, 2002 obtained from EP/TOMS observations . . . 417
6.15 Time series of effective temperature for winters 2003–2007 from ECMWF and daily muon rate . . . 418
6.16 Temperature profiles observed on February 16, 2001 during a major SSW event as well as data from before and after the event over Lerwick, Scotland . . . 420
6.17 Arrhenius plots for temperature-dependent diffusion coefficients for various species in ice as obtained by the LRD depth-profiling technique . . . 422
6.18 Relationship between the factor D_0 of the diffusion coefficient and the activation energy E . . . 423

6.19 Maximum day area covered by the ozone hole, lowest total column ozone, and ozone mass deficiency 426
6.20 O_3MD from pre-1976 averages integrated for 105 days (September 1–December 15) for the southern polar region and middle latitudes in the period 1979–2006 428
6.21 Long-term variation of ozone amount at Syowa 430
6.22 Size of the 2007 Antarctic ozone hole compared with the average size for the previous 12 years. 430
6.23 Average total ozone departures from pre-1976 values for the period 1990–2000 432
6.24 Integrated O_3MD for a period of 105 days 433
6.25 Total O_3MD north of 35°N (January 1–April 15) and south of 35°S (September 1–December 15) 434
6.26 Integrated ozone loss as a function of V_{PSC} for 1992/1993 to 2007/2008. . . . 436
6.27 Seasonal changes of 10-day averaged TOC of each month at the South Pole, Arrival Heights, Syowa Station, and Macquarie Island averaged over the period between 1993 and 2005 438
6.28 Seasonal change in the deviation of TOC from the polar night mean at the South Pole, Arrival Heights, and Syowa Station in the periods 1980–1992 and 1993–2005. 439
6.29 Geographical location of Antarctic stations and the Arctic station chosen for comparison of satellite EP-TOMS and ground-based measurements of total ozone content in 1996–2005 439
6.30 An artificial time series of events of a geophysical parameter in conventional time and in natural time 444
6.31 The entropy in natural time for various window lengths of 3 to 15 years sliding each time by one year through the whole MDOHA time series 445
6.32 The entropy in natural time under time reversal for various window lengths of 3 to 15 years sliding each time by one year through the whole MDOHA time series 446
6.33 The entropy change in natural time under time reversal for various window lengths of 3 to 15 years sliding each time by one year through the whole time series of MDOHA. 447
6.34 Time series of deseasonalized monthly means and the linear regression lines of TOC and tropopause height for 1984–2002 at Athens deduced from *in situ* observations 475
6.35 Time series of deseasonalized monthly means and the linear regression lines of tropopause temperature deduced from reanalysis data and TOC for 1984–2002 at Athens 476
6.36 Emissions of ODSs and their substitutes 480
6.37 Stratospheric EESC derived for mid-latitude and polar stratospheric regions relative to peak abundances, plotted as a function of time 482
7.1 Calculated monthly concentration of ozone in the surface layer for June 1850 494
7.2 Calculated monthly concentration of ozone in the upper troposphere for June 1850. 494
7.3 Calculated change in annual mean tropospheric ozone in Dobson units between 1850 and 1990. 495
7.4 RF due to the change in tropospheric ozone since pre-industrial times according to the chemistry–transport model 513
7.5 As Figure 7.4, but for the OsloRad model. 514
7.6 Meridional and seasonal distribution of the change in surface air temperature,

STL temperature, and STH temperature resulting from STL ozone reduction of 50% 518
7.7 Meridional and seasonal distribution of the adjusted change in net radiation at the tropopause; initial change in short-wave radiative divergence in the STL; and adjusted change in long-wave radiation at the tropopause 519
7.8 Meridional distribution of the change in annual mean surface air temperature for the various scenarios 520
7.9 Meridional distribution of the change in annual mean short-wave, long-wave, and net radiation at the tropopause for the various scenarios 521
7.10 Meridional and seasonal distribution of the change in STL temperature for the TOT(z) scenario and the TOT(φ, z, t) scenario 522
7.11 Trends in December to February zonal mean zonal wind 543
7.12 Stretched-out and cowboy solutions of the double spherical pendulum problem 552

Tables

1.1	Results of intercomparison of Dobson instruments (July 22 to August 10, 1990) at Arosa, Switzerland	13
1.2	Δ_{ETC}, the error in determination of the extraterrestrial constant for various stray light error models	27
1.3	The results from some stray light experiments taken by the Dobson D118 instrument in Athens between 1992 and 1994	32
1.4	Layers used for Umkehr ozone profile retrievals	53
1.5	An approximate correction for the air pump temperature	64
2.1	Instruments on board the PRIRODA module of the Mir Space Station	86
2.2	TOMS main characteristics	89
2.3	General description of OMI	91
2.4	Satellite instrumentation for BUV ozone observations	95
2.5	SBUV/2 wavelengths	96
2.6	TOC trends for January 1979 to March 1995 for five regions of northern mid-latitudes, the Arctic, and Antarctica	103
2.7	Explanation of acronyms for ozone-monitoring instruments found on ozone-monitoring satellites	131
2.8	Chronological summary of ozone-monitoring satellites (or Space Shuttle missions)	134
2.9	Inventory of the most "reliable" datasets with time records longer than 5 years of Ozone Vertical Profile	140
3.1	NDACC and WOUDC ground-based stations contributing to MIPAS O_3 validation	166
3.2	Results of the comparison between MIPAS v4.61 ozone profiles and NDACC ground-based measurements	175
3.3	Statistical means (MRD) and standard deviations (SD) of the relative differences (X-FTIR)/mean (FTIR) in percent of the O_3 partial columns defined by the given pressure limits	182
3.4	Results of the comparison of MIPAS data with mid-latitude ozone soundings	190
3.5	Summary of the coincidences between MIPAS-ENVISAT and MIPAS-B2	192

3.6 Best spatiotemporal coincidences selected for MIPAS-ENVISAT ozone validation with M-55 *Geophysica* aircraft measurements 203
3.7 Statistics over all comparisons of MIPAS to SAGE II 212
3.8 Statistics over all comparisons of MIPAS to HALOE. 228
3.9 Overview of lidar sites used by Van Gijsel *et al.* (2009). 233
4.1 The characteristics of total ozone anomalies in January in the equatorial region during negative Southern Oscillation Index anomalies 261
4.2 SBUV(/2) trends in %/decade by seasons and latitudinal zone over the period from January 1979 to May 1994. 264
4.3 Short-term Dobson trends in %/decade using data from January 1979 to February 1994. 265
4.4 Long-term Dobson trends in %/decade using data from January 1964 to February 1994. 265
4.5 Representative TOC trends in %/decade 267
4.6 Long-term total ozone trends derived from ground-based data 292
4.7 Important reactions for odd oxygen balance 302
4.8 Species in photochemical models 304
4.9 Multifractal DFA exponents for reconstructed TOC data at Arosa and Poluy 316
4.10 Calculated percentage change of surface ozone (at four stations) throughout the eclipse event of March 29, 2006 at Athens, Greece 325
4.11 Sites used for detection of the relationship between TOC and tropopause properties 330
5.1 Chronology of events leading up to the Montreal Protocol 345
5.2 Values of TOC (DU) from Spitsbergen and Halley in 1956 and Halley in 1996 357
5.3 Ratification of the Montreal Protocol 372
5.4 Halons used in various applications 373
5.5 Substances originally included in the Montreal Protocol 374
6.1 Scientific assessment reports 380
6.2 Lifetimes, relative fractional halogen release factors, and ozone depletion potentials for halocarbons 384
6.3a Comparison of annual changes in stratospheric column ozone and total ozone content in India before and after the implementation of the Montreal Protocol 398
6.3b Comparison of annual changes in total ozone content in different Indian latitude belts before and after the implementation of the Montreal Protocol 398
6.4 Linear trends at Belsk in selected Umkehr layers and corresponding TOC for the periods 1980–1995 and 1996–2007 399
6.5 Trend results for TOC computed by Yang *et al.* (2009) for 45°N, 45°S, and integrated 50°N–50°S. 400
6.6 Total ozone linear trends and their deviation from zero significance for periods I, 1973–1992 and II, 1993–2005 405
6.7 Warming and cooling at different heights in some SSWs. 420
6.8 Summary of rate parameters for the diffusion coefficients of selected compounds in ice 422
6.9 Number of days when the Antarctic ozone hole area covered a sunlit surface greater than 10, 15, 20, and 25 km^2 from 1995 to 2006 427
6.10 Linear trend from 1987 to 2007 springtime periods in each Umkehr layer at the Japanese Antarctic research station at Syowa. 429
6.11 Geographical co-ordinates of the high-latitude Dobson stations considered by Kravchenko *et al.* (2009) 440

6.12 Thirteen chemistry–climate models (CCMs) 460

7.1 Gaseous phase chemical reactions included in the model. 488

7.2 Aqueous phase chemical reactions included in the model 490

7.3 Net chemical production of ozone in the planetary boundary layer and the upper troposphere and net upward flux of ozone from the PBL 496

7.4 Radiative forcing due to changes in ozone content for the period 1972–1990 499

7.5 The relative roles of contributions of ozone and other greenhouse gases to the formation of RF 500

7.6 Decadal stratospheric column ozone change, stratospheric ozone forcing, and stratospheric RF gradient for standard tropical profiles 501

7.7 Same as Table 7.6, except for the mid-latitudes 502

7.8 Atmospheric composition changes 503

7.9 SAT and TOC changes 504

7.10 Minor gas concentrations used in RF calculations 505

7.11 Global and annual RF. 505

7.12 Adjusted annual global average net RF from various ozone changes and greenhouse gas changes over different time periods. 507

7.13 January and July clear-sky and average cloudiness sensitivities of ozone and sulfate aerosol and consequent RF for the industrial period 1850–1990 509

7.14 Emission rates applied in the model study of Berntsen *et al.* (1997) 512

7.15 Simulated global budget of tropospheric ozone 523

7.16 Changes in NO_x emissions due to a 20% reduction in the emissions in each region and the calculated changes in lifetime and steady-state concentration of methane 524

7.17 Total RF due to ozone changes for different regions 525

7.18 Results from perturbing the climate according to paleoclimate data. 529

7.19 Results from perturbing the climate according to projected greenhouse gas emission data 530

7.20 Climate impacts of 20% reduction in NO_x 539

7.21 Long-wave, short-wave, and net RF due to stratospheric ozone; global, Northern Hemispheric, and Southern Hemispheric averages, for the four seasons and the yearly average 540

Abbreviations and acronyms

AAO	Antarctic Oscillation
AAOE	Airborne Antarctic Ozone Experiment
AASE 1	Airborne Arctic Stratosphere Expedition
ABL	Atmospheric Boundary Layer
AC&C	Atmospheric Chemistry & Climate
ACCENT	Atmospheric Composition Change: the European Network of Excellence
ACD	Apparent Column Density
ACE	Aerosol Characterization Experiment
ACE-FTS	ACE Fourier Transform Spectrometer
ACVT	Atmospheric Chemistry Validation Team
AD	A and D wavelengths
ADEOS	ADvanced Earth Observing Satellite
AEM-2	Applications Explorer Mission-2
AEMET	Agencia Estatal de METeorología
AES	Atmospheric Environment Service
AF	Adjusted Forcing
AGCM	Atmospheric General Circulation Model
AIRS	Atmospheric InfraRed Sounder
ALA	Aerosol Limb Absorption
ALOMAR	Arctic Lidar Observatory for Middle Atmosphere Research
AM2	Atmosphere Model-2
AMAS	Advanced Millimeter-wave Atmosphere Sounder
AMF	AirMass Factor
AMSR	Advanced Microwave Scanning Radiometer
AMSU	Advanced Microwave Sounding Unit

AO	Arctic Oscillation
AOS	Acousto-Optical Spectrometer
APE	Airborne Polar Experiment
API	Advanced Pollution Instruments
AR	Assessment Report
ARC	ARCtic
ARCAS	All-purpose Rocket for Collecting Atmospheric Soundings
ARGOS	Advanced Remote Gaseous Oxides Sensor
amsl	Above mean sea level
ASOPOS	ASsessment of Operating Procedures for Ozone Sondes
ASSET	ASSimilation of EnvisaT data
ASUR	Airborne SUbmillimeter Receiver
ATLAS	ATmospheric Laboratory for Applications and Science
ATMOS	Atmospheric Trace MOlecule Spectroscopy
ATMS	Advanced Technology Microwave Sounder
ATOVS	Advanced TOVS
AURAMS	A Unified Regional Air quality Modeling System
AVE	Aura Validation Experiment
AWI	Alfred Wegener Institut
BADC	British Atmospheric Data Centre
BAPoMON	Background Air Pollution MOnitoring Network
BAS	British Antarctic Survey
BASCOE	Belgian Assimilation System for Chemical ObsErvations
BD	Brewer–Dobson circulation
BESOS	Balloon Experiment on Standards for Ozone Sondes
BIC	Balloon Intercomparison Campaign
BIRA	Belgisch Instituut voor Ruimte-Aëronomie
BLISS	Balloonborne Laser In-Situ Sensor
BM	Brewer–Mast (ozonesonde)
BMLS	Balloon Microwave Limb Sounder
BOMEM	BOMEM model FTIR
BP	Band Pass
BSMILES	Balloon-borne Superconducting subMIllimeter-wave Limb Emission Sondes
BUV	Backscattered UltraViolet
CALIPSO	Cloud–Aerosol Lidar and Infrared Pathfinder Satellite Observations
Caltech	California Institute of Technology
CAM	Community Atmosphere Model
CANDAC	CAnadian Network for the Detection of Atmospheric Change
CAO	Central Aerological Observatory

CARIBIC	Civil Aircraft for the Regular Investigation of the atmosphere Based on an Instrument Container
CC	Cross Correlation
CCD	Convective Cloud Differential; Charge Coupled Device
CCM	Chemistry–Climate Model
CCMVal	Chemistry–Climate Model Validation activity
CCO3L	Coordination Committee on Ozone Layer
CCOL	Coordinating Committee on the Ozone Layer
CCSM	Community Climate System Model
CCSR/NIES	Center for Climate System Research/National Institute for Environmental Studies
CDOM	Colored Dissolved Organic Matter
CEOS	Committee on Earth Observation Satellites
CERES	Clouds and the Earth Radiant Energy System
CF	Cloud cover Fraction
CFC	Chlorofluorocarbon
CFL	Courant–Friedrichs–Levy
CGE	Centro de Geofísica de Évora
CHEM	CHEMistry model
CHEOPS	CHEmistry of Ozone in the Polar Stratosphere
CHRONOS	Canadian Hemispheric and Regional Ozone and Nitrous Oxide System
CHMI	Czech HydroMeteorological Institute
CIRRIS	Cryogenic InfraRed Radiance Instrument for Shuttle
CLAES	Cryogenic Limb Array Etalon Spectrometer
CMAM	Canadian Middle Atmosphere Model
CMDL	Climate Monitoring and Diagnostics Laboratory
CMIP5	Coupled Model Intercomparison Project Phase 5
CNR	Italian National Research Council
CNRS	Centre National de la Recherche Scientifique
CONTRAIL	Comprehensive Observation Network for TRace Gases by AIrLiner
COP-7	Conference of Parties to the Vienna Convention
COPES	Core Project
CPV	CircumPolar Vortex
CrIS	Cross-track Infrared Sounder
CRISTA	CRyogenic Infrared Spectrometers and Telescopes for the Atmosphere
CTM	Chemistry–Transport Model; Chemical Tracer Model
CTMK	Continuous Time MarKov chain model
CUE	Critical Use Exemption
CUSUM	CUmulative SUM
CWT	Continuous Wavelet Transform
DAAC	Distributed Active Archive Center
DAO	Data Assimilation Office

DCC	Dobson Calibration Center
DE-1	Dynamics Explorer-1 (spacecraft)
DFA	Detrended Fluctuation Analysis
DIAL	DIfferential Absorption Lidar
DKRZ	Deutsches KlimaRechenZentrum
DLR	Deutsches Zentrum für Luft und Raumfahrt
DMI	Danish Meteorological Institute
DMSP	Defense Meteorological Satellite Program
DOAS	Differential Optical Absorption Spectroscopy
DoD	Department of Defense
DS	Direct Sun
D-TOC	Data Table Of Contents
DU	Dobson Unit
DWD	German Weather Service (Deutscher Wetterdienst)
DWSS	Defence Weather Satellite System
E	East
E39C	Chemistry–climate model (ECHAM4.L39)
EASOE	European Arctic Stratospheric Ozone Experiment
EC	Environment Canada
ECC	Electrochemical Concentration Cell (ozonesonde)
ECD	Electron Capture Detector
ECHAM	ECMWF HAMburg
ECMWF	European Centre for Medium Range Weather Forecasts
ECOC	ElectroChemical Ozone Cell
EDR	Environmental Data Record
EECl	Effective Equivalent Chlorine
EESC	Equivalent Effective Stratospheric Chlorine
emf	Electromotive force
ENSCI	ENvironmental SCIence Corporation
ENSO	El Niño Southern Oscillation
ENVISAT	ENVIronmental SATellite
EOS	Earth Observing System
EP	Earth Probe
EPA	Environmental Protection Agency
EPP	Emissions Prediction and Policy
EPPA	Emissions Prediction and Policy Analysis (model)
EQ	EQuator
EQUAL	ENVISAT QUality Assessment with Lidar
ERA	ECMWF Re-Analyses
ERB	Earth Radiation Budget
ERBS	Earth Radiation Budget Satellite
ERI	European Research Infrastructure
ERS	Earth Resources Satellite
ESA	European Space Agency

ESABC	ENVISAT Stratospheric Aircraft and Balloon Campaigns
ESL	Equivalent Solar Latitude
ESMOS	European Stratospheric MOnitoring Stations
ESRIN	European Space Research INstitute
ETC	ExtraTerrestrial Constant
EUMETSAT	EUropean Organisation for the Exploitation of METeorological SATellites
EV-PYR	Everest-Pyramid
EVA	A visible absorption spectrometer
ExMOP-1	Extraordinary Meeting of Parties to the Montreal Protocol-1
ExMOP-2	Extraordinary Meeting of Parties to the Montreal Protocol-2
FB	FeedBack
FD	Fixed Dynamic Heating
FFSL	Flux-Form Fermi–Lagrangian
FFT	Fast Fourier Transform
FIRS-2	Far InfraRed Spectrometer-2
FMI	Finnish Meteorological Institute
FOZAN	Fast OZone ANalyzer
FRESCO	Fast REtrieval Scheme for Cloud Observables
FT	Fixed air Temperature
FT-FIR	Fourier Transform Far-InfraRed
FTIR	Fourier Transform InfraRed
FTS	Fourier Transform Spectrometer
FU	Free University (Berlin)
FURM	FUll Retrieval Method
FVGCM	Finite Volume General Circulation Model
FWHM	Full Width Half Maximum
FZK	ForschungsZentrum Karlsruhe
GACM	Global Atmosphere Composition Mission
GAMETAG	Global Atmospheric Measurements Experiment on Tropospheric Aerosols and Gases
GAMS	Gas and Aerosol Monitoring Sensorcraft
GAS	Get Away Special (canister)
GASP	Global Air Sampling Program
GAW	Global Atmospheric Watch
GBMCD	Ground-Based Measurements and Campaign Database
GC	Governing Council
GCM	General Circulation Model
GCMAM	Global Climate/Middle Atmosphere Model
GCOM	Global Change Observation Mission
GDP	GOME Data Processor
GEOS CCM	Goddard CCM (labeled)

GFS	Global Forecast System
GGC	Grand Governing Council
GHG	GreenHouse Gas
GISS	Goddard Institute for Space Studies
GKSS	Gesellschaft für Kernenergieverwertung in Schiffbau und Schiffahrt mbH
GloPac	Global Hawk Pacific Mission
GMCC	Geophysical Monitoring for Climatic Change
GMI	Global Modeling Initiative
GMTR	Geo-fit Multi Target Retrieval
GO_3OS	Global Ozone Observing System
GODFIT	GOME Direct FITting
GOES	Geostationary Operational Environmental Satellite
GOME	Global Ozone Monitoring Experiment
GOMETRAN	GOME TRANsport model
GOMOS	Global Ozone Monitoring by Occultation of Stars
GPH	Geo-Potential Height
GRIPS	GCM-Reality Intercomparison Project for SPARC
GROMOS	GROund-based Millimeter-wave Ozone Spectrometer
GSFC	Goddard Space Flight Center
GUV	Ground-based Ultraviolet Radiometer
GWP	Global Warming Potential
HADAM2a	HADley Centre AtMospheric GCM-2a
HadGEM1	Hadley Centre Global Environmental Model-1
HALOE	HALogen Occultation Experiment
HAMMONIA	HAMburg Model Of the Neutral and Ionized Atmosphere
HCFC	Hydrochlorofluorocarbon
HFC	Hydrofluorocarbon
HH	Hitch Hiker (payload)
HIRDLS	HIgh Resolution Dynamics Limb Scanner
HIRS	HIgh Resolution Infrared Sounder
HIRS/2	High-resolution Infrared Radiation Sounder-2
HIRS/3	High-resolution Infrared Radiation Sounder-3
HIRS/4	High-resolution Infrared Radiation Sounder-4
HITRAN	HIgh-resolution TRANsmission molecular absorption database
IAGA	International Association of Geomagnetism and Aeronomy
IAGOS	In-service Aircraft for a Global Observation System
IAM	Interface Adapter Model; Integrated Assessment Model
IAMA	International Aerosol Modeling Algorithm
I-ARC	Izaña Atmospheric Research Center
IASB	Institut d'Aéronomie Spatiale de Belgique

IASI	Infrared Atmospheric Sounding Interferometer
IBEX	Infrared Balloon EXperiment
IFAC	Institute For Applied Physics Nello Carrara
IFE	Institut für Femerkundug (Institute of Remote Sensing)
IFOV	Instantaneous Field Of View
IGBP	International Geosphere–Biosphere Program
IGS	Infrared Grating Spectrometer
IGY	International Geophysical Year
IIP	Instrument Incubator Program
ILAS-1	Improved Limb Atmospheric Sounder-1
ILAS-2	Improved Limb Atmospheric Sounder-2
ILS	Instrumental Line Shape
IMG	Interferometric Monitor for Greenhouse Gases
IMK	Institut für Meteorologie und Klimaforschung
IMOS	Inadvertent Modification of the Stratosphere
INPE	National Institute for Space Research
INTA	Instituto Nacional de Técnica Aerospacial
INTEX	INtercontinental Chemical Transport Experiment
IOC	International Ozone Commission
IONS	INTEX Ozonesonde Network Study
IPCC	Intergovernmental Panel on Climate Change
IPSL	Institut Pierre Simon Laplace
IPY	International Polar Year
IR	InfraRed
IRA	Instrumentation et Réactivité Atmosphérique
IRCI	IntegRated Cloud Imager
IRF	Institutet för Rymdfysik
IRFS-1	InfraRed Fourier-transform spectrometer
IRFS-2	InfraRed Fourier-transform spectrometer
IRHS	InfraRed Heterodyne Spectroscopy
IRIS	InfraRed Interferometer Spectrometer
IRLS	InfraRed Limb Sounder
IRS	InfraRed Sounder
ISAC	Institute of Atmospheric Sciences and Climate
ISAMS	Improved Stratospheric And Mesospheric Sounder
ITCZ	Inter Tropical Convergence Zone
IZO	IZaa Observatory
JAL	Japan AirLines
JAXA	Japan Aerospace eXploration Agency
JEM	Japanese Experiment Module
JMA	Japan Meteorological Agency
JOSIE	Jülich Ozone Sonde Intercomparison Experiment
JPL	Jet Propulsion Laboratory
JPSS	Joint Polar Satellite System

KIMRA	KIruna Millimeter wave RAdiometer
KMI	Koninklijk Meteorologisch Instituut (Royal Meteorological Institute)
KNMI	Royal Netherlands Meteorological Institute
KOPRA	Karlsruhe Optimized Precise Radiative transfer Algorithm
LAGRANTO	LAGRangian ANalysis TOol
LAS	Limb Atmospheric Infrared Spectrometer
LFB	Lightning–ozone FeedBack
LHS	Laser Heterodyne Spectrometer
LID	LIDar Systems
LIDAR	LIght Detection And Ranging
LIDORT	LInearized Discrete Ordinate Radiative Transfer
LIMS	Limb Infrared Monitor of the Stratosphere
LKO	Licht-Klimatisches Observatorium; Light Climatic Observatory
LMDZ	Laboratoire de Météorologie Dynamique Zoom
LNO_x	NO_x production by lightning
LOP	Low-Ozone Pocket
LORE	Limb Ozone Retrieval Experiment
LOS	Line Of Sight
LPCE-CNRS	Laboratoire de Physique et Chimie de l'Environment (Orléans, France)
LRD	Laser Resonant Desorption
LRIR	Limb Radiance Inversion Radiometer
LS	Lower Stratosphere
LSC	Lyndon State College
LSST	Land and Sea Surface Temperature
LT	Local Time
LW	Long Wave
MAESTRO	Measurement of Aerosol Extinction in the Stratosphere and Troposphere Retrieved by Occultation
MAHRSI	Middle Atmosphere High Resolution Spectrographic Investigation
MANTRA	Middle Atmosphere Nitrogen TRend Assessment
MAP	Middle Atmosphere Program
MAS	Millimeter-wave Atmospheric Sounder
MASI	Model Assimilation and Satellite Intercomparison
MASTER	MODIS/ASTER airborne simulator
MDI	Metered Dose Inhaler
MDOHA	Maximum Daily Ozone Hole Area
MDR	Maximum Detectable Rigidity
MGC	Minor Gaseous Component
MINOS	Main Injector Neutrino Oscillation Search

MIPAS	Michelson Interferometer for Passive Atmospheric Sounding
MIR	Millimeter-wave Imaging Radiometer
MIRA	MIcrowave RAdiometry
MLF	MultiLateral Fund
MLO	Mauna Loa Observatory
MLS	Microwave Limb Sounder
MLTI	Mesosphere and Lower Thermosphere/Ionosphere
MOGUNTIA	Model Of the Global UNiversal Tracer Transport In the Atmosphere
MOHp	Meteorological Observatory of Hohenpeissenberg
MOMS	Modular Optoelectronic Multispectral Scanner
MOP	Meetings of Parties
MOS	Modular Optoelectronic Scanner
MOZAIC	Measurements of OZone and Water Vapor by Airbus In-service AirCraft
MOZART	Model for OZone And Related chemical Tracers
MR	Microwave Radiometer
MRD	Mean Relative Deviation
MRI	Meteorological Research Institute
MSC	Meteorological Service of Canada
MSU	Microwave Sounding Unit
MT	Middle Troposphere
MUTOP	Multi-sensor Upper Tropospheric Ozone Product
MWLS	Millimeter-Wave Limb Sounder
MWR	MicroWave Radiometer
NAM	Northern Hemisphere Annular Mode
NAPMN	National Air Pollution Monitoring Network
NASA	National Aeronautics and Space Administration
NASDA	NAtional Space Development Agency of Japan
NAT	Nitric Acid Trihydrate
NCAR	National Center for Atmospheric Research
NCDC	National Climatic Data Center
NCEP	National Centers for Environmental Prediction
NDACC	Network for the Detection of Atmospheric Composition Change
NDSC	Network for the Detection of Stratospheric Changes
NFB	No FeedBack
NGOs	Non Governmental Organization
NH	Northern Hemisphere
NHGOS	Northern Hemisphere Ground Ozone Stations
NIES	National Institute for Environmental Studies
NILU	Norsk Institutt for LUftforskning
NIRS	Near-InfraRed Spectrometer
NIST	National Institute of Standards and Technology

NIVR	Nederlands Instituut voor Vliegtuigontwikkeling en Ruimtevaart
NIWA	National Institute of Water and Atmospheric Research, Ltd. (New Zealand)
NLTE	Non-Local Thermal Equilibrium
NMC	National Meteorological Center
NMHC	Non-Methane HydroCarbons
NNORSY	Neural Network Ozone Profile Retrieval
NOAA	National Oceanic and Atmospheric Administration
NOZE	National OZone Expedition
NPOESS	National Polar-orbiting Operational Environmental Satellite System
NPP	NPOESS Preparatory Project
NTF	No Tropospheric Feedbacks
O_3MD	Ozone Mass Deficiency
O_3s	O_3 originating in the stratosphere
O3S	Ozonesonde Stations
O_3t	O_3 originating in the troposphere
OCO	Orbiting Carbon Observatory
OD	Ozone Depletion
ODP	Ozone Depletion Potential
ODS	Ozone Depleting Substance
ODUS	Ozone Dynamics Ultraviolet Spectrometer
OGOS	Operational Ground-based Ozone Sensor
OHP	Observatoire d'Haute Provence
OLME	Ozone Layer Monitoring Experiment
OM	Ozone Meter
OMI	Ozone Monitoring Instrument
OMPS	Ozone Mapping and Profiler Suite
OMS	Observations of the Middle Stratosphere
OOAM	Orbiting Ozone and Aerosol Monitor
ORACLE	Ozone layer and UV Radiation in A Changing cLimate Evaluated during IPY
OSIRIS	Optical Spectrograph and Infra-Red Imager System
OSN	Ozone Sounding Network
OSORT	Ozone and SOlar Radiation research Team
OVP	Ozone Vertical Profile
OZAFS	Ozone and Aerosol Fine Structure
PARASOL	Polarization and Anisotropy of Réflectances for Atmospheric Sciences coupled with Observations from a Lidar
PBL	Planetary Boundary Layer
PCMDI	Program for Climate Model Diagnosis and Intercomparison
PD	Present Day

PEM-West	Pacific Exploratory Mission-West
PERPA	National Service for Air Pollution Monitoring
PI	Pre Industrial
PION	(spectrometer–ozonometer)
PIRAMHYD	Passive InfraRed Atmospheric Measurements of HYDroxyl
PNJ	Polar Night Jet
POAM	Polar Ozone and Aerosol Measurement
POAM III	Polar Ozone and Aerosol Measurement-III
POES	Polar Operational Environmental Satellites
POLARIS	Photochemistry of Ozone Loss in the Arctic Region in Summer
POLECAT	POlar stratospheric clouds, LEe waves, Chemistry, Aerosols and Transport
PREMIER	PRocess Exploration through Measurements of Infrared and millimeter-wave Emitted Radiation
PREMOS	PREcision Monitor for OScillation measurement
PROFFIT	PROFile FITPSCs Polar Stratospheric Clouds
PTFE	Polytetrafluoroethylene
PV	Potential Vorticity
QBO	Quasi-Biennial Oscillation
QTO	Quasi-Triennial Oscillation
QUANTIFY	QUANTIFYing the Climate Impact of Global and European Transport Systems
RAL	Rutherford Appleton Laboratory
RAM	Radiometer for Atmospheric Measurements
RCC	Regional Calibration Center
RCP	Representation Concentration Pathway
RD	Relative Deviation
RDCC/A	Regional Dobson Calibration Centers for Asia
RDCC/E	Regional Dobson Calibration Centers for Europe
REFIR	Radiative Explorer in the Far-InfraRed
REPROBUS	REactive Processes Ruling the Ozone BUdget in the Stratosphere
RF	Radiative Forcing
RFD	Radiative Flux Divergence
RIS	Retro-reflector In Space
RIVM	National Institute for Public Health and the Environment
RMS	Root Mean Square
ROCOZ	ROCket OZonesonde
ROI	Reynolds Optimally Interpolated
RRS	Royal Research Ship
RSAS	Rayleigh Scattering Attitude Sensor
SA	Solar Activity

SABE	Solar Absorption Balloon Experiment
SABER	Sounding of the Atmosphere using Broadband Emission Radiometry
SAFARI	Southern Africa Fine Atmosphere Research Initiative
SAFIRE	Spectroscopy of the Atmosphere using Far IR Emission
SAGE	Stratospheric Aerosol and Gas Experiment
SAM	Southern Annular Mode
SAM I	Stratospheric Aerosol Measurement
SAMS	Stratosphere and Mesosphere Sounder
SAOZ	System for Analysis of Observation at Zenith
SAT	Surface Air Temperature
SAUNA	Sodankylä Total Ozone Intercomparison and Validation Campaign
SAWS	South African Weather Service
SBUS	Solar Backscattering UV Sounder
SBUV	Solar Backscatter Ultraviolet
SBUV/2	Solar Backscatter UltraViolet radiometer-2
SCIAMACHY	SCanning Imaging Absorption SpectroMeter for Atmospheric CartograpHY
SCIAVALUE	SCIAMACHY VALidation and Utilization Experiment
SCISAT	SCIentific SATellite
SCO	Stratospheric Column Ozone
SCOUT	Stratospheric–Climate links with emphasis On the Upper Troposphere and lower stratosphere
SD	Standard Deviation
SEM	Standard Error on the Mean
SES	Seismic Electromagnetic Signals
SESAME	Second European Stratospheric Arctic and Mid-latitude Experiment
SH	Southern Hemosphere
SHADOZ	Southern Hemisphere ADditional OZonesondes
SHIS	Scanning High-resolution Interferometer Sounder
SIC	Sea Ice Concentration
SIRHS	Stratospheric Sounding by InfraRed Heterodyne Spectroscopy
SIRIS	Stratospheric InfraRed Interferometer Spectroscopy
SLIMCAT	Single Layer Isentropic Model of Chemistry And Transport
SME	Solar Mesosphere Explorer
SMILES	Superconducting SubMIllimeter-Wave Limb-Emission Sounder
SMM	Solar Maximum Mission
SMR	Sub-Millimeter Radiometer

SOCOL	SOlar Climate Ozone Links
SOI	Southern Oscillation Index
SOLSE	Shuttle Ozone Limb Sounding Experiment
SOLSTICE	SOLar-STellar Irradiance Comparison Experiment
SOLVE	SAGE-III Ozone Loss and Validation Experiment
SOMORA	Stratospheric Ozone MOnitoring RAdiometer
SOPRANO	Sub-millimetric Observations of PRocesses in the Atmosphere Noteworthy for Ozone
SOZ	Surface OZone
SPARC	Stratospheric Processes And their Role in Climate
SPAS	Shuttle PAllet Satellite
SPATRAM	SPectrometer for Atmospheric TRAcers Monitoring
SPC	Science Pump Corporation
SPE	Solar Proton Event
SPEAM	SunPhotometer Earth Atmosphere Measurement
SPIRALE	SPectroscopie InfraRouge par Absorption de Lasers Embarqués
SPOT-3	Système Pour l'Observation de la Terre
SPS	Star Pointing Spectrometer
SPURT	Spurengastransport in der Tropopausenregion
SRES	Special Report on Emissions Scenarios
SRFG	Stratospheric RF Gradient
SSBUV	Shuttle SBUV
SSF/IC	Stiga Ski Field InterComparison
SIC	Sea Ice Concentration
SSBUV	Shuttle Solar Backscatter UltraViolet radiometer
SSH	Special Sensor H
SST	Sea Surface Temperature
SSTs	SuperSonic (jet) Transport aircraft
SSU	Stratospheric Sounding Unit
SSW	Sudden Stratospheric Warming
STE	Stratosphere–Troposphere Exchange
STEP	Stratosphere and Troposphere Exchange Project
STH	Higher stratosphere
STL	Lower stratosphere
STP	Standard Temperature and Pressure
STRATOC	STRATOspheriC Ozone Experiment
STREAMR	Stratosphere–TRoposphere Exchange And climate Monitor Radiometer
STS	(Space Shuttle mission)
SUMAS	SUbMillimeter Atmospheric Sounder
SUSIM	Solar Ultravioliet Spectral Irradiance Monitor
SUVR	Solar UltraViolet Radiation
SUVS	Solar UltraViolet Spectrophotometer
SUV-100	Solar UltraViolet (spectroradiometer)

SW	Short Wave
SWIFT	Stratospheric Wind Interferometer For Transport Studies
SYMOCS	SYstem for the Monitoring Of Stratospheric Compounds
SZA	Solar Zenith Angle
TANSO-FTS	Thermal And Near ir Sensor for climate Observation-Fourier Transform interferometer Spectrometer
TDL	Tunable Diode Laser
TDLHS	Tunable Diode LHS
TEAP	Technical and Economic Assessment Panel
TES	Tropospheric Emission Spectrometer
TG	Trace Gas
THESEO	THird European Stratospheric Experiment on Ozone
THIR	Temperature Humidity Infrared Radiometer
THT	Trajectory Hunting Technique
TIMED	Thermosphere Ionosphere Mesosphere Energetics and Dynamics
TIROS	Television InfraRed Observational Satellite
TOC	Total Ozone Content
TOM	Total Ozone Mapper; Tertiary Ozone Maximum
TOMS	Total Ozone Monitoring Spectrometer
TOMS/EP	Total Ozone Monitoring Spectrometer/Earth Probe Satellite
TOPS	Total Ozone Portable Spectrometer
ToR, TOR	Terms Of Reference
TOTE	Tropical Ozone Transport Experiment
TOU	Total Ozone Unit
TOVS	TIROS Operational Vertical Sounder
TOZ	Total OZone Column
TRACE A	TRansport and Atmosphere Chemistry Near the Equatorial Atlantic
TROPOZ	TROPospheric OZone Experiment
TROZ	TRoposphärische OZon
TrOZ	Tropospheric OZone
TRT	TRopospheric brightness Temperature
TTL	Tropical Tropopause Layer
UARS	Upper Atmospheric Research Satellite
UCI	University of California, Irvine
UGAMP	Universities Global Atmospheric Modeling Program
UGCM	UGAMP General Circulation Model
U.K.	United Kingdom
UKMO	United Kingdom Meteorological Office
ULAQ	University of L'Aquila
UMAS	UnManned Aerial mapping Systems

UMETRAC	Unified Model with Eulerian TRAnsport
UMSLIMCAT	Unified Model SLIMCAT
UNEP	United Nations Environmental Program
UPAS	Universal Processor for UV/VIS Atmospheric Spectrometers
U.S.A.	United States of America
U.S.S.R.	Union of Soviet Socialist Republics
UT	Upper Troposphere
UTC	Coordinated Universal Time
UT/LS	Upper Troposphere/Lower Stratosphere
UV	UltraViolet
UV-B	UltraViolet-B
UVISI	UV and Visible Imagers and Spectrographic Imagers
UVN	Ultraviolet Visible and Near-IR sounder
UVS	UltraViolet Spectrometer
UVSP	UltraViolet Spectrometer Polarimeter
VFD	Very Fast Delivery
VIIRS	Visible/Infrared Imager Radiometer Suite
VINTERSOL	Validation of INTERnational Satellites and study of Ozone Loss
VIS	VISible
VMR	Volume Mixing Ratio
VOC	Volatile Organic Compound
VOTE	Vortex Ozone Transport Experiment
VSLS	Very Short-Lived Substance
W	West
WACCM	Whole Atmosphere Community Climate Model
WCC	World Calibration Center
WCRP	World Climate Research Program
WG	Working Group
WIND	WIND Imaging Interferometer
WMO	World Meteorological Organization
WODC	World Ozone Data Center
WOUDC	World Ozone and Ultraviolet Data Center
YAG	Yttrium Aluminum Garnet
ZS	Zenith Sky
ZSW	Zentrum für Sonnenenergie- und Wasserstoff-Forschung

1

The traditional measurement of ozone concentration in the atmosphere

1.1 INTRODUCTION

Almost a century after Tyndall's discovery in 1859 that changes in the concentration of the atmospheric gases could bring about climate change, and almost 50 years after Arrhenius' calculation in 1896 of the global warming from human emissions of CO_2, a small group of scientists got together to increase international cooperation to a higher level in all areas of geophysics. They aimed to coordinate their data gathering and to persuade their governments to spend extra money on research. The result was the International Geophysical Year (IGY) of 1957–1958.

Under the IGY umbrella, the United States and the former Soviet Union officially announced plans to launch artificial satellites as cooperative scientific experiments. The era of satellite remote sensing began on October 4, 1957, when the former Soviet Union launched Sputnik 1, the world's first artificial satellite (a 55 cm diameter sphere that weighed 83 kg with four antennae coming off it) (Figure 1.1). It circled the Earth once every 96 min and transmitted radio signals that could be received on Earth, providing the first space views of our planet's surface and atmosphere. On November 3, 1957, a second satellite, Sputnik 2, was launched which carried a dog named Laika. The United States launched its first satellite, Explorer 1, on January 31, 1958, and its second, Vanguard 1, on March 17, 1958.

Among the principal priorities of IGY efforts was the study of atmospheric ozone (*http://nsidc.org/data/docs/noaa/g02184_station_alpha/index.html*). Ozone (an allotrope of oxygen) is produced naturally in the stratosphere and drifts downward by mixing processes, producing a maximum in its concentration near 25 km. It is also destroyed naturally by absorbing ultraviolet (UV) radiation (0.2–0.3 μm) and through chemical reactions with other atmospheric components. The natural balance between production and destruction leads to maintenance of the stratospheric ozone layer which, however, is now subject to substantial anthropogenic impacts. Much

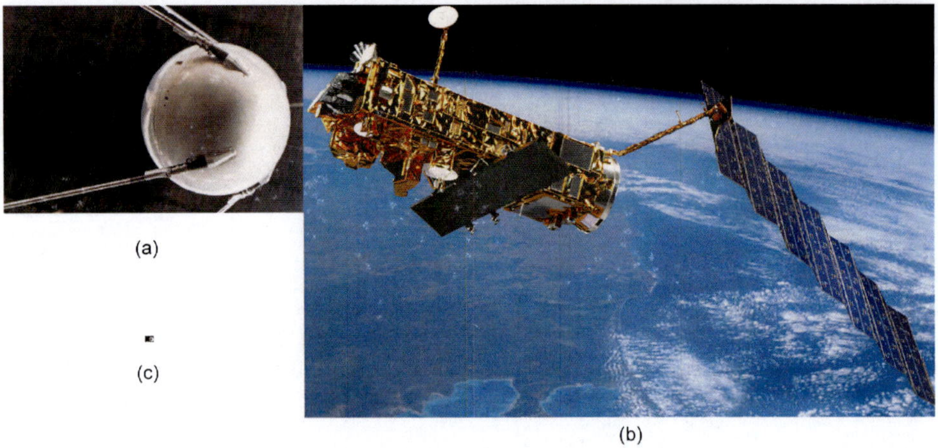

Figure 1.1. (a) Sputnik 1 (a basketball in diameter with mass 83 kg) (*source*: *http:// www.honors.umd.edu/HONR269J/projects/castell.html*). (b) ENVISAT with various instruments on board. In its orbit configuration, its dimensions are 26 m × 10 m × 65 m, and its total mass is 8,211 kg (*source*: ESA). (c) Sputnik on the same scale as ENVISAT in (b).

smaller amounts of ozone, of the order of 10%, are generated at the Earth's surface and enter the troposphere (see Figure 1.2).

Ozone is so rare in the atmosphere (there are only about three ozone molecules per 10 million air molecules) that if it was brought down to the Earth's surface and compressed, it would form a layer about as thick as a small coin. The thickness of this layer at standard temperature and pressure (STP) is used as a measure of total

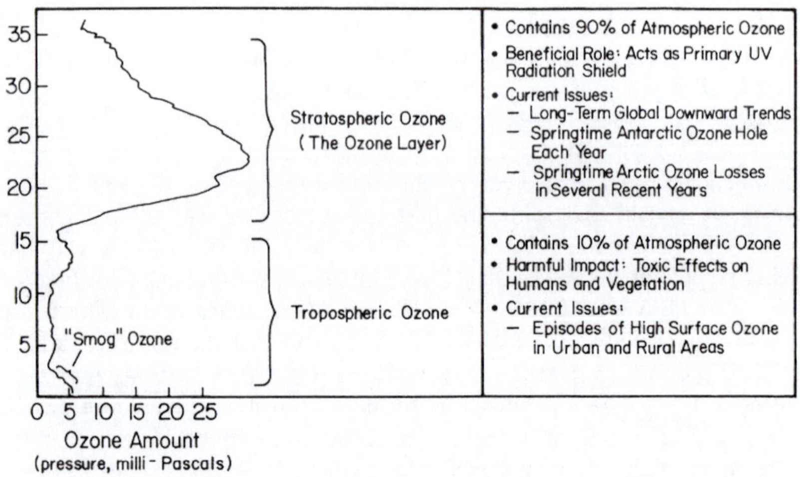

Figure 1.2. The vertical distribution of atmospheric ozone: partial pressure of ozone (mPa) and height (km) (*source*: Earth Observing System (EOS) Science Plan, Chapter 7, Ozone and Stratospheric Chemistry, *http://eospso.gsfc.nasa.gov/science_plan/*).

ozone content or total ozone column (TOC, or sometimes referred to as TOZ). The value of the TOC is usually expressed in Dobson units (DU) and one Dobson unit corresponds to a thickness of ozone of 10^{-2} mm, or 10 μm, at STP. Typical values of TOC are of the order of a few hundred Dobson units.

The ozone molecules in the stratosphere and those in the troposphere are, of course, chemically identical. However, they have very different roles in the atmosphere and very different effects on humans and other living beings. Although it is thin, the stratospheric ozone layer prevents harmful solar energy (the biologically damaging ultraviolet radiation) from reaching the Earth's surface. The absorption of ultraviolet radiation by ozone in the stratosphere provides an important source of heat, so that ozone thus plays a key role in the temperature structure of the Earth's atmosphere. It also strongly absorbs infrared thermal radiation (heating the stratosphere) and plays a major role in photochemistry (involved in the nitrogen, carbon, hydrogen, oxygen, and halogen chemical cycles). Various human activities, on the one hand, decrease the amount of ozone in the stratosphere, as was first pointed out in the early 1970s. Stratospheric ozone therefore is sometimes referred to as "good ozone", since it plays a beneficial role by absorbing most of the biologically damaging ultraviolet sunlight (called UV-B), allowing only a small amount to reach the Earth's surface.

On the other hand, tropospheric ozone, in cases of high concentration, is toxic to living systems. At the Earth's surface, ozone comes into direct contact with life-forms and displays its destructive side; hence, tropospheric ozone is often called "bad ozone". Because ozone reacts strongly with other molecules, high levels of ozone are toxic to living systems. Several studies have documented the harmful effects of ozone on crop production, forest growth, and human health. The substantial negative effects of surface-level tropospheric ozone from this direct toxicity contrast with the benefits of the additional filtering of UV-B radiation that it provides.

The dual role of ozone leads to two separate environmental issues. There is concern both about decreases of ozone in the stratosphere and there is concern about increases of ozone in the troposphere. Near-surface ozone is a key component of photochemical "smog", a familiar problem in the atmospheres of many cities around the world. Higher amounts of surface-level ozone are increasingly being observed in rural areas as well. There is some transport of ozone between the troposphere and the stratosphere (see Section 4.3.1), but this is quite limited and it does not lead to significant compensation between these two effects.

Observations conducted in recent years have led to a deeper understanding of the anthropogenic and natural causes of changes in the chemical composition of the atmosphere and their association with ozone depletion in the stratosphere. These observations and laboratory investigations indicate that the by-products of human-made chlorofluorocarbons (CFCs) and halons are the main causes of stratospheric ozone decline, especially in the Antarctic. This global-scale decline started in the 1970s and continued with statistically significant rates except over the 20°N–20°S tropical belt. Realization of the importance of this decline led to the formulation of the Montreal Protocol in 1987 (see Chapter 5).

Global stratospheric ozone levels are now no longer declining as fast as they were from the late 1970s until the mid-1990s, and some increases in ozone have recently been observed. Specifically, averaged between 60°S and 60°N, total column ozone is projected to increase in all models between 2000 and 2020, with most of the increase of 1 to 2.5% occurring after 2010. These improvements in the ozone layer have occurred during a period when stratospheric halogen abundances reached their peak and started to decline. These declining halogen abundances clearly reflect the success of the Montreal Protocol and its Amendments and Adjustments in controlling the global production and consumption of ozone-depleting substances (ODSs).

The reliable assessment of ozone depletion in the stratosphere and ozone increase in the troposphere depends on accurate monitoring of long-term changes in global TOC and in ozone vertical concentration profiles (OVPs). Therefore, there is a need for continuous and long-term collection of reliable ozone data. Ground-based measurements and ozonesonde data constitute a key component of the Global Ozone Network, both on their own account and by providing ground-based calibration for satellite-based instruments. However, the uneven geographical distribution of the existing ground-based network gives rise to a spatial sampling error when attempts are made to determine the global distribution of the atmospheric ozone content. The benefit of ground-based instruments is that it is easy to maintain them in good condition, including regularly checking their calibration, while satellite-based instruments provide better temporospatial coverage and resolution.

This chapter describes the instrumentation and measurement techniques that have been used for the measurement of the TOC or the OVP. Additional information in this respect has been presented elsewhere (WMO, 1990, 1995, 1999, 2007, 2010).

1.1.1 Observations of the total ozone column

M.A. Cornu in 1879 suggested that the cause of the rather sharp limitation of the UV part of the solar spectrum (as observed at the ground) was absorption by the atmosphere. In 1880 W.N. Hartley suggested that ozone was responsible for this absorption. C. Fabry and H. Buisson, in 1912, measured the absorption coefficients of ozone and concluded that there was about 5 mm (500 DU) of ozone in the vertical thickness of the atmosphere. The same researchers in 1920 made measurements of the amount of ozone at Marseilles by using a special spectrograph (Fabry and Buisson, 1913).

In the early 1920s G.M.B. Dobson initiated the first regular measurements of atmospheric ozone using spectrographic instruments (Dobson, 1931; Walshaw, 1989). Today, ozone observations are carried out daily by using different kinds of instrumentation, from early ground-based devices and balloons to the more advanced devices launched on rockets and flown on aircraft, rockets, and satellites (see Sections 1.2–1.8, 2.1–2.6).

At the time of the International Geophysical Year (in 1957/1958) the WMO assumed responsibility for establishing standard international procedures to ensure the uniformity and high quality of ozone observations conducted at ground-based ozone stations of which there are more than 160 at present (WMO, 2008). These standard procedures have been used in the WMO Global Ozone Observing System (GO_3OS) as well as the Background Air Pollution Monitoring Network (BAPoMON), which mainly constitute the WMO Global Atmospheric Watch (GAW). Since 1989 the latter also serves as an early-warning system to detect changes in atmospheric concentrations of greenhouse gases (WMO, 1990).

In connection with the International Geophysical Year, the World Ozone Data Center (WODC) was initiated and first located at the WMO Secretariat. From 1961 it has been run in Toronto for WMO, first by the Meteorological Service of Canada and, since 1972, by its successor the Atmospheric Environment Service (AES) of Environment Canada.

More than 40 GO_3OS stations (using filter instruments) making long-term observations supplemented by another 100 stations and satellites over the last few decades (data held by the World Ozone and Ultraviolet Data Centre, WOUDC), show that TOC decline has been observed over mid-latitudes and polar latitudes, which is approximately at a 10% level relative to the TOC of 1950s and 1960s. This will be discussed in detail in Section 4.1

We shall now briefly describe the basic instrumentation and the basic measurement techniques used for the measurement of TOC.

1.2 GROUND-BASED INSTRUMENTATION FOR TOC OBSERVATIONS

TOC was measured from the ground at a few sites in the late 1920s and early 1930s by using the Dobson spectrophotometer (initially consisting of a filter and a Fery prism (Miller *et al.*, 1949) and later on a photoelectric sensor), the development of which was based on earlier research with the Fabry–Buisson device which used a double-spectrograph with photographic plates (Dobson, 1968; Fabry and Buisson, 1921). In the 1920s, TOC was measured mainly by the two research groups around Fabry and Dobson at several sites in Europe with the aim to reveal the association between TOC variability and atmospheric circulation (Dobson and Harrison, 1926). A few years later, Dobson re-distributed his observational sites at more widely separated sites in order to study the geographical variability of TOC. Figure 1.3 shows seasonal TOC variability measured at Table Mountain, California, for the 1920s and 1930s and data from different instruments (Dobson spectrophotograph, spectrobolometer, stellar spectra from nearby Mount Wilson; see Bronnimann, 2005; Griffin, 2006).

The majority of the ground-based techniques that were used to measure TOC during the 20th century were mainly based on observations in the 300–340 nm band, where TOC is the most important absorber. Nowadays, three main types of primary instruments are used for routine TOC observations from the ground: the Dobson spectrophotometer (Section 1.2), the Brewer spectrophotometer (Section 1.3), and the M-83/124/134 filter ozonometer (Section 1.4). All of them use the principle of differential absorption by ozone of solar UV radiation within the 300–460 nm band. Determination of TOC by these three instruments becomes feasible through measur-

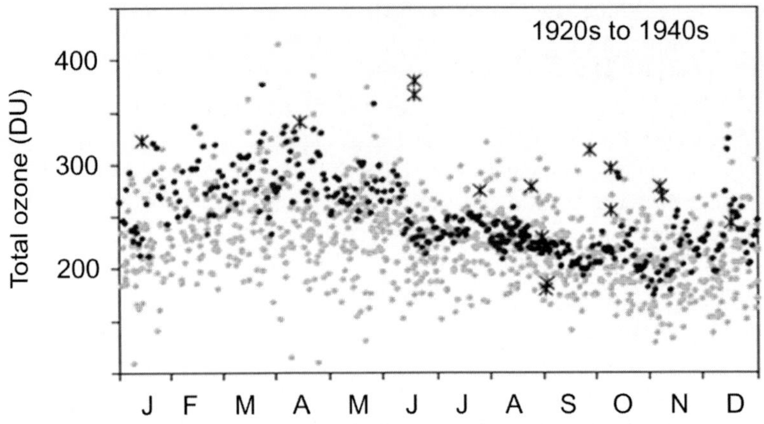

Figure 1.3. TOC at Table Mountain, California, as a function of season from three different TOC data sets from the 1920s to 1940s (black dots = Dobson Fery spectrograph, 1928–1929, Bronnimann, 2005; gray dots = spectrobolometer at 574 nm, 1925–1948, Bronnimann, 2005; stars = stellar spectra from nearby Mt. Wilson, 1935–1941, Griffin, 2006).

ing scattered solar radiation in the Hartley or Hartley–Huggins bands in the UV spectral region (Figure 1.4). Dobson and Brewer spectrophotometers can measure TOC with up to 99% accuracy, while filter ozonometers are less accurate (Bojkov *et al.*, 1988). Observations carried out with these instruments over longer time periods can also be used to measure OVP with approximately 5 km resolution and TOC with fairly high accuracy, by looking directly at the Sun or Moon.

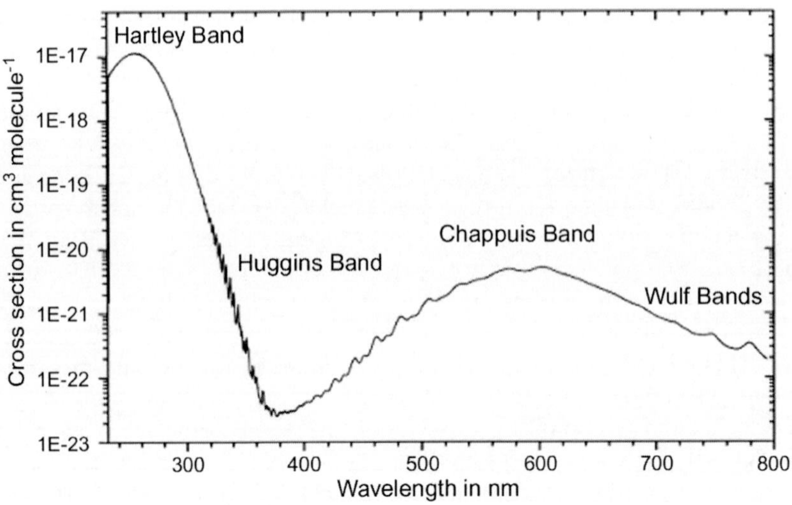

Figure 1.4. The O_3 spectrum at room temperature (Orphal, 2003).

In the following Sections 1.2, 1.3, 1.4 more detailed information about these basic instruments is presented. Secondary ground-based instruments will be described in Sections 1.5 and 1.6.

1.2.1 The Dobson ozone spectrophotometer

Dobson, Gordon Miller Bourne (1889–1976)
(*http://rammb.cira.colostate.edu/dev/hillger/modern.htm #dobson*)

The Dobson Spectrophotometer is the oldest device for measuring atmospheric TOC; it was developed in 1924 by G.M.B. Dobson. Today, the observations of TOC by this instrument constitute one of the longest geophysical measurement series in existence (TOC has been continuously observed over Arosa, Switzerland, since 1926). In addition, knowledge of the ozone layer prior to satellite measurements is derived from Dobson spectrophotometer records for more than 80 years.

The absorption spectrum of ozone in the UV region shows a large continuum from 200 to 300 nm, with a maximum near 250 nm, and this is called the Hartley band; there are some small oscillations which can be seen near the maximum (see Figure 1.4). At longer wavelengths (310–350 nm) the spectrum is dominated by more intense but also diffuse vibrational bands known as the Huggins bands. The instrument makes use of the rapid variation of the absorption cross-section of ozone for solar ultraviolet radiation, as a function of wavelength, in the Huggins bands. The instrument uses a quartz double-monochromator (see Figure 1.5). One monochromator is used to disperse the radiation and the second is used to reject interfering scattered radiation (stray light) (Dobson, 1931, 1957a, b; Dobson and Normand, 1957a, b, 1962).

The operation of the instrument involves measuring the difference between the intensity of solar UV at certain wavelength pairs in the Huggins bands. These pairs are selected in such a manner that the difference between the ozone absorption coefficient for each pair is as large as possible, while the difference between the corresponding wavelengths is as small as possible. By using one pair of wavelengths a differential measurement is made. One wavelength is significantly absorbed by ozone and the other passes through a variable optical attenuator (wedge). The idea of this is to produce a null measurement (i.e., the absorption of the second wavelength by the optical attenuator within the instrument is equal to the absorption in the atmosphere of the first wavelength of the pair).

The measured absorption combined with the extraterrestrial constant and the ozone absorption spectrum indicates the total ozone amount. Consider a case when the Sun is at an angle θ from the vertical (i.e., the zenith angle is θ). If we assume a horizontally stratified atmosphere and neglect scattering then the solar spectral

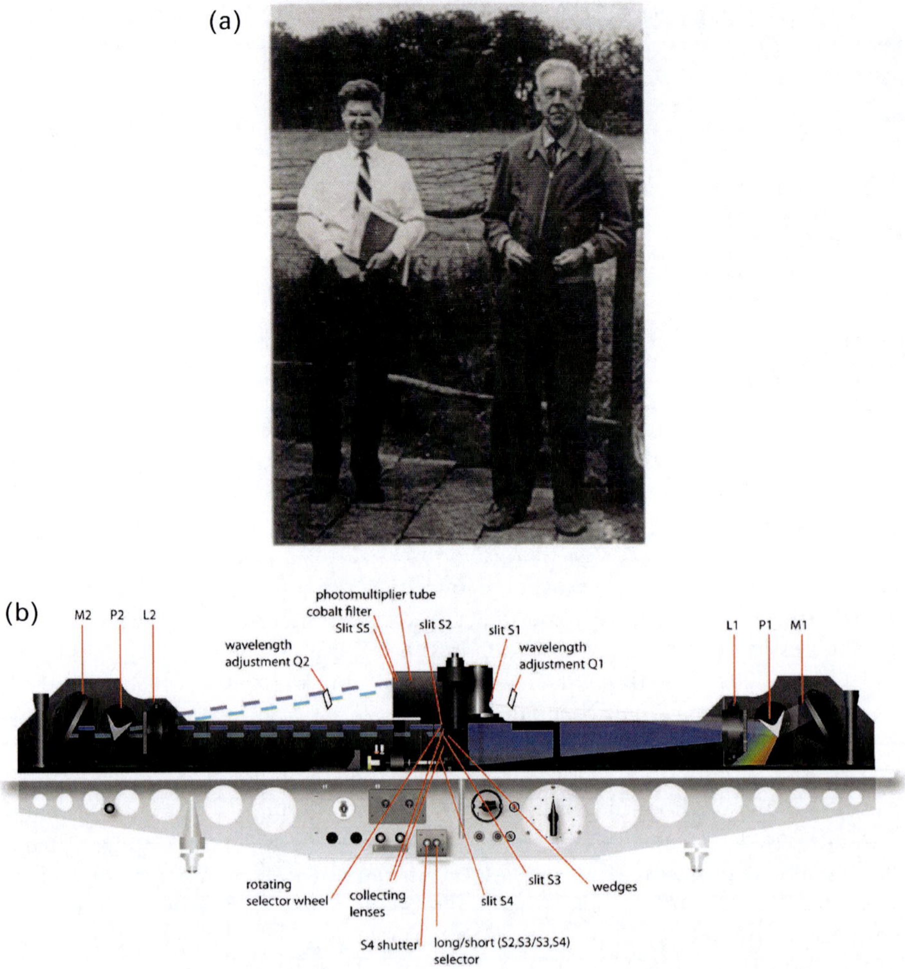

Figure 1.5. (a) Prof. G.M.B. Dobson (on the right) with his co-worker Dr. D. Walshaw around 1970 (Walshaw, 1989). (b) Dobson spectrophotometer optical diagram, L stands for lens, P stands for prism, and M stands for mirror (WMO, 2008).

radiance at frequency ν at the ground is:

$$L_\nu = L_{\nu\infty} \exp(-\chi_{\nu 0} \sec \theta) \tag{1.1}$$

where

$$\chi_{\nu 0} = k_\nu \int_0^\infty \rho(z')\, dz' = k_\nu m_0 \tag{1.2}$$

is the optical path between the top of the atmosphere and the ground; k_ν is the absorption coefficient; and m_0 is the total mass of absorbing material (i.e., the ozone) in a vertical column of unit area. From these two equations we obtain

$$L_\nu = L_{\nu\infty} \exp(-k_\nu m_0 \sec\theta) \tag{1.3}$$

Similarly, for a second frequency at the same zenith angle:

$$L_{\nu'} = L_{\nu'\infty} \exp(-k_{\nu'} m_0 \sec\theta) \tag{1.4}$$

Then, if we divide and take logs we obtain

$$\ln\left(\frac{L_\nu}{L_{\nu'}}\right) = \ln\left(\frac{L_{\nu\infty}}{L_{\nu'\infty}}\right) + (k_{\nu'} - k_\nu) m_0 \sec\theta \tag{1.5}$$

m_0 can be determined from this equation since the two quantities on the left-hand side are measured, $\sec\theta$ is measured, and all the other quantities on the right-hand side of the equation are known. In practice it is better to take measurements for a number of different solar zenith angles and then plot the measured values of the left-hand side against $\sec\theta$; the best fit to the total mass of ozone in the vertical column can then be found from the slope, which is $(k_{\nu'} - k_\nu)m_0$.

Equation (1.3) forms the basis of the Langley (extrapolation) method, which is a method for measuring the Sun's radiance with ground-based instrumentation, thereby removing the effect of the atmosphere. In the present context it is used to determine $L_{\nu\infty}$ and $L_{\nu'\infty}$ in equation (1.5). The Langley method is based on repeated measurements with a Sun photometer operated at a given location for a cloudless morning or afternoon, as the Sun moves across the sky. It is named after the American astronomer and physicist Samuel Pierpont Langley. If we take (natural) logarithms of equation (1.3) and revert to the total radiances I_0 and I integrated over all frequencies, we obtain

$$\ln(L_\nu) = \ln(L_{\nu\infty}) - k_\nu m_0 \sec\theta \tag{1.6}$$

If one assumes that the atmospheric conditions are constant (i.e., $k_\nu m_0$ is constant) and measurements of L_ν are taken at various times of day (i.e., at various solar zenith angles θ), one can plot $\ln(L_\nu)$ against $\sec\theta$ and then the intercept on the y axis will be the value of $\ln(L_{\nu\infty})$. This method is very widely used in processing data from Dobson spectrophotometers (and other ground-based instruments) for determining TOC.

The main limitation of the Langley method is the requirement of atmospheric stability and therefore a constant value of $k_\nu m_0$. This requirement can be fulfilled only under particular conditions, since the atmosphere is continuously changing. The required conditions are in particular the absence of clouds and the absence of variations in the atmospheric aerosol along the optical path. Since aerosols tend to be more concentrated at low altitude, Langley extrapolation is often performed at high mountain sites. Data from the NASA Glenn Research Center indicates that Langley plot accuracy is improved if the data are taken above the tropopause. Stray light is dealt with by allowing $L_{\nu\infty}$ and $L_{\nu'\infty}$ to be instrument dependent (see Section 1.2.4).There are very few places on the globe where the requirement of atmospheric stability is satisfied sufficiently often. Therefore the original version of the Langley method is not applicable for most observation stations. An attempt was made to overcome the disadvantages of the Langley method by taking a statistical approach and using the hypothesis that the effect of daily atmospheric variability may be minimized by averaging over a sufficiently large number of daily observations. However, a variational method, which is an alternative to the Langley method, has been proposed and tested with data from the Minsk ozone station (Krasouski and Liudchik, 2004; Liudchik, 2009; Liudchik and Krasouski, 2002, 2010).

One of the main factors affecting the accuracy of the determination of TOC measurements using a Dobson spectrophotometer arises from the stray light contribution, mainly to the shorter wavelength band of the wavelength pairs measured by this instrument. By using two pairs of wavelengths the scattering effects can essentially be filtered out.

The wavelength pairs most frequently used for precise observations are the standard double-pair wavelengths designated AD, while the pair CC′ is used less frequently for less precise observations (A: 305.5 and 325.4 nm; C: 311.4 and 332.4 nm; D: 317.6 and 339.8 nm; C′: 332.4 and 453.6 nm). Direct Sun AD observations generally provide the most precise measurements if the secant of the zenith angle, $\sec\theta$, is less than 3.

Dobson spectrophotometers can be operated manually, semi-automatically, or automatically (Evans *et al.*, 1994); they can also determine the vertical ozone distribution (OVP) by using the Umkehr technique (described in Sections 1.6.2 and 1.6.3). For instance, Dobson instruments D015 and D101 at the Light Climatic Observatory (LKO) in Arosa, Switzerland, are partially automated and located inside a spectrodome performing Umkehr observations in a completely automated way (Hoegger *et al.*, 1992).

1.2.2 Intercomparison of Dobson spectrophotometers

Nowadays, over 100 Dobson spectrophotometers constitute the global primary ground-based network of stations for monitoring atmospheric concentrations of ozone. This network provides the most reliable source of ozone data and is a key component of the GO_3OS, which is now part of the GAW of the WMO.

The Dobson network also serves a valuable function in providing ground-based calibration and validation data for satellite-based measurements and various remote-sensing instruments, which are used to measure atmospheric ozone (see Section 3.2).

The GAW/WMO has identified a World Calibration Center (WCC) and Regional Calibration Centers (RCCs) for ozone. The World Calibration Center is located at the World Dobson Spectrophotometer Central Laboratory (operated by the Climate Monitoring and Diagnostics Laboratory (CMDL) of the U.S. National Oceanic and Atmospheric Administration, NOAA). Since 1960 this Laboratory has organized intercomparisons of Dobson instruments, in order to maintain calibration within the network. These field intercomparisons (jointly organized by WMO and the International Ozone Commission) are usually accompanied by laboratory assessments and internal adjustments.

The Laboratory at Boulder also serves as the Regional Calibration Center for North America (*http://www.cmdl.noaa.gov/ozwv/dobson*). Other Regional Calibration Centers are:

— Regional Dobson Calibration Center for Africa
South African Weather Bureau, Pretoria, South Africa
(*coetzee@sawb.gov.za*)
— Regional Dobson Calibration Center (RDCC) for Asia
Japan Meteorological Agency (JMA), Tokyo, Japan
(*http://gaw.kishou.go.jp/wcc.html*)
— Regional Dobson Calibration Center for Australia
Bureau of Meteorology, Melbourne, Australia
(*j.easson@bom.gov.au*)
— Regional Dobson Calibration Center for Europe
Meteorological Observatory, Hohenpeissenberg, Germany
(*http://www.dwd.de/ozone*)
(*http://www.chmi.cz/meteo/ozon/dobsonweb/welcome.htm*)
— Regional Dobson Calibration Center for South America
Observatorio Central, Buenos Aires, Argentina
(*maximo@meteofa.mil.ar*).

In 1962 the Dobson spectrophotometer No. 83 (D083) was established as the U.S. standard for TOC measurements. It is maintained at the World Dobson Spectrophotometer Central Laboratory in Boulder, Colorado. This instrument was initially calibrated relative to Dobson spectrophotometer D072 (Launder, New Zealand) and additional calibrations were performed nine times between 1972 and 1987, using direct Sun observations (Komhyr *et al.*, 1989a). Since the early 1960s, almost all Dobson instruments have been calibrated either directly or indirectly, relative to instrument D083. Since 1974, most of the existing Dobson instruments have been modernized and calibrated by direct intercomparison with the WMO-designated World Primary Standard Dobson spectrophotometer D083. A secondary standard D065 is also maintained by the Laboratory in Boulder.

It should be noted that Komhyr (1980) suggested that traveling standard lamps, properly calibrated and adjusted, could also be used for basic error detection and stability testing of Dobson instruments. These kinds of systematic intercomparisons of Dobson instruments against traveling standard lamps have been undertaken since 1984 (Komhyr *et al.*, 1989a). In this context, satellite instruments providing ozone observations (see Chapter 2) are often used as the traveling reference instrument, in order to identify those field Dobson instruments that are systematically in error (Bojkov *et al.*, 1988). According to Basher (1995), WMO traveling standard lamp tests showed that more than 75% of Dobson instruments differ from the standard lamp by less than $\pm 1.5\%$.

To extend the calibration further, each participant in the international intercomparison gathers together a set of regional instruments, as well as the World Primary Standard Dobson Instrument or Regional Secondary Standard Instrument (calibrated in Boulder) for a period of 3 weeks in summer at a site that has relatively clear skies. The Meteorological Observatory of Hohenpeissenberg (MOHp) in Germany was designated to act as the European Ozone Center taking on the role of a Dobson Calibration Center (DCC). Several Dobson spectrophotometers have been used for total ozone observations at Arosa, Switzerland, beginning in 1926 and providing the world's longest series of such observations. From the mid-1950s to 1992 Dobson instrument D015 was calibrated by the statistical Langley plot method. In 1986 the calibration of another Dobson spectrometer at Arosa, D101, was changed by intercomparison with the primary world Dobson instrument D083. A statistical model based on simultaneous measurements of D101 and D015 for the period from 1987 to 1990 was then used to obtain a TOC series of data in line with the primary Dobson spectrophotometer. The normal MOHp (DCC) support in its role as the European Ozone Center includes among other things:

- Maintenance of each neighboring Dobson station will be guaranteed within the bounds of the usual DCC duties: regular (at least once every 4 years) check of the instrument's state and calibration as well as a kind of "first aid", if some trouble occurs with any instrument.
- Re-evaluation of a station's Dobson record should be done by its own staff, because they know the history of the instrument and have easy access to all necessary information, like test records and documents.

The DCC can give recommendations and advice on what to do and how to do it. The procedure and methods of re-evaluation are given in the NOAA Technical Report NESDIS 74 *Dobson Data Re-evaluation Handbook* and in WMO-GAW Report No. 29 *Handbook for Dobson Ozone Data Re-evaluation* (1993).

An intercomparison campaign was carried out at Arosa in Switzerland from July 22 to August 10, 1990 (see Figure 1.6). The main tasks of the campaign were the following:

- Technical inspection of the instrument
- Checking the electronic equipment
- Inspection, cleaning, and adjustment of the optical parts
- Wedge calibration test
- New Q-table (lever settings according to instrument temperature)
- New N-table (conversion of dial readings R into N for total ozone calculations).

The results of this intercomparison are shown in Table 1.1. An initial intercomparison of three reference instruments was carried out in summer 1999 at MOHp in collaboration with Hradec Kralove (Czech Republic). Since 2004 a series of intercomparisons has been held to evaluate the calibration of Dobson instruments in the international network. These intercomparisons include WMO-sponsored international campaigns and the ongoing intercomparisons organized by the Regional Dobson Calibration Centers for Europe (RDCC/E) at Hohenpeissenberg, and for Asia (RDCC/A) by the Japan Meteorological Agency's Aerological Obser-

Table 1.1. Results of intercomparison of Dobson instruments (July 22 to August 10, 1990) at Arosa, Switzerland. ΔX_{AD} is the percentage deviation from the U.S.A. standard (X = total amount of ozone expressed in Dobson Units; Basher, 1995).

Instrument No.	*Country–Location*	*Last intercomparison*	*Initial* ΔX_{AD} (%)	*Final* ΔX_{AD} (%)
13	Portugal–Lisbon	1987	0.28	−0.16
14	Norway–Tromso	1977	−0.99	−0.19
15	Switzerland–Arosa		-4.96	0.82
40	Belgium–Uccle	1986	−0.6	−0.15
41	United Kingdom	1985	−0.59	0.15
50	Iceland–Reykjavik	1977	−0.16	0.05
64	Germany–Potsdam	1977	0.22	−0.09
74	Czech–Hradek Kralove	1986	−0.66	−0.07
84	Poland–Belsk	1986	0.35	−0.05
92	Denmark–Greenland	1986	−0.5	−0.05
101	Switzerland–Arosa	1986	0.37	0.09
104	Germany–Hohenpeissenberg	1986	−2.4	−0.04
107	U.S.S.R.–Moscow	1988	−0.95	
110	Hungary–Budapest	1988	0.93	−0.24
118	Greece–Athens	Never[a]	−2.27	0.03
120	Spain	1989	−1.92	
121	Rumania–Bucharest	1988	−0.06	0.04
65	U.S.A.–standard	1990	0	0

[a] Has never been used or calibrated before this intercomparison.

vatory, Tokyo. In January 2006, four instruments, including the Australian and WMO region V standard D105, were compared with D083 at the Australian Bureau of Meteorology's facility near Melbourne. In March 2006, five instruments were compared with the Region II standard D116 at Tsukuba, Japan. In November–January 2006–2007, eight instruments were compared with D065 at the Argentine National Weather Service Villa Ortuzar Observatory in Buenos Aires. In April 2008, nine instruments were compared with D065 at the South African Weather

Service's facilities at Irene. During this period the RDCC/E and RDCC/A held intercomparisons for their respective regions. The instruments involved in these various campaigns comprise the majority of the operational Dobson spectrophotometers in the WMO Global Ozone Observing System. In general, the calibration of the instruments was good, within 1% of the standard.

As an example of the intercomparisons of Dobson spectrophotometers, we consider in some detail the case of Dobson spectrophotometer D118, which is based at Athens University, Greece, with various world standard instruments. More information about the Dobson spectrophotometer, in general, and the Athens Dobson Station (WMO Station No. 293), in particular, is given by Varotsos and Cracknell (1993, 1994). Intercomparison of the daily column ozone observations obtained by satellite instruments and Dobson spectrophotometer D118 has shown that satellite data compare well with the D118 measurements (Varotsos and Cracknell, 1994, 2007). It should be noted that the Athens location has already been suggested in the literature as representative for column ozone variations not only of the Mediterranean region but also of the entire mid-latitude belt in the Northern Hemisphere (Chandra and Varotsos, 1995). The Athens Dobson spectrophotometer, D118, has participated in three international intercomparison campaigns since it was installed at the Athens University campus; the first was the campaign in Arosa in 1990 mentioned above, the second was in Kalavryta (Greece) in 1997, and the third was in Hohenpeissenberg in 2006. The first two campaigns were organized by the World Dobson Calibration Center at the Climate Monitoring and Diagnostics Laboratory (CMDL), National Oceanic and Atmospheric Administration (NOAA) in Boulder, Colorado, which maintains the world primary D083 and secondary Standard D065 Dobson spectrophotometers. The third campaign was organized by the Regional Dobson Calibration Center, Europe (RDCC/E) at Hohenpeissenberg, Germany, which maintains the Regional Standard D064 Dobson spectrophotometer.

The whole experiment was divided into two parts. In the first part, after some initial mercury and standard lamp tests, the necessary maintenance of the optical, mechanical, and electronic sub-systems was performed. In the second part the Dobson D118 was checked against the World Secondary Standard Dobson Instrument D065 to determine the existing calibration level. During this part, both spectrophotometers made observations of total ozone at the same time, under clear-sky conditions. For each instrument the quantity N_i was determined where

$$N_i = \log\left[\frac{I_{0i}}{I'_{0i}}\right] - \log\left[\frac{I_i}{I'_i}\right] \tag{1.7}$$

I_{0i} and I'_{0i} are the intensities outside the atmosphere of solar radiation at the short and long wavelengths, respectively, of the A, C, and D wavelength pairs; I_i and I'_i are the measured intensities at the ground of solar radiation at the short and long wavelengths, respectively; and $i = \text{A, C, D}$. The parameter $\mu = \cos\theta$ expresses the relative path length of sunlight, through the ozone layer, assumed to be on average at the 22 km altitude from the Earth's surface. When $z = 0$ then $\mu = 1$.

The experimental results obtained from the intercomparison campaign at Arosa showed that TOC observations made by Dobson spectrophotometer D118 had a deviation of 0.03% compared with TOC observations made by the Standard Dobson D065 (see photo in Figure 1.6) while μ ranged from 1.15 to 3.2. It should be clarified here that X_{AD} stands for the total ozone amount (X) that was observed by using the A and D wavelengths. ΔX_{AD} is the difference in X_{AD} between two observations on the same morning (or afternoon) taken on high and low Sun, defining ΔX_{AD} as positive when the value nearer noon is the greater. During a calibration campaign the instrument before its calibration measures the "Initial ΔX_{AD}", whilst after its calibration it measures the "Final ΔX_{AD}".

The second WMO International Intercomparison of the Dobson spectrophotometers took place in the Stiga Ski Field, Kalavryta, Greece, in July–August 1997. This intercomparison of Dobson instruments (SSF/IC) was sponsored by the WMO and the University of Athens (Ozone and Solar Radiation Research Team, OSORT) and assisted by the NOAA Climate Monitoring and Diagnostics Laboratory's World Dobson Calibration Center. The SSF/IC consisted of the intercomparison of Dobson Instruments with the World Secondary Standard Dobson Instrument D065. The procedure during this campaign was the same as at Arosa. Instruments and personnel from the Czech Republic, Egypt, Germany, Greece, Romania, Russia, and the United States attended the intercomparison. In this context, long-term quality-controlled ozone data, obtained within the framework of the WMO/GAW, are essential for assessment of the state of the global ozone layer.

After the necessary checks and maintenance, the Athens Dobson D118 was ready for intercomparison with the secondary Standard Dobson D065. The final experiment took place on August 7, 1997. According to the results of this experiment the differences between N for Dobson D065 and Dobson D118 versus μ were similar to those presented in Figure 1.7. The fractions N_C/μ_C versus μ_C, N_D/μ_D versus μ_D, and N_A/μ_A versus μ_A were found to be similar to those presented in Figure 1.8. Specifically, the highest difference against the Standard Dobson (D065) for ADDSGQP observations, in the μ range 1.15 to 3.2, was −0.4% in total ozone. Here, ADDSGQP refers to the mode of measurement, where AD means that A and D pairs of wavelength were used, DS stands for Direct Sun ($\mu < 3.5$), and GQP means that the Ground Quartz Plate was placed on the inlet window.

A general recommendation was that total ozone observations and ozone vertical profile measurements made by the University of Athens (WMO station 293) should be continued, as these data play an important part in the understanding of both ozone climatology and ozone trends in the Eastern Mediterranean and southeastern Europe.

The third WMO International Intercomparison of Dobson spectrophotometers took place at the European Ozone Center at the Meteorological Observatory of Hohenpeissenberg, Germany (MOHp) from June 18 to July 11, 2006. During the first phase of the campaign, new U.S.-type electronics were installed, but normally this does not affect the calibration level. In addition, the shutter motor and its gear mechanism, responsible for adjusting the speed of the sector wheel, were replaced

Figure 1.6. (Top) Photo from the Dobson Intercomparison held in Arosa (July–August 1990). From left to the right, the observers are (1) U. Feister, (2) M. Degorska, (3) A. Ashbridge, (4) R. Hartmannsgruber, (5) J. Staehelin, (6) B. Thorkelson, (7) R. Evans, (8) C. Varotsos, (9), Observer (10), Z. Nagy (11) B. Rajewska-Wiech, (12) W. Komhyr, (13) K. Vanicek, (14) D. De Muer, (15) B. Grass, (16) M. Frimescu, (17) Observer, (18) V. Dorokhov, (19) A. Lapworth, (20) U. Kohler, (21) Observer, (22) D. Henriques, (23) Paul Eriksen, (24) Cacho A. Diaz, (25) G. Loska, (26) S. Trono, (27) H. De Backer, (28) K. Aeschbacher, (29) P. Plessing, (30) K. Henriksen, (31) Jim Kerr. (Bottom) Dr. D. Walshaw with Dr. C. Varotsos around 1985.

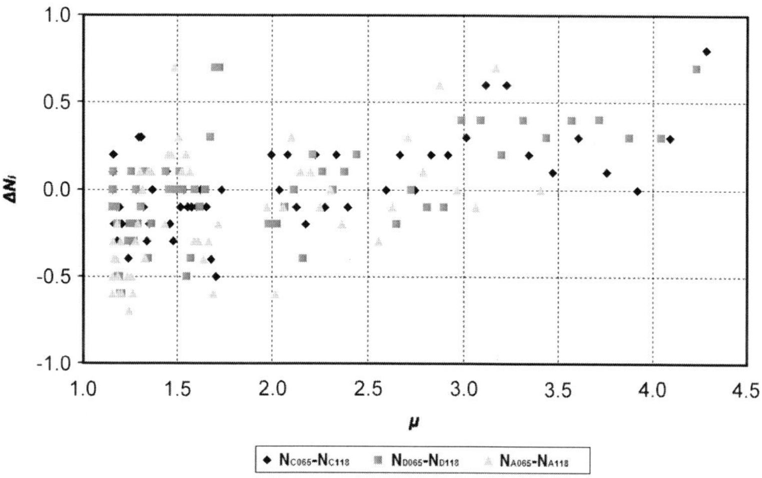

Figure 1.7. $\Delta N_i = (N_i$ for Dobson D065 $- N_i$ for Dobson D118), $i = $ A, C, D, on August 5, 1990, versus $\mu = \sec \theta$.

with a new motor free of the gear mechanism. The optical sub-system was checked, cleaned, and adjusted and the wedge calibration test was carried out.

In the next phase, the intercomparison of D118 against the Standard Dobson spectrophotometer of this station (D064) took place. The final experiment was conducted on July 11, 2006. Figure 1.9 shows the differences between N for Dobson D064 and Dobson D118 that arose from this experiment. Figures 1.10a–c illustrate the fractions N_C/μ_C versus μ_C, N_D/μ_D versus μ_D, and N_A/μ_A versus μ_A, respectively. From these figures it is evident that the highest difference against the Standard Dobson No. 064 ADDSGQP observations, in the μ range 1.15 to 3.2, was 0.6% in total ozone (Christodoulakis *et al.* 2008).

As mentioned already, the Dobson spectrophotometer is a mainstay instrument of the GO_3OS, and the Dobson stations are among those selected for trend analyses by the International Ozone Trends Panel. Thus, the accuracy and precision of the Dobson instruments used in the Dobson WMO Network must be known in order to perform trend analysis. According to the current experience, the calibration levels of many Dobson spectrophotometers have been changed at different times and therefore their records cannot be treated as entirely consistent. Therefore, the data from several Dobson stations were provisionally revised on a month-by-month basis. Nevertheless, the regional trends in TOC are non-uniform during a given time period.

As far as total ozone variability over the mid-latitude belt of the Northern Hemisphere (the most populated region of the planet) is concerned, it is well represented by total ozone variability over Athens (Chandra and Varotsos, 1995), which is very important for the validation of satellite observations. Thus maintenance of the calibration of the Dobson instrument in Athens is of great importance. In this context, the experimental results from the intercomparison campaign at Arosa

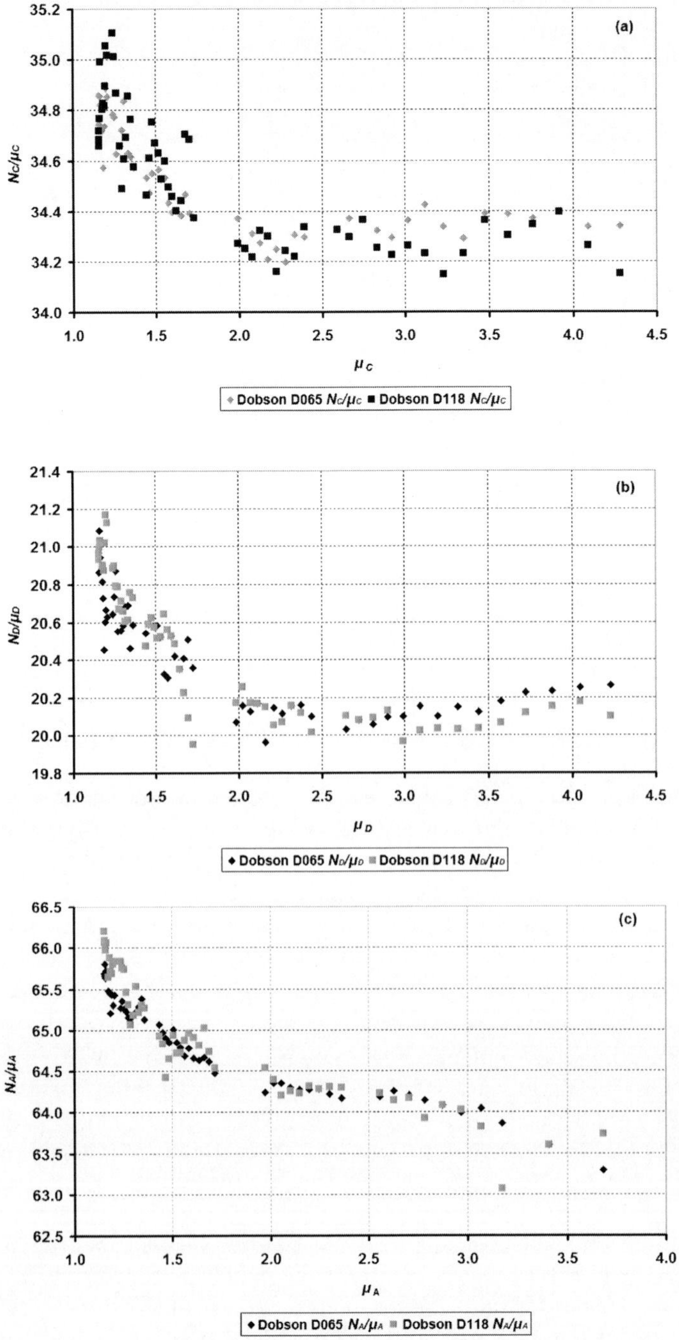

Figure 1.8. (a) N_C/μ_C versus μ_C for both D065 and D118, (b) and (c) the same but for the pairs D and A, respectively (Tzanis *et al.*, 2009b).

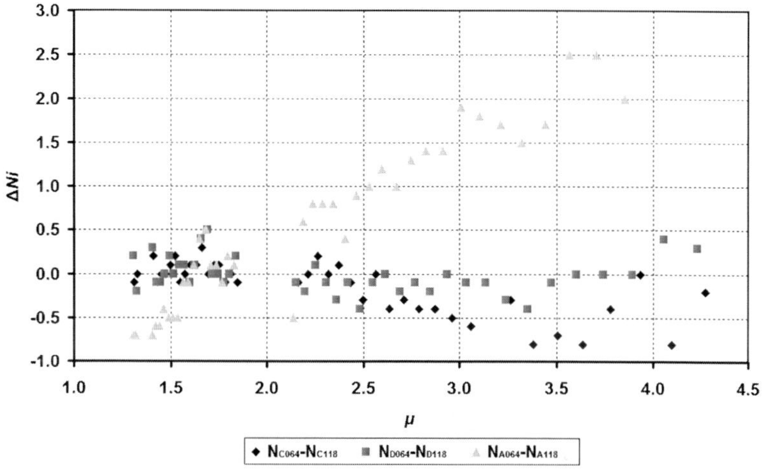

Figure 1.9. $\Delta N_i = (N_i$ for Dobson D064 $- N_i$ for Dobson D118), where $i =$ A, C, D on July 11, 2006, versus μ (Tzanis *et al.*, 2009b).

showed that TOC observations made by Dobson spectrophotometer D118 had a deviation of 0.03% from the TOC observations made by Standard Dobson D065, when μ ranged from 1.15 to 3.2. During the intercomparison campaign at Kalavryta, the highest difference against the Standard Dobson (D065) for ADDSGQP observations, in the μ range from 1.15 to 3.2, was −0.4% in total ozone. Finally, at Hohenpeissenberg the highest difference against the secondary Standard Dobson (D064), in the μ range from 1.15 to 3.2, was 0.6% in total ozone.

It should be noted that AD observations are regularly performed at noon to avoid μ values greater than 2.0 (where, according to Figure 1.10, the plausible deviation increases). It should be stressed that, according to the WMO, if the difference is less than 1%, then the test instrument is considered to be at the same calibration level as the Standard. Consequently, according to these results, Dobson spectrophotometer D118 installed at Athens, Greece, can be regarded as a reliable spectrophotometer and its measurements accurate and appropriate for further analysis.

Intercomparisons of Dobson spectrophotometers with other instruments will be discussed in some detail in Chapter 3.

1.2.3 Interference of SO_2 and NO_2 in Dobson TOC measurements

The precision of a well-calibrated Dobson spectrophotometer is about 0.5% and its accuracy is of the order of 2–4%. The instrument is also subject to additional errors in the presence of SO_2, NO_2, and aerosols (Komhyr and Evans, 1980; Varotsos *et al.*, 1995b). The sources of error in Dobson TOC measurement become significant where the Dobson instrument is placed in an area surrounded by a highly polluted atmo-

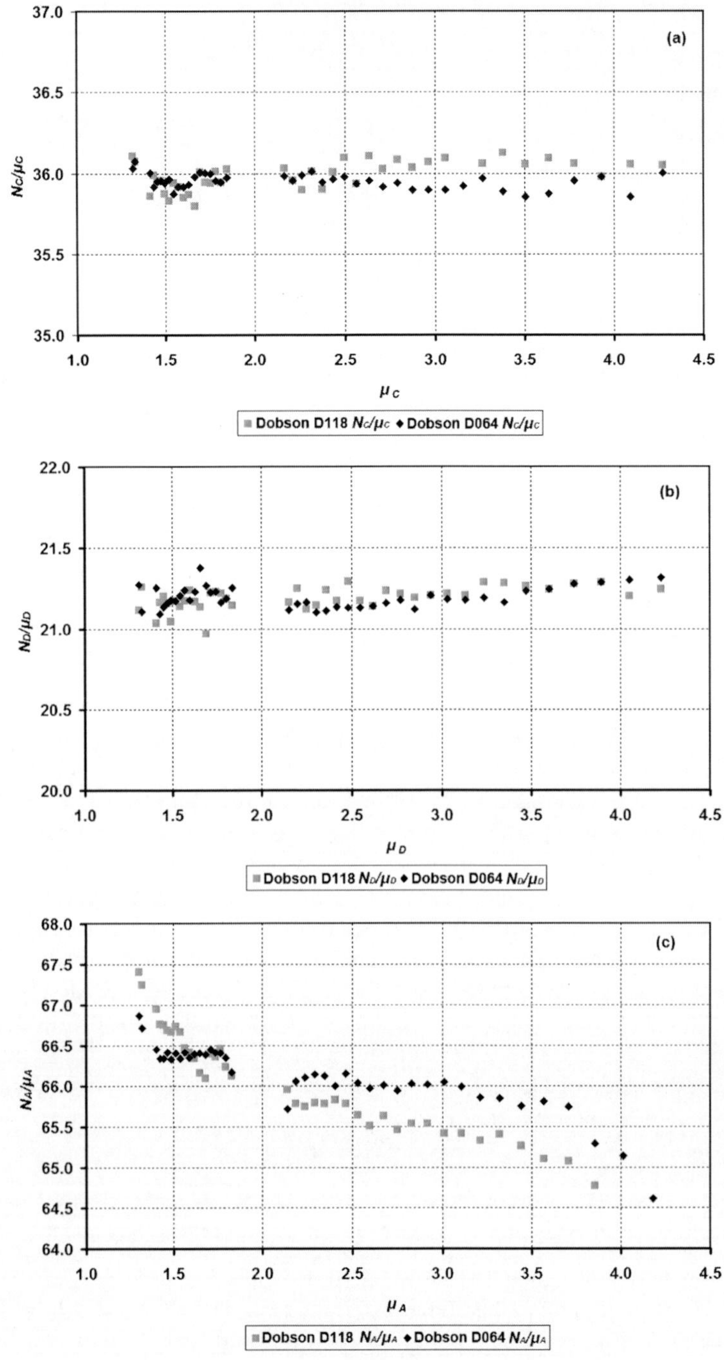

Figure 1.10. (a) N_C/μ_C versus μ_C for both D064 and D118, (b) and (c) the same but for the pairs D and A, respectively (Tzanis *et al.*, 2009b).

sphere. Two methods of identifying and correcting for SO_2 interference in Dobson measurements were suggested by Evans *et al.* (1980).

In the greater Athens area, where Dobson spectrophotometer D118 is installed, more than half of the industrial activities of the country are concentrated. Athens is situated in a relatively small basin of 450 km^2 surface area, surrounded by three mountains with heights of 100, 1,000 and 1,500 m. This creates ventilation problems, especially during meteorological conditions of low synoptic winds accompanied by high sunshine (anticyclonic conditions). Very high concentrations of urban pollutants, especially the photochemical ones (NO_2, O_3), are frequently observed under these conditions.

In order to find out the extent of the error in TOC measurements both in wintertime and summertime because of the presence of SO_2 and NO_2 in the Athenian atmosphere, the SO_2 and NO_2 concentration data of PERPA (National Service for Air Pollution Monitoring) at 13:00 LT were used. These data were obtained at the Patissia station located in the center of Athens and refer to the 2-year period 1989–1990.

Because the PERPA SO_2 and NO_2 concentration data are given in μg m^{-3}, the former must be multiplied by the factor 3.82×10^{-4} and the latter by the factor 5.32×10^{-4} to convert them into parts per million by volume (ppmv). The expression relating ppmv to μg m^{-3} is the following:

$$1\ \text{ppmv} = 10^{-3}[M/(M_i \rho)]\ \mu\text{g m}^{-3} \tag{1.8}$$

where $M = 28.97$ is the average molecular weight of the air; M_i is the molecular weight of the specific pollutant; and $\rho = 1.184\ \text{kg m}^{-3}$ is the air density for atmospheric pressure 1.013 hPa and 298.15 K.

Radiosonde temperature profiles as well as surface temperatures (obtained from the National Meteorological Service of Greece) were also used. All these data resulted from measurements made on a daily basis at 12:00 UTC at Hellinikon Airport, Athens, over 17 years (1974–1990).

The error in TOC measurement caused by the presence of absorbing species other than ozone is given by equation (1.9) (Komhyr and Evans, 1980):

$$\Delta x = x_1' - x_1 = \frac{\alpha_2 m}{\alpha_1 \mu} x_2 \tag{1.9}$$

where x_1' is the apparent TOC amount; x_1 is the true TOC amount, without the error owing to the presence of the specific absorbing species; and x_2 is the total amount of the absorbing species (all in atm-cm); α_1 is the ozone double-pair wavelength decimal absorption coefficient; α_2 is the corresponding coefficient for the absorbing species; m is the relative optical airmass; and μ is the relative optical ozone mass. As an approximation we can take $\mu = m$.

Thus the percentage error in true TOC amount x_1 because of the presence of the absorbing species will be:

$$\%\ \text{error in}\ x_1 = \frac{100\alpha_2}{x_1 \alpha_1} x_2 \tag{1.10}$$

The total amount x_2 of the absorbing species can be estimated in atm-cm (the Dobson Unit is defined to be 0.001 atm-cm) by the following equation, if we know its mixing ratio C in ppmv, as well as the height of the pollution mixing layer:

$$x_2 = \frac{1 - \exp[-(z/H)]}{1.251} C \tag{1.11}$$

where H is the scale height of the pollution mixing layer. It is given by the expression $H = R_d T/g_0$, where R_d is the gas constant for dry air ($= 287\,\mathrm{J\,kg^{-1}\,K^{-1}}$); T is the mean temperature of the mixing layer; and $g_0 = 9.80665\,\mathrm{m\,s^{-2}}$ is the global average of the acceleration due to gravity at mean sea level. $H = 8.49$ km for a layer at 17°C and z is the mixing layer's height.

Mixing heights were calculated from radiosonde temperature profiles and surface temperature measurements by the simple thermodynamic method of Holzworth (1972). This method has been widely used in several countries all over the world and seems to be appropriate for countries such as Greece, especially for the afternoon mixing height.

Potential temperature was calculated at the surface as well as in the layers at which the radiosonde data were available using the Poisson equation. If the potential temperature of a layer was greater than that of the surface, then the mixing height was defined at the bottom of that layer.

Mean total SO_2 amounts in the atmosphere over Athens for the periods December–March and May–August of the 2-year period 1989–1990 at local noon were estimated using equation (1.11). Afterwards, by using equation (1.10), the corresponding percentage errors in true TOC amount because of the presence of SO_2 for a value of true TOC amount of 0.300 atm-cm (300 DU) were calculated.

The results are shown in Figure 1.11. From this figure it is evident that the errors in true TOC amount due to the presence of SO_2 in Athens were higher between December and March than between May and August. This is due to the fact that SO_2 concentration is greater in winter than in summer since in winter the Athenian atmosphere is supplied with SO_2 from the central heating systems of buildings.

In order to estimate the percentage error in true TOC because of the presence of NO_2 the same method as used for SO_2 was followed. The results are presented in Figure 1.12. This figure shows that the errors in true TOC amount because of the presence of NO_2 in Athens are higher from May to August (not shown here), than from December to March. The explanation of this is that NO_2 concentration is greater in summer than in winter, since in summer there is greater photochemical pollution.

Svendby and Dahlback (2002) described the effect of SO_2 on measurements of TOC level in Oslo (59.9°N, 10.7°E), Norway, over a 20-year period from 1978 to 1998 using the Dobson spectrophotometer D056. When D056 ozone observations started in 1978 the SO_2 level in Oslo was significant and failure to make appropriate corrections may lead to false determinations of trends in TOC. For these Oslo data, without correcting for the effects of SO_2 the Oslo Dobson data showed a year-round TOC decrease of (-5.22 ± 0.63)% per decade from 1978 to 1998, while after correcting for the effect of the SO_2 the decrease was reduced to (-4.68 ± 0.63)% per decade.

(a)

(b)

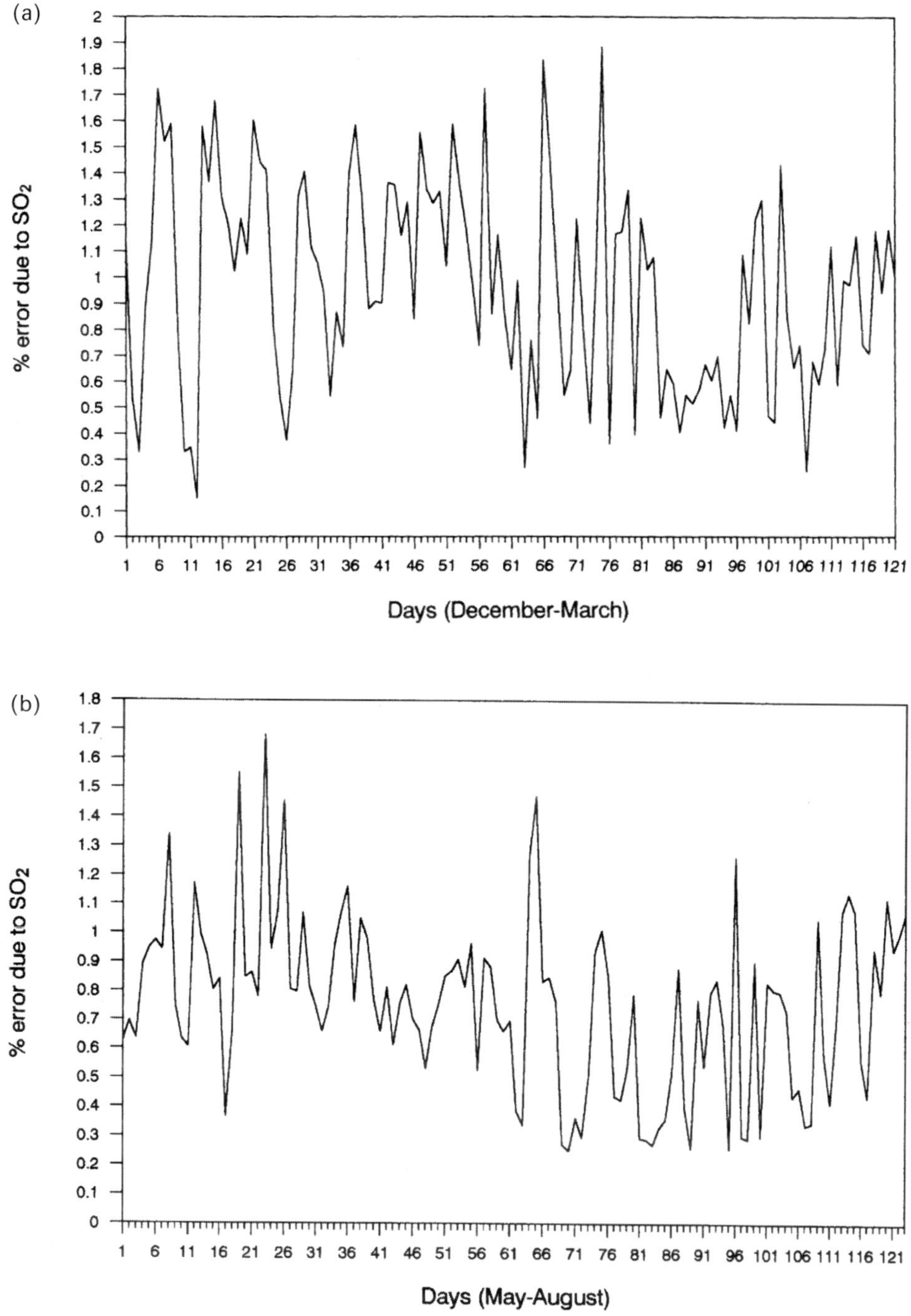

Figure 1.11. Mean daily percentage error in TOC caused by SO_2 for (a) December to March and (b) May to August of the 2-year period 1989–1990 (Varotsos *et al.*, 1995b).

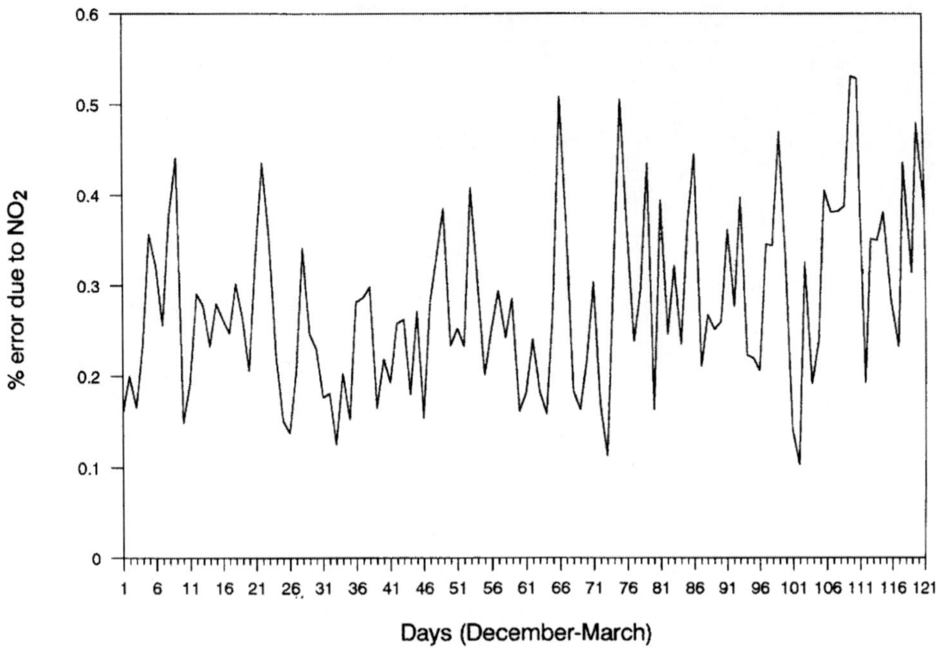

Figure 1.12. Mean daily percentage error in TOC caused by NO_2 for December to March of the 2-year period 1989–1990 (Varotsos *et al.*, 1995b).

1.2.4 Influence of stray light on Dobson TOC measurements

One of the main factors affecting the accuracy of determination of TOC measurements using a Dobson spectrophotometer arises from stray light contributions. The term "stray light" is considered here as being any radiation which contaminates the measurements made with a Dobson spectrophotometer. Therefore it includes both unwanted radiation scattered in the atmosphere by air molecules, aerosols, and clouds that enters the instrument's field of view as well as unwanted radiation scattered within the instrument. It should be distinguished from the internal stray light within the instrument which limits its ability to make accurate measurements at high total ozone amounts and high solar zenith angles (Evans *et al.*, 2009). The problem of stray light applies mainly to the shorter wavelength band of the wavelength pairs measured by this instrument. This means that the problem is principally evident in the lower intensity band of the Dobson bandpairs. The influence of stray light on Dobson TOC measurements is manifested by the characteristic effect of apparently declining ozone amounts with airmass above 2.5. For direct sunlight AD measurements, errors of 1%, 3%, and 10% may be present at airmasses 2.5, 3.2, and 3.8, respectively (Basher, 1982a).

A summary of early work on the effect of stray light on measurements made with Dobson spectrophotometers is given by Basher (1982a). According to Basher (1995) the error in direct Sun observations for a well-calibrated and well-operated instrument is about 1% (1σ) and for zenith sky measurements it is about 3%.

Radiation is scattered by the atmosphere into the instrument's field of view and contributes to stray light intensity which is approximately proportional to the field of view and to the airmass. This light contributes homochromatic (i.e., same wavelength) and heterochromatic (i.e., different wavelength) components and will include single and multiple scattered components. Its effect on ozone measurement is usually negative but may be positive and will depend on the relative intensity of the bandpair's intensities, the amount of the scattering material, the material's spectral and angular (phase function) scattering properties, and the optical depth along the beam's path.

Rayleigh scattering's inverse fourth-power dependence on wavelength and broad $\cos^2\theta$ phase function results in a nearly spatially uniform and bright (i.e., very hazy) sky radiance at 300–320 nm. Its main effect is the contribution of homochromatic stray light to the shorter wavelength bands and hence a reduction in the measured ozone amount. The attenuation by aerosol scattering (due to particles typically from 0.05 to 5 μm in diameter) has a weak spectral dependence. The homochromatic component and the internally scattered heterochromatic component both serve to reduce the measured ozone amount. The attenuation by individual cloud droplets (typically 10 to 1,000 μm in diameter) is essentially spectrally uniform, but coloring effects, as shown by colored rings about the Sun and the Moon, can also arise. For all but the thinnest of clouds multiple scattering becomes dominant. The effect of these factors on stray light is unclear (Basher, 1982a).

Thomas and Holland (1977) carried out a computer modeling experiment to determine ozone errors due to atmospheric scattering. They used a Monte Carlo simulation of radiative transfer through an atmosphere of fifty 2 km layers which included a Haze-C aerosol size distribution of 0.11 decimal optical depth. According to their model, which considered only homochromatic light, the ozone errors arising from this extra radiation received were less than 0.5% for single bandpairs and less than 0.2% for double-bandpair combinations. Calculations were limited to an airmass of 3, but extrapolation beyond this to an airmass of 4 suggested that the errors for single bandpairs may be about 2%. This is a larger estimate than that given by Dobson's earlier work as noted by Basher (1982a), which may be due in part to the greater density of the aerosol model used by Thomas and Holland. These authors showed that the errors are approximately linearly dependent on aerosol amount. The aerosol optical depth of 0.11 used by them may be compared with visible (500 nm) optical depths of 0.02 for very clear conditions to 0.5 for extremely hazy conditions.

We now turn to the stray light from internal scattering. Radiation entering the Dobson instrument must pass through either 28 or 32 optical surfaces and undergo three reflections before reaching the photomultiplier. This radiation is affected by undesired stray reflections at these components, by the usual optical aberrations of lenses and prisms, and by the diffraction effects of finite slit widths. There is also the possibility of the attenuation of short-wavelength bands by the poor-quality quartz optics of some instruments.

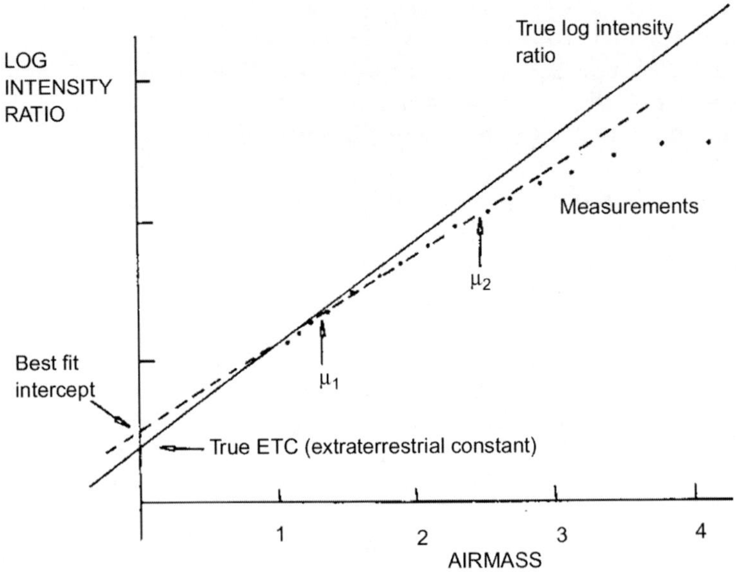

Figure 1.13. Illustration of the effect of stray light on log intensity ratios and on the determination of extraterrestrial constants (Basher, 1982a).

Some of the stray radiation will be homochromatic but the heterochromatic component, particularly from about 350 to 400 nm, is far more important because the incoming energy in this wavelength range is several orders of magnitude greater than that of the narrow and highly attenuated bands being measured. It is the second monochromator in the instrument that provides the means for reducing the effect of heterochromatic stray radiation.

The non-linearity of the log intensity ratio versus airmass function, stemming from the contribution of stray light to the short wavelength pair, induces the two errors shown in Figure 1.13. The first concerns the obvious reduction of the log intensity ratio (and thus of ozone measurement) with airmass. The second error concerns overestimation of the extraterrestrial constant (ETC) which can be calculated from the log intensity versus airmass plot. The latter will make the ozone measurements appear more constant with airmass and the residual curvature in the data seem small.

Basher (1982a) presented a simple homochromatic/heterochromatic stray light model for investigating non-linearity errors of the log intensity ratio versus airmass function. That model is characterized by two parameters. The first one, R_0, is the ratio of the energy of the monochromatic stray light source to that of the desired band as would be measured under the imaginary conditions of zero airmass. The second, a, is the atmosphere's relative attenuation coefficient of the stray light band to the desired band. If we consider the monochromatic and heterochromatic stray light at 370 nm then, according to Basher, the value of R_0 will be between 10^{-5} and 10^{-3}, depending on the quality of the instrument, and the parameter a would range from about 0.7 to 1.2, depending on ozone amount. The contribution of stray light to the longer wavelength band of a pair, in this example the 325.4 nm band, is negligible, and the errors are not very dependent on the assumption of monochromaticity or the center wavelength of the stray light band. The resulting equations for Δ_{ETC}, the error in the extraterrestrial constant, and for ΔX the total error in the ozone measurement are:

$$\Delta_{ETC} = \frac{1}{\mu_2 - \mu_1} \log \frac{(1 + R_0 10^{\mu_2 a})\mu_1}{(1 + R_0 10^{\mu_1 a})\mu_2} \tag{1.12}$$

and

$$\Delta X = \frac{-1}{\mu \Delta\alpha}(\Delta_{ETC} + \log(1 + R_0 10^{\mu a})) \tag{1.13}$$

where μ_1 and μ_2 are the airmass values at which the straight line used to determine the ETC intersect the log intensity versus airmass curve; and $\Delta\alpha$ is the bandpair's ozone absorption coefficient. Values of Δ_{ETC} for various values of R_0 and a are given in Table 1.2.

The more able the instrument to reject stray light, the smaller the R_0 value. The application of this model to total ozone X_{AD} measurements made by Basher in 1977 with the Wallops Island Dobson spectrophotometer showed that the value of the R_0 parameter at that time was about $10^{-3.8}$ to $10^{-3.6}$. Application of the same stray light model to the Belsk Dobson instrument D084 (Poland) gave R_0 values of 10^{-4} to $10^{-3.8}$ for all the analyzed time series (Degórska and Rajewska-Wiech, 1989). The value of 10^{-5} is not unattainable. This claim is verified by the airmass dependences taken with the Canadian Dobson D077 and shown by Olafson and Asbridge (1981).

A selection of results for ozone error, ΔX_{AD}, versus airmass μ for various values of R_0, μ_1, μ_2, and a were given by Basher (1982a). In the calculation of ΔX_{AD} it was assumed that the relative attenuation, a, and therefore the ozone amount is the same

Table 1.2. Δ_{ETC}, the error in determination of the extraterrestrial constant for various stray light error models (Basher, 1982a).

Airmasses		R_0, a				
μ_1	μ_2	10^{-5}, 1.0	10^{-4}, 0.8	10^{-4}, 1.0	10^{-4}, 1.2	10^{-4}, 1.0
1.25	2.0	0.0003	0.0012	0.0035	0.0094	0.0327
1.25	2.5	0.0008	0.0024	0.0083	0.0264	0.0724
1.25	3.0	0.0021	0.0050	0.0200	0.0717	0.1440
1.25	3.5	0.0053	0.0102	0.0471	0.1640	0.2417
1.25	4.0	0.0137	0.0209	0.0998	0.2870	0.3414

at the time of measurement as at the time of calibration of the ETC, and that the error source affects only the A bandpair. The following points were noted by Basher (1982a):

(i) The dependence of the ETC error, and therefore any ETC determination, on the airmass range used, and on the value of the parameter a, and therefore on the ozone amount. This may be a main cause for the difficulty of obtaining consistent ETCs for an instrument and for the lack of agreement between independently calibrated instruments.
(ii) The relative uniformity of the ozone error in the operating range of 1 to 3 in airmass, and the way this tends to conceal what may be a significant underestimation of ozone amount.
(iii) The increased error at low airmasses, which is principally due to ETC error. In middle latitudes this will result in noon values that in summer are lower than morning and afternoon values, and that in winter are higher. This sort of behavior has been observed in the Wallops Island Dobson instrument data (Geraci and Luers, 1978).
(iv) The strong dependence of the error on the parameters R_0, a, μ_1, and μ_2. An instrument whose R_0 is 10^{-4} might agree very well with another instrument whose R_0 is 10^{-5} during a summer intercomparison when the ozone level is low (i.e., when a is about 0.8), but it would disagree by 5% or more under winter–spring high-ozone conditions.

The measurements of any instrument can be compared to the calculated values to estimate the stray light levels present. The fact that for most Dobson instruments the X_{AD} measurement is not satisfactory beyond an airmass of 3 suggests that R_0 is typically about 10^{-4}. Basher made measurements of X_{AD} beyond this airmass using the Wallops Island Dobson instrument in 1977 (the average of the results for a range of ozone values is compared with two models in Figure 1.14). This suggests that the instrument's R_0 value at that time was about $10^{-3.8}$ to $10^{-3.6}$. The data of Figure 1.15 show that a value of 10^{-5} is the maximum desirable value. The airmass dependences shown by Olafson and Asbridge (1981) indicate that the Canadian D077 instrument is of this higher quality, which in turn indicates that the goal of $R = 10^{-5}$ is not unattainable.

A series of total ozone X_{AD} measurements for a large range of airmasses, extending to the value of 3.0, were taken over many days from 1992 to 1994 with Dobson spectrophotometer D118 in Athens, Greece. Among this measurement series were cases where the dependence of Dobson TOC measurements on airmass were first observed higher than about the value of 3.0, while in other cases they were observed higher than about the value of 2.5. The averages of the results for a range of ozone values were compared with some of the curves from Basher's stray light model. For that purpose, a reproduction of some of the stray light model's curves is given. Figures 1.16 and 1.17 show two clusters for various a values of some of the stray light model's curves. The first corresponds to the value $R_0 = 10^{-4.0}$ and

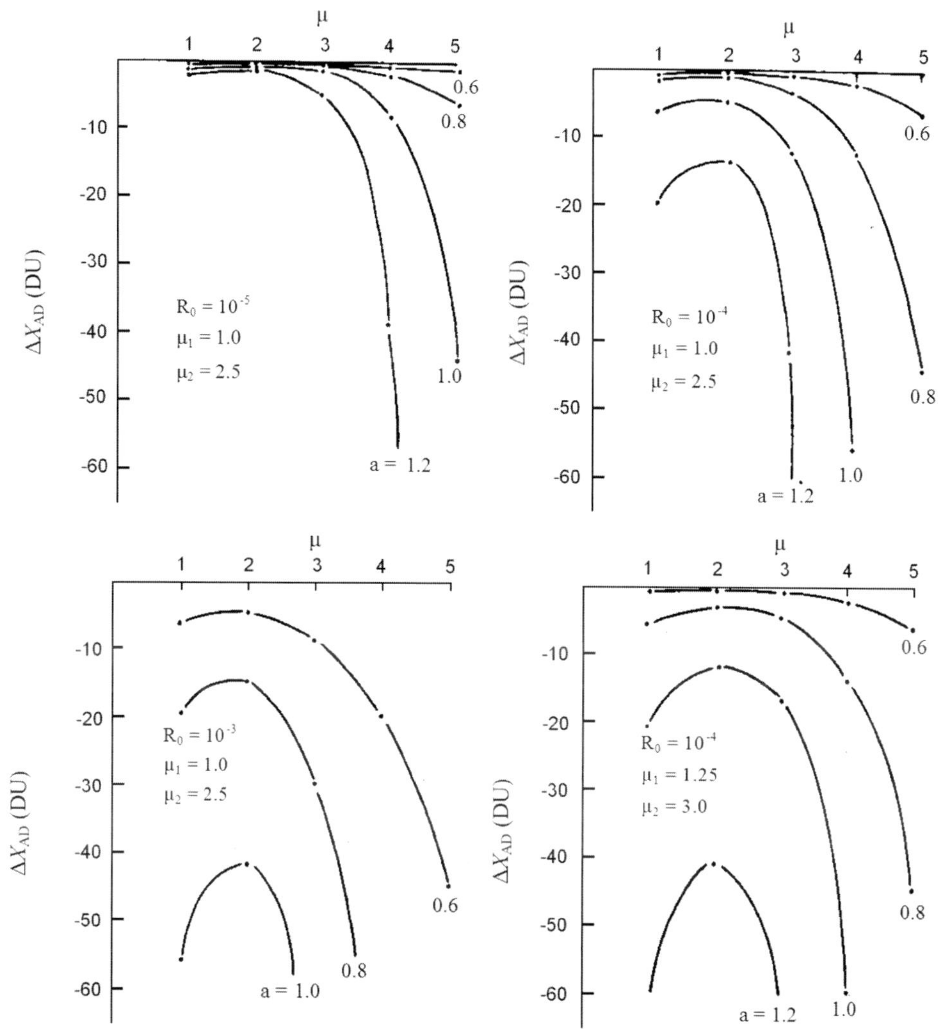

Figure 1.14. Variation of ozone error X_{AD} as a function of airmass μ for various values of model parameters, as calculated by the stray light model (Basher, 1982a).

the second to the value $R_0 = 10^{-3.0}$. From these figures it is obvious that the smaller the parameter R_0 the more able the specific Dobson instrument to reject the interfering stray light. Indeed, comparing any of these curves corresponding to the same value of the parameter a but to different values of the parameter R_0, we can observe that in the case of $R_0 = 10^{-3.0}$ the curves start to decline with much smaller airmass than for the corresponding curves with $R_0 = 10^{-4.0}$. Also these are more abrupt (i.e., the rate of increase in error of total ozone measurement, because of the presence of

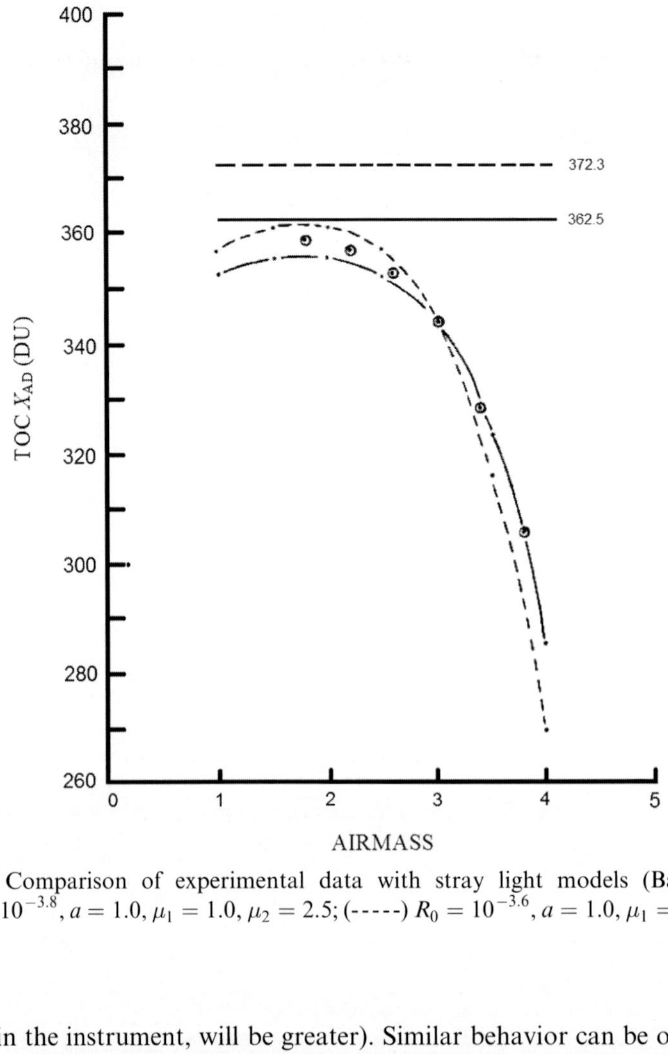

Figure 1.15. Comparison of experimental data with stray light models (Basher, 1982a): (——) $R_0 = 10^{-3.8}$, $a = 1.0$, $\mu_1 = 1.0$, $\mu_2 = 2.5$; (-----) $R_0 = 10^{-3.6}$, $a = 1.0$, $\mu_1 = 1.0$, $\mu_2 = 2.5$.

stray light in the instrument, will be greater). Similar behavior can be observed with the model curves for different values of a. The greater the value of a, the greater the influence of stray light on Dobson total ozone measurements.

The results of the comparisons are given in Table 1.3. The first column shows the dates the measurements were made. These are dates between 1992 and 1994. The second and third columns contain the values of R_0 and a, the stray light model parameters of the model curves fitting the data series. The fourth column contains the true total ozone amount as calculated from the model, while the last column shows the measured total ozone value (see also Figure 1.18).

The general conclusion from the abovementioned series of TOC measurements is that, in the period in which these measurements were made, the value of the R_0 of the Basher stray light model for the Dobson D118 instrument was in the range between $10^{-3.5}$ and $10^{-3.4}$. This means that this particular instrument does not

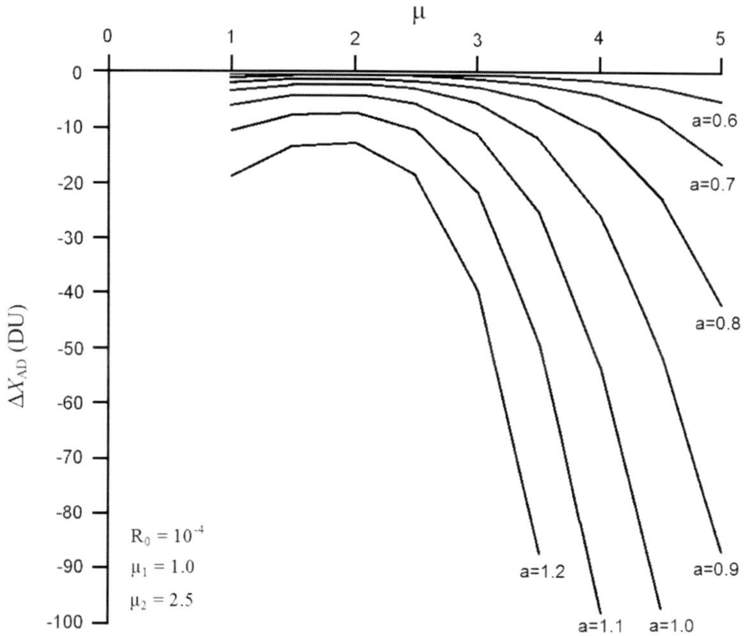

Figure 1.16. Variation of ozone error X_{AD} as a function of airmass μ for $R_0 = 10^{-4.0}$ as calculated by Basher's stray light model (Varotsos *et al.*, 1998).

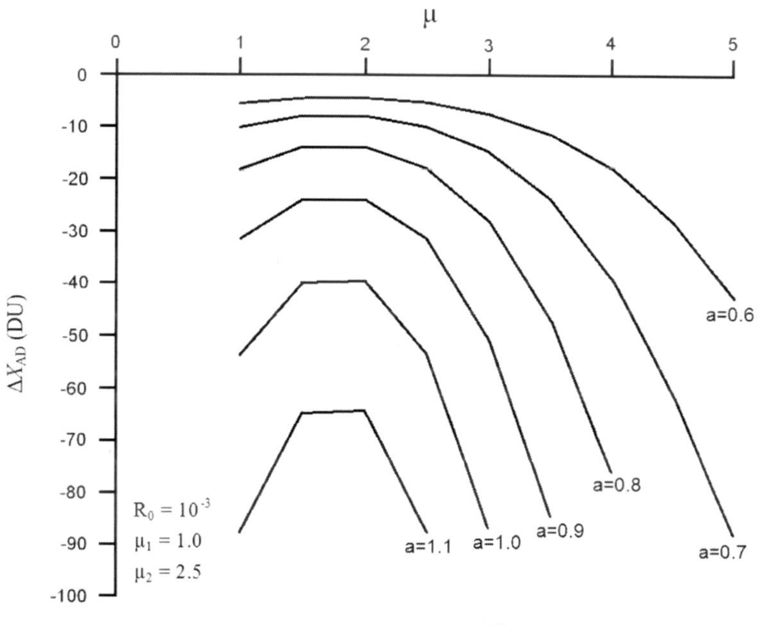

Figure 1.17. As Figure 1.16, but for $R_0 = 10^{-3.0}$ (Varotsos *et al.*, 1998).

Table 1.3. The results from some stray light experiments taken by the Dobson D118 instrument in Athens between 1992 and 1994 (Varotsos *et al.*, 1998).

Date	R_0	*a*	*True TOC* (DU)	*Measured TOC* (DU)
November 27, 1992	$10^{-3.4}$ $10^{-3.5}$	0.6	285 285	282
December 7, 1992	$10^{-3.4}$ $10^{-3.5}$	0.6	286 285	291
March 23, 1993	$10^{-3.4}$ $10^{-3.5}$	0.6	303 303	306
November 16, 1992	$10^{-3.4}$ $10^{-3.5}$	0.7	278 277	281
January 14, 1993	$10^{-3.4}$ $10^{-3.5}$	0.8	312 309	305
January 19, 1994	$10^{-3.4}$ $10^{-3.5}$	0.8	330 327	327
March 24, 1993	$10^{-3.4}$ $10^{-3.5}$	0.8	307 305	307
December 21, 1992	$10^{-3.4}$ $10^{-3.5}$	0.9	304 301	293
March 18, 1993	$10^{-3.4}$ $10^{-3.5}$	0.9	340 337	336

suffer from stray light interference when TOC measurements are made with low values of airmass μ (regularly obtained observations). Another result from the same series of measurements is that there is only a weak relation between the other Basher model parameter a and the TOC content because of the interference arising from the presence of SO_2 and NO_2, which absorb at the selected wavelengths of solar UVB radiation (Varotsos *et al.*, 1998).

Stray light effects are apparent when Dobson spectrophotometers are used to produce vertical ozone profiles using the Umkehr effect (see Section 1.6.2). Adjusting measurements for stray light components removes much of the bias between Umkehr profiles and other ozone-measuring datasets (Petropavlovskikh *et al.*, 2009). Evans *et al.* (2009) presented a method of making a measurement of the internal stray light, as well as other approaches evaluating the approximate level of internal stray light. These techniques can also be used to improve the results of direct Sun observations at high latitudes and high ozone.

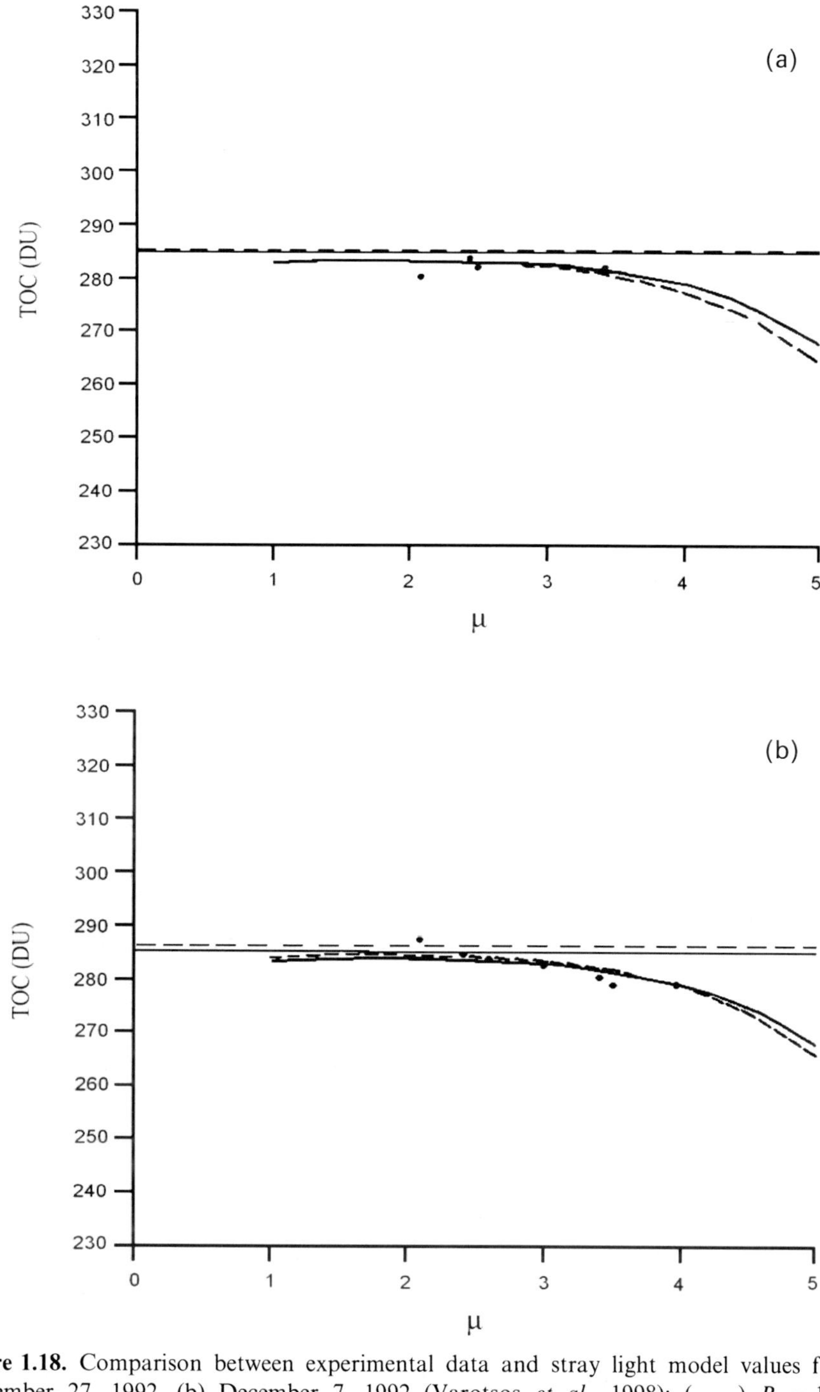

Figure 1.18. Comparison between experimental data and stray light model values for (a) November 27, 1992, (b) December 7, 1992 (Varotsos *et al.*, 1998): (——) $R_0 = 10^{-3.5}$, $a = 0.6$, $\mu_1 = 1.0$, $\mu_2 = 2.5$; (-----) $R_0 = 10^{-3.4}$, $a = 0.6$, $\mu_1 = 1.0$, $\mu_2 = 2.5$.

1.3 BREWER SPECTROPHOTOMETER

Brewer, Alan W. (1915–2007)
(*http://rammb.cira.colostate.edu/dev/hillger/modern.htm#brewer*)

Since the early 1980s, scientists around the world have been measuring TOC, SO_2, NO_2, with Brewer Spectrophotometers, which use a similar method to measure TOC to that of the Dobson instruments (Brewer, 1973).

These are grating spectrophotometers, developed in the early 1970s as an alternative to Dobson instruments in the World Ozone Network (Figure 1.19). They are modified Ebert-type spectrometers with focal length 16 cm, width 11 cm, and aperture ratio F/6. They use a single detector and are generally supplied with an automatic Sun tracker. Ground-based observations are made on both the solar and lunar disks as well as the zenith sky. They simultaneously measure the intensity of

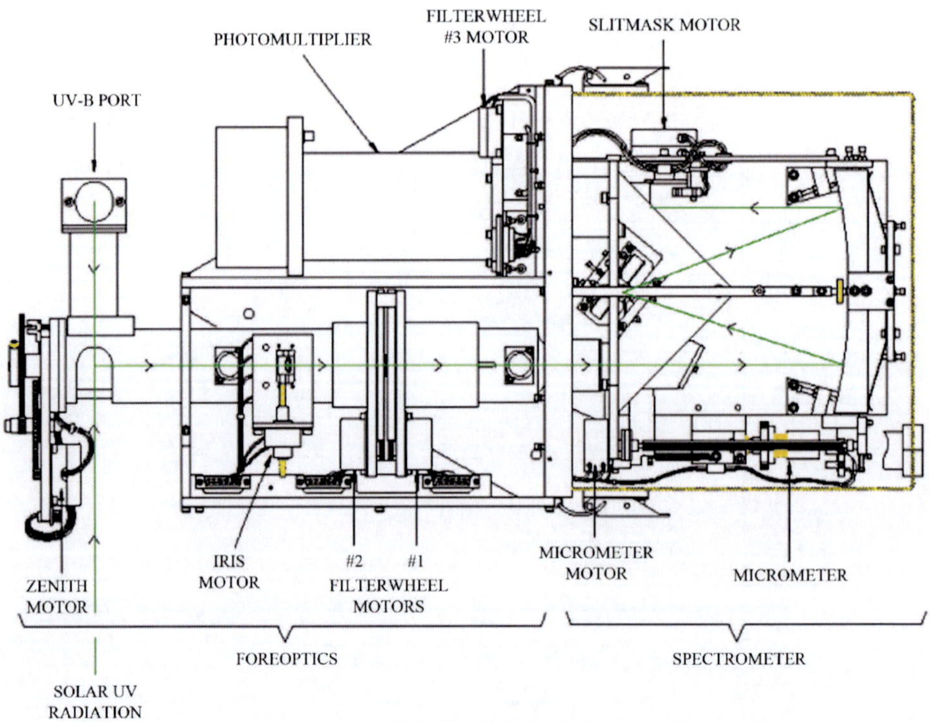

Figure 1.19. Brewer Mark IV mechanical assembly (top view) (*http://www.esrl.noaa.gov/gmd/grad/neubrew/MkIV.jsp*).

light at five wavelengths in the ultraviolet absorption spectrum of ozone with a resolution of 0.6 nm. The operational radiances at 306.3, 310.1, 313.5, 316.8, and 320 nm are monitored sequentially in 1.6 s. Measurements at these wavelengths allow correction for the effects of SO_2, which is a serious cause of error in ozone measurements in polluted air. Various intercomparisons of Brewer spectrophotometers, both between themselves and with Dobson instruments, have been carried out mainly by the Atmospheric Environment Service (AES) Canada. Periodic intercomparison of Dobson instrument D083 (World Primary Standard Dobson Spectrophotometer) with Canada's Standard Brewer Ozone Instrument D017 has achieved consistent results within 1% (Kerr *et al.*, 1988).

Differences between measurements of the TOC derived from Brewer and Dobson instruments are of the order of 3% (Köhler, 1999a, b; Vanicek, 1998). In 1997, over 130 Brewer spectrophotometers were in operation in 33 countries and in Antarctica.

To investigate the effects of scattered light on ozone retrieval a new double-monochromator Brewer instrument was built by AES. Stray light in the double-monochromator is several orders of magnitude less than that in a conventional single Brewer instrument. This instrument was made by linking two modified single Brewer optical frames together. The AES maintains a triad of calibrated Brewer instruments, as well as two traveling standards and one manufacturing standard. Each Brewer instrument is calibrated and maintained at WMO/AES ideal levels by means of intercomparison and using traveling standards (Kerr *et al.*, 1998).

Mark II and Mark IV model Brewer instruments are fully automated to use the direct Sun, zenith sky, or Moon as their light source. It is worth noting that Environment Canada (EC) recommends the Mark III Brewer Ozone Spectrophotometer as significantly superior to the Mark II and Mark IV Brewers for the measurement of solar radiation and ozone in the ultraviolet (UV) region of the spectrum. This is because of the much improved stray light performance of the double-monochromator optical system used in the Mark III. Instrument design and automation allow for nearly simultaneous observations of TOC, SO_2, and UV spectra (290–325 nm). The principle of measurements of total ozone and SO_2 column in the atmosphere with the Brewer spectrophotometer is similar to the method used for the Dobson spectrophotometer. The Brewer instrument measures spectral irradiances of solar radiation at five wavelengths selected by rotating a slit mask. From a series of zenith sky observations at twilight, OVP can be determined by utilizing the "Umkehr" inversion technique, which is described in Sections 1.6.2 and 1.6.3.

A new practical method for retrieving diurnal variability in tropospheric ozone OVP (Observations of ozone Vertical Profile) and day-to-day variability in stratospheric OVPs, using ground-based measurements of sky radiances obtained by a polarization-insensitive modified Brewer double-spectrometer and an optimal estimation retrieval method, was described by Tzortziou *et al.* (2008). Values of TOC derived by nearly simultaneous direct Sun Brewer observations were used to stabilize and constrain the retrieval. Using both simulated and measured radiances, their results demonstrated that the Brewer angular (0–72° viewing zenith angle) and spectral (303–320 nm) measurements of sky radiance in the solar principal plane provide sufficient information to derive the diurnal variability of the tropospheric OVP every 30 minutes. The Brewer measurements also provide stratospheric OVPs near sunrise and sunset each day. The OVP resolution was shown to give at least four points in the troposphere–low stratosphere, including good information in Umkehr layer 0 (0–5 km). Tzortziou *et al.* (2008) also showed that including a site-specific and time-dependent aerosol correction, based on Brewer direct Sun observations of aerosol optical thickness, was critical to minimize ozone retrieval residuals as a function of observing angle in the optimal estimation inversion algorithm, and improved the accuracy of the ozone profiles retrieved. Extending the technique to include longer wavelengths (>320 nm) could provide additional information on both OVPs and aerosol properties.

Just as we have seen in Section 1.2.2—that intercalibration of various individual Dobson spectrophotometers is important—so it is also important to carry out intercalibration of various Brewer spectrophotometers too. So within the GAW program of the WMO a World Calibration Center (WCC) for Brewer spectrophotometers has been established at the Meteorological Service of Canada (MSC), Toronto, Canada. The world Brewer calibration scale is defined by a set of three Brewer instruments maintained by MSC. Each of these instruments is calibrated absolutely at the Mauna Loa Observatory, Hawaii (MLO). Instruments in the world network are routinely calibrated by comparisons with traveling references. Outside Canada, these comparisons are mostly made with the Brewer No. 17 (B017) operated by the International Ozone Service, a Canadian commercial company. B017 is regularly checked against the three Brewer spectrophotometers maintained at the MSC.

In November 2003 the WMO/GAW Regional Calibration Center for the RA-VI region (Europe) (RBCC-E) was established by the Izaña Atmospheric Research Centre (I-ARC) at Izaña Observatory (IZO, Canary Islands), managed by the "Agencia Estatal de Meteorología" (AEMET, Spain). RBCC-E owns three Brewer spectroradiometers, a Regional Primary Reference (B157), a Regional Secondary Reference (B185), and a Regional Traveling Reference (B183) which can be transported for calibration campaigns outside IZO. These three spectrophotometers are absolutely calibrated at IZO and regularly compared against the three World Reference Brewer spectrophotometers maintained by the MSC in Toronto. As a Regional Brewer Calibration Center, the RBCC-E organizes regular intercomparison campaigns in Europe, mainly taking place in Central and South Europe (for details see the website *www.rbcc-e.org*).

The Sodankylä Total Ozone Intercomparison and Validation Campaign (SAUNA) took place in March–April 2006 and in February–March 2007 in Sodankylä, Finland (67.4°N, 26.6°E). The campaign involved a wide variety of ground-based systems including Brewer spectrophotometers, Dobson spectrophotometers, DOAS (Differential Optical Absorption Spectroscopy) and SAOZ (System for Analysis of Observation at Zenith) systems, and the NDSC (Network for the Detection of Stratospheric Change) traveling standard stratospheric ozone LIDAR; it also involved daily ozonesonde launches. The objective was to validate the performance of ground-based and satellite-flown ozone instruments at high latitude and low solar elevation. It was also to support the ongoing validation of the EOS–Aura OMI, the ERS–GOME, the ENVISAT–SCIAMACHY, SCISAT MAESTRO, and FTS instruments, by performing coincident satellite measurements.

In September 2007 the El Arenosillo Observatory (National Institute for Aerospace Technology, Spain) hosted an international calibration campaign for Brewer spectroradiometers, Dobson spectrophotometers, and UV-erythemal broadband radiometers in cooperation with several regional calibration centers and traveling reference instruments (*http://www.woudc.org*).

Stübi *et al.* (2008) described the analysis of 140 dual flights involving two types of ozonesondes—namely, the Brewer–Mast (BM) ozonesonde and the electrochemical concentration cell (ECC) ozonesonde. These dual flights were performed before the transition from BM to ECC as the operational ozonesonde for the Payerne Aerological Station, Switzerland. Different factors of ozonesonde data processing were considered and their influences on the OVPs were evaluated. The analysis showed good agreement between the data from the BM and the ECC ozonesondes. The profile of the ozone difference was found to be limited to ±5% (±0.3 mPa) from the ground to 32 km. The analysis confirmed the appropriateness of the standard BM data-processing method and the usefulness of normalization of the ozonesonde data. The conclusions of the extended dual flight campaigns were corroborated by analysis of the time series of the Payerne soundings for the periods of 5 years before and after the change from BM to ECC which occurred in September 2002. No significant discontinuity can be identified in 2002 attributable to the change of sonde.

1.4 FILTER OZONOMETERS M-83/124/134

The filter-type instrument is based on the same principle as the Dobson spectrophotometer in using differential absorption of ultraviolet radiation in the 300–350 nm Huggins band of ozone. The M-83 filter ozonometer was developed in 1963 (Gushchin, 1963). This instrument is equipped with optical light filters, which consist of a set of optical pigmented glasses. It uses 21 nm and 15 nm bandpass channels at central wavelengths 299 and 326 nm, respectively. Improvements were made in 1972 by using the stated bandwidths and a third filter for corrections due to aerosol effects. Improved versions, instruments M-124 and M-134, developed in the 1980s, were introduced into the former U.S.S.R.'s network of stations in 1985–1986 (Gushchin *et al.*, 1985b). This instrument measures ozone column content by looking directly at the Sun, using a peak spectral sensitivity at 299 nm (strong absorption by ozone) and another peak response at 326 nm (weak absorption by ozone). Like M-83, the improved instruments are intended to measure total ozone in the atmosphere, using both the direct Sun observation method and the zenith sky observation. Instead of two photoelectric receivers, the improved versions of M-83 employ only one sensing element (reducing substantially the weight and size of this device). The accuracy appears to be 2–4% and the precision of the measurement is approximately

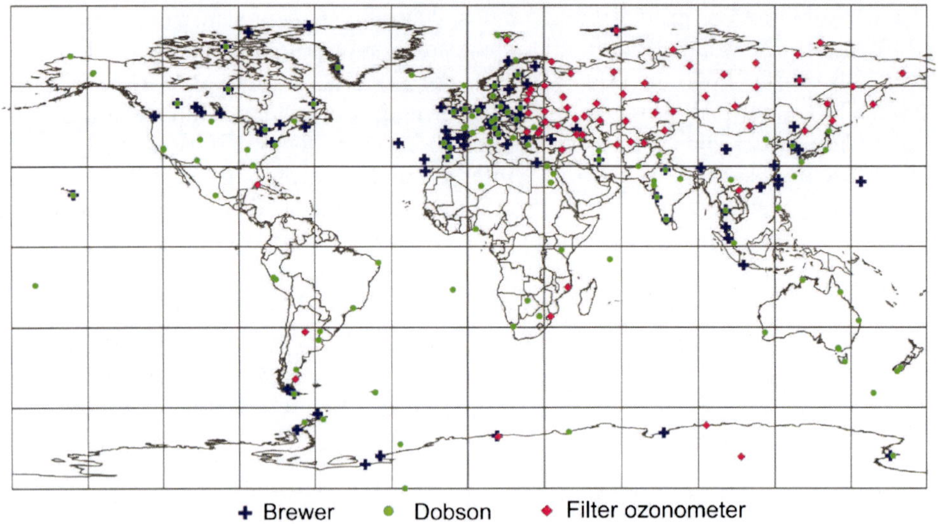

Figure 1.20. Map showing the location of total ozone stations where data were submitted to WMO within the period of 1979–2006 (Fioletov *et al.*, 2008).

5%. A network consisting of 45 stations has been operating for 30 years in the former U.S.S.R. and other countries (Figure 1.20).

The principle behind the operation of M-83/124/134 ozonometers (Figure 1.21) and, in general, filter-type instruments (Gushchin ozonometers) has often been criticized mainly because of the effect of finite wavelength intervals. As glass filters are used with broadband transmission characteristics for the selection of wavelength bands, their peak transmission shifts depend on the solar zenith angle and the amount of ozone. For instance, empirical curves have to be used to determine TOC from direct Sun or zenith sky measurements (Feister, 1994). Some years ago, a number of researchers (Bojkov, 1969) concluded that there is systematic disagreement between the M-83/124/134 ozonometers and Dobson spectrophotometers, which increases with increasing airmass μ (reaching 20–30%, when $\mu > 2$). This difference becomes even larger if visibility is below 5 km (in the presence of haze). Filter ozonometer data (covering more than one third of the northern mid and high latitudes from Eastern Europe to the Far East) were re-evaluated based on station instrument changes and calibrations (Bojkov *et al.*, 1994). It can be assumed that, in order to use filter-equipped instruments, careful study of the relevant errors, like the effective transmission factors of the filters, needs to be carried out in advance (Gushchin *et al.*, 1985a).

The GAW/WMO has designated one regional Calibration Center for Filter ozonometers, namely:

— Regional Calibration Center for Filter Ozone Instruments
Voeikov Main Geophysical Observatory, St. Petersburg, Russian Federation
(*ozon@peterlink.ru*)

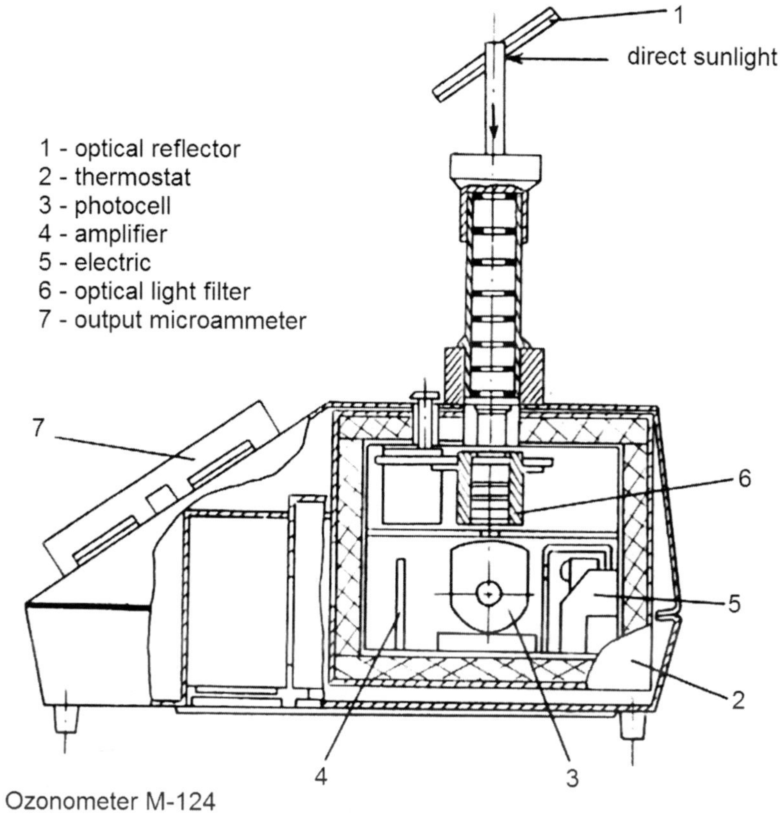

Figure 1.21. Schematic diagram of M-124 Ozonometer (Gushchin *et al.*, 1985a).

The total ozone observations conducted by Russia using filter ozonometers (M124/M83) cover a large area of Siberia. These instruments are regularly calibrated against the St. Petersburg Dobson instrument, which in turn is calibrated against the European reference instrument.

1.5 SECONDARY GROUND-BASED INSTRUMENTATION FOR TOC OBSERVATIONS

The Global Ozone Network of primary ground-based instruments is supplemented by several similar instruments such as the Canterbury ozonometer (New Zealand) and Sentran ozonometer (U.S.A.), which are based on various techniques. Among these is the differential optical absorption spectroscopy (DOAS) technique, which is one of the most powerful tools for observations of stratospheric trace gases (O_3, NO_2, NO_3, BrO, and ClO_2) (Balatsko *et al.*, 2010a, b; Platt and Stutz, 2008). The term "differential absorption spectrum" (simply) refers to studying $I(\lambda)$, the intensity of a beam of radiation with initial intensity $I_0(\lambda)$ after it has passed through an absorbing medium. To perform DOAS, two spectra are necessary; in one, referred to as the reference, the light has passed through little (ideally, of course, none) of the absorbing species in question and in the other the light has passed through a large amount of the absorber. The retrieved DOAS quantity is the apparent column density (ACD).

In the case of the study of trace gases in the atmosphere the wavelength range of interest is the UV and, to a lesser extent, the visible wavelength range. Each of the trace gases present contributes to the absorption of the radiation according to its own absorption spectrum (see Figure 1.22) and its concentration. The first step in DOAS applied to the determination of the concentrations of trace gases, including ozone, is to apply numerical filtering to the measured spectrum to separate the low-frequency (long wavelength) and high-frequency (short wavelength) parts of the spectrum. The low-frequency (i.e., long-wavelength) spectrum includes the broad-band gas absorptions and Rayleigh and Mie scattering, while the high-frequency (i.e., short wavelength) part of the spectrum contains the trace gas contributions. The short-wavelength intensity, $I(\lambda)$, can be written as:

$$I(\lambda) = I_0(\lambda) \exp\left(\sum \sigma_i'(\lambda) C_i L\right) \tag{1.14}$$

where $I_0(\lambda)$ is the extraterrestrial incoming radiation at the top of the atmosphere; $\sigma_i'(\lambda)$ are the absorption coefficients of the various trace gases labeled by i; C_i are the concentrations of these gases; and L is the path length. By writing

$$\sum \sigma_i'(\lambda) C_i L = \ln \frac{I(\lambda)}{I_0(\lambda)} = D'(\lambda) \tag{1.15}$$

and recalling that $D'(\lambda)$ is a (continuous) function of λ, then the C_i can be regarded as coefficients or variables to be determined by a least squares fitting procedure to equation (1.15). In the ground-based version $I_0(\lambda)$ is the incident sunlight and the instrument is mounted looking up vertically. In this common DOAS geometry the effective path along which the light travels is determined primarily by the solar zenith angle. As the Sun rises or sets the path length will change by an order of magnitude. The reference spectrum is obtained in the afternoon for solar zenith angles less than 80° and the retrieval is performed on spectra obtained at larger solar zenith angles, typically 88–90°. When the instrument is pointed at the Sun the measured signal comes from direct sunlight. In this case physical interpretation of the ACD is relatively straightforward; it represents the difference between the column density of the absorbing species along the line of sight and in the reference. However, if the viewing direction is away from the Sun, the source of the measured signal is scattered light. The ACD no longer has a clear physical interpretation because the path of the light through the atmosphere is complex and the measured sunlight may have been scattered in the atmosphere or reflected by the surface several times.

DOAS instruments have been employed not only on the ground but also on

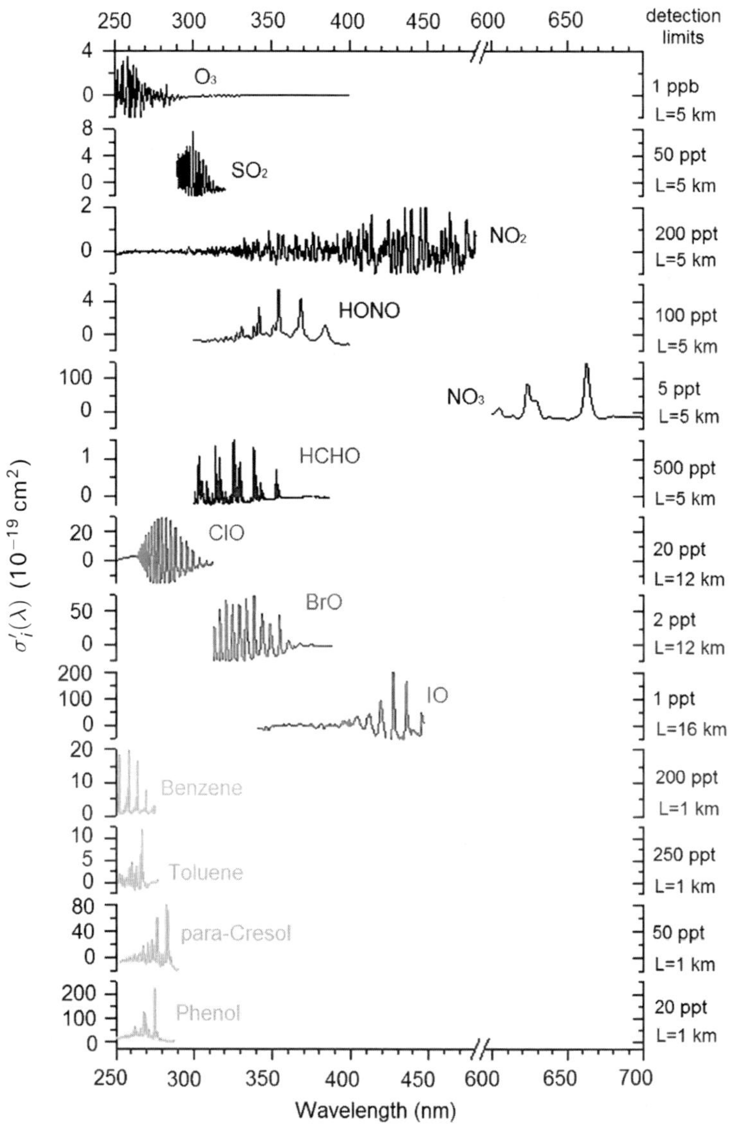

Figure 1.22. Differential absorption cross-sections of some trace gases absorbing in the UV/visible wavelength region. On the right axis the detection limits of the trace gases is listed together with the typical light path lengths used to measure them (*http://www.atmos.ucla.edu/~jochen/research/doas/DOAS.html*).

aircraft and in space in nadir and limb-sounding modes (see Section 2.5.1). For instance, DOAS is also used in some satellite-flown instruments (e.g., SCIAMACHY; see Section 2.5.2.5). In the following sections, brief information about some of the secondary ground-based instruments for TOC observations is given.

1.5.1 System for Analysis of Observation at Zenith (SAOZ)

SAOZ measures the ozone column by application of the DOAS method (described above) to zenith sky scattered light spectra in the UV–visible range, taken at twilight. This instrument allows TOC observations at polar sites in winter, when Dobson and Brewer measurements are strongly restricted.

In particular, SAOZ is a UV-visible diode array spectrometer (300–600 nm) developed for regular TOC and NO_2 monitoring at all latitudes up to the polar circle in winter. The instrument has a flat field grating, a 512 or 1,024 diode array detector and a 30° field-of-view limiter. Measurements are performed automatically from sunrise until sunset, through a solar zenith angle (SZA) of 84°. These are made by looking at the sunlight scattered at zenith in the visible wavelength range twice a day, just before sunrise and just after sunset, with a resolution of 0.6 nm. It should be mentioned that: (1) visible observations in the Chappuis bands (450–580 nm) at twilight have the advantages of being independent of stratospheric temperature, little contaminated by tropospheric ozone and multiple scattering, and permitting observations even in winter at the polar circle; (2) measurements in the UV (by Dobson and Brewer instruments) cannot be performed at SZA larger than 80° to 82° (i.e., at high latitudes in late autumn and winter).

In order to retrieve the constituent vertical total column the slant column is divided by the airmass, which corresponds to the enhancement of the optical path compared to the vertical. The resulting differential spectrum is filtered to leave features resulting from absorption by various species in the atmosphere: O_3, NO_2, H_2O, O_4. The absorber features are then correlated with laboratory cross-sections to obtain the total amount of each species present in the path.

Errors in TOC with this method are the sum of those of the spectral analysis of the measurements and those of the AMF used in the retrieval. The SAOZ standard AMF calculated for high latitude during winter is 16.59 for O_3 and 17.77 for NO_2 at 90° SZA. On average, the standard AMF is 3% smaller than that calculated from sondes with little seasonal dependence (Barlett and Vaughan, 1998; Sarkissian *et al.*, 1997). Ozone columns are therefore overestimated by the same factor. Overall accuracy is $\pm 10\%$ for ozone and between $+25\%$ and -45% for NO_2 at the polar circle in winter, where the column is smallest. In Section 1.7.3.9 a small balloonborne version of an SAOZ spectrometer is described.

1.5.2 MICROTOPS II (Total Ozone Portable Spectrometer)

This filter instrument is a continuation of a series of hand-held ozonometers starting with TOPS (Morys *et al.*, 1996). The new generation of these is MICROTOPS II, a five-channel Sun photometer with center wavelengths of 300, 305.5, 312.5, 940 and 1,020 nm for measurements of TOC, total water vapor, and aerosol optical thickness measurements (Morys *et al.*, 1996). With this instrument TOC is derived from measurements for three wavelengths in the UV region, given the site's latitude and longitude, universal time, altitude, and pressure. As in Dobson instruments the measurement at an additional third wavelength enables a correction to be made for particulate scattering and stray light. Total water vapor is determined by measurements at 940 nm and 1,020 nm. The angle of view of each optical channel is 2.5° and the resolution is 0.001 μW cm^{-2}. Typical agreement between various MICROTOPS II instruments (accuracy) is within 1–2% (Flynn *et al.*, 1996; Labow *et al.*, 1996). The repeatability of consecutive ozone measurements is better than 0.5%. Köhler (1999) recently made a 21-month intercomparison of the MICROTOPS II filter ozonometer with Dobson and Brewer spectrophotometers. The result is that MICROTOPS can measure total ozone with an accuracy comparable with that of conventional spectrophotometers (agreement is better than ±1%), over a reasonable range of μ. Adverse conditions (clouds, haze, and low Sun) result in deviations of more than ±2% or even ±3%.

1.5.3 High-Resolution Visible/Ultraviolet Absorption Spectroscopy

Absorption spectroscopy monitors the column abundance of various atmospheric species by measuring the absorption of light scattered by the sunlit zenith sky very precisely, typically 0.05% absorption (McKenzie and Johnston, 1982; Noxon *et al.*, 1979; Pommereau and Goutail, 1988a, b). An important feature of such measurements is that they can be carried out on cloudy as well as clear days. Observations of the stratospheric species that are achievable with UV/visible absorption spectroscopy have proved to be O_3, NO_2, NO_3, and ClO_2 (Megie *et al.*, 1998).

An example of this type of instrument is the crossed Czerny–Turner spectrometer (Mount *et al.*, 1987), with a grating chosen so that the appropriate color filters provide second-order light in the red region of the spectrum, approximately 605–685 nm (for NO_3 observations), and third-order light in the blue region, approximately 403 to 453 nm (for O_3, NO_2, and ClO_2 observations). The instrument is used at approximately 0.5 nm spectral resolution. The detection system is a Reticon diode array cooled by a refrigerator to about −70°C; the array contains 1,024 independent silicon diodes that simultaneously measure the spectrum over the indicated wavelengths. Eleven diodes cover the full width at half-maximum of the instrumental spectral profile. Thus, spectral lines are highly oversampled. Observations are usually made using the direct light from the Sun or Moon. Direct light observations have the advantage that weak sources (i.e., the Moon) can be observed.

The 1σ standard errors of the fitted column abundances are typically 2–5% for large solar zenith angles for NO_2. Ozone columns can be determined to 10–15% accuracy, using the Chappuis bands. Future development is expected to improve this considerably. The largest source of systematic error in the measurements is the evaluation of the airmass factors for scattered light that are used to convert measured slant column abundances to vertical column abundances. These errors largely affect the accuracy, but not the precision of the data.

1.5.4 Fourier transform spectrometer (FTS)

A Fourier transform spectrometer uses the Sun as a radiation source and measures the absorption by molecules in the stratosphere. A dispersive optical element such as

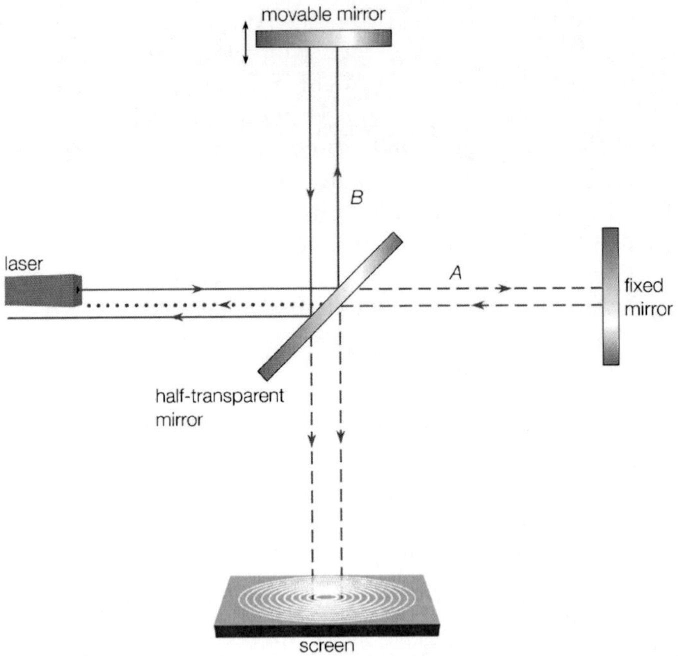

Figure 1.23. Optical diagram of a Michelson interferometer; in one of the arms the mirror can be moved to lengthen or shorten the optical path (*http://schools-wikipedia.org/wp/s/Speed_of_light.htm*).

a grating (as in the spectrometers described in the previous section) separates incident radiation into components with various wavelengths. A Fourier transform spectrometer (FTS) uses an interferometer. Radiation is not split into its component wavelengths but the beam is just split into two beams that follow different (optical) paths and are then recombined. The length of one of the two optical paths can be varied, as in a Michelson interferometer (see Figure 1.23), and the intensity of the output beam is recorded as the mirror is moved. By taking a Fourier transform of the output intensity as a function of mirror position one obtains the intensity as a function of frequency (i.e., the spectrum of the incident radiation). If $I(p)$ is the intensity as a function of the path difference p in the interferometer then the Fourier transform of $I(p)$, that is

$$I(\nu) = 4\int_0^{\infty} [I(p) - \tfrac{1}{2}I(p=0)] \cos(2\pi\nu p)\,\mathrm{d}p \tag{1.16}$$

is intensity as a function of frequency, ν. This can easily be converted into $I(\lambda)$, intensity as a function of wavelength, λ, instead of a function of frequency, ν, and this is the required absorption spectrum. This principle can be applied to radiation of various wavelengths—UV, visible, and infrared—let alone its use in radio astronomy.

In the present context the radiation incident on the FTS is solar radiation that has passed through the atmosphere. and the absorption spectrum is then used to determine the concentrations of trace gases, particularly ozone. With this technique all wavelengths are measured simultaneously. That is the main advantage of interferometers compared with most grating spectrometers (Grant, 1989).

The lack of highly accurate spectral line data and the overlaps within the instrument resolution of absorption lines of several species are the main factors that restrict measurement accuracy. Because of this a large spectral database is maintained in support of the Atmospheric Trace Molecule Spectroscopy (ATMOS) project (outlined in Section 2.5.1.2) that is updated continually with other field and laboratory measurements. In general, the FTS measurement accuracy of ozone is approximately 5% because of spectral parameter uncertainties. The resultant accuracy from further uncertainties (temperature and pressure profiles, instrument errors, and interference from other species) is from 10 to 20%. The instrument has been used in various experimental campaigns as a ground-based, balloonborne, or spaceborne device.

SIRIS, for instance, is a balloonborne FTS (see also Section 1.7.3.5). It is an emission mode remote-sensing instrument for measurement of the Earth limb from a balloon platform. Remote sensing of the limb allows vertical coverage from 5 to 40 km with a vertical resolution of 3 km. Another balloonborne-type instrument is the sub-millimeter high-resolution FTS.

Ground-based FTS (with a 0.01 cm^{-1} resolution) has been used by the Jet Propulsion Laboratory (JPL) group to measure stratospheric trace gases in the Antarctic at the time of the spring ozone hole and to establish that chlorine rather than nitrogen is responsible for the rapid decrease in ozone after the spring sunrise. A ground-based FTS instrument has also been operated in France (CNRS, Service d'Aéronomie/IPSL, Paris) to measure TOC making use of two spectral regions (2,084–2,085 cm^{-1} and 2,775–2,776 cm^{-1}) (Kramer, 2002).

The International Scientific Station of the Jungfraujoch located at 3,580 m altitude in the Swiss Alps is one of the alpine sites that constitute the primary (Network for Detection of Stratospheric Changes) station for the northern mid-latitude and is equipped with various instruments, such as infrared FTSs (Fourier Transform Infrared, FTIR, spectrometers), for TOC measurements. FTIR ozone data are retrieved from direct Sun absorption spectra in various spectral regions by two FTSs: a commercial Bruker IFSHR-120 spectrometer and a home-made FTS (De Mazière *et al.*, 1998).

A similar FTIR spectrometer has been used at Zugspitze/Garmisch (47.4°N, 11.0°E, 2,964 m above sea level, asl), since 1995. The main device is a Bruker IFS 120 HR spectrometer with a maximum resolution of $\Delta\nu = 0.00185\,cm^{-1}$. The optics of a 1.92 m Newtonian telescope (40 cm diameter) with a computer-controlled two-mirror solar tracker is matched to the spectrometer. For open path or laboratory measurements the radiation of three different internal sources can be detected through the telescope and solar tracker optics to an external retro-reflector positioned at a horizontal distance of 123 m. The reflected radiation is then directed back into the spectrometer through a beam splitter (Sussmann and Schaefer, 1997).

A ground-based Bruker 120 HR FTIR spectrometer is also used at Kiruna in Sweden (67.84°N, 20.41°E, 419 m asl) to derive the isotopic ratios of ozone from solar and lunar IR absorption spectra recorded in the 10 μm spectral region (Meier *et al.*, 1998; see also Section 3.8.5).

Finally, all thermal emission spectra have been measured also by a BOMEM M100 Michelson interferometer at Peterborough, Ontario, since 1991 (Evans *et al.*, 1994). The resolution of this instrument was of the order of $4\,cm^{-1}$. The results indicate that it provides useful information on TOC and on the tropo-stratospheric OVP (see also Section 2.5.3.3).

1.5.5 System for the Monitoring of Stratospheric Compounds (SYMOCS)

This instrument is a DOAS spectrometer (Mayer and Putz, 1998) developed by the Norwegian Institute for Air Research (Norsk institutt for luftforskning—NILU). It measures ozone, NO_2, BrO, and ClO_2 in the spectral region 332–482 nm (using the UV/visible spectroscopy method). This instrument was installed at Spitsbergen in the autumn of 1995 and participated in an NDSC (Network for the Detection of Stratospheric Change) validation campaign for ozone and NO_2 in the Observatoire d'Haute Provence (OHP) in June 1996. Just before winter 1997 SYMOCS moved to Andoya in Norway (ALOMAR) to participate in several satellite validation studies (GOME and SAGE III) (Fierli *et al.*, 1998).

1.5.6 Star Pointing Spectrometer (SPS)

The instrument consists of a telescope that focuses light from stars, planets, or the Moon onto a spectrometer,and two CCD array detectors. From the observations of atmospheric absorption, the atmospheric columns of several gases can be determined (Fish *et al.*, 1994). A novel version of this instrument (used for O_3, ClO_2, NO_2, and NO_3 retrieval) was deployed in Abisko (69°N) during the European Arctic Stratospheric Ozone Experiment (EASOE) (see Section 2.6.1). This instrument is particularly useful for polar measurements in winter, because it observes at night. By using an array detector as well, spectral noise from atmospheric flicker is avoided. Additional use of a two-dimensional array allows light from the sky adjacent to the star to be measured simultaneously and subtracted. This technique is very important at twilight at the poles during auroras or under hazy conditions near city lights.

1.5.7 MDR-23 (a Russian commercial device)

This is a grating scanning spectrometer with a 0.7 nm spectral resolution and a $51.2\,nm\,min^{-1}$ scanning rate. It measures TOC in the 303–333 nm spectral range at solar zenith angles lower than 70° and total NO_2 in the 435-450 nm spectral range during morning and evening twilights at solar zenith angles 84°–96°. TOC is derived using nomograms calculated using a single-scattering approach (Elokhov and Gruzdev, 1991, 1992). This kind of instrument was installed aboard the ship

Akademik Feodorov and TOC was measured in the 40°S–40°N latitude band in the Atlantic Ocean in May 1988.

1.5.8 Scanning spectrometer (EVA)

This operates in the range of 450–470 nm for TOC and 430–450 nm for total NO_2, in zenith view made at twilight. A fully automatic version, which was developed later, obtained a set of 30 spectra, 200 samples each, accumulated for each measurement to reduce photomultiplier noise. Instrumental errors are estimated to be $4–6 \times 10^{15}$ mol cm^{-2} for NO_2, and $1–2 \times 10^{19}$ mol cm^{-2} for O_3 (Yela *et al.*, 1998).

1.5.9 Solar IR spectroradiometer

This instrument is basically a double-monochromator on a heliostat to track the Sun. Solar radiation is focused on the entrance slit of the instrument by a Newtonian telescope and is converted to a square wave by the in-built chopper with frequency 163.4 Hz. Incoming radiation is dispersed by the gratings, centred blared at 1,600 nm and passed through the filter to a detector fixed at the exit slit. The electrical signal generated by the detector is amplified and the output is fed to a computer for data processing and analysis. The maximum error in TOC is less than 7%. The stray light is less than 0.1%, with maximum error 3% (Bose *et al.*, 1998).

1.5.10 Ground-based UV radiometer (GUV)

This is a multi-channel moderate-bandwidth UV filter instrument measuring absolute irradiance at 305 nm and 320 nm and used to derive total ozone. The optical part of the instrument consists of a Teflon diffuser, interference filters, and photosensitive detectors. The instrument is temperature stabilized at 40°C and the time resolution is 1 min. Details of the instrument and data analysis are given by van Roozendael *et al.* (1998a) and Dahlback *et al.* (1998). Ground-based UV–visible measurements of O_3, NO_2, ClO_2, and BrO were carried out over Ny-Ålesund (79°N, 12°E). The instrument operated in the near UV and visible spectrum (327–491 nm) and zenith sky spectra were measured every day for solar zenith angles smaller than 85°. Automatic spectral calibration with an HgCd lamp and measurements of the dark current and the detector were performed daily (Wittrock *et al.*, 1998). A UV spectroradiometer (absolutely calibrated) was also used for ground-based TOC determination by Huber *et al.* (1998). In this respect, a high-resolution double-monochromator with a full bandwidth of 0.72 nm at 50% maximum was used. Spectra were taken from 290 to 500 nm. The spectroradiometer was calibrated against a 1,000 W standard lamp. An automated solar UV spectrophotometer (SUVS) was described by Dorokhov *et al.* (1989) for measurements of TOC and UV radiation, using a similar technique to that of the Dobson spectrophotometer. Spectral measurements of solar radiation were also carried out in the Arctic with a double-monochromator H10D produced by Jobin-Yvon. The TOC was derived from direct solar measurements by applying Beer's law (Henriksen and

Stamnes, 1989). NILU UV ground-based multi-channel ultraviolet radiometers have been used since 2004 at Thessaloniki and Athens in Greece and the resulting values of TOC compared with values obtained from Brewer spectrophotometers at the same locations (Kazantzidis *et al.*, 2009). At both locations TOC was scarcely overestimated by these radiometers; the average difference being less than 1% (0.90% in Thessaloniki and 0.76% in Athens).

1.5.11 Spectrometer–Ozonometer PION

This instrument was constructed in 1985 in Belarus. It consists of a double-grating monochromator specially designed for reducing the level of internal scattering radiation. For measuring total ozone the spectral range was chosen within 295–320 nm. The instrument is equipped with a special input objective which forms a very narrow field of vision (about 40 angular minutes). The monochromator characteristics are very stable due to the absence of moving parts. The spectrometer–ozonometer PION has a precise automatic tracking system that tracks the Sun with an error not exceeding 2 angular minutes, a multi-wavelength method of total ozone calculation (up to 20 wavelengths in the 295–317 nm region), and a high speed of measurements (up to 1,000 measurements per day). The instrument has the ability to make total ozone observations at low Sun inclination angles ($\sim 8°$) as well as measuring the absolute intensities and spectral distribution of direct and scattered UV solar radiation in the spectral range of 280–450 nm. The processes of measurement and data storage are fully automated under the control of an external computer. A new modification of the instrument is being created to measure ozone concentration by means of zenith sky observation (WMO, 1996).

1.5.12 SPectrometer for Atmospheric TRAcers Monitoring (SPATRAM)

The SPATRAM (Spectrometer for Atmospheric TRAcers Monitoring) instrument is a multi-purpose UV/visible spectrometer for the measurement of radiation in the spectral range from 250 to 950 nm and has been installed at the Observatory of the Centro de Geofísica de Évora (CGE) in Évora (38°34′N, 7°54′W, 300 m amsl), in the south of Portugal since April 2004 (Figure 1.24).

In the SPATRAM, the spectrometer is installed inside a thermostatic box able to keep the internal temperature within the working range of the instrument (typically 15°C for the optical unit). The equipment allows for multiple input of radiation:

(i) from the primary input that is composed of a pair of flat and spherical mirrors that focus the light beam on the entrance slit (0.1 mm × 8 mm);
(ii) from the two lateral inputs, where the signal is carried to the entrance slit with an optical fiber connected to a very simple optical system; and
(iii) an additional input for spectral and radiometric calibration. A rotating mirror, driven by a stepper motor, allows the operator to choose between the primary input (pointing to the vertical), optical fiber inputs or additional input for spectral or radiometric calibration.

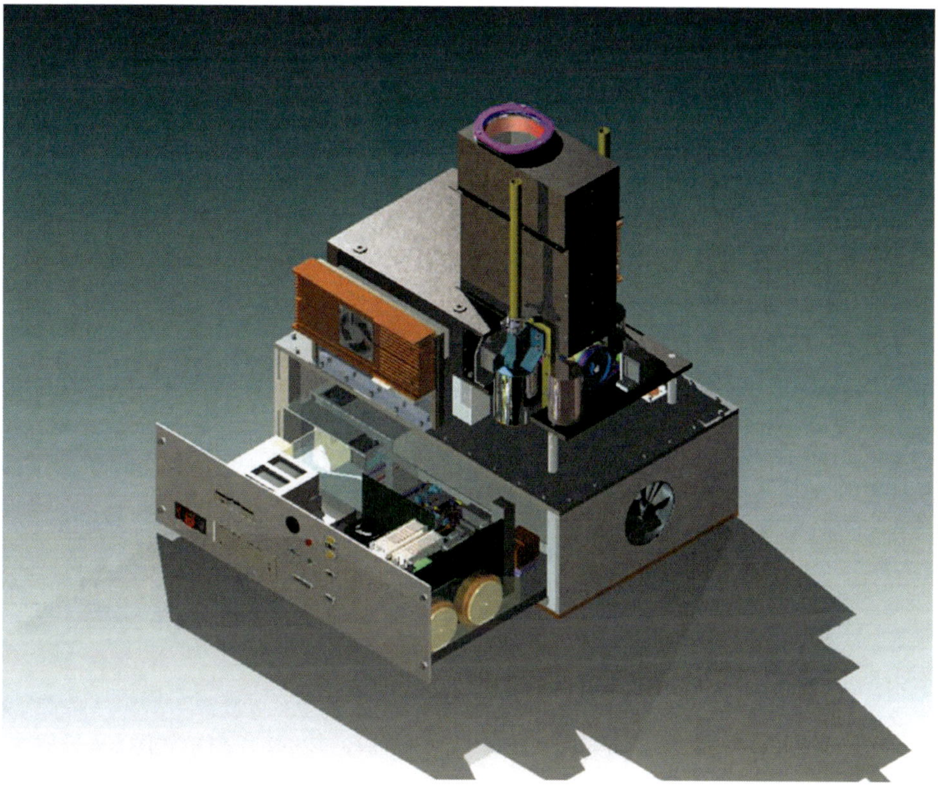

Figure 1.24. SPATRAM (SPectrometer for Atmospheric TRAcers Monitoring) installed since April 2004 at the University of Évora, Portugal (*source*: *http://www.dta.it/images/files/works/spatram.html*).

The spectrometer is composed of a holographic spherical diffraction grating of 1,200 grooves/mm and a Back-Illuminated CCD (Charge Coupled Device) sensor array both of which are customized in terms of cooling capability and AD signal conversion. In order to minimize the effects on measurements of the dark and read current, a two-stage Peltier circuit cools the CCD chip to −40°C. The grating is controlled by a stepper motor which, by varying the angle of incidence of light, allows the instrument to obtain measurements from 250 to 950 nm in spectral windows of 60 nm width each. The spectral dispersion is approximately 2.4 nm/mm (depending on the spectral region being investigated), and the overall typical spectral resolution is about 0.5 nm. Band-pass filters are used to remove the interference caused by the different order of dispersion and to reduce stray light inside the spectrometer. An internal mercury lamp is used for periodic checks of the diffraction grating positions, ensuring spectral accuracy better than 0.05 nm.

SPATRAM can provide the column content of all the compounds presenting absorption features in its spectral range. The standard products that have been obtained are O_3 and NO_2 in the UV and visible spectral region, respectively (specifically 320–340 nm for O_3 and 430–450 nm for NO_2), while the column contents of other species, such as SO_2, BrO, H_2CO were planned to be included in the retrieval (Bortoli *et al.*, 2009a, b).

1.6 OBSERVATIONS OF OZONE VERTICAL PROFILE (OVP)

In 1881 Hartley reached the important conclusion that ozone is a normal constituent of the higher atmosphere, being in larger proportion there than near the Earth's surface and being present in sufficient quantity to account for the ultraviolet limitation of the solar spectrum at ground level (Dobson, 1968). The first theoretical explanation of the ozone vertical profile (OVP) was given at a conference in Paris in May 1929 by Chapman (1930). However, in 1930 it was not yet possible to reach a general understanding because the ozone peak was misplaced at an average height of about 45 km.

The first measurements of the OVP were made by an indirect method (Umkehr) introduced by Götz in 1929 (Dobson, 1968). According to these observations the average height of the ozone layer is only of the order of 25 km and the main changes in OVP appear to be centered between 10 and 20 km. It should be stressed that these studies of OVP were made before direct soundings with the use of optical instruments (Dobson, 1968).

Since 1965 the use of chemical instruments has become the basis for the first analysis of the meridional pattern of the OVP (Griggs, 1964). Today, observations of OVP are made with the utilization of ground-based, airborne, and spaceborne instruments; we shall postpone the discussion of spaceborne instruments to the next chapter.

A number of instruments and techniques have been used to measure OVP from the ground. However, only a few have produced reliable and long-term data sets. In the following sections, brief information about the main instruments and techniques for OVP measurements from the ground is presented.

1.6.1 Primary ground-based instrumentation for OVP observations

OVPs are mainly retrieved from the ground using zenith sky measurements in the so-called Umkehr mode, which takes its name from the German word "umkehren" to turn back (round).

In 1929, F.W.P. Götz registered spectra of the zenith skylight, while the Sun was rising or setting while he was operating two spectrographs (a Fabry type and a Fabry–Buisson type) at Spitzbergen. He found that, when the Sun was fairly high, shorter wavelength radiation decreased in intensity more rapidly than longer wavelength radiation with increasing zenith distance of the Sun, because of the greater absorption of shorter wavelengths. He found, however, that when the Sun was very low the reverse occurred. The explanation he gave is that when the Sun is high most of the short-wavelength zenith skylight is scattered from the direct solar beam below the ozone region, but when the Sun is very low the absorption of the direct sunlight by the ozone is so great that the amount of light scattered above the ozone region and which passes down to the instrument by the short vertical path becomes predominant. Near the end of 1930 he wrote to G.M.B. Dobson in Oxford telling him of the effect and suggested that the ratio of two wavelengths should show a similar effect. He also suggested that this effect might be used as a means of estimating the OVP. Dobson, not really believing in Götz's suggestion, started to make measurements using his own instrument, the Dobson spectrophotometer (see Section 1.2.1) on the first clear day early in January 1931. He started before sunrise and he was surprised to find that the effect proposed by Götz was real. It was the first "Umkehr curve" (the term was suggested by Götz for "the turning back" of the curves, when the Sun is low) ever obtained. After that, in 1932 Götz and Dobson (joined later by A.R. Meetham) worked together for 6 weeks at Arosa and showed with certainty that the average height of the ozone in the atmosphere was about 22 km and not about 40–50 km as had been previously thought.

Umkehr observations may be made at ground level with both Dobson (WMO, 1991; Staehelin *et al.*, 1998b) and Brewer (Kerr *et al.*, 1988) spectrophotometers, measuring the ratio of diffusely transmitted zenith sky radiance at a wavelength pair in the ultraviolet, one wavelength strongly, the other weakly absorbed by ozone (e.g., for the C wavelength pair 311.5 nm is strongly absorbed and 332.4 nm is weakly absorbed). These wavelength pairs are measured in a series of zenith sky observations with the solar zenith angle changing from 60° to 90° during sunrise or sunset. During the Umkehr observation period, the Dobson or Brewer instrument also independently measures the total ozone column. The ground-based instrument performs observations of the blue zenith sky while the Sun is between the horizon and an elevation of 30° at sunrise or sunset.

1.6.2 Dobson Umkehr measurements and inversion

To reduce the time required for an Umkehr measurement, the Short method was developed, which uses the A, C, and D Dobson wavelength pairs (Mateer and DeLuisi, 1985; McElroy *et al.*, 1998). By using the triple-pair (A, C, D), OVP information is obtained, when measurements are made during a solar zenith angle change from 80° to 89°. However, for most of the past data and for current work, only the C pair is used for Dobson Umkehr inversion (Mateer and DeLuisi, 1992) and this requires measurements to be made over the range of solar zenith angle from 60° to 90°. The inversion algorithm of Mateer and DeLuisi uses new temperature-dependent ozone absorption coefficients (Bass and Paur, 1985). Umkehr inversions determine ozone concentrations in 10 Umkehr layers (layers 0 and 1 are combined into layer 1) divided into equal log-pressure vertical intervals starting at the surface (1 atm = 1,013 hPa) and extending to layer 10 (9.77×10^{-4} atm to the top of the atmosphere). These Umkehr layers are approximately 5 km thick and are centered roughly at the layer number times 5 km in height (e.g., layer 8 is centered at 40 km). Although standard Umkehr retrievals report 10 layers, because of a combination of the physical atmospheric scattering process, finite instrument spectral resolution, and real atmospheric vertical correlation, only data for layers 4–8 (19–43 km) are reliable. The vertical resolution of retrieval for these layers is from 11 to 14 km. Systematic uncertainty is estimated to be 15–20 and 5–12% (DeLuisi *et al.*, 1989).

The current Umkehr inversion algorithm due to Mateer and DeLuisi (1992) is based on the techniques of Rodgers (1990) and represents the first update of the original algorithm of Mateer and Dütsch (1964). Dobson zenith sky measurements may be written as

$$N(x,z) = 100\log_{10}\{[I(x,z,L_2)/F_0(L_2)]/[I(x,z,L_1)/F_0(L_1)]\} + C_0 \qquad (1.17)$$

where $N(x,z)$ is the relative logarithmic attenuation for the wavelength pair and is referred to as the N value. The quantity x refers to the ozone profile; z is the solar zenith angle; $I(x,z,L_1)$ is the 311.5 nm radiance and $I(x,z,L_2)$ is the 332.4 nm radiance of the C wavelength pair; F_0 is extraterrestrial flux; and C_0 is an instrumental constant. The measured $N(x,z)$ values are inverted to provide ozone amounts (in Dobson Units, DU) in the 10 Umkehr layers through the application of a radiative transfer code that takes into account primary and multiple atmospheric scattering, atmospheric refraction, and absorption by atmospheric ozone.

In the forward model, the N value comprises four components: the primary scattering component, N_{p}, the multiple scattering component N_{MS}, a refraction component N_{R}, and an instrumental parameter C_0

$$N_{\mathrm{p}} = N - N_{\mathrm{MS}} - N_{\mathrm{R}} - C_0 \qquad (1.18)$$

By using differences $Y_{\mathrm{i}} = N_{\mathrm{i}} - N_0$, where N_{i} is the N value at zenith angle z_{i} and N_0 is the lowest zenith angle, the instrumental constant C_0 and the extraterrestrial flux F_0 can be eliminated. The forward model uses the average of the Middle Atmosphere Program (MAP) 40°–50° North and South temperature profile with the temperature-dependent ozone cross-sections of Bass and Paur (1985). For the forward calculation, the atmosphere is divided into 61 layers in the vertical where each layer is 1/4 of the appropriate Umkehr layer. Table 1.4 shows the layer designations used in the Umkehr retrievals along with the pressure levels and approximate altitudes. The inverse model provides ozone content in 10 layers, where layer 10 includes all the ozone above layer 9, and layer 1 includes the retrieved ozone for both layers 0 and 1.

The inverse problem uses the observed N value corrected for multiple scattering and refraction with the value for the smallest solar zenith angle subtracted. Additionally, the integral of the retrieved ozone is constrained to have the value of the observed total ozone. The retrieval algorithm uses the optimal estimation method of Strand and Westwater (1968) as formulated by Rodgers (1976). At the nth iteration, retrieval $\mathbf{x}_{n+1}$ is obtained from $\mathbf{x}_n$ by the following expression:

$$\mathbf{x}_{n+1} = \mathbf{x}_{\mathrm{A}} + [\mathbf{S}_x^{-1} + \mathbf{K}_n^{\mathrm{T}}\mathbf{SE}^{-1}\mathbf{K}_n]^{-1}\mathbf{K}_n^{\mathrm{T}}\mathbf{SE}^{-1}[(\mathbf{y} - \mathbf{y}_n) - \mathbf{K}_n(\mathbf{x}_{\mathrm{A}} - \mathbf{x}_n)] \qquad (1.19)$$

where $\mathbf{x}_{\mathrm{A}}$ is the *a priori* ozone profile; $\mathbf{S}_x$ is the covariance uncertainty matrix for the first-guess profile; SE is the error covariance matrix for the measurements; $\mathbf{y}_n$ is the vector of calculated observations for $\mathbf{x}_n$; $\mathbf{K}_n$ is the averaging kernel; and the superscript T represents matrix transposition. The *a priori* profiles are significantly improved over the 1964 approximations. In layers above 5, the current *a priori*

Table 1.4. Layers used for Umkehr ozone profile retrievals (Mateer and DeLuisi, 1992).

Layer number	*Layer base pressure* (atm)	*Layer base approximate height* (km)
0	1.0	0.0
1	0.5	5.5
2	0.25	10.3
3	0.125	14.7
4	6.25×10^{-2}	19.1
5	3.12×10^{-2}	23.5
6	1.56×10^{-2}	28.0
7	7.81×10^{-3}	32.6
8	3.91×10^{-3}	37.5
9	1.95×10^{-3}	42.6
10	9.77×10^{-4}	47.9
11	4.88×10^{-4}	53.2
12	2.44×10^{-4}	58.3
13	1.22×10^{-4}	63.1
14	6.10×10^{-5}	67.8
15	3.05×10^{-5}	72.2

profiles are sinusoidal functions of Julian day and latitude in six latitude bands. The *a priori* amounts below layer 4 are quadratic functions of the measured total ozone amount only derived from ozonesonde climatology. The amounts in layers 4 and 5 result from a cubic fit to upper and lower layers.

Umkehr observations share most of the errors associated with measurements performed with ground-based instruments. However, these observations suffer some additional errors which do not significantly influence TOC measurements. Apart from instrument calibration problems, major sources of uncertainty include the inversion algorithm, the temperature dependence of the absorption cross-section, and the optical effects due to both stratospheric and tropospheric aerosols. The Umkehr inversion algorithm, which we have just described employing improved first-guess ozone profiles and improved ozone absorption coefficients (Mateer and DeLuisi, 1992; Reinsel *et al.*, 1994a, b), displays less error sensitivity in layer 9 to stratospheric aerosols normally existing in the Junge layer (stratospheric aerosol layer residing at 20 km altitude).

There are presently seven automated Dobson instruments that routinely make standard and short Umkehr measurements. These instruments were developed by NOAA/GMCC (Komhyr *et al.*, 1985) and are located at Perth (Australia), Lauder (New Zealand), Huancayo (Peru), Mauna Loa (Hawaii), Boulder (Colorado), Haute Provence (France), and Fairbanks (Alaska). The frequency of Umkehr measurements has significantly increased with the use of the automated Dobson, because observers are not required and a shorter observing time (for the Short Umkehr) decreases the chance of cloud interference.

1.6.3 Brewer Umkehr measurements

A Brewer Umkehr observation consists of zenith sky radiance ratio measurements for three wavelength pairs over the SZA range from 60° to 90°. The profile retrieval data cover the altitude range of layers 1–10 (0–53 km). A new five-wavelength version of the Brewer Umkehr inversion algorithm was suggested by McElroy *et al.* (1994). The performance of this new algorithm was assessed by comparing the OVP retrieved with those profiles generated using the six-wavelength algorithm, showing that there exists a much-reduced sensitivity of the five-wavelength algorithm. Until a complete error analysis has been carried out for the Brewer Umkehr system, the profiles for the standard Dobson Umkehr should be considered as only approximately applicable. Results of an intercomparison of Brewer Umkehr, Dobson Umkehr, and ozonesonde data indicated that the quality of the Brewer Umkehr technique is comparable to that of the Dobson "Short" Umkehr method (McElroy *et al.*, 1989).

An improved Brewer Umkehr retrieval algorithm produces ozone profiles directly from observations made by the Brewer spectrophotometer. It uses climatological ozone natural variability as a function of TOC, the season, and the year (Elansky *et al.*, 1998; Kosmidis *et al.*, 1998) to predict information about OVP (such as TOC). Furthermore, the estimation of measurement errors by using the new method leads to the determination of OVP by performing observations at one solar zenith angle (express Umkehr method). Note that the express Umkehr method allows for OVP measurements (1) between clouds and varying aerosol conditions and (2) at polar sites (the standard and short Umkehr are not applicable because of small changes in solar zenith angle).

Recently Petropavlovskikh *et al.* (2009, 2010) reached the following conclusions for the Umkehr observations obtained

(a) by the Dobson spectrophotometer:
  - the effect of band-pass (BP) shift (0.1 nm) error is small (wide BP), but the appropriate BP shape increases Umkehr retrieved ozone in the upper stratosphere;
  - the effect of temperature corections (climatology based) is small;
  - differences occur in the results from collocated instruments (these are

attributed to stray light effects and can be taken into account by including a stray light parameter in the individual Dobson instrument characterization);

(b) by the Brewer spectrophotometer:
   - the effect of band-pass shift and temperature corrections (climatology based) is small;
   - the stray light effect in single Brewers (1×10^{-4} level for Mark IV) is significant for ozone profile retrieval, but needs to be further evaluated;
   - the errors in Brewer Umkehr ozone profile retrievals related to the uncertainties in instrumental parameters are larger than errors due to X-section choice.

1.6.4 Secondary ground-based instrumentation for OVP observations

Besides direct *in situ* stratospheric techniques, remote-sensing observations have been made over a wide range of the electromagnetic spectrum, from the microwave to the ultraviolet. Of these, microwave and millimeter wave observations for rotational transitions offer certain advantages. First, calculations of spectral line intensities are simpler and more certain because the vibrational and electronic wave functions do not need to be known. Second, millimeter wave radiation, unlike ultraviolet and visible radiation, is unaffected by atmospheric aerosols. Third, the high spectral resolution permits the measurement of pressure-broadened line shapes, from which information can be obtained on ozone altitude distribution.

1.6.5 Lidar

Lidar uses the difference between two laser wavelengths (308 nm, 351 nm), one absorbed by ozone (308 nm) and the other not (351 nm). The two beams are sent out into the atmosphere, where they are scattered and reflected and are then collected by a telescope. The intensity of the scattered light from each laser beam is compared and a measure of the concentration of ozone in the range 10 km to 50 km is received.

The Differential Absorption Laser (DIAL) technique is the main basis of lidar measurements of the OVP in the troposphere and the stratosphere (from the ground up to 50 km). This technique requires measurements of the emission at two laser wavelengths with different ozone absorption cross-sections. The choice of the two wavelengths used in the experiment depends directly on the altitude range being monitored. The laser radiation that is scattered back to the surface is collected by a telescope. The derivative of the logarithm of each signal is computed and the ozone number density is obtained from the difference between the derivatives, divided by the differential ozone cross-section (Ancellet *et al.*, 1988; Eisele and Trickl, 1998; Godin *et al.*, 1989; McDermid *et al.*, 1990; Megie *et al.*, 1977; Pelon and Megie, 1982; Pelon *et al.*, 1986; Uchino *et al.*, 1983; Werner *et al.*, 1983; Zuev and Kabanov, 1987). The vertical profiles of ozone concentrations in the Planetary Boundary Layer (PBL) were measured at Schauinsland in the southern Black Forest with the Advanced Remote Gaseous Oxides Sensor (ARGOS) lidar (Bisling *et al.*, 1998).

Two types of error affect ozone lidar measurement: the random or statistical error (related to the signal-to-noise ratio) and the systematic error (related to errors in the value of ozone absorption cross-sections and to uncertainties in the Rayleigh and Mie scattering or the absorption by other species such as NO_2, SO_2). To obtain an ozone profile in the 15–50 km altitude range with resolution of 0.5–8 km, the corresponding uncertainty is 2 to 10%.

With respect to passive systems the lidar system has several advantages: (a) control of the emitting source and (b) higher altitude resolution of OVP.

However, in the presence of Polar Stratospheric Clouds (PSCs) conventional DIAL systems are strongly affected by the scattering by cloud particles. An alternative to reduction of the interference of particle scattering is to use the Raman DIAL technique (instead of the conventional Rayleigh DIAL). It should be clarified at this point that Raman scattering is an inelastic interaction of the optical beam involving excitation of the energy levels of a molecule and re-radiation at a different wavelength. In distinction, Rayleigh and Mie scattering are elastic interactions (i.e., there is no change of wavelength): the former is due to particles in the atmosphere, such as molecules or fine dust, that are much smaller than the optical wavelength λ of the laser, whilst the latter is associated with larger particles such as aerosols whose size is on the order of λ. Such an instrument was built and has been operated by the Forschungszentrum Geestacht GmbH (GKSS) at Esrange (near Kiruna, Sweden), since the winter of 1996/1997. This lidar uses two laser systems as emitters (Nd:YAG and XeCl excimer). The backscattered light is collected by an afocal Cassegrain telescope. It is separated by a filter polychromator with eight different channels. Both lasers are synchronized on a mechanical chopper wheel, which protects the photomultipliers against the very intense backscattered light from the lower atmosphere. The Meteorological Research Institute (MRI) in Japan developed an ozone lidar, the MRI Mark II, which consists of an XeCl excimer laser (8.8 W at 308 nm), a Nd:YAG laser (2.6 W at 355 nm, 2.0 W at 532 nm), and an 80 cm diameter receiving telescope (Fujimoto *et al.*, 1998).

The NIES ozone lidar system which is located at 36°N, 140°E, in Japan, is a typical UV differential absorption lidar (DIAL) system for low-stratospheric ozone measurements (Tatarov *et al.*, 2009). The system has been in operation since 1988, and several replacements and improvements have been made since it was first installed (Nakane *et al.*, 1992a, b). System parameters and improvements have been presented in detail by Park *et al.* (2006) and by Tatarov *et al.* (2009). Tatarov *et al.* applied an improved data-processing and retrieval algorithm (version 2) for ozone. This algorithm removed systematic errors in the signals (background signals, signal-induced noise, and dead time corrections) as well as errors due to the scattering and extinction of the atmosphere (air molecule extinction correction and aerosol corrections). The values of the TOC were calculated by means of the DIAL equation (Schotland, 1974).

1.6.6 Microwave radiometry

The Microwave Radiometer (MR) measures thermal emission from rotational transitions of the molecule under investigation. The frequency employed is typically below 300 GHz (1 mm wavelength). The sensor operates on the super-heterodyne principle, which provides very high spectral resolution, allowing determination of the exact shape of the emission line. The altitude range over which high-quality ozone profiles can be determined from MR measurements is typically 15 to 70 km (Zommerfeld and Kunzi, 1989) with an altitude resolution of ~10 km. By extending the bandwidth of the receiver from 200 MHz to 1,200 MHz it is possible to include the lower stratosphere (Rubin and Kunzi, 1989). The accuracy claimed for this sensor is 1 ppmv and the statistical error is of the order of 15%. The measuring time required for a complete profile is approximately one hour.

Note that accuracy can be improved by using observing sites at high altitudes, thus reducing attenuation by tropospheric water vapor. Model calculations (Kunzi and Rubin, 1988) indicate that for >90% of available observing time the sensor has to be at an altitude of 3,000 m in the tropics, whereas in polar regions the site can be at any altitude. Ozone profile data were also taken with microwave instruments in southern California during 1989 and 1995 (Parrish, A., *et al.*, 1998).

The operation and data retrieval of an MR can be fully automated and require only little maintenance (see also the MLS instrument and RAM described in Sections 1.7.3.3 and 1.6.9, respectively).

1.6.7 Ground-based Millimeter wave Ozone Spectrometer (GROMOS)

This spectrometer has been performing continuous measurements of the pressure-broadened millimeter wave line emission of ozone at 142.175 GHz at the University of Bern, Switzerland, since November 1994. Stratospheric and mesospheric ozone is measured on an operational basis from Bern as a part of ESMOS (European Strato-spheric MOnitoring Stations) using the GROMOS (Jost *et al.*, 1998). A similar instrument is the Operational Ground-based Ozone Sensor (OGOS) providing ozone profiles from 15 to 75 km (Zommerfeld and Kunzi, 1989). The collected ozone profiles are retrieved from microwave observations using the optimal estimation method. Retrieval of ozone profiles is carried out for the altitude range of 12–80 km, with an altitude resolution of 10–15 km and a time resolution of 1–2 h (Peter *et al.*, 1998). A ground-based millimeter wave radiometer observing pressure-broadened 204 GHz ClO transition lines is installed at the International Scientific Station at Jungfraujoch in the Swiss Alps. In 1995 the radiometer was modified and equipped with an additional front-end to measure the ozone transition line at 195 GHz (Siegenthaler *et al.*, 1998).

1.6.8 Stratospheric Sounding by Infrared Heterodyne Spectroscopy (SIRHS)

Trace constituents in the stratosphere may be detected by exploiting their infrared vibrational–rotational lines. Measurements of individual line shapes provide the density–altitude profile by finding an inverse solution of the radiative transfer equation. However, measurements of individual line intensity profiles are very difficult with conventional spectroscopy, because of pressure-broadening wings and Doppler cores in the line profiles.

Infrared heterodyne spectroscopy (IRHS) gives high specific detection sensitivity and ultrahigh spectral resolution ($\sim$5 MHz), and therefore individual line shapes can be measured without significant instrumental distortion (Abbas *et al.*, 1979). Stratospheric measurements may be performed either in absorption or in self-emission. In a heterodyne receiver, the infrared signal from the source is mixed with coherent radiation from a local oscillator (such as a CO_2 laser or a tunable diode laser), and the difference frequency signal is detected by radio-frequency techniques. The vertical mixing ratio profiles and total abundances of some constituents may be determined with high vertical resolution ($\sim$4 km) and diurnal variation may be followed by tracking the Sun or the Moon. The Rutherford Appleton Laboratory has developed a high-sensitivity airborne 500 GHz heterodyne receiver designed to measure the spectral line emission of ClO, BrO, O_3, N_2O, and HNO_3 (Siddans *et al.*, 1998).

1.6.9 Ground-based microwave radiometers

The MIRA Team (MIcrowave RAdiometry) from the IMK (Institute of Meteorology and Climate Research, Forschungszentrum und Universität Karlsruhe) has been operating a 278 GHz radiometer system, which is intended for long-term ground-based monitoring of the vertical profiles of stratospheric ClO, O_3, HNO_3, and N_2O since 1996. To improve the measurement of weak lines and to allow operation over a wider range of weather conditions an advanced balancing technique was implemented. Several measurement campaigns were carried out at Kiruna, Sweden, between February 1996 and March 2001, on the Zugspitze from February to July 2003 and at Ny-Ålesund, Svalbard, in March 1997. Since March 2004 the instrument has been operated on Pico Espejo at Mérida in Venezuela.

Another instrument, the ground-based KIruna Millimeter wave RAdiometer (KIMRA) has been operated almost continuously by the Swedish Institute of Space Physics (Institutet för Rymdfysik, IRF) at Kiruna (67.84°N, 20.41°E, 425 m asl), since 2002. KIMRA was developed as a cooperation between IMK and IRF. It was designed for the observation of thermal emission lines of stratospheric trace gases between 195 and 225 GHz to measure ClO, O_3, HNO_3, and N_2O in a continuous mode 24 hours a day throughout the year. Additionally, measurements of tropospheric transmission and tropospheric water vapor columns are routinely carried out. Data analysis of the measured spectra is performed at IMK in Karlsruhe using the same methods as for the MIRA measurements (*http://www.irf.se/program/afp/mm/*).

The radiometer at IRF has an obvious advantage in the location above the polar circle with only a short polar night. This is favorable for studying the evolution of polar stratospheric winter chemistry both inside and outside the polar vortex and for investigating early ozone loss. With a detailed study of the potential vorticity of the polar vortex using the Equivalent Latitude Method (Nash *et al.*, 1996) it is possible to identify whether measurements have been taken inside or outside the Arctic vortex. Since ground-based measurements over a longer time span like a week observe a substantial part of the inside vortex airmass, they provide some kind of vortex-averaged data.

The Radiometer for Atmospheric Measurements (RAM) is a ground-based passive microwave radiometer (installed at Ny-Ålesund; 78.9°N, 11.9°E) which detects the mixing ratio profiles of ozone as well as chlorine monoxide. It can be operated all year round independently of the weather, performing one to five profiles per hour. Its vertical resolution is 10–15 km and it is dependent on altitude. Its altitude range is 12–55 km with 0.1 ppm precision at a height of 20 km.

A similar radiometer was installed at the Bordeaux Observatory, France (45°N, 5°W) in 1995. This instrument is tuned to 110.836 GHz and ozone profiles are retrieved using the optimal estimation method in the altitude range 25–65 km.

Also, two ground-based microwave instruments, capable of measuring OVP between 20 and 70 km, are operated at the Jet Propulsion Laboratory. These measurements are considered accurate because the wavelength at which these instruments operate (2.7 mm) is very large compared with aerosol particle size (Parrish, A., *et al.*, 1998).

Since November 1994, the RAM is part of the permanent installation at the International Network for the Detection of Stratospheric Changes (NDSC).

1.6.10 Ground-based infrared solar spectroscopy

Ground-based high-resolution infrared (IR) solar spectroscopy is a powerful tool, used to measure OVP, in the study of the vertical distributions of various atmospheric trace gases (Pougatchev *et al.*, 1995). For instance, OVP in the troposphere and the lower stratosphere has been retrieved from a series of 0.0035–0.013 cm^{-1} resolution infrared solar spectra recorded by Fourier transform spectrometers at Kitt Peak in the U.S.A. (32°N, 112°W) and at Lauder in New Zealand (45°S, 169.7°E). Direct comparison of OVP from IR ground-based high-resolution solar spectra with ozonesonde, microwave, and Dobson data showed that IR total and O_3 columns above 20 km agree with Dobson and microwave measurements to better than 2% (see Section 3.3.3). For the lowest layer and 12–20 km zone the differences between IR and integrated ozonesonde columns are $(7 \pm 5)\%$ and $(2 \pm 4)\%$, respectively (Pougatchev *et al.*, 1998).

1.6.11 Stratospheric Ozone Monitoring Radiometer (SOMORA)

The stratospheric ozone monitoring radiometer (SOMORA) has operated in Switzerland since January 2000 and delivers data to the Network for the Detection of Atmospheric Composition Change (NDACC) (Maillard-Barras *et al.*, 2009). In 2002, SOMORA was moved from Bern to Payerne, about 50 km away.

SOMORA measures the ozone volume mixing ratio in the stratosphere and lower mesosphere (20–65 km) with a vertical resolution of 8–15 km and a time resolution of 30 min.

SOMORA is a triple-switched, total power microwave radiometer measuring the thermal emission line of ozone at 142.175 GHz (Figure 1.25). It essentially consists of a heterodyne receiver by which the broadband signal emitted spontaneously by the atmosphere is collected, filtered, and converted to a lower frequency prior to being analyzed by the detectors. The SOMORA detectors are two acousto-optical spectrometers (AOS). The first AOS is characterized by a detection bandwidth of 1 GHz distributed over 1,024 channels. Its spectral resolution is sufficient for the retrieval of ozone profiles in the stratosphere. The second AOS is used to monitor mesospheric ozone and focuses at the center of the observed emission line with a bandwidth of 50 MHz distributed over 2,048 channels. In the total power calibration mode, the dominant intrinsic system noise is separated from the very faint atmospheric signal by observing, besides the atmosphere, two calibration sources alternately: one cold (liquid nitrogen at 77 K) and one warm (electric heating plate at 300 K).The system noise temperature obtained in this way with the SOMORA instrument is of the order of 2,200 K (single-sideband, uncooled), whilst at the considered frequencies the typical background sky brightness temperature lies in the range of 50–250 K for an ozone line amplitude between 20 and 0 K. After 30 minutes of integration time, a sufficient signal-to-noise ratio is achieved with SOMORA spectral line measurements to retrieve ozone volume mixing ratio (VMR) profiles with low *a priori* information content in the altitude range of 20 to 60 km. Outside this altitude range, the measurement information content is low and the retrieved profiles are determined mainly by an *a priori* information guess (Barras *et al.*, 2009).

Figure 1.25. The SOMORA instrument (*http://www.issibern.ch/~yasmine/Publications/ndsc_application.pdf*).

1.7 AIRBORNE INSTRUMENTATION FOR OVP OBSERVATIONS

Airborne measurements of the vertical profile of ozone concentration are very accurate because the instruments are brought into direct contact with the atmosphere; they are not remote-sensing instruments. However, airborne measurements cannot provide a global picture of ozone distribution because of the local area they cover. The most common types of airborne measurements are from balloons, aircraft, and rockets using the instrumentation briefly described in the following sections.

Ozonesondes are balloon or rocketborne instruments, flown with standard meteorological radiosondes, performing *in situ* measurements of ozone vertical profiles up to altitudes of about 35 km. Ozonesonde measurements, after correction for residual ozone at altitudes above the balloon burst level, are normalized in order to reach agreement between the integrated ozone of the sonde and the TOC observed by a nearby TOC-measuring instrument (e.g., Dobson Spectrophotometer). The current ozonesonde network is geographically uneven with the large majority of stations in Europe and North America. The two most commonly used devices are the electrochemical and optical ozonesondes, which are now briefly discussed.

1.7.1 Electrochemical ozonesondes

Electrochemical ozonesondes utilize iodine iodide redox electrodes and have been used for many years to measure the spatiotemporal variation of the vertical distributions of ozone in both the troposphere and lower stratosphere.

Various versions of these ozonesondes have so far been constructed. The most commonly used are the electrochemical concentration cell (ECC) ozonesonde (Komhyr, 1969, 1987; Komhyr and Harris, 1971) and the Brewer–Mast (BM) electrochemical cell ozonesonde (Brewer and Milford, 1960; Mast and Saunders, 1962; Griggs, 1964). Both types measure ozone concentration via its reaction with an aqueous solution of potassium iodide.

Electrochemical concentration cell (ECC) ozonesondes are launched more or less regularly on board small meteorological balloons at a variety of stations from pole to pole. They yield the vertical distribution of ozone (expressed in volume mixing ratio (VMR)) from the ground up to burst point, the latter occurring typically around 30 km. Ozone VMR recorded at a typical vertical resolution of 100–150 m is converted to an ozone number density using pressure and temperature data recorded on board the same balloon. Errors in the ozone profiles of ozonesondes depend on a large number of parameters. For ECC sondes the important parameters are: the manufacturer of the sonde (SPC or ENSCI), the percentage of the sensing solution used in the electrochemical cell, and the type of correction applied for pump efficiency. Unfortunately, this information is not always given or well identified in the data files (see also Section 3.8.6).

The ozone sensor used in the Electrochemical Concentration Cell (ECC) ozonesonde is an iodine iodide redox electrode concentration cell that consists of two bright platinum electrodes immersed in potassium iodide solutions of different concentrations, contained in separate cathode and anode chambers. The chambers are fabricated from polytetrafluoroethylene (PTFE) and are linked together with an ion bridge that serves as an ion pathway and retards mixing of the cathode and anode electrolytes. The electrolyte in each chamber also contains potassium bromide and a buffer. The concentrations of these chemicals are the same in each half-cell. A similar technique is currently used in the model 2Z ECC ozonesonde produced by ENSCI Corporation. The ECC sensor does not require application of an external electromotive force (EMF) for operation (as in the BM sensor). Power for the cell is derived from the difference between the potassium iodide concentrations present in the two half-cells (Komhyr, 1969). A small pump is used to pass air through the ozonesonde cells. When air containing ozone is passed through these cells a chemical reaction occurs. The reaction is an iodide redox reaction. The platinum electrodes do not take part in the chemical reaction, since they are inert; they only carry electrons between cells of the sensor. Electrochemical reactions take place in the boundary layers of the electrodes. Each ozone molecule releases an iodine (I_2) molecule from the electrolyte, which is transformed by the cell to iodide ($2I^-$) resulting in current flow through the external circuit because of the flow of two electrons. An electrical current is generated in the proportion to the rate at which ozone enters the cell. The electronic interface relays the electric current data to the radiosonde. Ozone and meteorological data are telemetered to a receiving station where, using a microcomputer and appropriate software, they are transformed into readable information. The ozone concentration in the air is determined by an equation that also takes into account the airflow rate, air pressure, and pump temperature. The precision of an ECC for ozone measurement in the troposphere is 5–13% and the accuracy ±10%. In the stratosphere at 10 hPa the accuracy is ±5% but varies from ±5% to ±20% at pressure levels lower than 10 hPa. The Brewer–Mast sondes perform a little less well. ECC ozonesondes yield TOC that agree well with the Dobson spectrophotometer ozone values derived from observations on AD wavelengths (Komhyr *et al.*, 1989b).

In this respect, the EN-SCI Corporation conducted laboratory tests to check the performance of sample ECC sensors using a Thermo Environmental Instrument Incorporated Primary Standard Ozone Calibrator with calibration standards set by the U.S. National Institute of Standards and Technology (NIST). This quality control indicates that ECC sondes, when charged with standard 1% KI cathode electrolyte, overestimate ozone by 5–7% in the region of the atmospheric ozone maximum. The problem is corrected by using a 0.5% KI concentration of ECC sensor cathode electrolyte instead of the 1% KI solution concentration currently used. Additionally, ECC ozonesonde soundings made using the less concentrated KI cathode electrolyte yield integrated ozone amounts that are in excellent agreement with Dobson TOC observations (Komhyr *et al.*, 1995).

The use of balloonborne ozonesondes is a well-established technique used in the Global Atmosphere Watch (GAW) program of the World Meteorological Organization (WMO) to detect long-term changes of ozone and to validate satellite measurements up to 35 km altitude. Ozonesondes are also widely used in strategically designed networks such as SHADOZ in the tropics, IONS in the mid-latitudes, and MATCH in the polar regions to study atmospheric processes (see Sections

6.6.3.3, 6.6.3.2, and 6.6.3.1, respectively). Since 1996 several JOSIE (Jülich Ozone Sonde Intercomparison Experiment: *http://www.fz-juelich.de/icg/icg-2/josie*) ozonesonde simulation experiments to investigate and evaluate the performance of different ozonesonde types and to guarantee consistency in ozonesonde data have been conducted at the ozonesonde simulation facility at the Forschungszentrum Jülich (*http://www.fz-juelich.de/icg/icg-2/esf/*). In addition, in April 2004 the WMO/BESOS (Balloon Experiment on Standards for Ozone Sondes) field campaign at the University of Wyoming at Laramie was held to test the representativeness of JOSIE results in the real atmosphere. The results of JOSIE and BESOS have demonstrated that in studying trends in ozonesonde data it is necessary to exercise caution when there have been changes in instruments or in preparing/operating procedures. Under the auspices of WMO/GAW the Assessment of Standard Operating Procedures for Ozone Sondes (ASOPOS) has been conducted.

The Brewer–Mast (BM) ozonesonde uses a patented electrochemical sensing cell developed by Dr. Alan W. Brewer with one platinum electrode and one silver electrode. As with the ECC ozonesonde, the output of the bubbler cell is a current proportional to the rate of sampling of the ozone. The following relationship is used:

$$P_3 = 4.31 \times 10^{-3} ITt \tag{1.20}$$

where P_3 is the partial pressure of the ozone (mb); I is the current (μA); T is the air pump temperature (K); and t is the time (s) to pump 10 ml of air sample.

The calibration curve assumes a constant air pump temperature of 300 K (De Backer *et al.*, 1998). Table 1.5 gives an approximate correction for the air pump temperature. The percentage increase has to be added to the partial pressure (add) depending upon the altitude (ambient pressure).

Versions of the instrument are manufactured in the U.S.A., Germany, and India (e.g., the Wyoming digital electrochemical ozonesonde). Improvement in data quality is achieved by normalization of all ozonesonde ascents to Dobson spectrophotometer TOC. Recently, the EN-SCI Corporation (Environmental Science Corporation) produced the Model KZ-ECC which is a portable atmospheric ozone-sounding system.

In Japan model RSII-KC79 ozonesondes manufactured by Meisei Electric Company (Ibaraki) have been used for weekly observations at four sites since 1989 (Fujiwara *et al.*, 1998). The Japanese KC79 ozonesonde is a modified version of the KC68 sonde-type developed by Kobayashi and Toyama (1966). Both sonde types are based on the carbon–iodine ozone sensor. This is a single electrochemical cell containing a platinum gauze as the cathode and an activated carbon anode immersed in an aqueous neutral potassium bromide solution. As ozone is passed through the sensing solution it generates free bromine molecules (Br_2). At the Pt cathode, bromine is re-converted into two bromide ($2Br^-$) ions by the uptake of two electrons, while correspondingly at the activated carbon anode the following reaction takes place:

$$C + 2OH^- \rightarrow CO + H_2O + 2e \tag{1.21}$$

Table 1.5. An approximate correction for the air pump temperature. The percentage increase has to be added to the partial pressure (add) depending upon the altitude (ambient pressure) (De Backer *et al.*, 1998).

Ambient pressure (hPa)	*Add* (%)
1013–250	2
250–120	1
120–70	0
70–55	−1
55–38	−2
38–25	−3
25–17	−4
17–11.5	−5
11.5–7.7	−6
7.7–5.2	−7

releasing two electrons. Accordingly, one ozone molecule produces an electrical current of two electrons in the external circuit. The gas-sampling pump and the electrochemical cell are made of methacrylate resin. The pump flow rate is about $400\,cm^3/min$ with the pump motor speed being held constant by a governor. The sonde is enclosed in a Styrofoam flight box (see Figure 1.26).

The ozonesonde used in India is a hybrid Brewer–Milford-type ozone sensor made of Plexiglas and a non-reactive Teflon pump. All parts of the sonde are manufactured and assembled by the Indian Meteorological Department. The sonde is mounted in a polystyrene flight box. A schematic diagram of the sonde is shown in Figure 1.27. The typical flow rate of gas sampling is about $200\,cm^3/min$. The anode is a silver wire and the cathode a cylindrical Pt gauze in a neutrally buffered aqueous KI solution (0.25%).

1.7.2 Optical ozonesondes

This instrument consists of a multi-filter UV radiometer launched on board an ARCAS or a Super-Loki rocket to a height of about 70 km and permitted to fall back to Earth with a parachute. Direct solar irradiance in several channels is measured as the instrument descends on the parachute and ozone concentration is

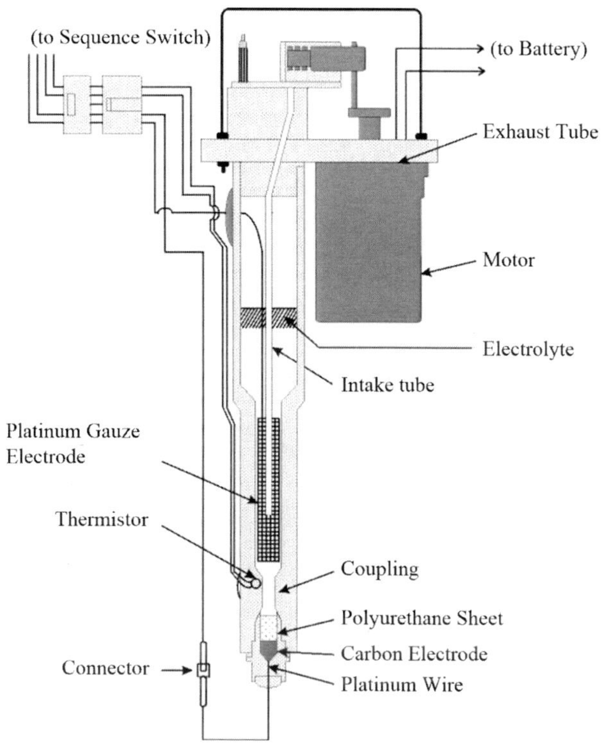

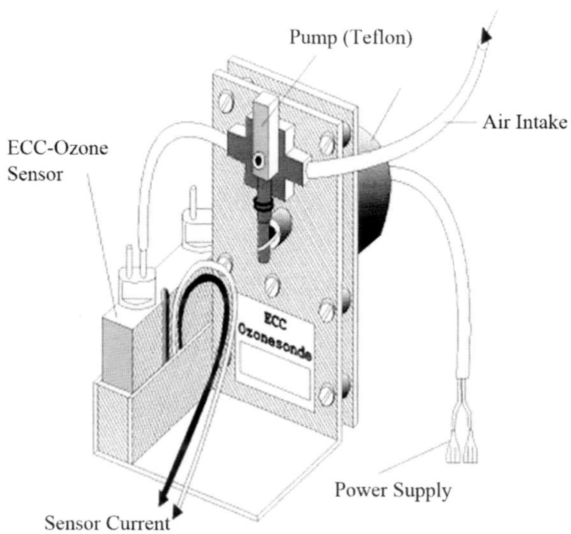

Figure 1.26. A schematic diagram of the Japanese ozonesonde (ICG, 2006) (top) and of an electrochemical concentration cell (ECC) ozonesonde (bottom) (*http://www.fz-juelich.de/icg/icg-2/josie/ozone_sondes/ecc/*).

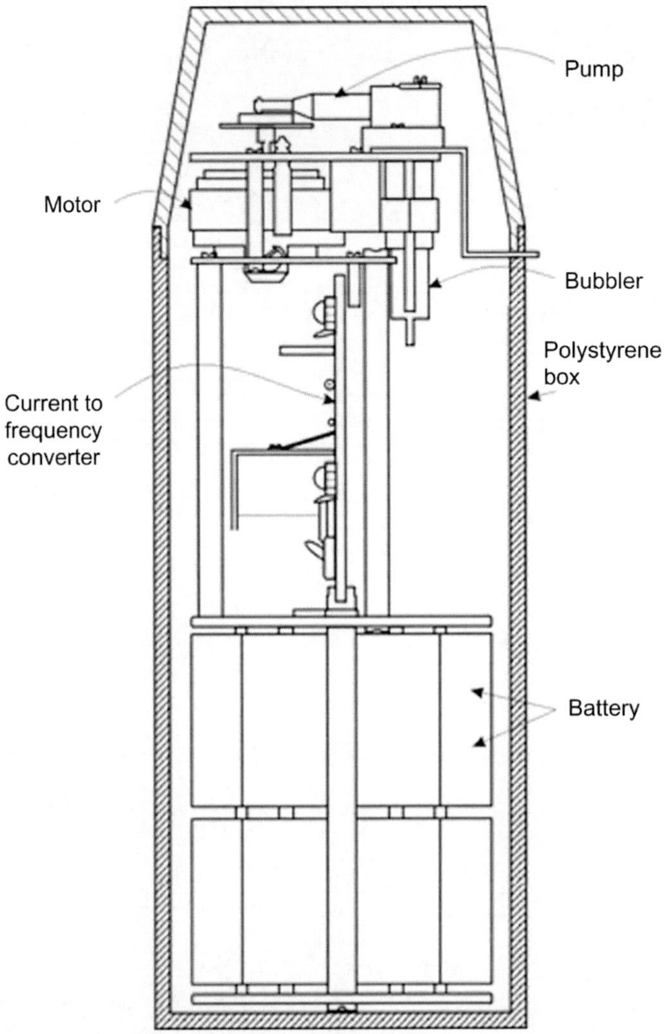

Figure 1.27. A schematic diagram of the Indian ozonesonde (ICG, 2006).

determined using the Beer–Lambert–Bouguer law. The precision and accuracy depend on the accuracy of the ozone absorption cross-sections and knowledge of the filter transmission. Precision is given as approximately 2.5% and accuracy is estimated at 5–7%. An example of this instrument is the optical Rocket Ozonesonde (ROCOZ), which uses four filters from 267 to 309 nm. The shortest wavelength is used to measure ozone from 39 to 54 km altitude and the longer from 20 to 47 km altitude. Samples are taken every 0.6 s, giving measurements every 80 m at 55 km and every 9 m at 25 km. The improved rocket ozonesonde (ROCOZ-A) includes electronics that provide essentially drift-free outputs throughout ozone soundings. It should be stressed that the fundamental ozone value measured by ROCOZ-A is the vertical ozone overburden as a function of geometric altitude (Barnes and Simeth, 1986). Measurements by ROCOZ-A ozonesonde were initiated by NASA in 1983.

1.7.3 Other balloon instrumentation

In the previous two sections, the basic versions of ozonesondes that can be carried into the troposphere and stratosphere using either balloons or rockets were discussed. Other instruments that perform *in situ* measurements of atmospheric ozone, mostly balloonborne, are presented in this section.

1.7.3.1 Chemiluminescence ozonometer

Regener (1960, 1964) developed a very fast response chemiluminescent ozonesonde that often exhibited considerable variations in sensitivity to ozone; this was later corrected. The chemiluminescence ozonesonde or rocketsonde exploits the chemiluminescence reaction between ozone and the dye Rhodamine B in order to measure the stratospheric ozone profile (20–65 km) with a vertical resolution of 0.5–1 km as the instrument ascends by balloon or descends by parachute. Atmospheric air is brought into a reaction chamber that is shielded from sunlight (Rhodamine B fluoresces in the 500–615 nm region and is sensitive to sunlight). In the chamber ozone reacts with the dye and the resulting luminescence is monitored (Krueger, 1986). Sometimes gas phase chemiluminescence is also used by employing such gases as 2-methyl-2-butene and NO (see also FOZAN described later in Section 1.7.3.8). A novel chemiluminescence sonde was described by Speuser *et al.* (1989) which can measure ozone up to 1.5 mb.

The accuracy of the chemiluminescence rocketsonde/ozonesonde is about 12% and the precision approximately 6%. Note that three types of ozonometric apparatus were used on the meteorological rockets M-100B: heterophase chemiluminescent, gas phase chemiluminescent, and optical (Perov and Tishin, 1989).

1.7.3.2 Mass spectrometer

This instrument has been used to measure vertical distribution of ozone in the stratosphere and in the mesosphere. For instance, a liquid helium-cooled mass spectrometer was developed and demonstrated in the 1990s for measuring ozone (and other species) when balloonborne in the stratosphere.

This instrument uses liquid helium to trap most of the gases entering it and ionizes the molecules, which travel along certain pathways. Then by using magnetic fields, various ions are separated into different paths in the measurement chamber. Given that the gas flow rate is known, thus the fractional composition of the gas can be identified. The measurement accuracy for ozone is about 2% (Mauersberger *et al.*, 1985).

1.7.3.3 Balloon Microwave Limb Sounder (BMLS)

The balloonborne version of this instrument operates in the 200–300 GHz range and was the first use of a microwave remote-sensing technique for limb sounding. The brightness temperatures at various frequencies within emission lines from various molecular species in the stratosphere can be measured using microwave local oscillators and a filter bank (35 filters per local oscillator). From these measurements and by looking at various limb angles the altitude profiles of concentrations of species can be inferred (ClO, O_3, H_2O, NO_2, HNO_3, N_2O, and HCN). For OVP measurements the O_3 spectral lines at 206.132 GHz were chosen because of the insensitivity of O_3 emission to stratospheric temperature variations. For instance, a temperature variation of 10°C changes the O_3 emission by only 1% (Waters *et al.*, 1993).

An important version of the MLS has been used on aircraft and in the Upper Atmospheric Research Satellite (UARS) launched in 1991. Another MLS version with additional frequencies between 63 and 640 GHz is currently used on the Earth Observing System (EOS) platform, which is in a 98° Sun-synchronous orbit (permitting measurements in polar regions).

1.7.3.4 Far-infrared emission measurements

The instruments used to study molecular species in the stratosphere (emission spectral region from 7 to 200 cm^{-1}) are Fourier transform spectrometers. To reduce detector noise cryogenically cooled detectors are used and then the resolution varies from 0.5 to 0.0033 cm^{-1}. Direct comparison of atmospheric emission with the blackbody signal at liquid nitrogen temperature leads to determination of the emissions made by atmospheric molecular species. These instruments are pointed outward by a few degrees above the horizon. An "onion-peeling" technique is used to separate signals into components from various altitudes.

SAFIRE (Spectroscopy of the Atmosphere using Far IR Emission) was launched on 2004 and is used for the Earth Observing System (EOS). The main aim of the SAFIRE experiment is to improve understanding of mid-atmosphere ozone distribution by performing global-scale measurements of the most important chemical, radiative, and dynamical processes that influence ozone changes. The instrument will measure ozone in the 10–100 km altitude range, with an accuracy of 10%. It operates day and night and covers the latitude belt 86°N-86°S (Megie *et al.*, 1989).

REFIR (radiative explorer in the far-infrared, Palchetti *et al.*, 2004) is a broadband FT spectro-radiometer capable of measuring atmospheric spectral radiance in the 100–1,100 cm^{-1} wave number region, with a resolution of 0.5 cm^{-1}, using room temperature detectors. A combined package of the REFIR and SAFIRE instruments, simultaneously operating in nadir geometry aboard the same platform (either a stratospheric balloon or high-altitude aircraft), is ideally suited for an airborne mission devoted to the study of atmospheric radiative properties and could make a significant contribution in improving our understanding of the processes that regulate the ERB (Earth Radiation Budget), including the distribution of water vapor and clouds (Cortesi *et al.*, 2005).

A Fourier transform infrared spectrometer built at JPL in 1984 is the Mk IV balloon interferometer (high resolution, 0.01 cm^{-1}) operating in solar absorption mode (Ostermann, 1999).

1.7.3.5 Stratospheric Infrared Interferometer Spectrometer (SIRIS)

Infrared emission spectroscopy using Fourier transform spectroscopy (Section 1.5.4) can be employed to measure the vertical distributions of ozone and other molecular species in the stratosphere. It is necessary to measure the vertical profiles of both temperature and absolute radiance to conduct such measurements. The stratospheric IR Interferometer Spectrometer using dichroic filters divides the 700–1,920 cm^{-1} region into four channels. The region between 1,125 and 1,425 cm^{-1} has been used to perform observations for O_3 and other species in the 12–40 km altitude region. The instrument, which is balloonborne, allows (by scanning below the horizon) measurements with a vertical resolution of 3 km (Grant, 1989).

The total error associated with SIRIS varies from 14% at 38 km to 42% at 16 km.

1.7.3.6 Balloonborne lasers

The Balloonborne Laser In Situ Sensor (BLISS) is a microprocessor-controlled high-resolution absorption spectrometer operating in the mid-infrared region using tunable diode laser (TDL) sources.

It performs observations of ozone and other stratospheric trace species by measuring the absorption from balloonborne diode lasers and reflected back from a retro-reflector suspended ~500 m below the balloon. For each gas of interest in the stratosphere, a narrowband tunable diode laser is scanned over a few selected infrared absorption lines and the mean gas concentration over the fixed path length is determined from the depth of the observed absorption lines through Beer's law ($dI/dx = -kI$) which states that there is a logarithmic dependence between the transmission of light through a substance and the product of the absorption coefficient of the substance and the distance the light travels through the material (i.e., the path length). This instrument's calibration uncertainty is 4.8%, overcoming various factors of uncertainty like temperature, pressure, and laser half-width at half-maximum. As far as accuracy is concerned, this is governed by precision in combination with uncertainty in the absorption cross-section of the ozone line and its temperature dependence and broadening coefficient, which lead to a total 10% accuracy in the 28–38 km altitude range (Grant, 1989). Another balloonborne *in situ* ozonometer based on the absorption of a laser beam by ozone in the Hartley band was built in France (Moreau *et al.*, 1989).

The balloonborne instrument SPIRALE (Spectroscopie Infrarouge par Absorption de Lasers Embarqués) has been developed (Moreau *et al.*, 2005; Pirre *et al.*, 2008) and used for *in situ* measurements of several tracer and chemically active species, including ozone, in the stratosphere. Laser absorption takes place in an open Herriott multipass cell located under the balloon gondola, with six lead salt diode lasers as light sources. One mirror is located at the extremity of a deployable mast

3.5 m below the gondola, enabling the measurement of very low abundance species throughout a very long absorption path. Fast measurements (every 1.1 s) allow a vertical resolution of 5 m for the profiles to be obained. A detection limit of a few tens of parts per trillion in volume has been demonstrated. Uncertainties of 3%–5% are estimated for the most abundant species rising to about 30% for the less abundant ones, mainly depending on the laser linewidth and the signal-to-noise ratio.

1.7.3.7 Ultraviolet Absorption Photometer

This is an *in situ* instrument that performs observations of ozone concentrations by measuring the absorption at 253.65 nm (near the UV ozone peak). It is actually a modified version of the Dasibi ozone monitor which is often used for ozone measurements at the ground level of air pollution. Dasibi-like devices use a mercury lamp as the spectral source, a narrowband filter, a silicon photodiode as the detector and a pump to pass air through the sample volume. For half of the measurement time, an ozone scrubber removes the ozone thus providing a reference atmosphere (Sections 1.7.4.3 and 1.8.3). In general, UV photometers have good precision (3%) and accuracy (5%). They are typically used on large balloons with other instruments and during return to the ground observe the fine structure in ozone distribution. For instance, a dual-beam UV absorption photometer (or balloon dual-beam UV *in situ* O_3 photometer) for balloonborne measurements of ozone has a response time of 1 s and an uncertainty at the ozone maximum of 3.6%, where 2% of this is the accuracy of the ozone cross-section. This photometer has two identical absorption chambers, each alternating between reference mode (ozone free) and sample mode by means of a four-port valve and ozone scrubber. The ratio of the absorption signals, along with the known lengths and ozone absorption cross-section, yields the ozone concentration (Proffitt and McLaughlin, 1983). The Deutsches Zentrum für Luft und Raumfahrt (DLR) research aircraft *Falcon* has also been used for ozone measurement using a UV absorption photometer (Ziereis *et al.*, 1998). A balloonborne Sun-tracking photometer operating at six wavelength bands centered at 210, 215, 290, 750, 860 and 950 nm was developed and used to measure the OVP in the 10 to 35 km altitude region (Lat *et al.*, 1989).

In the past this instrument has flown aboard gondolas carrying other instruments to provide data for the validation of ozone measurements of several satellite instruments including UARS MLS, HALOE, CLAES, and ADEOS ILAS. In 1996–1998 a suite of *in-situ* instruments was assembled and flown as the Observations of the Middle Stratosphere (OMS) gondola to expand the database obtained with the ER-2 (NOAA) instruments. Suggestions for improved operation practices of the ultraviolet O_3 monitors and recommendations for future testing were made by Dunlea *et al.* (2006).

Murata *et al.* (2009) described the development of a balloonborne optical ozone sensor and its operation since 1994 to observe the vertical distribution of upper stratospheric ozone in summer using a thin-film high-altitude balloon at Sanriku, Japan (39.2°N, 141.8°E). The sensor measures solar ultraviolet radiation in the

Hartley band at a wavelength of 300 nm. The optical system is described by Okano *et al.* (1996) and Murata *et al.* (2009). The balloon used in the observation is a thin-film high-altitude balloon developed by the Institute of Space and Astronautical Science (Japan Aerospace Exploration Agency) and it can attain an altitude of about 42 km. Vertical ozone distributions from 15 to 42 km were obtained with 1 km resolution from the altitude variation of the solar UV intensity. The temporal variations of ozone concentrations above 30 km from 1994 to 2007 showed correlations with the 11-year solar cycle and there was no evidence of decreasing or increasing trends.

1.7.3.8 Fast Ozone Analyser (FOZAN)

This is a chemiluminescent sensor based on the heterophase luminescence of ozone in a solid-state matrix filled with gallic acid dye. Inside a chemical reactor, the sensor is placed along a line where the airflow to be analyzed is provided by a pump. The luminescence light produced by the sensor is modulated by an optical liquid crystal chopper, regulated by a controller unit, and registered by the photomultiplier. A double-entry valve connects to the chemical reactor, either with a built-in calibrating ozone generator or with the airflow to be analyzed. The calibrating ozone generator supplies an airflow, with a strictly specified ozone concentration, to the chemical reactor. The generator produces ozone because of the effect of UV radiation from a mercury lamp on the airflow. The instrument can operate in two modes: (a) the measurement mode in which the airflow (after being thermostatted) is passed through the valve to the reactor to undergo the ozone–sensor chemiluminescence reaction; (b) the calibration mode in which the valve passes the airflow to the reactor through the ozone calibration cell and the calibrating ozone generator. The two modes are switched automatically. This instrument was used on board the high-altitude aircraft M-55 *Geophysica* (Figure 1.28) at the beginning of 1997 (Georgiadis *et al.*, 1998).

1.7.3.9 SAOZ sonde

This is a small balloonborne version of the SAOZ UV–visible spectrometer (Pundt and Pommereau, 1998). It measures vertical profiles of O_3, NO_2, H_2O, and ClO_2 by solar occultation in the UV and visible range (Pommereau and Piquard, 1994a, b). Measurements are performed at an altitude of 28–31 km and during the balloon ascent. One complete measurement cycle is made every 30 s for every 200 m in altitude during the balloon ascent and every 300–1,000 m during solar occultation depending on the altitude and latitude.

1.7.4 Aircraft instrumentation

Two of the balloonborne instruments—the UV photometer (Section 1.7.3.7) and the diode laser *in situ* (Section 1.7.3.6)—are also used on aircraft. Among other devices providing measurements of airborne ozone vertical profile are the Infrared Grating

Figure 1.28. The M55 *Geophysica* high-altitude aircraft (*source*: *http://www.phmi.uni-mainz.de/Bildergalerie/geophysica_I.JPG*).

Spectrometer and the Differential Absorption Lidar Technique, which are discussed in more detail in the following two sections.

1.7.4.1 Infrared Grating Spectrometer (IGS)

This instrument performs measurements of ozone and other trace species in the stratosphere. It observes absorption in the long optical path by the instrument, while observing the Sun at sunrise or sunset.

A modified version of this device, the grille spectrometer, is a correlation instrument with large-area geometrical patterns (grilles), instead of slits, at both the entrance and the exit of the spectrometer. The favorite arrangement has a set of hyperbolic curves in four quadrants, with alternate shaded bands adjacent to quadrants of mirror images. This specific arrangement scans rapidly ($0.2\,cm^{-1}\,s^{-1}$) over a limited spectral range (5–$10\,cm^{-1}$) (Grant, 1989).

This instrument was used to measure latitudinal variations in the abundances of ten stratospheric species including ozone on board aircraft between 62°N and 62°S. It can also be used on a balloon to measure H_2O, NO_2, HNO_3, and O_3 in the altitude range 23–40 km.

In this group belongs GOMOS, which measures the vertical profiles of NO_2, NO_3, H_2O, aerosol (extinction), and high-resolution temperature profiles, the latter being derived from atmospheric scintillations (Wehr, 2002). The spectrometer is a medium-resolution grating spectrometer operating in the ultraviolet–visible (at 250–675 nm with 1.2 nm resolution) and near-infrared (at 756–773 nm with 0.2 nm resolution for O_2 for the derivation of atmospheric density and temperature and at 926–952 nm with 0.2 nm resolution for H_2O).

1.7.4.2 Differential Absorption Lidar Technique

The instrument consists of (1) a telescope to capture backscattered photons; (2) a filter to reject out-of-band radiation; (3) a photomultiplier tube and (4) signal chain electronics (see Section 1.6.5). The conventional differential absorption lidar (DIAL) consists of a light source (e.g., a XeCl gas laser) emitting pulses (25 Hz) at two wavelengths, one of which lies in the ozone absorption band. Backscattered radiation (Rayleigh scattering) is observed by the telescope with a photon-counting detector. High resolution is achieved by measuring the time between the emission and detection of light pulses. For ozone measurements, wavelengths of 308 nm (ozone absorbent) and 353 nm (ozone almost transparent) are chosen (Godin *et al.*, 1989; Krueger *et al.*, 1989). The accuracy for ozone-mixing ratios is estimated to be 10–15% in the mid-stratosphere, with a decrease above and below this region. Airborne lidar provides OVP from aircraft cruise altitudes up to 25 km. The vertical and horizontal resolutions are 0.5 km and 50–100 km, respectively (Megie *et al.*, 1989).

1.7.4.3 MOZAIC, CARIBIC, IAGOS

An alternative to commissioning special flights on research aircraft to carry instruments for scientific work is to make use of the goodwill of commercial airlines. Weather services routinely use commercial aircraft to gather pressure, temperature, and wind data, but not generally for research. However, some projects have been developed that involve flying more sophisticated instruments on commercial aircraft for research purposes (Feder, 2008). The use of commercial aircraft allows the collection of highly relevant observations on a scale and in numbers impossible to achieve using research aircraft and where other measurement methods (e.g., satellites) have technical limitations.

Since the mid-1990s MOZAIC (Measurements of Ozone and Water Vapor by Airbus In-service Aircraft), CARIBIC (Civil Aircraft for the Regular Investigation of the atmosphere Based on an Instrument Container), and the predecessor to the current Japanese project, CONTRAIL (Comprehensive Observation Network for TRace Gases by AIrLiner), have been flying scientific instruments on commercial jet aircraft. NASA ran a similar project in the 1970s. Among the things the current experiments measure are ozone, water vapor, carbon monoxide, carbon dioxide, reactive nitrogen species, and aerosols. In the case of CONTRAIL, the main purpose is to understand the global carbon cycle. Data from all three projects are used in climate change and pollution studies. Other uses include validating satellite measurements and guiding computer models of various atmospheric chemical and physical processes.

The MOZAIC program was initiated in 1993 by European scientists, aircraft manufacturers, and airlines to better understand the natural variability of the chemical composition of the atmosphere and how it is changing under the influence of human activity, with particular interest in the effects of aircraft. The program was coordinated by the Laboratoire d'Aérologie (CNRS, Toulouse). MOZAIC consists of automatic and regular measurements of reactive gases by five long-range passenger airliners. A large database of measurements (about 30,000 flights since 1994) allows studies of chemical and physical processes in the atmosphere, validations of global chemistry transport models, and satellite retrievals. MOZAIC data provide detailed climatologies of trace gases at 9–12 km where sub-sonic aircraft emit most of their exhaust and which is a very critical domain (e.g., radiatively and as a result of stratosphere–troposphere exchanges) still imperfectly described in existing models. MOZAIC data also provide frequent vertical profiles over a large number of airports. MOZAIC has provided data from more than 100 million flight kilometers in the upper troposphere and lower stratosphere and 40,000 vertical profiles in the troposphere. The data are open to researchers worldwide.

CARIBIC began operation in 2004 and involves a mobile laboratory in an airfreight container, with instruments for 15 experiments that sample more than 40 different gases and which weighs 1.5 t. Once a month the container is loaded at Frankfurt airport onto a Lufthansa Airbus A340-600 aircraft with a built-in air inlet system. A month later it is unloaded and the data and samples are removed and analyzed and another month later the whole operation is repeated. The container organization is handled by the Max-Planck-Institute for Chemistry's Atmospheric Chemistry Division, Mainz, Germany (*http://www.caribic-atmospheric.com/*).

The Japanese CONTRAIL project began in 1993 and used the Boeing 747 aircraft of JAL (Japan Airlines). The program continues up to the present time. When the original aircraft were replaced the program was extended with new scientific equipment in new aircraft and over extended routes. The main purpose of the program is to obtain basic data on the concentration of greenhouse gases such as carbon dioxide (CO_2) that have been linked to global warming and human activity since the industrial revolution (*http://www.jal-foundation.or.jp/shintaikikansokue/Contrail_index (E).htm*).

MOZAIC and CARIBIC have now been merged into IAGOS (In-service Aircraft for a Global Observation System). IAGOS is one of the new European Research Infrastructures (ERIs) and is preparing the transition from a scheme of individual research projects into a sustainable infrastructure with enhanced measurement capabilities and global coverage. Several airlines are already involved (Lufthansa, British Airways, Air France, Austrian Airways, and Air Namibia) and several other airlines are in the process of becoming involved. It will establish and operate a distributed infrastructure for long-term observations of atmospheric composition, aerosol, and cloud particles on a global scale from a fleet of initially 10–20 long-range aircraft of various airlines (*http://www.iagos.org/*).

In a similar program, the ozone-mixing ratio of ambient air around a Transall C-160 aircraft was detected *in situ* using a UV absorption monitor (Horiba APOA-350E) (Zahn *et al.*, 1998). During the Arctic winters of 1991/1992 and 1994/1995, airborne UV-DOAS (Differential Optical Absorption Spectroscopy) on board a Transall aircraft measured the stratospheric trace gases NO_2, O_3, BrO, and ClO (Grund *et al.*, 1998). Note that on the Solar Absorption Balloon Experiment (SABE-2) the NASA LSC and NASA Goddard "Dasibi" photometers measured lower ozone than SBUV instruments for altitudes above 35 km (Weinstock and Anderson, 1989).

1.7.4.4 Electrochemical Ozone Sensor (ECOC)

This instrument performs *in situ* ozone measurements on board aircraft. For instance, during the APE/POLECAT campaign the M55 *Geophysica* (high-altitude aircraft) (Figure 1.28) sampled vortex filaments with sudden and intense tracer variation in *in situ* ozone data from the ECOC instrument on board the aircraft, while flying on an isentropic surface. This sensor consists of a Teflon sampler with cell, two pumps, and control electronics. The cell is filled with a KI–water solution similar to standard ozonesondes, but specially designed for use on high-altitude research aircraft. Both pumps are identical. The second pump is used to eliminate possible contamination of the electrochemical cell by the air leaving the pressure sensor during ascent. This pump also shortens the time taken by air samples to get from the air intake to the cell. The time resolution of the cell is 5 s and the accuracy of individual ozone readings is better than 5%. The instrument is temperature controlled and is assembled in a metal box and placed in an unpressurized compartment (Steffanutti *et al.*, 1999).

1.7.4.5 Airborne Submillimeter Receiver (ASUR)

ASUR and its predecessor SUMAS (Submillimeter Atmospheric Sounder) have been used for several campaigns such as the European Arctic Stratospheric Ozone Experiment, EASOE (Urban *et al.*, 1998). It is a passive heterodyne radiometer operating at sub-millimeter wavelengths (in the frequency range of 620–650 GHz) and employing novel superconductor–insulator–superconductor detector technology. Vertical concentration profiles are derived from pressure–broadened emission lines of the species under examination. It uses the German Air and Space Research Organization (DLR) research aircraft *Falcon* as a platform. Among the molecules measurable (which show rotational/vibrational transitions in emission) by ASUR are the chlorine species HCl and ClO, O_3, and N_2O (Kleipool *et al.*, 1998).

1.7.4.6 The Gas and Aerosol Monitoring Sensorcraft (GAMS)

The Gas and Aerosol Monitoring Sensorcraft (GAMS) concept is an outgrowth of a SAGE technology development study and a successful Instrument Incubator Program (IIP) proposal. The goal for GAMS is to make essential climate measurements in a cost-effective, sustainable manner. The term "sensorcraft" designates a spaceflight, remote-sensing system that does not distinguish between the instrument that performs the remote measurement and the spacecraft. The system does, however, provide the operational resources needed by the instrument in orbit. The instrument measures the vertical distribution of aerosols, ozone, and other trace species. The GAMS sensorcraft integrates the resource requirements of both the instrument and the spacecraft to minimize size, power consumption, and cost.

The GAMS instrument (Figure 1.29) measures direct solar transmittance along the LOS (line of sight) between the aircraft and the Sun in 1,024 spectral channels from 430 to 1,030 nm. The instrument package includes spectrometer, solar imager, and telescope assemblies that are pointed at the Sun by a two-axis rotating mirror.

Figure 1.29. Photograph of the GAMS spectrometer (*http://www.archive.org/details/1998-L-00864*).

The grating spectrometer and detectors were developed under NASA's Instrument Incubator Program and have exceptionally high performance. The spectrometer disperses the spectral region from 430 to 1,030 nm onto two 512-element hybrid diode–CCD arrays using a dichroic beam splitter. Spectral coverage is nearly continuous except for a gap of approximately 20 nm centered near 730 nm. The pixel-to-pixel spacing in wavelength is a uniform 0.58 nm/pixel across both detectors. Spectral resolution varies across each detector between values of 0.7 and 1.0 nm on the short-wavelength detector and 1.0 and 2.0 nm on the long-wavelength detector. The combination of a diode array, for primary light gathering and conversion of incident photon into electrons, with a pair of CCD-based serial 20-shift registers endows this detector with a 20 million electron full well and a small 10 μm pixel pitch. The entire detector assembly is referred to as the photon-to-bits board and achieves a shot noise-limited signal-to-noise ratio of 4,000:1 as it is digitized to 16 bits. The spectra are acquired at a rate of 180 Hz with a dynamic range above the noise floor in excess of 14 bits. The spectrometer is fed with an off-axis parabolic telescope and has a 0.5 arcmin vertical FOV. The solar imager is a commercially available 1,024 × 1,024 CCD camera with each pixel having a FOV of 0.3 arcmin. A narrow band filter at 865 nm is used in both the imager and the Sun position sensor in the gimbal servo-pointing system and allows pointing errors to be measured and modeled relative to the imager data and enables a means of evaluating scene homogeneity (Pitts *et al.*, 2005).

1.8 SURFACE OZONE MEASUREMENTS

Three methods are widely used for detecting surface ozone in clean air: the chemiluminescence method, the KI method, and the UV absorption method—all of which are briefly discussed in the following sections.

1.8.1 Chemiluminescence method

Several commercial instruments are available and, in general, these consist of an air-intake filter, detector cell, photomultiplier, amplifier, recorder, sample pump, ethylene supply, and flowmeter. Ambient air and ethylene are delivered simultaneously to a mixture zone of a detection cell, where the ozone reacts with the ethylene thus emitting light which is detected with a photomultiplier tube, after amplification of the photocurrent. The sensitivity is approximately 2 μg $O_3\ m^{-3}$. These instruments are calibrated against artificially generated test atmospheres of ozone (WMO, 1978).

1.8.2 Electrochemical potassium iodide method

The apparatus mainly consists of the measuring cell and air pump (integrated in one unit). The principle of the method makes use of the oxidation of the neutral buffered KI solution by ozone. Two electrons per reacting ozone are produced on a platinum electrode at the cathode. The electric current from the measuring cell is amplified by a DC amplifier. The instrument has a sensitivity of 2 to 5 μg $O_3\ m^{-3}$ and a range from this detection limit to clean air ozone concentrations of about 200 μg m^{-3} (Feister *et al.*, 1989). The method is considered accurate to within ±5% (WMO, 1978).

1.8.3 UV absorption method

The principle of "ultraviolet absorption" is based on the fact that ozone molecules absorb ultraviolet energy at a known wavelength of 254 nm. The UV light available

Figure 1.30. Advanced Pollution Instruments (API) Model 400E Ultraviolet Absorption Ozone analyzer (*source*: *http://www.anr.state.vt.us/air/monitoring/htm/InstrumentInfo.htm #ozone*).

for detection is proportional to the amount of ozone in the sample chamber. This method is used by several instruments like the Advanced Pollution Instruments (API) Model 400E Ultraviolet Absorption Ozone analyzer (Figure 1.30). The API unit is set up and operated to monitor ambient ozone concentrations up to 500 parts per billion.

The main instruments that use the UV absorption method are the Dasibi-1003–1008, the ThermoElectron 49 (TEI-49C), and the Environment S.A. In contrast to the Dasibi types, the ThermoElectron is a dual-beam instrument and has the fastest response (Helas *et al.*, 1989). Another instrument that is based on the UV absorption technique is the Dylec Model 1006 AHJ produced under licence from Dasibi Inc. The cycle time of measurements is about 12 s (Ogawa and Komala, 1989).

However, optical meters with gas vessels (e.g., ozone analyzers from Dasibi, Monitor Labs, Thermo Environmental Instruments, etc.) exhibit significant errors associated with the sampling of the air being analyzed, where *a priori* ozone destruction occurs (Balatsko *et al.*, 2008). To avoid this disadvantage an optical track analyzer TrIO-1 for ozone has been developed in the National Research Center for Monitoring the Ozonosphere of the Belarus State University, operating by the principle of DOAS equipment (Balatsko *et al.*, 2008). This instrument has some advantages over optical meters in the way it samples (e.g., it provides average values of ozone concentration over the probing track, etc.).

2

Satellite systems for studies of atmospheric ozone

It is now slightly over 50 years since the era of satellite remote sensing began on October 4, 1957, when the former Soviet Union launched Sputnik 1, the world's first artificial satellite (a 55 cm diameter sphere that weighed 83 kg with four antennae attached to it) (Figure 1.1). In the intervening years, satellites have revolutionized the Earth sciences, including the study of trace gases, in general, and ozone, in particular. With the passage of time, satellite-flown instruments became more complicated, while larger satellites carrying more and more instruments were launched. One of the largest of these was ENVISAT (ENVIronmental SATellite) which was launched on March 1, 2002 by the European Space Agency (ESA), with a mass of 8,211 kg and dimensions in orbit of 26 m × 10 m × 5 m. It was a Sun-synchronous satellite flown at an altitude of 800 km, with an inclination of 98° and an orbital period of 101 min. It carried instruments for monitoring the Earth's land, atmosphere, oceans, and ice caps (Figure 1.1) (*http://envisat.esa.int/*) and three of these were particularly relevant to atmospheric ozone (see Section 2.3.6). It successfully reached its nominal 5-year mission lifetime, having orbited the Earth more than 26,000 times. Figure 1.1c shows Sputnik on the same scale as ENVISAT in Figure 1.1b.

Recently, we have been seeing a trend to move away from large multifunctional Earth-observing satellites, like ENVISAT, to small satellites dedicated to one particular observational task. The advantages of this include simple and speedy design and manufacture, many more launch opportunities, low cost, non-competition between the requirements of different instruments, risk reduction, etc. Small satellites also provide opportunities for developing countries to become involved in technology transfer and the development of indigenous space-related capabilities. Thus we foresee a role for both large and small Earth-observing satellites in the future (Cracknell and Varotsos, 2007a).

However, before considering satellite remote sensing of atmospheric ozone we should also consider the use of aircraft. There have been a few studies of the atmosphere using instruments flown on aircraft to study ozone and related chemicals directly, most notably in the early study of the Antarctic ozone hole, using a modified U-2 spy plane called ER-2 (see Section 5.6). However, many of the satellite-flown instruments we discuss in this chapter have similar counterparts flown on aircraft as well. There have usually been two reasons for this. The first is for calibrating and evaluating prototypes of the instruments before flying them in space. The second is for simultaneous flying of similar (nominally identical) instruments in aircraft and in space for validation of the spaceborne instrument. Some instances of airborne versions of instruments will be given in later sections in this chapter. However, there is another rather new aspect of airborne remote sensing which we should mention.

Just as remote-sensing satellites for environmental studies can be regarded as a spin-off development from military spy satellites, so we are now seeing a similar spin-off from unmanned military aircraft in a new mission called Global Hawk Pacific, or GloPac, and one or two other unmanned aircraft NASA projects. NASA has staged environmental (manned) research flights from aircraft previously, but none has had the reach and duration of Global Hawk, a high-altitude, unmanned aircraft. The aircraft, which is distinguished by its bulbous nose and 35.4 m wingspan, can travel about 18,500 km in up to 31 hours, carrying almost 1 t of instrument payload. By contrast, NASA's manned research aircraft, the ER-2, can fly for only eight hours. Furthermore, researchers do not have access to their data until after the ER-2 aircraft lands (*http://gsfctechnology.gsfc.nasa.gov/GlobalHawk.htm*).

NASA has taken over two Global Hawk unmanned airraft that were originally built by Northrop Grumman for the U.S. Defense Advanced Research Projects Agency. The first two test flights were made on April 6 and April 13, 2010. These two flights were part of the five-flight GloPac mission to study the atmosphere over the Pacific and Arctic Oceans—NASA's first Earth science project to make use of unmanned aircraft, two of which were transferred from the U.S. Air Force to be based at NASA Dryden Flight Research Center on Edwards Air Force Base in California.

The Global Hawk aircraft used in the GloPac mission carried a suite of 12 instruments—provided by scientists from the Goddard Space Flight Center, the Jet Propulsion Laboratory, the National Oceanic and Atmospheric Administration (NOAA), Denver University, and the Ames Research Center. The first operational flights of the Global Hawk for GloPac were conducted in support of the Aura Validation Experiment (AVE). This was planned to encompass the entire offshore Pacific region with four to five 30 hour flights. Aura is one of the A-Train satellites (Figure 2.1) supported by the NASA Earth Observation System (*http://www.espo.nasa.gov/glopac/*). The NASA A-Train is a convoy of Earth-observing spacecraft following one another in polar orbit. Initially the A-Train comprised four satellites—Aqua, CloudSat, Cloud-Aerosol Lidar and Infrared Pathfinder Satellite Observations (CALIPSO) and Aura—which pass in sequence across the equator each day at around 1:30 PM local time each afternoon and carry between them 15 separate scientific instruments. From time to time further satellites are added to the A-Train (e.g., PARASOL and OCO satellites). Having several individual satellites in

Figure 2.1. A NASA Global Hawk lands at Dryden (top) and the A-Train (bottom) (*source*: NASA).

the A-Train avoids some of the engineering problems that would arise if one attempted the impossible by assembling 15 or more instruments on one large spacecraft like ENVISAT. The Global Hawk flights were designed to address various science objectives:

1. validation and scientific collaboration with NASA Earth-monitoring satellite missions, principally the Aura satellite;
2. observations of stratospheric trace gases in the upper troposphere and lower stratosphere from the mid-latitudes to the tropics;
3. sampling of polar stratospheric air and the break-up fragments of the air that move into the mid-latitudes;
4. measurements of dust, smoke, and pollution that cross the Pacific from Asia and Siberia;
5. measurements of streamers of moist air from the central tropical Pacific that move onto the West Coast of the United States (atmospheric rivers).

2.1 SATELLITE REMOTE SOUNDING OF TOC

We now turn to the satellite remote sensing of ozone. A number of instruments measure ozone vertical profiles and TOC from space by measuring backscattered UV solar radiation. The spectral distribution of backscattered radiation to space by the Earth's atmosphere depends on the presence of gases with selective absorption bands and molecular (Rayleigh) as well as aerosol scattering. The instruments can use spectrometers or filters for various spectral channels and they can observe in the nadir direction—as for solar-backscattered UV) or through scanning cross-tracks (as with total ozone-mapping spectrometry (Grant, 1989). These instruments operate best above 15–20 km (where there is strong absorption by ozone in the Hartley band).

During the second half of the 1960s several attempts at satellite remote sounding of ozone and other minor gas components from Soviet manned spacecraft were first made, using a hand-held spectrograph, and a complex of solar spectrometers functioning in a regime of occultation geometry (Kondratyev, 1972; Kondratyev *et al.*, 1996).

Since 1970, appropriate instrumentation launched on board various satellites (e.g., Nimbus-4), has provided values of the global distribution of TOC with very good spatiotemporal coverage. In the early days, TOC monitoring was carried out with NOAA's TOVS and its predecessors, with NASA's and METEOR's TOMS (Total Ozone Monitoring Spectrometer) and with the SBUV (Solar Backscatter UV), all instruments that were first flown on satellites in 1978. These measurements continued with SBUV/2 instruments on board the NOAA-9, -11, -14, -16, and -17 satellites, and TOMS instruments on the Russian Meteor-3, Earth Probe, and Japanese ADEOS satellites until the year 2003 (when instrument degradation was observed). The European Space Agency's GOME on the ERS-2 satellite, also performing BUV measurements, complemented the U.S. efforts. In addition, the Shuttle SBUV (SSBUV) experiment (conducting eight missions between October 1989 and January 1996) provided regular checks on the calibration of individual satellite instruments. Hilsenrath *et al.* (1997) summarized the contributions of the SSBUV (eight Space Shuttle missions from October 1989 to January 1996) in support of the U.S. long-term ozone-monitoring program (mainly from the viewpoint of validation of satellite observations). A significant decrease in Northern Hemisphere TOC from the winter of 1992 to the following winter was observed. However, the TOMS time series was interrupted during May 1993–July 1996 when TOMS/EP (Total Ozone Monitoring Spectrometer/Earth Probe Satellite) was launched. This gap was filled by continuous SBUV observations on various NOAA satellite missions and some GOME ozone observations since July 1995 aboard ERS-2, the first European satellite instrument measuring total ozone from the UV–VIS. An important source of minor gas components data is the TERRA satellite launched in December 1999. An instrumental problem in 2003 reduced the global coverage of GOME. Follow-up European satellite instruments are SCIAMACHY launched in 2002 on the ENVISAT platform of the European Space Agency (ESA) and OMI, a Dutch–Finnish instrument on the NASA EOS–Aura platform launched in 2004 and then later on the first European polar-orbiting meteorological satellite MetOp (Kaye, 1997). The first MetOp satellite (MetOp-A) was launched on October 19, 2006 and was declared fully operational in mid-May 2007. In addition, an interesting aspect of the OMI is also its synergy with the ENVISAT and GOME missions (Dobber *et al.*, 2006). The UV–VIS spectrometer GOME-2 was launched in 2006 on the first of a series of three operational EUMETSAT MetOp missions, which should allow continuous ozone monitoring until about 2020.

Multiple intercomparisons with ground-based instruments (e.g., Brewer, Dobson, and filter instruments) improved data-retrieval algorithms and therefore satellite observations of TOC became compatible with ground-based ones (Fioletov *et al.*, 2008; McPeters and Labow, 1996; see Chapter 3).

In the U.S. ozone research plans include both a total ozone mapper and an ozone-profiling instrument for the National Polar-orbiting Operational Environmental Satellite System (NPOESS; see Section 2.3.7).

A brief description of the satellite instrumentation providing TOC observations is given in Sections 2.3.1, 2.3.2, and 2.3.6, grouped by type of measurement (direct absorption, indirect absorption, and emission). It should be noted that several instruments provide not only TOC, but also OVP measurements.

2.2 DIRECT ABSORPTION MEASURING INSTRUMENTS

In this section brief information is given for a few instruments that are used to measure TOC by directly measuring the absorption features of ozone, either within a path defined by the instrument or by using an extraterrestrial source such as the Sun.

2.2.1 TIROS Operational Vertical Sounder (TOVS); GOES

The Television InfraRed Observational Satellite (TIROS)-N and its successors in the NOAA series of polar-orbiting meteorological satellites carried the TIROS Operational Vertical Sounder (TOVS) on board; later satellites in the program carried an

Advanced TOVS or ATOVS with more spectral channels. The TOVS is a set of three instruments—the High Resolution Infrared Sounder (HIRS/2), the Stratospheric Sounding Unit (SSU), and the Microwave Sounding Unit (MSU)—providing 27 spectral channels which are all, except one, in the infrared and microwave region of the spectrum with one panchromatic channel. The three instruments are used together to provide the vertical profiles of pressure, temperature, and humidity in the atmosphere. On later spacecraft in the NOAA series of polar orbiters there was the improved ATOVS (Advanced TOVS) with improved versions of the HIRS (HIRS/4) and the MSU (the AMSU, Advanced MSU) (Kramer, 2002).

HIRS has 19 infrared channels (3.8–15 μm) and 1 visible channel. The swath width is 2,160 km with 10 km resolution at nadir. The HIRS/4 instrument provides multispectral data from one visible channel (0.69 μm), 7 shortwave channels (3.7 to 4.6 μm) and 12 longwave channels (6.7 to 15 μm) using a single telescope and a rotating filter wheel containing 20 individual spectral filters. A rotating scan mirror provides cross-track scanning of 56 steps in increments of 1.8 degrees. The mirror steps rapidly (<35 ms), then holds at each position while optical radiation, passing through the 20 spectral filters, is sampled. This action takes place every 0.1 s. The instantaneous field of view for each channel is approximately 0.7° which, from a spacecraft altitude of 837 km, encompasses a circular area of 10 km at nadir on the Earth. Three detectors are used to sense optical radiation. A silicon photodiode, at nominal instrument temperature (15°C), detects the visible radiation. An Indium Antimonide detector and Mercury Cadmium Telluride detector (mounted on a passive radiator and operating at 95 K) sense the shortwave and longwave IR radiation, respectively. The shortwave and visible optical paths have a common field stop, while the longwave path has an identical but separate field stop. The size and registration of the optical fields of view in all channels is determined by these stops. IR calibration of the HIRS/4 is provided by programmed views of two radiometric targets: the warm target mounted on the instrument baseplate and a view of deep space. Data from these views provides sensitivity calibrations for each channel at 256 s intervals if commanded. It uses CO_2 absorption bands for temperature sounding (CO_2 is uniformly mixed in the atmosphere). In addition, it measures water vapor, ozone, N_2O, and cloud and surface temperatures.

Channel 9 of the HIRS, at a wavelength of 9.7 μm, is particularly well suited for monitoring stratospheric ozone concentration; this is a (general) window channel, except for absorption by ozone. Thus, the radiation received by the HIRS instrument was emitted from the Earth's surface and not from various levels in the Earth's atmosphere, but it is attenuated by the ozone in the atmosphere. The less ozone, the greater the amount of radiation reaching the satellite. A 3 DU drop in lower-stratospheric ozone produces a measurable (circa 0.2°C) increase in the brightness temperature in this channel. It appears that, strictly speaking, what is measured is lower-stratospheric ozone and a correction has to be applied to obtain total column ozone (TOC). Images are now regularly produced from TOVS data giving hemispherical daily values of total ozone. This means that there is now a long time series of such data available. TOVS data have been used to determine atmospheric ozone concentration from 1978 to the present (Neuendorffer, 1996; Kondratyev, 1997,

1998b) and the standard deviation is approximately 7%. Nowadays, TOVS data are also used when the more reliable TOMS data are not available (*http://www.theozonehole.com/2010oct.htm*). Since TOVS ozone data are only sensitive to variations in the lower stratosphere, long-term TOVS total ozone trends are, at best, only indicative of lower-stratospheric ozone trends. An advantage of TOVS over other systems that use solar UV radiation is that TOVS data are available at nighttime and in the polar regions in winter. The drawbacks are that when the Earth's surface is too cold (e.g., in the high Antarctic Plateau), too hot (e.g., the Sahara desert), or too obscured (e.g., by heavy tropical cirrus clouds) the accuracy of this method declines.

Li *et al.* (2001) discussed the potential for using Geostationary Operational Environmental Satellite (GOES) Sounder radiance measurements to monitor total atmospheric ozone with a statistical regression using GOES Sounder spectral band 1–15 radiances. The advantage is the high temporal frequency of the availability of the data. Hourly GOES ozone products have been generated since May 1998. GOES ozone estimates were compared with TOMS TOC data and ozone measurements from ground-based Dobson spectrometer ozone observations. The results showed that the percentage root-mean-square (rms) difference between instantaneous TOMS and GOES ozone estimates ranged from 4% to 7%. Also, daily comparisons for 1998 between GOES ozone values and ground-based observations at Bismarck (North Dakota), Wallops Island (Virginia), and Nashville (Tennessee) show that the rms difference is approximately 21 DU.

2.2.2 Laser Heterodyne Spectrometer (LHS)/Tunable Diode LHS (TDLHS)

This spectrometer measures TOC by performing observations of the ozone profiles in the 0.9 μm region in the infrared. A laser is used as a local oscillator to direct a narrow bandwidth beam onto a detector or photomixer. Usually a CO_2 laser is used as the local oscillator, but occasionally a tunable diode laser is employed. The Sun or the Moon acts as the source of radiation beyond the atmosphere. Radiation is selectively absorbed by ozone or other molecular species in the atmosphere and is detected by the spectrometer. A TDLHS system was used as a ground-based station instrument at the University of Denver to determine TOC (McElroy *et al.*, 1991). A CO_2 and TDLHS can also be used to measure the OVP. Such measurements can be made with a vertical resolution of 4 km, which is limited by the rate of change in the spectral weighting functions with altitude or with lower resolution, if an insufficient number of channels are used. The accuracy of the measurements is approximately 5–10% for column content and 10–30% for profile measurements. The precision is approximately 2%. The advantages of the LHS over other techniques are its ultra-high spectral resolution, high spatial resolution, high quantum detection efficiency, and very good signal-to-noise ratio. The high resolution makes the system very selective as interference problems due to overlapping lines or bands are minimized (Fogal *et al.*, 1989).

Table 2.1. Instruments on board the PRIRODA module of the *Mir* Space Station.

Instrument	*Spectral range*	*Channels/ Bandwidth*	*Swath*	*Spatial resolution*	*Remarks*
MSU-SK	0.5–12.5 μm	5/100 nm	350 km	120×300 m	Conical scanner
MSU-E	0.5-0.9 μm	3/100 nm	45 km	25 m	Pushbroom scanner
MOS-A	757–767 nm	4/1.4 nm	80 km	2.7 km	Imaging spectrometer
MOS-B	408–1,010 nm	13/10 nm	80 km	650 m	Imaging spectrometer
ALISA	532 nm	—	3′	150 m (vertical)	LIDAR Pulse Power (40 mJ)
TV camera	400–750 nm	—	15°	300 m	Color
MOMS-2P	400–750 nm	4 + 3/80 nm	44/88 km	5/16 m	Stereo/MS
OZON-MIR	0.26–1 μm	4	$2' \times 25'$	—	Occultation

2.2.3 OZON-MIR

The PRIRODA module of the *Mir* Space Station was launched in mid-1996. This module has two instruments on board, known as OZON-MIR and ISTOK-1. These instruments use the occultation technique to measure the concentrations of a number of trace gases and aerosols. ISTOK-1 also has a limb emission mode, although retrieval studies suggest that the errors in this mode are considerably larger than in the occultation mode (WMO, 1999). The instruments and their specifications are given in Table 2.1.

2.3 INDIRECT ABSORPTION MEASURING INSTRUMENTS

For a decade or two the principal instruments for studying TOC from satellites were the TOMS (Total Ozone Monitoring Spectrometer) and the SBUV (Solar Backscatter UV radiometer), although the SBUV was also capable of profiling (i.e., studying the OVP). TOMS is a scanning instrument that enables total ozone, TOC, to be determined. The SBUV is a nadir-pointing profiling instrument which enables both TOC and the vertical profile, OVP, to be determined. We consider the TOMS first.

2.3.1 Total Ozone Mapping Spectrometer (TOMS)

At about the same time that the first TOVS was flown, the Nimbus-7 satellite was launched and this carried, among other instruments, the first Total Ozone Mapping

Spectrometer (TOMS). A schematic diagram of the Nimbus-7 TOMS is shown in Figure 2.2.

The TOMS uses a similar principle to that of the Dobson spectrophotometer: it utilizes the wavelength dependence of the Earth's ultraviolet albedo in the Huggins band of the ozone absorption spectrum. The TOMS uses a single monochromator and scans across the sub-orbital track, sampling radiation backscattered from the underlying surface and atmosphere. It uses an infrared cloud cover photometer to avoid problems from clouds. It has six UV wavelength bands from 312.5 to 380 nm (see Table 2.2), and TOC is inferred by utilizing the wavelength dependence of the Earth's ultraviolet albedo in the Huggins band of the ozone absorption spectrum. The first four wavelength regions are used in pairs to provide three estimates of ozone concentration by the differential absorption method, while the other two (free of ozone absorption) are used to determine the effective background albedo. The Temperature Humidity Infrared Radiometer (THIR) on board the satellite is used to determine cloud locations and heights. Precision is quoted as 2% or better (with a small drift in diffuser plate reflectance of 0.4% per year). The TOMS instrument flown on the Nimbus-7 satellite and its successors have been used to measure the global distribution of ozone. This instrument was followed by the TOMS flown on the Meteor-3 satellite. The time that had elapsed since the Nimbus-7 TOMS was designed and built enabled some improvements to be incorporated in the Meteor-3 TOMS.

The basic differences between the Meteor-3 TOMS (engineering model) and the Nimbus-7 TOMS instrument were the following:

- An Interface Adapter Model (IAM) was added to make the TOMS instrument compatible with Meteor-3, which was launched on August 15, 1991 in an orbit with an 82.5° inclination. The orbit precessed relative to the Earth–Sun line with a period of 212 days, unlike Nimbus-7 which had a Sun-synchronous orbit. For two periods within this cycle the instrument was close to the terminator and this made ozone retrieval difficult.
- Replacement of the single diffuser with a three-diffuser carousel allowed calibration by comparison of diffusers with different rates of exposure.
- The higher altitude of Meteor-3/TOMS (1,200 km compared to 996 km for Nimbus-7/TOMS) meant greater overlap between successive orbits for the Meteor-3/TOMS than for Nimbus-7/TOMS.
- The Meteor-3 instrument was further improved when the diffraction grating was replaced, the mirrors resurfaced, flight-qualified electronics added, and a solid-state data recorder replaced the tape recorder, which had been used for Nimbus-7 and so almost no data were lost during the 3-year mission.

To obtain daily high-resolution global maps of atmospheric ozone, Meteor-3/TOMS measured solar irradiance and the radiance backscattered by the Earth's atmosphere in six (1 nm) selected wavelength bands in the ultraviolet. The experiment used a single monochromator and scanning mirror to sample backscattered

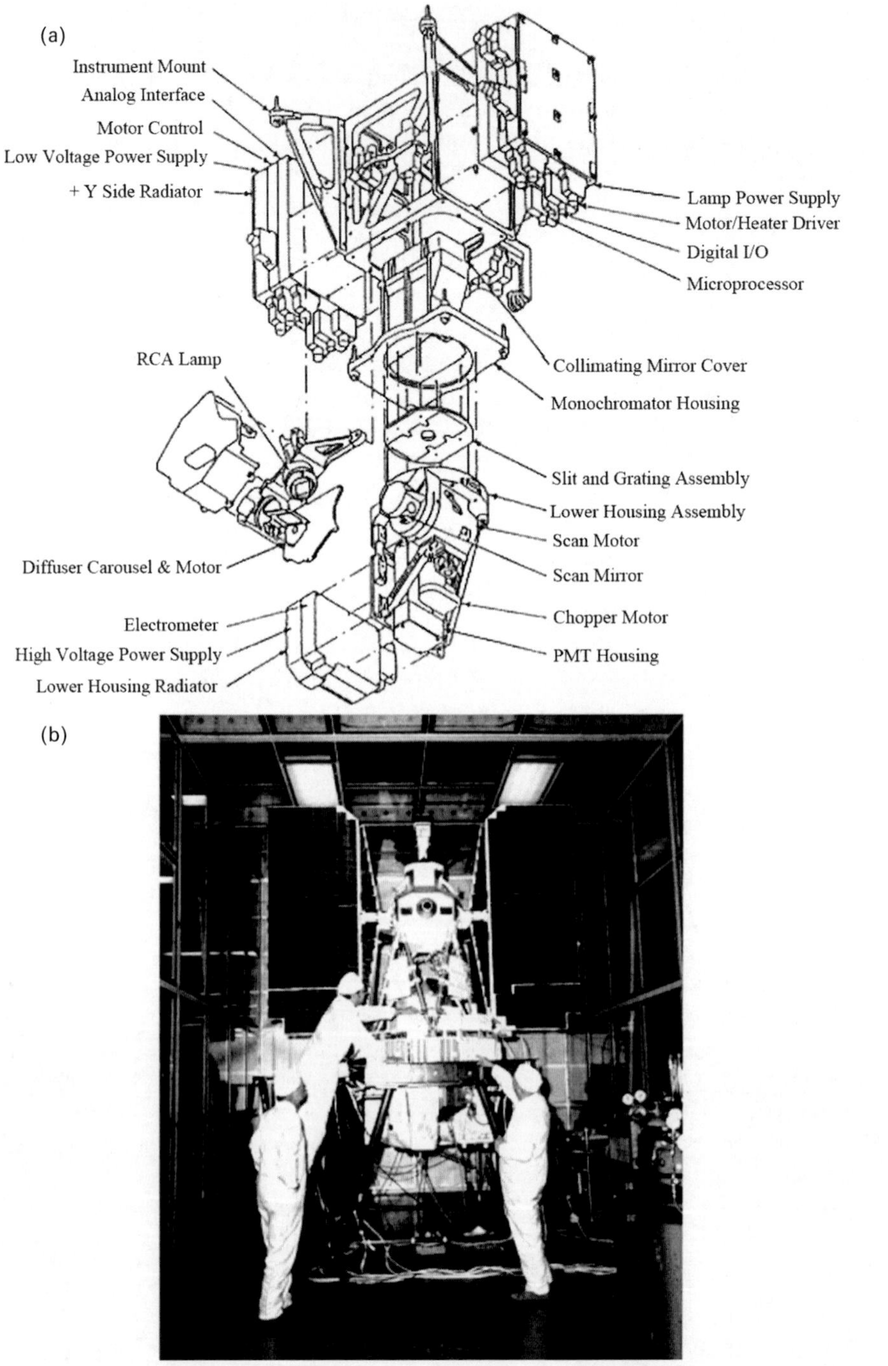

Figure 2.2. (a) TOMS optical diagram and (b) the Nimbus-7 spacecraft which carried the first TOMS instrument, before its launch in 1978 (NASA).

Table 2.2. TOMS main characteristics (McPeters and Labow, 1996).

Spectral bands	6
Wavelengths	312.3, 317.4, 331.1, 339.7, 360, 380 nm
Bandwidth	1.0 nm
Total ozone accuracy	<2%
Ozone trend accuracy (goal)	1% per decade
Sulphur dioxide accuracy	±25%
Scan time	6.3 s
Data rate	approx. 700 bps
Mass	<33 kg
Power	<20 W

SUVR at 35 sample points at 3° intervals. It scanned the Earth in 3° steps to 51° on each side of the sub-satellite point in a direction perpendicular to the orbital plane.

In normal operation, the scanner measured 35 scenes, one for each scanner view angle stepping from right to left. It then quickly returned to the first position, not performing measurements on the retrace. Eight seconds after the start of the previous scan, another scan would begin. Observations used for ozone retrieval were made during the sunlit portions of the orbit.

The measurements made using the TOMS on board the Meteor-3 satellite began on August 22, 1991 and ended on December 28, 1994. This system provided daily global coverage of the Earth's TOC by measuring backscattered Earth radiance. These data were archived at the Goddard Space Flight Center (GSFC), Distributed Active Archive Center (DAAC) and have been processed independently by the Central Aerological Observatory (CAO) of the Russian State Committee on Hydrometeorology. There was a one-and-a-half-year overlap between Nimbus-7 and Meteor-3 permitted an intercalibration that allows the data from the two satellites to be used to form a continuous 16-year dataset, which can be used to study ozone trends from November 1978 to December 1994.

A detailed description of the adaptations to TOMS for Meteor-3 is provided by Herman *et al.* (1996) and pre-launch and post-launch calibrations are described by Jaross *et al.* (1995). As a result of post-launch calibration it was found that there was severe degradation of an aluminum diffuser plate deployed to reflect sunlight into the instrument. On board Nimbus-7 severe degradation of the diffuser plate was also observed as time passed (Wellemeyer *et al.*, 1996). The three-diffuser system aboard Meteor-3 reduced the exposure and degradation of the diffuser and allowed calibration by comparing signals reflected by diffusers with different rates of exposure. As mentioned above, diffusers were arranged along the sides of an equilateral triangle and mounted on a carousel, so that a given diffuser could be placed into view on demand. For Meteor-3 TOMS TOC the absolute error is ±3% and the drift for 40 months is less than ±1%. Meteor-3 TOMS agrees with both Nimbus-7 TOMS and ground-based measurements to better than 1% (McPeters *et al.*, 1996).

Of the five TOMS instruments that were built, four entered successful orbit. Nimbus-7 and Meteor-3 provided global measurements of total column ozone on a daily basis and together provided a complete dataset of daily ozone from November 1978–December 1994. After an 18-month period when the program had no on-orbit capability, ADEOS TOMS (see Section 2.3.3) was launched on August 17, 1996 and provided data until the satellite which housed it lost power on June 29, 1997. Earth Probe TOMS was launched on July 2, 1996 to provide supplemental measurements, but was boosted to a higher orbit to replace the failed ADEOS. The transmitter for the Earth Probe TOMS failed on December 2, 2006. Since January 1, 2006 data from the Ozone Monitoring Instrument (OMI, see Section 2.3.2) has replaced Earth Probe TOMS.

Their historical and current daily and monthly data are widely available at the NASA website *http://toms.gsfc.nasa.gov* Spatially, a good global coverage in TOMS (OMI) data is combined with a high resolution of 1° by latitude and 1.25° by longitude (McPeters *et al.*, 1998).

In addition to ozone, the TOMS instrument measures sulfur dioxide released in volcanic eruptions. These observations are of great importance in the detection of volcanic ash clouds that are hazardous to commercial aviation.

2.3.2 Ozone Monitoring Instrument (OMI)

The OMI instrument mentioned above is a nadir-viewing near-UV/visible CCD spectrometer aboard NASA's Earth Observing System's (EOS) Aura satellite. Aura flies in formation about 15 minutes behind Aqua, both of which orbit the Earth in a polar Sun-synchronous pattern. Aura was launched on July 15, 2004 and the OMI has collected data since August 9, 2004. The other instruments flown on Aura include HIRDLS (see Section 2.5.3.8), MLS (see Section 2.5.3.4), and TES (see Section 2.5.3.5).

OMI measurements cover a spectral region of 264–504 nm with a spectral resolution between 0.42 nm and 0.63 nm and a nominal ground footprint of 13 km × 24 km at nadir (see Table 2.3). Essentially complete global coverage is achieved in one day. The significantly improved spatial resolution of OMI measurements as well as the vastly increased number of wavelengths observed—as compared to TOMS, GOME (Global Ozone Monitoring Experiment), and SCIAMACHY (SCanning Imaging Absorption Spectrometer for Atmospheric Chartography)—sets a new standard for trace gas and air quality monitoring from space. OMI observations provide the following capabilities and features:

- mapping of ozone columns at 13 km × 24 km and profiles at 13 km × 48 km (a continuation of TOMS and GOME ozone column data records and the ozone profile records of SBUV and GOME);

Table 2.3. General description of OMI (OMI, 2009).

Parameter	*Value*
Wavelength range:	UV-1: 264–311 nm UV-2: 307–383 nm VIS: 349–504 nm
Spectral resolution (FWHM)	UV-1: 0.63 nm UV-2: 0.42 nm VIS: 0.63 nm
Spectral sampling (FWHM)	UV-1: 1.9 px UV-2: 3.0 px VIS: 3.0 px
Telescope FOV	115° (2,600 km on ground)
IFOV	12 km × 6 km (flight direction × cross-flight direction)
Detector	CCD: 780 × 576 (spectral × spatial) pixels
Mass	65 kg
Duty cycle	60 minutes on daylight side 10–30 minutes on eclipse side (calibration)
Power	66 W
Data rate	0.8 Mbps (average)

- measurement of key air quality components: NO_2, SO_2, BrO, HCHO, and aerosol (a continuation of GOME measurements);
- ability to distinguish between aerosol types, such as smoke, dust, and sulfates;
- the ability to measure aerosol absorption capacity in terms of aerosol absorption optical depth or single-scattering albedo;
- measurement of cloud pressure and coverage;
- mapping of the global distribution and trends in UV-B radiation;
- combining processing algorithms including TOMS Version 8, DOAS (Differential Optical Absorption Spectroscopy, see Section 1.5), hyperspectral BUV retrievals, and forward modeling to extract the various OMI data products;
- near real-time measurements of ozone and other trace gases.

OMI's scientific mission objectives (discussed in detail by Levelt *et al*., 2006) were directly related to the Aura mission objectives. The OMI mission seeks answers to the following questions:

- Is the ozone layer recovering as expected?
- What are the sources of aerosols and trace gases that affect global air quality and how are they transported?
- What are the roles of tropospheric ozone and aerosols in climate change?
- What are the causes of surface UV-B change?

The OMI employs hyperspectral imaging to observe solar backscatter radiation in the visible and ultraviolet improving the accuracy and precision of the TOC amounts and also allow for accurate radiometric and wavelength self-calibration over the long term. The instrument observes Earth's backscattered radiation with a wide-field telescope feeding two imaging grating spectrometers. It is a contribution of the NIVR (Netherlands Institute for Air and Space Development) of Delft, in collaboration with the Finnish Meteorological Institute (FMI), to NASA's Aura mission. The Dutch industrial efforts focused on the optical bench design and assembly, thermal design, and project management. The detector modules and the readout and control electronics were provided by Finnish industrial partners. An extensive discussion of the OMI can be found in the *OMI User's Guide* (OMI, 2009). It can distinguish between aerosol types, such as smoke, dust, and sulfates, and measures cloud pressure and coverage, which provide the data necessary to derive tropospheric ozone. OMI continues the TOMS record for total ozone and other atmospheric parameters related to ozone chemistry and climate (*http://www.itc.nl/research/products/sensordb/AllSensors. aspx*).

Ozone profiles have been derived from the BUV/SBUV series of instruments since the 1970s. However, vertical ozone information is limited to above about 25 km because of the selection of the 12 wavelengths used by these instruments. Using hyperspectral OMI measurements (270–330 nm), ozone vertical profiles from the surface to about 60 km are derived at high spatial resolution (13 km × 48 km at nadir) and with daily global coverage. The retrieved ozone profiles agree very well with Microwave Limb Sounder (MLS) (see Section 2.5.3.4) measurements; the global mean biases are within 2.5% above 100 hPa and 5–10% below 100 hPa (for further details see Liu *et al.*, 2010).

Veefkind *et al.* (2006) described an algorithm for deriving total column ozone from spectral radiances and irradiances measured by the Aura OMI; this algorithm is based on the differential optical absorption spectroscopy (DOAS, see Section 1.5) technique. The main characteristics of the algorithm and an error analysis are described. The algorithm was successfully applied to the first available OMI data and comparisons with data from ground-based instruments were very encouraging and clearly showed the potential of the method.

The Aura satellite that carries the OMI has the capability to directly broadcast measurements to ground stations at the same time as measurements are being stored in the spacecraft's memory for later transmission to Earth. The Finnish Meteorological Institute's very fast delivery (VFD) processing system utilizes this direct broadcast to produce maps of total ozone and ultraviolet radiation over Europe within 15 minutes of the satellite overpass of the Sodankylä ground station in northern Finland. VFD products include maps of total ozone, ultraviolet index, and ultraviolet daily dose. The aim of this service is to provide up-to-date information on the ozone and ultraviolet situation for the general public and snapshots of the current situation for scientists. The accuracy of VFD products compares well with standard off-line OMI ozone products and ground-based Brewer measurements (Hassinen *et al.*, 2008; Ialongo *et al.*, 2008).

2.3.3 Advanced Earth Observing Satellite (ADEOS I, II)

The Japanese ADEOS carried the fourth TOMS into orbit on August 17, 1996. It remained in range between 10:45 AM to 11:45 AM throughout the ozone data record from September 11, 1996 to June 29, 1997. Its orbital inclination was 98.6° and the nominal orbit altitude was 800 km with a period of 100.9 min.

The approach taken to calibrate the three diffuser plates (by examining the differences in degradation of diffuser reflectivity resulting from the different rates of exposure) was first used with Meteor-3 TOMS and proved to be successful. In addition the Japanese ADEOS/TOMS is equipped with UV lamps for monitoring the reflectivity of the solar diffusers.

For ADEOS/TOMS TOC, the absolute error was ±3%, the random error is ±2%, and the drift over the 9-month data record was less than ±0.5%.

The ADEOS/TOMS observations of TOC are approximately 1.5% higher than a 45-station network of ground-based measurements (Herman *et al.*, 1997; Krotkov *et al.*, 1998; McPeters and Labow, 1996; Seftor *et al.*, 1997; Torres and Bhartia, 1999; Torres *et al.*, 1995).

The complementary instrumentation of ADEOS includes the Improved Limb Atmospheric Spectrometer (ILAS), the Interferometric Monitor for Greenhouse Gases (IMG), and the Retro-reflector In Space (RIS) instrument. ILAS is described in Section 2.5.3.10. IMG is a nadir-observing Michelson-type FTS (see Sections 1.5.4 and 2.3.10) designed to measure the density profiles of CO_2 and H_2O, total ozone column, and mixing ratios of CH_4, N_2O, and CO in the troposphere. The RIS, which is used together with laser ground stations, supports vertical profile and/or column measurements of a small number of gases.

ADEOS-II (a Sun-synchronous orbit satellite) launched by NASDA, NASA, and CNES in December 2002 was developed to include sensors like the Advanced Microwave Scanning Radiometer (AMSR), an advanced microwave radiometer developed by the National Space Development Agency of Japan (NASDA). It was equipped with the Improved Limb Atmospheric Spectrometer-II (ILAS-II), an improved spectrometer for measuring infrared radiation at the edge of the atmosphere, which was developed by the Environmental Agency of Japan (Kondratyev *et al.*, 1998). The mission ended in October 2003 after the satellite's solar panels failed (*http://www.jaxa.jp/projects/sat/adeos2/index_e.html*).

2.3.4 Solar Backscattered Ultraviolet Radiometer (SBUV)

We now move on to consideration of the SBUV instruments. The SBUV instruments are nadir-viewing instruments that determine total column ozone and the ozone vertical profile by measuring sunlight scattered from the atmosphere in the ultraviolet spectrum. As we did much earlier on in the case of temperature sounding, we assume that the atmosphere is infinitely deep so that the ground can be neglected.

The Nimbus series of satellites, starting with Nimbus-4 in 1970, was used to measure global distributions of ozone for seven years, using the backscattered ultraviolet (BUV) technique, which is briefly described here. Incident solar radiation at 317.5 and 360 nm passes through the ozone within the upper atmosphere and is backscattered largely within the troposphere at altitudes below the peak in ozone concentration. This backscattered radiation passes once again through the ozone to the satellite. The radiation at 317.5 nm is attenuated by ozone during both passages through the atmosphere. However, measurements at the 317.5 nm BUV alone cannot be used to determine TOC without knowledge of the varying reflectivity of the backscattering region. In order to determine this reflectivity, the 360 nm BUV (which is not attenuated by ozone), is measured with the BUV technique. The first synoptic global images of TOC were obtained in the sunlit hemisphere, using the imaging instrumentation on board the *Dynamics Explorer-1* (DE-1) spacecraft. The first instrument in this series was the Backscatter Ultraviolet (BUV) instrument which was flown on NASA's Nimbus-4 satellite, which was launched in 1970. It was in operation until 1977, but not continuously due to spacecraft power problems. An improved version, the Solar Backscatter Ultraviolet radiometer (SBUV) was flown on Nimbus-7 which was launched in 1978. This was followed by the SBUV/2 on the NOAA-9, -11, -14, -16, and -17 satellites (see Table 2.4).

The BUV instrument is a downward-viewing double monochromator with a 1 nm bandpass, which is stepped through 12 wavelength increments every 32 s through the 255–340 nm spectral region. In addition, there is a separate filter photometer with a 5 nm bandpass centered at 380 nm, which is used to measure backscattered radiance. This instrument was designed to determine the vertical ozone profile, while measuring backscattered solar UV radiation directed vertically upwards from the ground. Two wavelength pairs are used for measuring total ozone, and eight short wavelengths are used to profile the ozone. The precision of this instrument is better than 2% for zenith angles less than about 60° and clear conditions. The accuracy is limited to approximately 6%.

The SBUV instruments—starting with the SBUV on Nimbus-7—are all of similar design: nadir-viewing double-grating monochromators of the Ebert–Fastie type. The instruments use 12 narrow bands in the wavelength range 0.25–0.34 μm, using sunlight or moonlight, while viewing the Earth in the fixed nadir direction with an instantaneous field of view (IFOV) on the ground of approximately 180 km × 180 km. The spectral resolutions for the SBUV(/2) monochromators are all approximately 1.1 nm. The wavelength channels used for Nimbus-7 SBUV were very similar to those of the SBUV/2 (see Table 2.5), except that on Nimbus-7 channel 1 had been at 256 nm. This was moved to 252.0 nm on the SBUV/2 in order to avoid emission in the nitric oxide gamma band that contaminated the SBUV channel 1 measurement.

The total ozone algorithm uses the four longest wavelength bands (312.57, 317.56, 331.26, and 339.89 nm), whereas the profiling algorithm uses the shorter

Table 2.4. Satellite instrumentation for BUV ozone observations (Hilsenrath *et al.*, 1997).

Instrumentation	*Satellite*	*Observation period*
BUV	Nimbus-4	1970–1975
SBUV	Nimbus-7	1979–1990
SBUV/2	NOAA-9	1985–1998
SBUV/2	NOAA-11	1989–2003
SBUV/2	NOAA-14	1995–present
SBUV/2	NOAA-16	2000–present
SBUV/2	NOAA-17	2002–present
SSBUV	Shuttle	8 flights, 1989–1996
TOMS	Nimbus-7	1978–1993
TOMS	Meteor-3M	1991–1994
TOMS	Earth Probe	1996–2005
TOMS	ADEOS	1996–1997
GOME	ERS-2	1995–present
Aura	OMI TO3	2004–present
IASI	MetOp	2006–present
GOME 2	MetOp	2006–present

wavelengths. Ozone profiles and total column amounts are derived from the ratio of observed backscattered spectral radiance to incoming solar spectral irradiance. This ratio is referred to as the "backscattered albedo". The only difference in the optical components between radiance and irradiance observations is the instrument diffuser used to make the solar irradiance measurement; the remaining optical components are identical. Therefore, a change in diffuser reflectivity will result in an apparent trend in ozone.

The Shuttle Solar Backscatter Ultraviolet Spectral Radiometer (SSBUV) was designed and developed at Goddard Space Flight Center to calibrate the Nimbus and NOAA instruments. It carries out measurements of ozone concentrations by comparing solar ultraviolet radiation with that scattered back from the atmosphere. In late 1989, the Space Shuttle *Atlantis* carried the instrument for the first time, in an appropriate orbital flightpath to assess performance by directly comparing data from identical instruments (SBUV) on board the NOAA spacecraft and the Nimbus-7. The Shuttle SBUV (SSBUV) was flown on eight missions between October 1989 and January 1996 and provided regular checks on individual satellite instrument calibrations. Multiple intercomparisons with ground-based instruments have improved data retrieval algorithms and, therefore, satellite ozone measurements have become compatible with those of the network of ground-based measurements. The principal purpose was to compare observations from several ozone-measuring instruments on board NOAA-9, NOAA-11, Nimbus-7, and UARS; this was because of the degradation of SBUV and SBUV/2 instruments in space. The quality of in-orbit calibration depends on the flight-to-flight calibration repeatability in SSBUV. Given accurate measurements of backscattered radiance, it is necessary to account for differences in solar zenith angle and effective surface reflectivity (Heath *et al.*, 1993) (see Section 3.2 for a further discussion of the intercomparison of different instruments).

Table 2.5. SBUV/2 wavelengths (NOAA, 2009).

Channel	*Wavelength* (nm)
1	252.00
2	273.61
3	283.10
4	287.70
5	292.29
6	297.59
7	301.97
8	305.87
9	312.57
10	317.56
11	331.26
12	339. 89

A UV spectrometer was also flown aboard the EXOS-C/Ohzora satellite (launched in February 1984) to measure backscattered UV in the wavelength region 250–320 nm with a spectral resolution of 1 nm for the period March 1984–September 1987 (Ogawa *et al.*, 1989). Complete global coverage of total ozone was obtained. However, an anomaly in the Earth's magnetic field in the South Atlantic affects the distribution of Van Allen radiation belts and, therefore, these radiation levels send too much noise to the detectors. This was overcome in the Solar Backscattered UV instrument (SBUV) by adding a 50 Hz optical chopper.

2.3.5 Global Ozone Monitoring Experiment (GOME)

GOME was launched in 1995 aboard the ESA's Earth Resources Satellite (ERS-2) in a polar Sun-synchronous orbit. It is a nadir-viewing multichannel spectrometer measuring solar irradiance and upwelling radiance (backscattered from the atmosphere and earthshine radiance) in the wavelength range 240–790 nm at moderate spectral resolution (0.2–0.4 nm). The four main channels provide continuous spectral coverage of the wavelengths between 240 and 790 nm (channel 1: 237–314 nm,

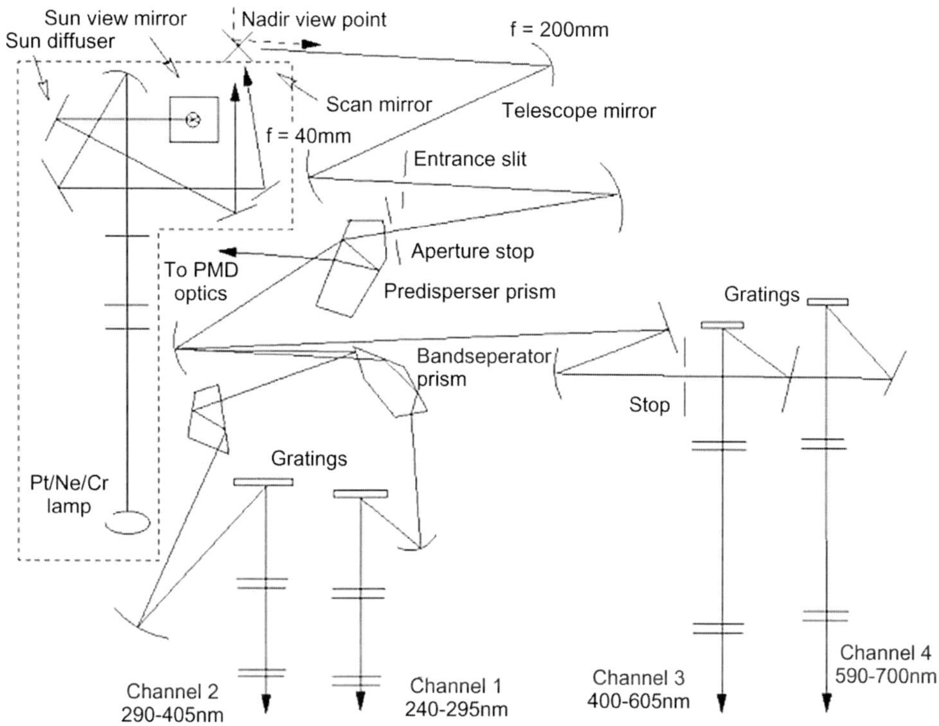

Figure 2.3. Schematics of GOME Optics. The GOME instrument is a four-channel spectrometer. Adjacent to the spectrometer is a calibration unit housing a Pt/Cr/Ne hollow cathode discharge lamp and the fore optics for solar viewing (Weber *et al.*, 1998).

channel 2: 311–405 nm, channel 3: 410–600 nm, channel 4: 610–790 nm) (see Figure 2.3). The instrument exploits the fact that near 260 nm the penetration depth of solar radiation into the atmosphere strongly increases with increasing wavelength, because at this specific wavelength the ozone absorption cross-sections have their maximum. When radiation has a wavelength greater than about 310 nm, it penetrates the tropopause reaching the surface. Therefore, the measurement of backscattered radiation in the UV–visible region provides information about the vertical ozone concentration profile. In general, GOME measures the TOC, ozone vertical profile, and total column of several trace constituents, including BrO, NO_2, and ClO_2 and also obtains information on clouds, aerosols, and surface spectral reflectance. It can measure TOC at a higher horizontal resolution than TOMS and, thus, complements TOMS observations. Retrieval of OVP from backscattered UV–visible spectra is based on the "eigenvector method", which uses *a priori* information (climatological mean profile with the corresponding covariance matrix) as the original optimal estimation approach.

Alternatively, the Full Retrieval Method (FURM), which was developed at the Institute of Remote Sensing of the University of Bremen, can be used. This is mainly based on the optimal estimation approach and contains the radiative transfer code GOMETRAN as an essential tool (Rozanov *et al.*, 1997, 1998). An explicit treatment of clouds in the retrieval algorithm based on a new GOMETRAN cloud parameterization has been developed by Kurosu *et al.* (1997). An advanced radiative transfer model, SCIATRAN, has been developed by Rozanov *et al.* (2002) mainly for the retrieval of atmospheric constituents from global nadir radiance measurements of the SCIAMACHY satellite spectrometer. This is a further development of the successful GOMETRAN. SCIATRAN solves the radiative transfer equation using the Finite Difference Method for a plane-parallel vertically inhomogeneous atmosphere taking into account multiple scattering.

A second GOME-2 instrument was flown on MetOp-1, ESA's first polar-orbiting meteorological satellite which was launched on October 16, 2006. GOME-2 on the satellites ERS-2 and MetOp 1 operates in four bands—band 1: 240 to 295 nm, band 2: 290 to 405 nm, band 3: 400 to 605 nm, band 4: 590 to 790 nm; these are slightly different from the bands on GOME-1 (launched on April 20, 1995) (*http://rammb.cira.colostate.edu/dev/hillger/ozone-monitoring.htm*). Some early work on comparison of its TOC data with reliable ground-based measurements recorded by five Brewer spectroradiometers in the Iberian Peninsula was described by Antón *et al.* (2009). A similar comparison for the predecessor instrument GOME/ERS-2 is also described. The period of study was a whole year from May 2007 to April 2008. The results show that GOME-2/MetOp ozone data are already of very good quality. TOC values were found to be on average 3.05% lower than Brewer measurements. This underestimation is higher than that obtained for GOME/ERS-2 (1.46%). However, the relative differences between GOME-2/MetOp and Brewer measurements show significantly lower variability than the differences between GOME/ERS-2 and Brewer data. The dependences of these relative differences on satellite solar zenith angle (SZA), satellite scan angle, satellite cloud cover fraction (CF), and ground-based total ozone measurements were analyzed. For both GOME instruments, the differences show no significant dependence on solar zenith angle. However, GOME-2/MetOp data show a significant dependence on satellite scan angle (+1.5%). In addition, GOME/ERS-2 differences present a clear dependence on CF and ground-based total ozone; such differences are minimized for GOME-2/MetOp. The comparison between the daily total ozone values provided by both GOME instruments shows that GOME-2/MetOp ozone data are on average 1.46% lower than GOME/ERS-2 data without any seasonal dependence. Finally, deviations of the *a priori* climatological ozone profile used by the satellite retrieval algorithm from the true ozone profile are analyzed. Although excellent agreement between *a priori* climatological and measured partial ozone values is found for the middle and high stratosphere, relative differences greater than 15% are common for the troposphere and lower stratosphere.

2.3.6 ESA ENVISAT, GOMOS

Among the many instruments on ENVISAT launched by the European Space Agency (ESA) the Global Ozone Monitoring by Occultation of Stars (GOMOS) instrument is relevant to stratospheric ozone measurements (Megie *et al.*, 1989). It is a stellar occultation instrument onboard the ENVISAT satellite (see Bertaux *et al.*, 1991, 2000a, b, 2004, 2010; ESA, 2003, *http://envisat.esa. int/handbooks/gomos/*; Kyrölä *et al.*, 2004) (see Section 1.7.4.1). GOMOS measurements start at an altitude of 130 km and the first few measurements are used to determine a star's undisturbed spectrum (the reference spectrum). Horizontal transmission spectra are calculated from the star spectra measured through the atmosphere and the reference spectrum. The integration time is 0.5 s, which gives an altitude-sampling resolution of 0.5–1.6 km depending on the tangent altitude and the azimuth angle of the measurement. GOMOS measures both during the day and night (Kyrölä *et al.* 2006).

The spectral ranges of GOMOS detectors are 248–690 nm, 755–774 nm, and 926–954 nm, which make it possible to retrieve the vertical profiles of O_3, NO_2, NO_3, H_2O, O_2, and aerosols. For O_3 they are retrieved from the UV–visible spectral range 248–690 nm. Retrieved ozone profiles have 2 km vertical resolution below 30 km and 3 km above 40 km. Retrieval algorithm details and data quality are discussed by Kyrölä *et al.* (2010b) and Tamminen *et al.* (2010), *http://envisat. esa.int/dataproducts/gomos*. Retrieved ozone distributions for 2002–2008 are described by Kyrölä *et al.* (2006, 2010a).

2.3.7 The Ozone Mapping and Profiler Suite (OMPS) and the NPOESS

The Ozone Mapping and Profiler Suite (OMPS) is one of the five instruments to be included in the U.S. National Polar-orbiting Operational Environmental Satellite System (NPOESS). For many years two parallel and rather similar, but not identical, polar-orbiting meteorological satellite programs were run by the National Oceanic and Atmospheric Administration (NOAA) and the DoD (Department of Defense). NASA had also been involved, in various ways, in a variety of polar-orbiting environmental remote-sensing programs.

NPOESS was planned as a tri-agency program with the Department of Commerce (specifically NOAA), the Department of Defense (DoD, specifically the Air Force), and NASA. It was designed to merge the civil and defense weather satellite programs of the NOAA Polar Operational Environmental Satellites (POES) series and the U.S. Department of Defense's Defense Meteorological Satellite Program (DMSP), respectively, in order to reduce costs and to provide global weather and climate coverage with improved capabilities over the earlier systems. NPOESS was to be the United States' next-generation satellite system that would monitor the Earth's weather, atmosphere, oceans, land, and near-space environment. NPOESS satellites were to host proven technologies and operational versions of sensors that were under operational prototyping by NASA.

The NPOESS Preparatory Project (NPP) program aimed to bridge the gap between the old and the new systems by flying new instruments on a satellite originally to be launched in 2005. However, the program encountered technical, financial, and political problems leading to a delay of the launch of NPP until 2011 (NPP was successfully launched on October 28, 2011) and of the launch of the first NPOESS platform (C-1) until late 2014. These would each constitute delays of about five years from the original plan. This led to a review of the NPOESS, and as a result of this review the White House announced on February 1, 2010 that the NPOESS satellite partnership was to be dissolved. Two separate lines of polar-orbiting satellites to serve military and civilian users would be pursued instead:

- The NOAA/NASA portion is called the "Joint Polar Satellite System" (JPSS). The JPSS-1 spacecraft will be constructed by Ball Aerospace & Technologies Corp. under a fixed price contract of $248 million with a performance period to February 1, 2015.
- The Defense Department's portion is called the "Defense Weather Satellite System" (DWSS).

The existing partnership with Europe through the European Organization for the Exploitation of Meteorological Satellites (EUMETSAT), which operates the MetOp polar-orbiting weather satellite program, would continue.

The five-instrument suite for NPOESS includes the Visible/Infrared Imager Radiometer Suite (VIIRS), the Cross-track Infrared Sounder (CrIS), the Clouds and the Earth Radiant Energy System (CERES), the Advanced Technology Microwave Sounder (ATMS), and the Ozone Mapping and Profiler Suite (OMPS). OMPS was designed to monitor ozone from space. It is comprised of two sensors, a nadir sensor and a limb sensor. Measurements from the nadir sensor are used to generate total column ozone measurements, while measurements from the limb sensor generate the ozone profiles of along-track limb-scattered solar radiance. OMPS will make measurements used to generate estimates of total column and vertical profile ozone data. These will continue the daily global data produced by the current ozone-monitoring systems, the Solar Backscatter Ultraviolet radiometer (SBUV)/2, and the Total Ozone Mapping Spectrometer (TOMS), but with higher fidelity. Collection of these data contributes to fulfilling the U.S. treaty obligation to monitor ozone depletion for the Montreal Protocol to ensure there are no gaps in ozone coverage.

OMPS consists of two telescopes, a nadir one and a limb one, feeding three detectors. The instruments use a configuration of working and reference solar diffusers to maintain calibration. The nadir sensor uses a wide-field-of-view push-broom telescope to feed two separate spectrometers. The nadir total column spectrometer measures scene radiance from 300 to 380 nm with a resolution of 1 nm sampled at 0.42 nm. Measurements from this spectrometer will be used to generate TOC with daily coverage of the sunlit Earth to continue heritage records from the Total Ozone Mapping Spectrometer (TOMS). The performance specifications for total column ozone Environmental Data Records (EDRs) from OMPS Nadir Mapper measurements are identified in table 2 of Flynn *et al.* (2009). The NPOESS algorithm is an extension of the TOMS Version 7 total ozone algorithm (McPeters and Labow, 1996). It uses multiple triplets of measurements: one chosen for ozone sensitivity, a second chosen to give an estimate of cloud and surface reflectivity, and a third to estimate the variation of reflectivity with wavelength. The triplets used in the final ozone estimates are selected so that the ozone-sensitive channels maintain sensitivity to the column as the solar zenith angle and column amount vary.

The OMPS nadir telescope also feeds a nadir profile grating spectrometer. It will measure from 250 to 310 nm with a 1.1 nm FWHM bandpass and 0.4 nm sampling and a 250 by 250 km^2 nadir-only FOV. These measurements will be used to continue the SBUV/2 ozone profile record. Ozone profile estimates will be obtained from the Version 8 SBUV/2 ozone profile retrieval algorithm with adaptations to use the increased number of wavelengths present in OMPS Nadir Profiler measurements. The OMPS limb sensor measures along-track limb-scattered solar radiance with 1 km vertical sampling in the spectral range from 290 to 1,000 nm. Three vertical slits sample the limb spaced at 250 km cross-track intervals to provide for weekly global coverage. The limb instrument will be flown as an experimental component on the NPP mission but is not currently manifested for future NPOESS satellites. The capabilities and complexities of retrieving ozone profiles from limb measurements are analyzed by Loughman *et al.* (2005).

2.3.8 Ozone Dynamics Ultraviolet Spectrometer (ODUS)

This is a Japanese ultraviolet instrument measuring ozone and several other species (e.g., O_3, SO_2, and NO_2 in urban polluted air, BrO, and OClO, aerosols) in the troposphere and stratosphere. ODUS uses techniques similar to those used by TOMS and GOME (see Sections 2.3.1 and 2.3.5) launched on the GCOM-A1 platform (2007–2012).

2.3.9 Ozone Layer Monitoring Experiment (OLME)

This instrument consists of UV cameras using both charge-coupled devices and UV photodiode detectors. It was flown aboard Chile's FaSAT Bravo satellite in 1997 and measured ozone in the Antarctic and sub-Antarctic regions of Chile. Another instrument, the OM-2 (developed in Israel), uses filters for the measurement of total ozone and the ozone vertical profile aboard the Israeli Techsat-1 satellite (Parnes, 1992).

2.3.10 Interferometric Monitor for Greenhouse Gases (IMG)

This is a Michelson-type interferometer which was launched on August 17, 1996 on the Japanese ADEOS platform. This instrument monitors IR spectra (3.3–14 μm) from the Earth's surface and the atmosphere, observing with a nadir view. For data inversion the atmospheric temperature profile is needed in order to retrieve the total amounts of O_3, CO, and CH_4, and a few points of the vertical profile. For these data an inversion algorithm based on neural network techniques is currently used (Hadji-Lazaro *et al.*, 1998).

2.3.11 Infrared Atmospheric Sounding Interferometer

This instrument is a Michelson-type interferometer providing spectra between 3.4 and 15.5 μm, with an apodized resolution of $0.5\,cm^{-1}$ (Hadji-Lazaro *et al.*, 1998). The total amount of ozone under cloud-free conditions is measured with a horizontal resolution of 25 km and an accuracy of 5%. It was launched on October 19, 2006 on the satellite MetOp-1.

Let us briefly discuss some results illustrating TOC observations (some more results will be considered in Chapter 3).

2.4 OBSERVED VARIABILITY IN TOTAL OZONE COLUMN

The TOC is characterized by significant spatiotemporal variability. As far as temporal variability is concerned, it consists of large periodic and aperiodic components. Periodic variations have timescales ranging from day-to-day changes, through seasonal and annual variations, to an 11-year solar cycle. Aperiodic variations include the irregular Quasi-Biennial Oscillation (QBO) with a period of roughly 26–30 months, the irregular El Niño Southern Oscillation (ENSO) with a period of 4–7 years, and other interannual signals. Ozone trends are important phenomena that will be considered in Chapters 4 and 6.

As far as spatial variability is concerned, it is characterized by latitudinal and longitudinal dependence. Satellite systems for studying atmospheric ozone have the advantage that they provide frequent (near) gobal coverage, whereas traditional ground-based and ozonesonde measurements only provide data at a small number of locations. As we shall see in Chapter 3, the algorithms that are used in processing satellite data need to be validated and refined by comparison with the results of more conventional measurements. In the meantime we discuss the spatial variations in TOC that are revealed by satellite data.

2.4.1 Latitudinal variation of TOC

A very important feature of global TOC distribution is the strong latitudinal gradient with lower values over the equator and tropics and higher values over mid and high latitudes. This gradient is characterized by a well-pronounced annual cycle, reaching a maximum in spring and a minimum in autumn. The amplitude of this annual cycle is a function of latitude, with a maximum at about 60° north and south latitude (Varotsos *et al.*, 2000). In the tropics seasonal variations are small, with ozone maxima in summer. Such a latitudinal distribution results from the relatively long lifetime (months to years) of ozone in the lower stratosphere and the Brewer–Dobson circulation that transports stratospheric ozone from the tropics toward the poles and downwards at high latitudes.

Bojkov and Fioletov (1995), based on re-evaluated TOC observations from over 100 Dobson and filter radiometer stations from pole to pole, presented the following TOC trend results:

— Up to 1995 TOC continued its decline (which started in the 1970s), with statistically significant year-round and seasonal trends except over the equatorial belt.
— The cumulative year-round TOC reduction over the 35°–60° belts of both hemispheres since the early 1970s is close to 8%. In the southern mid-latitudes, it is difficult to distinguish the seasonal dependence of TOC trends, while the cumulative decline in the northern mid-latitudes in winter and spring is about 9% and 4–6% for summer and autumn.
— Observations from 12 Dobson polar stations have demonstrated that the northern polar region shows the same ozone decline as northern mid-latitudes or even a slightly stronger one (the cumulative decline is about 7% year-round and 9% for winter and spring). In the southern polar region the December–March trend is also nearly the same as the mid-latitude trend reaching about 10%, while during September–November the decline reaches 40%.
— The eruption of Mount Pinatubo in 1991 did not significantly influence overall declining TOC trends.
— TOC trends are likely to continue until the peak levels of chlorine are reached in the stratosphere.

It appears that more recent volcanic eruptions than Mount Pinatubo have no long-term effects on TOC (Clarisse *et al.*, 2008; Flentje *et al.*, 2010).

The TOC trends for various geographic regions for January 1979 to March 1995 are shown in Table 2.6 (Bojkov and Fioletov, 1998).

The afore-mentioned results provide the observed TOC variability which marked the beginning of the TOC depletion phenomenon. Nowadays, according

Table 2.6. TOC trends for January 1979 to March 1995 for five regions of northern mid-latitudes, the Arctic, and Antarctica (Bojkov and Fioletov, 1998).

	Trend (%/decade)			
	Year	*Dec.–Mar.*	*May–Aug.*	*Sep.–Nov.*
North America (36°–60°N)	-4.2 ± 1.4	-4.5 ± 2.2	-4.5 ± 1.9	-2.6 ± 1.4
Europe (40°–60°N)	-4.8 ± 1.6	-7 ± 3.2	-4.3 ± 1.9	-1.7 ± 2.0
European Russia (45°–60°N)	-4.4 ± 1.6	-6.3 ± 3.0	-2.8 ± 1.9	-2.1 ± 1.8
Western Siberia (50°–60°N)	-5.8 ± 1.7	-8.2 ± 3.1	-3.6 ± 1.9	-3.6 ± 2.1
Eastern Siberia and Far East (43°–62°N)	-6.4 ± 1.7	-7.8 ± 2.3	-5.2 ± 2.4	-5.7 ± 2.2
Arctic (60°–80°N)	-5.3 ± 1.7	-7.7 ± 3.2	-3.8 ± 1.9	-2.5 ± 2.2
Antarctica (65°–90°S)		-5.4 ± 3.0		-22 ± 8

to the recent scientific ozone assessment (WMO, 2010a), the average total ozone values in 2006–2009 remain at roughly 3.5% and 2.5% below the 1964–1980 averages, respectively, for 90°S–90°N and 60°S–60°N. Mid-latitude (35°–60°) annual mean TOC amounts in the Southern Hemisphere (Northern Hemisphere) over the period 2006–2009 have remained at the same level as observed during 1996–2005: at ~6% (~3.5%) below the 1964–1980 average. The TOC decline in Arctic winter and spring between 2007 and 2010 has been variable, but has remained in a range comparable to the values prevailing since the early 1990s. In addition, springtime Antarctic TOC losses continue to occur every year. Although ozone losses exhibit year-to-year variations that are primarily driven by year-to-year changes in meteorology, the October mean TOC within the vortex has been about 40% below 1980 values for the past 15 years. In other words, the springtime Antarctic TOC does not yet show a statistically significant increasing trend.

As far as the future projection is concerned, the global annually averaged TOC is expected to return to 1980 levels before the middle of the century and before stratospheric halogen loading returns to 1980 levels. Global (90°S–90°N) annually averaged TOC will likely return to 1980 levels between 2025 and 2040, well before the return of stratospheric halogens to 1980 levels between 2045 and 2060. Simulated changes in tropical TOC from 1960 to 2100 are generally small.

Chandra *et al.* (1996) and Varotsos *et al.* (1997a, b), using measurements of TOC by Nimbus-7 TOMS version 7, suggested that trends over the latitudes centered at 40°N in the Northern Hemisphere vary from −3 to −9% per decade during winter and within −2 to −3% per decade during summer.

To account for the effects of dynamical perturbations on TOC Chandra *et al.* (1996) used tropospheric or stratospheric temperature as an index of dynamical variability. Figure 2.4 depicts the relative changes in regional and local trends, with respect to zonal mean trends. The regional and zonal trends chosen correspond to the Mediterranean region (35–45°N, 0–30°E) and Athens, Greece (38°N, 24°E). This figure also shows the ranges (shaded area) in seasonal trends over the longitude circle at 40°N derived from TOMS data. Figure 2.4a is based on the linear regression model which includes the QBO signal (based on the 30 hPa Singapore zonal wind) and the solar cycle signal (based on F10.7 radio emission flux at the wavelength 10.7 cm). Figure 2.4(b) shows the results of using the lower stratosphere (50–150 hPa) temperature from channel 4 of the Microwave Sounding Unit (MSU) on the NOAA series of polar-orbiting meteorological satellites (Randel and Cobb, 1994). Figure 2.4(c) shows the results of using 500 hPa data from the National Meteorological Center (NMC) analyses.

From Figure 2.4 it can be seen that, for most of the data, the Athens and Mediterranean values are very close to one another (see also Section 3.2). They also generally follow the 40°N zonal mean derived from TOMS data. Thus, taking the effects of dynamical perturbations on TOC trends into account, it seems plausible to use tropospheric or stratospheric temperature as an index of dynamical variability. The use of temperature has an additional advantage, since the effects of the QBO and to some extent the F10.7 (or UV radiation) are implicitly included in temperature changes (Chandra *et al.*, 1996).

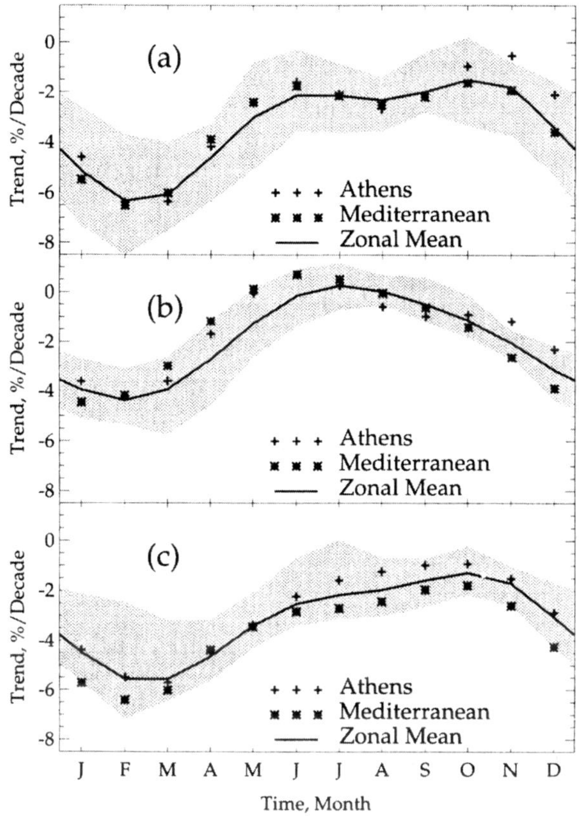

Figure 2.4. Monthly trends in total ozone in the Mediterranean region and at Athens compared with the trends derived from zonally averaged data. Also shown in this figure is the range of variability in seasonal trends over the latitude circle centered at 40°N (shaded). The regression models in the three panels contain (a) QBO and solar cycle terms, (b) lower-stratospheric temperature (MSU) term, and (c) the NMC 500 hPa temperature (Chandra *et al.*, 1996).

Inspection of Figure 2.4 suggests a significant difference in both zonal and regional trends according to season even though relative differences between the Mediterranean and the zonal mean trend do not vary significantly. In general, TOC trends are reduced by 1 to 3%, as a result of using the MSU temperature as a proxy for dynamical variability. This is mainly due to negative trends in lower-stratospheric temperature, which vary from −1.5 to −2 K per decade during winter (Chandra and Stolarski, 1991; Randel and Cobb, 1994). Similar analysis using the National Meteorological Center (NMC) temperature results in a smaller change in TOC trends of 1% or less. This is probably due to the positive trend of about 0.4 to 0.8 K during the same period in the 500 hPa temperature time series.

The afore-mentioned phase change in the trend of temperature from the troposphere to the stratosphere is consistent with the changes in stationary wave patterns suggested by Hood and Zhou (1998).

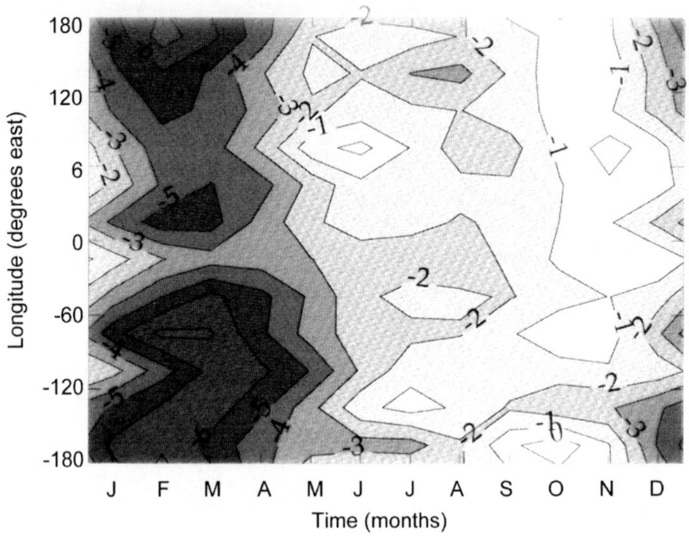

Figure 2.5. A cross-section of monthly trends in column ozone over the latitude belt 35°N–45°N derived from the 13 years of Nimbus-7 TOMS measurements (version 7). The trend values are given in % per decade. Negative and positive longitudes correspond to west and east longitudes, respectively (Varotsos *et al.*, 1997a).

2.4.2 Longitudinal variation of TOC

Re-examination of the influence of interannual variability on satellite-derived trends in total ozone during winter and spring confirms that dynamical perturbations in the atmosphere cause large longitudinal spread in total ozone trends at mid-latitudes (Varotsos and Kondratyev, 1998a).

Figure 2.5 is a cross-section of TOC trends (% per decade) as a function of the month of the year and longitude from 180°W to 180°E in the region covering the latitude band from 35°N to 45°N (centered at 40°N). This figure shows negative trends in all seasons over the entire latitude belt centered at 40°N and is generally consistent with the pattern discussed by Chandra *et al.* (1996). The trends vary from −1 to −2% per decade during summer to −2 to −6% per decade during winter. The 2σ uncertainties associated with these trends are in the range of 2 to 3% in summer and 3 to 6% in winter. Chandra *et al.* (1996), using the same analysis but with version 6 TOMS data, found that the trends vary from −2 to −3% per decade during summer to −3 to −9% per decade during winter.

2.5 SATELLITE INSTRUMENTATION FOR OVP OBSERVATIONS

Satellite-based techniques and instruments for OVP observations are presented in this section. Note that the satellite-based solar occultation technique and satellite-based nadir measurements have important advantages over ground-based measurements, because they provide near-global distributions, as opposed to point measurements from ground-based *in situ* techniques. It is self-calibrating because unattenuated sunlight is measured during each event allowing data to be normalized to extra-atmospheric values. The longer absorption path in the limb geometry provides higher sensitivity than that from nadir-looking geometry. Moreover, higher vertical resolution is inherent in this technique because most of the absorption occurs at the tangent altitude. Finally, this technique simplifies data interpretation because the Sun is used as a constant background source.

Note that the first instrument used to measure ozone in the 55–85 km region was a UV spectrometer and polarimeter onboard the Solar Maximum Mission spacecraft launched in February 1980. This used the technique of solar occultation in the UV region by the Earth's limb. The first attempts at satellite occultation sounding were conducted from Soviet manned spacecraft (Kondratyev, 1972, 1998b; Kondratyev *et al.*, 1996). The Russian Ozone and Aerosol Fine Structure (OZAFS) experiment obtained a few ozone profiles from satellite occulation sounding aboard the Soviet Salyut-7 orbiting station in 1965.

2.5.1 Direct-absorption measuring instruments

A few instruments use the Beer–Lambert–Bouguer law to determine ozone concentration by directly measuring the absorption features of ozone either within a path defined by the instrument or by using an extraterrestrial source (i.e., the Sun). The main advantage of direct absorption measuring instruments is that in the data-processing algorithm only a few assumptions have to be made regarding quantities that are not directly measurable. However, their main disadvantage is the uncertainty of the absorption cross-section and also the limited signal-to-noise ratio at the extreme measurement range.

In general, direct absorption measuring instruments are divided into two groups according to the source of the radiation they use. Those instruments that rely on an external radiation source (i.e., the Sun or the Moon) are called "passive" instruments (i.e., optical ozonesondes, IR spectrometers, and the laser heterodyne spectrometer). Passive instruments usually measure an average molecular density along the atmospheric path (observing at sunrise or sunset from above the atmosphere or descending through the atmosphere).

Instruments with their own radiation sources are called "active" instruments and include the source, the atmospheric path length, the collector, the aperture, the spectral analyzer, the detector, the signal chain, the data storage unit, and the instrument controller (i.e., DIAL, UV photometer).

2.5.1.1 Stratospheric Aerosol and Gas Experiment (SAGE)

The SAGE instruments are satelliteborne multichannel radiometers that measure gas extinction profiles using solar occultation. SAGE I was launched aboard the dedicated Applications Explorer Mission-2 (AEM-2) in February 1979 and operated for 34 months until November 1981, when the spacecraft's electrical system failed.

SAGE II was launched on the Earth Radiation Budget Satellite (ERBS) in October 1984 and globally monitored the vertical distribution of stratospheric aerosols, ozone, water vapor, and nitrogen dioxide, by measuring the extinction of solar radiation through the atmosphere during ERBS solar occultations (Mauldin *et al.*, 1985, 1986). Both SAGE I and SAGE II are in approximately 600 km circular orbits with inclination angles of 56 and 57°, respectively. They measure the absorption of sunlight by ozone with four channels (SAGE I), at 0.385, 0.45, 0.6 and 1.0 μm wavelengths at spacecraft sunrise and sunset (solar occultation geometry). The 0.6 μm channel is used to measure stratospheric ozone by using the Chappuis band. The 0.45 μm channel is used to measure the concentration of stratospheric NO_2, and the other two channels are used to determine aerosol optical depth. Vertical resolution is approximately 1 km for altitudes above 10 km. SAGE II is a seven-channel instrument and—in addition to ozone, NO_2, and aerosols—it measures water vapor concentrations. It makes measurements up to 65 km as against 55 km for SAGE I. The extra channels enable better corrections for aerosols to be made. SAGE III is an improved version with higher spectral resolution, greater spectral wavelength coverage, and a lunar occultation capability, which allows for a broader range of latitudes (than that of solar occultation) and for the measurements of NO_3 and ClO_2. It relies upon the flight-proven designs used in the Stratospheric Aerosol Measurement (SAM I) and the first and second SAGE instruments. In SAGE III (for the Earth Observing System Program) the filters are replaced by a spectrometer, a diode array is used for detecting radiation in the various wavelength regions, and the spectral range is extended to 1.55 μm for observation of aerosols with larger particles and aerosols near the Earth's surface. This configuration enables SAGE III to make multiple measurements of absorption features of target gaseous species and multiwavelength measurements of broadband extinction by aerosols from the middle troposphere through the stratosphere. By making these measurements in the correct color region, SAGE III produced accurate profiles of ozone or water vapor (NASA). One SAGE-III instrument was launched on board the Russian spacecraft METEOR 3M in 1998 (polar Sun-synchronous orbit), and another was launched from Baikonur, Russia, on December 10, 2001 for installation on the International Space Station (51.5° inclination orbit).

Regarding precision and accuracy, the major uncertainty in the ozone concentrations measured by either SAGE I or SAGE II is Chappuis band ozone absorption cross-sections and their temperature dependences, which are estimated to give a 6% uncertainty. The measurement precision of SAGE II is 4–7% in the 15–52 km altitude range, becoming 20% at 60 km and 50% at 10 km. When statistical and systematic uncertainties are combined, accuracy is better than 10% for heights between 15 and 52 km. Figure 2.6 shows the differences between ozone measurements made by various different systems and by SAGE-II (WMO, 1999)..

2.5.1.2 *Atmospheric Trace Molecule Spectroscopy (ATMOS)*

The first chemically comprehensive set of space-based measurements was made by the ATMOS instrument, an infrared occultation interferometer. This instrument

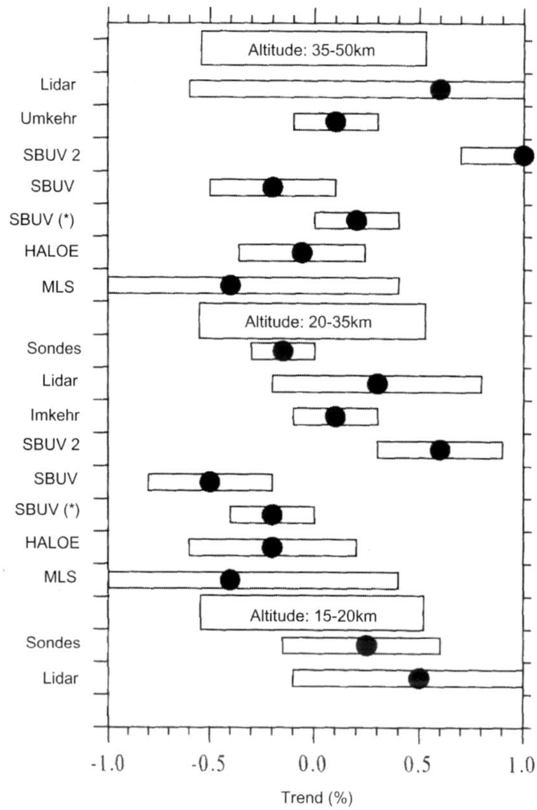

Figure 2.6. Trends of differences (i.e., drifts) between ozone measurements made by various ozone-profiling instruments and SAGE-II, in % per year [(Sounding – SAGE-II)/SAGE II]. The trends of ozonesondes, lidar, and Umkehr differences are presented as averages for eight northern mid-latitude sounding stations. The trends of satellite differences are presented as global means. Average differences are indicated by dots, and bars represent the 99% confidence interval of drift estimation. The entries for SBUV are for SBUV/SAGE II comparison; the entries for SBUV* are for SBUV/combined SAGE I and SAGE II comparison (WMO, 1999).

uses a Fast Fourier Transform (FFT) interferometer, which operates in the near and mid-infrared region of the spectrum to detect many molecular constituents in the upper atmosphere and to measure their vertical and horizontal distributions. It was first flown aboard the Shuttle *Challenger* (as a part of *Spacelab 3*) in April–May 1985 and measured solar radiation at sunsets and sunrises (as seen from the orbiter). Twenty occultations were recorded from this experiment and these produced simultaneous concentration profiles for a large number of minor upper-atmosphere species, several of which had not previously been observed. Simultaneity of observations is a very important aspect in the study of the highly interactive nature of photochemical processes. Subsequent flights of the ATMOS instrument (March 1992, April 1993, November 1994) extended ATMOS coverage to the tropics and high latitudes and were used for trend determination by comparison of the mid-1985 and late-1994 observations (Grant, 1989).

2.5.1.3 Polar Ozone and Aerosol Monitor (POAM)

The POAM II instrument is a satelliteborne multichannel radiometer similar to SAGE II. It was carried onboard the Système Pour l'Observation de la Terre (SPOT-3) satellite in an orbit similar to that of the SAM-II sensor on the Nimbus-7 satellite (i.e., a Sun-synchronous 98° inclination, 833 km orbit with a 10:30 LT equatorial crossing time). In normal operation, it provided for Southern Hemisphere observations much closer to the South Pole, extending to 88°S at the vernal equinox from October 15, 1993 until November 14, 1996. It is a solar occultation instrument with nine narrowband channels between 352 and 1,060 nm that retrieve stratospheric ozone, aerosol extinction, NO_2, and water vapor (Glaccum *et al.*, 1996). Its ozone profiles were used to study the breakup of the 1993 Antarctic polar vortex (Bevilacqua *et al.*, 1995), polar vortex dynamics during the 1994 and 1995 northern winters (Randall *et al.*, 1995), and polar mesospheric clouds in the Southern Hemisphere (Debrestian *et al.*, 1997; Fromm *et al.*, 1997).

One POAM instrument called the "Orbiting Ozone and Aerosol Monitor" (OOAM) flew on a 45° inclination orbit in the summer of 1997. Another POAM III instrument flew in early 1998 on the French SPOT-4 satellite in a polar Sun-synchronous orbit observing ozone at the wavelength of 0.603 μm.

2.5.1.4 Halogen Occultation Experiment (HALOE)

This instrument (Figure 2.7) was flown on the Upper Atmosphere Research Satellite (UARS) covering two longitude sweeps each day (15 profiles each): one at the latitude of sunsets and one at the latitude of sunrises. Each sunrise or sunset observation is separated by 1.6 hr and 24° longitude. It has eight infrared channels(-from 2.45 to 10.01 μm): four are broadband (60–120 cm^{-1}) filter radiometer channels for CO_2, H_2O, O_3, and NO_2 and four are gas filter correlation radiometer channels for HF, HCl, CH_4, and NO.

The HALOE instrument uses the principle of the satellite solar occultation technique, and the measurement principle is based on both gas filter correlation radiometry and broadband filter radiometry (Figure 2.7). In the first technique incoming radiation is split into different channels for each species and passed through a narrowband filter; it is then further split into two paths for each species. One path includes a cell containing the gas to be measured.

As the atmosphere passes in front of the Sun, the path containing the cell will see little change in absorption, because the appropriate portion of the Sun's spectrum is already absorbed by the gas in the cell. The conventional broadband technique is described in other sections describing other instruments (e.g., LIMS in Section 2.5.3.1). The vertical resolution for the retrievals of aerosol, NO_2, H_2O, and O_3 is 2 km. The accuracy for ozone-mixing ratios is estimated to be 10–15% in the mid-stratosphere, with a decrease in accuracy above and below this region (Grant, 1989).

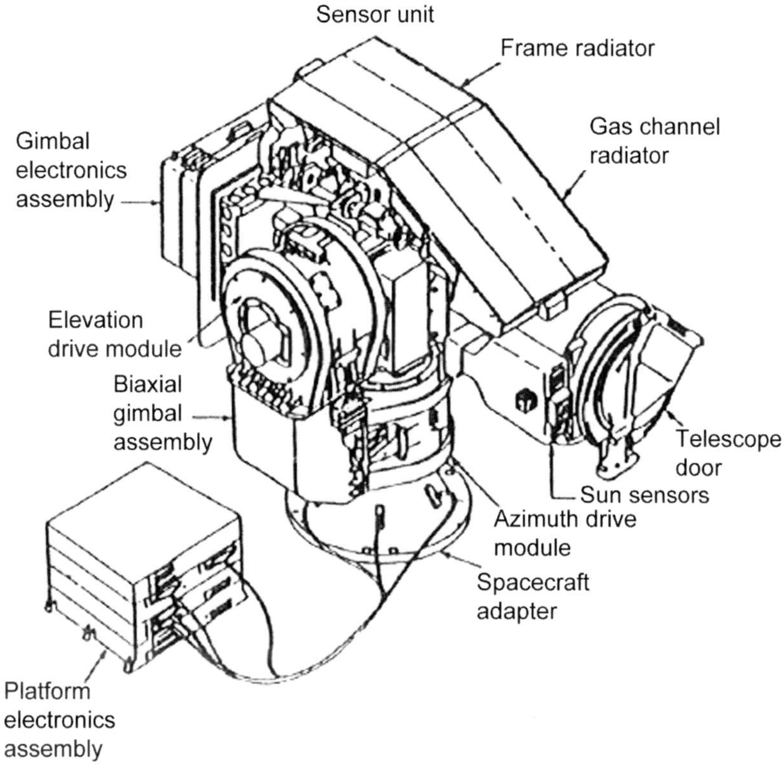

Figure 2.7. HALOE instrument (NASA, 1986).

Data obtained from the Halogen Occultation Experiment (HALOE) on board the Upper Atmospheric Research Satellite (UARS) during the period 1992–2004 have been analyzed by Fadnavis and Beig (2010) to study the effect of the 11-year solar cycle using a multifunctional regression model. The effects on ozone were analyzed over the 0–30°N and 0–30°S belts. The solar effect on ozone was found to be significant in most of the stratosphere: 2 ± 1.1 to $4 \pm 1.6\%/100$ sfu (solar flux unit). It is negligible in the lower mesosphere whereas it is of the order of 5%/100 sfu in the upper mesosphere. These observed results were in reasonable agreement with model simulations.

Stellar occultations obtained by the Solar–Stellar Irradiance Comparison Experiment (SOLSTICE) on board UARS are also used to study the ozone concentration in the mesosphere between 55 and 85 km (de Toma *et al.*, 1998).

2.5.1.5 SunPhotometer Earth Atmosphere Measurement (SPEAM)

This Canadian instrument was flown on the U.S. Space Shuttle *Challenger* in October 1984 and provided observations at 315 and 324 nm at 63.34°S, 81.96°E, from which OVPs are inferred. Observations started several minutes before the occultation time. The instrument was manually pointed at the Sun through the quartz window, and this was maintained until the occultation was over. The small size of the instrument, and the fact that it can be returned to the laboratory for calibration between flights, make this technique very useful. It was also flown on the Space Shuttle in 1992 making measurements of O_3, NO_3, and NO_2. It was actually a dual-wavelength interference filter sunphotometer, similar to the commercial single-channel device, formerly used in the Canadian solar radiation network. The flight version of it (STS 41-G) differs from its conventional ground-based version in that it can observe simultaneously at two different wavelengths (McElroy *et al.*, 1991). SPEAM made measurements as a part of a Canadian payload aboard two Space Shuttle flights: STS-41G in October 1984 and STS-52 in October 1992.

2.5.1.6 *The Atmospheric Chemistry Experiment (ACE)*

One of the newest satellites for solar occultation studies is the Atmospheric Chemistry Experiment (ACE) developed for remote sensing of the Earth's atmosphere. This Canadian-led satellite mission, also known as SCISAT-1, was launched on August 12, 2003. There are two instruments on board the spacecraft that provide vertical profiles of ozone and a range of trace gas constituents, as well as temperature and atmospheric extinction due to aerosols. The ACE Fourier Transform Spectrometer (ACE-FTS) measures in the infrared (IR) region of the spectrum and the Measurement of Aerosol Extinction in the Stratosphere and Troposphere Retrieved by Occultation (ACE-MAESTRO) operates in the UV/visible/near-IR.

The main objective of the ACE mission was to understand the global-scale chemical and dynamical processes that govern the abundance of ozone from the upper troposphere to the lower mesosphere, with an emphasis on chemistry and dynamics in the Arctic. SCISAT, the platform carrying the ACE-FTS and ACE-MAESTRO, is in a circular low-Earth orbit, with a 74° inclination and an altitude of 650 km. From this orbit, the instruments measure up to 15 sunrise (hereinafter SR) and 15 sunset (hereinafter SS) occultations each day. Global coverage of the tropical, mid-latitude, and polar regions (with the highest sampling in the Arctic and Antarctica) is achieved over the course of one year and the ACE measurement latitude pattern repeats each year.

The primary instrument for the ACE mission, the ACE-FTS (Figure 2.8), is a successor to the Atmospheric Trace MOlecule Spectroscopy (ATMOS) experiment, an infrared FTS that operated during four flights on the Space Shuttle (in 1985, 1992, 1993, and 1994). ACE-FTS measures high-resolution (0.02 cm^{-1}) atmospheric spectra between 750 and 4,400 cm^{-1} (2.2–13 μm). This sensor uses the solar occultation technique to determine the profiles of atmospheric trace gases, temperature pressure, and aerosols. A feedback-controlled pointing mirror is used to target the center of the Sun and track it during the measurements. Typical signal-to-noise ratios are more than 300 from 900 to 3,700 cm^{-1}. From the 650 km ACE orbit, the instrument's field of view (1.25 mrad) corresponds to a maximum vertical resolution of 3–4 km. The vertical spacing between consecutive second ACE-FTS measurements depends on the satellite's orbital geometry during occultation and can

Figure 2.8. The ACE-FTS instrument (interferometer side) (*http://www.ace.uwaterloo.ca/instruments_acefts.html*).

vary from 1.5 to 6 km. The altitude coverage of the 20 measurements extends from the cloud tops to 100–150 km.

The Measurement of Aerosol Extinction in the Stratosphere and Troposphere Retrieved by Occultation (ACE-MAESTRO) is a dual-grating diode array spectrophotometer that extends the wavelength range of ACE measurements into the near-IR to UV spectral region (Figure 2.9). It records over a nominal range of 400–1,010 nm with a spectral resolution of 1.5–2 nm for its solar occultation measurements. The forerunner of the ACE-MAESTRO is the Sun Photo Spectrometer instrument which was used extensively by Environment Canada as part of the NASA ER-2 stratospheric chemistry research program. ACE-MAESTRO uses the same Sun-tracking mirror as the ACE-FTS, receiving 7% of the beam collected by the mirror. The ACE-MAESTRO instrument's vertical field of view is 1 km at the limb (Figure 2.9). Observation tangent altitudes range from the surface to 100 km with a vertical resolution estimated at better than 1.7 km.

The processing of ACE-MAESTRO version 1.2 occultation data is done in two stages and is described by McElroy *et al.* (2007). In summary, the raw data are

Figure 2.9. Photograph of the MAESTRO instrument, showing occultation (solar) and backscatter viewing ports (McElroy *et al.*, 2007).

converted to wavelength-calibrated spectra, corrected for stray light, dark current, and other instrument parameters in the first step. The corrected spectra are then analyzed by a non-linear least squares spectral-fitting code to calculate slant path column densities for each spectrum, from which the vertical profiles of O_3 and NO_2 volume-mixing ratios are subsequently derived. The retrieval algorithm does not require any *a priori* information or other constraints. The inversion routine uses the pressure and temperature profiles and tangent heights from the ACE-FTS data analysis to fix the tangent heights for ACE-MAESTRO. Vertical profiles for the trace gases are determined by adjusting an initial guess (high vertical resolution model simulation) using a non-linear Chahine relaxation inversion algorithm. The final profiles are provided both on the tangent grid and interpolated onto a 0.5 km spacing vertical grid (Dupuy *et al.*, 2009).

2.5.1.7 Upcoming ESA Mission

PREMIER (PRocess Exploration through Measurements of Infrared and millimeter wave Emitted Radiation) is one of the three candidate Earth Explorer mission concepts to progress to the next phase of consolidation. Following a further down-selection, this will lead to ESA's seventh Earth Explorer mission envisaged to be launched around 2016.

The PREMIER candidate mission aims to quantify the processes that control the composition of the mid to upper troposphere and lower stratosphere—which equates to 5–25 km above the surface of the Earth. This region of the atmosphere is particularly important for climate studies because it is where the atmosphere cools to space and where it is most sensitive to changes in the distribution of radiative gases and clouds. Through the transport of water vapor, ozone, methane, clouds, and aerosols, this region is host to numerous important interactions between the composition of the atmosphere and climate.

PREMIER will address these issues by measuring temperature, water vapor, and a large number of trace gases such as ozone and methane at high spatial resolution. The proposed mission concept employs two innovative instruments: an infrared-imaging spectrometer and the first millimeter wave sounder optimized for the upper troposphere. These instruments are based on proven heritage used on earlier missions such as the millimeter wave radiometer on the Swedish Odin satellite, which is now an ESA Third Party mission, and the MIPAS (Michelson Interferometer for Passive Atmospheric Sounding) instrument on ENVISAT.

The payload complement of PREMIER consists of two instruments observing the limb of the Earth: (a) the Millimeter Wave Limb Sounder (MWLS) and (b) the Infra-Red Limb Sounder/Cloud Imager (IRLS/IRCI).

The Millimeter Wave Limb Sounder (MWLS) concept description is based on the STREAMR instrument concept being developed by the Swedish Space Corporation within the framework of the dedicated Swedish National Program.

The STREAMR instrument is designed to provide spatially well-resolved (1–2 km vertically and 50 km horizontally) information on the distribution of Upper Troposphere/Lower Stratosphere (UT/LS) key constituents such as water vapor, ozone, and carbon monoxide on a global scale using bands in the 310–360 GHz spectral region. The STREAMR measurement concept is based on tomographic multibeam limb sounding in the orbital plane using Schottky diode heterodyne detectors. The instrument observes the limb at 14 tangent altitudes simultaneously with a staring view concept.

An offset antenna system receives thermal radiation from the atmospheric limb. Additional optical elements fold the path and refocus the beams on the focal plane. Calibration devices can be viewed by rotating a switch mirror close to the secondary aperture stop. A second rotating mirror selects the calibration source from one of two cold-sky views: a temperature-controlled warm load or a sideband filter to calibrate the sideband ratios.

Individual mirror–horn combinations couple the signals to the waveguides of the 14 sub-harmonically pumped Schottky mixers integrated with low-noise amplifiers. The down-converted signals in the 9 to 21 GHz spectral region are distributed after amplification to a set of autocorrelation spectrometers and divided into six separate units to simplify thermal control.

The Infra-Red Limb Sounder and Cloud Imager (IRLS/IRCI) instrument is a Fourier transform spectrometer combining the functions of spectrometry and imagery. The IRLS provides two mutually exclusive measurement modes: the atmospheric Chemistry Mode, featuring the highest spectral resolution to observe minor trace gases, and the atmospheric Dynamics Mode, providing a higher spatial resolution in order to resolve finer atmospheric structures. The IRCI must work continuously, acquiring images at a high spatial resolution in the same spectral ranges as the IRLS but with a reduced spectral resolution (10–20 cm^{-1}).

The spectral range of the IRLS covers two bands: band A (770–980 cm^{-1} or 10.2–13 μm) and band B (1070–1650 cm^{-1} or 6.0–9.4 μm). The spectral resolution is 0.2 cm^{-1} and 1.25 cm^{-1} in the chemistry and dynamic modes, respectively.

The IRLS measures limb radiance at vertical spatial-sampling intervals of 2 km and 0.5 km in the chemistry and dynamic modes, respectively. Radiance from the limb enters the instrument via a pointing mirror, which is also used to view cold space, and a flat or cavity blackbody for calibration. After beam size adaption in the afocal telescope, a Michelson interferometer produces interferograms, which are acquired by two detector arrays after band splitting by a dichroic filter. A laser metrology system, using a diode laser or a Nd:YAG laser, monitors the path difference between the two arms. The detectors are cooled to 55 K with Stirling cycle or pulsed tube coolers, with the possibility of operating the band B detectors at a slightly higher temperature (ESA, 2008).

2.5.2 Scattering-measuring instruments

2.5.2.1 Shuttle Ozone Limb Sounding Experiment (SOLSE)

The SOLSE is an ozone-measuring payload which was flown on board the Space Shuttle in the autumn of 1997 and included two UV instruments (Section 2.5.2.2). One was the Shuttle Ozone Limb Sounding Experiment (SOLSE). This is a spectrometer that uses UV wavelengths consisting of a two-dimensional detector (altitude wavelength). It measures ozone concentration in the stratosphere and upper troposphere (Hilsenrath *et al.*, 1991).

2.5.2.2 Limb Ozone Retrieval Experiment (LORE)

LORE and SOLSE were the two UV instruments of the ozone-measuring payload flown aboard the Space Shuttle in the autumn of 1997. LORE is a multiwavelength version of the Rayleigh Scattering Attitude Sensor (RSAS), a single-channel (355 nm) instrument that flew with SSBUV on STS-72 in January 1996. It uses optical filters and covers the UV, visible, and infrared channels. It obtains ozone profiles as low as the 10–15 km altitude range (Hilsenrath *et al.*, 1991).

2.5.2.3 Solar Mesospheric Explorer-Ultraviolet Spectrometer (SME-UVS)

The ultraviolet spectrometer (UVS) on the Solar Mesosphere Explorer (SME) satellite is designed to measure the ozone density of the mesosphere in the 1.0–0.1 hPa (48–70 km) region at sunlit latitudes. To achieve this, the instrument measures altitude profiles of Rayleigh-scattered solar photons in the 200–340 nm spectral region.

This instrument has an off-axis parabolic telescope, an Ebert–Fastie grating spectrometer, and a pair of photomultiplier tubes. The spectrometer is used at a pair of fixed wavelengths separated by approximately 30 nm (265.0 and 296.4 nm) and operates with a spectral resolution of 1.5 nm. The spacecraft rotates at 5 rpm and observes the Earth's limb, leading and trailing the SME ground track. It measures the altitude profiles of Rayleigh-scattered UV. Random errors vary from 3.4% at a height of 48 km to 10.2% at 68 km. Temperature and pressure errors vary from 4.5% at 48 km to 10.1% at 68 km. The SME-UVS provided a number of ozone density datasets in the measured ozone concentration, which help to explain ozone photochemistry (Grant, 1989). The Solar Maximum Mission (SMM) spacecraft carried a high-resolution ultraviolet spectrometer polarimeter (UVSP), employed since 1980 to measure the ozone profile by the technique of solar occultation (Aikin, 1989).

2.5.2.4 *Solar Mesosphere Explorer (SME)-Near-IR Spectrometer*

This instrument uses an off-axis parabolic mirror and an Ebert–Fastie spectrometer. There are two channels with a resolution of 12 nm and a separation of 0.6 μm. The 1.27 μm channel is used for emission measurements and the 1.87 μm channel is used for reference purposes (i.e., to correct data for radiances that are due to oxygen emission). Random error sources for ozone density include temperature (3% at 50 km, 1% at 90 km); systematic errors include calibration errors (10% in the laboratory before launch and 20% after launch).

The SME Near-Infrared Spectrometer has generated a number of datasets on spatiotemporal variations in ozone concentration since its launch in October 1981. A similar instrument was developed in Japan, the Infrared Air-glow (IRA), and flew on board the EXOS-C satellite (Ogawa *et al.*, 1989).

2.5.2.5 *SCanning Imaging Absorption Spectrometer for Atmospheric Chartography (SCIAMACHY)*

This is a multichannel spectrometer measuring radiation in the spectral range 240–1,750 nm at relatively high resolution (0.2 nm–0.5 nm) on board the European satellite ENVISAT-1, launched in 1999. This instrument was designed to measure the column and profile distribution of a number of gases, including high horizontal resolution measurements of ozone. It is a combination of two spectrometers also operating in the UV/visible/near-infrared part of the spectrum to observe transmitted, reflected, and scattered light. It provides measurements from nadir (scattered and reflected solar photons), limb (scattered and reflected solar photons), occultation (transmitted solar or lunar photons), and solar reference (extraterrestrial solar photons), achieving global coverage within three days.

From the limb measurements, the stratospheric profiles of pressure, temperature, and several trace gases (O_3, O_2, O_4, NO, NO_2, NO_3, N_2O, CO, CO_2, CH_4, H_2O, BrO, ClO, OClO, HCHO, SO_2, and aerosol) can be retrieved (Kramer, 2002; Rozanov *et al.*, 1997).

2.5.2.6 FY-3 Total Ozone Unit (TOU)

FY-3 satellites are Chinese second-generation polar orbit meteorological satellites. The Total Ozone Unit (TOU) is one of the main payloads on FY-3 satellites and the first instrument for daily global coverage of total ozone monitoring in China. The main purpose of TOU is to measure the Earth's backscatter ultraviolet radiation for retrieving a daily global map of atmospheric ozone. TOU will provide the important parameters for environmental monitoring, climate forecasting, and global climate change research.

TOU is an instrument used to measure incident solar radiation and backscattered ultraviolet radiance such that total ozone can be retrieved. Considering the characteristics of the atmospheric ozone profile, the measuring wavelength should be selected so that its effective scattering layer is situated in the troposphere in order to ensure the accuracy of measuring total ozone from the satellite.

The effective scattering layer for 302.5 nm is situated above the height of the ozone maximum and that for the 307.5 nm has two effective scattering layers. Therefore, the shortest wavelength for measuring total ozone should be longer than 308 nm.

Using the single-wavelength method makes it difficult to obtain total ozone precisely. Therefore, a wavelength pair method is used to eliminate the effects of atmospheric molecular scattering and aerosol scattering. In this method, a non-ozone-absorbing wavelength is used to obtain the reflectivity of underlying surfaces and the heights of cloud tops (Wang *et al.*, 2009).

2.5.3 Emission-measuring instruments

A few instruments measure ozone in emissions, observing features from the near-IR to millimeter wavelengths. These instruments observe the Earth's limb from a space-borne platform, and some are used in nadir-viewing directions (Grant, 1989; Kramer, 2002). The signals in emissions are generally weaker than those in absorption and, therefore, the instruments are generally cooled, except the millimeter wave ones (they can observe in very narrow spectral regions using heterodyne detection techniques).

2.5.3.1 Limb Infrared Monitor of the Stratosphere (LIMS)

The first space-based measurements of stratospheric trace constituents (besides ozone) were made from the Limb Infrared Monitor of the Stratosphere (LIMS) and the Stratosphere and Mesosphere Sounder (SAMS) on board Nimbus-7.

LIMS is a cryogenically cooled limb-scanning infrared filter radiometer, launched on Nimbus-7 in 1978 with a lifetime of 7 months. It was developed from the earlier Limb Radiance Inversion Radiometer (LRIR) with a limited number of channels. It has six channels (centered between 6.2 and 15.0 μm) that permit observation of emissions by CO_2, HNO_3, O_3, H_2O, and NO_2.

Two channels near 15 μm can be used to retrieve vertical temperature profiles. The other channels are used for molecular species measurements. Ozone, for example, is observed in the 926–1,141 cm^{-1} spectral region, with a vertical field of view of 1.8 km and a horizontal field of view of 18 km from 15 to 64 km altitude. Because of its geometry, this technique has an inherently high vertical resolution and can sound at high altitudes. The instrument scans from above the atmosphere to below the Earth's horizon and back every 24 s. The precision of this instrument was found to vary from 2–4% at 30 hPa to 1% at 0.4 hPa. Absolute accuracy was about 10%. Many experiments provided data about the distribution and variability of temperature and several minor constituents (O_3, H_2O, HNO_3, and NO_2) over the latitude range from 84°N to 64°S. Some data on ozone and aerosol profiles in the stratosphere were also obtained by the Aerosol Limb Absorption (ALA) instrument which flew on board the Japanese Ohzora satellite, launched in February 1984. ALA was a two-channel sunphotometer (6,500, 1,000 nm) that used the solar occultation technique. Another occultation instrument, the Limb Atmospheric Infrared Spectrometer (LAS) also flew aboard Ohzora (Ogawa *et al.*, 1989).

2.5.3.2 Solar Mesosphere Explorer Near-Infrared Spectrometer (IRS/SME)

The near-infrared spectrometer (NIRS) on board the SME satellite was designed to carry out ozone measurements between 50 and 90 km over most latitudes at 3:00 PM local time. The instrument measures emission from O_2 ($^1\Delta g$) at 1.27 μm that is primarily due to the photodissociation of ozone. Vertical resolution is better than 3.5 km. Note that at the time of the El Chichón eruption in 1982, there was no SAGE instrument operating and data about the eruption were obtained from SME (Grant, 1989).

2.5.3.3 Michelson Interferometer (IRIS, MIPAS)

This instrument (IRIS-D) was designed to record the infrared emission spectrum of the Earth and its atmosphere between 400 cm^{-1} and 1,600 cm^{-1} in order to provide the vertical profiles of temperature, humidity, and ozone concentration. IRIS-D launched on board Nimbus-4 is an advanced version of the IRIS-B launched on board Nimbus-3 a year earlier (on April 14, 1969). The subsequent improvement resulted in an increase in spectral resolution from 5 cm^{-1} to 2.8 cm^{-1}.

The MIPAS (Michelson Interferometer for Passive Atmospheric Sounding) is a middle-infrared Fourier transform spectrometer operating on board the ENVISAT platform and acquiring high-resolution spectra of atmospheric limb emission in five spectral bands within the frequency range from 685 to 2,410 cm^{-1} (14.6 to 4.15 μm) (Fischer *et al.*, 2007) during both daytime and nighttime with global coverage. It measures temperature, ozone, and other chemical species (especially nitrogen and chlorine-containing species) (Carlotti *et al.*, 1995; Guld *et al.*, 1994). It also measures aerosols and PSCs. In its standard observation mode, the instrument scanned 17 tangent altitudes for each limb sequence, viewing in the rearward direction along the orbit with a sampling rate of approximately 500 km along track and with a horizontal resolution across track of about 30 km. The vertical scanning grid ranges between 6 km and 68 km, with steps of 3 km from 6 to 42 km, 5 km from 42 to 52 km, and 8 km from 52 to 68 km. On a daily basis, MIPAS covered the Earth with 5° latitude by 12.5° longitude spacing. Complete global coverage was obtained approximately every three days by 73 scans per orbit and 14.3 orbits per day scanning the latitudinal range from 87°S to 89°N. MIPAS operation was temporarily halted at the end of March 2004 because of excessive anomalies observed in the interferometric drive unit and resumed in January 2005 in a new operational mode at a reduced spectral resolution (0.0625 cm^{-1}) and on a finer vertical grid. In addition to being flown on ENVISAT, a MIPAS instrument has also been used on a Transall C-160 aircraft (Guld *et al.*, 1994).

2.5.3.4 *Microwave Limb Sounder (MLS)*

The MLS was originally an airborne millimeter wavelength (1.0–1.5 mm) radiometer. The present version operates in the 200–300 GHz range. Using local oscillators and filter banks (35 filters per local oscillator), brightness temperatures (day and night) at various frequencies, along emission lines from various molecular species, are measured with a vertical resolution of about 5 km. The MLS was found to have a precision of 4% and an accuracy of the order of 9%. MLS was used on board the Upper Atmospheric Research Satellite (UARS) and proposed for the Earth Observing System (EOS). It was also used in the Balloon Intercomparison Campaign (BIC). MLS has been acquiring millimeter wavelength emission measurements of the stratosphere in both hemispheres since late September 1991. Measurements are made as the instrument's field of view is vertically scanned through the atmosphere limb in a plane perpendicular to the UARS trajectory. MLS pointing geometry together with the inclination of the UARS orbit leads to measurement latitudinal coverage that extends from 80° on one side of the equator to 34° on the other (Froidevaux *et al.*, 1996).

A new MLS instrument, which was a significantly enhanced version of the UARS MLS, was flown on NASA's Aura satellite (originally known as EOS-CHEM-1) which was launched on July 15, 2004 to measure temperature, ozone, and other chemical species in the lower stratosphere and the upper troposphere. Note that, given the length of the MLS dataset, its use in studies of long-term trends and shorter term phenomena such as the Quasi-Biennial Oscillation (QBO) is becoming possible (de Toma *et al.*, 1998).

The upcoming Global Atmosphere Composition Mission (GACM) (planned launch in 2016–2020) will include a Microwave Limb Sounder along with IR and UV spectrometers. The objectives of the GACM mission are to measure ozone and related gases for intercontinental air quality (NO_2, SO_2, HCHO, CO, and aerosols) and stratospheric ozone. A second-phase mission featuring an active differential absorption lidar (DIAL) system will be launched in 2022 or shortly thereafter. It will measure O_3 with a vertical resolution better than 2 km and aerosols and atmospheric structure at resolutions better than 150 m.

2.5.3.5 *Tropospheric Emission Spectrometer (TES)*

The Tropospheric Emission Spectrometer (TES) was the first satellite instrument to provide simultaneous concentrations of carbon monoxide, ozone, water vapor, and methane throughout Earth's lower atmosphere. TES is an imaging infrared Fourier Transform Spectrometer designed especially to measure the state and composition of the troposphere (i.e., the layer of the atmosphere that extends from the Earth's surface to about 16 km in altitude). However, in principle, TES is capable of observing trace gases at any altitude and in practice its use extends from the ground up to an altitude of about 32 km so that it includes the important regions of the upper troposphere and the lower stratosphere. While the instrument can detect and measure many components of the troposphere, one of its main purposes is to study ozone. TES is flown on the Aura satellite which, as we have mentioned before, was launched on July 15, 2004. TES has both nadir and limb-viewing capability and covers the spectral range 650–2,250 cm^{-1} (4.5–15.4 μm) at either 0.08 cm^{-1} or 0.02 cm^{-1} spectral resolution.

TES has four co-aligned focal plane detector arrays of 1×16 elements (pixels), with each array optimized for a different spectral region. Each pixel's instantaneous field of view (IFOV) is 0.075 mrad high by 0.75 mrad wide. At the limb, this corresponds to about 2.3 km altitude by 23 km parallel to the horizon. TES observes both in nadir view and in limb view. Limb viewing provides a much longer path through the atmosphere, and looking through a larger mass of air improves the chances of observing sparsely distributed substances that might be missed in the nadir view. Limb viewing's sideways angle also makes it easier to determine the altitudes of observed substances. But limb viewing is very susceptible to interference (only rarely does the line of sight reach the surface). Nadir viewing is less affected by clouds, but looking straight down makes it more difficult to determine altitudes. A complete description of TES is given by Beer *et al.* (2001) (see also Beer, 2006 and Schoeberl *et al.*, 2006).

Felker *et al.* (2011) described the use of TES data in combination with other information to generate a Multi-sensor Upper Tropospheric Ozone Product (MUTOP). Determination of ozone concentration in the upper troposphere (UT) layer is especially difficult without direct *in situ* measurement. However, it is well understood that UT ozone is correlated with dynamical tracers like low specific humidity and high potential vorticity. Thus the approach of Felker *et al.* (2011) was to create map view products of upper-troposphere (UT) ozone concentrations through the integration of TES ozone measurements with the two synoptic dynamical tracers of stratospheric influence: specific humidity (derived from the GOES geostationary satellite) and potential vorticity (obtained from an operational forecast model). This approach results in the spatiotemporal coverage of a geostationary platform, which is a major improvement over the much less frequent individual polar overpasses, while retaining the ability of TES to determine UT ozone concentration.

Felker *et al.* (2011) note that there are several advantages in this multisensor-derived product approach: (1) it is calculated from two operational fields (GOES specific humidity and Global Forecast System (GFS) potential vorticity (PV)), so layer average ozone can be created and used in near-real time; (2) the product provides the spatial resolution and coverage of a geostationary platform as it depicts the distribution of dynamically driven ozone in the UT; and (3) the 6-hour temporal resolution of the imagery allows for the visualization of rapid movement of this dynamically driven ozone in the UT (see the paper by Felker *et al.*, 2011 for details on this method).

2.5.3.6 Cryogenic Limb Array Etalon Spectrometer (CLAES)

This instrument measures the concentrations of members of the nitrogen and chlorine families—as well as ozone, water vapor, methane, and carbon dioxide—by observing infrared thermal emissions at wavelengths from 3.5 to 12.7 μm. In normal operation it utilizes a telescope, a spectrometer, and a linear array of 20 detectors to perform simultaneous observations for 20 altitudes ranging from 10 to 60 km. To avoid interference from the thermal emissions of the detectors and optics the CLAES cryogenic system is used, although it limits the lifetime of the instrument. This system consists of two subsystems: a block of solid neon at −260°C (cooling the detectors) and a surrounding block of solid carbon dioxide at −150°C (reducing emissions from the optical system). It has been used on board the UARS research satellite (Grant, 1989).

2.5.3.7 Improved Stratospheric and Mesospheric Sounder (ISAMS)

This is a filter radiometer that observes, with eight detectors, infrared molecular emissions by means of a movable off-axis reflecting telescope from 4.6 to 16.6 μm. It utilizes a Stirling cycle refrigerator to cool its detectors to −195°C and carries samples of some of the gases to be measured in cells within the instrument. Installing these cells in front of the detectors, the atmospheric radiation collected by the telescope passes through them, allowing the spectra of the gases in the cells to be matched with the spectra observed in the atmosphere. As some gases cannot be confined in cells (because of their chemical activity) ISAMS employs broadband filters to isolate portions of the spectrum and, thereby, permits measurements of the gases. The instrument flown on board UARS is an improved version of that aboard Nimbus-7 and measures the concentrations of nitrogen chemical species, as well as ozone, water vapor, methane, and carbon monoxide. CLAES and ISAMS data, through their excellent spatial and comprehensive chemical information content, have provided significant information on both chemical and transport processes in the stratosphere, including issues such as the tropical mid-latitude transport of tracers (Koutoulaki *et al.*, 1998).

2.5.3.8 High Resolution Dynamics Limb Scanner (HIRDLS)

This is a multichannel limb-scanning infrared radiometer developed jointly by the U.S. and the U.K. It is carried by the EOS-CHEM platform (Figure 2.10). HIRDLS has measured temperature, ozone, and a number of trace gases with a vertical resolution of 1 km from the tropopause to the mesopause since January 23, 2005. It provides information on small-scale variability in the atmosphere for use in transport studies (Dials *et al.*, 1998).

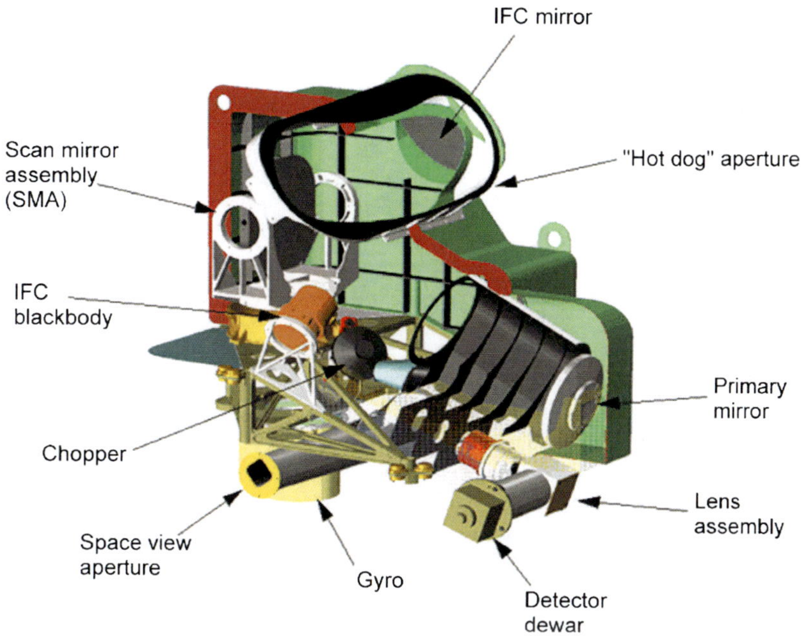

Figure 2.10. HIRDLS components (*http://www.ucar.edu/communications/staffnotes/0511/hirdls.html*).

The general objectives of HIRDLS are twofold: to provide information to assess the role of the stratosphere, especially the lower stratosphere and the upper troposphere (UT/LS) in climate processes; and to observe the processes that affect the stratospheric ozone layer at a time when the concentrations of active chlorine have reached a maximum and are beginning to decrease.

2.5.3.9 Cryogenic Infrared Spectrometers and Telescopes for the Atmosphere (CRISTA)

This is an infrared limb sounder designed to measure the densities of mid-atmospheric trace gases with high spatiotemporal resolution. It contains four limb-scanning liquid helium-cooled spectrometers and three telescopes that are simultaneously operated to acquire global maps of temperature and atmospheric trace gases. The telescopes (used to sense three atmospheric volumes 600 km apart) are each followed by infrared grating spectrometers; the center telescope is additionally equipped with a spectrometer for the far-infrared spectrum. A complete spectrum is measured in about 1.2 s, while a complete altitude scan is achieved in less than 1 min, providing trace gas profiles on a grid lower than 500 km × 600 km and with vertical resolution of about 2–3 km.

CRISTA was successfully used aboard the CRISTA-SPAS (Shuttle Pallet Satellite) free flyer together with ATLAS-3 (Kaye and Miller, 1996) during a

Space Shuttle mission (STS-66) in November 1994. A second flight of CRISTA and the Middle Atmosphere High Resolution Spectrographic Investigation (MAHRSI) for NO and OH observations took place in August 1997, as part of the STS-85 Space Shuttle mission. The U.S. Cryogenic Infrared Radiance Instrument for Shuttle (CIRRIS), which was designed mainly for airglow studies, was flown on board the Space Shuttle in April–May 1991 (STS-39) and relayed information on OVP and other traces.

2.5.3.10 Improved Limb Atmospheric Spectrometer (ILAS)

ILAS is an occultation instrument, getting vertical profile measurements at high northern and southern latitudes. It consists of two spectrometers, one using the infrared for trace constituent measurements and the other visible wavelengths to derive temperature and pressure profiles.

This spectrometer was developed by the Environmental Agency of Japan and installed on board the ADEOS spacecraft, which was launched on August 17, 1996. ILAS measures the vertical profiles of O_3, NO_2, HNO_3, N_2O, CH_4, H_2O, CFC-11, CFC-12, N_2O5, and aerosols in the altitude range of 10–60 km with an altitude resolution of 1–2 km. It performs 14 measurements per day on an Arctic and Antarctic latitude circle (58–73°N, 65–90°S). ILAS O_3 profiles are in good agreement with ozonesonde profiles between 20 and 24 km. Higher ILAS ozone below 20 km may be attributed to the presence of aerosols, while lower ILAS O_3 at 24–32 km look like being due to a systematic difference between ILAS and ozonesondes, probably owing to the ILAS operational algorithm (Yushkov *et al.*, 1998). A modified version of the ILAS-II is scheduled to fly on board the Japanese ADEOS-2 satellite (see Section 2.3.3).

2.5.3.11 Millimeter wave Atmospheric Sounder (MAS)

This is a Shuttleborne limb-scanning radiometer that measures the molecular emission spectra of O_3, H_2O, ClO, and O_2 in the Earth's atmosphere, in order to provide their geographical and vertical distributions. MAS belongs to the Atmospheric Laboratory for Applications and Science (ATLAS) Shuttle payload package. It measured OVP at altitudes between 20 and 80 km over more than 90% of the Earth's surface during the three ATLAS missions (in 1992, 1993, 1994).

The MAS instrument uses a single 1 m diameter off-axis parabolic antenna (pointed toward the Earth's limb) and observes radiation along paths through the Earth's atmosphere at tangent heights from 10 to 140 km. The O_3 spectrum is measured with 50 RF filters, arranged in three banks (wide, intermediate, and narrow having widths of 40, 2, and 0.2 MHz, respectively).

The precision of mixing ratio results is estimated at 2–4% at altitudes of 29–62 km, increasing at both higher and lower altitudes to about 8%. Taking measurement errors into account the total uncertainty is estimated to be about 5% near the O_3 peak (Daehler *et al.*, 1998).

2.5.3.12 *Advanced Millimeter wave Atmospheric Sounder (AMAS)*

This instrument was designed as a limb-viewing, heterodyne spectrometer with two sub-millimeter receivers to measure ClO and BrO at 500 GHz and HCl at 625 GHz and receivers to measure O_3, N_2O, and O_2 in the 300 GHz region, H_2O near 325 GHz and CO and HNO_3 close to 345 GHz. AMAS is designated for launch alongside NASA's TOMS on the Russian Meteor-3M/2 and will monitor global distributions of important constituents in the stratosphere and upper troposphere (Reburn *et al.*, 1998). The European Space Agency is investigating several scenarios for future satelliteborne limb sounders in the millimeter and sub-millimeter spectral range. For instance, MASTER is planned to be an instrument in the millimeter range focusing on the upper troposphere and on exchange processes between the troposphere and the stratosphere tracking CO, H_2O, O_2, SO_2, N_2O, and HNO_3 in the 200–350 GHz band (STAR, 2001). SOPRANO was planned to focus on stratospheric chemistry with a sub-millimeter receiver observing O_3, ClO, HCl, O_2, BrO, HOCl, CH_3Cl, H_2O, N_2O, HNO_3 in the 500–950 GHz band. PIRAMHYD is planned to measure OH at even smaller wavelengths (2.5 and 3.5 THz) than SOPRANO (Bühler *et al.*, 1996). These instruments are planned to operate in the millimeter and sub-millimeter spectral range, allowing the retrieval of atmospheric mixing ratio profiles of several molecular species (STAR, 2001).

2.5.3.13 *Infrared Atmospheric Sounding Interferometer (IASI)*

The Infrared Atmospheric Sounding Interferometer (IASI) is probably the most advanced instrument carried on the MetOp satellite. The sophisticated IASI instrument is a Fourier Transform Spectrometer based on a Michelson Interferometer coupled to an integrated imaging system that observes and measures infrared radiation emitted from the Earth (Figure 2.11).

The optical interferometry process offers fine spectral samplings of the atmosphere in the infrared band between the wavelengths of 3.4 and 15.5 μm. This enables the instrument to establish temperature and water vapor profiles in the troposphere and the lower stratosphere, as well as measure quantities of O_3, CO, CH_4, SO_2, and other compounds (Clarisse *et al.*, 2008). The IASI system aims at observing and measuring the infrared spectrum emitted by the Earth. One of its missions is to provide information on the total amount of ozone with an accuracy of 5% and a horizontal sampling of typically 25 km. It may also provide information on ozone vertical distribution with an accuracy of 10% and a vertical resolution that provides two or three pieces of independent information.

The Infrared Atmospheric Sounding Interferometer (IASI) scanning mirror directs emitted infrared radiation from one swath into the uncovered interferometer (Figure 2.12). One swath covers 30 scan positions towards the Earth and two calibration views. One calibration view is into deep space and the other is via an internal blackbody. Each scan starts on the left with respect to the flight direction of the satellite.

The sequence concludes with the interferometer (housed in the MetOp satellite) showing incident radiation being directed through the long split opening to make

Figure 2.11. The MetOp satellite carrying the IASI instrument (European Space Agency).

observations through the atmosphere, the calibration process then quickly takes place. The first flight model of IASI was launched on October 15, 2006 on board the first European meteorological polar-orbiting satellites, MetOp-1. IASI delivers temperature, moisture, and ozone profile information for the upper atmosphere.

2.5.3.14 Stratospheric Wind Interferometer for Transport Studies (SWIFT)

The Stratospheric Wind Interferometer for Transport Studies (SWIFT) is an instrument carried on the Japanese GCOM-A1 satellite designed to measure wind profiles in the stratosphere and simultaneously provide profiles of ozone density. Ozone concentration and distribution is not only based on chemical reactions but is also greatly affected by the movement of air in the stratosphere. Good knowledge of the dynamics in the stratosphere and troposphere is indispensable for the correct scientific interpretation of observed ozone distributions.

SWIFT is aimed at providing much needed information about global wind distributions in the stratosphere by measuring the thermal emission of ozone mol-

Figure 2.12. The scanning mirror of IASI directing emitted infrared radiation from one swath into the uncovered interferometer. One swath covers 30 scan positions towards the Earth and two calibration views. One calibration view is into deep space and the other is via an internal blackbody. Each scan starts on the left with respect to the flight direction of the satellite (*source*: European Space Agency).

ecules in the infrared at a wavelength of 9 μm. Because ozone molecules are moving in the atmosphere by the wind, the wavelength of their emission line appears to be shifted by the Doppler effect. SWIFT is able to measure this tiny wavelength shift and, consequently, stratospheric wind can be calculated to an accuracy of $5\,\mathrm{m\,s^{-1}}$. Although this concept has already been successfully employed for wind measurements in the mesosphere with the Canadian WIND II instrument (WIND Imaging Interferometer) on the UARS satellite, this is the first time that this technique will have been used to take direct stratospheric wind measurements from space. WIND II measures winds in the upper mesosphere and lower thermosphere using Doppler shifts in visible airglow features.

Since SWIFT will observe an ozone emission line, it will also be able to measure ozone concentration simultaneously with the wind and, as such, will offer further possibilities for ozone transport studies.

2.5.3.15 Sounding of the Atmosphere using Broadband Emission Radiometry (SABER)

The Sounding of the Atmosphere using Broadband Emission Radiometry (SABER) instrument is one of four instruments mounted on the Thermosphere Ionosphere Mesosphere Energetics and Dynamics (TIMED) satellite. SABER sounds the MLTI (the mesosphere and lower thermosphere/ionosphere, 60–180 km altitude) region of the Earth's atmosphere by measuring infrared limb emission. The instrument measures accurate ground-processed values of atmospheric limb radiance to retrieve temperature, ozone, water vapor, carbon dioxide, and key parameters describing the energetic procedures of the high atmosphere. SABER results are being used to study the chemistry and dynamics of the mesosphere and the effects on the atmosphere due to major solar storm events (Cooper, 2004).

SABER (launched in 2001 and expected to be out of service by 2020) is an infrared radiometer which observes emission from the Earth's limb in 10 relatively narrow spectral intervals. It has a 2 km instantaneous field of view which scans the Earth limb from the Earth's surface to 400 km tangent altitude. Profiles of emitted radiance are recorded simultaneously through one telescope in all 10 spectral channels. These profiles are being analyzed using a variety of models in order to derive the profiles of kinetic temperature, minor species concentration, energy loss rates, solar heating rates, chemical heating rates, and radiative cooling rates. The SABER radiometer will be accurately calibrated to better than 5% (3% is the goal) in all spectral channels.

Daytime ozone is being determined by SABER by means of two different techniques. Emission by the fundamental asymmetric stretch bands of ozone in the vicinity of 9.6 μm will be observed in order to derive the ozone concentration using emission directly from the ozone molecule itself. The SABER spectral bandpass for this channel is 1,010–1,140 cm^{-1} and has been chosen to minimize the contribution from the hot bands of ozone and also from the laser bands of carbon dioxide which provided a considerable signal to the Limb Infrared Monitor of the Stratosphere (LIMS) experiment. Daytime ozone will also be inferred from measurements of the emissions from molecular oxygen dayglow at 1.27 μm. This technique was applied in the SME experiment and will provide a second measure of daytime ozone on the SABER experiment.

Nighttime ozone concentrations will be provided by measurements at 9.6 μm. The SABER instrument has sufficient sensitivity to record ozone emissions up to 100 km altitude (day and night). The choice of the spectral interval and the specifics of the non-LTE problem make the retrieval of nighttime ozone under non-LTE conditions relatively straightforward (Mlynczak, 1997).

2.5.3.16 Superconducting Submillimeter-Wave Limb-Emission Sounder, SMILES

The Superconducting Submillimeter-Wave Limb-Emission Sounder (SMILES) was successfully launched and attached to the Japanese Experiment Module (JEM) on the International Space Station (ISS) on September 25, 2009 (Kikuchi *et al.*, 2010). It has been making atmospheric observations since October 12, 2009 with the aid of a

4 K mechanical cooler and superconducting mixers for sub-millimeter limb emission sounding in the frequency bands of 624.32–626.32 GHz and 649.12–650.32 GHz. On the basis of the observed spectra, data processing has been retrieving vertical profiles for atmospheric minor constituents in the middle atmosphere, such as O_3 (with isotopes), HCl, ClO, HO_2, BrO, and HNO_3. Results from SMILES have demonstrated its high potential to observe atmospheric minor constituents in the middle atmosphere. Unfortunately, SMILES observations have been suspended since April 21, 2010 owing to the failure of a critical component (Kikuchi *et al.*, 2010). Previous versions of the instrument (BSMILES) were flown on balloons in 2003, 2004, and 2006.

2.5.3.17 *Odin satellite*

Odin is a small Swedish satellite built jointly by Canada, France, and Finland; it was launched on February 20, 2001 to a Sun-synchronous 600 km orbit. It has two instruments, a sub-millimeter radiometer using microwave wavelengths to measure ozone, chlorine monoxide, water vapor, and other constituents, and an optical spectrograph and infrared-imaging system (OSIRIS) using UV–visible and near-infrared wavelengths for studying O_3, NO_2, aerosols, and other constituents. The SMR is a limb-sounding instrument that employs four tunable heterodyne receivers in the range 486–581 GHz and a 1 mm wave receiver at 119 GHz, to observe atmospheric thermal emission spectra for determination of the vertical distribution of trace species relevant to stratospheric and mesospheric chemistry and dynamics (Frisk *et al.*, 2003; Murtagh *et al.*, 2002).

Odin was the first satellite to employ sub-millimeter (480–580 GHz) radiometry of atmospheric thermal emission in a limb-sounding mode to measure the global distributions of several species important for ozone chemistry in the stratosphere (O_3, ClO, N_2O, and HNO_3). These observations were performed simultaneously, while another mode is focused on odd hydrogen chemistry, giving H_2O, O_3, HO_2, and CO. The retrieval method is the Optimal Estimation Method (von Scheele, 1997).

2.5.3.18 *Neural Network Ozone Profile Retrieval (NNORSY)*

A variety of national and international-funded projects enabled the Center for Solar Energy and Hydrogen Research (ZSW) in Stuttgart, Germany (*http://nnorsy.zsw-bw.de/NNORSY-Intro.htm*) to develop the Neural Network Ozone Profile Retrieval (NNORSY) to provide very fast total ozone and ozone profile retrieval from different satellite instruments. NNORSY uses neural networks that are trained on GOME UV/VIS spectra by using collocated ozone profiles. These measured data include data from ozonesondes (e.g., SHADOZ, WOUDC) and from satellites (e.g., SAGE, HALOE, POAM, NOAA). The full database is divided into the training dataset and a smaller test dataset that is not used as input information for the neural nets. The whole training process was carried out in a huge number of training steps called "epochs", and training progress could be monitored by comparing neural network outputs with the independent test dataset. Once trained, NNORSY was ready for operational use. Since NNORSY is able to deal with instrument degradation, it can be used for near-real time applications even when the training database is slightly outdated.

2.5.4 Summary of ozone-monitoring satellites

Table 2.7 explains the acronyms used for ozone-monitoring instruments found on ozone-monitoring satellites and Table 2.8 gives a chronological summary of ozone-monitoring satellites (or Space Shuttle missions).

2.6 OBSERVED VARIABILITY IN VERTICAL OZONE DISTRIBUTION

In the troposphere, the ozone concentration falls (on average) with increasing altitude until the tropopause is reached. In the stratosphere, ozone concentration increases rapidly with altitude to a maximum near 5 hPa, with an often appearing secondary maximum in the lower stratosphere near 100 hPa. A long-term decrease in stratospheric ozone has occurred at mid-latitudes in both hemispheres. In the Northern Hemisphere the decrease is larger in winter and spring (11% since 1979) than in summer or autumn (4% since 1979). The most dramatic changes have been seen at high latitudes in wintertime vortices over the Arctic and Antarctic. Analyses of measurements of ozone in the 1994/1995, 1995/1996, and 1996/1997 winters have shown chemical losses of up to 50% at some altitudes in each winter. Current ozone datasets came from various ozone-measuring platforms using different techniques with different spatiotemporal coverage. WMO (1999) provided the most reliable datasets with time records longer than 5 to 10 years (Table 2.9).

The Arctic vortex has been colder than usual in recent winters with record low temperatures observed during the three winters leading up to the 1997/1998 winter. It should be noted that no major mid-winter warming of the stratosphere has occurred since 1990/1991. Prior to that time, they had occurred every couple of years, and the longest gap in the 40-year record had been 4 years. Under cold conditions (temperatures below approximately 195 K or −78°C), polar stratospheric clouds (PSCs) can form.

Chemical reactions on PSC particles convert chlorine compounds into forms that can rapidly destroy ozone in the presence of sunlight.

It is worth noting that nacreous (mother of pearl) clouds are a rare form of stratospheric cloud which occur high up in the stratosphere. These kinds of clouds appeared, for instance, on November 30, 1999 and were widely seen across Scotland and the far north of England (Figure 2.13). They require unusually cold conditions of about −80°C in the stratosphere to form and are usually associated with a strong northwesterly airflow over the U.K. The spectacular colors are caused by diffraction effects in particles which are very uniform and have sizes comparable with the wavelength of light. As mentioned above these clouds also provide a platform for the chemical reactions that deplete ozone, and the ozone layer crippled shortly after the November display before recovering to normal levels again. There was briefly an

Table 2.7. Explanation of acronyms for ozone-monitoring instruments found on ozone-monitoring satellites (Colorado State University, *http://rammb.cira.colostate.edu/dev/hillger/satellites.htm*).

Instrument acronym	*Instrument name*	*Satellite(s) or Space Shuttle mission(s)*
ACE-FTS	Atmospheric Chemistry Experiment-Fourier Transform Spectrometer	SciSat
ATLAS	ATmospheric Laboratory for Applications and Science	STS-45, STS-56, STS-66 (3 missions)
BUV	Backscatter UltraViolet spectrometer (and photometer)	OV1-1, OV1-10, Kosmos-45, -65, -92, -122, Nimbus-4, Explorer-55
CLAES	Cryogenic Limb Array Etalon Spectrometer	UARS
GAS	Get Away Special (canister)	STS-34, STS-41, STS-43, STS-45, STS-56, STS-62, STS-66, STS-72 (8 missions)
GOME	Global Ozone Monitoring Experiment	ERS-2, MetOp-2 (MetOp-A), MetOp-B, MetOp-C
GOMOS	Global Ozone Monitoring by Occultation of Stars	Envisat
HALOE	HALogen Occultation Experiment	UARS
HH	Hitch Hiker (payload)	STS-87, STS-107 (2 missions)
HIRDLS	HIgh Resolution Dynamics Limb Sounder	EOS-Aura
HIRS/2	High-resolution Infrared Radiation Sounder/2	NOAA-6–NOAA-14 (9 satellites)
HIRS/3	High-resolution Infrared Radiation Sounder/3	NOAA-15–NOAA-18 (4 satellites)
HIRS/4	High-resolution Infrared Radiation Sounder/4	NOAA-19, MetOp-2 (MetOp-A), MetOp-B, MetOp-C
ILAS-1	Improved Limb Atmospheric Sounder-1	ADEOS-1
ILAS-2	Improved Limb Atmospheric Sounder-2	ADEOS-2

(continued)

Table 2.7 (*cont.*). Explanation of acronyms for ozone-monitoring instruments found on ozone-monitoring satellites (*source*: Colorado State University, *http://rammb.cira.colostate.edu/dev/hillger/satellites.htm*).

Instrument acronym	*Instrument name*	*Satellite(s) or Space Shuttle mission(s)*
IRFS-1	InfraRed (sounder)-1	Meteor-M1
IRFS-2	InfraRed (sounder)-2	Meteor-M2
IRIS	InfraRed Interferometer Spectrometer	Nimbus-3, Nimbus-4
IRS	InfraRed Sounder	MTG-I, MTG-S
ISAMS	Improved Stratospheric and Mesospheric Sounder	UARS
LIMS	Limb Infrared Monitor of the Stratosphere	Nimbus-7
LORE	Limb Ozone Retrieval Experiment	STS-87, STS-107 (2 missions)
LRIR	Limb Radiance Inversion Radiometer	Nimbus-4
MLS	Microwave Limb Sounder	UARS, EOS-Aura
MTVZA	Microwave Imaging/Sounding Radiometer	Meteor-M1, Meteor-M2
OLME	Ozone Layer Monitoring Experiment	FASat-Alfa, FASat-Bravo
OM	Ozone Meter	Techsat-1a, Techsat-1b/OSCAR-32
OMI	Ozone Monitoring Instrument	EOS-Aura
OMPS	Ozone Mapping and Profiler Suite	NPP/JPSS series (formerly NPOESS series)
OOAM	Orbiting Ozone and Aerosol Measurement	STEP-4
OSIRIS	Optical Spectrometer and InfraRed Imaging System	Odin
POAM	Polar Ozone and Aerosol Measurement	SPOT-3, SPOT-4
PREMOS	PRecision Monitor for OScillation measurement	Picard

SAGE	Stratospheric Aerosol and Gas Experiment	Explorer-60, ERBS, Meteor-3M-1
SBUS	Solar Backscattering UV Sounder	FY-3A, FY-3B, FY-3C
SBUV	Solar Backscatter UltraViolet radiometer	Nimbus-7
SBUV/2	Solar Backscatter UltraViolet radiometer/2	NOAA-9, NOAA-11, NOAA-13, NOAA-14, NOAA-16, NOAA-17, -18, -19
SMR	Sub-Millimeter Radiometer	Odin
SOLSE	Shuttle Ozone Limb Sounding Experiment	STS-87, STS-107 (2 missions)
SOLSTICE	SOLar STellar (Inter-)Comparison Experiment	UARS, SORCE
SSBUV	Shuttle Solar Backscatter UltraViolet radiometer	STS-34, STS-41, STS-43, STS-45, STS-56, STS-62, STS-66, STS-72 (8 missions)
SSH	Special Sensor H	DMSP-5D1-F1–DMSP-5D1-F5 (5 satellites)
SUSIM	Solar Ultravioliet Spectral Irradiance Monitor	UARS
TANSO-FTS	Thermal And Near ir Sensor for climate Observation-Fourier Transform interferometer Spectrometer	GOSAT
TES	Tropospheric Emission Spectrometer	EOS-Aura
TOM	Total Ozone Mapper	FY-3A, FY-3B, FY-3C
TOMS	Total Ozone Mapping Spectrometer	Nimbus-7, Meteor-3-5, TOMS-EP, ADEOS-1, QuikTOMS
TOU	Total Ozone Unit	FY-3A, FY-3B, FY-3C
UV	UltraViolet (radiometer, or telescope/spectrophotometer)	SAMOS-9, OAO-2, OAO-3, Explorer-64
UVISI	UV and Visible Imagers and Spectrographic Imagers	MSX
UVN	UltraViolet Visible and Near-IR sounder	MTG-S

Table 2.8. Chronological summary of ozone-monitoring satellites (or Space Shuttle missions) (Colorado State University, *http://rammb.cira.colostate.edu/dev/hillger/satellites.htm*).

Launch date	*Satellite (or Space Shuttle mission)*	*Instrument acronym(s) or name(s)*
12-08-1960	Echo-1	Ground-based analysis of sunlight reflected by the balloon (passive method, inaccurate results; solar occultation; sunlight from limb was reflected to ground from the balloon)
18-07-1962	SAMOS-9	UV radiometer (first experiment with active sensor; solar occultation limb scanning)
27-03-1964	Ariel-2	Broadband filter photometer, simple prism spectrometer (solar occultation limb scanning)
13-09-1964	Kosmos-45 (experimental weather satellite)	BUV radiometer (first known BUV or backscatter UV instrument; downward looking)
21-01-1965	OV1-1	BUV radiometer (downward looking, 30° left of nadir)
17-04-1965	Kosmos-65 (experimental weather satellite)	BUV radiometer
16-10-1965	Kosmos-92 (experimental weather satellite)	BUV radiometer
25-06-1966	Kosmos-122 (pre-Meteor weather satellite)	BUV radiometer
11-12-1966	OV1-10	BUV spectrophotometer
28-07-1967	OGO-4	Ebert–Fastie scanning spectrometer
07-12-1968	OAO-2	UV telescope/spectrophotometer (starlight occultation limb scanning)
13-04-1969	Nimbus-3	IRIS
08-04-1970	Nimbus-4	BUV spectrometer, IRIS

21-08-1972	OAO-3	UV telescope/spectrometer (starlight occultation limb scanning)
12-06-1975	Nimbus-6	LRIR (limb scanning)
20-11-1975	Explorer-55 (AE-5 or AE-E)	BUV (solar occultation limb scanning rather than downward looking)
11-09-1976	DMSP-5D1-F1	SSH (multichannel filter radiometer)
04-06-1977	DMSP-5D1-F2	SSH (multichannel filter radiometer)
01-05-1978	DMSP-5D1-F3	SSH (multichannel filter radiometer)
24-10-1978	Nimbus-7 (longest lasting mission; data 1978–1993)	LIMS (limb scanner, an update of the LRIR), TOMS-1, SBUV
18-02-1979	Explorer-60 (AEM-B)	SAGE-1
06-06-1979	DMSP-5D1-F4	SSH (multichannel filter radiometer)
27-06-1979	NOAA-6	HIRS/2 (atmospheric sounder with 9.7 μm ozone channel)
14-07-1980	DMSP-5D1-F5 (launch failed)	SSH (multichannel filter radiometer)
23-06-1981	NOAA-7	HIRS/2
06-10-1981	Explorer-64/SME	UV ozone experiment, Four-Channel Infrared Radiometer, Airglow Instrument
28-03-1983	NOAA-8	HIRS/2
05-10-1984	ERBS (launched from STS-41G)	SAGE-2
12-12-1984	NOAA-9	SBUV/2 (based on the SBUV/TOMS-1 flown on Nimbus-7), HIRS/2
17-09-1986	NOAA-10	HIRS/2

(*continued*)

Table 2.8 (*cont.*). Chronological summary of ozone-monitoring satellites (or Space Shuttle missions) (*source*: Colorado State University, *http://rammb.cira.colostate.edu/dev/hillger/satellites.htm*).

Launch date	*Satellite (or Space Shuttle mission)*	*Instrument acronym(s) or name(s)*
24-09-1988	NOAA-11	SBUV/2, HIRS/2
18-10-1989	STS-34	SSBUV-1 (GAS canister)
06-10-1990	STS-41	SSBUV-2 (GAS canister)
14-05-1991	NOAA-12	HIRS/2
02-08-1991	STS-43	SSBUV-3 (GAS canister)
15-08-1991	Meteor-3-5 (Meteor-TOMS)	TOMS-2
12-09-1991	UARS (launched from STS-48)	CLAES, ISAMS, HALOE, SUSIM, MLS, SOLSTICE
24-03-1992	STS-45	SSBUV-4 (GAS canister, ATLAS-1 payload)
08-04-1993	STS-56	SSBUV-5 (GAS canister, ATLAS-2 payload)
09-08-1993	NOAA-13	SBUV/2, HIRS/2
26-09-1993	SPOT-3	POAM-2 (solar occultation limb scanning)
03-04-1994	STS-62	SSBUV-6 (GAS canister)
03-11-1994	STS-66	SSBUV-7 (GAS canister, ATLAS-3 payload)
30-12-1994	NOAA-14	SBUV/2, HIRS/2
20-04-1995	ERS-2	GOME-1 (BUV technique)

28-03-1995	TechSat-1a microsatellite (failed)	OM-2 UV radiometer (SBUV technique)
31-08-1995	FASat-Alfa microsatellite (failed to separate from Sich-1)	OLME
11-01-1996	STS-72	SSBUV-8 (GAS canister)
24-04-1996	MSX	UVISI
02-07-1996	TOMS-EP	TOMS(-3)
17-08-1996	ADEOS-1	TOMS(-4), ILAS-1
23-10-1997	STEP-4	OOAM
13-11-1997	STS-87	SOLSE-1 (HH payload), LORE-1
24-03-1998	SPOT-4	POAM-3 (solar occultation limb scanning)
13-05-1998	NOAA-15	HIRS/3
10-07-1998	FASat-Bravo microsatellite	OLME
1999	Meteor-3M-2 (this Meteor-3M-1 follow-up was canceled in 1999)	SAGE-3
10-07-1998	TechSat-1b/OSCAR-32 microsatellite	OM-2 UV radiometer (SBUV technique)
21-09-2000	NOAA-16	SBUV/2, HIRS/3
20-02-2001	Odin	OSIRIS, SMR
21-09-2001	QuikTOMS (launch failed)	TOMS-5

(*continued*)

Table 2.8 (*cont.*). Chronological summary of ozone-monitoring satellites (or Space Shuttle missions) (*source*: Colorado State University, *http://rammb.cira.colostate.edu/dev/hillger/satellites.htm*).

Launch date	*Satellite (or Space Shuttle mission)*	*Instrument acronym(s) or name(s)*
10-12-2001	Meteor-3M-1	SAGE-3
01-03-2002	Envisat	GOMOS (starlight occultation limb scanning)
24-06-2002	NOAA-17	SBUV/2, HIRS/3
14-12-2002	ADEOS-2	ILAS-2
16-01-2003	STS-107 (lost on entry)	SOLSE-2 (HH payload), LORE-2 (limb-viewing spectrometers)
25-01-2003	SORCE	SOLSTICE
13-08-2003	SciSat	ACE-FTS
15-07-2004	EOS-Aura (formerly EOS-Chem)	OMI, HIRDLS, TES, MLS (OMI-SBUV technique; data similar to earlier TOMS data, but higher horizontal resolution; HIRDLS and MLS limb scanners; TES limb and nadir scanner)
20-05-2005	NOAA-18	SBUV/2, HIRS/4
19-10-2006	MetOp-2 (MetOp-A)	GOME-2, HIRS/4
27-05-2008	FY-3A	TOM or TOU, and SBUS
06-02-2009	GOSAT (formerly GCOM-A1)	TANSO-FTS (similar to SciSat's ACE-FTS). ODUS (Ozone Dynamics UV Spectrometer), a SBUV instrument, was planned but not used
06-02-2009	NOAA-19	SBUV/2, HIRS/4

17-09-2009	Meteor-M1 (or Meteor-M) (replacement for Meteor-3M)	MTVZA imager/sounder; IRFS-1
15-06-2010	Picard	PREMOS
04-11-2010	FY-3B	TOM or TOU, and SBUS
Future satellites with ozone equipment		
201?	FY-3C	TOM or TOU, and SBUS
201?	Meteor-M2	MTVZA imager/sounder; IRFS-2 advanced IR sounder
201?	CX-1 microsatellite (University of Colorado)	(BUV?) spectrophotometer
201?	DSCOVR (formerly Triana, or "Gore Sat") (to rest at Lagrange point)	BUV?
201?	NPP/JPSS series (formerly NPOESS series)	OMPS (nadir-viewing UV sensor and limb-viewing UV/visible sensor)
2012?	MetOp-B	GOME-2, HIRS/4
2016?	MetOp-C	GOME-2, HIRS/4
2016–2030	MTG-I	IRS
2018–2026	MTG-S	IRS, UVN

Table 2.9. Inventory of the most "reliable" datasets with time records longer than 5 years of OVP (Ozone Vertical Profile) (WMO, 1999).

Platform / *Satellites*	*Middle/Upper stratosphere* ($Z > 25$ km)	*Low stratosphere* ($Z < 25$ km)	*Troposphere*
SAGE I	Feb. 1979—Nov. 1981	Feb. 1979–Nov. 1981	
SAGE II	Oct. 1984—present	Oct. 1984–present	
SBUV + SBUV/2	1978–present		
HALOE	Oct. 1991–present	Oct. 1991–present	
MLS	Oct. 1991–present	Oct. 1991–present	
Ground-based microwave	1989–present	1989–present	
Umkehr/Dobson	1957–present		
Lidar	1990–present	1985–present	1990–present
Balloonborne sondes		1965–present	1965–present

Figure 2.13. Cloud detail view from an image taken in Aberdeen (Scotland) with a digital camera (© George Soja and Brian Ward, Aberdeen).

ozone hole over Europe arising from the accelerated ozone-depleting reactions that occur on cloud particles. There was another display of nacreous clouds on January 29, 2000 which was visible from the south of England (*http://www.nezumi.demon.co.uk/astro/nacreous/3/nacreous3.htm*).

In general the Arctic experiences highly extreme cold as well as sudden stratospheric warmings (SSWs) at times. As a result the degree of ozone loss is mostly controlled by the strength of the vortex and magnitude of air temperature within. For instance, according to Kuttippurath *et al.* (2010a, b) the winters 1995, 1996, 2000, and 2005 were very cold and cumulative total ozone loss was as high as ~25–35% (WMO, 2007). On the other hand, the winters 1997–1999, 2001, 2002, 2006, and 2009 were relatively warm and the loss was minimal about 10–15%, while the winters 2003, 2007, and 2008 were moderately cold and, hence, the loss was on an average scale of about 15–20% (Goutail *et al.*, 2010; WMO, 2007).

As far as the upper atmosphere is concerned the temperature shows an annual variation in the upper mesosphere above about 65 km with the largest values in winter, but this general behavior is interrupted by SSW events resulting in a cooling of the upper-mesosphere/mesopause region (Sonnemann *et al.*, 2007). The chemical reaction rates are temperature dependant and entail an ozone decrease if the temperature increases. This should reduce the wintertime enhancement of ozone, but to a different degree between daytime and nighttime ozone. However, the effect is insignificant close to the turning point of annual temperature variation at 65 km where the annual variation of temperature is very small. The influence of SSWs on the chemistry was investigated by Sonnemann *et al.* (2006). They found that an SSW is able to modulate ozone concentrations. During major SSWs the west wind changes into a summery east wind system, prolonging the time of sunset and reducing the nighttime ozone level. However, the decrease in temperature has the tendency to enhance the ozone level so that the resulting effect is damped or even reversed. As the zonal wind during daytime does not strongly affect ozone, the ozone increases during daytime when the temperature decreases.

Sonnemann *et al.* (2007) also suggested that at sunset the wintry west wind system conveys an air parcel more quickly through this period and shortens the time of effective odd oxygen destruction by catalytic cycles. This effect results from the imbalance between net odd oxygen production by the dissociation of molecular oxygen and the ozone dissociation rate transforming only one odd oxygen constituent into another. Atomic oxygen is then involved in the odd oxygen-destroying catalytic cycle.

We shall now briefly discuss some outstanding field campaigns.

2.6.1 EASOE

A major European campaign, the European Arctic Stratospheric Ozone Experiment (EASOE) (Section 6.6.3.1) was organized to study the polar regions during the winter of 1991/1992. Particular goals were:

- to measure the change in ozone concentration with altitude throughout the winter;
- to measure the concentrations of other trace chemical species (especially chlorine and nitrogen);
- to investigate the role of polar stratospheric clouds, and in particular to study dehydration and denitrification of the Arctic stratosphere;
- to study the meteorological processes that move chemically perturbed air southward.

Measurements were made using ground-based instruments at 16 sites ranging from the Arctic Circle to southern Europe from large stratospheric balloons and from three research aircraft. Forty-three stratospheric balloons, capable of carrying scientific payloads of up to 500 kg directly into the ozone layer, were launched from the Swedish Space Corporation rocket range at ESRANGE (European Space Range) near Kiruna, Sweden. In addition, over 1,000 ozonesondes were launched on small balloons from more than 20 sites. A wide variety of meteorological information was provided by the various European meteorological services, including some information provided especially for EASOE scientists.

Much new information was gained. Despite this activity, many intriguing questions remain: What causes the mid-latitude loss? How are the losses over the poles linked to those at mid-latitudes? (see Chapters 3 and 4).

2.6.2 SESAME

In 1994 and 1995 European scientists conducted SESAME, the Second European Stratospheric Arctic and Mid-latitude Experiment, to investigate the processes occurring at both high and mid-latitudes and to see how they are linked (see Section 6.6.3.1). At the same time a U.S.-led expedition was looking at similar processes in the Southern Hemisphere.

2.6.3 THESEO

The Third European Stratospheric Experiment on Ozone (THESEO) was organized to investigate long-term ozone loss over northern high and mid-latitudes during 1998 and 2000. The processes involved include both those that occur over mid-latitudes and those resulting from the exchange of air between the mid-latitude stratosphere and neighboring parts of the atmosphere. In the first few months field measurements were made mainly at mid and high latitudes and focused on a limited number of specific processes. From the end of November until the beginning of February (1997/1998) the temperature within the vortex fell below the critical temperature for the formation of polar stratospheric clouds (PSCs). However, because of the occurrence of a series of mid-stratospheric warmings the extent of the PSCs was not large. Moreover, outside the vortex in the upper stratosphere unusually high temperatures were observed. This winter was finally warmer than the previous three (exceptionally cold) winters and, therefore, TOC values were higher than those in the past few winters. These higher TOC values masked any chemical ozone loss (estimated to be 15–20% inside the vortex).

2.6.4 SOLVE

The SOLVE campaign was designed to examine the processes that control polar to mid-latitude stratospheric ozone levels. The mission was conducted during the 1999–2000 northern winter from Kiruna (Sweden). The SOLVE campaign employed NASA ER-2, NASA DC-8, and OMS *in situ* and remote-sensing balloon payloads, ground station observations, and an extensive theory team. The mission also acquired correlative measurements needed to validate the Stratospheric Aerosol and Gas Experiment (SAGE) III satellite mission and used these satellite measurements to help quantitatively assess high-latitude ozone loss. The results of SOLVE both expanded our understanding of polar ozone processes and provided greater confidence in our current ozone-monitoring capabilities. This knowledge provides the basis for setting sound public policies which will help to preserve the Earth's ozone layer.

During the 1999/2000 winter, the THESEO 2000 campaign sponsored by the European Union (consisting of a core of 12 major EU-funded projects) and the NASA-sponsored SAGE III SOLVE campaign obtained measurements of ozone and other atmospheric gases and particles using satellites and aircraft; large, small, and long-duration balloons; and ground-based instruments. Scientists from Europe, the United States, Canada, Russia, and Japan joined forces in mounting the biggest field measurement campaign yet to measure ozone amounts and changes in the Arctic stratosphere. The total amount of information collected by the THESEO 2000/SOLVE campaign was more extensive than any information collected by past polar measurement campaigns. Most of the measurements were made near Kiruna, Sweden, with additional measurements being taken from satellites and through a network of stations at mid and high northern latitudes.

2.6.5 ORACLE-O_3

The acronym comes from the title "Ozone layer and uv RAdiation in a changing CLimate Evaluated during the International Polar Year (IPY).

The IPY themes that were addressed are as follows:

1. Current state of the environment
2. Change in the polar regions
3. Polar–global linkages/tele-connections
4. Exploring new frontiers
5. The polar regions as vantage points
6. The human dimension in polar regions.

The project was divided into seven main activities: (1) ozone loss (detection and impact on UV radiation); (2) PSCs (polar stratospheric clouds) and cirrus; (3) atmospheric chemistry; (4) UV radiation; (5) ozone, climate change, and feedback; (6) data management; and (7) education, outreach, and communication. The project implied precise quantification of polar ozone losses in both hemispheres achieved with concerted international campaigns during which hundreds of ozonesondes were launched in real time coordination from station networks in the Arctic and Antarctic. Satellite coverage of ozone and ozone-depleting substances were made from satellites such as ENVISAT, Aura, ACE, Odin, POAM III, and SAGE III along with ground-based station data.

2.6.6 SCOUT-O3

SCOUT-O3 was a European Commission Integrated Project with €15 million of funding. The project had 59 partner institutions and over 100 scientists involved from 19 countries. The project commenced on May 1, 2004 and ran until April 31, 2009.

SCOUT-O3's aim was to provide predictions about the evolution of the coupled chemistry/climate system, with emphasis on ozone change in the lower stratosphere and the associated UV and climate impact, to provide vital information for society and policy use.

SCOUT-O3 was structured to investigate eight scientific activities, as follows:

- ozone, climate, and UV predictions
- SCOUT-O3 tropical
- extratropical ozone and water vapor
- UV radiation
- chemistry and particles: studies of kinetics, photochemistry, heterogeneous processes, and nucleation
- global diagnostic chemical transport modeling
- SCOUT-O3 database and public outreach
- management.

2.6.7 Match

The Match campaigns are part of the ORACLE-O_3 project (see Section 2.6.5) within the International Polar Year (IPY), which actually covers two years—March 2007 to March 2009 (see also Section 6.6.3.1).

The two main aims of the Match campaigns are:

1. to measure chemical ozone loss in polar regions; and
2. to check our understanding of the underlying processes.

As mentioned above, during the North Hemisphere winter 1991/1992 a concerted campaign funded by the EU and national agencies (EASOE) took place. The activities included the launch of a large number ($\sim$1,400) of ozonesondes in the Arctic and in the mid-latitudes. This dataset was used afterwards to determine Arctic ozone losses for the first time with the Match Method. Since the launches were not coordinated only those matches that happened by chance were used. Fortunately, because of the large number of soundings enough matches were found for a successful analysis. It was in this way that the method was established and for the Second European Stratospheric Arctic and Mid-latitude Experiment (SESAME) in the winter 1994/1995 the method was advanced to its active form with coordinated launches as described above. In nearly every Arctic winter since then a Match Campaign has been carried out. The only exceptions were the winters 2001/2002 and 2003/2004, which were relatively warm stratospheric winters with minor ozone losses. In the meantime, the winters 1992/1993 and 1993/1994 were analyzed by the passive Match approach, too. In total, analyzed ozonesonde Match data exist for 12 Arctic winters and one Antarctic winter (2003).

2.6.8 ARC_IONS

The Arctic Intensive Ozonesonde Network (ARC-IONS)—which covered Canada, Alaska, and the mid–upper U.S.—consisted of two field phases with 18 sites, most launching daily (square symbols in Figure 2.14) (*http://croc.gsfc.nasa.gov/arcions*).

ARC-IONS is one of the three coordinated ozonesonde networks, the IONS series, that operated over North America in support of satellite validation and multiple aircraft and ground sites during field campaigns (Figure 2.14). During IONS-04, the 250+ profiles obtained over eastern North America represented the largest single set of free tropospheric ozone measurements ever compiled for that region. In IONS-06 and ARC-IONS (2008), multiple phases (two 3 to 4-week campaigns) and a larger set of sites expanded the IONS dataset to a total of 1,400 ozone

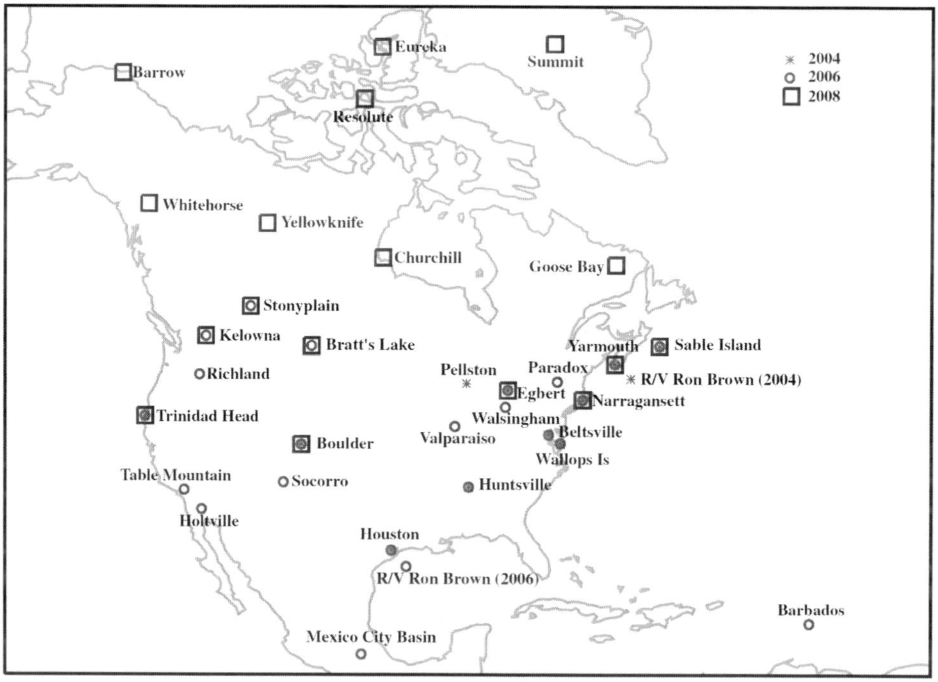

Figure 2.14. Locations of IONS sites: IONS-04 (INTEX Ozonesonde Network Study), where INTEX = Intercontinental Chemical Transport Experiment, conducted in July–August 2004; IONS-06, March–May 2006, supporting INTEX-B, and August–September 2006. ARC-IONS, Arctic Intensive Ozonesonde Network, April 2008 and June–July 2008, shifted to higher latitudes (*source*: Thompson *et al.*, 2011).

and pressure–temperature–humidity profile sets (IONS data are available at *http://woudc.org*).

IONS data are used for forecasting and flight planning during the field phase to determine ozone budgets, satellite validation, and evaluation of chemical transport models at various scales. During field phases, the ozone profiles are used in a manner similar to the Match concept.

The scientific goals of IONS-04 were to study continental outflow patterns and variability during INTEX-NA, particularly major regions of pollution flow above the boundary layer and the convective contribution to ozone in the middle to upper troposphere. IONS-04 profiles tracked pollution from Alaskan and Canadian forest fires and from New York City. In addition to the sonde coordination with aircraft and surface measurements in INTEX-A, the large number of IONS-04 profiles enabled new air quality model validation exercises. Thompson *et al.* (2011) suggested that, whereas the Canadian air quality forecast models AURAMS and CHRONOS show considerable skill at predicting ozone in the planetary boundary layer and just above (Figure 2.15), they have large errors in the free troposphere, owing largely

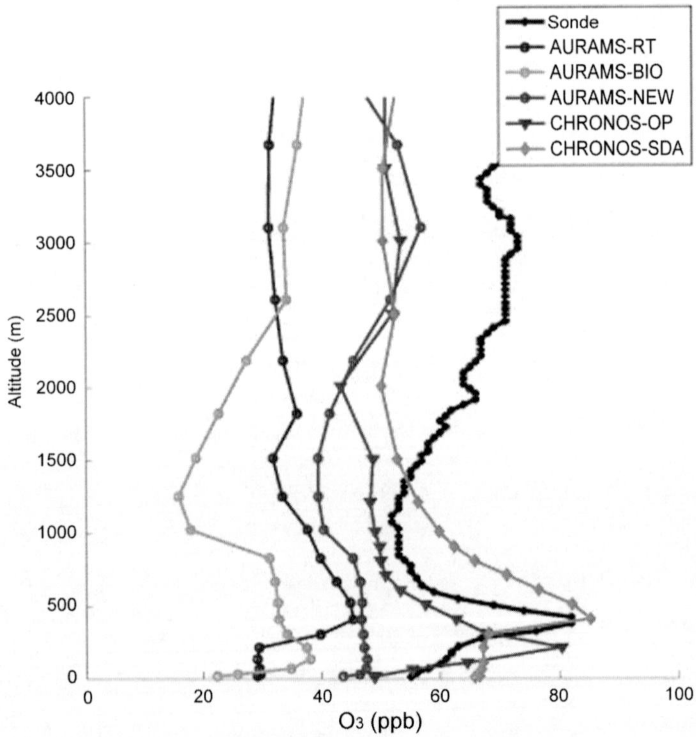

Figure 2.15. Comparisons of several Canadian air quality models with IONS-04 sondes (line without symbol) at the surface to 4 km. AURAMS = A Unified Regional Air Quality Modeling System; CHRONOS = Canadian Hemispheric and Regional Ozone and NO_x System (Thompson *et al.*, 2011).

to the inadequate treatment of model domain boundaries (including the upper boundary).

The scientific goals of IONS-06 were: the determination of intercontinental transport, primarily the flux of ozone from Asia to and across the North American continent; the evaluation of tropospheric ozone budgets, including the stratospheric component; and the validation of Aura satellite tropospheric ozone. IONS-06 data provided surprising new insights on tropospheric processes and their effect on the ozone budget. Sub-tropical ozone over Mexico City and Houston in spring revealed robust wave activity in the free troposphere and tropopause region. In summer, although lightning and convection influences over these sites were intense, 39% of the soundings exhibited stratospheric signatures. Over the central to eastern U.S. and Canada during the summer phase of IONS-06, a similar mixture of interleaved sources in ozone profiles to those in IONS-04 was observed (Thompson *et al.*, 2011).

3

Intercomparisons between various atmospheric ozone datasets

3.1 INTRODUCTION

Ozonesondes provide direct measurements of ozone concentrations as a function of height in the atmosphere (OVP) and, therefore, also provide estimates of TOC. But these measurements are only made at a few geographical locations and often also somewhat infrequently. One important reason for studying atmospheric ozone in great detail is to identify long-term trends in ozone concentration and, in particular, to study the rate of recovery of the ozone layer in response to the implementation of the Montreal Protocol. Dobson spectrophotometers, and other ground-based instruments, have been used for many years and provide quite long-term datasets. The question of intercomparison of data from the various early laboratories is very important and has been discussed by Staehelin *et al.* (2009). These laboratories between them provided data for more than four decades before the beginning of anthropogenic ozone depletion; they included Arosa (Switzerland), Oxford (United Kingdom), Lerwick (United Kingdom), Tromsø (Norway), Svalbard (Norway), and Vigna di Valle (Italy). However, Dobson spectrophotometers and other ground-based instruments only provide data for rather few geographical locations (see Figure 1.20). On the other hand satellite-flown instruments provide frequent data over enormous areas, approaching total global coverage, but they do not provide direct measurements of ozone concentrations; algorithms have had to be developed to determine TOC and OVP from the spectral data recorded by these instruments. Ground-based measurements constitute a key component of the Global Ozone Network, both on their own account and by providing ground truth for satellite-based instruments. However, the uneven geographical distribution of the existing ground-based network gives rise to a spatial sampling error, when attempts are made to determine the global distribution of ozone content. The benefit of ground-based instruments is that it is easy to maintain them in good condition, while the benefit of satellite-based instruments is that they provide better spatiotemporal coverage and resolution.

Satellite measurements since the 1970s have provided data on the daily global distribution of total ozone content (TOC) and favor the study of the spatiotemporal variations and long-term trends in TOC fields and time series (Eskes, 2004; Lambert *et al.*, 2000; Levelt *et al.*, 1996; Schoeberl *et al.*, 1986). Continuous validation of satellite data using the ground-based ozonometric network has been undertaken and is ongoing and is discussed in this chapter (see also Balis *et al.*, 2007a; Bramstedt *et al.*, 2003; Cortesi *et al.*, 2007; Evtushevsky *et al.*, 2008a, b; Fioletov *et al.*, 2008; Grytsai *et al.*, 2000; ; Ionov *et al.*, 2008; Kulinich *et al.*, 2005; McPeters and Labow, 1996; Weber *et al.*, 2005). Total Ozone Mapping Spectrometer (TOMS) instruments were in operation from 1978 to 2005, and the Ozone Monitoring Instrument (OMI) since 2004. Their historical and current daily and monthly data are widely available at the National Aeronautics and Space Administration (NASA) website *http://toms.gsfc.nasa.gov*. Spatially, a good global coverage of TOMS (OMI) data is combined with a high resolution of 1° by latitude and 1.25° by longitude (McPeters *et al.*, 1998).

The continuous monitoring of the quality of satellite-based products through validation with long-term ground-based measurements provides the means to instrument intercomparison and algorithm development. Despite data gaps in between satellite missions, the ozone layer can be globally monitored in a consistent way ensuring that the trends in ozone will be properly identified.

In Chapter 2 we described a large variety of satellite-flown instruments for measuring atmospheric ozone concentrations, both OVP and TOC. However, like so many other remote-sensing techniques, they do not directly measure the quantity one is trying to study, in this case OVP and TOC. All these instruments need some algorithm to calculate the OVP or TOC from the actual measurements made by the instruments. We may loosely describe this as calibration of the remote-sensing instrument. Then, once an algorithm has been established for a given instrument the results obtained using the algorithm need to be validated. Ozonesonde data and data from ground-based spectrophotometers therefore play a key role in algorithm development and validation.

One thing of interest and importance, as we shall see later (in Chapter 4 onwards), is the question of temporal changes in atmospheric ozone concentration. Another crucial parameter for the reliability of ozone measurements is selection of the ozone absorption coefficients for in-field ozone observations. Usually, knowing the ozone absorption cross-sections in the ultraviolet and infrared spectral range, with an accuracy better than 1%, is of the utmost importance for atmospheric remote-sensing applications. For this reason, various ozone intensity intercomparisons and measurements have been published in recent years (e.g., Gratien *et al.*, 2010). However, the corresponding results proved not to be consistent and, thus, have raised a controversial discussion in the community of atmospheric remote sensing. In this context, Gratien *et al.* (2010) have reported a new laboratory intercomparison of the ozone absorption coefficients in the mid-infrared (10 μm) and ultraviolet (300–350 nm) spectral regions. This intercomparison confirmed that the

IR and UV cross-sections recommended in the literature are in disagreement by about 4%.

Undoubtedly, the Earth's atmosphere has indeed changed markedly since the industrial revolution, with several changes being of anthropogenic origin. The consequences of the quantitative and qualitative changes in the atmospheric system have become serious, especially in the case of ozone following the industrial production of CFCs and other ozone-destroying chemicals. However, as is well known, these changes are of both natural (due to internal processes and external forcings) and anthropogenic origin (i.e., the atmosphere possesses both internal and external variability). In order to quantify this variability, the reliability of observational data is fundamental and constitutes a major priority in the international research community.

In the case of ozone, therefore, in looking for long-term trends in atmospheric ozone concentration one needs to have accurate long-term datasets. The problem of instrumental bias has been discussed throughout the whole history of ozone measurements, but it acquired special importance after satellite observations started. Even at the early stages of satellite TOC observations very serious attention was paid to this problem. It is important, therefore, to reconcile the various different datasets obtained from ground-based measurements, ozonesondes, and satellite data. The most reliable information about the spatiotemporal variation of the concentration of atmospheric ozone will be obtained by making use of all sources of data from ozonesondes, from ground-based instruments, and from satellite-flown instruments. Not only is it necessary to intercalibrate different instruments that are operating at any given time, but to construct a long-term archive it is necessary to intercalibrate different versions of the same instrument, such as Dobson spectrometers, that have been operated over different periods. The question of the intercalibration of different Dobson spectrophotometers was discussed in Sections 1.2.2 and 1.2.3 and of Brewer spectrophotometers in Section 1.3.

In an early comparison of Nimbus-7 TOMS data from 1978 to 1985 with surface observations, Bojkov *et al.* (1988) found a significant bias. According to TOMS data, an inhomogeneous trend of TOC depletion was observed of about 3.5% during the period 1978–1985 compared with Dobson data, the cause of which was probably ageing of the TOMS diffuser plate. While there were still comparatively few satellite systems observing atmospheric ozone, the existing long-term ground-based observations were very important. The history of ground-based observations was discussed by Farman (1989).

Eventually, after filtering out the drift of TOMS sensitivity, it became possible to use satellite information for verifying the reliability of the global net of 92 ground-based ozone stations. The result was that there was close agreement (within ±2% for about two-thirds of the stations) between satellite and Dobson data, whereas in the case of M-83 filter ozonometers much stronger deviations were observed.

Bojkov *et al.* (1990) carried out a statistical seasonal trend analysis of revised Dobson TOC data for 29 Northern Hemisphere stations located between 15 and 64°N latitude using data for 1986. For explanation of solar cycle and Quasi-Biennial Oscillation-induced ozone variation, the 10.7 cm solar flux series and the 50 hPa equatorial wind series were used. The following statistical model for monthly ozone series y_t was applied:

$$y_t = \sum_{i=1}^{12} \mu_i I_{i;t} + \sum_{i=1}^{12} \beta_i I_{i;t} R_t + \gamma_1 (Z_{1;t} - 120) + \gamma_2 Z_{2;t-k} + N_t \qquad (3.1)$$

where μ_i is the ozone mean in month i ($i = 1, \ldots, 12$); $I_{i;t}$ is the indicator series for the ith month of the year: 1 if month t corresponds to month i of the year, and 0 otherwise; β_i is the trend in Dobson units/yr after December 1969 in month i of the year; R_t is the linear ramp function, beginning in January 1970, $(t - t_0)/12$ if $t > t_0$ (where t_0 corresponds to December 1969) and 0 if $t \leq t_0$; $Z_{1;t}$ is the solar 10.7 cm flux series; γ_1 is the associated coefficient; $Z_{2;t-k}$ is the QBO series lagged k months; γ_2 is the associated coefficient; N_t is the residual noise series.

Rathman *et al.* (1997) considered the results of a preliminary comparison between TOC data from the TIROS Operational Vertical Sounder (TOVS) and the Global Monitoring Experiment (GOME). Two comparisons were made: (a) TOVS ozone analysis maps versus GOME level-2 data; (b) TOVS data located at Northern Hemisphere Ground Ozone Stations (NHGOS) versus GOME data. Rathman *et al.* (1997) emphasized that both analyses consistently showed an offset in the TOC values between the datasets: 35 DU and 10 DU for (a) and (b) comparisons, respectively, despite good correlation between the spatiotemporal features of the datasets. However, a noticeably poor correlation was observed in the latitudinal belts 10–20°N and 10–20°S. The smallest region that was statistically representative of the ozone value correlation dataset of TOVS data at NHGOS and GOME level-2 data was determined to be a region enclosed by an effective radius of 0.75 arc-degrees (83.5 km).

In this chapter we shall discuss the validation of some of the satellite-flown instruments that we discussed in Chapter 2. We shall also consider the intercomparison of Dobson, ozonesonde, and satellite data and—as an example—consider in some detail the case of the Dobson spectrophotometer of the University of Athens (Section 3.2). Then we shall, as an example of the validation of ozone data obtained from a satellite-flown instrument, consider in some detail (Sections 3.3–3.7) the validation of data from MIPAS (Michelson Interferometer for Passive Atmospheric Sounding) which was flown on ENVISAT (Section 2.5.3.3). This involves data from a wide variety of other instruments. Finally, we shall present other intercomparisons between datasets obtained recently from various ozone-monitoring systems (Section 3.8).

3.2 TOTAL OZONE MEASUREMENTS OVER ATHENS: INTERCOMPARISON BETWEEN DOBSON, TOMS (VERSION 7), SBUV, AND OTHER SATELLITE MEASUREMENTS

Comparison between total ozone measurements from TOMS and ground-based measurements has shown significant differences between different ground stations

(Heese *et al.*, 1992). These authors presented the relative mean difference in TOMS data between 1986 and 1989. This is as high as +6.9% and +8.3% at Oslo and Longyearbyen, respectively, but lower at other stations and within the accuracy of ±3.6% for this comparison. Tromsø and Resolute have a mean difference of +2.9% and +1.2% respectively, while Reykjavik is the only station with a negative mean difference of −2.5% (Heese *et al.*, 1992). The Norwegian stations in higher northern latitudes have solar zenith angles often higher than 80° over a significant part of the year. The relative difference between TOMS and Dobson total ozone results shows a higher scatter at large solar zenith angles, so that daily ozone values in particular may be in error. The problem of high scatter at large solar zenith angles does not arise for measurements at lower latitudes.

In this section we describe an illustrative example of the intercomparison of total ozone data from various satellite systems and from Dobson measurements. This example is the case of a study that was done for Athens (37.59°N, 23.44°E), Greece, where the Dobson spectrophotometer has been installed since 1989 (Varotsos and Kondratyev, 1998b). It should be noted that, because of the sunny weather that characterizes that location, the correlation obtained between the observations of the various instruments is good and Athens can be regarded as a representative ground-based station for the mid-latitudes. Similar validation problems (homogenization of ozone records) were discussed by Köhler and Claude (1998) in connection with the interpretation of the long TOC time series (since 1967) of Dobson observations in Hohenpeissenberg.

We first consider a period when overlapping TOMS, SBUV, and Dobson measurements were available was chosen. This period covering March 1, 1991 until December 31, 1991 is the only suitable period when all the data were available simultaneously. TOMS data were obtained in several versions, version 7 being the latest version available at that time. It should be noted that the Athens TOC data are assumed to be representative of the column ozone variations not only for the Mediterranean region but also for the entire mid-latitude belt in the Northern Hemisphere (Efstathiou *et al.*, 2003). Version 7 data are corrected for drift, caused by the degradation of the instrument's diffuser plate, using the internal pair justification technique. Version 7 data are obtained after careful intercomparison with ground-based measurements. In principle, better agreement between version 7 data and ground-based measurements should be expected when compared with version 6 data.

The plausible SO_2 and NO_2 interference in Dobson total ozone data has been considered (Varotsos *et al.*, 1995b). The estimated SO_2 and NO_2 contribution is not very large (<1%), so that the Dobson total ozone correction with regard to these trace gases is very slight and therefore can be neglected (Section 1.2.3).

In Figure 3.1 scatter plots for TOMS (version 7) and Dobson observations are shown and in Figure 3.2 for SBUV and Dobson measurements. The slope of the regression line for Figure 3.1 (TOMS version 7) is 1.006 and that for Figure 3.2 (SBUV) is 1.076 indicating that both TOMS (version 7) and SBUV measurements (NASA Goddard Space Flight Center), in general, broadly agree very well with Dobson measurements and that, based on the slopes of the regression line, the agreement for version 7 TOMS data and the Dobson data is slightly better than

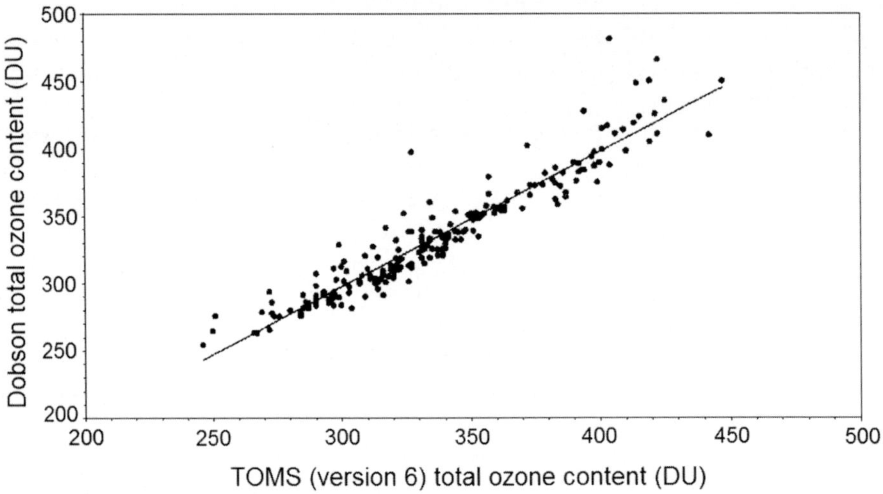

Figure 3.1. Scatter diagram for TOMS (version 7) and Dobson observations at Athens, Greece (March 1, 1991–December 31, 1991). All days (cloudy and cloud-free) were included (Varotsos *et al.*, 1997a).

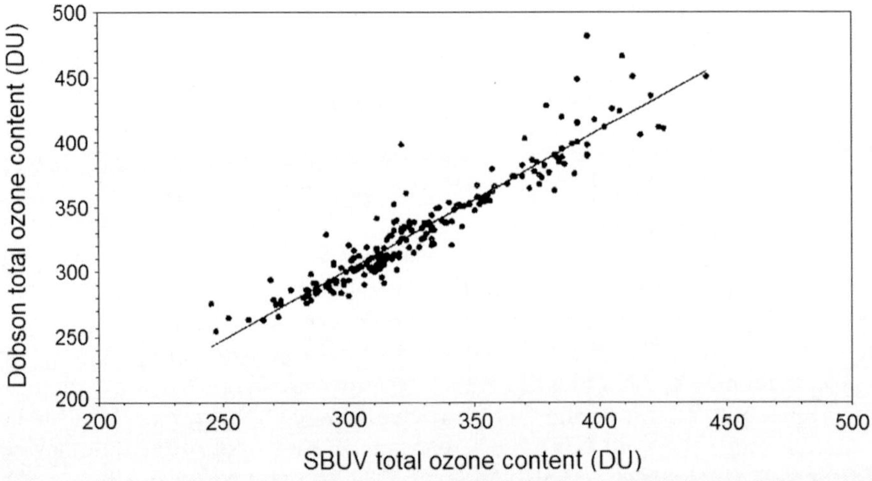

Figure 3.2. Scatter diagram for SBUV (version 7) and Dobson observations at Athens, Greece (March 1, 1991–December 31, 1991). All days (cloudy and cloud-free) were included (Varotsos *et al.*, 1997a).

for the SBUV. A closer inspection of these figures, however, shows considerable scatter when the total ozone content exceeds 380 DU (see the far right of both Figures 3.1 and 3.2). In particular, the scatter is greatest when the total ozone content is between 370 and 420 DU. In view of this, the data were plotted again, with the exclusion of all the cloudy days (see Figures 3.3 and 3.4 corresponding to

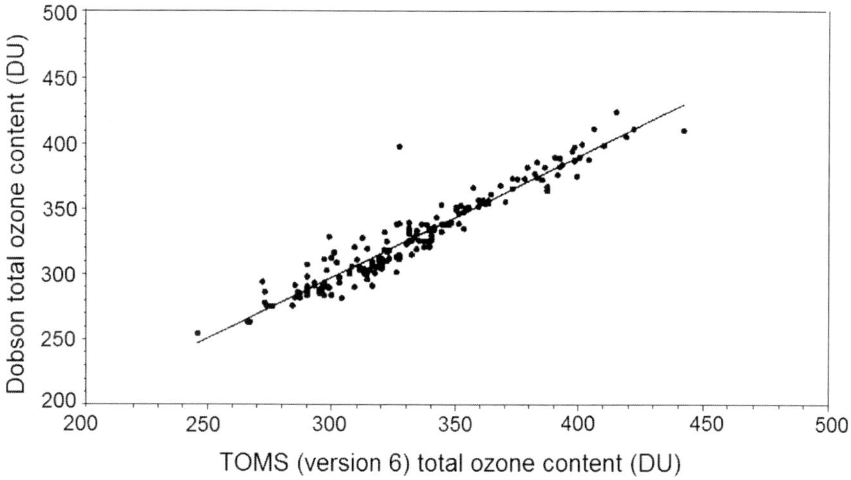

Figure 3.3. Scatter diagram for TOMS (version 7) and Dobson observations at Athens, Greece (March 1, 1991–December 31, 1991). Only cloud-free days were included (Varotsos *et al.*, 1997a).

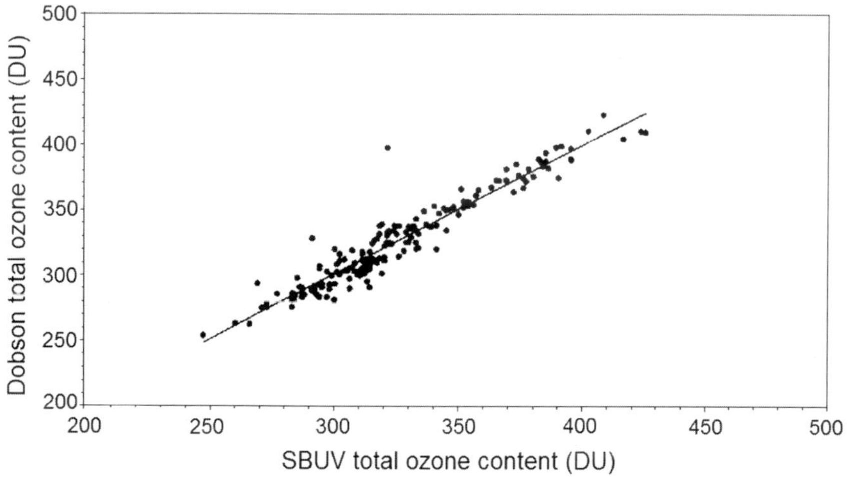

Figure 3.4. Scatter diagram for SBUV (version 7) and Dobson observations at Athens, Greece (March 1, 1991–December 31, 1991). Only cloud-free days were included (Varotsos *et al.*, 1997a).

TOMS version 7 and SBUV data, respectively). While the slope of the regression line is 0.932 in Figure 3.3 (TOMS version 7 versus Dobson), it is 0.994 in Figure 3.4 (SBUV versus Dobson), which clearly indicates that on sunny days the agreement between SBUV and Dobson measurements is much better than the version 7 TOMS measurements.

Note that on cloudy days the Dobson measurements are higher than the TOMS (version 7) and SBUV measurements. This observation points to a possible source of error in the correction of both TOMS and SBUV measurements for tropospheric ozone, which is only partially measured, particularly so on cloudy days, because the effective reflecting surface for backscattered radiance lies in the upper troposphere. So, the amount of tropospheric ozone below this layer depends on the standard ozone profiles used in the TOMS and SBUV algorithm. These standard profiles are taken from climatological annual means. Thus, seasonal and day-to-day variations of up to 50% in tropospheric ozone and a long-term trend of 0.5–1.0% per year observed by sonde measurements at several stations in the Northern Hemisphere (Bojkov *et al.*, 1988) will not be recognized by TOMS and SBUV.

The above-mentioned influence of thick clouds on column ozone content deduced from TOMS observations has been thoroughly studied by Varotsos (1995a, b). In particular, Varotsos (1995b) has examined qualitatively the expected errors in satellite total ozone measurements as a function of the frequency of appearance of the cut-off low systems over Greece. These systems are isolated vortices for at least a few days combined with tropopause foldings over mid-latitudes contributing to stratosphere–troposphere exchange. This mechanism increases the "hidden" ozone quantity under the cloud top, thus exceeding the standard tropospheric ozone content used in the TOMS and SBUV algorithm.

The result obtained was that underestimation of the column ozone by the satellite instrumentation due to the cut-off low events can reach 20%. In addition, Varotsos (1995b) pointed out that underestimation of the total ozone amount by the TOMS instrument, due to cut-off low systems over Athens, shifts with the season resulting in an underestimation of the total ozone negative trend in this region.

In this context, Varotsos and Kondratyev (1995) studied mid-latitude total ozone trends in the Northern Hemisphere. Varotsos and Kondratyev (1995) confirmed the above-described results obtained by Varotsos (1995b), notably that, although TOMS measurements are corrected for total ozone content to account for the hidden ozone beneath the cloud top, nevertheless satellite observation underestimates the total ozone under the cloud top since the amount of ozone within mixed stratospheric air exceeds the background tropospheric value. Thus, the air within cut-off low vortices is isolated for at least a few days and, hence, any reduction in its ozone value (measured from a satellite) can be attributed to the vertical transport of ozone within the troposphere under the cloud (Ebel *et al.*, 1991; Fleig *et al.*, 1982; Varotsos and Kondratyev, 1995).

From the preceding discussion it is clear why there is a better correlation between SBUV ozone measurements and ground-based Dobson measurements on sunny days. Cloudy days associated with cut-off lows over Athens are also included in the analysis despite the noise interference deduced from the clouds' presence. The latter is of great importance in order to gain information on Dobson performance on cloudy days.

We now turn to an intercomparison with data from a longer period for which data from some other satellite systems were available. In the following we also describe the intercomparison of column ozone observations over Athens that were derived from ground-based and satelliteborne instrumentation just before International Polar Year 2007–2008 activity (see Cracknell and Varotsos, 2007a, b).

The first set of column ozone data used in this study was deduced from the clear-sky observations of the Dobson spectrophotometer D118 installed in Athens on the university campus. The second dataset consists of observations made by the SCanning Imaging Absorption SpectroMeter for Atmospheric CHartographY (SCIAMACHY), which is an instrument on board the European Space Agency's ENVIronmental SATellite (ENVISAT) (see Section 2.5.2.5). By using the nadir and limb processes, SCIAMACHY observes the whole atmosphere, while the occultation process confirms the measurements taken by the limb method. The SCIAMACHY total column ozone (TM3DAM-SCIA level-4 overpass) is provided at *http://www.temis.nl/protocols/o3field/overpass_scia.html* The TOC measurements derived from the SCIAMACHY observations that refer to the greater Athens area are discussed later. The third dataset comprises the column ozone observations obtained by the Earth Probe TOMS (Total Ozone Mapping Spectrometer), along with the Ozone Monitoring Instrument (OMI) on board Aura (see Section 2.3.2).

To validate statistically the comparison between ground-based and satellite TOC data, Spearman's and Wilcoxon's tests were used. The Wilcoxon signed rank test is a non-parametric statistical hypothesis test for the case of two related samples or repeated measurements on a single sample. It is used as an alternative to the paired Student *t*-test when the population cannot be assumed to be normally distributed. The test is named after Frank Wilcoxon (1892–1965) who, in a single paper, proposed both it and the rank sum test for two independent samples (Wilcoxon, 1945). Like the *t*-test, the Wilcoxon test involves comparisons of differences between measurements, so it requires that the data are measured at set intervals. However, it does not require assumptions about the form of the distribution of the measurements. It should therefore be used whenever the distributional assumptions that underlie the *t*-test cannot be satisfied. Spearman's rank correlation coefficient, named after Charles Spearman (Spearman 1904), is a non-parametric measure of correlation; that is, it assesses how well an arbitrary monotonic function can describe the relationship between two variables, without making any assumptions about the frequency distributions of the variables.

Varotsos and Cracknell (2007) presented an updated comparison between the Dobson spectrophotometer D118 operating in Athens, Greece, and other ground-based and satelliteborne instrumentation for column ozone observations. More specifically, Figure 3.5 shows the daily column ozone observations above the area of greater Athens, as deduced from the satelliteborne (SCIAMACHY, TOMS) and ground-based instrumentation (Dobson) from October 2002 to September 2006. In an attempt to validate statistically the above comparison between ground-based and satellite total ozone data, Spearman's and Wilcoxon's tests were employed. The application of the Spearman's test to the SCIAMACHY and Dobson observations reveals that the correlation coefficient between these datasets is 0.91 (at the 95% confidence level), while between the observations of SCIAMACHY and TOMS, the correlation coefficient is 0.93 (at the 95% confidence level). To investigate deviations between the column ozone observations obtained by the different observing

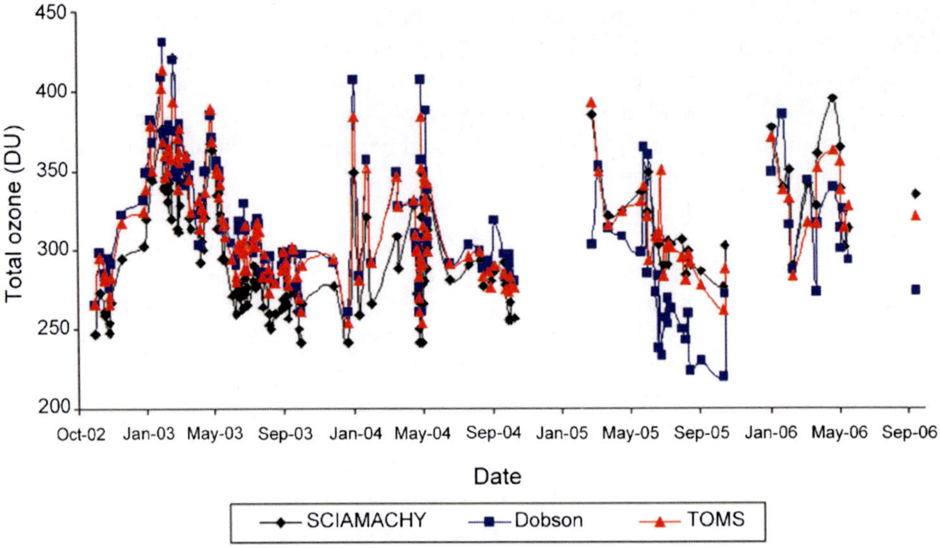

Figure 3.5. Intercomparison of column ozone observations performed by satelliteborne and ground-based instrumentation in Athens, Greece (2002–2006) (Varotsos and Cracknell, 2007).

instruments, Wilcoxon's test was applied. The latter revealed that SCIAMACHY underestimates the column ozone with respect to Dobson by 23.4 DU, while TOMS exhibits an overestimation with respect to SCIAMACHY of 20.3 DU. In other words, at the beginning of the International Polar Year 2007–2008 campaign, the percentage difference for the column ozone observations obtained by Dobson, SCIAMACHY, and TOMS over Athens was of the order of 6%.

Tzanis (2009b) has intercompared the daily column ozone observations made by the Dobson spectrophotometer, SCIAMACHY, Earth Probe-TOMS and OMI over Athens, Greece, during 2002–2008. The results obtained by Tzanis (2009b) are presented in Figure 3.6, according to which the column ozone data obtained by SCIAMACHY seem to agree fairly well with the corresponding data deduced from the observations of the other three instruments. To validate statistically the comparison between the column ozone observations made by SCIAMACHY and the other observing instruments (Dobson, Earth Probe-TOMS, OMI), Tzanis (2009b) employed Spearman's and Wilcoxon's tests. Using Spearman's test for the SCIAMACHY and Dobson observations it was found that the correlation coefficient between these datasets was 0.87 (statistically significant at the 99% confidence level) while between the column ozone observations of SCIAMACHY, Earth Probe-TOMS, and OMI, the correlation coefficient was 0.85 and 0.93 (statistically significant at the 99% confidence level), respectively. Application of Wilcoxon's test revealed that SCIAMACHY overestimates the column ozone with respect to Dobson, Earth Probe-TOMS, and OMI by 10, 15, and 3 DU, respectively.

Comparisons have also been made by Tzanis (2009b) with data from some other systems, namely Earth Probe, ADEOS, and OMI (see Sections 2.3.1, 2.3.3, and

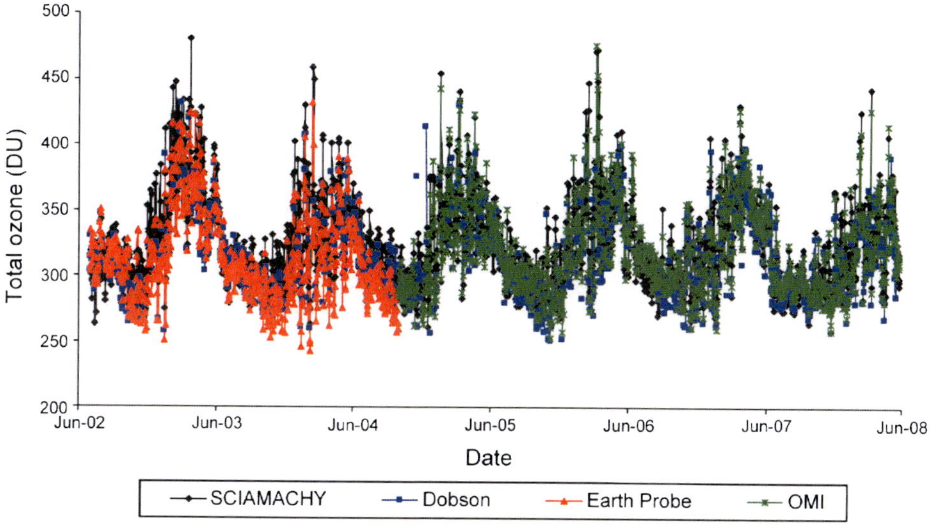

Figure 3.6. Comparison of column ozone observations performed by satelliteborne and ground-based instrumentation in Athens, Greece during 2002–2008 (Tzanis, 2009b).

2.3.2). Figure 3.7 shows the monthly mean TOC observations above the greater Athens area as deduced from the satelliteborne (TOMS on board Nimbus-7, Meteor-3, Earth Probe, ADEOS, and OMI on board Aura), and ground-based instrumentation (Dobson) from 1991 to 2008. Inspection of Figure 3.7 shows that very good agreement exists between the TOC data obtained by the Dobson

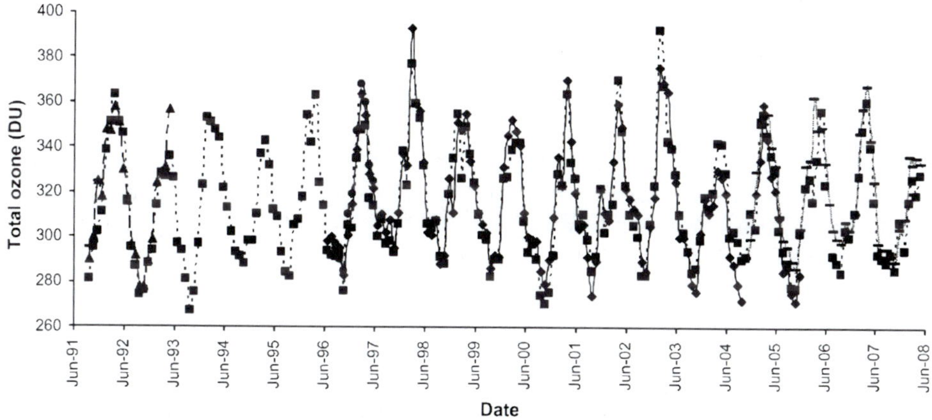

Figure 3.7. Comparison of monthly mean column ozone observations performed by satelliteborne and ground-based instrumentation in Athens, Greece during 1991–2008 (Tzanis, 2009b).

instrument and the corresponding data derived from the observations made by these satelliteborne systems.

To obtain a measure of this agreement between satellite and ground-based TOC data, Spearman's and Wilcoxon's statistical tests were used. Application of the Spearman test to the Dobson and Nimbus-7, ADEOS, Earth Probe, and OMI observations reveals that the correlation coefficients between these datasets is 0.95, 0.96, 0.94, and 0.93 (statistically significant at the 99% confidence level), respectively. In the following, to explore deviations between the TOC observations obtained by the aforementioned instruments, the Wilcoxon test was applied to the TOC data and revealed that Dobson underestimates the column ozone with respect to Nimbus-7, ADEOS, and OMI by 5, 10, and 8 DU, respectively, while the difference between Dobson and Earth Probe observations was found to be zero.

Applying Spearman's test to the Dobson and Nimbus-7, ADEOS, Earth Probe, OMI, and SCIAMACHY observations, it was found that the correlation coefficient between these datasets is 0.95, 0.96, 0.94, 0.93, and 0.87, respectively, while the correlation coefficient between TOC observations of SCIAMACHY/Earth Probe-TOMS and OMI is 0.85 and 0.93, respectively. Application of Wilcoxon's test revealed that Dobson underestimates the column ozone with respect to Nimbus-7, ADEOS, and OMI by 5, 10, and 8 DU, while SCIAMACHY overestimates the column ozone with respect to Dobson, Earth Probe-TOMS, and OMI by 10, 15, and 3 DU, respectively. From the analysis presented in these studies from Athens it is evident that the TOC observations derived from the Dobson instrument compare well with the total ozone observations from various satelliteborne instruments and, hence, the Athens Dobson station may be used as a ground data station for satellite column ozone observations. This is an illustration of a conclusion that can be made for any one of the many ground stations that has been involved in the international calibration activities described in Section 1.2.2.

Total ozone measurements using the Dobson spectrophotometer D056 were performed on a regular basis in Oslo (59.9°N, 10.7°E), Norway, from 1978 to 1998. The dataset for the entire period of observations has been critically examined by Svendby and Dahlback (2002). Considerable effort was expended to optimize the zenith sky charts used for CC′ zenith observations. The charts were constructed using a semiempirical method, where a large number of quasi-simultaneous observations of direct Sun and zenith observations were used in combination with a radiative transfer model. The same radiative transfer model was used for constructing AD and CD zenith sky charts for various optical thicknesses of cloud cover. When the D056 ozone observations started up in 1978 the SO_2 level in Oslo was significant. The significance of the effect of SO_2 on Dobson spectrophotometer measurements is discussed in Section 1.2.3. Without the SO_2 corrections the revised Oslo Dobson data showed a year-round ozone decrease of $(5.22 \pm 0.63)\%$ per decade from 1978 to 1998, while after SO_2 corrections the decrease is reduced to $(4.68 \pm 0.63)\%$ per decade. The Dobson data from Oslo were also compared with ozone data from the Brewer spectrometer B042 and TOMS data. The mean and standard deviation of the monthly percentage difference between Dobson and TOMS amount to $(-0.91 \pm 1.41)\%$.

3.3 GEOPHYSICAL VALIDATION OF MIPAS-ENVISAT OPERATIONAL OZONE DATA

The continuous monitoring of the quality of satellite-based products through validation with long-term ground-based measurements provides the means to achieving instrument intercomparison and algorithm development. Despite data gaps in between satellite missions, the ozone layer can be globally monitored in a consistent way ensuring that the trends in ozone will be properly identified.

In addition to comparisons between ground-based data and satellite data it is also important to consider intercomparisons between observational data from various satellites. One such intercomparison, made for Nimbus-7 and NOAA-11 data over 18 months, indicated (Hilsenrath *et al.*, 1997) that TOC value differences did not exceed 1% (except high latitudes where they reached 4%). In Section 3.2 we consider the comparison of ozone concentration derived from ground-based instruments with ozone data derived from various satellite systems, by taking as an example the case of ground-based data from Athens. In a similar way, in this section we consider the validation of ozone data from satellites by considering the case of data from MIPAS (see Section 2.5.3.3) as an example of validation carried out for other satellite systems (e.g., POAM III—Randall *et al.*, 2003 and GOME—Weber *et al.*, 2005). MIPAS data have been the subject of a particularly detailed and extended validation program. In considering the validation of MIPAS data we shall, inevitably to some extent, be involved with the validation of data from other satellite systems too. A further brief consideration of the validation of some other satellite systems will be found in Section 3.8.

3.3.1 Introduction to MIPAS

Ozone is one of the six atmospheric trace gases (H_2O, O_3, HNO_3, CH_4, N_2O, and NO_2) that, along with temperature, constitute the set of target products of the Michelson Interferometer for Passive Atmospheric Sounding (MIPAS) (Fischer and Oelhaf, 1996) on board the European ENVIronment SATellite (ENVISAT) and ozone plays a pivotal role in the majority of the research areas covered by the scientific mission of the instrument (Fischer *et al.*, 2000). MIPAS is described in Section 2.5.3.3.

MIPAS operational products are important in several studies including chemistry, transport processes, and the energy budget in the upper atmosphere, as well as the stratosphere–troposphere exchange of ozone. A crucial step towards the exploitation of MIPAS O_3 operational products in quantitative studies investigating these scientific issues is, of course, a thorough validation process, based on comparison with a comprehensive suite of ground-based and other satellite datasets and capable of deriving an overall assessment of the reliability and quality of MIPAS ozone measurements. This aim has been accomplished—for the set of ozone data obtained by MIPAS during the period from July 6, 2002 to March 26, 2004 (i.e., during the instrument's nominal spectral resolution mission) see Section 2.5.3.3—through a series of dedicated experiments executed by different teams and providing results that were subsequently combined into a general and consistent picture (Cortesi *et al.*, 2007).

The coordinated effort for validation of MIPAS operational ozone data v4.61/v4.62 involved the comparison with collocated measurements of the O_3 vertical distribution from a variety of observation platforms and techniques and the combination of the resulting pieces of information into coherent and quantitative statements about the validity of the selected products. This comparison involved different categories of correlative data, obtained from ground-based stations, from high-altitude aircraft and balloon campaigns, and from other satellite missions as well as from assimilated O_3 fields by ECMWF. This took advantage of the redundancy and complementarity of reference datasets to strengthen statistical confidence in the results and to achieve the widest spatial (vertical and geographical) and temporal (diurnal and seasonal) coverage. To this aim, and within the practical limits posed by the large number of validation measurements, special attention was paid to the selection of uniform criteria and methods for individual comparison. With reference to the general guidelines proposed by Fischer *et al.* (2007) for the validation of MIPAS operational products, baseline criteria of 300 km and 3 h were adopted as the ideal for maximum spatiotemporal separation, respectively, between MIPAS and correlative ozone profiles. Departure from these criteria was allowed in a number of specific cases and under suitable conditions, up to a maximum of 500 km and 10 h, in order to increase the statistical value of the comparison. A validation approach relying on the terminology and methodology described by von Clarmann (2006) for the statistical bias and precision determination with matching pairs of O_3 VMR (volume mixing ratio) measurements was followed (cp., for instance, Section 2.5.3.3) and in some cases rigorously applied to evaluate the effects of coincidence errors or horizontal smoothing (cp. Section 2.5.3.3). Comparisons were mostly performed between profiles of O_3 VMR using pressure as the vertical coordinate. With the objective to reduce systematic and random comparison errors associated with the MIPAS vertical smoothing error, correlative profiles measured at much higher vertical resolution than that of MIPAS were transformed using the method described in Section 3.3.2, which uses both the averaging kernels and the *a priori* profiles associated with MIPAS retrievals. This operation was generally performed by using a common routine. Trajectory-hunting techniques were applied to calculate Lagrangian coincidences whenever direct matching did not provide sufficient statistics for the comparison (particularly in the case of comparison with balloonborne measurements, cp. Section 2.6.7).

The present section represents the final outcome of this activity, which involved scientists from the sub-groups of the ENVISAT Atmospheric Chemistry Validation Team (ACVT) contributing to the geophysical validation of MIPAS ozone profiles (i.e., the GBMCD or Ground-Based Measurements and Campaign Database, the ESABC or ENVISAT Stratospheric Aircraft and Balloon Campaigns, and the MASI or Model Assimilation and Satellite Intercomparison sub-groups). The activity started three months after the launch of ENVISAT (March 1, 2002) with the calibration and validation experiments of the commissioning phase and continued during the 12 months of the main validation phase (September 1, 2002 to September

1, 2003) and the first part of the long-term validation program. The preliminary results of the geophysical validation of MIPAS ozone measurements were presented during the First and the Second ENVISAT Validation Workshops (e.g., Sembhi *et al.*, 2006; Taddei *et al.*, 2006) held at ESA's European Space Research Institute (ESA-ESRIN, Frascati, Italy), in December 2002 and May 2004, respectively (ESA, 2003, 2004). A first attempt was made there to achieve quantitative evaluation of the quality of MIPAS near-real time (produced within three hours of the measurement time) and off-line (produced with a less stringent constraint for the processing time and using an extended retrieval range) O_3 data products, by combining the results of comparisons with ozonesonde, lidar, and microwave measurements from individual ground-based stations and networks (Blumenstock *et al.*, 2004) with remote-sensing and *in situ* observations from balloon and aircraft field campaigns (Cortesi *et al.*, 2004), as well as with profiles from concurrent satellite sensors (Kerridge *et al.*, 2004). As a further and closing step in the process of gradual merging and integration of individual validation results, a coordinated effort—focusing on MIPAS O_3 data versions v4.61 and v4.62, to homogenize criteria and strategies of the comparison with different correlative datasets and to update the pre-launch estimates of precision and accuracy of the selected MIPAS ozone products—has been made. In the following sub-sections an overview of the latter phase, with presentation of final results and conclusions, is given.

3.3.2 MIPAS ozone data

The MIPAS instrument was described in Section 2.5.3.3. The data obtained during the MIPAS instrument's full spectral resolution mission, from July 6, 2002 to March 26, 2004, have been processed by using v4.61 and v4.62 of ESA's level-1b and level-2 (based on an unconstrained non-linear least squares fit procedure) operational algorithms (as described in detail by Kleinert *et al.*, 2007 and Raspollini *et al.*, 2006 respectively), and provide a self-consistent set of quasi-continuous measurements for temperature and the six target species. For the purposes of MIPAS ozone validation, the two versions of the ESA operational processor are substantially equivalent; as a baseline v4.61 data were adopted, using v4.62 only for those cases where v4.61 ozone profiles in coincidence with the selected validation measurements were not available. Retrieval of ozone VMR vertical distribution for v4.61/v4.62 data products was carried out using three microwindows: two microwindows (1,122.800–1,125.800) cm^{-1} and (1039.375–1040.325) cm^{-1} (the latter are used in the altitude interval 52–68 km) in MIPAS band AB (1,020–1,170 cm^{-1}), associated with the ozone fundamental modes υ_1 and υ_3; and the third microwindow (763.375–766.375) cm^{-1} in MIPAS band A (685–970 cm^{-1}), close to the center of the O_3 υ_2 band. The total error budget of the ozone vertical distribution retrieved from individual MIPAS scans can be evaluated by combining the random contribution due to the mapping of radiometric measurement noise into retrieved profiles (expressed by the square root of the diagonal elements of the error variance–covariance matrix included in ESA level-2 data products) and *a priori* estimates of systematic components (Dudhia *et al.*, 2002) derived from the analysis carried out at Oxford University (see data available for five different atmospheric scenarios at *http://www-atm.physics.ox.ac.uk/group/mipas/err*, hereafter indicated as the "Oxford University error dataset"). In the case of ozone retrievals, the dominating sources of systematic uncertainty come from the propagation of pressure and temperature retrieval error, from spectrocopic errors, from the effects due to atmospheric horizontal gradients, and from radiometric gain and calibration errors. Further systematic components, such as those due to interfering species (H_2O, CO_2, N_2O_5) or non-local thermal equilibrium (NLTE) effects contribute less than 1% to the total error budget. NLTE can have a larger effect above 55 km.

3.3.3 Comparison of MIPAS data with WMO/GAW ground-based measurements

3.3.3.1 Comparison with NDACC and WOUDC ozonesondes, lidar, and microwave networks

The necessity for careful monitoring of the endangered ozone layer was considered to be of prime importance after the discovery of the Antarctic hole (see Chapter 5). Therefore, a ground-based global network was established in 1991 to monitor not just the ozone but also the chemical and physical parameters that influence the ozone budget. The Network for the Detection of Stratospheric Changes (NDSC) relied on worldwide measurement stations equipped with multiple instruments for parallel monitoring of the variety of atmospheric parameters involved in the ozone depletion issue. In recent years, new developments in the measurement techniques broadened the scope of the network to the monitoring of atmospheric composition in the free and upper troposphere (Godin-Beekmann, 2009). The name of the network was changed to Network for the Detection of Atmospheric Composition Changes (NDACC) in order to reflect these new priorities. At present, the NDACC includes more than 70 research stations throughout the world, equipped with active and passive remote-sensing instruments to monitor the atmospheric composition and its link to climate change. Godin-Beekmann (2009) provides an overview of the network implementation and operation, with particular emphasis on data quality issues. It includes examples of recent results on the long-term evolution of atmospheric parameters relevant to ozone depletion and describes the new development in network-observing capabilities.

A comprehensive intercomparison between MIPAS ozone measurements and correlative data obtained from extensive ground-based networks contributing to the WMO (World Meteorological Organization) Global Atmosphere Watch (GAW) program was carried out at the Belgian Institute for Space Aeronomy (BIRA-IASB). The comparison dataset included ozone profiles from 39 ozonesonde stations (O3S), 8 lidar systems (LID), and 7 microwave radiometers (MWR) associated with the Network for Detection of Atmospheric Composition Change (NDACC), formerly the NDSC (Kurylo and Zander, 2001), and/or the World Ozone and Ultraviolet Data Center (WOUDC). The WOUDC has its origin in the International Polar Year (IPY) of 1957/1958, which also brought about introduction of the International Geophysical Year (IGY), when a concerted effort was made to gather together the plethora of geophysical data, including ozone datasets. The World Meteorological Organization (WMO) coordinated the early years of this data collection which included surface, tropospheric, and stratospheric measurements made from both ground-based instruments as well as balloonborne payloads. In 1961, Canada, working under the auspices of the WMO, became the permanent home for this data archive, which was previously known as the World Ozone Data Center (later to be renamed the World Ozone Data and Ultraviolet Radiation Centre or WOUDC). The stations involved in this intercomparison are identified in Table 3.1.

Prior to using data uploaded routinely to the WOUDC archive, their quality was investigated carefully on statistical and climatological grounds. As discussed in Section 1.7 there are several different types of ozonesondes in use. However, as shown during the JOSIE (Jülich Ozone Sonde Intercomparison Experiment) chamber comparison (Smit and Sträter, 2004), if ozonesondes are operated in a specific way, a similar level of precision and accuracy is achievable from the different sonde types. Typical error estimates are:

— systematic error from 3% (0–20 km) to 5% (20–35 km);
— precision from 5% (0–20 km) to 7% (20–35 km).

Differential absorption ozone lidar (DIAL) systems (see Section 1.6.5) provide the vertical distribution of nighttime ozone number density at altitudes between 8–15 km and 45–50 km. Actual operation depends on cloud cover and other measurement conditions. Typical vertical resolution ranges from 300 m up to 3 km depending on the altitude. The accuracy of the lidar ozone profile depends on the duration of the measurement and on the vertical resolution chosen to process the data. Individual error bars are given in each ozone file. Typical accuracy estimates range from 3 to 7% from 15 to 40 km. At 40–45 km and above, due to the rapid decrease in the signal-to-noise ratio, the error bars increase and significant bias reaching 10% may exist (Godin *et al.*, 1999; McDermid *et al.*, 1998).

Millimeter wave radiometers (MWR) (see Section 1.6.7) operate night and day, providing ozone VMR from 20–25 to 70 km, with a vertical resolution of 8 to 12 km. Ozone VMR is converted into number density using ECMWF or NCEP meteorological analyses of pressure and temperature. Individual error bars are usually given in each ozone data file. Typical accuracy ranges from 5% at 20 km to 20% at 70 km where the information content is smaller leaving a larger weight to *a priori* constraints (Connor *et al.*, 1995; Tsou *et al.*, 1995, 2000). Its low vertical resolution poses additional problems for comparisons, for which dedicated methods have been developed (Calisesi *et al.*, 2005).

Taking into account the ground-based error contribution does not change the total error budget dramatically; this contribution is small compared to the contribution of both MIPAS errors and horizontal smoothing differences in the presence of large horizontal inhomogeneities in the ozone field. As comparisons are based on profiles convoluted with MIPAS averaging kernels, for the ground-based error

Table 3.1. NDACC and WOUDC ground-based stations contributing to MIPAS O_3 validation (Cortesi *et al.*, 2007).

Ozonesondes					
Station	*Location*	*Latitude* (°)	*Longitude* (°)	*Institute*	*Sonde type*
Alert	Canada	82.50	−62.33	MSC	ECC
Eureka	Canada	80.05	−86.42	MSC	ECC
Ny-Ålesund	Svalbard	78.91	11.88	AWI	ECC
Thule	Greenland	76.51	−68.76	DMI	ECC
Resolute	Canada	74.72	−94.98	MSC	ECC
Scoresbysund	Greenland	70.48	−21.97	DMI	ECC
ESRANGE	Sweden	67.88	21.06	NIES	ECC
Sodankylä	Finland	67.37	26.67	FMI	ECC
Keflavik	Iceland	63.97	−22.60	INTA	ECC
Orland	Norway	63.42	9.24	NILU	ECC
Jokioinen	Finland	60.82	23.48	FMI	ECC
Churchill	Canada	58.75	−94.07	MSC	ECC
Edmonton	Canada	53.55	−114.1	MSC	ECC
Goose Bay	Canada	53.32	−60.38	MSC	ECC
Legionowo	Poland	52.40	20.97	INWM	ECC
De Bilt	Netherlands	52.10	5.18	KNMI	ECC
Valentia	Ireland	51.93	−10.25	ME	ECC
Uccle	Belgium	50.80	4.35	KMI	ECC
Praha	Czech Republic	50.02	14.45	CHMI	ECC
Hohenpeissenberg	Germany	47.80	11.02	DWD	Brewer–Mast
Payeme	Swiss Alps	46.49	6.57	MCH	ECC
Tsukuba	Japan	36.05	140.13	IMA	Carbon–Iodine
Paramaribo	Surinam	5.81	−55.21	KNMI	ECC
San Cristóbal	Galapagos	−0.92	−89.60	CMDL	ECC
Nairobi	Kenya	−1.27	36.80	MCH	ECC
Malindi	Kenya	−2.99	40.19	RPSM	ECC
Natal	Brazil	−5.42	−35.38	INPE	ECC
Watukosek	Java	−7.50	112.6	JAXA	ECC
Ascension Island	Congo	−7.98	−14.42	NASA	ECC
Tutuila	Samoa	−14.23	−170.56	CMDL	ECC
Fiji	Fiji	−18.13	178.42	CMDL	ECC

Ozonesondes (*cont.*)					
Station	*Location*	*Latitude* (°)	*Longitude* (°)	*Institute*	*Sonde type*
Saint-Denis	Réunion	−21.05	55.47	CNRS	ECC
Irene	South Africa	−25.25	28.18	SAWS	ECC
Lauder	New Zealand	−45.03	169.68	NIWA	ECC
Marambio	Antarctica	−65.28	−56.72	INTA	ECC
Dumont d'Urville	Antarctica	−66.67	140.01	CNRS	ECC
Syowa	Antarctica	−69.00	39.58	JMA	Carbon–Iodine
Neumayer	Antarctica	−70.65	−8.25	AWI	ECC
Belgrano	Antarctica	−77.87	−34.63	INTA	Undefined
Lidar					
Station	*Location*	*Latitude* (°)	*Longitude* (°)	*Institute*	
Eureka	Canada	80.05	−86.42	MSC	
Ny-Ålesund	Svalbard	78.91	11.88	AWI	
Alomar, Andoya	Norway	69.28	16.02	NILU	
Hohenpeissenberg	Germany	47.80	11.02	DWD	
Haute Provence	French Alps	43.94	5.71	CNRS	
Tsukuba	Japan	36.05	140.13	NIES	
Table Mountain	California	34.23	−117.41	JPL	
Mauna Loa	Hawaii	19.54	−155.58	JPL	
Lauder	New Zealand	−45.03	169.68	RIVM	
Microwave radiometers					
Station	*Location*	*Latitude* (°)	*Longitude* (°)	*Institute*	
Ny-Ålesund	Svalbard	78.91	11.88	IFE	
Kiruna	Sweden	67.84	21.06	IMK	
Bremen	Germany	53.11	8.86	IFE	
Zugspitze	German Alps	46.49	6.57	MCH	
Mauna Loa	Hawaii	19.54	−155.58	UMAS	
Lauder	New Zealand	−45.03	169.68	UMAS	

(according to Calisesi *et al.*, 2005), the term:

$$\mathbf{AK}^{\mathrm{T}}\mathbf{W}^{\mathrm{T}}\mathbf{S}_{\mathrm{GR}}\mathbf{WAK} \tag{3.2}$$

was considered, where **AK** is the MIPAS averaging kernel matrix; **W** is the interpolation matrix from the ground-based grid to the MIPAS grid; and $\mathbf{S}_{\mathrm{GR}}$ is ground-based error covariance (for details see Calisesi *et al.*, 2005).

When it comes to the error budget of ground-based comparisons, it is necessary to realize that MIPAS and ground-based instruments offer a different perception of atmospheric ozone. Such differences must be considered to interpret comparison results properly. To evaluate the comparison error budget, one takes into account, along with the measurement and retrieval error of MIPAS and of the correlative instrument, the contributions associated with the vertical and horizontal smoothing differences and with the spatial separation of the two ozone profiles. Expanding the theory and formalism of Rodgers (1990), the following expression was used for the total comparison error covariance S:

$$\begin{aligned} S = S_M + S_N + (A_{M,V} - A_{N,V})S_V(A_{M,V} - A_{N,V})^{\mathrm{T}} \\ + (A_{M,H} - A_{N,H})S_H(A_{M,H} - A_{N,H})^{\mathrm{T}} + S_{\Delta\mathrm{O}_3} \end{aligned} \tag{3.3}$$

where S_M = MIPAS error (measurement, retrieval, and retrieval parameters);
S_N = correlative instrument error (measurement, retrieval, and retrieval parameters);
A_M = MIPAS averaging kernels, vertical (V index) and horizontal (H index);
A_N = correlative instrument averaging kernels, vertical (V) and horizontal (H);
S_V = atmospheric variability covariance (vertical);
S_H = atmospheric variability covariance (horizontal);
$S_{\Delta\mathrm{O}_3}$ = spatial distance error.

The effect of differences in vertical resolution can be estimated by means of the vertical averaging kernels (**AK**) associated with the MIPAS retrieval of the ozone profile. First, the **AK**s of the low-resolution data are used to map the high-resolution profile to the low-resolution perception. The *a priori* profile used in optimal estimation retrievals is also included as it may introduce an additional bias. Second, the smoothing difference error is estimated as the difference between smoothed and original profiles. For MIPAS comparison with high vertical resolution measurements (O3S ozone sonde or lidar):

$$\Delta x_V = x_a^M + A_M(x_N - x_a^M) - x_N \tag{3.4}$$

where Δx_V = vertical smoothing error;
x_N = high-resolution profile (O3S or lidar);
x_a^M = MIPAS ozone profile used to compute vertical averaging kernels

and for MIPAS comparison with lower vertical resolution measurements (MWR):

$$\Delta x_V = x_a^N + A_N(x_M - x_a^N) - x_M \tag{3.5}$$

where x_M = high-resolution profile (MIPAS);
x_a^N = MWR *a priori* ozone profile.

As the MIPAS processor retrieves only one-dimensional profiles, no **AK**s are available for the study of horizontal smoothing. The MIPAS uncertainties associated with horizontal smoothing are calculated rather as an estimate of the ozone gradient interfering with the MIPAS line of sight (LOS) (i.e., the horizontal component of atmospheric noise associated with the MIPAS measurement). The horizontal smoothing error is written as

$$\Delta x_H = \pm \text{abs}(\nabla X_{\text{MEDIAN}} \cdot I_{\text{ENVISAT}})|\text{MIPAS}|_{90\%} \tag{3.6}$$

where Δx_H = horizontal smoothing error (or horizontal component of atmospheric noise);
∇X_{MEDIAN} = ozone gradient at the median point of MIPAS LOS;
I_{ENVISAT} = ENVISAT direction (MIPAS LOS is backward along track);
$|\text{MIPAS}|_{90\%}$ = LOS extension of 90% information airmass.

The ozone gradient was estimated from four-dimensional ozone fields generated by the Belgian Assimilation System of Chemical Observations from ENVISAT (BASCOE; Errera and Fonteyn, 2001; Fonteyn *et al.*, 2003). BASCOE is a data assimilation system of stratospheric chemistry using the four-dimensional variational (4D-VAR) method. In the course of a run, BASCOE can ingest satellite observations. The resulting assimilated field is an estimate of the chemical composition of the stratosphere based both on the set of observations and on the physical laws describing the evolution of the system synthetized into the model. They are defined at 37 hybrid pressure levels from 0.1 hPa down to the surface. The horizontal resolution of BASCOE standard outputs is 3.75° in latitude by 5° in longitude. In this work the off-line version v3d24 of BASCOE fields was used.

Finally, to complete the comparison error budget, the ozone partial column difference induced by the spatiotemporal separation of the two ozone profiles can be estimated by:

$$\Delta O_3 = O_3(|\boldsymbol{X}^{\text{MIPAS}}_{\text{MEDIAN}}|) - O_3(|\boldsymbol{X}_{\text{STATION}}|) \tag{3.7}$$

where $|\boldsymbol{X}^{\text{MIPAS}}_{\text{MEDIAN}}|$ = the estimated geolocation of the median point of MIPAS LOS;
$|\boldsymbol{X}_{\text{STATION}}|$ = the ground-based station geolocation; and
$O_3(X)$ = ozone partial column at the corresponding location and time estimated using BASCOE-assimilated ozone fields.

The along-orbit distribution median position and 90% extension as a function of tangent altitude of the MIPAS information content was estimated by De Clercq and Lambert (2006) using their two-dimensional radiative transfer model of the MIPAS full limb scanning sequence. It is important to note that BASCOE absolute ozone fields have been shown to compare reasonably well with HALOE, CRISTA, and MLS and, more important here, that relative fields are accurate (Errera and Fonteyn, 2001; Fonteyn *et al.*, 2003).

The first part of the study concentrated on analysis of the time series of the differences between MIPAS and ground-based ozone partial column data (i.e., the integral of ozone partial pressure as a function of height). The analysis included assessments of the different contributions to total comparison error, as defined above (Section 3.3.2). Comparison results vary significantly between the lower stratosphere, where dynamics and chemistry interfere with the clear influences of tropospheric dynamics, and the higher stratosphere, where photochemistry dominates. Consequently, a classification based on regularities in the pattern of O_3 partial column differences emerges: in the lower stratosphere (75–35 hPa), results regroup around synoptic and regional systems and the systems linked to stratospheric transport; reaching into the middle stratosphere (35–15 hPa), we move from large synoptic groups to more zonal behavior and we can extend these described synoptic systems to group more stations; in the middle and upper stratosphere (15–7 hPa, 7–3 hPa, 3–0.8 hPa), zonal symmetry becomes dominant and comparison results follow this behavior. Deviations from zonal symmetry nevertheless exist and must be taken into account. Typical output of the comparison carried out for each of the aforementioned groups of measurement sites is displayed in Figure 3.8, presenting the results obtained at western and central European stations. The plot shows (black dots) the percentage relative difference in ozone partial column (73–35 hPa) between MIPAS and correlative ozonesonde data at western and central European stations for 2003, and smoothing and collocation errors (running mean in plain and standard deviation in dashed) estimated by the aforementioned methods. Gray rectangles identify monthly means (central line) and standard deviations of the differences.

In general, the comparison error is dominated by the effect of differences in horizontal smoothing of atmospheric variability. While ground-based instrumentation captures only a portion of the airmass probed by MIPAS, MIPAS smoothes atmospheric inhomogeneities over several hundred kilometers. The red curves in Figure 3.8 give the range of atmospheric variability smoothed by the MIPAS measurement (i.e., the upper limit of the expected difference between MIPAS and ground-based ozone column data). It was concluded from this plot that differences in horizontal smoothing can account for the observed standard deviation of the comparisons in most of the cases, but not for systematic differences like those appearing in Figure 3.8 in the summer of 2003. Horizontal smoothing differences are followed in magnitude by errors associated with geolocation differences. The latter also correlate with the standard deviation of comparisons, but their amplitude is dominated by MIPAS horizontal smoothing effects. Errors associated with vertical smoothing differences are smaller. Their effect could account for the small, constant offset in the comparisons. In most cases, comparison results can be interpreted by considering the different error contributions. However, in some cases, they cannot account fully for the difference noticed between MIPAS and correlative partial column data. MIPAS reports larger partial columns than ground-based instruments:

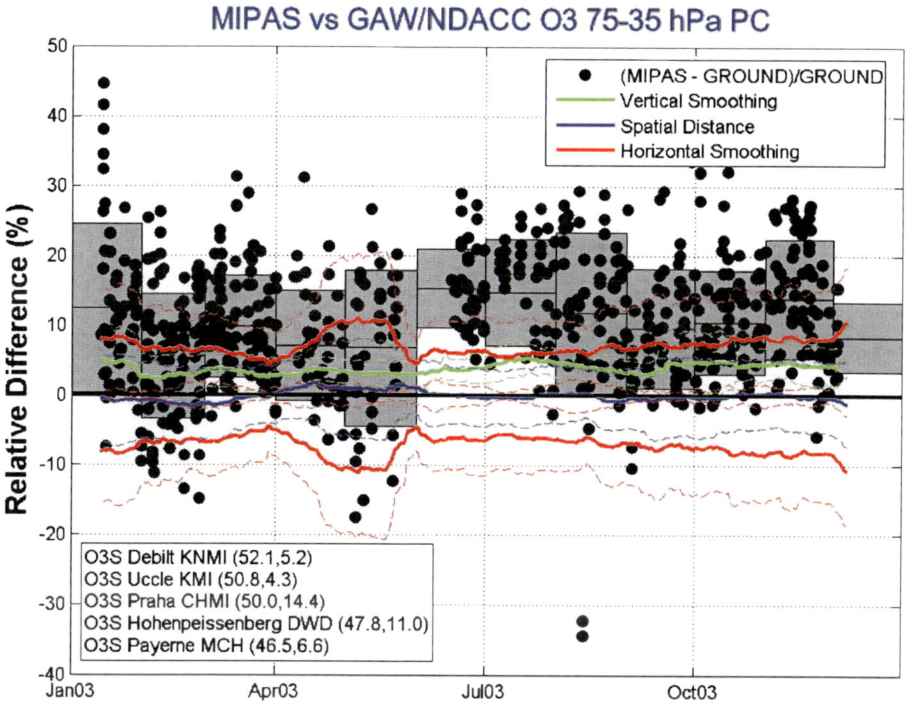

Figure 3.8. Time series of the percentage relative difference in ozone partial column (75–35 hPa) between MIPAS and correlative ozonesonde data at five western and central European stations for 2003, and estimated smoothing and collocation errors (running mean in plain and 1σ standard deviation in dashed). Gray-shaded rectangles identify monthly means (central line) and standard deviations of the differences (Cortesi *et al.*, 2007).

1. In the 75–35 hPa layer at stations from northern (see Figure 3.8) and southern mid-latitudes, the equator, and the tropics.
2. At 35–15 hPa over stations at the equator, in the tropics, and in Antarctica during ozone hole events.
3. In the 3–0.8 hPa layer at European stations.

At 7–3 hPa, MIPAS partial columns underestimate correlative observations in Hawaii. The comparison error budget cannot account for these observed differences. In all other analyzed situations, MIPAS partial column data agree well with those reported by the ground-based instrumentation, and the observed differences fit well within the comparison error budget.

In the comparison of MIPAS O_3 vertical profiles with ground-based data the first step was instrumental in getting an overall view of the agreement between MIPAS and WMO/GAW ground-based data, and also in determining periods and groups of stations where comparison results are sufficiently consistent to allow the meaningful derivation of statistical values. The second step involved derivation of the vertically resolved statistics of comparisons between MIPAS v4.61 ozone profiles and correlative data obtained at NDACC and WOUDC stations. The comparisons were performed at each individual station listed in Table 3.1 and summary plots were computed for stations belonging to the same synoptic system/zonal region and showing mostly identical comparison results.

At Arctic and Northern Hemisphere and Southern Hemisphere mid-latitude sites, the results can be separated between October 1 to March 31 and April 1 to September 30. At tropical and equatorial stations, the weak seasonal variation allows us to draw annual plots. At Antarctic stations results can be separated between "ozone hole" (i.e., for 2003, August 21 to October 15) and "normal ozone" periods (i.e., for 2003, October 16 to August 20).

A few examples of the results obtained for the absolute and relative differences of MIPAS O_3 vertical profiles with ozonesonde and lidar data are shown in Figures 3.9a and b, respectively. Each plot of Figure 3.9 shows, for each collocated pair of profiles, absolute differences between MIPAS and correlative measurements (light gray lines). To eliminate vertical smoothing differences, high-resolution correlative measurements have been previously convoluted with MIPAS averaging kernels and biased by the first-guess profile, following the method proposed by Rodgers and Connor (2003).

Black lines depict the statistical values (mean and 1σ standard deviation) of the absolute or relative differences between MIPAS and ground-based data. Red lines depict the total systematic error of the comparison. The mean difference between MIPAS and ground station data should be compared to these lines. The total systematic error of the comparison is calculated as the sum of MIPAS systematic error and the systematic bias due to non-perfect collocation (spatiotemporal distance, as explained above). The yellow block delimited by dashed red lines depicts the total random error of the comparison. This value should be compared with the 1σ standard deviation of the differences. This total random error of the comparison is calculated as the quadratic sum of MIPAS random error, ground-based random error, random contribution of spatiotemporal distance, and line-of-sight inhomogeneity.

Figures 3.10a, b show the results of comparison between ozonesonde and lidar data, respectively, in terms of relative differences. These results are similar to those obtained from absolute difference comparisons, but should be considered carefully:

- The total error budget of the comparison is, first, calculated for absolute difference and, second, a percentage is estimated.
- Low ozone concentrations lead to large relative differences although absolute differences are small. In these cases, the mean and standard deviation of relative differences are not relevant. The percentages obtained below 12–15 km at middle and high latitudes, below 20 km at tropical and equatorial stations, and during "ozone holes" in Antarctica should not be considered.

An overall summary of the results obtained from the comparison of O_3 vertical profiles is presented in Table 3.2, with a detailed assessment of the quality of the

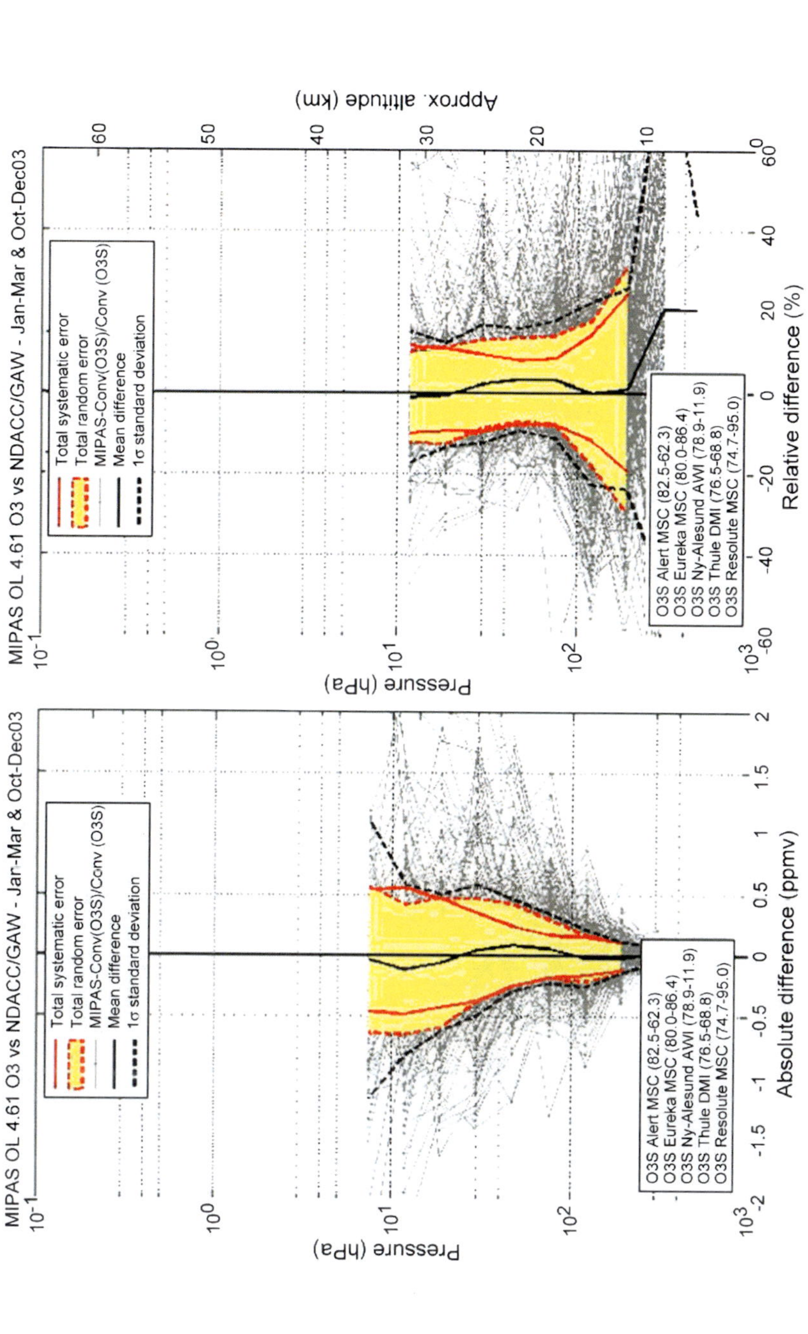

Figure 3.9. Vertically resolved statistics of the absolute differences between MIPAS O_3 data and NDACC and WOUDC measurements in the Arctic (see main text for explanations) (Cortesi *et al.*, 2007).

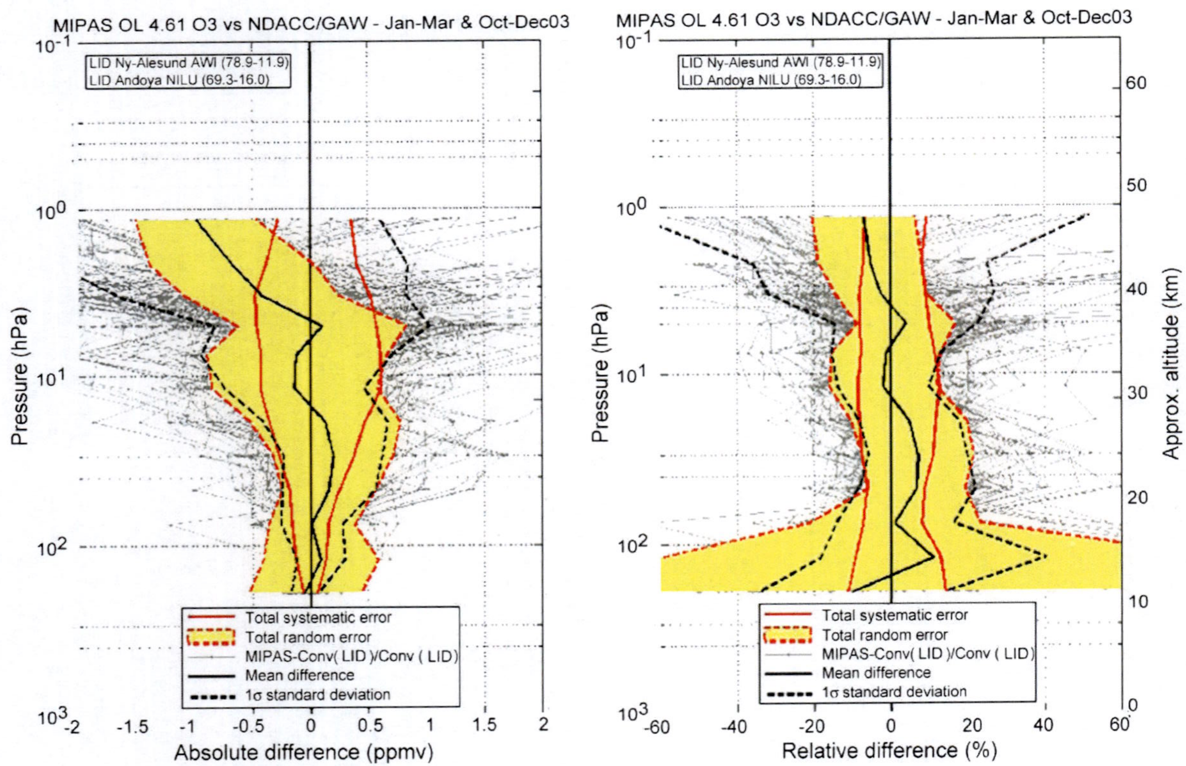

Figure 3.10. Vertically resolved statistics of the relative differences between MIPAS O_3 data and NDACC and WOUDC measurements in the Arctic (see main text for explanations) (Cortesi *et al.*, 2007).

Table 3.2. Results of the comparison between MIPAS v4.61 ozone profiles and NDACC ground-based measurements (Cortesi *et al.*, 2007).

	Arctic and Northern Atlantic	*Northern mid-latitude/ Europe*	*Tropic of Cancer/ Hawaii*	*Equator*	*Tropic of Capricorn/ Polynesia*	*Southern mid-latitude/ New Zealand*	*Antarctica*
Lower stratosphere	Observed differences fit within the total error budget of the comparison at all stations (mean within ±5%). Scatter of the the differences follows the estimated horizontal uncertainty, with large variability in winter, weaker in summer	Mean positive bias of ±10% at European stations in the 12 to 25 km altitude range. Larger bias in June and July (±15%). Scatter of the differences fits within the estimated horizontal inhomogeneity effect. Similar results at other northern mid-latitude locations	Comparisons with Mauna Loa lidar show a positive mean difference below 25 km. Large variability before May could explain part of this, but not after the decrease of variability after May	Positive mean differences +10% to +25% depending on stations. Relative differences are larger than in Europe due to lower ozone columns. Part of the bias may be accounted for in the vertical smoothing effect, but not fully	Scattered results with a majority of large positive differences (+15–20%)	Results at Lauder are similar to those obtained at European stations and show a mean positive difference of +10% in the 15–20 km altitude range	Every station has a unique behavior and provides different results reflecting the influence of the polar vertex and circumpolar belt. The large range of the partial column induces use of absolute differences. Differences fit within comparison error budget
Middle stratosphere	Similar to lower stratosphere, observed differences fit within the total error budget of the comparison at all stations. Scatter of the difference follows the estimated horizontal uncertainty	Above 20–25 km, differences at European stations fit within the total error budget of the comparisons. Still show a positive mean difference (weaker than at lower altitudes) in June and July. More scattered results in southern Canada	Good agreement with lidar from 25 to 40 km	Positive mean differences at all equatorial stations weaker than in lower stratosphere (+10%)	Positive mean differences of about +10%. Similar to equatorial results	Good agreement, differences fit within comparison error budget (±5%)	Good agreement during normal ozone periods. Standard deviation explained by large atmospheric variability associated with polar vortex. MIPAS ozone larger than ozonesonde during ozone hole events
Upper stratosphere	Good agreement with lidar data and MWR data at higher altitudes	Good agreement with lidar data. At altitude higher than 40 km a mean positive difference is observed with MWR data	A mean negative difference is observed in comparisons with lidar data above 40 km. At higher altitude, good agreement with MWR data is observed			Good agreement with lidar data and MWR data at higher altitudes	

agreement between MIPAS and ground-based measurements (O3S, LID, and MWR) for each altitude region and synoptic or regional system.

3.3.3.2 Comparison of MIPAS data with NDACC/EQUAL lidar network

A purely statistical analysis of the differences between MIPAS O_3 vertical profiles and lidar data was carried out by the groups involved in the EQUAL (ENVISAT QUality Assessment with Lidar) project, based substantially on the same NDACC dataset adopted by the Belgian Institute for Space Aeronomy team for the pseudo-global intercomparison described in Section 3.3.3.1—that is, measurements from the ground-based lidar stations listed in Table 3.1 with the addition of the Eureka (80.05°N, 86.42°W) site. The selection of collocated pairs of MIPAS and lidar observations was based on matching criteria that were slightly relaxed with respect to the agreed baseline, in order to obtain a sufficient number of coincident profiles for a statistically meaningful comparison: useful matches were chosen within a 400 km, 10 h window. A total of 627 matching pairs was identified and was used to validate MIPAS O_3 level-2 off-line data v4.61 and v4.62 in the period from July 6, 2002 to March 26, 2004. The comparison was based on statistical analysis of the differences between profiles of O_3 number density measured as a function of altitude by MIPAS and by lidar stations in the range from 10 km to 50 km. The vertical co-ordinate for MIPAS profiles was transferred from pressure to altitude by using ECMWF data: we interpolated ECMWF pressure and geo-potential height (GPH) to the MIPAS retrieval pressure grid and converted the resulting GPH values to geometric altitude.

The results of the comparison for the whole set of collocated pairs are summarized in Figure 3.11. In the left panel, the mean profiles of O_3 number density measured by MIPAS and by lidars are displayed, along with the corresponding 1σ standard deviations. The mean and the median of the percentage differences between MIPAS and lidar O_3 profiles relative to the lidar values are plotted in the middle panel. On the same graph, we show the mean relative difference $\pm 1\sigma$ standard deviation (light-green profiles) and indicate, for some of the altitude levels, the number of MIPAS and lidar pairs taken into account by the statistics at that level. In the right panel, the standard deviation of the relative differences is compared with the standard deviations of the selected MIPAS and lidar profiles. The mean relative difference is lower than $\pm 5\%$ between 15 and 40 km, whilst slightly larger values of positive and negative bias (up to $\pm 15\%$) are obtained outside this altitude range, respectively, above 40 km and below 15 km. The quality of the agreement in the lower and middle stratosphere is confirmed by the substantial match between the mean and the median of the differences at these altitudes. The occurrence of outliers in the distribution of the relative differences leads to an increase of the standard deviation and, when asymmetric, introduces a discrepancy between the mean and the median values, as happens, in our case, at altitudes below 20 km and to a lesser extent above 35–40 km. To better identify possible sources of the observed discrepancies, we have extended the statistical analysis of MIPAS and lidar O_3 collocated profiles by investigating their latitude dependency. No distinction was found

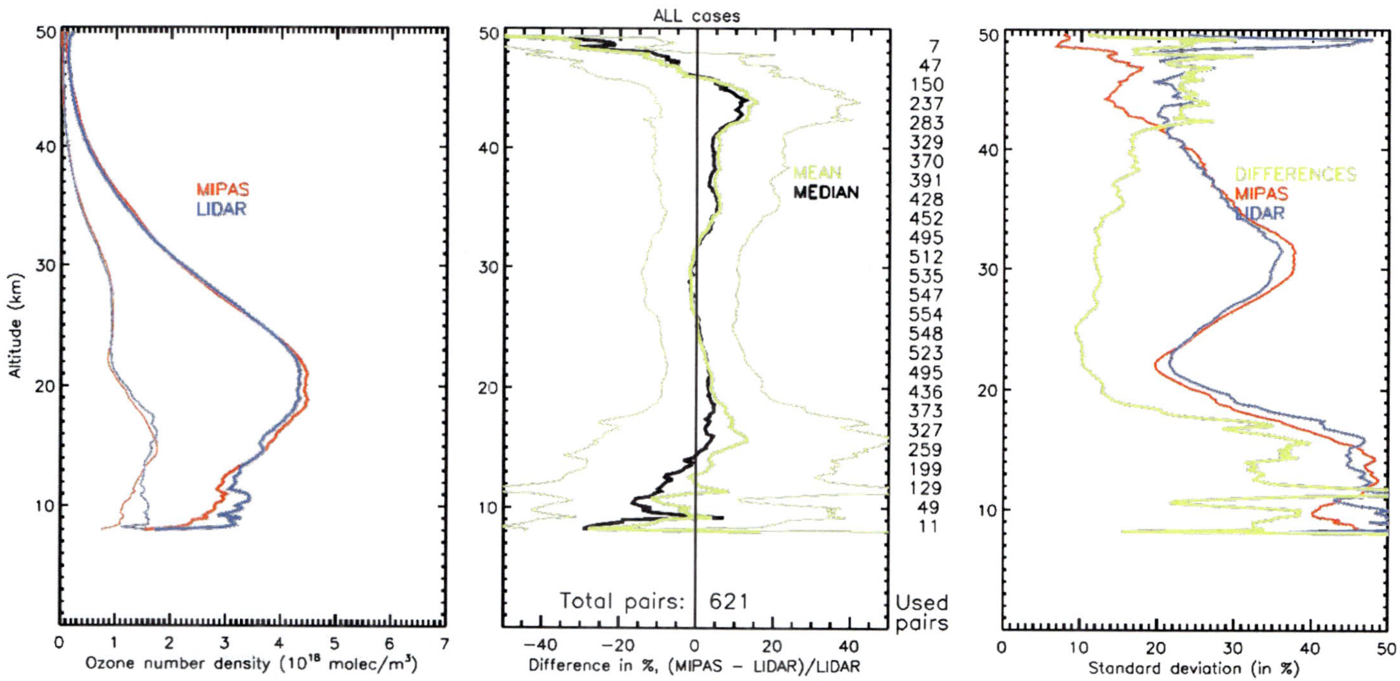

Figure 3.11. Results of the comparison between MIPAS O_3 profiles and ground-based lidar measurements matching the coincidence criteria of 400 km and 10 h. On the left, MIPAS and lidar mean profiles are shown by bold red and blue lines, respectively; the same color code is used for the thin lines representing the 1σ standard deviations. The plot in the middle panel shows the mean (bold green) and the median (bold black) of the relative differences, with the thin green lines indicating the $\pm 1\sigma$ standard deviations from the mean difference; on the right side of the plot, the number of coincident pairs that have been used in the calculations are reported for some of the altitude levels. On the right panel, we display the standard deviations of the relative differences (bold green) and of MIPAS (bold red) and lidar (bold blue) O_3 profiles (Cortesi *et al.*, 2007).

between the Southern and Northern Hemisphere. We calculated the mean and the median of the relative differences, as well as their standard deviations, for three latitude bands corresponding to the tropical (from the Equator to latitude 23.5°), to mid-latitude (from latitude 23.5° to 66.5°) and to the polar (from latitude 66.5° to the North Pole) regions (the results are displayed in Figure 3.12). A small positive bias (less than 5%) is generally found between 20 and 40 km both in the mid-latitude and in the tropical regions, with the exception of the 21–24 km range in the latter, where the mean difference increases up to 10%. In the tropics larger values of mean relative differences (up to 50%) are found below 20 km, associated with a standard deviation of the differences that exceeds those of the individual instruments. At high latitudes, MIPAS O_3 data are biased low with respect to lidar measurements, with differences that remain always below 7% from 15 km up to 40 km altitude. Once again, the discrepancy increases at the lowest tangent altitude of MIPAS (below 12 km), with a negative bias up to −20% and a standard deviation of the mean relative differences comparable to the ones of MIPAS and lidar profiles. Notably, the larger differences between the mean and the median of the distribution observed below 20 km are mostly localized at the mid-latitude, while elsewhere they remain either small (less than a few percent in the polar region) or negligible (in the tropics) for the whole altitude range.

3.3.3.3 Comparison of MIPAS data with NDACC FTIR network

MIPAS v4.61 ozone data for the period July 6, 2002 to March 26, 2004 were compared with ground-based Fourier Transform InfraRed (FTIR) measurements at five stations: Kiruna, Sweden (67.8°N, 20.4°E) and Jungfraujoch, Switzerland (46.5°N, 8.0°E) in the Northern Hemisphere; and Lauder, New Zealand (45.0°S, 169.7°E), Wollongong, Australia (34.4°S, 150.9°E), and Arrival Heights, Antarctica (77.5°S, 166.4°E) in the Southern Hemisphere (see Table 3.3). These instruments are all operated within the NDACC. Quality control is applied according to the NDACC guidelines. In addition to column amounts of O_3, low vertical resolution profiles are obtained from solar absorption spectra by using the Optimal Estimation Method of Rodgers (2000) in the inversion programs: PROFFIT (PROFile FIT) for the Kiruna station (described by Hase, 2000; Hase *et al.*, 2004) is based on the forward model KOPRA (Karlsruhe Optimized Precise Radiative transfer Algorithm; Höpfner *et al.* 1998); and SFIT2 (Pougatchev *et al.*, 1995; Rinsland *et al.*, 1998) for the other stations. The SFIT2 and PROFITT codes have been cross-validated successfully by Hase *et al.* (2004). The retrieval process in both codes involves the selection of retrieval parameters: spectral microwindows, spectroscopic parameters, *a priori* information, and model parameters. The choice of these retrieval parameters has been optimized independently at each station. An exception was made for the spectroscopic database: all stations agreed on using the HITRAN 2004 database (Rothman *et al.*, 2005) in order to avoid biases due to different spectroscopic parameters. For 49 infrared bands of O_3 the line positions and intensities have indeed been updated in the HITRAN 2004 database following those of the MIPAS database (*mipas-pf-3.1* for the v4.61 products) (Raspollini *et al.*, 2006).

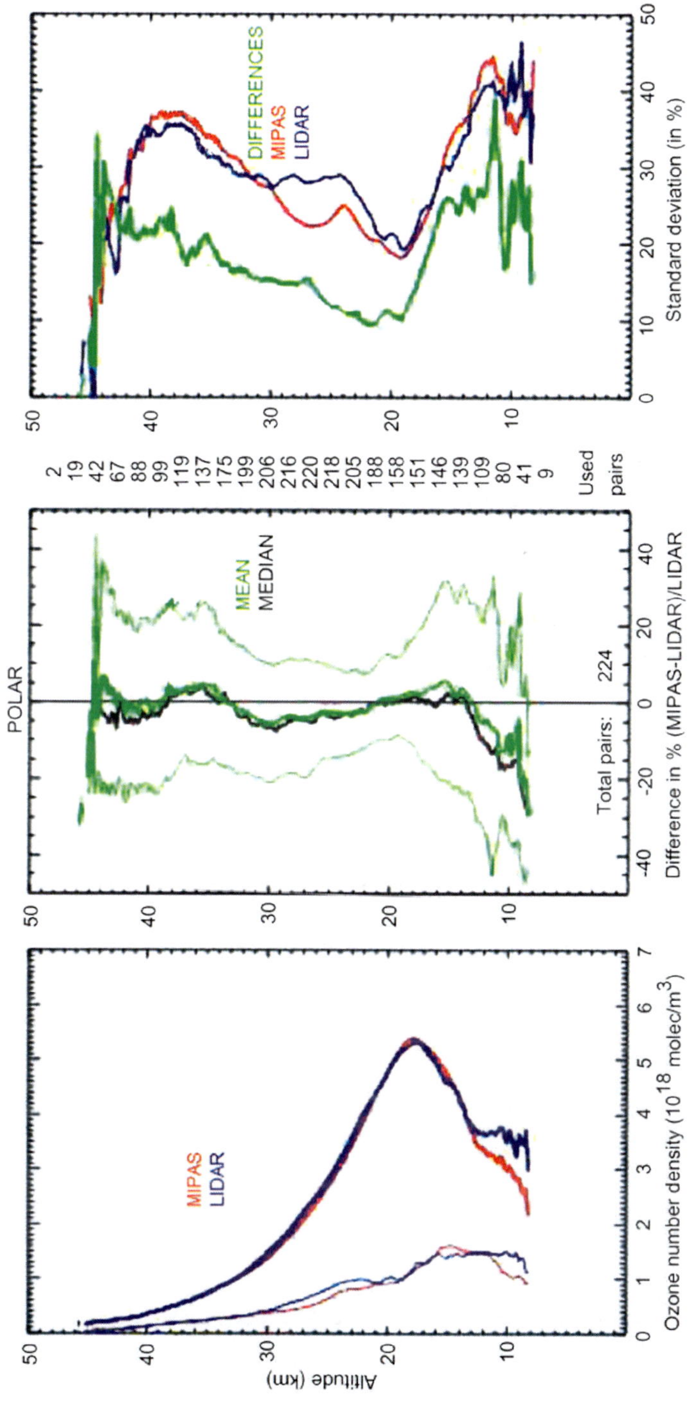
POLAR
MIPAS
LIDAR
Altitude (km)
Ozone number density (10^{18} molec/m^3)
MEAN
MEDIAN
Total pairs: 224
Difference in % (MIPAS-LIDAR)/LIDAR
Used pairs
DIFFERENCES
MIPAS
LIDAR
Standard deviation (in %)

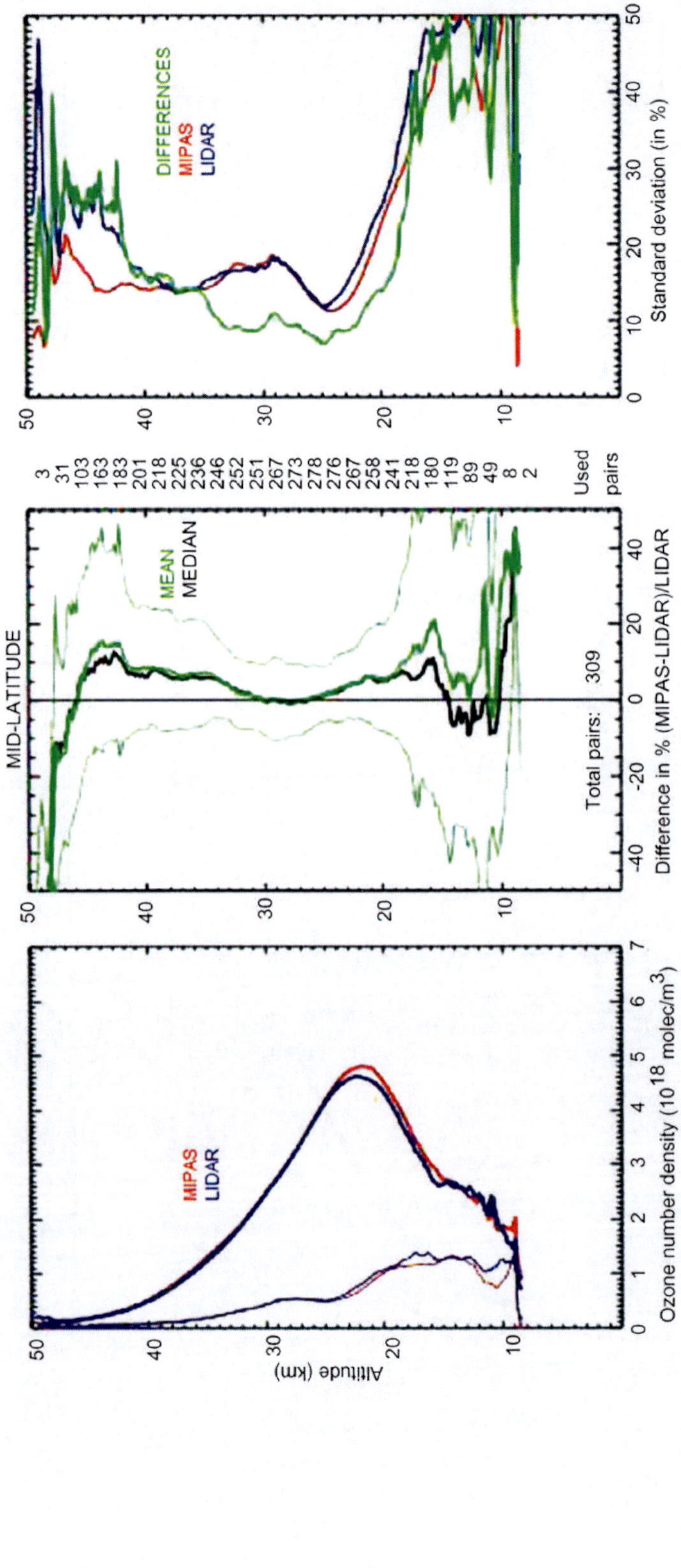
MIPAS
LIDAR
Altitude (km)
Ozone number density (10^{18} molec/m^3)
MID-LATITUDE
MEAN
MEDIAN
Total pairs: 309
Difference in % (MIPAS-LIDAR)/LIDAR
Used pairs
3 31 103 163 183 201 218 225 236 246 252 251 267 273 278 276 267 258 241 218 180 119 89 49 8 2
DIFFERENCES
MIPAS
LIDAR
Standard deviation (in %)

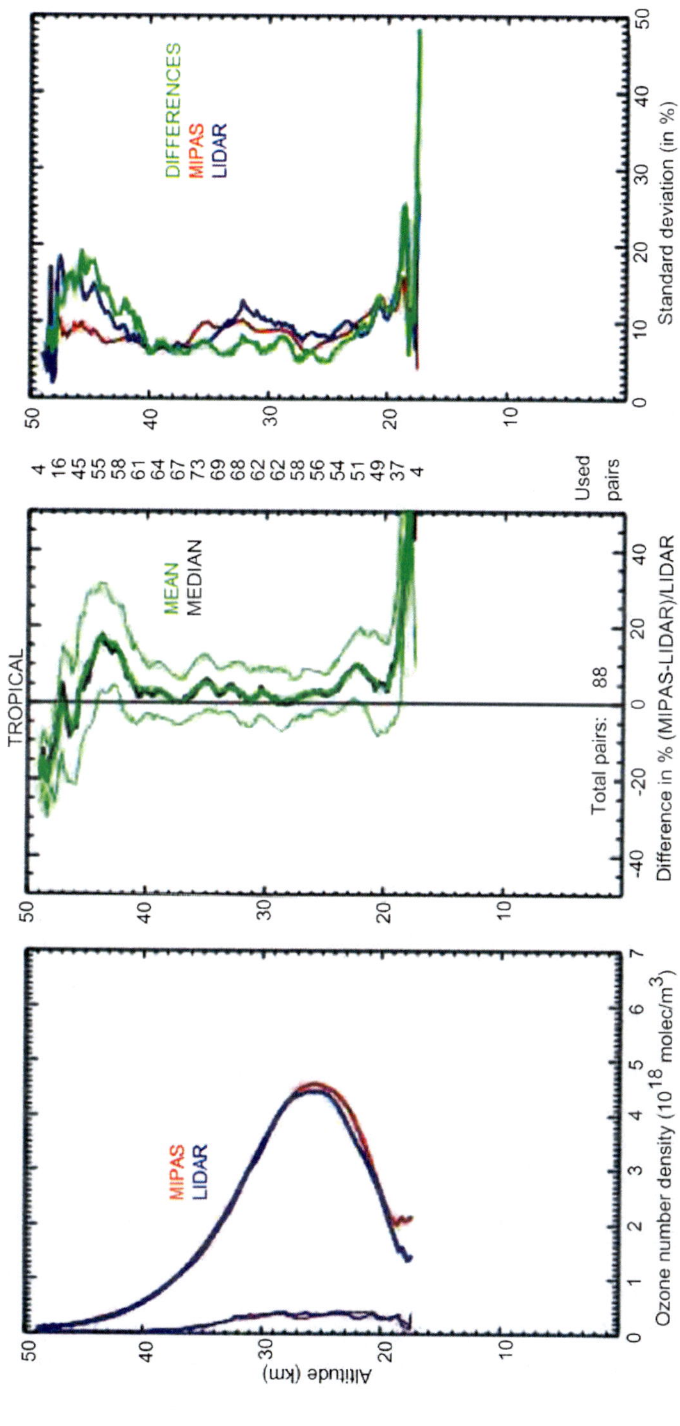

Figure 3.12 (*cont.*). Results of the comparison between MIPAS O_3 profiles and ground-based lidar measurements: zonal averages. The same format is used as for the plots in Figure 3.11 (Cortesi *et al.*, 2007).

Table 3.3. Statistical means (MRD) and standard deviations (SD) of the relative differences (X-FTIR)/mean (FTIR) in percent of the O_3 partial columns defined by the given pressure limits. X is the MIPAS O_3 partial column collocated within 3 h and 300 km from the ground-based FTIR measurements. The number N of comparison pairs for each station, the combined random error, and the 3σ standard error on the mean (SEM) are also reported (Cortesi *et al.*, 2007).

Station	*Pressure range* (hPa)	*O_3 partial column* MRD $\pm$ SD (%)	*Random error* (%)	*N*	*SEM* (%)
Kiruna	2–168	$+1.3 \pm 6.3$	5.6	24	3.9
Jungfraujoch	2–214	-3.5 ± 6.1	5.5	12	5.3
Wollongong	1–196	-0.4 ± 2.3	6.1	4	3.5
Lauder	3–185	-5.6 ± 2.9	5.5	17	2.1
Arrival Heights	2–163	-7.1 ± 8.1	7.1	16	6.1

Pairs of coincident ozone profiles from MIPAS and from each of the five FTIR stations were selected for comparison according to the baseline criteria ($\pm$3 h, 300 km), with spatial separation between satellite and ground-based observations evaluated at the MIPAS nominal tangent height of 21 km (Table 3.3). Each spatially collocated MIPAS scan is compared with the mean of the FTIR measurements recorded within the chosen temporal coincidence criterion. The comparison is made on a pressure grid. The MIPAS profiles are degraded to the lower vertical resolution of the ground-based FTIR measurements, following:

$$x_s = x_a + A(x_m - x_a) \tag{3.8}$$

where x_m and x_s are the original and the smoothed MIPAS profiles; and x_a and A are the FTIR *a priori* profile and averaging kernel matrix, respectively.

For the sake of homogeneity, a common approach was agreed for the calculation of O_3 partial columns and vertical profile differences in the comparisons.

The absolute differences (MIPAS – FTIR) between MIPAS-smoothed profiles and the low vertical resolution FTIR measurements were calculated. The mean relative difference in percent and the associated 1σ standard deviation were then obtained by dividing the mean absolute differences and standard deviation, respectively, by the mean of the FTIR O_3 profiles.

The boundaries of partial columns, defined by pressure levels as indicated in Table 3.4, were chosen taking into account:

— the ground-based FTIR sensitivity, which is reasonable up to around 40 km for O_3;
— the lowest altitudes of valid MIPAS profiles which have a mean of about 12 km over the dataset selected for comparison.

As for the vertical profiles, the absolute differences between MIPAS and FTIR O_3 partial columns were first calculated and then divided by the mean of the FTIR partial columns to obtain the relative differences. In Table 3.3, the mean and the standard deviation of partial column relative differences are reported for each station, along with the number N of coincident pairs and the estimated random error on O_3 partial column differences. The random error covariance matrix of the difference (MIPAS – FTIR) was calculated, using the work of Rodgers and Connor (2003) for the comparison of remote-sounding instruments and of Calisesi *et al.* (2005) for re-gridding between the MIPAS and the FTIR data (see Vigouroux *et al.* 2007, for more details). The FTIR random error budget was estimated for typical measurement at Kiruna. There are different contributions to the MIPAS random error covariance matrix. The error covariance matrix due to noise is given in the MIPAS level-2 products for each profile. The mean of the covariance matrices of the coincident MIPAS profiles was used as the noise contribution to the MIPAS random error matrix. Two coincident MIPAS profiles at Lauder have been removed from the comparisons, because their random errors were especially large. Following the approach adopted for MIPAS comparison with other satellite measurements, the systematic errors with random variability (i.e., errors due to the propagation of pressure and temperature random covariance into the ozone retrieval) were added to the MIPAS random error budget (as explained in detail in Section 3.3.3.1).

Time series of O_3 partial columns at the five ground-based stations are displayed in Figure 3.13. For each station, the upper panel in the plot shows the results of FTIR measurements and of collocated MIPAS data. In the lower panel, the mean relative differences between MIPAS and FTIR partial columns are plotted. In Table 3.3 the mean and the standard deviation of these relative differences for each station are given. The estimated random error on the relative difference of O_3 partial columns, combining the ground-based FTIR and MIPAS error budgets, is around 6% for all the stations except Arrival Heights (7%). The agreement is good for Kiruna, Jungfraujoch, and Wollongong, where there is no statistically significant bias, as can be seen in Table 3.3 by comparing the mean of the differences to the 3σ standard error on the mean (SEM) ($\mathrm{SEM} = 3.5D/N^{1/2}$). A small negative bias of MIPAS O_3 partial column is observed in the comparison with Lauder and Arrival Heights data, which is presently not explained by known contributions to the systematic error budget of the comparison. It must be noticed, however, that a spectral microwindow region at 2,100 cm^{-1} was selected for ozone retrieval at Lauder and Arrival Heights and that a high bias in ozone total column (on average, 4.5%) was observed when comparing these results with those obtained from the analysis of Arrival Heights spectra in retrievals employing microwindows in the 1,000 cm^{-1} region. Differences of up to 4% have been observed in retrievals of total column O_3 when employing different microwindow spectral regions (Rinsland *et al.*, 1996). This suggests that different choices of spectral microwindows might explain the different biases observed at different stations.

For all the stations, except Arrival Heights, the standard deviations are within 6%, which is comparable to the estimated random error on the difference. For Arrival Heights, the standard deviation (8.1%) is larger than the estimated random error of 7.1%. This is not surprising considering the potential vorticity differences between the observed MIPAS and ground-based airmasses that can

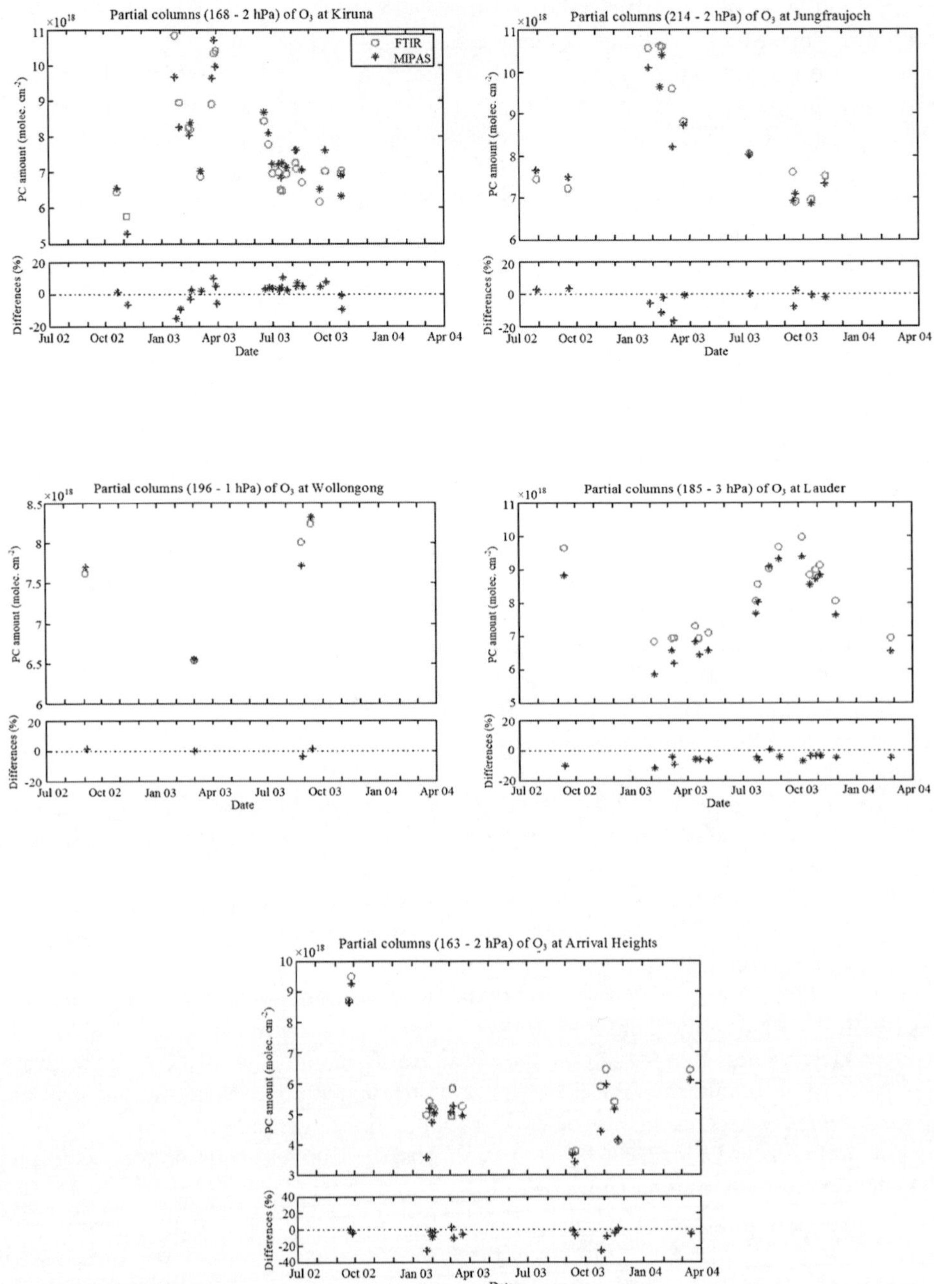

Figure 3.13. Time series of ozone partial columns. (Upper panel) Ground-based FTIR (circles) and MIPAS v4.61 (stars) O_3 partial columns for collocated measurements at the five stations. (Lower panel) Relative differences between MIPAS and ground-based FTIR O_3 partial columns (Cortesi *et al.*, 2007).

occur at the pole during the spring. The stronger atmospheric gradient at the poles during spring not only affect the error due to the collocation of airmasses, it also increases the horizontal smoothing error (as already seen in Section 3.3.1). For comparison with Kiruna measurements, a potential vorticity (PV) criterion has been applied, so that critical coincidences with relative differences in potential vorticity larger than 15% have been neglected. For Arrival Heights, tests performed by applying the same criterion resulted in a reduction of the standard deviation, but showed no influence on the bias.

Results of the comparison between O_3 vertical profiles retrieved from collocated measurements of MIPAS and each of the five ground-based FTIR stations are displayed in Figure 3.14. The individual plots show the mean and 1σ standard deviation of the relative differences (MIPAS − FTIR) in the ozone volume mixing ratio versus pressure. The combined random error associated with the O_3 mean difference is represented by the shaded gray area. The 3σ standard error on the mean is also reported to facilitate discussion of the statistical significance of the observed bias. The black solid lines in each plot mark the pressure levels adopted as the lower and upper limits for the calculations of ozone partial columns. We notice in Figure 3.14 that, except for Kiruna, the profile differences are oscillating. First, one should remember that the retrieval of vertical profiles from ground-based FTIR solar absorption spectra is an ill-posed problem. Therefore, the inversion needs to be constrained by some *a priori* information and the inversion results depend on this information and on some additional retrieval parameters (as mentioned in Section 3.3.3.1). The number of degrees of freedom for the signal of retrieved profiles between 12 and 40 km is only about 3.5. In the present exercise a common retrieval strategy for the five stations has not been defined. Only for Lauder and Arrival Heights have similar retrieval parameters been used. This probably explains the fact that we observe similar oscillations in the difference profiles at the Lauder and Arrival Heights stations.

The bias is below 10% at Kiruna in the whole altitude range and is usually not significant when taking into account the 3σ standard error on the mean. The bias is below 10% for Jungfraujoch and is 15% for Lauder and Wollongong at pressures lower than 80 hPa. The bias is below 25% at Arrival Heights in the whole altitude range. The error can be statistically significant at some pressure levels but, as previously pointed out, the FTIR profiles have to be interpreted with care considering their small degrees of freedom. Regarding the standard deviations, in Figure 3.14 we can see that they are roughly in agreement with the combined random error in the middle stratosphere, whereas they are greater than the random error in the lower stratosphere, especially at Arrival Heights where the variability of O_3 is expected to be larger.

3.3.3.4 *Comparison of MIPAS data with ozone soundings at individual mid-latitude stations*

A statistical analysis of the differences between coincident O_3 profiles obtained by MIPAS and by mid-latitude ozonesondes was conducted using the methodology

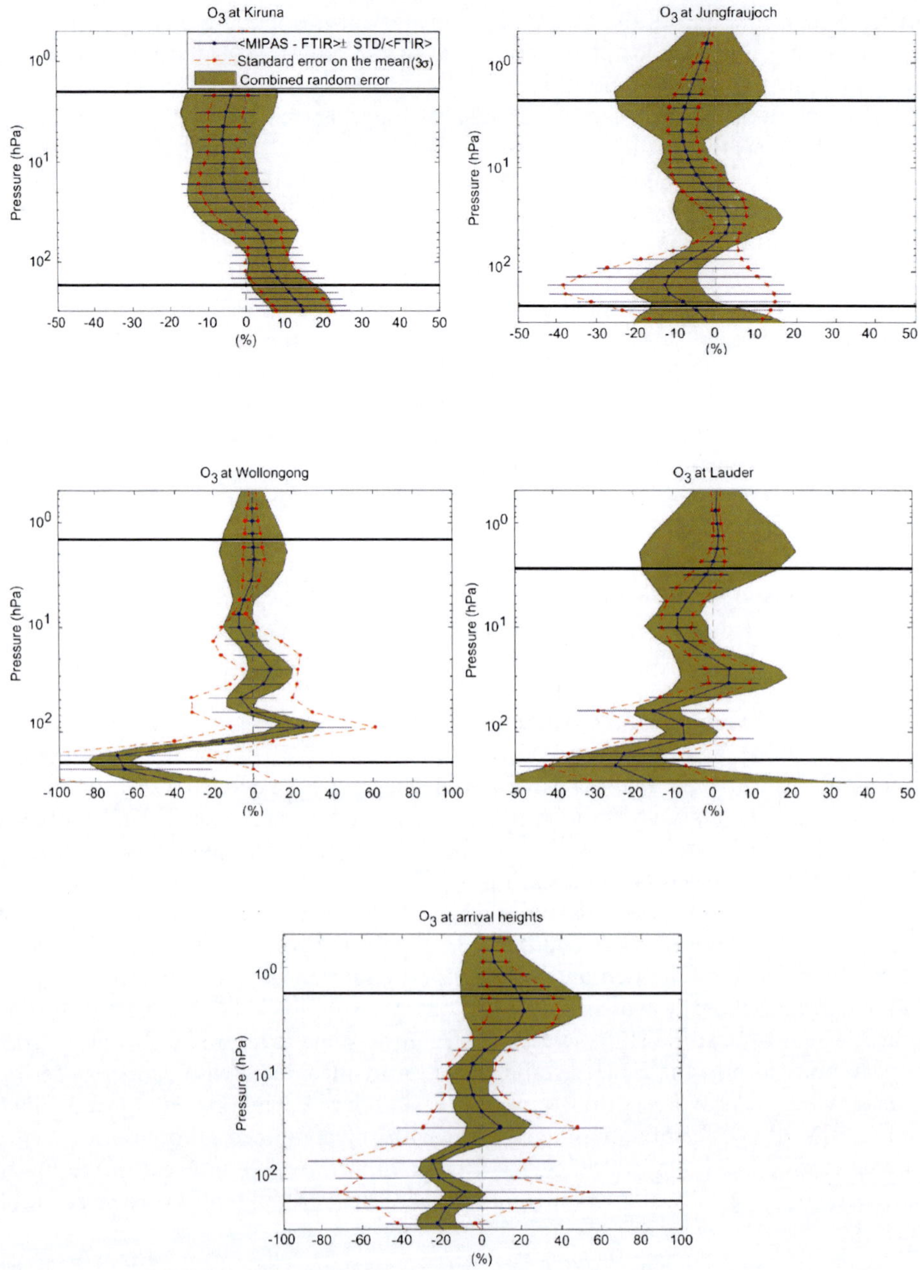

Figure 3.14. Statistical means (blue line) and standard deviations (error bars) of the relative differences between MIPAS and FTIR O_3 profiles (MIPAS – FTIR)/mean(FTIR), with red dots indicating the 3σ standard error on the mean; the shaded areas correspond to the estimated random error on the relative differences. The two black horizontal bars show the pressure ranges used for the partial columns of Table 3.4 (Cortesi *et al.*, 2007).

suggested by von Clarmann (2006) for bias and precision determination with matching pairs of measurements. The correlative data considered here consisted of ozone soundings from four sites that were not included as part of the NDACC datasets selected in Section 3.3.3.1 and that were provided by:

— the team of the University of L'Aquila who contributed to MIPAS validation activity by operating a Väisälä balloon sounding system from L'Aquila, Italy, (42.38°N, 13.31°E), with ECC ozonesondes having a precision of 4–12% in the troposphere and 3–4% between 100 and 10 hPa. The various sources of systematic errors are also altitude dependent and are between ±12% (Komhyr *et al.*, 1995);
— the team of the University of Athens who performed measurements of the O_3 vertical profiles for the location of Athens, Greece, (37.60°N, 23.40°E), by using electrochemical concentration cells (ECC, EN-SCI, Inc.), with corrections based on observations of the total ozone content made with the Dobson spectrophotometer D118 installed at the campus of Athens University;
— the team from Environment Canada and the University of Toronto who obtained O_3 profiles in coincidence with MIPAS overpasses from ozonesonde launches in Vanscoy, Canada (52.02°N, 107.05°W) during the MANTRA (Middle Atmosphere Nitrogen TRend Assessment) balloon campaign in 2002;
— the team of the Institute of Atmospheric Physics of the Chinese Academy of Science who provided results of the ozone soundings from Beijing, China (39.48°N, 116.28°E) in the period 2002–2004.

Coincident pairs of MIPAS and ozonesonde profiles were selected by applying the baseline criteria of 300 km and 3 h for maximum spatiotemporal separation. The comparison was then carried out according to the procedure employed by Ridolfi *et al.* (2007) to validate MIPAS temperature data against radiosonde measurements from L'Aquila and Potenza. Here below we briefly summarize the basic steps of this approach, while referring to the above-mentioned papers for a precise definition of the terminology and validation strategy (von Clarmann, 2006) and for a more detailed explanation of the individual steps of the comparison and of the underlying approximations (Ridolfi *et al.*, 2007).

First of all, the effects of MIPAS vertical smoothing on the comparison were taken into account. Correlative ozone data on the same pressure grid of the MIPAS matching profile were obtained by considering together the original high vertical resolution measurement of the ozonesonde $\mathbf{x}_{\text{ref,hires}}$, with the MIPAS averaging kernels, and the *a priori* profile:

$$\mathbf{x}_{\text{ref,smoothed}} = \mathbf{x}_0 + \mathbf{A}(\mathbf{x}_{\text{ref,hires}} - \mathbf{x}_0) \tag{3.9}$$

where $\mathbf{x}_{\text{ref,smoothed}}$ is the smoothed ozonesonde profile; $\mathbf{A}$ is the MIPAS averaging kernel matrix; and $\mathbf{x}_0$ is the *a priori* profile that was used as the linearization point for calculation of the averaging kernels. Both $\mathbf{A}$ and $\mathbf{x}_0$ in equation (3.9) were represented over the vertical grid of the matching MIPAS profile by using the shrinking/stretching and interpolation methods described by Raspollini *et al.* (2006). $\mathbf{x}_0$ is the ozone vertical distribution retrieved from MIPAS measurements when the true state of the atmosphere is equal to the *a priori* profile ($\mathbf{x}_{\text{ref,hires}} = \mathbf{x}_0$).

In order to correct for the spatiotemporal mismatch between MIPAS and the ozonesonde measurement of each comparison pair, equation (15) of von Clarmann (2006) was followed using assimilated ozone fields from the ECMWF:

$$\mathbf{x}_{\text{ref}} = \mathbf{x}_{\text{ref,smoothed}} + \mathbf{x}_{\text{mipas}}^{\text{ecmwf}} - \mathbf{x}_{\text{ref}}^{\text{ecmwf}} \tag{3.10}$$

where $\mathbf{x}_{\text{ref}}^{\text{ecmwf}}$ is the ECMWF ozone field interpolated at the location and time of the ozone sounding, whilst the term $\mathbf{X}_{\text{mipas}}^{\text{ecmwf}}$ is the ECMWF field at the location and time of the MIPAS scan (see below).

In order to include the effects of MIPAS horizontal smoothing, the following expression was used for the calculation of $\mathbf{X}_{\text{mipas}}^{\text{ecmwf}}$:

$$\mathbf{x}_{\text{mipas}}^{\text{ecmwf}} = \text{diag}|\mathbf{A}\mathbf{X}_{\text{mipas}}^{\text{ecmwf}}| \tag{3.11}$$

where $\mathbf{X}_{\text{mipas}}^{\text{ecmwf}}$ is a matrix whose columns represent ECMWF ozone values interpolated at the time of each MIPAS scan and at the points along the MIPAS line of sight that we used to calculate $\mathbf{A}$. A detailed description of the procedure adopted for the calculation of $\mathbf{X}_{\text{mipas}}^{\text{ecmwf}}$ is given by Ridolfi *et al.* (2007).

MIPAS O_3 measurements and ozonesonde-corrected values from the selected pairs of coincident profiles were binned in pressure according to the vertical grid defined by MIPAS nominal retrieval levels, so that no more than a single entry per profile could be associated with each pressure bin. This made it possible to discard vertical correlations between values of the individual profiles and to perform statistical analysis over the binned pairs, in line with the hypothesis that horizontal correlation between measurements are negligible after debiasing (as suggested in Section 8 of the paper by von Clarmann, 2006).

The bias b_i at the ith pressure bin was computed from the expression:

$$b_i = \frac{1}{n_i}\sum_{k=1}^{n_i}[x_{\text{mipas},i}(k) - x_{\text{ref},i}(k)] \tag{3.12}$$

with the associated standard deviation given by:

$$\sigma_{b_i} = \sqrt{\frac{\sum_{k=1}^{n_i}[x_{\text{mipas},i}(k) - x_{\text{ref},i}(k) - b_i]^2}{n_i(n_i - 1)}} \tag{3.13}$$

where the sums extend over the n_i comparison pairs that provide a valid entry for the ith bin. The validation of our current estimate of MIPAS systematic error $\sigma_{\text{mipas,sys}}$ obtained from the *a priori* values provided by Oxford University requires that the bias b_i is equal to zero within its total uncertainty $\sigma_{b_i,\text{tot}}$ expressed by:

$$\sigma_{b_i,\text{tot}} = \sqrt{\sigma_{b_i}^2 + \sigma_{b_i,\text{sys}}^2} \tag{3.14}$$

where $\sigma_{b_i,\text{sys}}$ is the systematic error in the bias that we evaluated from the root sum square of $\sigma_{\text{mipas,sys}}$ and of the ozonesonde systematic error $\sigma_{\text{ref,sys}}$ (associated with the corrected value $\mathbf{x}_{\text{ref}}$ and calculated from the estimated bias of the ozonesonde):

$$\sigma_{b_i,\text{sys}} = \sqrt{\sigma^2_{\text{mipas},i,\text{sys}} + \sigma^2_{\text{ref},i,\text{sys}}} \tag{3.15}$$

The precision p_i of the result of the comparison at each pressure bin:

$$p_i = \sigma_{b_i}\sqrt{n_i} \tag{3.16}$$

was calculated and compared with the random error of the difference $d_i(k) = x_{\text{mipas},i}(k) - x_{\text{ref},i}(k)$ given by:

$$\sigma_{d_i,\text{rnd}} = \sqrt{\sigma^2_{\text{mipas},i,\text{rnd}} + \sigma^2_{\text{ref},i,\text{rnd}}} \tag{3.17}$$

where $\sigma_{\text{mipas},i,\text{rnd}}$ and $\sigma_{\text{ref},i,\text{rnd}}$ are the random errors of MIPAS and of the ozonesonde, respectively. In order to validate MIPAS random error, it is necessary to verify that the precision p_i is consistent with the random error of the comparison $\sigma_{d_i,\text{rnd}}$.

The statistical analysis described above was applied to a validation dataset consisting of 22 matching pairs of MIPAS and ozonesonde profiles. The results obtained from the application of equations (3.9) to (3.17) are presented in Table 3.4, where we report for each altitude bin the bias b_i and its standard deviation σ_{b_i}, the systematic error $\sigma_{b_i,\text{sys}}$ on the bias, the precision p_i, and the random error $\sigma_{d_i,\text{rnd}}$ on the difference d_i.

The quantifiers $\chi^2_{R,i}$ and L_i in the last two columns of Table 3.4 characterize the significance levels of these results. The reduced chi-square $\chi^2_{R,i}$, with expectation value equal to 1.0, is defined by:

$$\chi^2_{R,i} = \frac{1}{(n_i - 1)} \sum_{k=1}^{n_i} \frac{[x_{\text{mipas},i}(k) - x_{\text{ref},i}(k) - b_i]^2}{\sigma^2_{d_i,\text{rnd}}} \tag{3.18}$$

and tests the consistency of the differences $d_i(k)$ with their expectation value b_i within their random error $\sigma_{d_i,\text{rnd}}$. L_i is the probability that a new comparison might yield a smaller value of the reduced chi-square $\chi^2_{R,i}$.

In the left panel of Figure 3.15, the vertical profile of the bias b_i is shown as a function of the approximate center altitude of each pressure bin (solid line), with error bars corresponding to the 95% confidence interval derived from the t-statistics for each altitude bin (see Ridolfi *et al.*, 2007, and references therein). For comparison, the curves $\pm\sigma_{b_i,\text{sys}}$ of the systematic error of the bias (dashed lines) are overplotted (Figure 3.15). A statistically significant bias (i.e., a bias that is different from zero beyond the 95% confidence interval defined above) is found for most of the altitude bins. This bias is, however, consistently lower than the combined systematic error of the comparison, as expected to validate the current estimate of MIPAS systematic uncertainties.

In the right panel of Figure 3.15, the precision p_i (solid line) is compared with the random error $\sigma_{d_i,\text{rnd}}$ on the difference d_i (dashed line); here, the error bars represent the 95% confidence interval computed from the chi-square statistics of each altitude

Table 3.4. Results of the comparison of MIPAS data with mid-latitude ozone soundings (Cortesi *et al.*, 2007).

Altitude range (km)	n_i	b_i (ppmv)	σ_{b_i} (ppmv)	$\sigma_{b_i,\text{sys}}$ (ppmv)	p_i (ppmv)	$\sigma_{d_i,\text{rnd}}$ (ppmv)	$\chi^2_{R,i}$ (ppmv)	L_i
0.0–7.5	5	−0.04	0.01	0.17	0.03	0.02	1.69	0.890
7.5–10.5	9	−0.01	0.02	0.20	0.06	0.03	2.77	0.973
10.5–13.5	19	−0.07	0.03	0.16	0.13	0.06	7.97	0.987
13.5–15.5	21	−0.09	0.04	0.20	0.18	0.13	5.01	1.000
16.5–19.5	19	−0.05	0.05	0.15	0.24	0.21	1.71	1.000
19.5–22.5	19	0.15	0.11	0.22	0.49	0.23	3.70	1.000
22.5–25.5	18	0.15	0.08	0.33	0.33	0.35	0.84	1.000
25.5–28.5	17	−0.37	0.12	0.41	0.49	0.47	1.13	1.000
28.5–31.5	15	−0.42	0.17	0.55	0.67	0.63	1.35	1.000
31.5–34.5	9	−0.18	0.34	0.70	1.04	0.73	1.91	0.993
34.5–37.5	6	−0.01	0.43	0.78	1.07	0.82	1.75	0.941
37.5–40.5	4	−0.53	0.50	0.74	0.99	0.70	1.98	0.793

bin. We observe a reasonable agreement between the two curves over the whole range of the comparison, with significant discrepancies found for the altitude bins at 21, 15, and 12 km, where in any case the precision value never exceeds the combined random error by a factor larger than 2.

3.4 COMPARISON OF MIPAS DATA WITH STRATOSPHERIC BALLOON AND AIRCRAFT MEASUREMENTS

3.4.1 MIPAS-B2

A balloonborne version of the MIPAS-ENVISAT instrument, MIPAS-B2, operated by a team of Forschungszentrum Karlsruhe (IMK-FZK), was flown during mid-latitude (Aire sur l'Adour, France, September 24, 2002) and Arctic (Kiruna, Sweden) validation flights between March 20–21, 2003 and July 3, 2003 and obtained a set of correlative data in very good spatiotemporal coincidence with satellite measurements (Oelhaf *et al.*, 2003). The high quality of the collocations, combined with several features of the MIPAS-B2 instrument configuration that closely match those of MIPAS-ENVISAT (spectral coverage, spectral resolution, sensitivity, and radio-

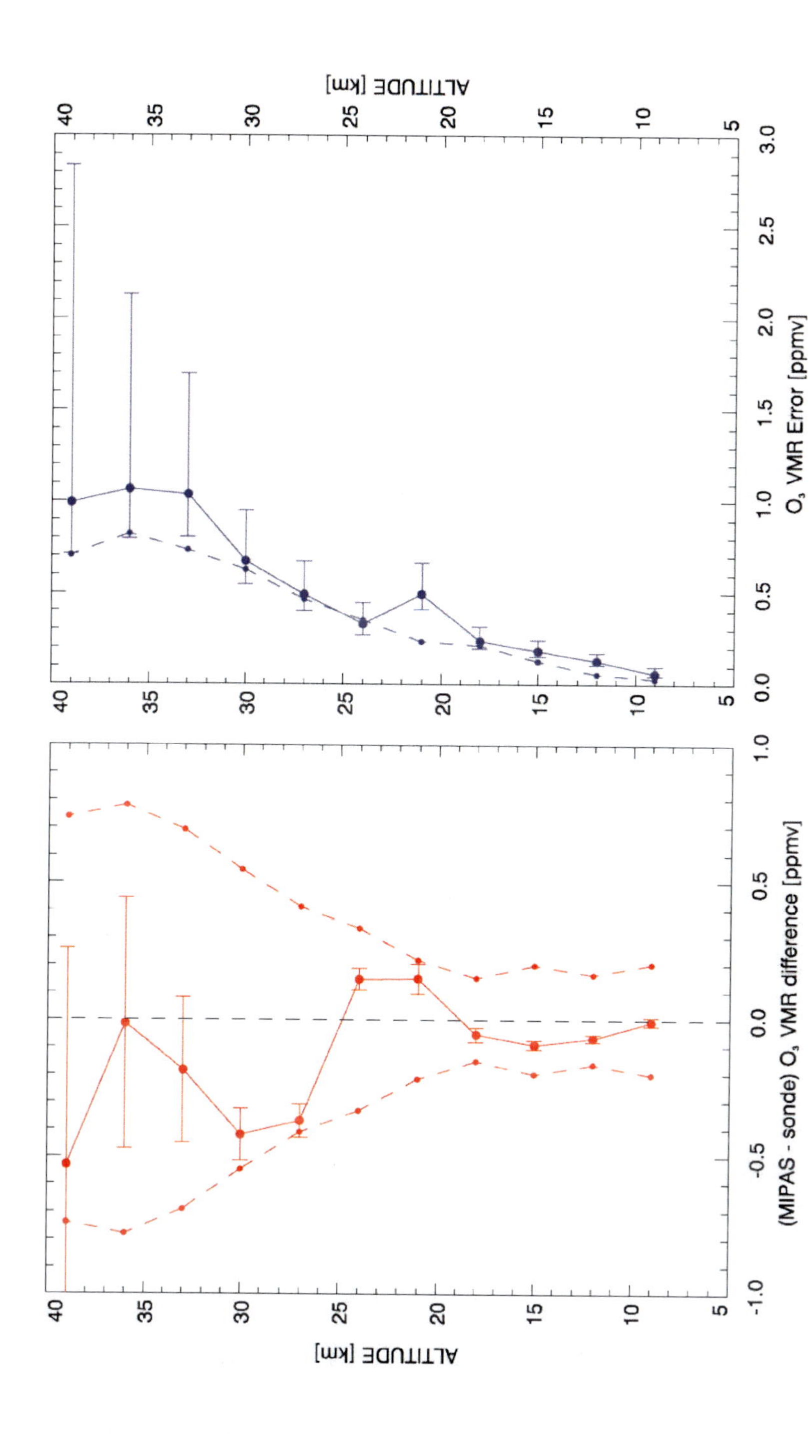

Figure 3.15. Results of statistical analysis for MIPAS O_3 bias and precision determination by comparison with matching measurements from mid-latitude ozone soundings (see text in Section 3.3.3.1 for explanations) (Cortesi *et al.*, 2007).

metric accuracy, etc.), offered a unique opportunity for validation of the vertical profiles of ozone and other MIPAS target species. A detailed description of the MIPAS-B2 spectrometer is given by Friedl-Vallon *et al.* (2004). The limb-sounding observations acquired during the ENVISAT validation flights were processed using a least-squares-fitting algorithm based on the forward model KOPRA (Karlsruhe Optimized and Precise Radiative transfer Algorithm) together with a Tikhonov–Phillips regularization procedure (Hopfner *et al.*, 2002). A total of 34 ozone micro-windows were chosen in the mid-infrared spectral region to infer vertical ozone profiles from the measured spectra. The resulting vertical resolution of the profiles lies typically between 2 and 3 km and is therefore comparable to MIPAS. Error estimation includes random noise, temperature errors, line-of-sight inaccuracies, and spectroscopic data errors. A detailed description of level-2 MIPAS-B2 data analysis is given by Wetzel *et al.* (2006) and references therein. Table 3.5 provides an overview of the coincidences used in this for the comparison of MIPAS-B2 and MIPAS-ENVISAT ozone measurements. For MIPAS-B2 flights 11 and 13, a close-to-perfect coincidence with MIPAS-ENVISAT could be reached in time and space. For flight 14, this is true only for the coincidence in space while the time difference amounts to several hours. However, both observations were carried out in the same airmass. We used MIPAS-ENVISAT ozone data version 4.61 exclusively for our comparison.

In Figure 3.16 we present the results of the comparison between all the available pairs of O_3 matching profiles listed in Table 3.5. Each panel shows on the left side the MIPAS-ENVISAT and MIPAS-B2 ozone VMR profiles, retrieved from the coincident limb-scanning sequences and on the right their absolute difference with over-plotted combined random and total errors. The MIPAS-B2 measurements have been cross-checked with ozonesondes launched shortly after the launch of the

Table 3.5. Summary of the coincidences between MIPAS-ENVISAT and MIPAS-B2. The spatiotemporal separation between MIPAS-ENVISAT and MIPAS-B2 scans for each of the available comparison pairs are given in the last two columns. The distance between the coincident scans is calculated at 20 km (Cortesi *et al.*, 2007).

Location	*Date*	*MIPAS-ENVISAT* (orbit, scan)	*MIPAS-B2* (flight, scan)	*Distance* (km)	*Time difference* (min)
Aire sur l'Adour	Sep. 24, 2002	2,975, scan 14	F11, scan S	207	14
Aire sur l'Adour	Sep. 24, 2002	2,975, scan 15	F11, scan S	358	15
Aire sur l'Adour	Sep. 24, 2002	2,975, scan 16	F11, scan N3	79	14
Kiruna	March 20, 2003	5,508, scan 20	F13, scan N3a	78	14
Kiruna	March 21, 2003	5,515, scan 30	F13, scan D15c	28	20
Kiruna	July 3, 2003	7,004, scan 6	F14, scan 3	2	506

MIPAS-B2 instrument. These comparisons have shown a general good agreement between MIPAS-B2 and the ozonesondes (see, e.g., Wetzel *et al.*, 2006). In general, an excellent agreement is obtained both for mid-latitude and for high-latitude measurements over the whole range of vertical overlap, with significant discrepancies occasionally observed at the lowest levels (below ~100 hPa) or in proximity of the peak of the O_3 vertical distribution (above ~10 hPa, where MIPAS-ENVISAT overestimates the ozone content). The absolute difference between MIPAS-ENVISAT and MIPAS-B2 ozone values is mostly within the combined total error, often remaining below its random component. An overall statistical analysis of the comparison, showing the mean profiles of the O_3 absolute difference and corresponding total, random, and systematic errors is displayed in Figure 3.17. Average values have been calculated over all the pairs of coincident profiles: the mean absolute difference is shown (solid red line), along with the standard error of the mean (error bars). A bias between MIPAS-ENVISAT and MIPAS-B2 ozone values, which is marginally higher than the combined systematic errors, is only observed at some pressure level below 100 hPa. Moreover, the standard deviation never exceeds the combined random error value, except for a few levels above 10 hPa.

3.4.2 FIRS-2 and IBEX

Two balloonborne high-resolution Fourier Transform Far-Infrared (FT-FIR) spectrometers were deployed in field campaigns for the validation of the ENVISAT chemistry payload: the Far InfraRed Spectrometer (FIRS-2) of the Harvard–Smithsonian Center for Astrophysics, Cambridge, MA (Johnson *et al.*, 1995) and the Infrared Balloon Experiment (IBEX) operated by the Institute for Applied Physics "Nello Carrara" (IFAC-CNR), Firenze, Italy (Bianchini *et al.*, 2006).The FIRS-2 and IBEX instruments are capable of retrieving the vertical distributions of a number of trace gases from float altitude (approximately 35–40 km) down to the tropopause, with vertical resolutions of ~2–3 km, from limb-sounding observations of the atmospheric emission spectrum. FIRS-2 measurements cover the spectral region of 80 to 1,220 cm^{-1}, while IBEX operates under photon noise-limited conditions and acquires spectra in narrow bands (typically 2 cm^{-1} wide) within the interval 10–250 cm^{-1}. FIRS-2 observations of O_3 concentrations use transitions both in the rotational band between 80 and 130 cm^{-1} and the v_2 band between 730 and 800 cm^{-1}. The former lend the most weight above 25 km, while the latter contributes almost entirely below 20 km. In addition, a comparison has been made between MIPAS O_3 data v4.61 and the ozone profiles retrieved from FIRS-2 measurements during flights from the National Scientific Balloon Facility balloon launch site at Fort Sumner, NM (34°N, 104°W) on October 20, 2002 and on July 19–20, 2003 and with those obtained by IBEX in the trans-Mediterranean flight from Trapani, Italy (38°N, 12°E) to Spain on July 29–30, 2002. In both cases, useful coincidences between MIPAS observations and measurements of the two FT-FIR spectrometers could be obtained only after substantial relaxation of spatiotemporal matching criteria (as shown, for instance, in previous analyses carried out for MIPAS O_3 validation; Cortesi *et al.*, 2004). No matching pair is available for comparison if

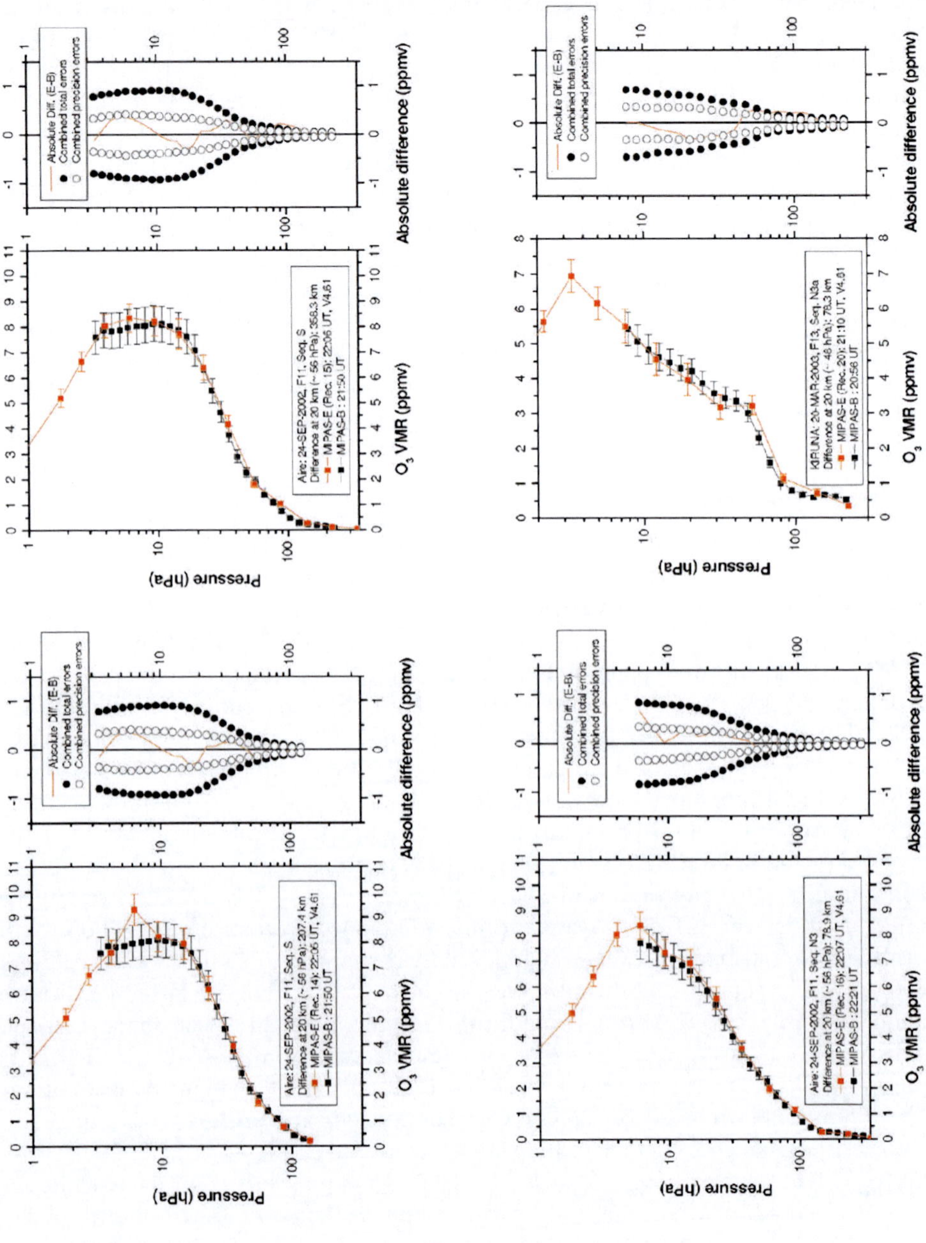
Pressure (hPa)
O_3 VMR (ppmv)
Absolute difference (ppmv)
Absolute Diff. (E-B)
Combined total errors
Combined precision errors
Aire: 24-SEP-2002, F11, Seq. S
MIPAS-E
MIPAS-B
KIRUNA: 20-MAR-2003, F13, Seq. N3a
Aire: 24-SEP-2002, F11, Seq. N3

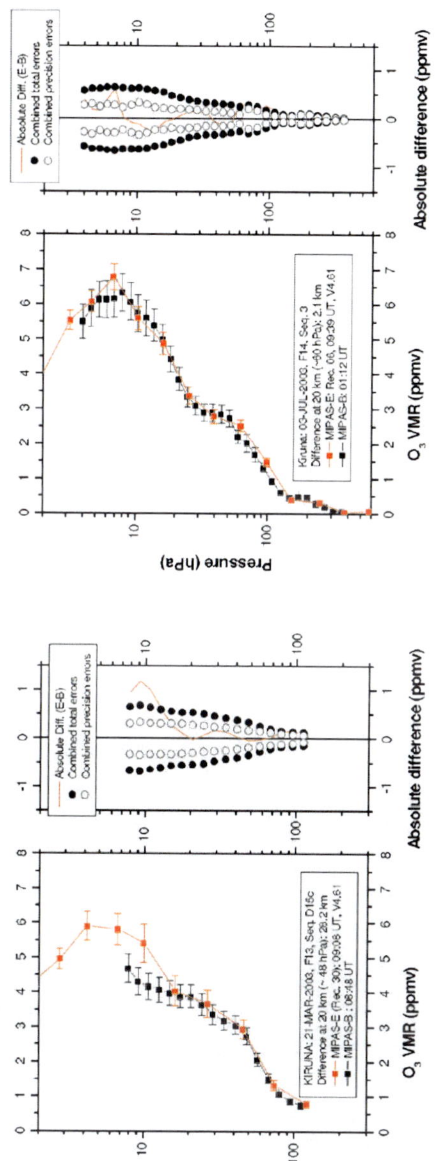

Figure 3.16. Results of the comparison between coincident MIPAS-ENVISAT and MIPAS-B2 ozone measurements at mid-latitudes and in the Arctic region (Cortesi *et al.*, 2007).

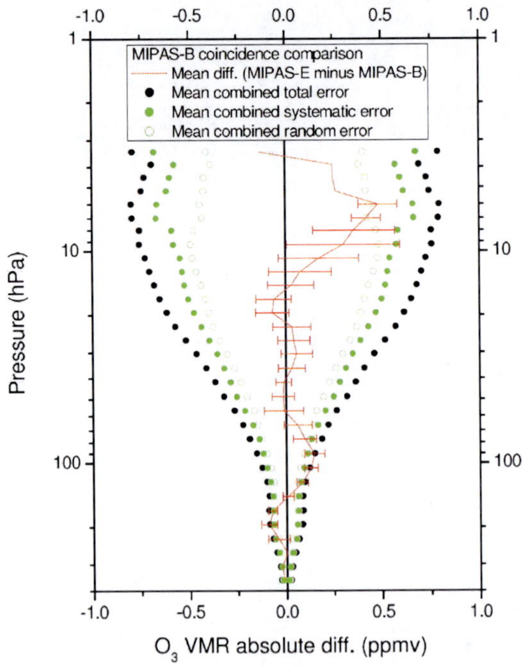

Figure 3.17. Absolute difference between MIPAS-ENVISAT and MIPAS-B2 ozone volume mixing ratios averaged over all the available collocations (Cortesi *et al.*, 2007).

one applies the baseline criteria for maximum spatiotemporal separation. As a consequence, it was decided to exploit the two sets of correlative balloon data, using a trajectory-hunting technique (THT) (Danilin *et al.*, 2002) that launches backward and forward trajectories from the locations of measurements and finds air parcels sampled at least twice within a prescribed match criterion during the course of several days (see also Section 6.6.3.1).

A similar procedure was applied for comparison of MIPAS ozone profiles with both FIRS-2 and IBEX measurements, relying on isentropic trajectories calculated using the University of L'Aquila Global Trajectory Model (Dragani *et al.*, 2002; Redaelli, 1997), based on the meteorological fields produced by ECMWF. Four days backward and forward isentropic trajectories, departing from the geolocations of FIRS-2 and IBEX retrieved profiles were calculated and MIPAS O_3 profiles at locations within 2 degrees in longitude, 2 degrees in latitude and 2 h in time along these trajectories were identified and vertically interpolated in potential temperature, to obtain the O_3 volume mixing ratio value to be compared with the corresponding FT-FIR measurements. The resulting comparison pairs were then binned by altitude, in steps of $\Delta h = 1.5$ km, and averaged, and 1σ RMS values of the differences (MIPAS FT-FIR data) in O_3 volume mixing ratios were calculated. Preliminary results of a so-called self-hunting analysis of MIPAS data that matches satellite observation with themselves—providing a test for the precision of the instrument's

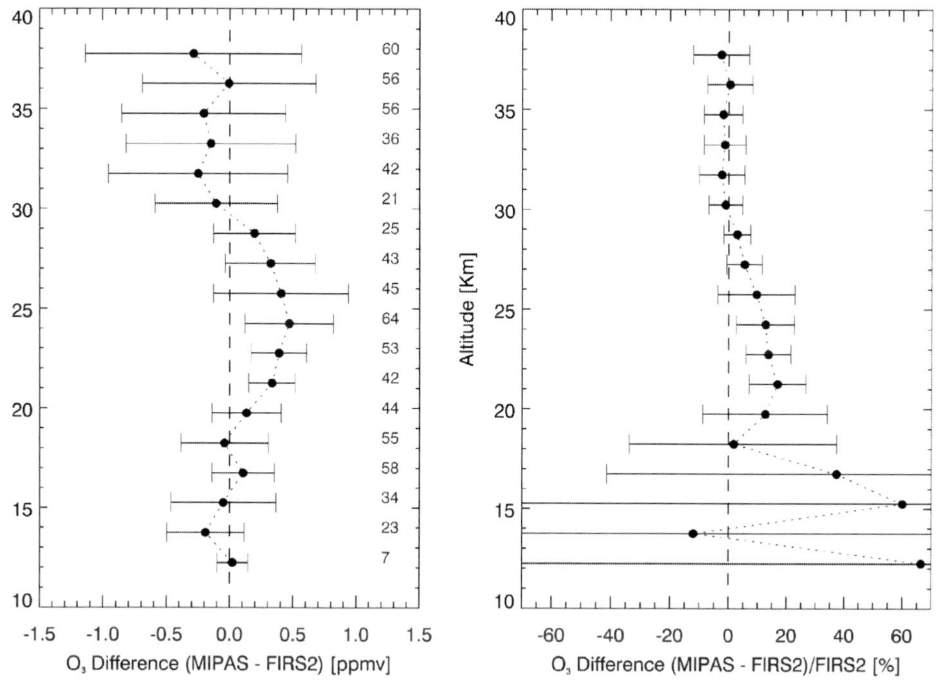

Figure 3.18. Comparison between MIPAS v4.61 and FIRS-2 (October 30, 2002 and September 19/20, 2003) ozone measurements. Mean absolute and relative differences between MIPAS and FIRS-2 O_3 VMR reconstructed using trajectory analysis and averaged in altitude bins of $\Delta h = 1.5$ km are shown in the left and right panel, respectively. Error bars represent 1σ standard deviations. The number of elements per altitude bin is also displayed (Cortesi *et al.*, 2007).

products and the quality of the calculated trajectories and thus assessing noise in the technique and providing estimates of its possible extension to multi-platform comparison for the selected time period—are given by Taddei *et al.* (2006).

Results of the comparison between MIPAS O_3 measurements and data from the FIRS-2 flights on October 20, 2002 and on July 19–20, 2003 are shown in Figure 3.18. Mean absolute and relative differences between MIPAS v4.61 and FIRS-2 O_3 VMR calculated with THT and binned by altitude values ($\Delta h = 1.5$ km) are displayed in the left and right panel, respectively; 1σ error bars and total number of reconstructed data in each bin are also indicated. Very good agreement within 1σ error bars, with relative differences within $\pm 10\%$, is found down to about 24 km. At lower levels the mean relative difference increases, mainly resulting from the small values of the ozone mixing ratio at these altitudes, although the absolute difference remains reasonably small.

Mean absolute and relative differences between MIPAS v4.61 and IBEX O_3 data obtained during the trans-Mediterranean flight of July 29–30, 2002 are presented in Figure 3.19. MIPAS measurements agree reasonably well with the balloon profile down to approximately 27 km (mean relative differences within $\pm 10\%$). At lower

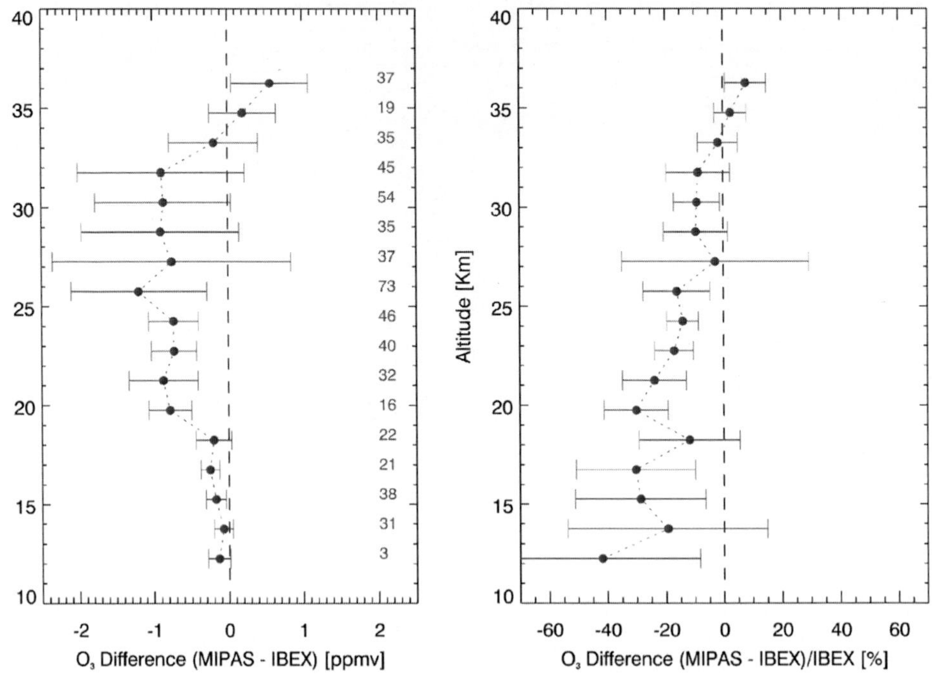

Figure 3.19. Comparison between MIPAS v4.61 and IBEX (July 29–30, 2002) ozone measurements. Mean absolute and relative differences between MIPAS and IBEX O_3 VMR are plotted as in Figure 3.18 (Cortesi *et al.*, 2007).

altitudes, MIPAS appears to underestimate the ozone content by up to 30–40% with respect to IBEX.

3.4.3 SPIRALE

SPIRALE (SPectroscopie InfraRouge par Absorption des Lasers Embarqués) is a balloonborne instrument operated by LPCE-CNRS (Laboratoire de Physique et Chimie de l'Environnement, Orléans, France) and employing the technique of tunable diode laser absorption spectroscopy to perform simultaneous *in situ* measurements of several minor atmospheric constituents (Moreau *et al.*, 2005). The instrument contributed to the ESABC program with a mid-latitude and a high-latitude flight, carried out, respectively, from Aire sur l'Adour on October 2, 2002 and from Kiruna on January 21, 2003 to measure O_3, CH_4, N_2O, CO, NO, NO_2, HNO_3, and HCl volume mixing ratio (VMR) profiles. MIPAS ozone data versions 4.61 and 4.62 have been compared with SPIRALE O_3 profiles obtained during the descent phases of the October 2002 flight and during the ascent phase of the January 2003 flight. For the Arctic flight, direct coincidences with two MIPAS scans (orbit 4,677, scan 20, v4.62 and orbit 4,678, scan 6, v4.61), whose temporal separation from the SPIRALE measurements satisfied the baseline matching criterion $\Delta t < 3$ h, were available. The location of this flight was close to the vortex edge and, although the spatial separation does not satisfy the baseline criterion $\Delta s < 300$ km (300–500 km for scan 20, 600–800 km for scan 6), MIPAS and SPIRALE measurements were made at locations close in potential vorticity (PV) (5 to 25% for scan 20, 5 to 35% for scan 6). Direct coincidences were not possible in the case of the mid-latitude flight. For the latter, comparison was carried out by means of trajectory analysis with MIPAS profiles from orbit 3,019, scans 14 and 15 (v4.61) on September 27 at 23:52:50 UT and 23:54:11 UT, respectively.

Estimations of the uncertainties of SPIRALE measurements have been previously described in detail (Moreau *et al.*, 2005). In brief, random errors mainly come from the signal-to-noise ratio and from fluctuations of the laser emission signal, which have more important effects at lower altitudes (6% below 18 km) than at higher altitudes (2%). Systematic errors originate from the laser line width (increasing from 1% at lower altitudes to 3% at higher altitudes) and the spectroscopic parameters which are well determined (5%) at the wave numbers used (2,081.7–2,082.5 cm^{-1}). Adding random errors and systematic errors quadratically results in total uncertainties of 6% at altitudes above 18 km ($p < 80$ hPa) and 8% below 18 km ($p > 80$ hPa). MIPAS systematic errors have been computed at Oxford University (Cortesi *et al.*, 2007). Polar winter nighttime conditions and day and night mid-latitude conditions have been used, respectively, for the Arctic case and the mid-latitude case.

In Figure 3.20, the O_3 profile obtained by SPIRALE during the Kiruna 2003 flight is compared with coincident MIPAS O_3 profiles from orbit 4,678, scan 6 and from orbit 4,677, scan 20. Both the SPIRALE original high vertical resolution profile and its smoothed version after the application of MIPAS averaging kernels are displayed. In general, good agreement is observed in both cases, with MIPAS O_3 data from orbit 4,677, scan 20 mostly matching SPIRALE smoothed values within the error bars (with the only notable exception of the level above 100 hPa, where

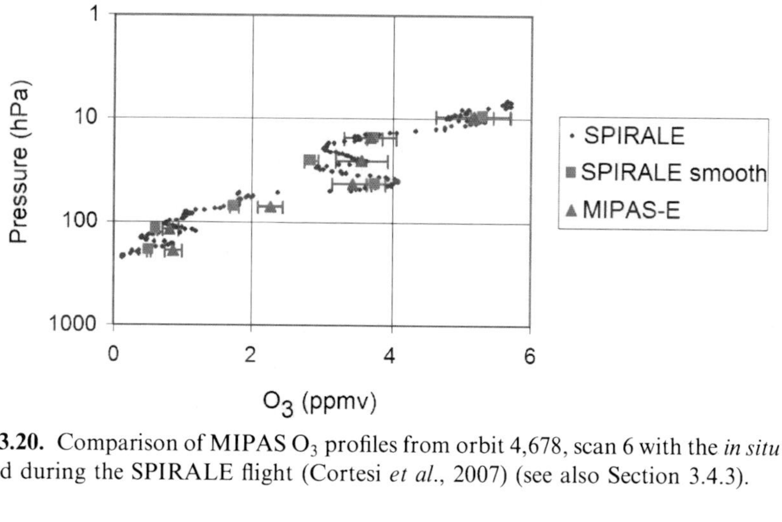

Figure 3.20. Comparison of MIPAS O_3 profiles from orbit 4,678, scan 6 with the *in situ* profiles acquired during the SPIRALE flight (Cortesi *et al.*, 2007) (see also Section 3.4.3).

MIPAS O_3 is closer to SPIRALE raw data). Slightly larger discrepancies are found in the comparison with MIPAS orbit 4,678, scan 6, possibly due to increased comparison errors introduced by the greater spatial separation (600–800 km, PV differences up to 35%).

The feasibility of using long trajectories for MIPAS validation by comparison with data of the SPIRALE flight on October 2, 2002 at Aire sur l'Adour was investigated by means of PV analysis of sets of trajectories ending close to each point of the SPIRALE profile. For each point of the SPIRALE profile (with potential temperature steps of $\Delta\theta = 25$ K), seven backward trajectories have been calculated:

— the trajectory ending at the point of the SPIRALE profile;
— four trajectories ending close to this point on the same isentropic surface ($\pm 0.5°$ in latitude and $\pm 0.5°$ in longitude);
— two trajectories ending ± 6.25 K (about 250 m) above and below the point of the SPIRALE profile.

For each trajectory, PV at 00:00 UT on 28 September was then computed, along with mean PV and standard deviation for each set of seven trajectories. Finally, the difference between the mean value and PV at the end of the trajectories (SPIRALE profile) was calculated as a function of potential temperature. It was found that between 400 K and 600 K and between 700 K and 900 K, the standard deviation is very low (<2–3%) and PV is conserved relatively well on 4.5-day trajectories (differences are less than 10%). This is not the case below 400 K, between 600 K and 700 K, and above 900 K. Airmass mixing probably occurs on these isentropic surfaces. SPIRALE data are therefore no longer representative of the measurements made by MIPAS on the same isentropic surface. Moreover, by comparing the PV values of SPIRALE and MIPAS profiles, we found that PV differences are lower than 10% between 400 K and 600 K for both profiles and above 700 K for profile 14. It was concluded, therefore, that SPIRALE data may be used to validate:

— MIPAS profile 14 of orbit 3,019 on potential surfaces between 400 K and 600 K and between 700 K and 900 K, which corresponds to the retrieval nominal MIPAS altitudes of 18, 21, 24, 30, and 33 km;
— MIPAS profile 15 of orbit 3,019 on potential surfaces between 400 K and 600 K, which corresponds to the nominal MIPAS altitudes of 18, 21, and 24 km.

The results of the comparison are shown in Figure 3.21, highlighting an almost perfect overlapping between MIPAS and SPIRALE O_3 measurements.

3.4.4 MIPAS-STR, SAFIRE-A, and FOZAN on board the M-55 *Geophysica* aircraft

Simultaneous measurements of the ozone vertical distribution in strict coincidence with MIPAS-ENVISAT overpasses were obtained by the *in situ* and remote-sensing instruments of the M-55 *Geophysica* high-altitude aircraft during dedicated

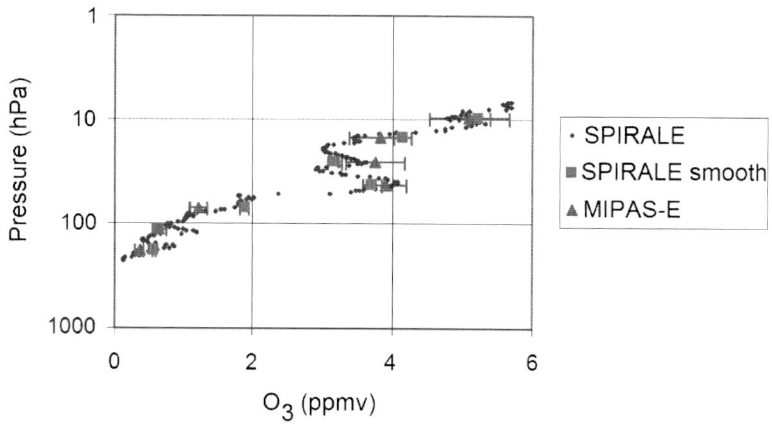

Figure 3.21. Comparison of MIPAS O_3 profiles from orbit 4,677, scan 20 and scan with the *in situ* profiles acquired during the SPIRALE flight (Cortesi *et al.*, 2007) (see also Section 3.4.3).

mid-latitude flights (Forlì, Italy, July and October 2002) and flights in the Arctic region (Kiruna, Sweden, February–March 2003), aimed at the validation of satellite chemistry sensors (as reported in detail by Cortesi *et al.*, 2004).

The remote-sensing payload embarked on board the M-55 stratospheric platform during these missions consisted of two FT spectrometers operating in limb-sounding geometry and capable of retrieving the ozone VMR profile from the upper troposphere up to the flight altitude and the total ozone column above: these spectrometers were the MIPAS-STR (MIPAS STRatospheric aircraft, FZK-IMK, Karlsruhe, Germany) and SAFIRE-A (Spectroscopy of the Atmosphere by using Far-InfraRed Emission–Airborne, IFAC-CNR, Firenze, Italy). MIPAS-STR is an aircraft version of the satellite spectrometer and operates in the middle-infrared spectral region with similar characteristics and performance (Piesch *et al.*, 1996). SAFIRE-A is a high-resolution FT instrument, performing limb emission measurements in narrow bands ($\Delta s \sim 1$–$2\,\mathrm{cm}^{-1}$) within the far-infrared spectral region (10–250 cm^{-1}) (as described by Bianchini *et al.*, 2004). Both instruments obtain ozone profiles with a vertical resolution (approximately 1–2 km) that is slightly better, but still comparable with the one of MIPAS-ENVISAT v4.61/v4.62 data and are, therefore, directly compared with satellite measurements without correcting for vertical smoothing effects.

The chemiluminescent ozonesonde FOZAN (Fast Ozone ANalyzer), jointly operated by ISAC-CNR (Bologna, Italy) and CAO (Central Aerological Observatory, Moscow, Russia) teams, provides *in situ* measurements of ozone concentration at the balloon height (Yushkov *et al.*, 1999) with a sampling rate of 1 Hz and precision and accuracy equal to 8% and 0.01 ppmv, respectively. High-resolution vertical profiles (typically, a vertical resolution of about 10 m is obtained during the ascent and descent phases of the flight) of O_3 are reconstructed from FOZAN measurements acquired during take-off and landing, as well as during occasional dives performed by the aircraft close to the geolocation of MIPAS-ENVISAT scans.

MIPAS averaging kernels are applied to FOZAN high-resolution O_3 data to obtain the smoothed profile to be compared with satellite retrieved values. The results reported are based on the use of both high-resolution and smoothed FOZAN data.

A total of 11 flights and about 45 flight hours were undertaken with the M-55 *Geophysica* aircraft for the validation of the ENVISAT chemistry payload within the framework of the 2002–2003 ESABC field campaigns. The results of these airborne measurements have been stored and are now accessible at the ENVISAT Cal/Val database of the Norwegian Institute for Air Research (NILU, *http://nadir.nilu.no/calval/*). Using these data, multiple coincidences can be identified based on the agreed matching criteria ($\Delta s < 300$ km, $\Delta t < 3$ h) between MIPAS-ENVISAT and remote-sensing and *in situ* aircraft observations, thus obtaining a comprehensive set of collocated O_3 profiles to be considered for validation purposes. Here, we have selected a sub-set of the above comparison pairs including only those flights for which at least two sensors of the M-55 *Geophysica* payload provided useful ozone measurements (for mutual data quality check) and choosing, for each MIPAS scan, the O_3 profiles measured with the best spatiotemporal coincidence by MIPAS-STR, SAFIRE-A, and FOZAN. The resulting validation dataset is shown in Table 3.6, illustrating the combinations of MIPAS-ENVISAT, MIPAS-STR, SAFIRE-A, and FOZAN profiles that have been used for the comparison. All the comparisons with correlative data provided by the M-55 *Geophysica* payload have been carried out using MIPAS-ENVISAT data v4.61. In the case of aircraft remote-sensing measurements, the O_3 vertical distribution retrieved from the individual MIPAS-ENVISAT scans was compared with the mean VMR profile of MIPAS-STR (or SAFIRE-A) obtained by averaging over all the limb-scanning sequences collocated with the selected satellite overpass.

SAFIRE-A mean profiles have been calculated over fixed pressure levels, corresponding approximately to a regular altitude grid with steps of 1.0 km. MIPAS-STR O_3 profiles have been retrieved on a fixed altitude grid. The VMRs of one altitude have been averaged to get the mean profile (Hopfner *et al.*, 2001; Keim *et al.*, 2004). The UTC time interval covered by SAFIRE-A and MIPAS-STR averages is indicated in Table 3.6.

In situ vertical profiles, measured by FOZAN during ascent or descent phases of the flight, are compared with collocated MIPAS-ENVISAT measurements and with the remote-sensing data recorded on board the aircraft when flying at a flight altitude between 17 and 20 km immediately before/after the M-55 ascent/descent. As previously stated, the comparison is made using both high vertical resolution *in situ* data and the smoothed profile obtained by convolution with MIPAS averaging kernels.

The comparisons cover the altitude range between ~25 km (slightly above the maximum flight altitude) and MIPAS-ENVISAT lowest tangent altitude. Aircraft measurements conducted in the polar region were aimed at validating MIPAS-ENVISAT products in the presence of strong vertical and horizontal gradients. Consequently, the corresponding dataset (February–March 2003 data) generally includes data acquired at the border of the polar vortex, with vertical and horizontal inhomogeneities much larger than those encountered at mid-latitudes (July and

Table 3.6. Best spatiotemporal coincidences selected for MIPAS-ENVISAT ozone validation with M-55 *Geophysica* aircraft measurements. For each MIPAS scan, we report the interval of the scans used to calculate the MIPAS-STR and SAFIRE-A collocated mean profiles and the corresponding UTC time window, along with the flight segment of the closest FOZAN profile. N.A. indicates cases for which either no data or no coincidence within 300 km and 3 h are available (Cortesi *et al.*, 2007).

Date	*MIPAS-ENVISAT*		*MIPAS-STR*	*SAFIRE-A*	*FOZAN*
	Orbit	*Scan* (UT)	*Scan interval* (UT)	*Scan interval* (UT)	*Flight segment* (UT)
July 22, 2002	2,051	12 (09:19)	11-22 (07:28–07:59)	N.A.	Landing (09:30–10:12)
		13 (09:20)	39–47 (09:06–09:27)	N.A.	Take-off (06:01–06:28)
October 24, 2002	3,403	14 (21:22)	N.A.	5–9 (19:36–20:26)	Dive (19:50–20:28)
		15 (21:23)	N.A.	10–15 (20:33–21:22)	Take-off (18:41–19:06) Landing (21:58–22:36)
March 2, 2003	5,250	19 (20:34)	21–26 (20:28–20:40)	13–14 (20:36–20:45)	Dive (19:43–20:26)
		20 (20:36)	45–50 (21:18–21:30)	14–19 (20:45–21:25)	Take-off (18:40–19:10)
		21 (20:37)	64–65 (22:06–22:09)	20–23 (21:34–21:53)	Landing (22:24–23:05)
March 12, 2003	5,386	27 (08:46)	42–47 (07:47–08:01)	15–16 (09:21–09:30)	N.A.
		28 (08:47)	31–36 (08:45–08:58)	9–14 (08:32–09:13)	N.A.
		29 (08:49)	6–11 (09:16–09:30)	2–8 (07:36–08:24)	Take-off (07:13–07:44) Landing (11:00–11:47)
	5,387	21 (10:27)	42–47 (07:47–08:01)	15–16 (09:21–09:30)	N.A.
		22 (10:28)	54–59 (09:53–10:06)	17–20 (09:39–10:02)	N.A.

October 2002 data). To avoid strong gradients along the line of sight of the remote-sensing instruments, which decrease the quality of the measured profiles, the flights were planned with long north–south legs. The aircraft measurements were performed in the west–east direction, while the MIPAS-ENVISAT measured north–south along the gradients. Very high-quality coincidences, both in the spatial and in the temporal domain, characterize the correlative dataset available from the M-55 *Geophysica* campaigns, particularly for remote-sensing measurements, considering that the time difference between MIPAS-STR/SAFIRE-A and MIPAS-ENVISAT is on average less than 1 h (see Table 3.6).

Results of the comparison between MIPAS-ENVISAT ozone profiles and the M-55 correlative measurements obtained during Northern Hemisphere mid-latitude flights (Forlì, Italy, July 22, 2002 and October 24, 2002) and during the Arctic campaign (Kiruna, Sweden, March 2, 2003 and March 12, 2003) are shown in Figure 3.22 and Figure 3.23, respectively. Each plot displays the ozone vertical distribution retrieved by MIPAS-ENVISAT for one of the selected overpasses and collocated O_3 profiles measured by the remote-sensing and *in situ* sensors of the aircraft. Ozone VMR values are plotted versus pressure in a range corresponding roughly to the 6–25 km interval, as indicated by the approximate altitude scale reported on the right axis of the plots. The error bars on the MIPAS-ENVISAT, MIPAS-STR, and SAFIRE-A profiles indicate the total uncertainty regarding the corresponding ozone values.

Very good agreement is found at mid-latitudes, with aircraft O_3 measurements and satellite data generally matching within their total error bars (with the only exception being the MIPAS-ENVISAT orbit 2,051/scan 12, which overestimates the O_3 VMR below 100 hPa compared to MIPAS-STR, but still matches, however, the *in situ* measurements acquired by FOZAN during landing). Reasonably good results are also found, on the other hand, from comparison of the ozone profiles from Arctic flights, despite the larger atmospheric inhomogeneities that characterize the measurement scenario at higher latitudes. The occurrence of strong vertical gradients is highlighted in the comparison with *in situ* measurements (see, for instance, plots of MIPAS-ENVISAT orbit 5,250/scan 19 and orbit 5,386/scan 29) and can account for the observed differences with remote-sensing data, whilst horizontal gradients encountered at the border of the polar vortex might at least partially justify the discrepancy in ozone values retrieved by airborne and satelliteborne limb sounders. We can notice from Figure 3.23 that MIPAS-ENVISAT normally tends to be in very good agreement with MIPAS-STR and only occasionally to show significant differences, mostly in terms of a slight over-estimate of ozone VMR. The latter trend is more pronounced in comparison with SAFIRE-A mean profiles, which are almost consistently lower than MIPAS-ENVISAT O_3 values.

In order to investigate the origin of the observed differences, we must remember that our selection of collocated ozone profiles was based on standard criteria for the maximum separation, in space and in time, between pairs of satellite and aircraft measurements and did not take into account any further requirement for the proximity of the observed airmasses. This implies, for observations performed

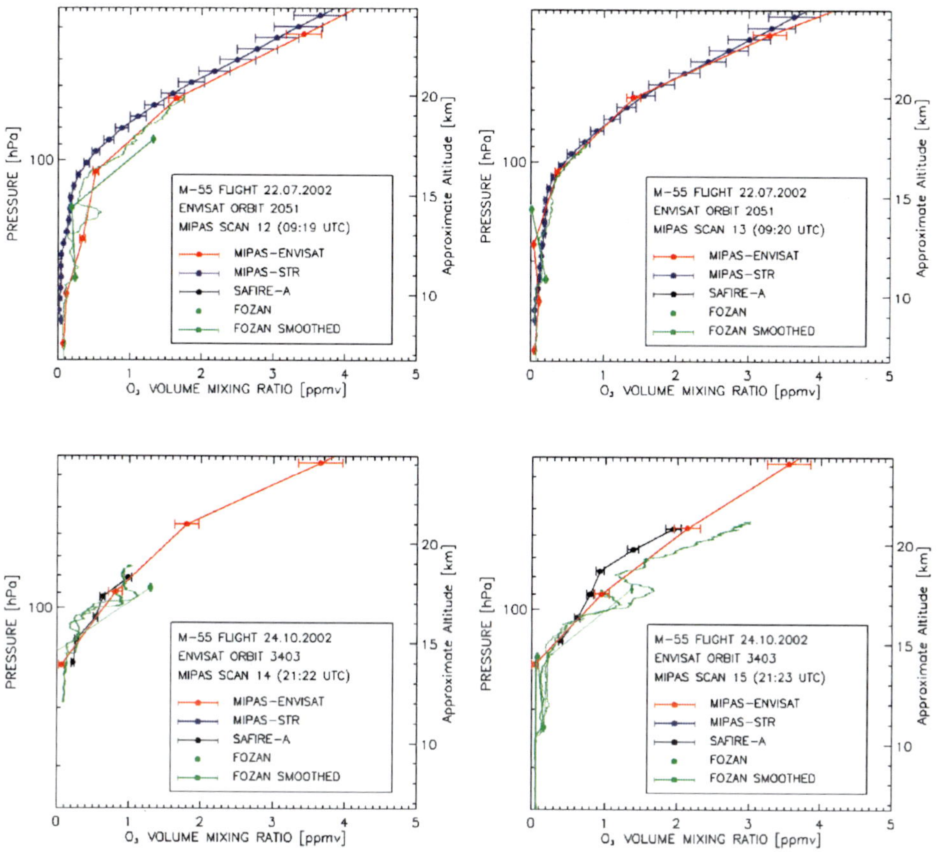

Figure 3.22. Results of the comparison between MIPAS-ENVISAT v4.61 ozone data and correlative measurements performed by the remote-sensing and *in situ* payload of the M-55 *Geophysica* aircraft during mid-latitude flights on July 22, 2002 and on October 24, 2002 from Forlì, Italy (42°N, 12°E) (Cortesi *et al.*, 2007).

across strong vertical and horizontal gradients, that matching measurements, satisfying the spatiotemporal coincidence criteria, can be associated with substantially different conditions and, thus, explain the observed discrepancy between the ozone mixing ratios retrieved from airborne and satellite data.

We can look, for instance, at the potential vorticity field on the isentropic surface $\theta = 420$ K (approximately 18 km) in the region covered by the M-55 flight on March 12, 2003 (from NCEP data at 12:00 UTC) (as displayed in Figure 3.24). Also noticeable in the figure is the geolocation of a particular set of collocated measurements from MIPAS-ENVISAT (orbit 5,386, scan 28), MIPAS-STR (scans 31–36), and SAFIRE-A (scans 9–14): MIPAS-ENVISAT and MIPAS-STR limb measurements mostly overlap on a region with PV values of about (25 ± 1) pvu (1 pvu $= 10^{-6}$ km^2 kg^{-1} s^{-1}), whilst the SAFIRE-A mean profile results from aver-

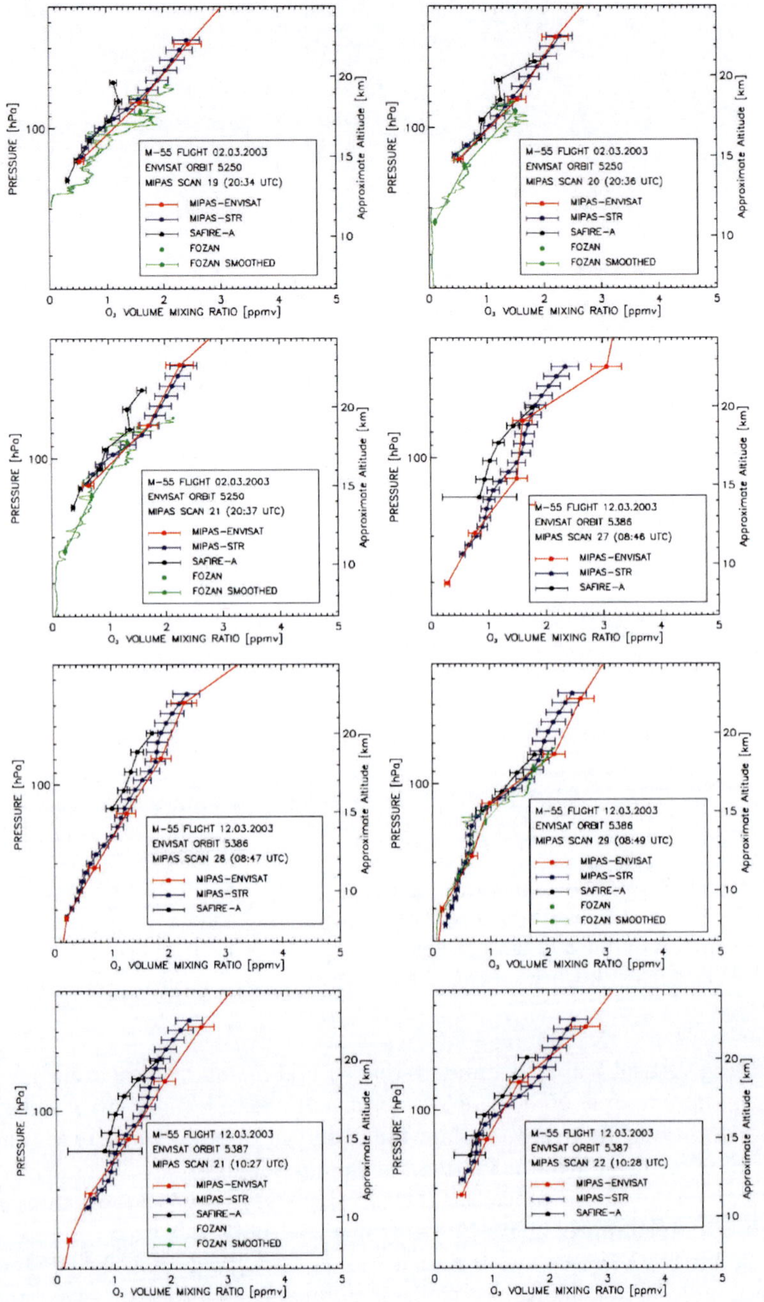

Figure 3.23. Results of the comparison between MIPAS-ENVISAT v4.61 ozone data and correlative measurements performed by the remote-sensing and *in situ* payload of the M-55 *Geophysica* aircraft during high-latitude flights on March 2, 2003 and March 12, 2003 from Kiruna, Sweden (68°N, 20°E) (Cortesi *et al.*, 2007).

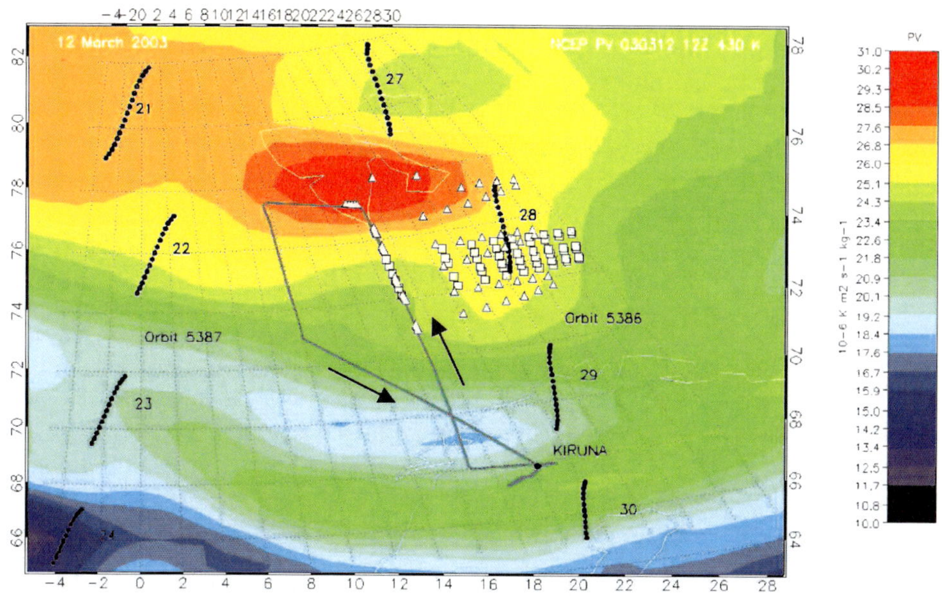

Figure 3.24. False-color map of potential vorticity on the isentropic surface $\theta = 420$ K and the M-55 *Geophysica* route during the ENVISAT validation flight from Kiruna on March 12, 2003. The geolocation of MIPAS-ENVISAT tangent points for the selected overpasses (orbits 5,386 and 5,387) is indicated (black circles). The geographical coverage of collocated aircraft measurements, in coincidence with MIPAS-ENVISAT orbit 5,386, scan 28, is also displayed, with the white triangles and white squares corresponding, respectively, to O_3 mean profiles measured by SAFIRE (scans 9–14) and MIPAS-STR (scans 31–36) (Cortesi *et al.*, 2007).

aging over a more extended area including airmasses with PV values as high as ~30 pvu. In the plot of Figure 3.24 we observe, correspondingly, matching ozone values retrieved at ~18 km by MIPAS-ENVISAT and MIPAS-STR (approximately 1.8–2.0 ppmv) and lower O_3 VMR measured by SAFIRE-A (approximately 1.6 ppmv). This example, as well as similar checks performed using different combinations of coincident data, confirm that whenever a significant difference is found between simultaneous ozone measurements of MIPAS and one of the M-55 *Geophysica* sensors this is mostly due to sampling of different airmasses across a region of strong horizontal (and vertical) gradients. However, a more comprehensive and quantitative analysis of the O_3 differences in the (PV, θ) space is needed.

3.4.5 ASUR

Measurements of the ozone VMR profile gathered by the Airborne Sub-millimeter Radiometer ASUR (Mees *et al.*, 1995) during the SCIAMACHY Validation and Utilization Experiment SCIAVALUE (Fix *et al.*, 2005) are used in this study to validate MIPAS ozone data products v4.61. ASUR is a passive heterodyne radiometer for middle-atmospheric sounding, operating in the frequency range

604–662 GHz and flying on board aircraft to avoid signal absorption due to tropospheric water vapor. The mixing ratio profiles of stratospheric trace gases O_3, ClO, HCl, HNO_3, N_2O, etc. are retrieved on a 2 km altitude grid using the optimal estimation method (Rodgers, 1990). The retrieved ozone profiles from 16 km to 50 km have a vertical resolution of 7–10 km, decreasing with altitude, and a horizontal resolution of about 20 km. An error in instrument calibration led to systematically high values in early ASUR publications. This error has been rectified and measurement accuracy is now better than 10% (Kuttippurath *et al.*, 2007). The collocated ozone profiles obtained by MIPAS and ASUR were compared within the baseline coincidence criteria $\Delta s < 300$ km and $\Delta t < 3$ h. The MIPAS ozone profiles were combined with the ASUR averaging kernels, to account for the lower vertical resolution of the ASUR measurements. Smoothed MIPAS values were used to calculate the absolute and relative differences from the collocated ASUR measurements. The mean profiles of the differences were finally obtained by averaging over the available coincidences in different latitude bands (the tropics, mid-latitudes, and the Arctic).

Mean profiles of the absolute difference between ASUR and MIPAS O_3 VMR and of their relative difference with respect to the ASUR values, calculated from the available dataset of direct coincidences, are reported in Figure 3.25 for three latitude bands, corresponding to the tropics (5°S–30°N), mid-latitudes (30°N–60°N), and the Arctic (60°N–80°N), as well as for all of these regions combined. Both absolute and relative differences are plotted as a function of altitude, with an approximate pressure scale derived from the U.S. standard atmosphere displayed on the right axis. The yellow-shaded area represents the 1σ standard deviation from the mean profile. The total number of coincidences is 50 with the majority (i.e., 22 instances) in the Arctic, 7 instances in mid-latitudes, and 21 instances in the tropics. The MIPAS-ASUR deviation is -0.9 to $+0.4$ ppmv or -40 to $+4$% in the tropics at 20–40 km, whereas at mid-latitudes the difference is within 0.9 ppmv or -15 to $+25$%.

The agreement between the profiles is very good in the Arctic between 20 and 40 km, where the difference is within ± 0.4 ppmv or -6 to $+4$%.

3.5 COMPARISON WITH SATELLITE MEASUREMENTS

Correlative measurements of OVPs are obtained by several satellite sensors operating simultaneously with the MIPAS-ENVISAT spectrometer and employing different observation modes. In this section we consider the validity of MIPAS O_3 data against coincident profiles retrieved by four solar occultation instruments (SAGE II, HALOE, POAM III, and ACE), by a nadir-viewing sensor (GOME) and by a limb emission sounder (Odin-SMR).

A common strategy was followed for the validation of MIPAS O_3 profiles by comparison with these spaceborne sensors, using the key concepts of the scheme for statistical bias and precision determination with matching pairs of measurements described by von Clarmann (2006) and based on the comparison:

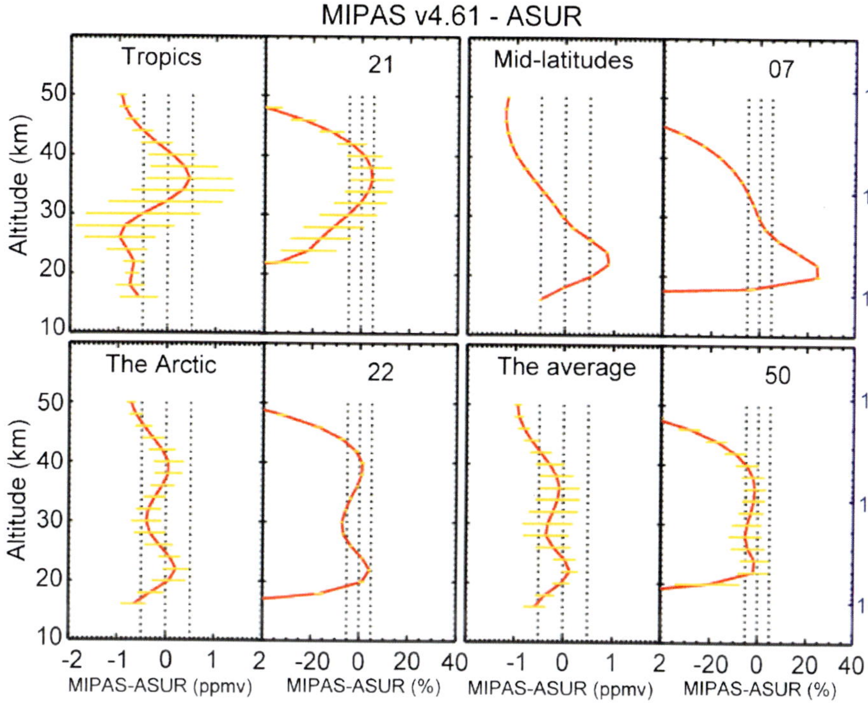

Figure 3.25. The absolute (MIPAS-ASUR ozone VMR in ppm) and percentage (ΔVMR/MIPAS VMR in %) difference between the MIPAS and ASUR ozone profiles in the tropics (top left), mid-latitudes (top right), the Arctic (bottom left), and the average of all these latitude sections (bottom right). The thick red line indicates the mean Δ profile at each section and the yellow-shaded area represents standard deviation from the mean profile. The dotted line stands for ±0.5 ppm or ±5%. The number of averaged Δ profiles in each climatic region is also noted in the plots (Cortesi *et al.*, 2007).

1. Between the mean percentage difference (MIPAS-REFERENCE) O_3 VMR and the combined systematic error of the two instruments, in order to identify unexplained biases in MIPAS ozone measurements.
2. Between the standard deviation of the mean relative difference and the combined random error, in order to validate the precision of MIPAS.

Details of the procedure for the implementation of this scheme were agreed and slightly adapted in individual cases to better exploit the specific features of each dataset. Unless otherwise noted, the standard criteria for maximum spatiotemporal separation of 300 km and 3 h with the reference measurements were strictly applied to select the comparison pairs available during the overlapping period of operation of MIPAS and the validating instrument.

For each of the selected pairs, both MIPAS and the reference instrument O_3 profiles were interpolated on a common pressure grid to enable statistical analysis of collocated measurements having different vertical resolutions; the interpolation grid was generally defined by averaging the pressure values of selected MIPAS scans (details about the interpolation of O_3 vertical profiles are provided in the relevant sub-sections, whenever a different choice has been made, such as, for instance, in the case of MIPAS/POAM comparison—see Section 3.5.2). With the only exception of the comparison with GOME observations, no averaging kernels have been applied, because of the similar vertical resolution of MIPAS and the reference instruments.

The interpolated profiles were used to calculate the relative deviation (RD) in ozone VMR values retrieved by MIPAS and by the correlative sensor at each pressure level (p) using equation (3.19):

$$\mathrm{RD}(p) = 100 \times \frac{\mathrm{MIPAS}[\mathrm{O}_3]_p - \mathrm{REFERENCE}[\mathrm{O}_3]_p}{\mathrm{REFERENCE}[\mathrm{O}_3]_p} \tag{3.19}$$

The mean relative deviation (MRD) and root mean square (RMS) of the relative deviation between all MIPAS and correlative sensor pairs were determined, along with corresponding quantities averaged over sub-sets of latitudinal or seasonal bands, whenever further investigation was required to isolate the source of discrepancies identified in the global average or to diagnose zonal and seasonal patterns in O_3 mean differences.

In all cases, beside the MRD over all the available coincidences, the mean profiles of both MIPAS and the reference instrument are displayed in the plots of the global average.

Combined random and systematic error estimates on the O_3 VMR difference between matching profiles were based on the expected uncertainties of MIPAS measurements and on the validated precision and accuracy of the correlative data.

As far as MIPAS errors are concerned, we refer, in general, to ESA level-2 products for random error due to propagation of instrument noise through retrieval and to the *a priori* estimate of systematic errors provided by Oxford University.

In the following an important effort has been made to evaluate properly the combined error budget associated with the mean relative difference of collocated O_3 profiles. This is because some of the components, listed in the Oxford University dataset as systematic error in individual profiles, show random variability over the longer timescale involved when averaging different MIPAS scans and/or orbits and tend to contribute to standard deviation of mean difference rather than to bias. Taking this into account for the purpose of our comparisons with concurrent satellite sensors, we have considered the error contribution due to the propagation of pressure and temperature (p, T) random covariance into the retrieval of O_3 VMR (taken from the Oxford University dataset) as a randomly variable component and combined it with measurement noise—using the root-sums-square method—to obtain the MIPAS random error. MIPAS systematic error was conversely calculated by subtracting the p, T propagation error from the overall systematic error given in the Oxford University files. In the following sub-sections, details of individual comparisons with various satellite sensors are provided.

A very brief description of the instrument and of the correlative dataset is given in each case, specifying the data version adopted for the comparison with MIPAS v4.61 and/or v4.62 profiles and referring to the most recent publications and updated information for details about their measurement validation and quality assessment (ESA, 2006). Results of the comparison with each of the validating sensors are presented and discussed.

3.5.1 Comparison of MIPS data with SAGE II O_3 profiles

The Stratospheric Aerosol and Gas Experiment II (SAGE II) (McCormick, 1987), which was launched on October 5, 1984 on board the Earth Radiation Budget Satellite (ERBS), is a seven-channel sunphotometer, at visible and near-infrared wavelengths ranging from 0.385 μm to 1.02 μm, which uses the solar occultation technique to measure the aerosol volume extinction coefficient and O_3, NO_2, and H_2O mixing ratios (Section 2.5.1.1). The limb measurements of absorption by trace gases are inverted (Chu *et al.*, 1989) to provide vertical profiles with a horizontal resolution of about 200 km and a vertical resolution of 1 km in the range 8–38 km and of 5 km in the range 38–50 km (Mauldin *et al.*, 1985a, b). The latitude coverage of SAGE II measurements is from 80°N to 80°S. SAGE II ozone concentration profiles are retrieved using spectra from the 0.60 μm wavelength channel. Validation of SAGE II data version 6.1 (Wang, H. *et al.*, 2002) shows agreement within 10% with ozonesonde measurements from the tropopause up to 30 km, with SAGE II slightly overestimating (<5%) the ozone content between 15 to 20 km. A former version of SAGE II (v5.96) had been extensively validated within 7% at 20 to 50 km (Cunnold *et al.*, 1989). Version 6.2 of SAGE II was improved by adjustment to the aerosol clearing and by the correction of channels 520 and 1,020 nm for absorption of the oxygen dimer.

In this work, SAGE II ozone data v6.2 were used to validate MIPAS data v4.61/v4.62 for the period of the instrument full spectral resolution mission. The baseline coincidence criteria (spatial separation <300 km and temporal separation <3 h) are applied to select the SAGE II and MIPAS ozone profiles to be compared. A total of 326 pairs of matching profiles were identified. To estimate the error budget of the comparison, the values for precision and accuracy of SAGE II ozone data given by Cunnold *et al.* (1989) were used: in the range from 16 to 53 km accuracy is between 5% and 7%, precision between 4.1% and 6.7%, and systematic error is between 1.5% and 6.2%.

In Table 3.7 and Figure 3.26 the statistics used to compare the MIPAS and SAGE II collocated ozone profiles over all the available collocations (total number = 326) are presented. Total error budget estimates are reported for both instruments, combining the random error contributions in Table 3.7. In Figure 3.26 the solid red line represents mean relative difference, with error bars indicating the standard error of the mean (i.e., 1σ standard deviation of the MRD divided by the square root of the number of matching pairs in the sample). The combined random and systematic errors are represented as dotted and dashed lines, respectively, while the shaded area corresponds to the MRD $\pm 1\sigma$ standard deviation. In

Table 3.7. Statistics over all comparisons of MIPAS to SAGE II: (a) zonal averages, (b) seasonal averages for the Southern Hemisphere and Northern Hemisphere, and (c) all collocations (Cortesi *et al.*, 2007).

(a)

Latitude band	*Pressure range* (hPa)	*Mean relative difference*	*Root mean square*	*N*	*Month of year*
90°S–60°S	70–0.75	+2% ↔ +9%	+5% ↔ +11%	69	Dec. 3, Feb. 4
60°S–30°S	70–0.75	−7% ↔ +11%	+9% ↔ +18%[a]	64	Jan. 3/4, April–May 3, July 3
30°N–60°N	70–0.75	−6% ↔ +4%	+5% ↔ +12%	29	Jan. + Mar. 3, Apr. 3, July 2/3 (22)
60°N–90°N	70–0.75	−4% ↔ +4%	+5% ↔ +7%	169	April + June 3, July 2/3, Sep. 3

(b)

Hemisphere	*Season*	*Pressure range* (hPa)	*Mean relative difference*	*Root mean square*	*N*
SH	Spring	70–0.75	0% ↔ +9%	+4% ↔ +12%	32
	Summer	70–0.75	−5% ↔ +14%	+5% ↔ +15%[a]	45
	Autumn	70–0.75	−5% ↔ +7%	+5% ↔ +15%	25
	Winter	70–0.75	−10% ↔ +12%	+6% ↔ +15%	26
NH	Spring	70–0.75	−4% ↔ +4%	+5% ↔ +10%	101
	Summer	110–0.75	−5% ↔ +3%	+5% ↔ +10%	95

(c)

Collocations	*Pressure range* (hPa)	*Mean relative difference*	*Root mean square*	*N*
ALL	100–0.75	−4% ↔ +4%	+5% ↔ +14%	326

[a] Except at $p > 45$ hPa close to 40%.

the entire pressure range from the lower stratosphere to the upper stratosphere (approximately from 100 hPa to 1 hPa), the mean deviation of MIPAS O_3 VMR relative to SAGE II is within ±5%, mostly reflecting the positive bias of MIPAS to SAGE II that never exceeds the combined systematic error. The good agreement between the two datasets is confirmed by the root mean square of their mean difference which is significantly larger than the combined random error only at lower altitudes for pressure values higher than ~30 hPa. Mean relative differences for different latitude bands are plotted in Figure 3.27, highlighting the fact that the main source of discrepancy is concentrated in the Southern Hemisphere mid and

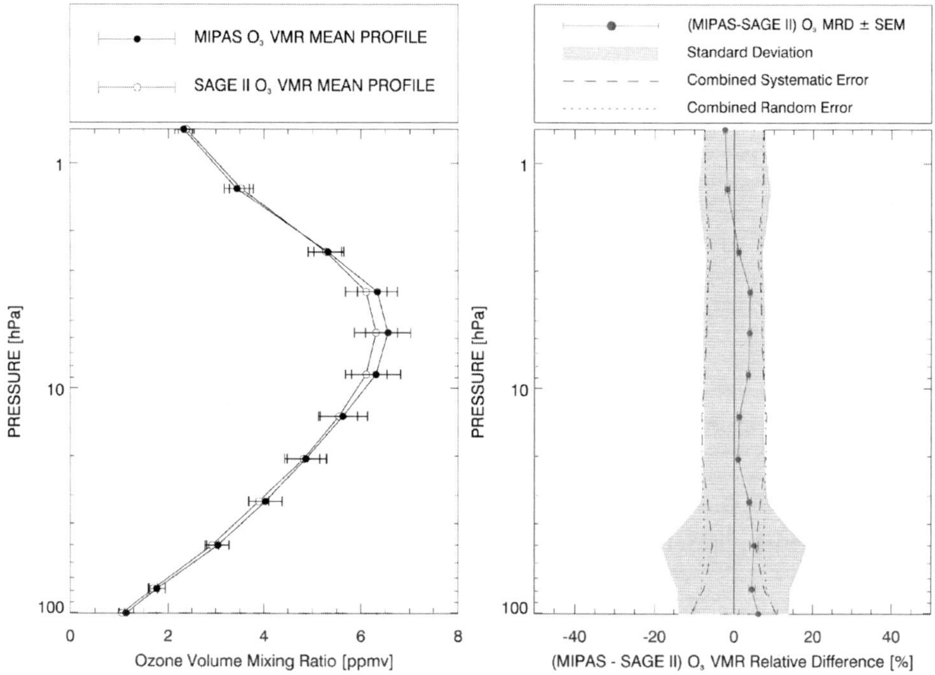

Figure 3.26. Comparison between MIPAS and SAGE II: statistics over all the collocated O_3 profiles. (Left panel) MIPAS and SAGE II O_3 mean profiles with total error bars are shown. (Right) Mean relative differences and standard deviations, along with combined random and systematic errors, are plotted (Cortesi *et al.*, 2007).

high latitudes with evidence of a significant high bias of MIPAS in the 60°–90°S above 10 hPa. The seasonal dependence of relative differences has also been investigated (as shown by the plots in Figure 3.28). Complete seasonal coverage is obtained only for the Southern Hemisphere, where the observed bias is (marginally) higher than the systematic uncertainties on the MRD between approximately 10 and 2 hPa and below ~30 hPa in autumn and winter, and for a peak centered around ~45 hPa in spring and summer. In the Northern Hemisphere, on the other hand, the available collocations provide smaller values of the mean difference with no evidence of significant biases throughout the whole stratosphere.

3.5.2 Comparison with POAM III O_3 profiles

The Polar Ozone and Aerosol Measurement III (POAM III) instrument (Lucke *et al.*, 1999), operating on the SPOT-4 spacecraft since March 23, 1998, is a nine-channel photometer that performs solar occultation measurements in selected bands from 0.354 to 1.018 μm to derive the profiles of O_3, NO_2, and H_2O, as well as temperature and wavelength-dependent aerosol extinction. Ozone profiles are primarily retrieved from spectra recorded by the channel centered at

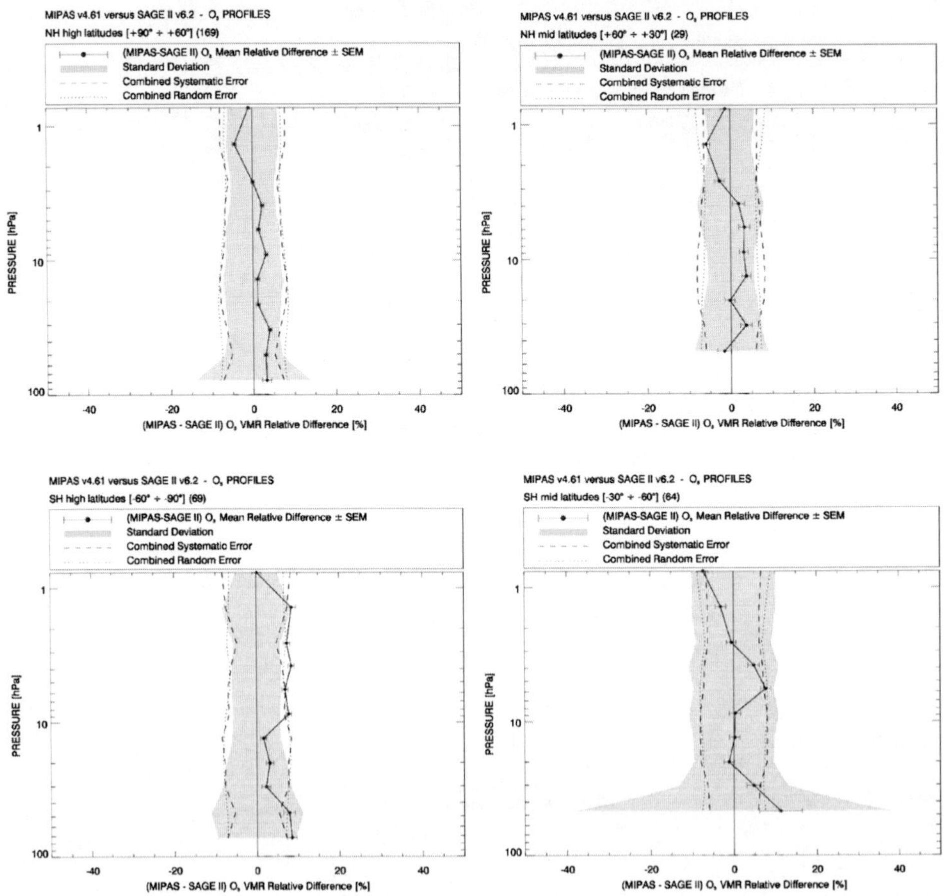

Figure 3.27. Comparison between MIPAS and SAGE II: zonal averages. Relative differences and comparison errors averaged over four latitude bands. The number of coincident pairs in each zone is indicated in parentheses (Cortesi *et al.*, 2007).

603 nm, near the peak of the Chappuis absorption band. The vertical resolution of ozone retrieval is 1 km throughout the stratosphere, but degrades rather quickly to 2–3 km in the upper troposphere. The horizontal resolution is estimated to be approximately 30 km perpendicular to the line of sight (i.e., parallel to the terminator) and about 200 km parallel to the line of sight. Details of the retrieval algorithm and error analysis for POAM III version 3.0 are given by Lumpe *et al.* (2002). Validation of POAM III ozone has been performed by Randall *et al.* (2003).

Ozone data version 4.0 from the POAM III instrument are used for comparison and validation of MIPAS data versions v4.61/v4.62. The selection of collocated MIPAS and POAM profiles was based on the standard criteria of 300 km and 3 h for the maximum spatiotemporal separation of matching measurements and resulted

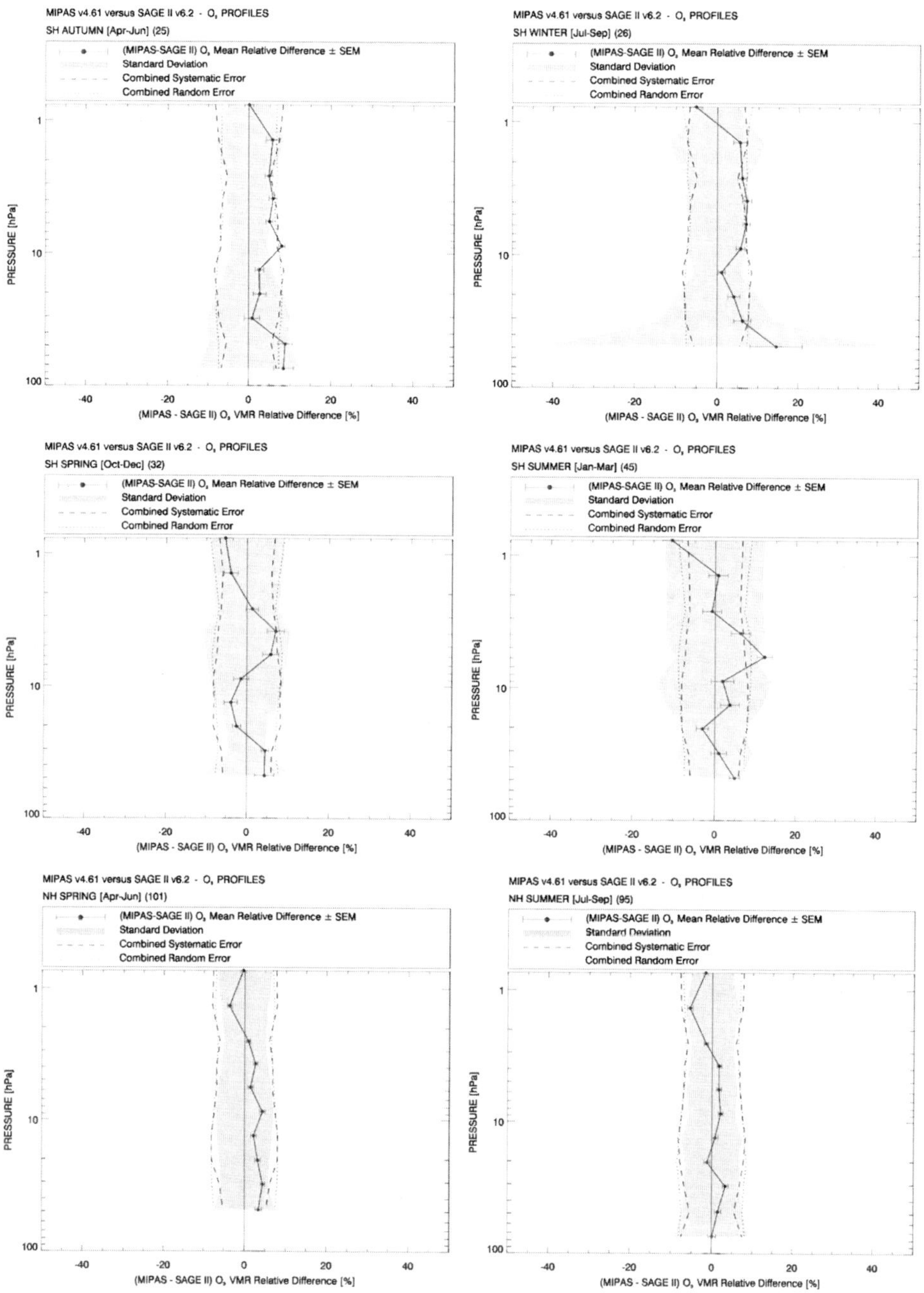

Figure 3.28. Comparison between MIPAS and SAGE II: seasonal averages. Relative differences and comparison errors averaged over different seasons in the Northern and Southern Hemisphere. The number of coincident pairs in each season/hemisphere is indicated in parentheses (Cortesi *et al.*, 2007).

in a total of 1,571 comparison pairs within the three latitude bands (90°N–60°N, 60°N– 30°N, and 60°S–90°S) in the period from June 1, 2002 to March 26, 2004. A fine vertical pressure grid, equidistant in logarithmic pressure, was selected and both POAM and MIPAS results were interpolated on to this common grid. POAM pressure is derived from UKMO (United Kingdom Meteorological Office) pressure values.

POAM III error analysis has been carried out by Lumpe *et al.* (2002). The random error is below 5% throughout the stratosphere with a minimum value of 1% at 20 km. In the troposphere the random error is rapidly increasing to values of more than 10%. According to Lumpe *et al.* (2002), POAM III ozone profiles are neither affected by improper removal of sunspot artifacts nor by aerosol feedback errors in gas retrieval, which means that systematic errors are negligible for ozone.

The global average of the relative differences of MIPAS O_3 profiles with respect to collocated POAM III measurements is less than ±5% between approximately 60 hPa and 0.2 hPa (see Figure 3.29). No evidence of unexplained biases is found within the whole range from the upper troposphere (~300 hPa) up to the lower mesosphere (~0.12 hPa), with the only exception of a localized peak around 100 hPa where the MRD exceeds the combined systematic error. The available dataset of MIPAS/POAM coincident ozone measurements provides only partial coverage for calculation of zonal means over different latitude bands, with no matching pairs satisfying the baseline criteria of 300 km and 3 h at mid-latitudes in the Southern Hemisphere (i.e., in the range 30°S to 60°S). Results of the comparison carried out over the other latitude bands are shown in Figure 3.30, where we can notice that the peak of the MRD mainly originates from the high-latitude data of both hemispheres. No correlation was found, however, with any seasonal cycle (e.g., polar stratospheric cloud climatology), as displayed in Figure 3.31, showing an MRD profile that exceeds the combined systematic error around 100 hPa in most of the seasons with peak values ranging between 10% and 20%.

3.5.3 Comparison with Odin-SMR O_3 profiles

Cortesi *et al.* (2007) compared Odin-SMR version 1.2 data from July 20, 2002 to March 26, 2004 with collocated MIPAS ozone profiles v4.61. By applying the standard coincidence criteria of $\Delta s < 300$ km and $\Delta t < 3$ h, they selected a total number of 1,270 matching profiles. Odin-SMR data used for this comparison (available at *http://www.rss.chalmers.se/gem/*) were obtained from the stratospheric mode band at 501.8 GHz. The O_3 line is at 501.5 GHz, allowing retrieval of the O_3 profile between 21 and 45 km with a vertical resolution of 3.5–4 km. The retrieval algorithm is based on the Optimal Estimation Method. Version 1.2 puts more weight on the *a priori* information with respect to previous versions and this leads to smoother and less noisy profiles with the drawback of slightly reduced resolution and altitude range. The Odin-SMR level-2 analysis uses temperature data from the ECMWF in the stratosphere as well as data from model climatology in the mesosphere (Hedin, 1991). Ozone retrieval in this band is dominated by spectroscopic error.

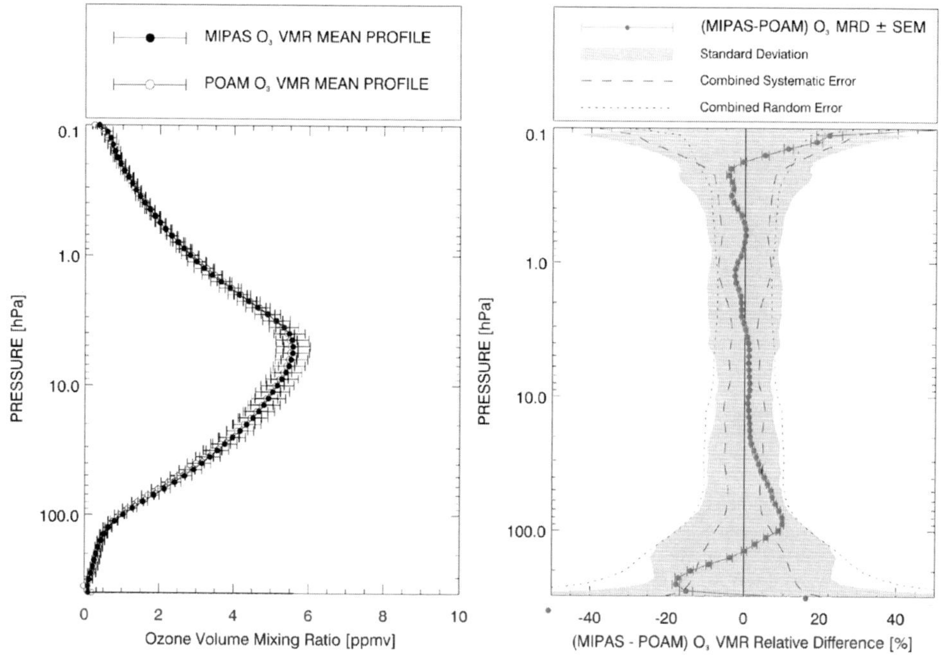

Figure 3.29. Comparison between MIPAS and POAM III: statistics over all the collocated O_3 profiles. Global mean profiles of O_3 VMR measured by MIPAS and POAM III, with error bars indicating the corresponding total uncertainties (left panel). Mean relative difference between MIPAS and POAM III ozone data and combined error budget (right panel) (Cortesi *et al.*, 2007).

Expected total systematic error is lower than 0.4 ppmv above 25 km and increases to ~0.75 ppmv at 20 km. In terms of relative units, the error is of the order of 5% above 30 km and increases to 35% at 20 km.

The comparison included all the matching pairs of measurements available in the test period. Only good quality Odin-SMR profiles were selected and a measurement response (defined by Urban *et al.*, 2005 as the sum of the averaging kernel at a given altitude and providing an estimate of the relative contribution to the information coming from measurements and from *a priori* data) larger than ~0.75 was used to ensure that the information comes from measurements and not from *a priori* data. For MIPAS, only profiles associated with a successful pressure/temperature and O_3 retrievals were considered.

The Odin-SMR systematic error results from individual instrumental errors—that is, calibration error (pointing uncertainty, antenna and sideband response knowledge, spectrometer resolution), model error (temperature knowledge), and spectroscopic error. The Odin-SMR random error for single-profile retrieval is due to intrinsic receiver noise.

On average, a typical systematic error profile was considered for both MIPAS (from the Oxford University error estimate) and Odin-SMR measurements (Wang *et*

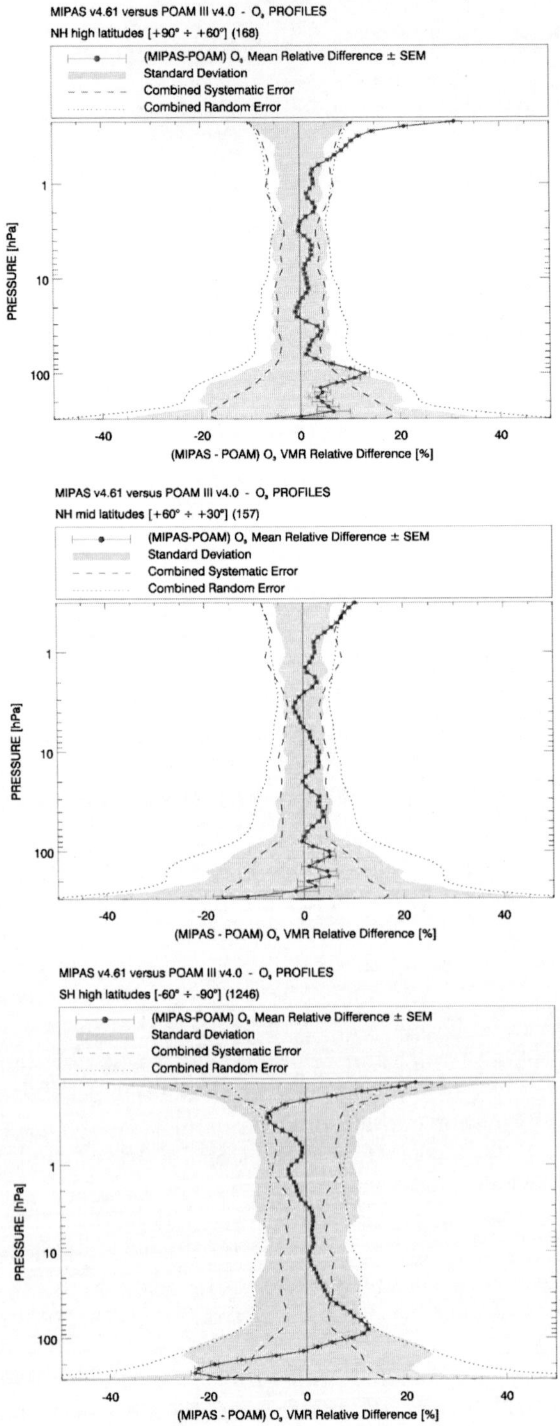
MIPAS v4.61 versus POAM III v4.0 - O$_3$ PROFILES
NH high latitudes [+90° ÷ +60°] (168)
(MIPAS-POAM) O$_3$ Mean Relative Difference ± SEM
Standard Deviation
Combined Systematic Error
Combined Random Error
PRESSURE [hPa]
(MIPAS - POAM) O$_3$ VMR Relative Difference [%]
MIPAS v4.61 versus POAM III v4.0 - O$_3$ PROFILES
NH mid latitudes [+60° ÷ +30°] (157)
(MIPAS-POAM) O$_3$ Mean Relative Difference ± SEM
Standard Deviation
Combined Systematic Error
Combined Random Error
PRESSURE [hPa]
(MIPAS - POAM) O$_3$ VMR Relative Difference [%]
MIPAS v4.61 versus POAM III v4.0 - O$_3$ PROFILES
SH high latitudes [-60° ÷ -90°] (1246)
(MIPAS-POAM) O$_3$ Mean Relative Difference ± SEM
Standard Deviation
Combined Systematic Error
Combined Random Error
PRESSURE [hPa]
(MIPAS - POAM) O$_3$ VMR Relative Difference [%]

◀ **Figure 3.30.** Comparison between MIPAS and POAM III O_3 profiles: zonal averages (Cortesi *et al.*, 2007).(Hedin, 1991). Ozone retrieval in this band is dominated by spectroscopic error. Expected total systematic error is lower than 0.4 ppmv above 25 km and increases to ~0.75 ppmv at 20 km. In terms of relative units, the error is of the order of 5% above 30 km and increases below up to 35% at 20 km.

al., 2007). These systematic error profiles are then multiplied by the respective mean O_3 profiles of the matching pairs of measurements. The combined systematic error is given by the root sum square of the two instruments' systematic errors. The combined random error is given by the root sum square of the averaged random error profiles of the two instruments.

The global average of the percentage difference between MIPAS and Odin-SMR ozone values, calculated over the full set of collocated measurements, is presented in Figure 3.32, where the mean profile of the relative difference between MIPAS and Odin-SMR with respect to the latter is plotted along with error bars representing the standard error on the mean (1σ). The MRD values are within $\pm 5\%$ of approximately 40 to 1 hPa, with MIPAS mostly overestimating the O_3 content. The resulting bias is always constantly lower than the combined systematic error in the full range (60–1 hPa). Outside this interval, both in the upper-stratospheric layers and in the upper troposphere/lower stratosphere (UT/LS), the average O_3 VMR values retrieved by Odin-SMR become increasingly higher than those measured by MIPAS. This discrepancy could be due to a lack of statistics. There are not so many points as can be seen from standard deviation at altitudes below 60 hPa. Moreover, the theoretical retrieval altitude grid for Odin-SMR O_3 at 501.5 GHz is between 21 and 45 km (60–1 hPa), therefore altitudes below 60 hPa might include mainly *a priori* information.

No significant variations in seasonal and latitudinal mean differences are present between MIPAS and Odin-SMR O_3; the global average of the differences is representative of the overall comparison between the different capabilities of the two instruments.

Kopp *et al.* (2007) measured the ozone profiles retrieved by Odin-SMR (level-2 data versions 2.1 and 2.0, respectively) and compared their results to measurements taken by ground-based millimeter wave radiometers in the Arctic, at Kiruna in Sweden, in the mid-latitudes on the Zugspitze, Germany, and in the tropics at Merida, Venezuela. The Kiruna Microwave Radiometer (KIMRA) covers the frequency range 195–224 GHz, and the Millimeter Wave Radiometer MIRA 2, which was operated on the Zugspitze and at Merida, measures in the frequency band 268–281 GHz. From the measurements, ozone profiles in the vertical range between approximately 15–65 km were retrieved using the Optimal Estimation Method. Since ground-based measurements have a lower vertical resolution than those of Odin the latter were degraded using the averaging kernels of ground-based retrievals. Comparison of the resulting profiles to the ground-based data enables the identification of biases in Odin measurements and their possible latitudinal variation. In general, good agreement between satellite and ground-based measurements for the 501.8 GHz band was found in the stratosphere, except for negative bias in the

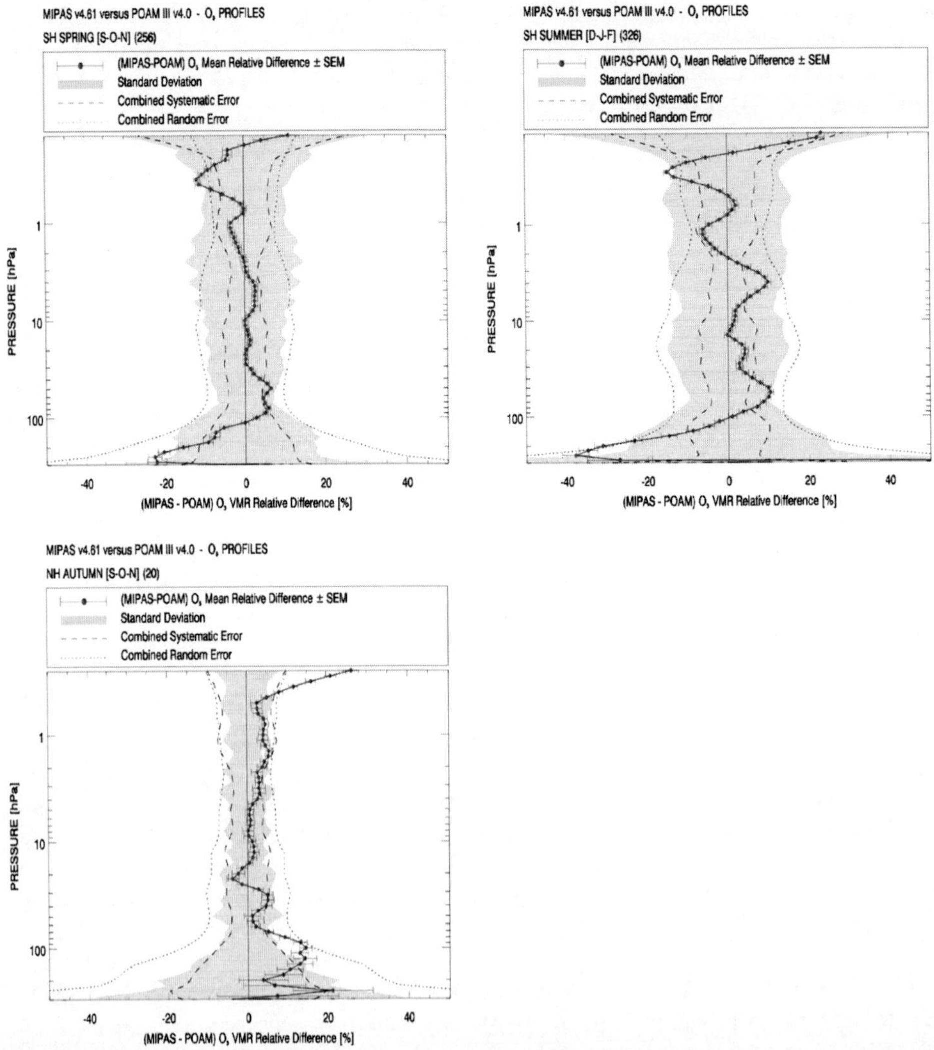

Figure 3.31. Comparison between MIPAS and POAM III: seasonal averages for the Northern Hemisphere (60°N–90°N) and Southern Hemisphere high latitudes (60°S–90°S) (Cortesi *et al.*, 2007).

Odin data of about 10–15% in tropical measurements. Odin measurements taken at 544.9 GHz yielded systematically 20–30% lower ozone mixing ratios in the middle stratosphere than ground-based measurements at all sites.

3.5.4 Comparison with ACE-FTS O_3 profiles

The ACE-FTS (Atmospheric Chemistry Experiment) instrument is a high-resolution (bandwidth $\Delta s = 0.02\,\text{cm}^{-1}$) Fourier transform spectrometer operating from 2 to

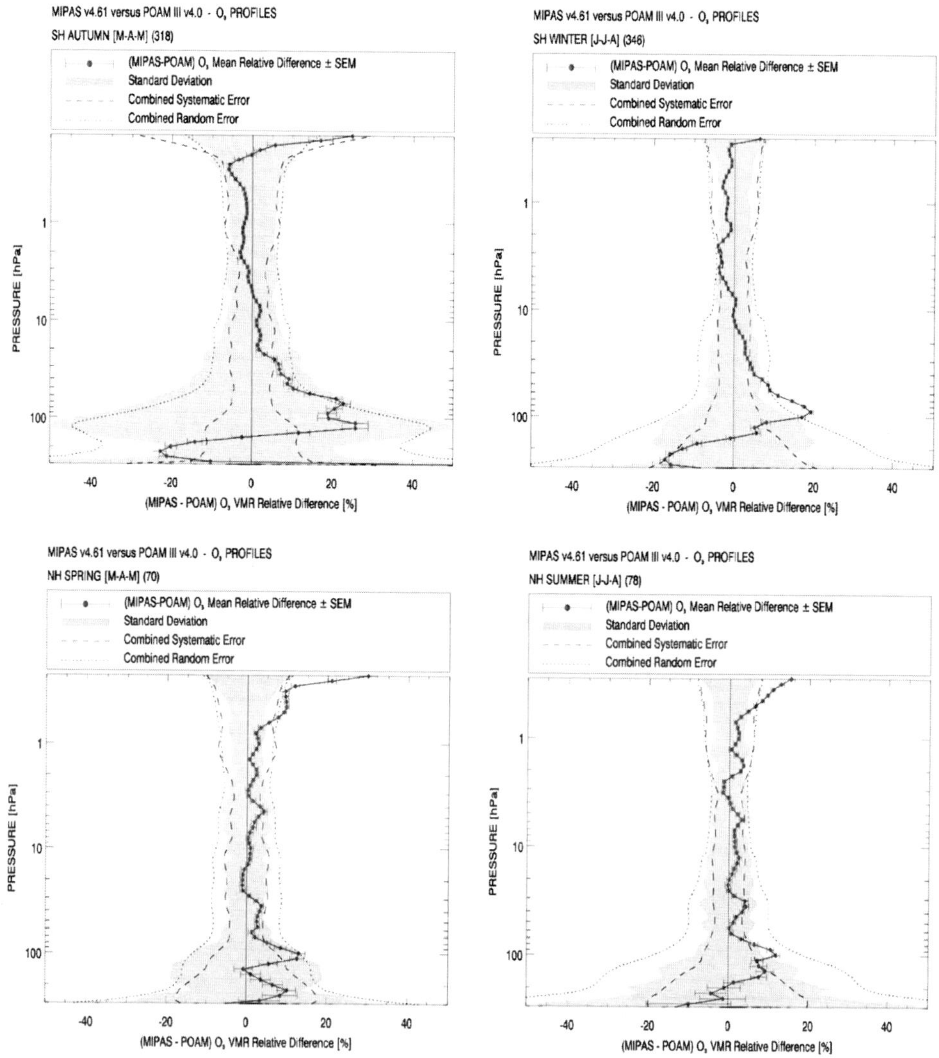

Figure 3.31 (*cont.*)

13 μm ($\sigma = 750$–$4{,}100\,\text{cm}^{-1}$). It performs solar occultation measurements of the vertical distribution of trace gases and temperature from the cloud top up to about 100 km. ACE-FTS measurements are recorded every 2 s. This corresponds to a measurement spacing of 2–6 km, which decreases at lower altitudes due to refraction. Latitude coverage is from 85°N to 85°S. The instrument was launched on August 12, 2003, as part of the ACE mission (Bernath *et al.*, 2005), on board the Canadian satellite SCISAT-1. A modified global fit approach (Boone *et al.*, 2005) is adopted for the retrieval of pressure, temperature, and volume mixing ratio profiles.

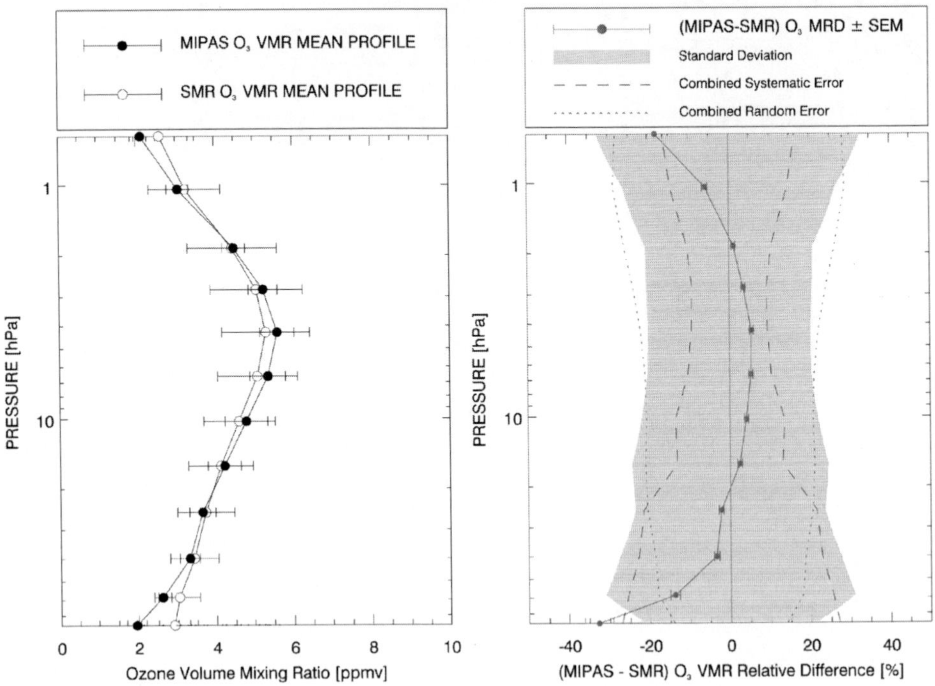

Figure 3.32. Comparison between MIPAS and Odin-SMR: statistics over all the collocated O_3 profiles (Cortesi *et al.*, 2007).

Results of ACE-FTS ozone measurements have been validated against ozonesondes and other satellite measurements (Fussen *et al.*, 2005; McHugh *et al.*, 2005; Petelina *et al.*, 2005; Walker *et al.*, 2005).

Here, MIPAS ozone data v4.62 are compared with ACE-FTS version 2.2 data in the period from February 4, 2004 to March 26, 2004. During the first five months of the mission, only sunsets were measured because of problems with spacecraft pointing at sunrise. Therefore, the latitude coverage for this comparison is limited to 20°N–85°N. The selected coincidence criteria were 300 km and 6 h. A slightly relaxed temporal mismatch, compared to the generally adopted 3 h, was chosen in order to increase the statistics of the comparison since the ozone does not vary substantially when relaxing the timescale from 3 to 6 h. A total of 152 matching pairs of profiles is available for the comparison of MIPAS and ACE O_3 data. The ACE operational retrieval employs a weighted non-linear least squares fit. *A priori* profiles are used only as a first guess and to constrain the shape of the profiles above the highest analyzed measurement.

The comparison included all the matching pairs of measurements available in the test period. Only retrieved ACE points of the O_3 profiles were used in the comparison according to the quality flags specified by the ACE team. The ACE profile above the highest analyzed measurement is given as a scaled initial guess profile and is not taken into account in the comparison. For MIPAS, only ozone profiles associated with successful pressure/temperature and O_3 retrievals were considered.

The estimated systematic error profile for ACE-FTS data version 2.2 is based on validation comparisons of ACE-FTS with different satellite instruments (POAM III, SAGE III, and HALOE) and balloonborne ozonesonde (McHugh *et al.*, 2005; Walker *et al.*, 2005). This estimated systematic error is up to 10% below 35 km and up to 35% above.

The results of the comparison between coincident O_3 measurements of MIPAS and ACE-FTS can be summarized by the plot shown in Figure 3.33, where the mean relative difference of MIPAS ozone VMR with respect to ACE values is displayed. A pronounced peak of the MRD, corresponding to the low bias of MIPAS, emerges above approximately 1.0 hPa, but appears fully justified by our estimate of the combined systematic errors of the two instruments (larger than $\pm 40\%$ above ~ 2 hPa). This is also the case throughout the whole profile from 0.1 hPa down to 250 hPa if we exclude a thin layer around 10 hPa, where a percentage mean relative difference of about 15–20% cannot be explained by estimated systematic uncertainties.

3.5.5 Comparison with HALOE O_3 profiles

The HALOE (HALogen Occultation Experiment) instrument, which operated from September 1991 to November 2005 on board the Upper Atmospheric Research Satellite (UARS), is a solar occultation infrared radiometer that obtains profiles of pressure, temperature, aerosol, and several trace gases. It uses broadband and gas filter radiometry to measure solar energy absorption over the 2.45 μm to 10.04 μm spectral range (Russell *et al.*, 1993). A single ozone profile is retrieved from 9.6 μm channel radiances during each of the 15 daily sunset and 15 daily sunrise events with an effective 2.5 km vertical resolution although data are oversampled at 300 m intervals. The UARS was in a circular orbit inclined at 57° at an altitude of 600 km from which vertical ozone profiles from 12 to 90 km were obtained nearly globally between 30°S to 70°N and 70°S to 30°N following a yaw maneuver every 36 days. The HALOE retrieval algorithm incorporates a modified onion-peeling approach with no *a priori* assumption and simulates the gas and broadband measurements using specific line-by-line forward models obtaining O_3 and interfering gas spectroscopic information from the HITRAN 1991–1992 database (SPARC, 1998). HALOE ozone measurements have been extensively validated (as described by Bruhl *et al.*, 1996) for the results obtained with version 17 of the retrieval software. Total error estimates associated with the HALOE O_3 channel were obtained and values range from 95% at 0.01 hPa and 11% at 0.1 hPa and gradually increase to 30% at 100 hPa. Significant systematic errors below 50 km are uncertainties in the retrieval algorithm's forward model (in particular, its spectral line parameters and approximations) and the instrument's altitude registration. Pointing errors increase rapidly in the lower stratosphere and below where cloud and aerosol interference start to dominate. In general, HALOE version 17 data were

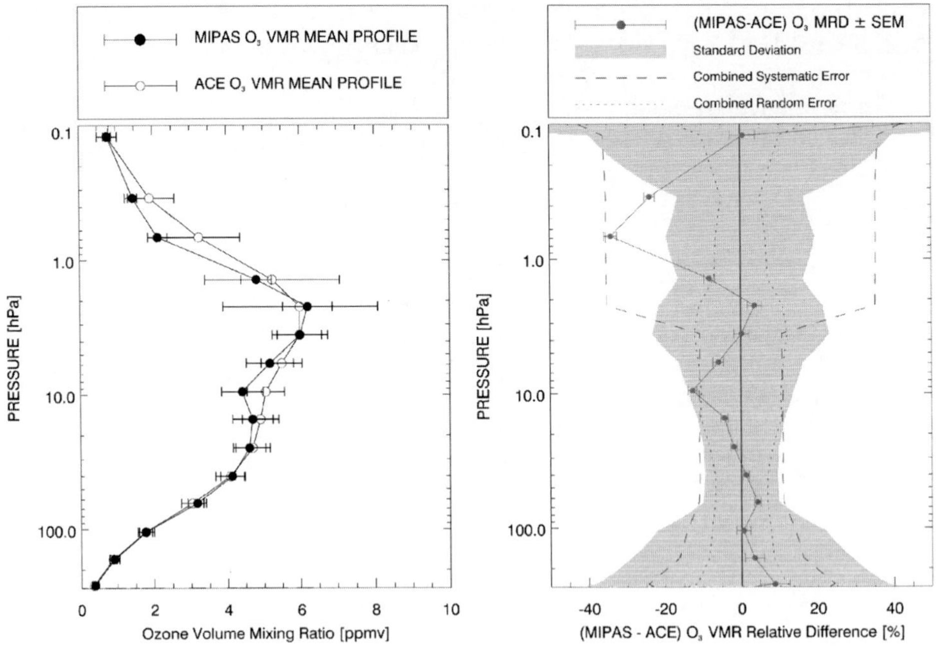

Figure 3.33. Comparison between MIPAS and ACE-FTS: statistics over all the collocated O_3 profiles (Cortesi *et al.*, 2007).

found to agree well within the errors associated with comparative sources, with a tendency to be low by 5% between 30 hPa and 1 hPa.

Off-line MIPAS version 4.61 and 4.62 level-2 ozone were compared to the HALOE version 19 ozone profiles. The coincidence criteria used for validation are a distance and time difference of 300 km and 3 h and are applied to MIPAS data from July 2002 to March 2004. Conditions fulfilled for the comparison were (a) only profiles corresponding to successful retrieval flags were selected, (b) the profile should exceed 45 km and reach 12 km and below, and (c) no additional cloud flagging was applied (i.e., $CI \leq 1.8$ is assumed from the v4.61 and v4.62 processing). The total number of matches for the above coincidence criteria were 156 with 141 (98 v4.61 and 43 v4.62) profiles fulfilling the conditions applied. The estimate of HALOE error budget was based partly on information contained in the data files, which provided—along with the ozone profile—the random error component (consisting of noise and aerosol) error. Table 2.2 of Bruhl *et al.* (1996) was consulted for the remaining random and systematic error components. All HALOE version 19 data were screened for cloud and aerosol effects in accordance with Hervig and McHugh (1999).

Results of the comparison between MIPAS and HALOE ozone measurements are shown in Figure 3.34, in terms of the mean relative difference obtained by averaging the deviation of MIPAS O_3 values relative to coincident HALOE profiles.

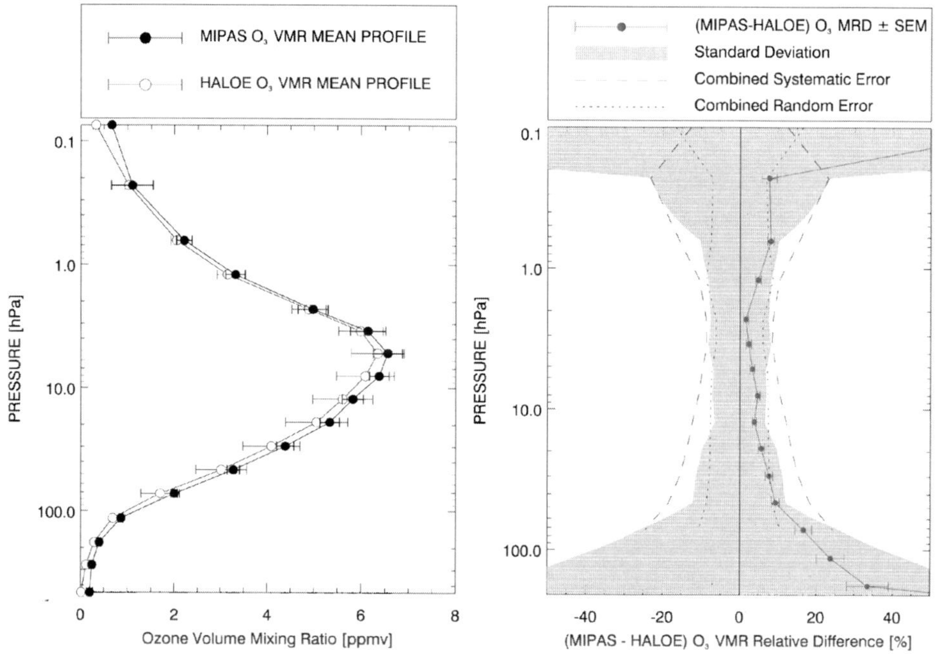

Figure 3.34. Comparison between MIPAS and HALOE: statistics over all the collocated O_3 profiles (Cortesi *et al.*, 2007).

From the global average, MIPAS data are found to show constantly higher O_3 concentrations from 0.1 to 100 hPa relative to HALOE, with MRD values less than 10% in the 0.2–50 hPa interval and increasing to 25% at 100 hPa. At pressures less than 100 hPa, where estimates of HALOE random and systematic uncertainties are available, the combined systematic error fully accounts for the observed positive bias of MIPAS. Combined random errors are fairly consistent with the root mean square of the relative differences from 0.1 down to 50 hPa. Zonal and seasonal averages of the relative difference (MIPAS – HALOE) are plotted in Figure 3.35 and Figure 3.36, respectively. No evidence for MRD higher than the expected systematic uncertainties emerges when limiting calculation of the average to the selected latitude bands or seasons. Table 3.8 summarizes the results of the comparison between MIPAS and HALOE.

In the MIPAS operational processor (Raspollini *et al.*, 2006), both v4.61 and v4.62 algorithms included a cloud detection algorithm to identify clouds in MIPAS spectra so that such spectra are not included in the retrieval of pressure/temperature and trace gases.

For both versions, the cloud index for band A (the most commonly used cloud flag is CI-A) is set with CI-A ≤ 1.8 as flagging cloud (Spang *et al.*, 2004); CI-A arises from the ratio between the integrated signal from 788 to 796 cm^{-1} and the integrated signal from 832 to 834 cm^{-1}.

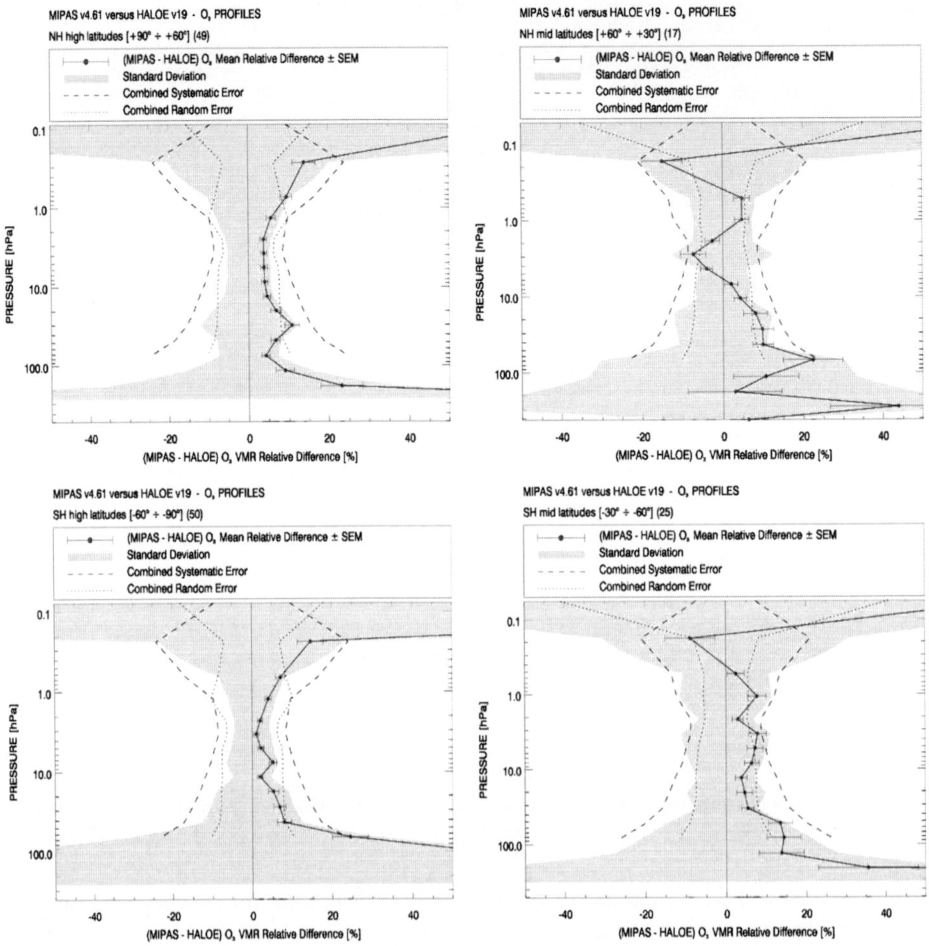

Figure 3.35. Comparison between MIPAS and HALOE O_3 profiles: zonal averages (Cortesi *et al.*, 2007).

A number of tests were performed to investigate whether the increased MIPAS–HALOE MRD below 50 hPa may be due to inefficiencies in the cloud detection algorithm. Two possible scenarios are:

— the cloud detection algorithm does not effectively identify and remove cloudy level-1b spectra allowing contaminated spectra to enter level-2 processing and resulting in anomalous ozone concentrations;
— the current CI-A threshold is not rigorous enough meaning that optically thinner clouds in the MIPAS FOV have a significant effect on lower altitude ozone concentrations and that this threshold should be raised.

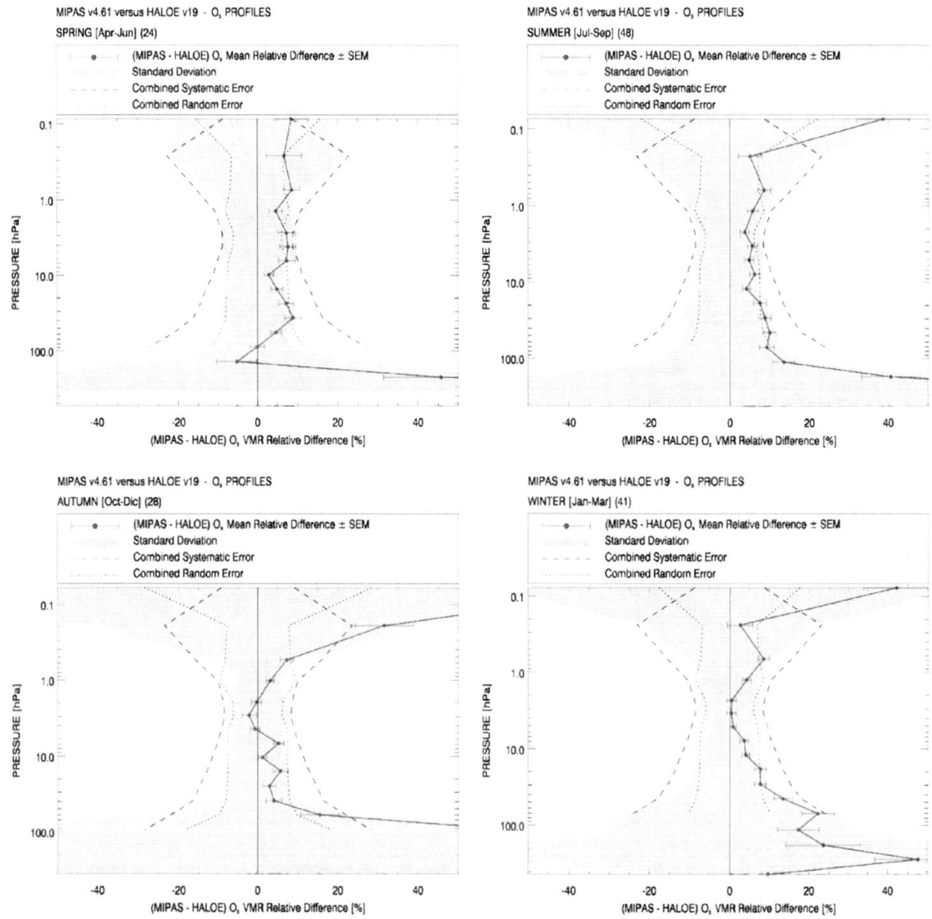

Figure 3.36. Comparison between MIPAS and HALOE O_3 profiles: seasonal averages (Cortesi *et al.*, 2007).

To test the above hypotheses, each MIPAS ozone profile used in the MIPAS versus HALOE comparisons was isolated and compared to its corresponding CI value. For the MIPAS data that were used in this MIPAS-HALOE analysis, ozone data corresponding to $CI \leq 1.8$ were successfully removed in both versions of MIPAS data. In general, it was found that the v4.61 processor did not always remove data corresponding to cloudy level 1b spectra. A sub-section of MIPAS data was then cloud-screened using a range of CI thresholds, including CI-A ≤ 2.2 (Sembhi *et al.*, 2006) and up to CI-A ≤ 3.0, and the analysis repeated. No significant change was found in the MRD between 50 and 100 hPa. Thus, we can verify that for these cases, which are mid-latitude/polar tropospheric clouds, the current CI can sufficiently remove cloud-corrupted ozone data and increasing the threshold does not improve the MRD. No coincidences were found in PSC-dominated seasons/latitudes or in the

Table 3.8. Statistics over all comparisons of MIPAS to HALOE: (a) zonal averages, (b) seasonal averages, and (c) all collocations (as in Table 3.7) (Cortesi *et al.*, 2007).

(a)

Latitude band	*Pressure range* (hPa)	*Mean relative difference*	*Root mean square*	*N*	*Month of year*
90°S–60°S	70–0.20	+4% ↔ +13%	+7% ↔ +20%	49	July 2, May–Jun.–July 3
60°S–30°S	68–0.10	−14% ↔ +22%	+20% ↔ +30%	17	Nov. 2, Nov. 3
30°N–60°N	81–0.10	−8% ↔ +14%	+28% ↔ +30%	25	May 3, Jan. 4
60°N–90°N	65–0.20	+14% ↔ +24%	+23% ↔ +31%	50	Jan. 3/4, July 2/3, Aug. 3, Nov. 3

(b)

Hemisphere	*Season*	*Pressure range* (hPa)	*Mean relative difference*	*Root mean square*	*N*
Both	Spring	88–0.20	−0.08% ↔ +6%	+8% ↔ +21%	24
	Summer	70–0.20	+5% ↔ +9%	+12% ↔ +20%	48
	Autumn	94–0.10	+31% ↔ +64%	+38% ↔ +94%	28
	Winter	70–0.20	+2.5% ↔ +22%	+20% ↔ +26%	41

(c)

Collocations	*Pressure range* (hPa)	*Mean relative difference*	*Root mean square*	*N*
ALL	70–0.20	+7% ↔ +16%	+21% ↔ +25%	141

tropics. It should be noted that the anomalous ozone concentration observed in MIPAS data in the tropical upper troposphere and lower stratosphere (UT/LS), where more frequent and higher tropical cirrus persist, are successfully removed when using a higher CI threshold (Sembhi *et al.*, 2006). From the tests described above, it is concluded that MIPAS-HALOE ozone comparisons observed between 50 and 100 hPa are not affected by cloud contamination and that the cloud detection algorithm efficiently removes corrupt v4.61 and v4.62 MIPAS data in these cases.

Comparisons of updated HALOE ozone data (versions 18 and 19) with correlative satellite instruments show that, generally, HALOE possesses a 5–10% negative bias at all altitudes below the ozone peak ($p \sim 10$ hPa), particularly in comparison to SAGE II versions 5.93 and 6.0 (Morris *et al.*, 2002), POAM III (Randall *et al.*, 2003), and ACE-FTS (McHugh *et al.*, 2005). The largest differences of greater than 30% usually occur at 15 km and below but also differences of up to 20% occur near 22 km in some regions (mostly the tropics and sub-tropics) with HALOE < SAGE II. These comparisons and also the results of Borchi and Pommereau (2007) show that HALOE has a tendency to be low near altitudes of 15 to 20 km and below and differences are largely due to the band model used to simulate ozone in the HALOE forward model and aerosol/cirrus effects that become dominant when the ozone signal reduces.

It is possible that systematic discrepancies in addition to inaccuracies in HALOE pointing at lower altitudes are likely to contribute to the increased MRD below 50 hPa. Forward model errors were improved in HALOE version 2.0 (Cortesi *et al.*, 2007). The remaining MRD is likely to be biased toward MIPAS forward model and instrumental factors such as assumptions of a horizontal homogeneous atmosphere and uncertainties in the apodized instrument line shape (ILS) that are most significant between 12 and 20 km.

3.5.6 Comparison with GOME O_3 profiles

The Global Ozone Monitoring Experiment (GOME) is a nadir-viewing backscatter UV–visible spectrometer measuring contiguously between 237–790 nm with a spectral resolution of 0.2–0.4 nm (Section 2.3.5). It has been operating on the second European Remote Sensing (ERS-2) satellite since 1995, with global coverage available up to May 2003. At the Rutherford Appleton Laboratory (RAL), a retrieval scheme has been developed to retrieve ozone profiles spanning the troposphere and stratosphere (Munro *et al.*, 1998), with vertical resolution of approximately 6 km in the stratosphere. The data produced by this retrieval scheme have been validated against ozonesondes and have been found to agree within 10% in the altitude range between 12 and 40 km. Larger biases have been identified in the tropical UT/LS below 50 hPa with GOME O_3 values up to 50% higher compared to ozonesondes.

In this work, MIPAS version 4.61 ozone data have been validated against GOME profiles for the period between November 2002 and May 2003. Matching MIPAS and GOME profiles were found using the specified coincidence criteria of 3 h and 300 km, with the best matched GOME profile used if there was more than one match to a given MIPAS profile.

GOME ozone data were available on a fixed pressure grid between 1,000 and 0.01 hPa. However, the comparison was restricted to altitudes below 1 hPa, as GOME values are not reliable at high altitudes. Additionally, any points where GOME *a priori* information was found to contribute significantly to the profile (using a cut-off where the reduction in error in the retrieved GOME data is less than 50% of the *a priori* error) were removed.

In order to compare the MIPAS data to GOME, the MIPAS profiles were first interpolated (linearly in log pressure) to the GOME pressure grid. Since GOME has a lower resolution than MIPAS, the GOME averaging kernels were applied to the MIPAS data to degrade its resolution to match that of GOME. As the GOME averaging kernels were only quoted in units of number density the retrieved MIPAS temperature profile was used to convert to units of VMR. In order to apply the averaging kernels the MIPAS profile was extended to cover the complete range of the GOME pressure grid (1,000–0.01 hPa) using ECMWF data below the lowest MIPAS level, and the GOME *a priori* profile at high altitudes. However, only values in the range of the original MIPAS data were used in the comparison, and a stringent check was applied to remove any points with a significant contribution from altitudes outside the MIPAS range.

Relative differences between the smoothed MIPAS profiles and the collocated GOME profiles were then determined by computing the mean absolute difference and dividing it by the mean GOME profile to obtain global, zonal, and seasonal MRD profiles.

For the estimate of GOME total error budget, random errors were obtained from the data file, while the systematic error was taken to be 10%. This dataset has been validated to have a bias better than 10% in the range from 12 to 40 km, although the errors may be greater than this at lower and higher altitudes. In the tropics below approximately 50 hPa, larger differences (up to 50%) are observed. Below 100 hPa, random errors in the GOME data can become large. MIPAS errors were interpolated to the GOME pressure grid and also had the GOME averaging kernels applied to give the appropriate errors for the smoothed MIPAS profile. In order to apply the averaging kernels, the random error profiles were extended with errors of 100% above and below the MIPAS amplitudes. The averaging kernels were applied to the random error using equation (3.20):

$$\mathbf{S}_{\mathbf{A}_{\text{rnd}}} = \mathbf{A}\mathbf{S}_{\text{rnd}}\mathbf{A}^{\text{T}} \tag{3.20}$$

where $\mathbf{S}_{\text{rnd}}$ is the MIPAS random covariance matrix (only the diagonal elements), whilst the systematic errors were derived from equation (3.21):

$$\mathbf{S}_{\mathbf{A}_{\text{sys}}} = \mathbf{A}\mathbf{S}_{\text{sys}} \tag{3.21}$$

where $\mathbf{S}_{\text{sys}}$ is the systematic error profile.

Results of the comparison between MIPAS and GOME O_3 measurements, averaged over the whole set of collocated profiles, are shown in Figure 3.37. Global means of GOME ozone-retrieved values and of MIPAS smoothed profiles are displayed in the left panel. On the right, the statistics of the relative differences and of the comparison error budget are presented. Only points at latitudes south of 60°N have been included in the global zonal mean, as there was found to be a problem with a number of GOME retrievals in Northern Hemisphere high latitudes in April and early May 2003. The mean relative difference between MIPAS and GOME ozone mixing ratios is within the combined systematic error in the pressure range between about 1.0 hPa and 200 hPa.

Moreover, as GOME only measures in sunlight, and the period of overlap between GOME and MIPAS was restricted, this comparison could achieve only a limited seasonal and latitudinal coverage. Calculation of seasonal mean relative differences for the periods December 2002–February 2003 and March–May 2003 showed that the resulting profiles (not shown here) do not exhibit any relevant features with respect to the global average. Zonal MRD profiles were obtained by averaging over the latitude bands (30°N–60°N), (30°S–30°N), (60°S–30°S), and. (90°S–60°S) (as shown in Figures 3.37 and 3.38). Peculiar behavior is found at

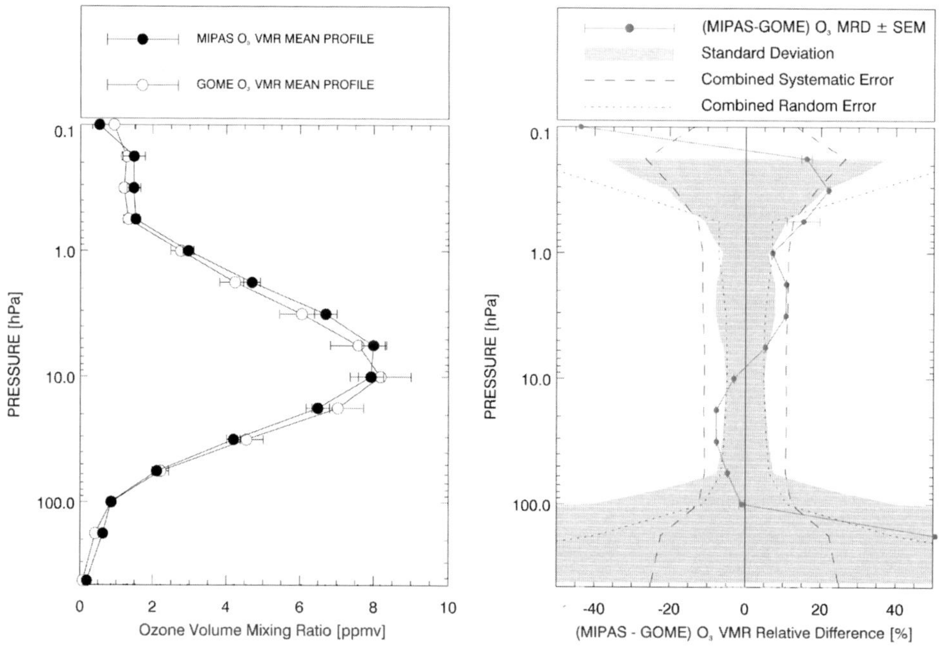

Figure 3.37. Comparison between MIPAS and GOME O_3 profiles. MIPAS and GOME O_3 VMR mean profiles calculated on all the collocations available from 90°S to 60°N (left panel). Corresponding statistics for the relative differences between MIPAS and GOME and associated comparison error budget (right panel) (Cortesi *et al.*, 2007).

higher pressure levels for low latitudes compared to mid and high latitudes: MRD values within ±5% are obtained in the belt from 30°S to 30°N, with MIPAS mostly underestimating the ozone content with respect to GOME around ~200 hPa; in the other bands, large positive values of the MRD are generally found below ~100 hPa. The latter can be explained by a few anomalously high values in the MIPAS profiles at these heights, possibly due to the presence of undetected clouds. In the tropics, most of the points at the lowest pressure levels are removed, either because the *a priori* contribution to the GOME retrieved value is more than 50% or because a significant area of the GOME averaging kernels lies below the bottom of the MIPAS profile (generally, 15 km at low latitudes). If we take into account that GOME retrieved profiles generally overestimate the ozone content in the tropical UT/LS, the negative bias observed for MIPAS data in comparison with GOME appears to be reasonably justified.

3.5.7 Comparison with SCIAMACHY and GOMOS

Van Gijsel *et al.* (2009) described the validation of ozone profiles obtained from three instruments on ENVISAT—the SCanning Imaging Absorption spectroMeter for

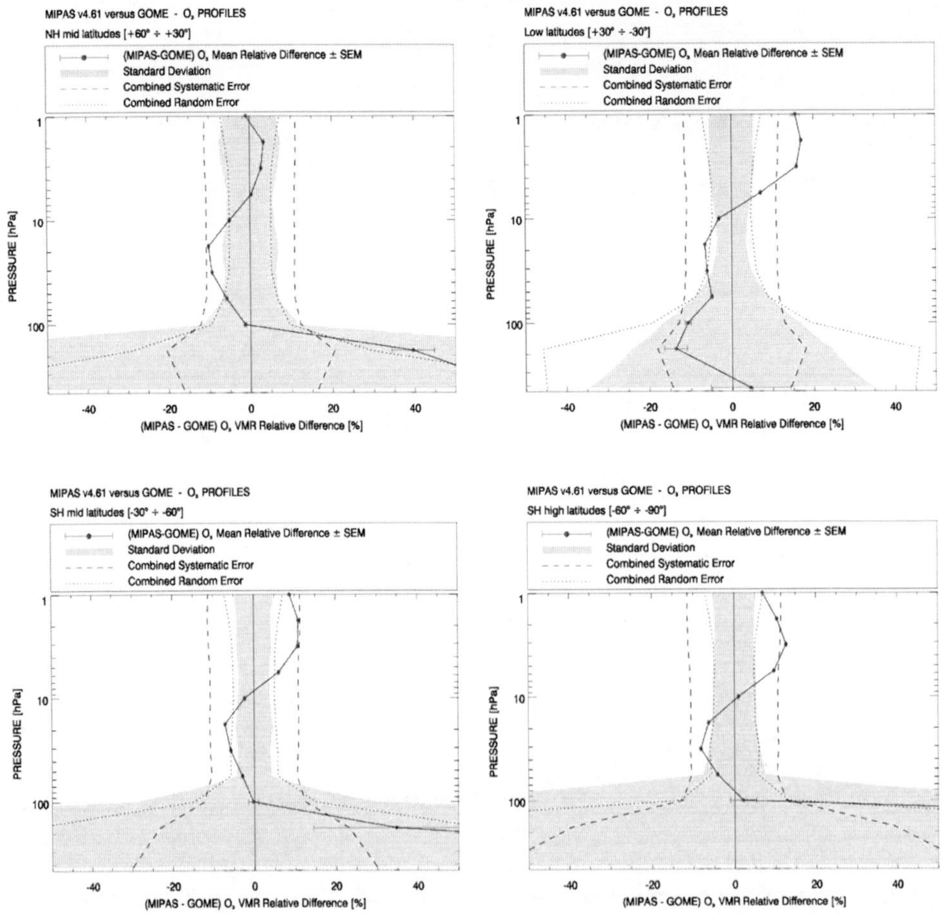

Figure 3.38. Comparison between MIPAS and GOME O_3 profiles: zonal averages (Cortesi *et al.*, 2007).

Atmospheric CHartographY (SCIAMACHY) (Section 2.5.2.5), the Michelson Interferometer for Passive Atmospherc Sounding (MIPAS), and the Global Ozone Monitoring by Occultation of Stars (GOMOS) (Section 2.2.8)—with data from ground-based lidar systems using DIAL (DIfferential Absorption Lidar). The ground-based measurements of stratospheric ozone were taken at 11 sites (see Table 3.9) with differential absorption lidars (DIALs), which make use of two wavelengths. Since the absorption by ozone is wavelength dependent, comparison of the return signals allows the derivation of ozone concentration as a function of altitude (see Section 1.6.5).

The validation approach followed by Van Gijsel *et al.* (2009) is similar to that of Meijer *et al.* (2004). Both satellite and lidar data were first interpolated using a linear

Table 3.9. Overview of lidar sites used by Van Gijsel *et al.* (2009).

Site	*Latitude* (°)	*Longitude* (°)
Alomar	69.3	16.0
Eureka	80.1	−86.4
Hohenpeissenberg	47.8	11.0
Réunion	−20.8	55.5
Lauder	−45.0	169.7
Mauna Loa	19.5	−155.6
Ny Ålesund	78.9	11.9
Observatoire Haute Provence	43.9	5.7
Río Gallegos	−51.6	−69.3
Table Mountain	34.4	−117.7
Tsukuba	36.1	140.1

spline to a common altitude grid and then compared in a quality assessment. The selected datasets were subsequently compared in terms of mean, median, and standard deviation of the differences for various subsets, such as the geographical region, measurement conditions, and collocation criteria. A lower confidence limit for the altitude was set for all the lidar data to 18 km and an upper confidence limit to 45 km. One prototype and three operational ESA satellite products from three atmospheric sensors have been compared with lidar: MIPAS reduced-resolution (RR) prototype data, SCIAMACHY level-2 processor version 3.01; MIPAS full-resolution (FR) level-2 processor version 4.61; and GOMOS processor version 5.00 (for details see Van Gijsel *et al.*, 2009). The SCIAMACHY ozone profiles (version 3.01) have been more precisely corrected for the altitude shift present in previous data versions and showed reasonable agreement with lidar. Ozone profile validation results showed good agreement of MIPAS (version 4.61) full-resolution data and a small positive bias was found for MIPAS reduced-resolution data throughout the stratosphere ranging from 0 to 20%. Nevertheless, individual comparisons showed very good agreement in the high vertical structures in the profiles. GOMOS data (version 5.00) from measurements under dark limb conditions agreed very well with the correlative data and no significant systematic errors are observed, except for the northern polar regions where GOMOS seems to underestimate ozone number densities slightly.

3.6 COMPARISON OF MIPAS DATA WITH ECMWF ASSIMILATED FIELDS

As part of the coordinated effort for the validation of MIPAS full spectral resolution measurements, we compared MIPAS O_3 profiles v4.61 with assimilated ozone fields obtained from ECMWF operational analysis data archived at the British Atmospheric Data Centre (BADC). The ozone mass mixing ratio was provided every 6 h on an N80 grid, and vertically on 60 model levels up to 0.1 hPa. This was converted to the volume mixing ratio, and spatially (in latitude and longitude) and temporally interpolated to the average geolocation and time of each MIPAS scan. The profiles were then interpolated vertically and had MIPAS averaging kernels applied. Prior to October 2003 the operational ECMWF system assimilated only data from SBUV/2 and GOME, which are limited in vertical resolution and restricted to daytime-only measurements. SBUV/2 data have been assimilated since April 2002 as six layers, with the lowest layer covering the altitude range between 16 hPa and the surface, and has been restricted to observations with solar zenith angles less than 84°. Total column ozone data from GOME were assimilated between April 2002 and June 2003, at latitudes between 40°N and 50°S, and for solar zenith angles less than 80°. Between October 7, 2003 and March 25, 2004 MIPAS (version 4.59) data were also assimilated. Therefore, the comparison is limited to the period from July 2002 to September 2003, during which ECMWF data represent an independent source for the validation of MIPAS-ENVISAT O_3 products.

Quantitative errors were not available for ECMWF ozone data, and no errors have been included for ECMWF in the plots shown in this section. In the analysis of ERA-40 ozone data quality conducted by Dethof and Hélm (2004), ozone profiles were generally found to compare well with independent observations, except for the case of high-latitude winter and spring profiles in both hemispheres, where large discrepancies in ECMWF data were observed. In addition, low bias was observed for peak ozone values in the tropics. The "Assimilation of ENVISAT Data" project (ASSET) compared ozone analyses including ECMWF for the period between July to November 2003 (Geer *et al.*, 2006). In general, it was found that ECMWF data agreed to within ±10% compared to sonde data throughout much of the stratosphere. However, larger biases were seen in the UT/LS, the troposphere, the mesosphere, and for profiles in the Antarctic region. Above 5 hPa there was observed to be a positive bias compared to HALOE, while in the mesosphere the model does not include diurnal variability. Low bias was observed in ECMWF data at the tropical tropopause, whilst in the lower stratosphere ECMWF data were generally biased high compared to the sondes.

The procedure adopted for comparing ECMWF and MIPAS ozone data is based on the same scheme described previously for comparison with concurrent satellite measurements, although data were averaged over pressure bins, rather than interpolated to a fixed pressure grid. All the MIPAS ozone profiles within the comparison period were selected, except those for which any of the quality flags were set as bad or that contained ozone VMR values greater than 100 ppmv, or equal to 10^{-10} ppmv, or where the associated variances were negative. For each

MIPAS profile, collocated values were obtained, as previously mentioned, by interpolation of ECMWF ozone VMR fields, both horizontally and temporally. MIPAS averaging kernels were applied vertically to ECMWF profiles using a modified version of the routine generally adopted in all other cases. In this procedure, nominal MIPAS averaging kernels were adjusted to match the true pressure levels of each individual MIPAS measurement, whilst correlative ECMWF data were interpolated to the fine-pressure grid on which the averaging kernels were supplied. These averaging kernels were then applied to the adjusted ECMWF data, providing correlative data on the same pressure grid as each of the individual profiles.

The absolute differences between MIPAS O_3 VMR and ECMWF values were computed for each of the individual MIPAS profiles. These were then binned into fixed pressure bins, defined by the midpoints between the nominal retrieval levels for the pressure profile from the mid-latitude reference atmosphere. The mean relative difference (defined by the ratio between the mean absolute difference and the mean ECMWF profile in percent) in each pressure bin was then determined, along with the corresponding mean pressure in each bin.

Global mean profiles of the relative differences have been calculated, along with zonal and seasonal averages over five latitude bands (90°N–65°N, 65°N–20°N, 20°N–20°S, 20°S–65°S, 65°S–90°S) and four seasons (JJA, including data for July–August 2002 and June–August 2003; SON, including data for September–November 2002 and September 2003; DJF, including data for December 2002 and January– February 2003; MAM, including data for March–May 2003). Random and systematic error estimates were allocated to each mean profile of the relative differences, taking into account only the contribution from MIPAS uncertainties.

Mean O_3 VMR profiles from global averages of MIPAS v4.61 and ECMWF data are shown in Figure 3.39, along with their mean relative difference and combined error estimates. MRD mostly falls within MIPAS systematic error and appears to be associated with a slight altitude shift between MIPAS and ECMWF profiles, which is reflected in significant biases (i.e. $|\text{MRD}| >$ MIPAS systematic error) around 2 hPa and 50 hPa. A closer insight can be gained by examining the latitudinal and seasonal dependency of the relative difference between MIPAS and ECMWF ozone profiles. This is shown in Figure 3.40, where zonal and seasonal averages, calculated over the 2002–2003 data, are displayed. A substantially good agreement is evident throughout all seasons at mid-latitudes both in the Northern and in the Southern Hemisphere, whilst major differences are clearly highlighted in the tropics and at high latitudes, particularly in the Antarctic region. In the latitude band between 20°N and 20°S, we observe that MIPAS constantly overestimates the O_3 mixing ratio relative to ECMWF by up to 100% at pressures higher than ~50–60 hPa (approximately 20–25 km). On the other hand, negative bias in the range of ~10 to ~25% characterizes the MRD at levels above ~5 hPa in Southern Hemisphere high latitudes, especially during summer and spring (and, slightly reduced, during winter, confirming the bias already observed by the ASSET results in ECMWF data). In the same latitude band large positive and negative differences are found in different seasons around ~100 hPa (up to +40% in winter and autumn and ~40% in summer) possibly connected to the presence of polar stratospheric clouds.

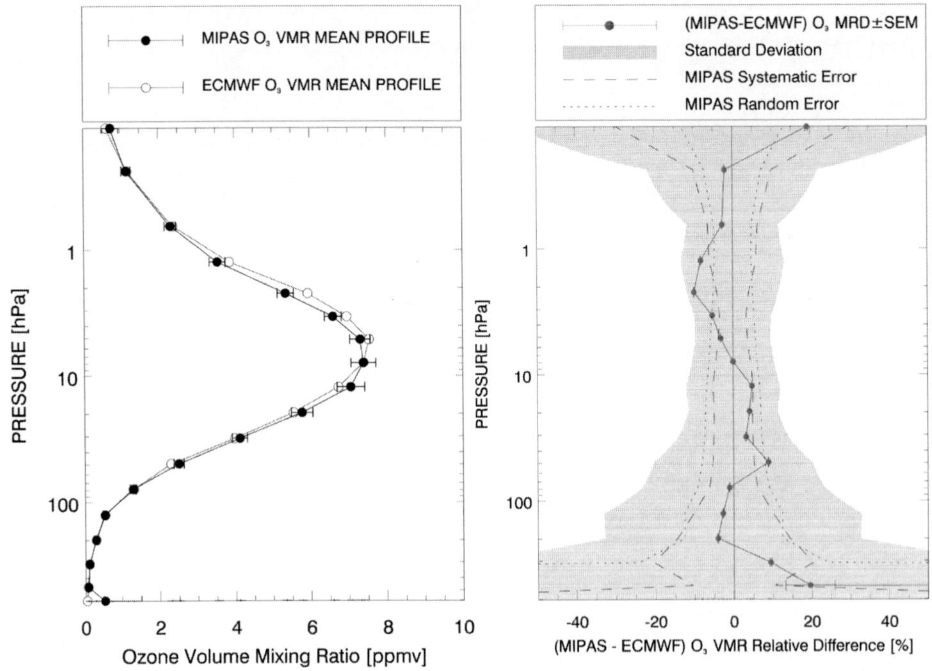

Figure 3.39. MIPAS v4.61 and ECMWF O_3 VMR mean profiles: global averages (on the left) and corresponding mean relative difference, standard deviation, and MIPAS errors (on the right) (Cortesi *et al.*, 2007).

Similarly, we assume that the discrepancy observed between MIPAS and ECMWF ozone values at the tropical tropopause might be caused by the presence of high-altitude cirrus clouds in the latitude range 20°N–20°S.

In summary, it was concluded that good agreement is found between MIPAS v4.61 and ECMWF ozone data, with the only notable exception of discrepancies observed at Southern Hemisphere high latitudes at about 100 hPa and in the tropical tropopause, which might be attributed, respectively, to the presence of polar stratospheric clouds and high-altitude cirrus (see Section 1.6.5). All other relevant differences found in seasonal and zonal averages can be explained by known effects due to the quality of ECMWF data.

3.7 SUMMARY OF MIPAS COMPARISONS

In this section we summarize the main points of comparison between the different categories of correlative data selected for the validation of MIPAS O_3 operational products and we will make an attempt to merge the key results obtained from each group of reference measurements into an overall assessment of MIPAS ozone data quality. We start our summary by focusing on the outcome of comparisons with other satellite sensors that—in view of the better spatiotemporal coverage—are capable of providing, in their own right, a general indication on the validity of MIPAS O_3 profiles. In Figure 3.41 we report the global average of the relative difference between MIPAS and collocated ozone profiles obtained by concurrent spaceborne instruments.

Excellent agreement is found in most of the comparisons at pressures ranging from approximately 50 hPa up to 1.0 hPa, with MRD values constantly within ±10% (with the only small exception of the value ~10 hPa for the comparison with ACE-FTS). The results of individual comparisons consistently show that, within this pressure range (roughly corresponding to the altitude interval between 50 km and 20–25 km), the observed bias is always lower than the combined systematic error. The slightly larger bias observed in the comparison with ACE-FTS measurements, marginally exceeding the combined systematic error around 10 hPa, can possibly be explained by the limited number of coincident profiles available for the validation of MIPAS measurements and by the coarse characterization of the ACE-FTS systematic error available for our comparison. Below 20–25 km and above 50 km, an increase in the absolute values of the global MRD of ozone VMR profiles is generally observed. At higher pressure levels, particularly around 100 hPa, MIPAS O_3 values are 5% to 25% larger compared to the majority of the validating satellite sensors. Only in the case of the comparison with POAM III data, however, is this positive bias larger than the combined systematic errors of the comparison. No coherent indications can be derived from the large differences observed at the lowermost levels, where the occurrence of stronger atmospheric gradients results in a significant enhancement of different components of the comparison error (primarily those due to time–space mismatch and to differences in vertical and horizontal smoothing). The effect of larger natural variability on the spatiotemporal scale of the selected coincidence criteria can also be highlighted by looking at the standard deviation of mean relative differences between MIPAS and other spaceborne sensors compared with combined random error. In general, SD and random uncertainty match well in the stratosphere down to approximately 20 km or ~60–80 hPa (25 km or ~30–40 hPa in the case of SAGE II and Odin-SMR), whilst standard deviations become increasingly larger than estimated random error in the lowermost stratosphere and in the upper troposphere. Much greater SD values are found throughout the full altitude range only in the case of the comparison with ACE (Figure 3.33). The remarkably good quality of MIPAS v4.61 and v4.62 ozone profiles in the pressure range 1–50 hPa, emerging from the results of satellite comparison, is amply confirmed by the extensive analysis which was conducted using a variety of ground-based correlative data. Ground-based validation and satellite measurements, on the other hand, also reflect a similar degradation in the outcome of the comparison for the UT and LS regions in the middle stratosphere. In particular, pole-to-pole validation—based on ozonesonde, lidar, and MWR data from the NDACC network—clearly indicates the variability of results for different synoptic regions below 25 km, with a prevalence of positive biases between 5% and 20%. Only in a few cases is a significant bias found between 25 and 40 km and the mean difference is always lower than 10%. Evidence of low bias in

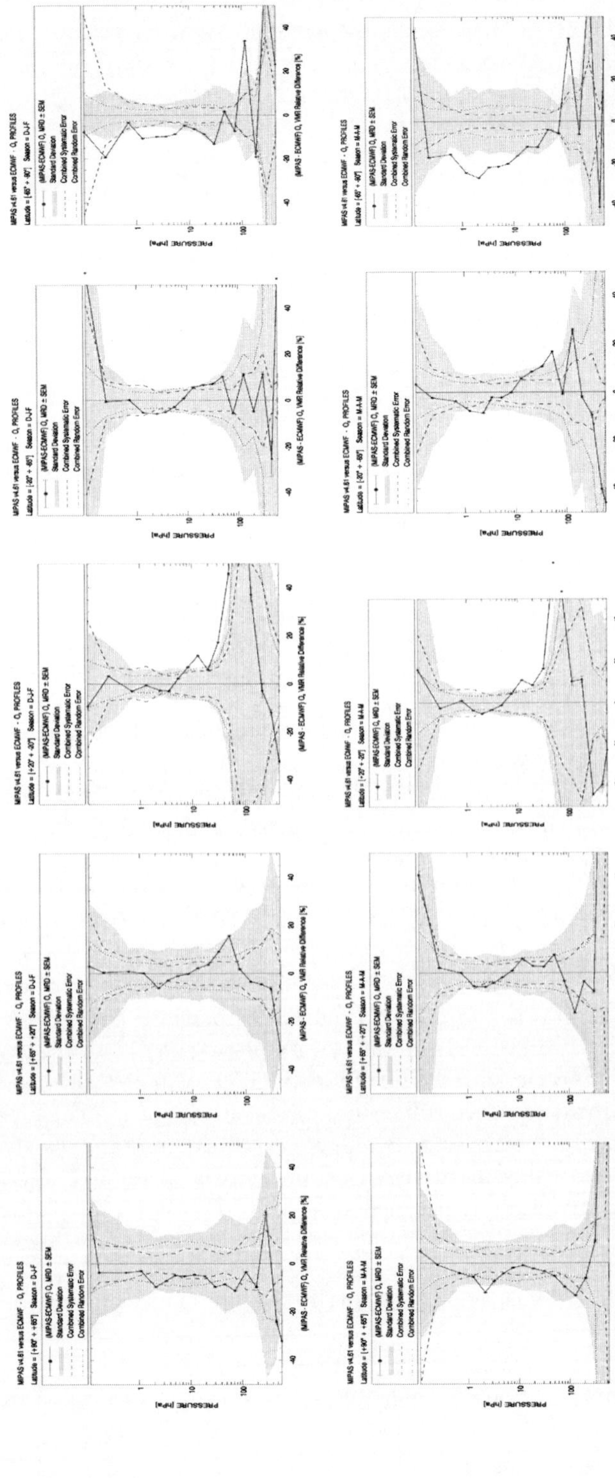

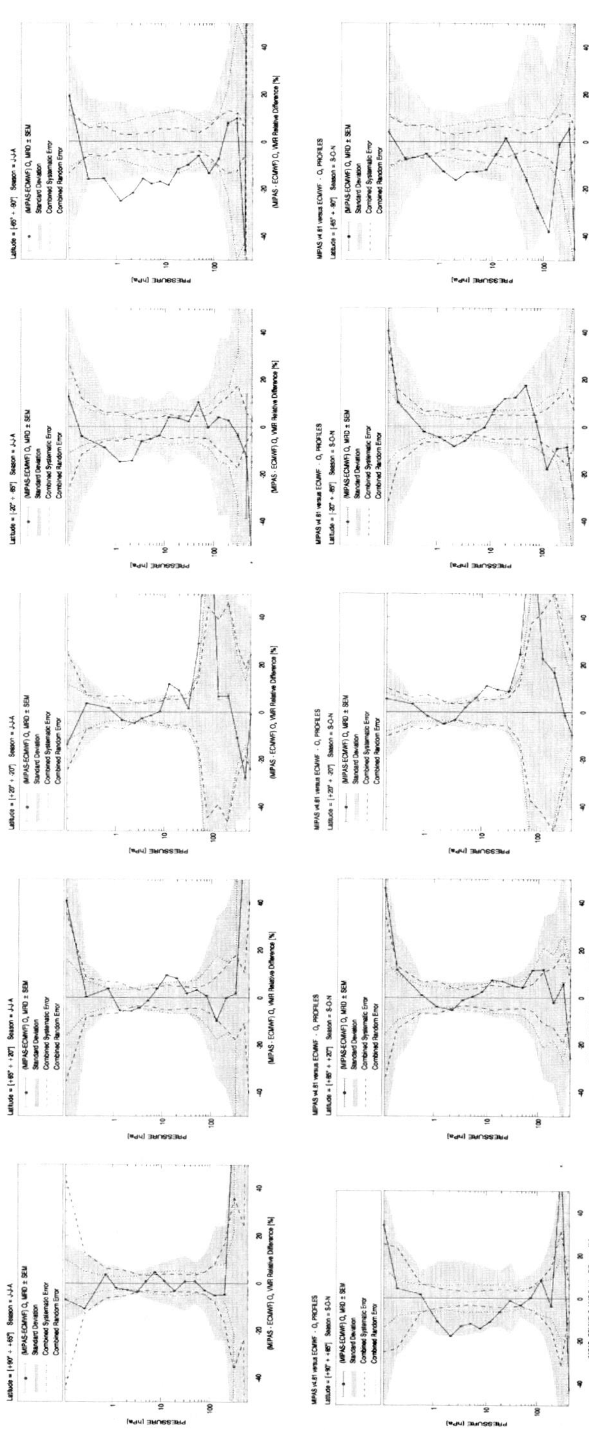

Figure 3.40. Comparison between MIPAS v4.61 and ECMWF O_3 VMR mean profiles: zonal and seasonal averages (Cortesi *et al.*, 2007).

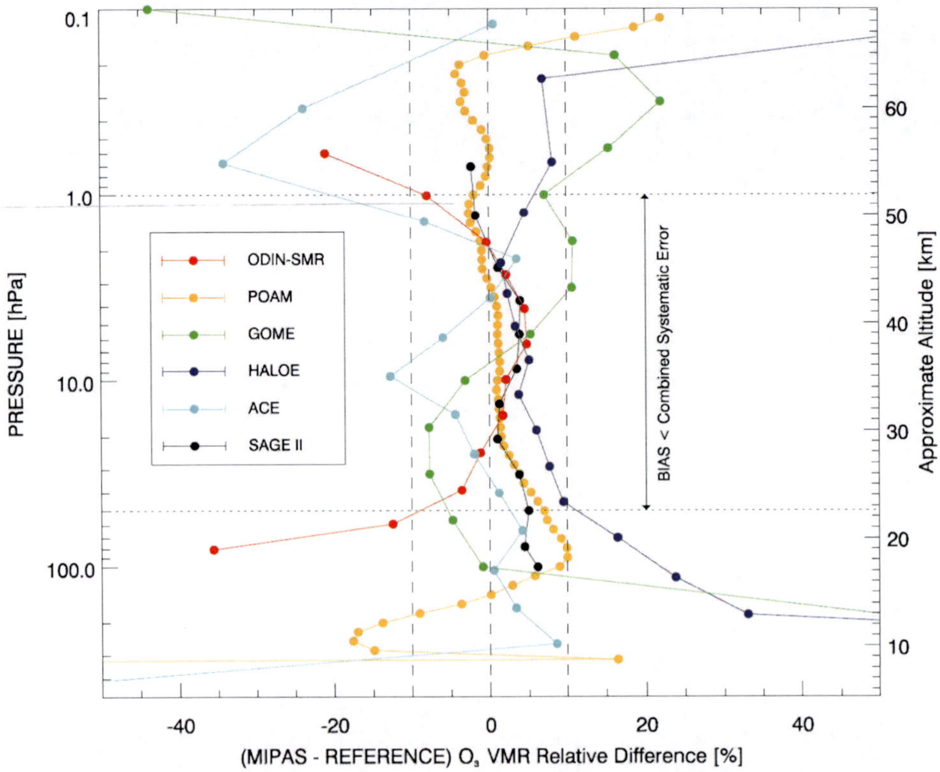

Figure 3.41. Summary plot of global mean relative differences between MIPAS O_3 VMR profiles and coincidence measurements by concurrent satellite sensors (Cortesi *et al.*, 2007).

MIPAS ozone measurements is also occasionally found, as in the case of the comparison with FTIR O_3 partial columns described in Section 3.3.3.3. Here, significant mean differences are obtained at two stations (Lauder and Arrival Heights), which could possibly be caused by the use of different microwindows for the retrieval of the O_3 profile.

In order to better investigate the source of the larger discrepancies found in the altitude range between 25 km and MIPAS lowest tangent heights, a valuable tool is offered by the comparison with coincident measurements acquired on board high-altitude platforms. The possibility of planning validation flights coinciding optimally with the satellite overpass and under the most favorable meteorological conditions makes it feasible (especially for aircraft payloads) to acquire correlative measurements with minimum spatiotemporal mismatch (often much lower than the required 300 km and 3 h) and under completely cloud-free conditions. Most of the balloon and aircraft data presented in this section do in fact satisfy the above-mentioned requirements, as in the case of MIPAS-B balloon data that were generally obtained in almost perfect spatiotemporal coincidence with MIPAS-ENVISAT or in the case of the M-55 *Geophysica* validation flights that were mostly executed under cloud-free conditions. In these cases, we obtain substantial agreement between MIPAS O_3 data and collocated reference profiles at lower-stratospheric and upper-tropospheric altitudes, where significant biases and lower precision had been generally found by ground-based and satellite validation experiments. Our results from balloonborne validation measurements (MIPAS-B, FIRS-2, and SPIRALE) typically provide a mean difference of O_3 mixing ratio within ± 0.5 ppmv for the full vertical range of the comparison (~10–35 km). The only exception is offered by the results of the comparison with O_3 profiles recorded during the trans-Mediterranean flight of the IBEX spectrometer. In this case, a low bias of MIPAS O_3 VMR was observed using trajectory analysis with mean relative differences as high as 30% between 15 and 20 km. The airborne dataset from the validation campaigns with the M-55 *Geophysica* aircraft provides further evidence of the good quality of MIPAS-ENVISAT ozone measurements in the range from 10 to 20 km, showing that a good match is normally found between satellite and aircraft profiles and that discrepancies exceeding our estimate of the total error budget can often be explained in terms of different airmasses measured by the satellite or aircraft sensors. Even though balloon and aircraft measurements can be used to derive information of limited statistical value—due to the sparse character of their geographical and temporal coverage—these results still suggest that the large discrepancies observed below 20–25 km must be partly ascribed to the influence of natural variability in the outcome of the comparison.

Additional sources of the discrepancies observed in O_3 VMR values at lower altitudes can be identified on the basis of complementary hypotheses that emerge from specific sub-sets of individual results. A critical issue is certainly represented by the current choice of thresholds for the cloud index value (Raspollini *et al.*, 2006), which may not be sufficiently stringent to enable the removal of all significant cloud contamination effects from MIPAS ozone retrievals at all latitudes (Glatthor *et al.*, 2006; Sembhi *et al.*, 2006). This might explain, for instance, the high MIPAS O_3 values responsible for the large differences observed in the comparison with GOME collocated profiles for $p > 100$ hPa in the extratropics (cp. Section 3.5.6); and this is reasonably the cause of the positive bias (MRD > 30% below 18 km) between MIPAS measurements and lidar profiles in the tropics (cp. Section 3.3.3.2), in ECMWF results (cp. Section 3.6), and in the poor agreement that is found in the same region with respect to ASUR data (cp. Section 3.4.5). However, no definite proof of cloud contamination has been established in this study.

Moreover, when evaluating the outcome of our validation exercise, we should properly take into account some of the limitations associated with our estimate of the systematic and random component of the comparison error budget. The primary limitation here is whether the choice of considering different *a priori* systematic errors as contributing either to bias (purely systematic errors) or to standard deviations (systematic errors with a random variability) should in principle be made according to the kind of spatiotemporal average of the individual comparison. On the contrary, the application of uniform criteria to a variety of time and space scales might result in under or over-estimation of both MIPAS systematic and random uncertainty. In addition to this, we must remember that the *a priori* error values we used for our estimates rely on linear approximation of the Oxford University reference forward model and tend to underestimate the actual contributions of systematic uncertainties to total error budget especially at the lower altitudes.

In this section, we have presented the results of an extensive analysis aimed at validation of MIPAS-ENVISAT O_3 vertical profiles obtained during the instrument's full spectral resolution mission (July 6, 2002–March 26, 2004) and retrieved using versions 4.61 and 4.62 of the ESA operational processor.

The validation strategy was based on the synergistic use of a variety of correlative datasets from independent sources, with complementary features in terms of the trade-offs between accuracy and spatiotemporal coverage.

It is of interest to present the comparison of MIPAS ozone partial columns and vertical profiles with collocated measurements from instruments at more than 50 NDACC ground-based stations (ozonesondes, lidar, FTIR, and microwave radiometers), from remote-sensing and *in situ* sensors aboard stratospheric aircraft (MIPAS-STR, SAFIRE-A, FOZAN, ASUR), from balloons (FIRS-2, IBEX, MIPAS-B2, SPIRALE), and from six concurrent satellite sensors (SAGE II, POAM III, ACE-FTS, Odin-SMR, HALOE, GOME), as well as with assimilated fields from ECMWF.

Special attention has been paid to rigorous selection of reference data, based on homogeneous criteria that were only slightly adapted from case to case to match the specific features of each validation dataset and of the selected comparison methodologies.

The overall picture that can be derived from the output of comparisons with individual groups of collocated ozone measurements provides a sound basis for the required assessment of the validity of MIPAS ozone profiles over a wide range of altitudes, latitudes, and seasons.

The very good agreement that was found between 50 hPa and 1 hPa with the majority of correlative datasets demonstrates the inherent high quality of MIPAS ozone measurements through most of the stratosphere. The mean relative differences with correlative data are within $\pm 10\%$ and no apparent bias was observed in this pressure range that could not be explained by known systematic effects already included in the comparison error budget. Similarly, the variability of global mean differences between MIPAS and coincident O_3 profiles appears to be fully consistent with the expected random error from 1 hPa down to at least 30–40 hPa.

We can conclude therefore that in the altitude interval between approximately 20–25 km and 50–55 km the existing estimate of MIPAS O_3 systematic error sources provided by Oxford University are substantially correct and that MIPAS O_3 precision error, as computed from level-2 data and from the Oxford University *a priori* estimate for p, T error propagation, is equally appropriate. We recall here that, according to the pre-launch calculations of the Oxford team, the systematic and the random components of the *a priori* error budget (evaluated for a single ozone profile at mid-latitudes and under daytime conditions) have an average value of $\sim 6\%$ and $\sim 5\%$, respectively, in the altitude interval between 20 km and 52 km. At lower and higher altitudes, a roughly linear increase in both random and systematic uncertainties is expected up to ~15–20% at 10 km and up to ~30–35% at 68 km.

Below 20 km, we generally observe a degradation of the agreement between MIPAS and most coincident data, with the appearance of statistically significant biases from 5% to approximately 25% at 100 hPa and standard deviation substantially larger than the combined random errors by a factor of 1.5 to 3.0 in the range ~50–100 hPa. Part of the discrepancy at pressure levels greater than ~100 hPa can reasonably be traced to the higher variability of airmasses in the lowermost stratosphere and upper troposphere (as clearly shown in Section 3.3.3.3 by the detailed analysis, based on NDACC data, demonstrating that atmospheric inhomogeneities and, particularly, horizontal gradients represent a major component of the comparison error budget).

Further sources of uncertainty, affecting the results of our comparisons, have been identified that can be more directly translated into specific recommendations for possible improvements of MIPAS ozone data quality and error estimate.

The positive biases associated with unrealistically large ozone values at the bottom end of MIPAS profiles and observed with respect to various sets of correlative data—cp. the results of the comparison with ground-based ozonesonde and lidar measurements (ASUR or GOME data are typical examples)—can be reasonably ascribed to residual cloud contamination. This interpretation would suggest that a more conservative choice, in terms of cloud-filtering capabilities, for the threshold value of the cloud index is needed. The questions of the effects of clouds and making corrections for them in ozone retrieval by various instruments have been considered for various satellite-flown instruments by Acarreta *et al.* (2004), Koelemeijer *et al.* (2001), Wang *et al.* (2008) among others.

The evidence for an underestimate of MIPAS random error in the lower stratosphere and upper troposphere is also consistent with the linear approximation of the forward model adopted by Oxford University for the *a priori* evaluation of systematic errors with random variability. As clearly demonstrated by the results of the comparison in the middle and upper stratosphere, this approximation properly describes *a priori* uncertainties down to 20–25 km, but might become inadequate at lower altitudes.

In general, we cannot avoid the intrinsic limit of the linear approach to calculation of *a priori* contributions to MIPAS random error, but some margin of improvement can still be identified in our estimate of specific components. A typical example is provided by p, T error propagation that we entered in our calculation of the overall random uncertainty of the comparison. This is an approximate value, both as a consequence of the linearization introduced to calculate p, T propagation matrices and of the assumptions made for the choice of the pressure and temperature error value to propagate. Since the latter values are mostly underestimated at lower altitudes, this leads to underestimation also of the contribution of p, T error propagation to the overall random error budget.

A more realistic estimate could be obtained by considering the actual values for the pressure and temperature retrieval error and propagating it by means of the pre-computed matrices included in ESA level-2 data products. A problem due to incorrect implementation of the p, T error propagation algorithm in MIPAS operational data v4.61 and v4.62 prevented the use of this procedure in the comparison, but will be corrected in future versions, thus making it possible to slightly improve the estimate of MIPAS random uncertainty.

At pressures lower than 1 hPa and, particularly, for the uppermost retrieval levels of MIPAS ozone profiles, a tendency to observe larger differences is generally shown by our analysis. However, fewer coincidences, mostly from correlative measurements provided by other satellite sensors, are available at these altitudes and the output of the comparison cannot achieve the same statistical value as for the rest of the profile. Moreover, the larger uncertainties of the reference data in this range and the relatively poor characterization of their random and systematic errors do not allow us to consider them, in a strict sense, as a useful dataset for validation purposes. As a consequence, we cannot derive any quantitative assessment for the quality of MIPAS ozone profiles for $p < 1$ hPa.

Taking into account the summary of the results and the recommendations and caveats expressed, it can be asserted that MIPAS ozone operational data v4.61 and v4.62 are validated in the vertical range from $p \sim 1$ hPa down to the lower stratosphere and, therefore, can be used in quantitative scientific studies.

In Section 3.2 we used the ground data from Athens as an example of the comparison of ground data and satellite measurements. So also in Sections 3.3–3.7 we have used MIPAS data as an example of the comparison of one satellite ozone dataset with various other remotely sensed ozone datasets.

3.8 OTHER INTERCOMPARISONS BETWEEN VARIOUS OZONE-MONITORING SYSTEMS

Very recently Zhang *et al.* (2010) have presented and analyzed three different methods to validate and intercompare satellite retrievals of atmospheric composition. The intercomparison methods were: (1) independent validation of the Tropospheric Emission Spectrometer (TES) and Ozone Monitoring Instrument (OMI) (both aboard the EOS Aura satellite) with ozonesonde measurements (*in situ* method); (2) use of a chemical transport model as the comparison platform (CTM method); and (3) comparison of OMI ozone profiles with TES profiles smoothed by OMI averaging kernels (averaging kernel smoothing method). Zhang *et al.* (2010) have applied each method to the analysis of differences between TES and OMI tropospheric ozone retrievals. For the *in situ* method using the global ozonesonde database, they found 528 TES/sonde coincidences and 2,568 OMI/sonde coincidences within 60°S–60°N for 2005–2007 (80% of these were between 20°N and 60°N). Both instruments showed an overall positive bias. The global mean bias at 500 hPa was 5.3 ± 12.3 ppbv for TES and 2.8 ± 6.6 ppbv for OMI. Application of the CTM method to the 180 TES/OMI/sonde coincidences for 2006 showed that it closely reproduces the results of the *in situ* method while providing a globally complete intercomparison perspective. The CTM method revealed that differences between TES and OMI were generally within ± 10 ppbv (18%). The largest differences were observed at summertime northern mid-latitudes where TES was higher than OMI (TES – OMI differences > 10 ppbv), and over tropical continents where TES was lower than OMI (TES – OMI differences < –10 ppbv).

Zhang *et al.* (2010) analyzed three different methods for the validation and intercomparison of satellite determinations of atmospheric ozone concentration and applied them to tropospheric ozone retrievals from TES and OMI. The first method (the *in situ* method) uses *in situ* vertical profiles for absolute instrument validation; this is limited by the sparseness of *in situ* data. The second method (the CTM method) uses a chemical transport model (hence, CTM) for the intercomparison; it provides a globally complete intercomparison with relatively small noise added by model error. The third method (the averaging kernel smoothing method) involves smoothing the retrieved profile from one instrument by the averaging kernel matrix of the other; it also provides a global intercomparison. Application to data from a full year (2006) of TES and OMI data showed mean positive biases of 5.3 ppbv (10%) for TES and 2.8 ppbv (5%) for OMI at 500 hPa relative to *in situ* data from ozonesondes. They showed that the CTM method (using the GEOS-Chem CTM) closely approximated results from the *in situ* method while providing global coverage (Fusco and Logan, 2003). It showed that differences between TES and OMI are generally less than 10 ppbv (18%), except for northern mid-latitudes in summer and over tropical continents. The CTM method allows for well-constrained CTM evaluation in places where the satellite observations are consistent. Zhang *et al.* found that GEOS-Chem underestimates tropospheric ozone in the tropics, reflecting a combination of possible factors, and overestimates ozone in the northern subtropics and southern mid-latitudes, likely because of excessive stratospheric influx.

3.8.1 TOMS, GOME, GOMOS, and SCIAMACHY data

Antón *et al.* (2010) described the validation of empirically corrected total ozone column (TOC) data provided by the Earth Probe TOMS using ground-based measurements recorded by a well-calibrated Brewer spectroradiometer located at El Arenosillo in Spain. In addition, satellite TOC observations derived from the OMI with the TOMS algorithm were also used in their work. The agreement between EP-TOMS TOC data and Brewer measurements was excellent ($R^2 \sim 0.92$) even for the period 2000–2005 when a higher EP-TOMS instrument degradation occurred. Despite its low magnitude, the EP-TOMS-Brewer relative differences depend on the solar zenith angle (SZA), showing a clear seasonal cycle with amplitude between ±2% and ±4%. Conversely, OMI-Brewer relative differences show a constant negative value around –1% with no significant dependence on SZA. No significant dependence on ground-based to satellite-based differences with respect to the EP-TOMS scene or to the OMI cross-track position is observed for either satellite retrieval algorithm. Finally, the TOC estimated by the two satellite instruments has also been compared, showing good agreement ($R^2 \sim 0.88$). Overall, we conclude that the empirical correction of the EP-TOMS data record provides a reprocessed set of high quality. However, according to Antón *et al.* (2010) the EP-

TOMS data after 2000 should not be used in calculations of global ozone trending due to remaining errors in the dataset and because it is no longer an independent dataset.

Comparison between GOME-2/MetOp and Brewer TOC data for one whole year of measurements in five locations in Spain has been performed by Antón *et al.* (2009). This comparison showed that satellite TOC underestimates ground-based measurements with a mean offset of 3.05%. Although the relative differences between GOME/ERS-2 data and Brewer measurements showed a smaller offset, these differences had a significantly higher variability than the differences obtained when GOME-2/MetOp and Brewer TOC data were compared. In addition, Antón *et al.* (2009) found that TOC data from the new GOME-2/MetOp instrument showed no dependence on solar elevation and cloudiness conditions. However, this instrument showed a slight dependence with respect to TOC values measured by Brewer spectroradiometers, and a significant dependence on the satellite scan angle. Direct comparison between both satellite instruments indicated that GOME-2/MetOp ozone data underestimates GOME/ERS-2 data by 1.46% on average and that relative differences showed no seasonal dependence. The *a priori* ozone profiles used in the GOME-2/MetOp TOC retrieval algorithm were compared with true ozone profiles from ozonesonde measurements. The results showed that there was excellent agreement in the middle and high stratosphere where the greater part of the ozone amount is located, while differences increased in the lower stratosphere and troposphere.

The combined total ozone measurements from the Global Ozone Monitoring Experiment (GOME), SCIAMACHY, and GOME-2 constitute a data record covering more than 13 years, and this record is expected to extend until 2020. Since the onset of GOME observations in 1995, total ozone has been operationally retrieved using algorithms based on the differential optical absorption spectroscopy (DOAS) technique. There have been a number of algorithm improvements since then and the high accuracy for the product has culminated in the latest version of the operational GOME Data Processor (GDP) 4.1 (Liu *et al.*, 2005; Van Roozendael *et al.*, 2006) used for GOME/ERS-2 since 2004, for SCIAMACHY/ENVISAT since 2006, and for GOME-2/MetOp since 2007. Although correlative studies show excellent overall agreement with ground-based reference observations as well as with complementary satellite data, persisting discrepancies have been observed, particularly for large solar zenith angles (Balis *et al.*, 2007b).

Lerot *et al.* (2010) described in considerable detail a direct-fitting retrieval algorithm (GODFIT: GOME Direct-FITting) in which simulated backscattered spectral radiances from a radiative transfer model (Linearized Discrete Ordinate Radiative Transfer, LIDORT, v3.3) are fitted to measured radiances in a physically consistent way (Spurr *et al.*, 2004, 2005, 2007). Lerot *et al.* compared the GOME total O_3 columns (collection 2) retrieved with GODFIT with the OMI columns (collection 2) retrieved with the TOMS v8.5 algorithm (Bhartia and Wellemeyer, 2002). Compared with results from GDP 4.1, Lerot *et al.* showed that using GODFIT leads to a reduction in seasonal dependences on ground-based satellite differences and to an improvement in the accuracy of measurements at high latitudes. This is particularly useful for the monitoring of ozone concentrations in polar regions and of the gradual recovery of the ozone layer, which is expected to happen in the next decades as a consequence of the Montreal Protocol and its amendments (see Chapters 5–7). GODFIT has been implemented in the operational Universal Processor for UV/VIS Atmospheric Spectrometers (UPAS) at the German Aerospace Center (DLR) to produce the new GDP 5.0 total ozone data product.

Fioletov *et al.* (2008) described the comparison between Dobson and Brewer spectrophotometer and filter ozonometer data available from the World Ozone and Ultraviolet Data Center (WOUDC) with satellite total ozone measurements from TOMS, OMI, and GOME. Five characteristics of the difference with satellite data were calculated for each site and instrument type: the mean difference, the standard deviation of daily differences, the standard deviation of monthly differences, the amplitude of the seasonal component of the difference, and the range of annual values. All these characteristics were calculated for five 5-year-long bins and for each site separately for direct Sun (DS) and zenith sky (ZS) ozone measurements. The main percentiles were estimated for the five characteristics of the difference and then used to establish criteria for "suspect" or "outlier" sites for each characteristic. About 61% of Dobson, 46% of Brewer, and 28% of filter stations located between 60°S and 60°N have no "suspect" or "outlier" characteristics. In nearly 90% of all cases, Dobson and Brewer sites demonstrated 5-year mean differences with satellites to be within ±3% (for DS observations). The seasonal median difference between all Brewer DS measurements at 25°–60°N and GOME and OMI overpasses remained within ±0.5% over a period of more than 10 years. The performance of the satellite instruments was also analyzed to determine typical measurement uncertainties. It was demonstrated that systematic differences between analyzed satellite instruments were typically within ±2% and very rarely outside the ±3% envelope. As the satellite instrument measurements appear to be better than ±3%, it was argued that ground-based instruments with precision values worse than ±3% are not particularly useful for the analysis of long-term changes and comparison with numerical simulations.

The validation of ozone distributions derived from GOMOS (see Section 2.3.6) flown on ENVISAT for the period 2002–2008 is described by Kyrölä *et al.* (2010a). GOMOS measurements provide a possibility of investigating nighttime global vertical distributions of ozone (and of other trace gases) from the tropopause to the mesosphere and even to the lower thermosphere.

As is the case with all new instruments, a considerable amount of effort has been devoted to the geophysical validation of GOMOS products and their comparisons with results from other satellite-flown instruments. GOMOS data validation has been conducted since the summer of 2002. Comprehensive validation of ozone profiles against ground-based and balloonborne instruments was described by Meijer *et al.* (2004). They showed that dark limb GOMOS ozone profile data agree very well with the results from other sources between 14 and 64 km altitude; the differences showed only a small (2.5–7.5%) negative bias in the GOMOS data and a standard deviation of 11–16%. We have already mentioned the comparison of

GOMOS data with MIPAS data in Section 3.5.7; the extensive validation work involving the comparison of GOMOS data with ACE (Section 2.5.1.6) data validation by Dupuy *et al.* (2009) should also be mentioned.

Mze *et al.* (2010) compared ozone profiles obtained by GOMOS with ozonesonde data from the SHADOZ network. They only used nighttime ozone profiles and selected eight Southern Hemisphere stations; 7 years of GOMOS datasets and 11 years of balloonborne sondes were used in their work. A monthly distribution of GOMOS ozone concentrations was performed in the upper troposphere and in the stratosphere (15–50 km). A comparison with SHADOZ was made for altitudes in the range between 15 km and 30 km. In the 21–30 km altitude range, satisfactory agreement was observed between GOMOS and SHADOZ, although some differences were observed depending on the station. The range for monthly differences generally decreases with increasing height and was found to be within $\pm 15\%$. The agreement declined below about 20 km. Median differences were almost within $\pm 5\%$, particularly above 23 km. However, a large positive bias was found below 21 km in comparison to SHADOZ.

Kyrölä *et al.* (2010a) used GOMOS dark limb measurements from 2002 to 2008 and restricted their work by requiring the solar zenith angle at the tangent point to be greater than 107°. Instrumental problems with GOMOS made it necessary to restrict the azimuth range of the pointing system from the original −10° to 90° (measured with respect to the antivelocity vector of ENVISAT) to a window of 20° with a variable center. These changes meant that the selection of stars was not constant during the timeframe of their study. To ensure the reliability of GOMOS data during the whole period considered, Kyrölä *et al.* paid special attention to the quality of retrievals from each star individually. Kyrölä *et al.* (2006) studied the quality of GOMOS measurements during 2003 and found that stars with a magnitude weaker than 1.9 and cooler than 7,000 K may easily fail to capture the whole ozone profile from 15 to 100 km. Over the course of time, the GOMOS signal-to-noise ratio decreased due to aging of the instrument. Kyrölä *et al.* (2010a) examined the data quality for the whole period 2002–2008 and compared time series formed by individual stars. In order to guarantee data quality and consistency over the whole period considered they removed all ozone measurements from 57 different stars. The temperatures and magnitudes of these stars fulfilled approximately the restrictions mentioned above. The remaining number of occultations considered was then 173,223. Extensive discussion of the method used and the results obtained for ozone can be found in sections 3 and 4 of Kyrölä *et al.* (2010a).

Van Gijsel *et al.* (2009) described the validation of ozone profiles obtained from three instruments on ENVISAT—SCIAMACHY, MIPAS, and GOMOS—with data from ground-based lidar systems using DIAL. This has already been discussed in Section 3.5.7 in connection with MIPAS.

3.8.2 MLS data

There have been several studies on the validation of Aura Microwave Limb Sounder (MLS) ozone data (see Section 2.5.3.4).

Froidevaux *et al.* (2008) compared MLS version 2.2 ozone data with coincident ozone profiles from other satellite instruments, as well as from aircraft lidar measurements taken during Aura Validation Experiment (AVE) campaigns. Ozone comparisons were also made between MLS data and balloonborne data. Froidevaux *et al.* found better agreement in the comparisons using MLS version 2.2 ozone data than the version 1.5 data. The agreement and MLS uncertainty estimates in the stratosphere were often of the order of 5%, with values closer to 10% (and occasionally 20%) at the lowest stratospheric altitudes, where small positive MLS biases can be found. There is very good agreement in the latitudinal distributions obtained from MLS and in the coincident profiles from other satellite instruments, as well as from aircraft lidar data along the MLS track.

Krzyścin *et al.* (2008) compared MLS data from 2004 to 2006 with the results of Umkehr measurements at Belsk (52°N, 21°E). MLS ozone vertical profiles from version 1.5 and 2.2 retrieval algorithms and various Umkehr retrievals were considered and there was good agreement between the ozone content in Umkehr layers. A mean difference of the order of ±5%, a standard deviation of the relative differences of less than 10%, and a correlation coefficient larger than 0.5 were found in the mid-stratosphere (Umkehr layers 5–7) and in the upper stratosphere (combined layers 8–10). In the lower stratosphere (combined layers 2–4) the mean difference and/or standard deviation were larger but the correlation coefficient was still high (of the order of 0.8). The ozone content in Umkehr layers obtained by integration of MLS version 1.5 and 2.2 ozone profiles behaved very similarly in comparisons with ground-based data.

The precision, resolution, and accuracy of the data produced by the MLS version 2.2 processing algorithm were discussed and quantified by Livesey *et al.* (2008). Ozone accuracy was estimated at ~40 ppbv +5% (~20 ppbv +20% at 215 hPa). Comparisons with expectations and other observations show good agreements for the ozone product, generally consistent with the systematic errors quoted above.

3.8.3 SAGE data

Fioletov *et al.* (2006) compared coincident ozone data from the SBUV and SAGE II instruments with data from ozonesondes and the Umkehr method to estimate the stratospheric ozone variability and standard uncertainties of these different measurements. Below 20 km over northern midlatitudes, estimated measurement uncertainties for SBUV(/2), ozonesondes, and Umkehr are similar (~8% for 0 to 20 km integrated ozone), although only ozonesondes have a high vertical resolution there. From 20 to 28 km, the estimated uncertainties (4–6%) for all systems were substantially smaller than the ozone variability in winter (10–15%), but they were comparable in summer (−5%). Above 28 km, ozonesonde uncertainties were larger than or comparable to ozone variability and much larger than the uncertainties with SBUV(/2), Umkehr, or SAGE II data. Umkehr measurement uncertainties at 24–32 km were about 5% and lower than ozonesonde uncertainties (7–13%) at these levels. SBUV(/2) data were used to evaluate differences between different types of ozonesondes and to show that correction by total ozone measurements noticeably reduces ozonesonde uncertainties. The latitudinal dependence of ozone variability and instrument uncertainties were studied using pairs of collocated SBUV(/2) and SAGE II measurements. There was good correlation between these measurements over middle and high latitudes. Over the tropical region, the correlation coefficients are modest (about 0.5) but significant in all layers except at 28–31 km. It was noted that, where ozone variability is comparable to instrument uncertainties, comparison with a climatology based on a large number of observations may provide a better insight into instrument performance than intercomparison of a small number of quasi-coincident measurements.

McLinden *et al.* (2009) presented a stratospheric vertically resolved, monthly, zonal mean ozone dataset based on SAGE and SBUV data from 1979 to 2005. Drifts in individual SBUV instruments and inter-SBUV biases were corrected using SAGE I and II by calculating the differences between coincident SAGE-SBUV measurements. In this way the daily, near-global coverage of SBUV(/2) is combined with the stability and precision of SAGE to provide a homogeneous ozone record suitable for trend analysis. The resultant SAGE-corrected SBUV dataset, showed, for example, a more realistic Quasi-Biennial Oscillation signal compared to the one derived from SBUV data alone. Furthermore, this methodology can be used to extend the dataset beyond the lifetime of SAGE II.

3.8.4 TES data

Shephard *et al.* (2008) compared TES nadir observations with nearly coincident spectral radiance measurements from the Atmospheric Infrared Sounder (AIRS) on Aqua and special scanning high-resolution interferometer sounder (SHIS) underflights. Modifications to the L1B calibration algorithms for TES version 2 data resulted in significant improvements for TES-AIRS comparisons. The comparison of TES with SHIS (adjusted for geometric differences) showed mean and standard deviation differences of less than 0.3 K at warmer brightness temperatures of 290–295 K. TES/SHIS differences were less than 0.4 K at brightness temperatures of 265–270 K. TES/AIRS comparisons showed mean differences of less than 0.3 K at 290–295 K and less than 0.5 K at 265–270 K with standard deviation less than 0.6 K for the majority of the spectral regions and the brightness temperature range. Results from long-term comparison of TES sea surface temperature (SST) observations with the Reynolds optimally interpolated (ROI) SST product was considered evidence of the radiometric stability of TES.

3.8.5 ACE and IASI data

Batchelor *et al.* (2010) have recently reported on the Canadian Arctic Atmospheric Chemistry Experiment (ACE) validation campaigns at Eureka, Nunavut (80.05°N, 86.42°W), which took place during International Polar Year (IPY) springs of 2007 and 2008. Within the framework of these campaigns there took place an intercomparison of measurements obtained by four Fourier transform infrared (FTIR) spectrometers during the highly variable Arctic polar sunrise period. Comparisons between the three ground-based FTIR spectrometers have shown small inter-instrumental differences well within the estimated uncertainties of measurements and consistent with side-by-side intercomparisons conducted around the world. TOC measurements, made with the higher resolution Canadian Network for the Detection of Atmospheric Change (CANDAC) Bruker 125HR (a high-resolution Fourier transform spectrometer; Batchelor *et al.*, 2009) were shown to compare with the similar resolution Environment Canada ABB Bomem DA8 (a FTIR spectrometer; Batchelor *et al.*, 2009) to within 3.5%. TOC measurements from both these instruments were shown to agree to within 6.5%. The mean biases between the ACE-Fourier Transform Spectrometer (FTS) and Bruker 125HR for 2007/2008 were, respectively, 1.1/−1.5% for TOC. These values were generally insignificant within the standard error (Figure 3.42).

Ozone data for the upper troposphere/lower stratosphere (UT/LS) measured by the Atmospheric Chemistry Experiment Fourier Transform Spectrometer (ACE-FTS) on Canada's SCISAT-1 satellite have been validated using aircraft and ozonesonde measurements by Hegglin *et al.* (2008). They found upper bounds to the relative differences in the mean found between the ACE-FTS and SPURT air-

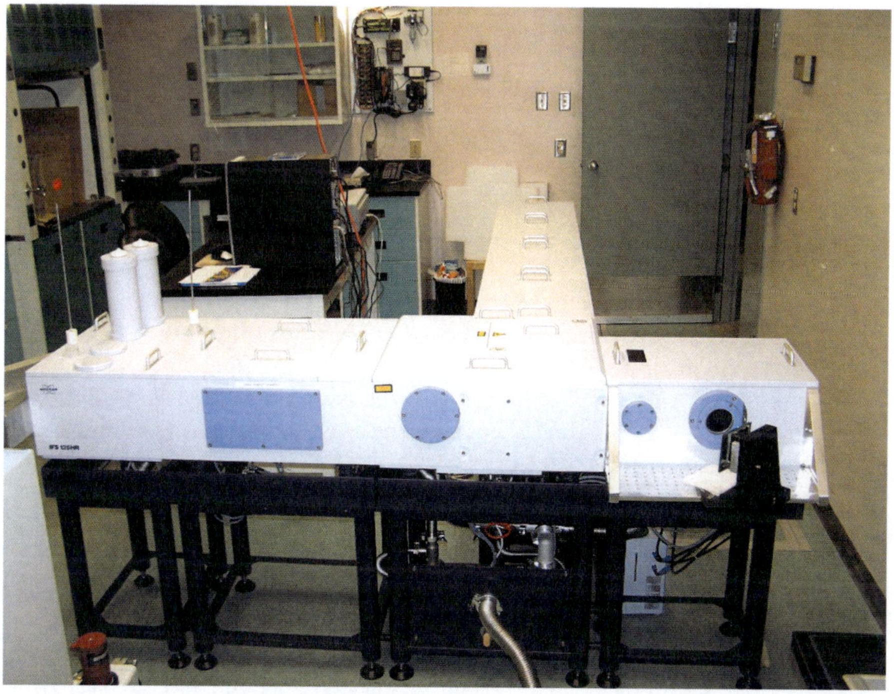

Figure 3.42. Bruker 125HR installed at Polar Environment Atmospheric Research Laboratory (PEARL), located on Ellesmere Island, Canada (*http://acebox.uwaterloo.ca/eureka/Eureka 2008/Our%20Instruments.html*).

craft measurements in the upper troposphere (UT) and lower stratosphere (LS), respectively, of ±25% and ±19% for ozone. These relative differences for ozone can be narrowed down by using a larger dataset obtained from ozonesondes, yielding a high bias in the ACE-FTS measurements of 18% in the UT and relative differences of ±8% for measurements in the LS. When taking into account the smearing effect of the vertically limited spacing between measurements of the ACE-FTS instrument, the relative differences decrease by 5–15% around the tropopause, suggesting a vertical resolution of the ACE-FTS in the UT/LS of around 1 km. Hence, the ACE-FTS offers unprecedented precision and vertical resolution for a satellite instrument, which will allow a new global perspective on UT/LS tracer distributions.

The ground-based—Fourier Transform Infrared spectrometer (FTIR) and Brewer—measurements performed at Izaña (Tenerife) between March and June 2009, were used by Viatte *et al.* (2010) to validate TOC observations from the Infrared Atmospheric Sounding Interferometer (IASI) sensor aboard the MetOp satellite launched in October 2006 (Clerbaux *et al.*, 2009). A scatter of only 0.7% between the two ground-based measurement methods was found, which documented the very good quality of the ground-based data. However, a systematic mean relative difference of 4.2% was observed. According to Viatte *et al.* (2010) this systematic difference may be due to systematic errors in the spectroscopic parameters. Furthermore, they implemented comparisons of TOC data over Izana, as they had been derived from two different IASI retrieval algorithms. According to their results excellent agreement of −2.0(±1.4)% and 1.5(±2.2)% was found when comparing FTIR and Brewer with IASI results derived from an analytical algorithm. By contrast, differences of −5.2(±1.9)% and −0.9(±2.5)% were found with the operational product of IASI compared to the FTIR and Brewer measurements. Finally, Viatte *et al.* (2010) also compared the TOC over Izana derived by FTIR and Brewer with TOC data derived from other satellite instruments (OMI, GOME-2). This comparison showed excellent agreement of −0.5(±0.7)% and 3.5(±1.2)% for OMI, and −2.4(±1.1)% and 1.5(±1.5)% for GOME-2 with FTIR and Brewer, respectively.

The coherence of stratospheric ozone time series retrieved from various observational records at Haute Provence Observatory (OHP, 43.93°N, 5.71°E) was discussed by Nair *et al.* (2011). The analysis took the form of an intercomparison of collocated ozone measurements by Light Detection and Ranging (lidar) with measurements from the SBUV/2, SAGE II, HALOE, MLS, and GOMOS, as well as with *in situ* ozonesondes and ground-based Umkehr measurements performed at OHP. A detailed statistical study of the relative differences of ozone observations was performed to detect any specific drifts in the data. On average, all instruments showed their best agreement with lidar at 20–40 km, where deviations are within ±5%. Discrepancies were found to be somewhat higher below 20 km and above 40 km. The agreement with SAGE II data is remarkable since average differences were within ±1% at 17–41 km. In contrast, Umkehr data systematically underestimated lidar measurements in the whole stratosphere, albeit a near-zero bias was observed at 16–8 hPa (~30 km). Drifts were estimated using simple linear regression for the long-term (more than 10 years long) datasets

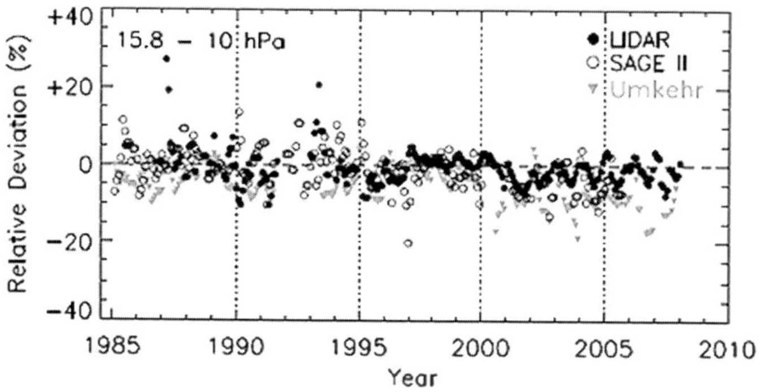

Figure 3.43. Monthly average of the relative differences of collocated ozone column measurements of SBUV(/2) with lidar, SAGE II, and Umkehr at 15.8–10 hPa. The dashed horizontal line represents 0% and the dotted vertical lines represent 1990, 1995, 2000, and 2005. Data are smoothed by a 3-month running mean (*source*: Nair *et al.*, 2011).

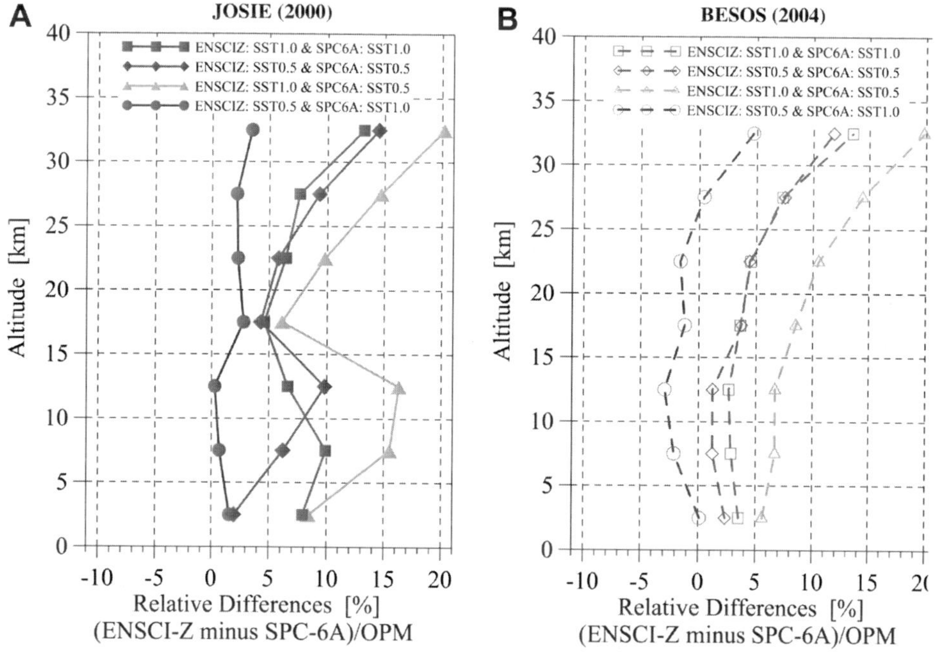

Figure 3.44. Comparison of ozone measurement, relative to a standard ozone photometer instrument, as measured by different combinations of ECC instrument type (ENSCI or SPC) and sensing solution types in intercomparisons: (A) JOSIE-2000 laboratory tests, Jülich, Germany; (B) a field experiment, BESOS, in Wyoming, 2004. The data are averaged over 5 km altitude bins (*source*: Thompson *et al.*, 2011).

analyzed, from monthly averaged difference time series. The derived values were less than ±0.5% yr^{-1} in the 20–40 km altitude range and most drifts were not significant at the 2σ level (Figure 3.43).

3.8.6 Ozonesonde intercomparisons

Several JOSIE (Jülich Ozone Sonde Intercomparison Experiment) simulation experiments to assess ECC sondes with various sensing solutions have been conducted at this facility. In addition, in 2004 the BESOS (Balloon Experiment on Standards for Ozone Sondes) field campaign at the University of Wyoming (Laramie) confirmed JOSIE results in the real atmosphere. The JOSIE and BESOS activities characterized systematic biases among different instrument types and techniques (Figure 3.44).

ECC sondes are manufactured by either the Science Pump Corporation (model type: SPC-6A) or the Environmental Science Corporation (model type: ENSCI-Z). When operated under the same conditions the differences between the two ozonesonde types are about 5–10% (see also Section 1.7.1).

When ozonesondes of the same type are filled with different sensing solutions, systematic differences are also observed. For each ECC type the use of 1.0% KI and full buffer (SST 1.0 in Figure 3.44) gives 5% larger ozone values compared to a 0.5% KI and half-buffer (SST 0.5) solution, and as much as 10% larger values compared with 2.0% KI without buffer (SST 2.0) (Thompson *et al.*, 2011).

4

The dynamics of atmospheric ozone

As mentioned in Chapter 1, the most characteristic feature of global total ozone dynamics is the presence of strong spatiotemporal variability, which has been documented and analyzed in detail in recent years. The development of various kinds of ozone observations constitutes a great step forward. However, a number of problems still remain unresolved. This is why there is a need for long-term continuity of ozone measurements (Kaye and Readings, 1998). Several important conclusions were made in this context by the Committee on Earth Observation Satellites (CEOS) Analysis Group (Kaye and Readings, 1998):

- There is a need to use all relevant observational techniques for ozone observations including space-based measurements of TOC and vertical ozone profile, ozonesondes, aircraft, and ground-based measurements (especially Dobson, Umkehr, microwave, and lidar remote-sounding techniques). Long-term trend studies should be a principal purpose of data interpretation.
- Simultaneous monitoring of various other quantities, such as meteorological parameters, solar forcing, aerosols, and other atmospheric trace constituents, is a prerequisite of adequate ozone data interpretation.
- Special attention must be paid to problems of calibration and data quality. In this context combined ground-based, balloon, aircraft, and satellite observations are important for data verification.

As mentioned in the *Stratospheric Processes and Their Role in Climate Report* (SPARC) (Harris and Hudson, 1998; Harris *et al.*, 1998) global ozone observations display both long-term and considerable interannual variability. Sources of long-term variability include changes in solar output, temporal changes in the abundance of stratospheric aerosols, and variability in stratospheric circulation leading to changes in ozone transport on seasonal, interannual, and decadal timescales. Several components of interannual variability have been identified in ozone observations: in particular, the Quasi-Biennial Oscillation (QBO) and the 11-year solar cycle (Kramarova and Kuznetsov, 2009; Sekiyama *et al.*, 2006). Moreover, numerous anthropogenic impacts (through ozone-depleting substances) on the ozone layer (see Chapter 5) have to be mentioned as well as very complex ozone–climate interactions.

Among the remarkable findings reported in the Report on the SPARC/IOC Workshop on Understanding Ozone Trends (SPARC, 2001) were the following:

- The overall quality and availability of ozone data are in reasonably good shape. Clarification is needed however on the possible satellite bias between the Northern and Southern Hemispheres and some apparent degradation in EP-TOMS in 2000 (see Sections 2.3.1 and 3.2). Several composite datasets for total ozone, in which instrument-to-instrument biases have been addressed, have been prepared. The aim is to intercompare these composite datasets and to have identified and understood discrepancies by December 2001, in time for the assessment. The need for continued high-quality ozone data from ground and space was clearly recognized.
- The 2D models can reproduce most of the northern mid-latitude ozone trends. In particular, several 2D models have simulated the observed drop in ozone following the Mt. Pinatubo eruption in 1991, as well as the more recent higher values, supporting the link between chlorine chemistry and aerosol loading. The agreement between measurements and models does not appear to be as good for the Southern Hemisphere. It has proved hard to accurately quantify the impact of polar chemical processes at mid-latitudes (a certain influence on mid-latitude trends) from measurements and from models. High-altitude loss should be the simplest to model.
- Ozone trends in the lowermost stratosphere are particularly sensitive to changes in dynamics, whether changes in Rossby wave events, in the frequency of low-ozone events (mini-holes and laminae) or changes in principally tropospheric phenomena such as the North Atlantic Oscillation/Arctic Oscillation. Diagnosing the links between these undoubtedly coupled phenomena is hard, but it is necessary in order to avoid "double-counting" when calculating ozone trends. In the long term, the magnitude of the dynamical influence on ozone trends depends critically on whether the short-term relations between dynamics and ozone stay constant (as currently assumed) or vary with time as the overall atmospheric system responds to the original forcing (whatever its origin). As a result, the dynamical influence depends on the time period considered. Current estimates of the influence are likely to be upper limits. It is important not to simply add ozone trends derived from individual statistical studies looking at the influence of one particular dynamical process, as the various dynamical processes in the atmosphere are clearly related. It is unclear which dynamical proxies should be used in statistical models. A better understanding of which dynamical processes affect ozone is clearly needed, as well as any radiative feedback or radiative forcings from the changes in stratospheric water vapor or greenhouse gases.

- Another approach, which takes account of atmospheric motions, is to consider the atmosphere as a number of regimes with boundaries chosen according to dynamic criteria rather than simple geographic criteria, which in ozone trend studies have typically been latitude bands. New work has been presented that uses meteorological tropopause fronts to separate total ozone into regimes. Narrow distributions of total ozone were found if the regions close to the fronts were excluded. Within the regimes, small trends in TOC were observed. Changes in the relative contributions of tropical, mid-latitude, and polar regimes to overall trends at mid-latitudes were also reported. Trends in the lowermost stratosphere, calculated relative to the tropopause height, have been used to distinguish stratospheric and tropospheric effects.
- The development of tools to investigate the turnaround in ozone depletion resulting from decreased halogen loading is progressing. An improved understanding of the processes affecting ozone trends is an essential element of this work. Most of the current statistical effort is being devoted to the further development of time series approaches rather than the introduction of different statistical techniques.

To summarize achievements made in the study of global TOC dynamics, we begin with the problem of TOC trends. As the end of the 20th century appoached, this problem was discussed in detail in several reports (Bojkov and Visconti, 1998; SPARC, 1997; WMO, 1999).

4.1 TOTAL OZONE TRENDS

The problem of filtering out regular trends in strong spatiotemporal variability is fairly complicated and includes two principal aspects; (1) analysis of instrumental biases and (2) developing an adequate statistical approach. We have already discussed the intercomparison of different ozone datasets in Chapter 3.

The results of TOC observations at various sites were considered in a number of papers (WMO, 1999). For instance, Denis *et al.* (WMO, 1999) discussed observations at Réunion Island (21°S) in the western Indian Ocean and at Tarawa (1°N) in the western Pacific which started in 1992–1993 in this intertropical area within the NDSC (Network for Detection of Stratospheric Change) program. Besides the signature of stratospheric transport in the annual cycle in the tropics and QBO modulation at the equator, the data indicate the presence of a significant tropospheric component in TOC variability. At both stations TOC minima occur during the season of maximum development of the ITCZ.

Dubrovský and Kalvová (1998) studied the mean annual TOC cycle and its correlation with meteorological conditions on the basis of a 29-year series (1962–1990) of Dobson measurements in Hradec Králové (Czech Republic).

Fioletov *et al.* (1998) analyzed the great difference between TOC trends derived from ground-based (Dobson measurements) and satellite (SBUV, SBUV/2, TOMS) observations in the Southern Hemisphere (the ratio of relevant TOC values is nearly

1:2 at 30°–55°S). It has been shown that the very strong austral winter TOC decline, calculated from SBUV-SBUV/2 data, was mostly related to instrument and/or algorithm errors for high solar zenith angles (>83°) as well as to specific features of the orbit of the NOAA-11 satellite. Actual ozone decline (winter and year-round) is approximately the same for both hemispheres in the mid-latitudes (except Antarctica).

Hare and Fioletov (1998) discussed typical problems with the use of TOC data from the World Ozone and Ultraviolet Radiation Data Center (WOUDC) located in Toronto, since the early 1960s (this data series is nearly 40 years long).

Analysis of re-evaluated Dobson TOC data, obtained at five stations on the U.S. mainland between 1979 and 1994, made by Komhyr *et al.* (1998) yielded average linear downward trends of −4.54, −5.26, −2.35, and −2.06% per decade, respectively, for winter, spring, summer, and autumn seasons, and −3.68% on an annual basis. The contribution of various causes of TOC change (QBO, Pinatubo eruption, etc.) were assessed.

Reinsel *et al.* (1989) carried out a trend analysis of volcanic aerosol-corrected Umkehr ozone profile data from 10 stations between 1977 and 1987. The results indicated a significant overall negative trend, exclusive of trend variations associated with solar flux variations, of the order of −0.5% per year in Umkehr layers 7–9 over the period mentioned, and a significant positive solar cycle association in all layers 4–9 with the maximum solar effect estimated to occur in layer 9. A comparison was made by Reinsel *et al.* (1989) between solar backscattered ultraviolet (SBUV) monthly average ozone profile data near the 10 Umkehr stations and corresponding Umkehr data, corrected for aerosol errors, to investigate possible drifts. The results showed that a substantial overall negative linear drift in SBUV data relative to corrected Umkehr data in layers 7–9 took place, with estimated values of the drift of the order of −1.0% per year for layers 8 and 9. The problem of volcanic aerosol correction of Umkehr observations of ozone profile for five northern mid-latitude Dobson stations was also considered by DeLuisi *et al.* (1989). Using corrected data they reached the conclusion that between 1979 and 1986 decreases in ozone concentrations in layers 8 and 9 were, respectively, 9% and 15%. According to the Dobson Umkehr data considered by Tiao *et al.* (1986), ozone trends for the 3A, 3B, and 4A layers (100–50 hPa) were about −0.5% per year.

Karol *et al.* (1992) studied TOC linear trends in the Northern Hemisphere on the basis of ground-based Dobson and filter observations between 1975 and 1986. They discovered strong changes of not only amplitude but even sign of mean monthly TOC trends dependent on season and time period. TOC trends also varied according to the phase of the 11-year solar activity cycle (rise and fall in 1975–1980 and 1981–1986, respectively) and QBO phases. Fioletov *et al.* (2002) analyzed long-term variations in TOC mean zonal values in the Northern Hemisphere as well as TOC trends for two latitudinal zones (40°–52°N and 53°–64°N) again using ground-based observation data. In the case of long-term variations, QBO-induced variations were very distinctly shown.

The results obtained by Bojkov *et al.* (1990) indicate significantly more negative trends during winter (December to March: −1.2%, −2.1%, and −3.0% per decade for latitudes 35°, 45°, and 55°N, respectively), than during summer (May to August: −0.6% per decade, without clear dependence on latitude), especially at higher latitudes, with trends in winter becoming more negative with increasing latitude. The mean annual trend for all latitudes is about (−0.84 ± 0.82)% per decade. The trends also display some regional variation, being considerably less negative in Japan than in North America and Europe. Bojkov *et al.* (1990) have concluded that trend results, based on unrevised published data, were on average less negative than trends from revised Dobson data for European stations by about 1.0% per decade across all seasons, with only small average differences for stations in North America and Japan. Later on, Bojkov *et al.* (1995) considered the revised Dobson TOC trend over the Northern Hemisphere for 1964 to 1994.

Staehelin *et al.* (1998a, b) conducted a trend analysis of the homogenized Dobson total ozone series at Arosa, Switzerland (47°N, 1,800 m amsl) between 1926 and 1996 (earlier results were considered by Dütsch and Staehelin, 1989). Analysis of this, the longest TOC series in the world, was based on the use of a statistical model. Ozone trends for annual means are (−2.3 ± 0.6)% per decade when a simple linear change from 1970 to 1996 is assumed. The strongest depletions (by about −4% per decade) were found between January and March. A different approach of applying multiple regression—when terms for the solar cycle, stratospheric aerosol loading, and high mountain temperature record were included—resulted in annually averaged ozone trends of the order of (−1.9 ± 0.6)% per decade. Staehelin *et al.* (1998a) demonstrated that the recent TOC decrease cannot be accounted for by the higher average aerosol content in the second half of the last century. Assessment of the levels of total and stratospheric ozone depletion led to the conclusion that the decrease of ozone content in the stratosphere was approximately 30% larger than that of total ozone (this assessment is based on a crude estimate of the increase in tropospheric ozone). It has been assumed that the acceleration observed in TOC trends between the 1970s and the 1980s over northern mid-latitudes might be partially attributed to the larger increase in tropospheric ozone in the 1970s.

Staehelin *et al.* (1998), applying the methodology of TOC signal processing similar to that of Stolarski *et al.* (1991), deduced total ozone trends from Nimbus-7 TOMS data. The linear trend obtained from the data averaged between 65°N and 65°S latitudes was (−0.26 ± 0.14)% per decade over the 11.6-year period from November 1978 to May 1990. The trend was near zero (0.0002 ± 0.2% per year) at the equator and increased towards both poles. At 50°N the annually averaged trend is (−0.5 ± 0.21)% per year, but there were strong seasonal variations at this latitude, from more than −0.8% per year in winter and early spring (February and March) to about −0.2% per year in summer (July and August). The solar cycle component of TOC variability (for global average data) was (3.7 ± 0.6) DU/100 units of F10.7 or 1.5% over the solar cycle, while the QBO contribution was (−0.8 ± 0.2) DU/10 m s^{-1} of 30 hPa wind or 1% over a QBO cycle (Dikty *et al.*, 2010b; Sekiyama *et al.*, 2006).

Special attention has been paid to QBO-induced TOC variability (Bowman, 1989; Kramarova and Kuznetsov, 2009; Yang and Tung, 1994, 1995; Zerefos *et al.*,

1992). Yang and Tung (1995) came to the following three conclusions:

- the QBO signal was strongest in the mid and high latitudes and was present mainly in the winter–spring season;
- the extratropical ozone QBO signal was out of phase with the equatorial ozone signal, the latter being in phase with the equatorial stratospheric zonal wind QBO signal;
- there was no gradual phase propagation. Instead, there were distinct regions, in each of which the ozone QBO signal had fairly constant phase and large phase shifts occurred in between the transition regions.

Kane *et al.* (1998) studied the latitude dependence of Quasi-Biennial Oscillation (QBO) and Quasi-Triennial Oscillation (QTO) characteristics using TOMS total ozone data between 1979 and 1992 in the form of 12-month moving averages. Comparison of these data was made with stratospheric low-latitude zonal wind and equatorial eastern Pacific sea surface temperature (SST). Equatorial ozone dynamics were characterized by a strong QBO component with a period of about 30 months, and its maxima tallied with westerly wind maxima. At other latitudes, TOC maxima spacing was often different over 30 months, more so in the Northern Hemisphere.

Spectral analysis carried out by Kane *et al.* (1998) showed that both hemispheres had one peak at 20 months and another peak at ~30 months, only up to ~50° latitude. At higher latitudes these peaks shifted to ~23 and ~36 months. For 0–50° latitude the ~30-month periodicity in the Northern Hemisphere showed abrupt changes, while the southern latitudes showed a roughly gradual phase shift. The Northern Hemisphere had an additional periodicity at ~4 years, roughly matching the sea surface temperature (SST). Analysis of the same data without 12-month averaging showed the above characteristics more clearly and revealed additional peaks at ~0.7 years (8–9 months) and ~1.3 years (16–17 months). All of these peaks showed shifts (mostly increases) at higher latitudes.

Dikty *et al.* (2010b) investigated the modulating nature of 27-day solar rotation forcing on stratospheric ozone using a new ozone profile dataset from SCIAMACHY for 2003–2008. Continuous wavelet transform (CWT), fast Fourier transform (FFT), and cross-correlations (CCs) were applied to SCIAMACHY ozone data in the tropics (<20° latitude) between 20 and 60 km altitude. The maximum correlation between the Mg II index and ozone was weaker during the maximum of solar cycle 23 than in the previous two solar cycles that were investigated in earlier studies using different datasets. The magnitude of the ozone signal is highly time dependent and may vanish for several solar rotations even close to solar maximum conditions. The FFT analysis reveals, besides the 27-day signal, several frequencies close to 27 days. Ozone sensitivity (ozone change in percent per percent change in 205 nm solar flux) is on average about 0.2%/% above 30 km altitude and smaller by about a factor of 2 compared to earlier studies. For selected 3-month periods the sensitivity may rise beyond 0.6%/% in better agreement with earlier studies. Analysis of the 27-day solar forcing was also carried out with stratospheric temperatures from the European Centre for Medium-Range Weather Forecasts operational analysis. Although direct radiation effects on temperature are weak in the upper stratosphere, temperature signals with statistically significant periods in the 25 to 35-day range similar to ozone could be found with the applied methods.

Powerful perturbations in general atmospheric circulation, connected with El Niño/Southern Oscillation (ENSO) phenomena, have a significant impact on the ozone layer (Shiotani, 1992; UNEP, 1994; WMO, 1995, 1999; Zerefos *et al.*, 1992). Using statistical analysis of TOMS data, Randel and Cobb (1994) studied in detail the global response of TOC to ENSO dynamics. The Southern Oscillation Index is commonly used for quantitative description of ENSO.

As is now known, (1) El Niño events result in a positive (negative) response in the extratropics (near the equator), which is characterized by strong longitudinal asymmetry; (2) this effect is more strongly pronounced in northern winter months. Chernikov *et al.* (1999) summarized ENSO observation results (Table 4.1). These point out that El Niño's influence on total ozone occurs by a strengthening of upward motions over the oceans, because of sea surface temperature increase with subsequent TOC decrease. Photochemical ozone destruction is possibly strengthened by an increase of cloudiness and albedo during El Niño events. The importance of these causes of ENSO-induced TOC change is confirmed by TOC increase in the extratropics (the closure of a meridional circulation cell here results in downward air flows and TOC increase) and the stability of local TOC minima positions during El Niño's development in different years (Table 4.1).

Using Nimbus-7 TOMS (1978–1993) and Earth Probe (1996–1998) TOC data as well as relevant global meteorological information, Chernikov *et al.* (2000)

Table 4.1. The characteristics of total ozone anomalies in January in the equatorial region during negative Southern Oscillation Index (SOI) anomalies (Chernikov *et al.*, 2000).

Time period	*SOI min*	*Characteristics of centers of total ozone anomalies*					
		Pacific Ocean			*Africa*		
		Latitude (°S)	*Longitude* (°E)	ΔO_3 (%)	*Latitude* (°N)	*Longitude* (°W)	ΔO_3 (%)
June 1982–April 1983	−4.6	15	120	−1.3	10	10	−1.8
November 1986–September 1987	−2.0	5	160	−2.2	5	0	−2.2
September 1991–April 1992	−3.4	0	135	−1.2	10	5	−2.7
March 1997	−3.2	10	130	−4.1	5	0	−2.5

ΔO_3 is the total ozone deficit from the mean value.

carried out an analysis of the interrelationships between global fields of TOC and meteorological parameters. The following conclusions were reached:

- The El Niño event of 1997/1998 was one of the most intensive phenomena in the history of observations, strongly affected the ozone layer, and resulted in a decrease of TOC. Basic regions with negative TOC anomalies in the equatorial zone were located over the central and eastern Pacific, as well as near the Greenwich Meridian.
- Negative total ozone anomalies were associated with regions of horizontal wind divergences in the upper troposphere (200 hPa level) over the eastern equatorial Pacific and eastern Africa. Positive total ozone anomalies were observed in regions with wind convergence in the upper troposphere over the western equatorial Pacific and eastern part of the Indian Ocean.
- Short-term TOC variations (of the order of a few days) were discovered for the first time with typical scales of about several thousand kilometers and with absolute values between 200 and 225 DU. Such variations were observed mainly over the El Niño region. The largest values of their characteristics (amplitude, areal extent, frequency, etc.) were registered in mid-January 1998 with a time lag of 2–3 weeks after a maximum in sea surface temperature (SST) in the El Niño region.
- Ozone mini-holes formed on both sides of the equator in the El Niño event over the Pacific during the second half of December 1997, the sizes of which increased with time reaching about 1 million km^2 by the end of December. TOC values decreased to 200–225 DU.
- It was found that the negative and positive components of the Southern Oscillation Index (SOI) are better predictors than net SOI. Both negative and positive SOI led to a decrease in total ozone in the equatorial region, but with different intensities. This may be an indication of the limited applicability of linear assumption in the parameterization of interrelationships between ENSO and total ozone variations.
- The unusually strong influence of the El Niño event in 1997/1998 on the ozone layer could have influenced global climate because of a chain of processes: ozone decrease, ocean warming, troposphere warming, stratosphere cooling, and the intense interaction of the last two. A similar situation existed after the strongest El Niño event in 1982/1983.

Considering both satellite and ground-based observational data available for March 1991, Stolarski *et al.* (1992) analyzed the relevant trends. In the case of ground-based observations the average trend for the 26–64°N latitude belt was −2.7% per decade in the winter, −1.3% per decade in the summer and, −1.0% per decade in the autumn. Stolarski *et al.* (1992) emphasized that all of these declines are statistically significant.

Chandra *et al.* (2009) studied the effects of the 2006 El Niño on tropospheric O_3 and CO at tropical and sub-tropical latitudes measured from the OMI (Section 2.3.2) and MLS (Section 2.5.3.4) instruments on the Aura satellite. The 2006 El Niño-induced drought caused forest fires (largely started in order to clear land) to burn out of control during October and November in the Indonesian region. Chandra *et al.* (2009) used a global model of atmospheric chemistry and transport (GMI CTM) to quantify the relative importance of biomass burning and large-scale transport in producing the observed changes in tropospheric O_3 and CO. Model results show that during October and November biomass burning and meteorological changes contributed almost equally to the observed increase in tropospheric O_3 in the Indonesian region. The biomass component was 4–6 DU, but was limited to the Indonesian region where the fires were most intense. The dynamical (meteorological) component was 4–8 DU but it covered a much larger area in the Indian Ocean extending from South East Asia in the north to Western Australia in the south. By December 2006, the effect of biomass burning was reduced to zero and the observed changes in tropospheric O_3 were mostly due to dynamical effects. Model results showed an increase of 2–3% in the global burden of tropospheric ozone.

Hilsenrath *et al.* (1997) pointed out that since 1978 there was a mean global TOC depletion of about 2–3%, while the ozone content in the upper stratosphere decreased by 7–8%. These trends are higher than trends predicted by numerical modeling. From the viewpoint of the quality of observational data, reliable ground-based and on-board calibration became very important. Another important aspect of the problem is adequate control of the sensitivity of instrumentation (in the case of Nimbus-7 such control reduced the drift of sensitivity by 1.0% for each 10 yr period).

Fioletov *et al.* (2002) estimated global and zonal total ozone variations from ground-based and satellite measurements. Six datasets of monthly average zonal total ozone were intercompared and then used to estimate latitudinal and global total ozone temporal variations and trends. The datasets were prepared by different groups and were based on TOMS, SBUV-SBUV/2, GOME, and ground-based measurements. Different approaches have been used to homogenize the records over the period 1979–2000. Systematic differences of up to 3% were found between different datasets for zonal and global total ozone area weighted average values. However, when these systematic differences were removed by deseasonalizing the data, the residuals agreed to within ±0.5% of long-term mean ozone values. All datasets show changes in the rate of total ozone decline in recent years. While global ozone was fairly constant during the 1990s, the average value of the 1990s was about 2–3% lower than those of the late 1970s. About 38% of global ozone is located between 25°S and 25°N where the data show no decline. The strongest decline and the largest variability occur over the 35°N–60°N zone during the winter–spring season with the largest negative deviations occurring in 1993 and 1995. The decline in autumn is much smaller at these latitudes. Over the 35°S–60°S zone the ozone decline shows less seasonal dependence, and the largest deviations there were observed in 1985 and 1997. Sliding 11-year trends were calculated to estimate ozone changes over different time intervals. The first interval was from 1964 to 1974, and the last interval was from 1990 to 2000. The steepest year-round trends, of up to −5% per decade, occurred in the 11-year periods ending between 1992 and 1997 over

Table 4.2. SBUV(/2) trends in %/decade by seasons and latitudinal zone over the period from January 1979 to May 1994, with 95% uncertainty limits (two standard errors, labeled 2se) (WMO, 1995).

Zone (°)	*Dec.–Feb.*	*2se*	*Mar.–May*	*2se*	*Jun.–Aug.*	*2se*	*Sep.–Nov.*	*2se*	*Year*	*2se*
65N	−5.6	4.2	−6.3	2.9	−3.5	1.4	−4.3	1.6	−5.0	2.0
55N	−6.0	3.4	−6.1	2.6	−3.0	1.6	−3.7	1.5	−4.8	1.9
45N	−6.4	2.9	−5.7	2.3	−2.8	1.5	−3.1	1.5	−4.6	1.8
35N	−4.9	2.4	−4.5	2.5	−3.1	1.5	−2.9	1.5	−3.9	1.8
25N	−3.2	2.0	−2.7	2.5	−2.7	1.6	−3.0	1.3	−2.9	1.6
15N	−2.0	1.6	−1.8	1.8	−2.0	1.9	−2.6	1.4	−2.1	1.4
5N	−1.3	1.8	−1.7	2.2	−1.8	1.5	−1.6	1.8	−1.6	1.6
5S	−1.5	1.3	−1.8	1.7	−2.5	1.3	−2.1	1.5	−2.0	1.2
15S	−0.7	1.1	−0.3	1.1	−1.5	1.8	−1.0	1.5	−0.9	1.1
25S	−3.1	0.9	−2.7	1.3	−3.6	2.6	−2.7	1.8	−3.0	1.4
35S	−4.4	1.0	−5.3	1.8	−6.5	2.6	−3.9	2.0	−5.0	1.5
45S	−4.4	1.4	−5.0	1.7	−6.6	2.5	−3.5	2.2	−4.9	1.5
55S	−4.6	1.6	−6.3	1.9	−10.7	3.0	−6.3	3.3	−7.0	2.1
65S	−5.8	1.4	−7.6	2.1	−14.3	3.8	−13.6	5.2	−10.4	2.4

the 35°–60°N zone and between 1985 and 1993 over the 35°–55°S zone. More recent 11-year trends have smaller declines.

The UNEP/WMO *Scientific Assessment of Ozone Depletion: 1994* (WMO, 1995) provides a survey of TOC trends from ground-based data from January 1964 to February 1994 and SBUV/2 satellite data from January 1979 to May 1994 (Tables 4.2–4.4). Figure 4.1 illustrates specific features of TOC trend annual variations. The representative trends (annual averages, in % per decade) for north and south mid-latitudes and the tropics are given in Table 4.5—uncertainties (±) are expressed at the 95% confidence limits with two standard errors.

The principal conclusions made by WMO (1995) were as follows:

- Northern Hemisphere mid-latitude trends are significantly negative in all seasons, but are much larger in winter/spring (about 6% per decade) than in summer and autumn (about 3% per decade).

Table 4.3. Short-term Dobson trends in %/decade using data from January 1979 to February 1994. Tabulated numbers are averages of individual trends within latitude zones, with 95% uncertainty limits (two standard errors, labeled 2se) (WMO, 1995).

Zone (°)	*N*	*Dec.–Feb.*	*2se*	*Mar.–May*	*2se*	*Jun.–Aug*	*2se*	*Sep.–Nov.*	*2se*	*Year*	*2se*
50–65N	7	−6.2	1.5	−6.9	0.5	−4.6	1.4	−2.2	1.1	−5.2	0.5
40–50N	9	−5.4	1.5	−6.1	0.6	−3.0	0.6	−2.0	0.9	−4.3	0.5
30–40N	8	−3.9	1.3	−3.5	1.4	−1.6	1.4	−0.9	0.8	−2.5	1.0
20–30N	5	−1.7	0.7	−1.8	0.2	−0.9	1.7	−0.5	0.9	−1.2	0.2
0–20N	3	0.5	1.1	−0.0	0.4	−0.7	0.6	−0.8	0.4	−0.3	0.2
30–0S	5	−1.1	0.7	−1.1	1.4	−2.0	0.7	−1.4	0.5	−1.4	0.6
55–30S	6	−3.6	1.9	−2.9	1.2	−3.6	1.7	−2.8	1.4	−3.2	1.3

Table 4.4. Long-term Dobson trends in %/decade using data from January 1964 to February 1994 (trends from January 1970). Tabulated numbers are averages of individual trends within latitude zones, with 95% uncertainty limits (two standard errors, labeled 2se) (WMO, 1995).

Zone (°)	*N*	*Dec.–Feb.*	*2se*	*Mar.–May*	*2se*	*Jun.–Aug*	*2se*	*Sep.–Nov.*	*2se*	*Year*	*2se*
50–65 N	7	−4.0	1.0	−3.4	0.5	−1.4	0.4	−1.2	0.5	−2.6	0.4
40–50N	9	−3.7	0.8	−3.6	0.9	−1.8	0.6	−1.3	0.4	−2.7	0.5
30–40N	8	−2.4	1.0	−1.8	0.7	−0.6	0.5	−0.7	0.5	−1.4	0.6
20–30N	5	−1.5	0.8	−1.1	0.5	0.0	0.3	−0.4	0.7	−0.7	0.4
0–20N	3	−0.4	0.8	−0.0	0.4	−0.8	0.7	−0.7	0.9	−0.3	0.3
30–0S	5	−1.2	0.7	−1.2	1.5	−1.7	0.3	−1.4	0.7	−1.4	0.7
55–30S	6	−1.8	1.0	−1.9	0.8	−2.5	0.6	−1.6	0.8	−2.0	0.7

- Tropical (approximately 20°S–20°N) trends are slightly negative, but not statistically significant, when suspected drift in the satellite data is incorporated into the uncertainty.
- Southern mid-latitude trends are significantly negative in all seasons and increase in magnitude for high latitudes.

Similar more recent data can be found (e.g., WMO, 1999).

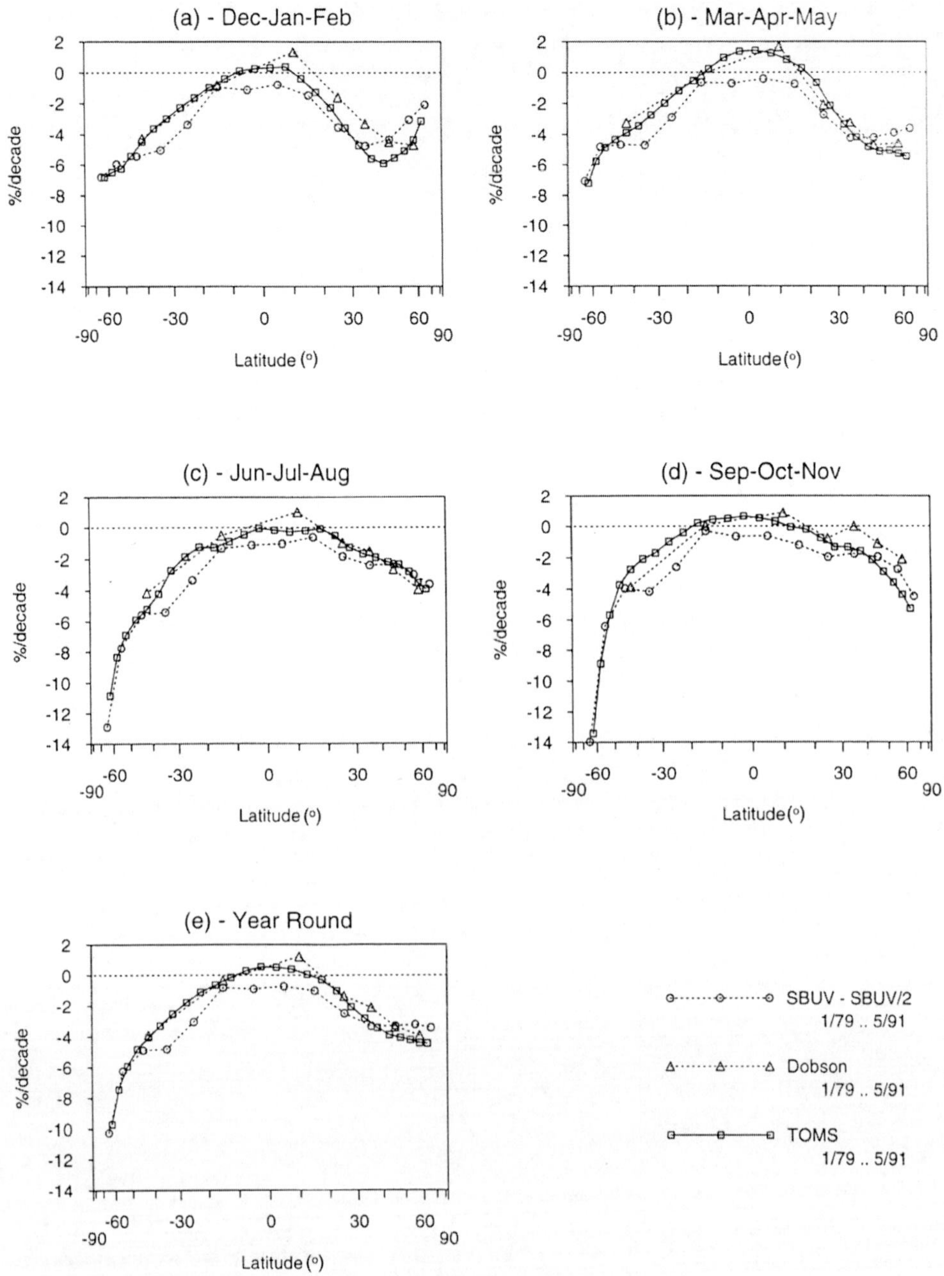

Figure 4.1. SBUV(/2), Dobson, and TOMS seasonal total ozone trends in percent per decade by latitude to May 1991. Open circles are SBUV(/2) trends over January 1979 to May 1991; open triangles are Dobson trends over January 1979 to May 1991; open squares are TOMS trends over January 1979 to May 1991. The Dobson trends are averages within the latitudinal zones of individual trends at 59 Dobson stations (WMO, 1995).

Table 4.5. Representative TOC trends (annual averages, in %/decade) (WMO, 1995).

		Latitude		
		Mid-south	*Equatorial*	*Mid-north*
Recent				
Jan. 1979–May 1994	SBUV + SBUV/2	−4.9 ± 1.5	−1.8 ± 1.4	−4.6 ± 1.8
Jan. 1979–Feb. 1994	Dobson Network	−3.2 ± 1.3	−1.1 ± 0.6	−4.8 ± 0.8
Jan. 1979–Feb. 1994	Ozonometer (former U.S.S.R.)	N.A.	N.A.	−4.9 ± 0.8
Pre-Pinatubo				
Jan. 1979–May 1991	SBUV + SBUV/2	−4.9 ± 2.3	−0.8 ± 2.1	−3.3 ± 2.4
Jan. 1979–May 1991	TOMS	−4.5 ± 2.1	+0.4 ± 2.1	−4.0 ± 2.1
Jan. 1979–May 1991	Dobson Network	−3.8 ± 1.3	+0.2 ± 2.1	−3.9 ± 0.7
Jan. 1979–May 1991	Ozonometer (former U.S.S.R.)	N.A.	N.A.	−3.8 ± 1.0
The corresponding ozone loss (in %) accumulated over 15.3 years for trends calculated through 1994 are:				
		Latitude		
		Mid-south	*Equatorial*	*Mid-north*
	SBUV + SBUV/2	−7.4 ± 2.3	−2.7 ± 2.2	−7.0 ± 2.7
	Dobson Network	−4.8 ± 2.1	−1.7 ± 0.9	−7.3 ± 1.3
	Ozonometer (former U.S.S.R.)	N.A.	N.A.	−7.5 ± 1.3

According to WMO (2007) the following developments should be mentioned:

- Global mean total column ozone values for 2002–2005 were approximately 3.5% below 1964–1980 average values. The 2002–2005 values are similar to the 1998–2001 values and this indicates that ozone is no longer decreasing. These changes are evident in all available global datasets, although differences of up to 1% between annual averages exist between some individual sets.
- Averaged for the period 2002–2005, total column ozone for Northern Hemisphere and Southern Hemisphere midlatitudes (35°–60°) are about 3% and 5.5%, respectively, below their 1964–1980 average values and are similar to their 1998–2001 values.

- Statistical analyses have been applied to the gridded monthly means of total ozone from combined TOMS and SBUV measurements (version 8 of the data) for the period 1978–2003 (Krzyścin, 2006). They focus on detecting a change in the trend pattern by searching for a turnaround in the previous downward trend. The results appear to be consistent among models using different formulations of the trend pattern. The 2003 level of total ozone after the removal of variations due to parameterized dynamical/chemical forcing on ozone is still below the long-term (1978–2003) mean level over extratropical regions. The deficit is $\sim$2–5% in the Northern Hemisphere and much larger in the Southern Hemisphere and exhibits clear seasonal variability: $\sim$15% in autumn, $\sim$10% in winter, and ~ -5% in spring and summer. The present total ozone level is higher outside the tropics than that in the mid-1990s but it is too early to announce that ozone recovery has started there because of trend uncertainties due to errors in the regression estimates for individual grid points and longitudinal variability of the trend pattern. A rigorous statistical test has shown a statistically significant turnaround for some grid points over the extratropical region and a deepening of the ozone negative trend has not been found for any grid point.

SPARC has recently established the Chemistry–Climate Model Validation Activity (CCMVal) for coupled chemistry–climate models (CCMs). SPARC released the final version of the ozone database within this framework on September 22, 2009. It covers the period 1850 to 2100 and can be used as ozone forcing for Coupled Model Intercomparison Project Phase 5 (CMIP5) models that do not include interactive chemistry. This database consists of the following two parts (Figures 4.2 and 4.3):

A. The historical part of the AC&C (Atmospheric Chemistry & Climate)/SPARC ozone database covering the period 1850 to 2009 is now available at the CMIP5 website (*https://pcmdi-cmip.llnl.gov/*) at the Program for Climate Model Diagnosis and Intercomparison (PCMDI). The historical ozone dataset combines separate stratospheric and tropospheric data sources:
    1. Stratospheric data are constructed using a multiple linear regression analysis of SAGE I and II satellite observations and polar ozonesonde measurements for the period 1979–2005 (as described by Randel and Wu, 2007). The regression includes terms representing equivalent effective stratospheric chlorine (EESC) and 11-year solar cycle variability.
    2. The stratospheric time series are extended backwards to 1850 based on regression fits combined with an extended proxy time series of EESC and solar variability.
    3. Tropospheric data are derived from model simulations. Tropospheric ozone simulations were performed using two chemistry–climate models: the Community Atmosphere Model (CAM) version 3.5 and the NASA-GISS PUCCINI model. Both models simulated tropospheric and stratospheric chemistry with feedback to the radiation and were driven by the recently

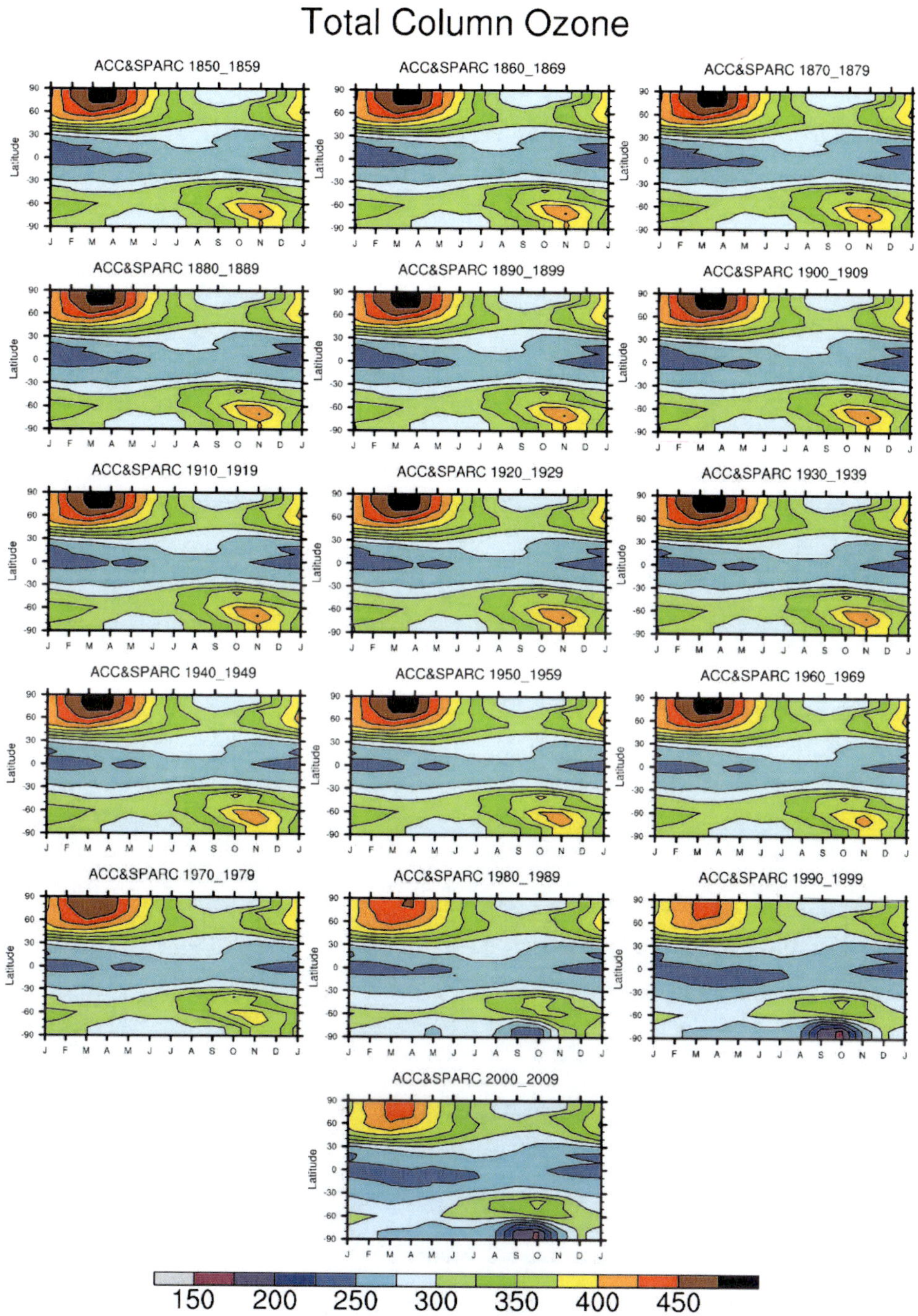

Figure 4.2a. Time series of decadal mean total column ozone (1850–1859 to 2000–2009) (*http://www.pa.op.dlr.de/CCMVal/SPARC_CCMValReport/SPARC_CCMValReport.html*).

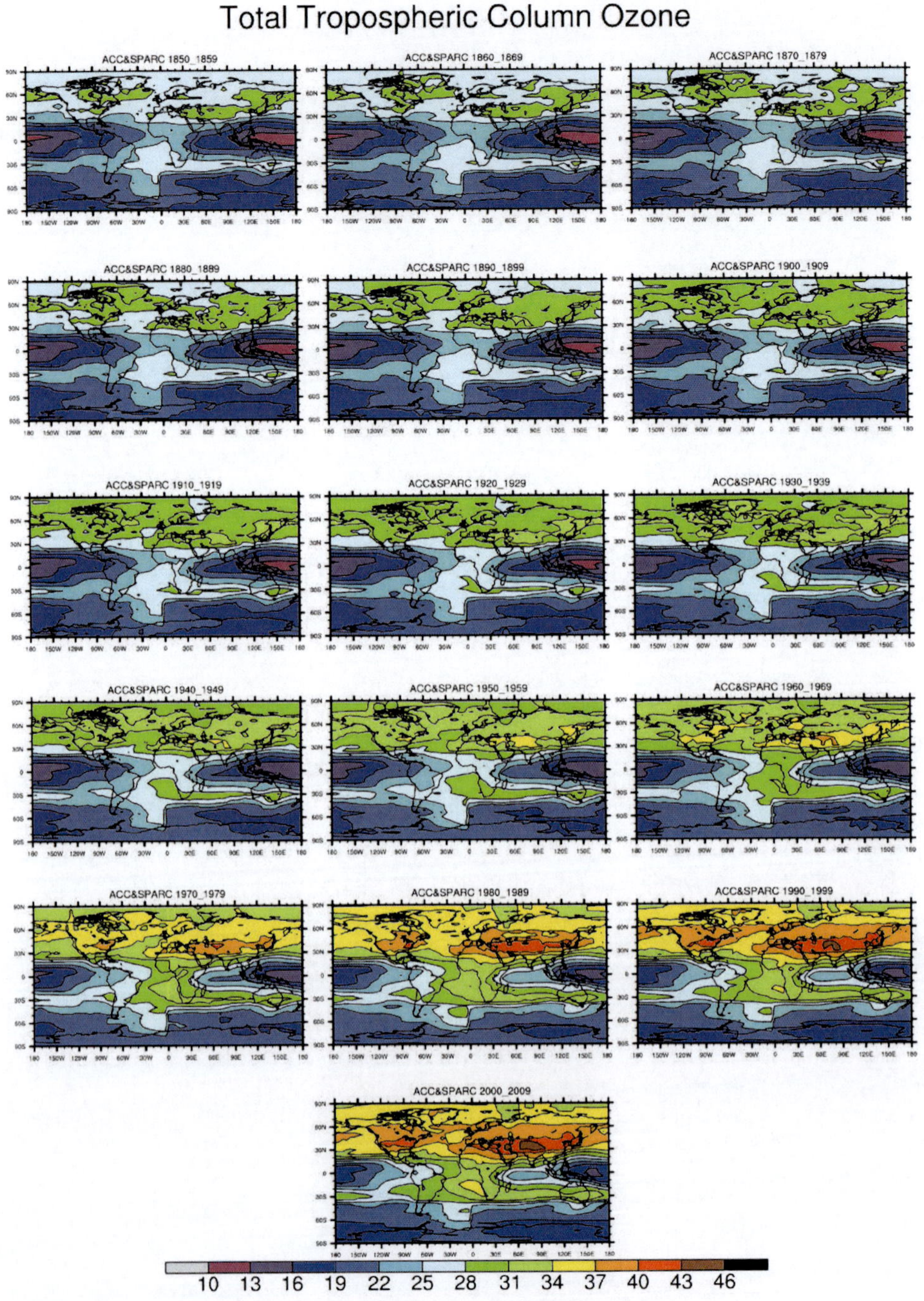

Figure 4.2b. Decadal annual means for tropospheric column ozone (using the chemical tropopause $O_3 = 150$ ppbv) (*http://www.pa.op.dlr.de/CCMVal/SPARC_CCMValReport/SPARC_CCMValReport.html*).

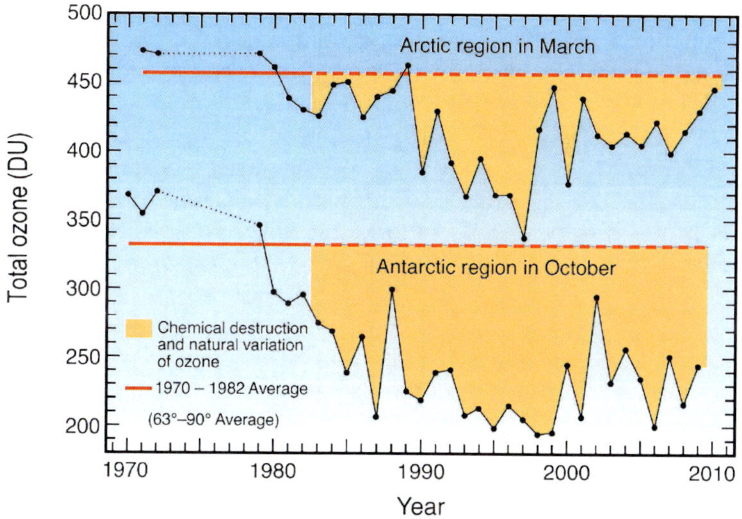

Figure 4.3. Average total ozone in polar regions (Fahey and Hegglin, 2011).

available historical (1850–2000) emissions succintly described by Lamarque *et al.* (2009). Emissions were kept constant from 2000 onwards in both simulations. In addition, CAM used sea surface temperatures (SSTs) from a previous CCSM3 simulation while the NASA-GISS model used observed SSTs. For the climatology, decadal averages were performed for each model, then the NASA-GISS results were interpolated to the CAM vertical grid and an average of both was taken. This average field represents the tropospheric ozone field in the climatology.

4. Stratospheric and tropospheric data are combined by simply merging the two datasets across the climatological tropopause, to produce a smooth final dataset.

B. The future part of the AC&C/SPARC ozone database covering the period 2010 to 2099. The future time series is a seamless extension of the historical database (merged in 2009), so that the entire time series (1850–2100) can be used as ozone forcing for climate models that do not include interactive chemistry. The future ozone time series combines separate stratospheric and tropospheric data sources:

1. Stratospheric ozone projections are taken from the future reference simulations (REF-B2) of the 13 CCMs that performed a future simulation until 2100 under the Special Report on Emissions Scenarios (SRES) A1b greenhouse gases (balanced on all energy sources) scenario and the A1 (of a more integrated world) scenario in CCM Validation Activity 2 (CCMVal-2) (*http://www.pa.op.dlr.de/CCMVal/SPARC_CCMValReport/SPARC_CCMValReport.html*). In the stratosphere, the multi-model mean of the REF-B2 simulations is used in all representation concentration pathway (RCP) scenarios (i.e., stratospheric ozone in this database does not vary among the RCPs).

2. The future tropospheric ozone time series continues the historical CAM3.5 simulation until 2100 using the four different RCPs that were generated by integrated assessment models (IAMs) and were harmonized with the historical emissions in both amplitude and geographical distribution. Since RCP-based simulations from an atmosphere–ocean climate model were not available when these runs were started, SSTs/SICs from SRES simulations closest to the RCP greenhouse gas scenarios were used.
3. As in the historical part, future stratospheric and tropospheric data are combined by simply merging the two datasets across the climatological tropopause, to produce a smooth final dataset (Cionni *et al.*, 2011).

According to the recent *Scientific Assessment of Ozone Depletion: 2010* (WMO, 2010a), the main characteristics of total ozone variability deduced from the global total ozone network are the following:

- Average total ozone values in 2006–2009 remain at roughly 3.5% and 2.5% below the 1964–1980 averages, respectively, for 90°S–90°N and 60°S–60°N. Mid-latitude (35°–60°) annual mean total column ozone amounts in the Southern Hemisphere (Northern Hemisphere) over the period 2006–2009 have remained at the same level as observed during 1996–2005, at ~6% (~3.5%) below the 1964–1980 average.
- The ozone loss in Arctic winter and spring between 2007 and 2010 has been variable, but has remained in a range comparable to the values prevailing since the early 1990s. Substantial chemical loss continues to occur during cold Arctic winters.
- Global ozone is projected to increase approximately in line with the decline of ozone-depleting substances, and the increase is accelerated by cooling of the upper stratosphere. Global ozone is not very sensitive to circulation changes, so high confidence can be placed in this projection.

Long-term changes in average total ozone are shown in Figure 4.3 for the Antarctic and Arctic for the respective regions defined by latitudes between 63° and 90° (Fahey and Hegglin, 2011). Total ozone is measured with satellite instruments. The reference values (red lines) are averages of springtime total ozone values available from observations between 1970 and 1982. Each point represents a monthly average for October in the Antarctic or March in the Arctic. After 1982, significant ozone depletion is found in most years in the Arctic and all years in the Antarctic. The largest average depletions have occurred in the Antarctic since 1990. Ozone changes are a combination of chemical destruction and natural variations. Variations in meteorological conditions influence year-to-year changes in ozone, particularly in the Arctic. Essentially all of the decrease in the Antarctic and usually about 50% of the decrease in the Arctic each year are attributable to chemical destruction by reactive halogen gases. In the Arctic, the other 50% is attributable to natural variations in the amounts of ozone transported to polar regions before and during winter. Average total ozone values over the Arctic are naturally larger at the beginning of each winter season because, in the preceding months, more ozone is transported poleward in the Northern Hemisphere than in the Southern Hemisphere.

4.2 OZONE VERTICAL PROFILE VARIABILITY

The variability of tropospheric and stratospheric ozone concentrations, along with special features in the vertical ozone structure, have received significant scientific attention in recent decades (e.g., Kondratyev and Varotsos 2001a, b, 2002; Varotsos *et al.*, 1994a–c, 2000, 2001a, b). Reliable information on ozone vertical profile (OVP) variability in the stratosphere is very important for solving a number of problems, such as ozone impact on climate change (see Chapter 7) and ozone depletion due to emissions of man-made chemicals (see Chapters 5 and 6). During the last 40 years many researchers have reported the existence of a laminar structure (layers with enhanced and depleted ozone amounts) in vertical ozone profiles (Reid and Vaughan, 1991; Varotsos *et al.*, 1994a, b, 1999). Dobson (1973) examined the occurrence of a laminar ozone structure in the stratosphere over a wide latitudinal and longitudinal range. In Dobson's analysis the criterion for detection of laminae within a certain height interval was the change in ozone partial pressure to a value greater than 3 mPa. In the same study, ozone profiles were separated into three groups: group 0 with profiles containing at least two laminae, group 1 with profiles exhibiting moderate lamination, and group 2 with profiles extensively laminated. One of Dobson's findings was that the features within a laminated structure vary with latitude and season. In particular, between January and April the lamination phenomenon was most frequently present; also at latitudes below about 20°N during spring and below about 30°N during autumn, a laminated ozone structure was very seldom found. Dobson (1973) also observed a characteristic ozone minimum in the vertical ozone distribution at 14–17 km, but no explanation was given for both the incidence of minima at the preferred height region and its constancy with latitude. It was stated though that there is a strong correlation between the existence of the characteristic ozone minimum at 14–17 km and the occurrence of the double tropopause, especially at latitudes around 4°N. Dobson also made the assumption that the laminar structure and the characteristic ozone minimum are of the same origin, since the variations in appearance frequency are very similar for both season and latitude.

Subsequently special features in vertical ozone structure, especially in the lower stratosphere, have received significant scientific attention in the past few decades. For example, many researchers have reported the existence of layers with enhanced and depleted ozone amounts in vertical ozone profiles (Hering, 1964; Hering and Borden, 1964; Reid and Vaughan, 1991; Varotsos *et al.*, 1992). Perliski and London (1989) discussed satellite-observed long-term averaged seasonal and spatial ozone variation in the stratosphere. They emphasized the presence of various kinds of temporal changes: annual, semi-annual, quasi-biennial, daily, and solar cycle induced variability as well as long-term trends. The analysis of TOMS data from October 1978 to September 1987 made it possible to identify the basic features of ozone annual and semi-annual variations in the lower, mid, and upper stratosphere for the latitude interval 65°S–65°N.

Reid and Vaughan (1991) examined the ozone laminar structure in the altitude range of 9.5–21.5 km. Laminae were classified according to their vertical extent (depth), as well as the difference in ozone partial pressure (magnitude). They also used criteria to separate genuine laminar features from instrumental noise or large-scale features in ozone profiles. The acceptable depth for laminae was chosen to be between 200 m and 2.5 km, while the ozone partial pressure anomaly was at least 2 mPa. They also showed that laminae are most frequently found below 18 km at high latitudes during winter and spring, which is in agreement with the findings of Dobson. But Reid and Vaughan (1991) disagreed with Dobson about the origin of laminae. Based on the observation that their magnitudes are greatest and their depths are least during winter and spring, they ruled out Dobson's suggestion that they all originate near the subtropical jetstream, although the possibility remains that a small number of laminae may be generated by stratosphere–troposphere exchange (as Begum, 1989 believes). Reid and Vaughan (1991) suggested that laminae may represent evidence for a process that can cause the exchange of air in and out of the polar vortex. This argument is supported by more recent polar ozone campaigns, when particularly sharp and deep laminae near the winter polar vortex were detected (Reid *et al.*, 1993). Analysis of data from 800 experimental ozonesondes for 18 stations during EASOE (European Arctic Stratospheric Ozone Experiment) showed that laminae are absent above the 450 K level within the polar vortex, and are most abundant near its edge. Additionally, the small amplitude of inertia gravity waves could not totally support the appearance of laminar features (Reid *et al.*, 1994).

Orsolini *et al.* (1995), using a high-resolution off-line transport model, demonstrated that large-scale isentropic advection can create the laminar structure of a vertical tracer profile. This is an additional indication that a lamination profile relates to the exchange of air in and out of the polar vortex.

Križan and Laštovička (2005, 2006) considered ozone laminations in the Southern Hemisphere in comparison with earlier studies of laminations in the Northern Hemisphere. They noted that there are large differences between the characteristics of ozone laminations in the Northern Hemisphere and the Southern Hemisphere. The overall ozone content (deficit) in the positive (negative) laminae per profile and the number of laminae were observed to be much smaller in the Southern Hemisphere compared with the Northern Hemisphere. No detectable trend in these two lamina characteristics was found in the Southern Hemisphere contrary to a strong negative trend with a reversal to a strong positive trend in the mid-1990s in the Northern Hemisphere. The pattern of trends in the ozone content in one lamina was found to be quite different; there was no trend except for a slight negative trend in the North American sector. Some dynamical factors, mainly the degree of stability of the polar vortex, were suggested as contributing (or not) to the long-term behavior of lamina characteristics.

The high vertical resolution and ability of HIRDLS (Section 2.5.3.8) to sound the tropopause region permit the study of the transport of upper-tropospheric air with low ozone across the tropopause break in thin layers that have not been observed before. Olsen *et al.* (2008) described HIRDLS observations and simulation of a lower-stratospheric intrusion of tropical air to high latitudes. On January 26, 2006 low mixing ratios of ozone and nitric acid were observed in a ~2 km layer near 100 hPa extending from the subtropics to 55°N over North America. Subsequent evolution of the layer was simulated with the Global Modeling Initiative (GMI) model and confirmed by HIRDLS observations. Air with low concentrations of ozone was transported poleward to 80°N. Although there was evidence of mixing with extratropical air, much of the tropical intrusion returned to the subtropics. This study demonstrated that HIRDLS and the GMI model can resolve thin intrusion events. The observations combined with simulation are a first step towards development of a quantitative understanding of the lower-stratospheric ozone budget.

Further studies of lamination were made using ozonesonde data from Athens by Varotsos *et al.* (1992, 1994a–c, 1995a). During the winter period of November 1996–April 1997 a total of 25 ozone soundings were performed at the Athens station (Varotsos, C.A. *et al.*, 1999). These ozone soundings were classified into five groups, according to their specific features.

1. The first group consists of profiles in cyclonic flow at 200 hPa, generally on the western or southern side of an elongated trough. The main characteristics of the profiles of the first group are (a) high tropopause; (b) small secondary maximum in ozone (<7 mPa); and (c) ozone minimum at 100 hPa.
2. The second group is similar meteorologically to the first, but consists of profiles with a rather larger secondary layer in the lower stratosphere and again an ozone minimum at 100 hPa.
3. The third group is mainly characterized by many thin layers in the lower stratosphere (laminated structure). The profiles of the third group were observed during fairly cyclonic flow or on the eastern side of the trough.
4. The fourth group contains profiles with one thick layer in the lower stratosphere and an ozone minimum at 14–17 km. The profiles of the fourth group were usually observed during cut-off low events.
5. The fifth group consists of profiles with an ozone depletion effect in the region, where the partial pressure of ozone takes the maximum value. The meteorological situation of the latter category is characterized by the movement of the polar vortex to mid-latitudes of the Northern Hemisphere, resulting in more intense northwestern circulation above the site.

A further set of ozone soundings made in 2002–2004 in Athens were compared with the OVPs derived from MIPAS (Michelson Interferometer for Passive Atmospheric Sounding) observations that referred to the greater Athens area (see Section 3.3.3.4). The derived OVPs were classified into the first five groups just described, but a new group was added to them: the sixth group includes profiles characterized by scalable ozone increments from the region of characteristic ozone minima at 100 hPa, up to

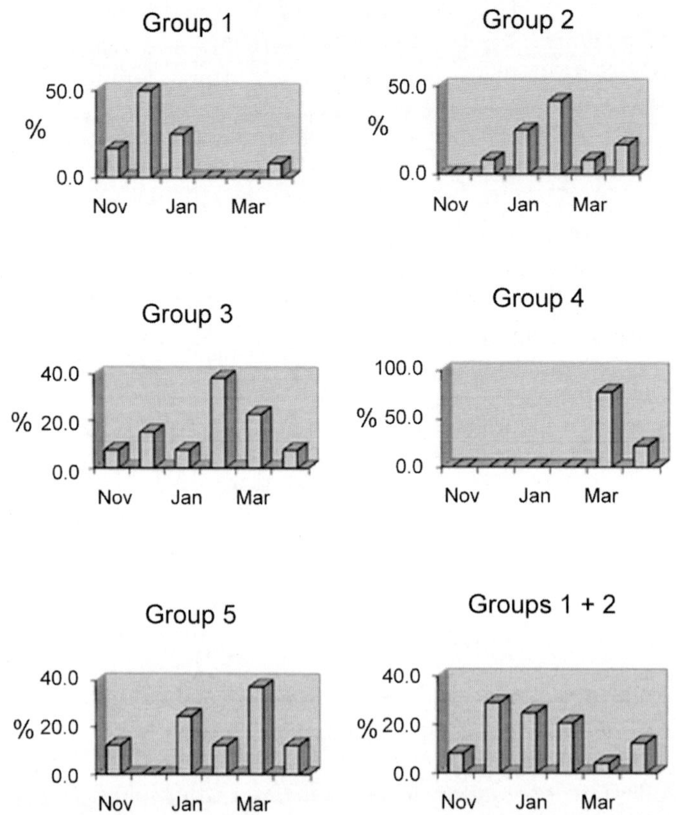

Figure 4.4. Monthly distribution of the occurrence of each group of ozone soundings for the period from November 1996 to April 1997 over Athens, Greece (Varotsos *et al.*, 1994b).

the ozone maximum concentration region. Such profiles were usually observed on the southern side of the trough during cyclonic flow.

From the 1996–1997 ozone-sounding dataset for the Athens station, the monthly distribution of the occurrence of each group is shown for the period from November to April in Figure 4.4. The soundings in the first group (ozone minima at the 100 hPa level and small secondary maximum) are more likely to occur in early winter, while those in the second group (same characteristics as the first group but with larger secondary maximum) are more likely to occur at the end of the winter. Both groups are characterized by an ozone minimum in the lower stratosphere. Therefore, in order to examine the occurrence of the characteristic minimum the two groups were merged into one (1 + 2). The characteristic ozone minimum shows maximum occurrence during the winter period. This is in agreement with the findings of Dobson (1973).

The soundings in the third group (laminated ozone structure) are more likely to occur at the end of winter and in early spring. This is also in agreement with Dobson

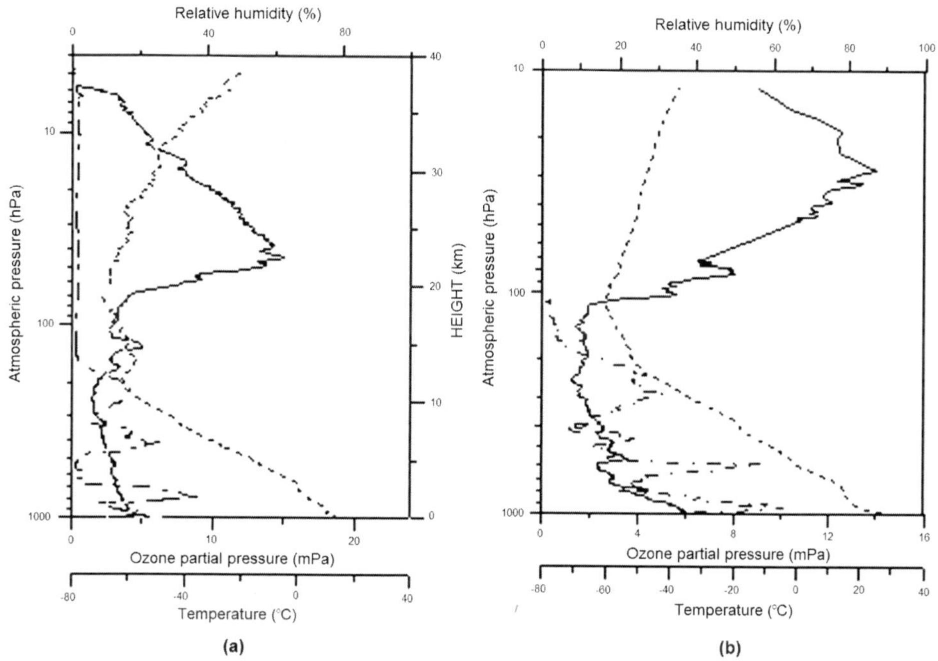

Figure 4.5. Ozone partial pressure (——), relative humidity (– · · –), and temperature (– – –) over Athens (a) January 24, 1997 (Varotsos *et al.*, 1999) and (b) November 23, 2004 (Varotsos *et al.*, 2008a).

(1973), who observed the maximum in lamination during February and March. The soundings in the fourth group (thick ozone layer in the lower stratosphere and ozone minimum at 14–17 km) mostly occur in early spring while those in the fifth group (ozone decline effect in the altitude range of maximum partial pressure) are more likely to occur during the late winter and early spring.

Examples of OVPs obtained from ozonesondes launched from Athens for the first five groups in 1996–1997 and 2002–2004 are shown in Figures 4.5-4.9.

On January 24, 1997 (Figure 4.5a) and November 23, 2004 (Figure 4.5b) characteristic ozone minima in the lower stratosphere (100 hPa) were observed. On April 6, 1997 (Figure 4.6a) and June 1, 2005 (Figure 4.6b) a characteristic ozone minimum was also observed, as well as a larger secondary maximum in the lower stratosphere. On these days Athens was influenced by the subtropical jetstream; this is in agreement with observations made by Dobson (1973) that the 15 km minimum is associated with air coming from the subtropical region. On January 10, 1997 (Figure 4.7a) and April 18, 2005 (Figure 4.7b) a laminated ozone structure was observed. This was strongly associated with the establishment of northwestern circulation over the region, so it could be concluded that the Athens station was under the influence of airmasses of polar origin. In the case of April 18, 2005 (Figure 4.29) an interesting feature in the vertical ozone structure is the ozone depletion effect that

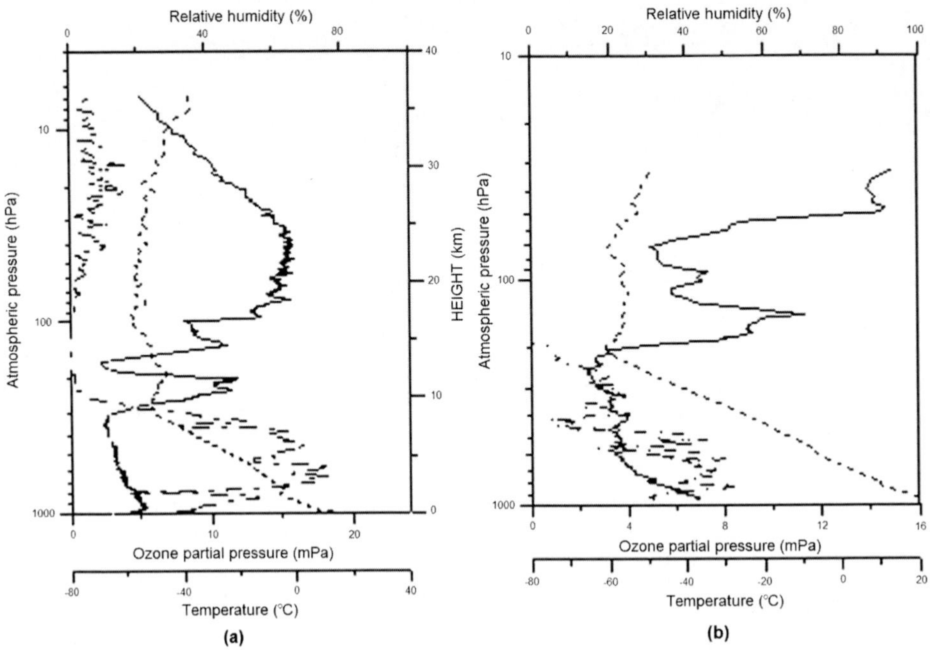

Figure 4.6. Ozone partial pressure (——), relative humidity (– · –), and temperature (– – –) over Athens (a) April 6, 1997 (Varotsos *et al.*, 1999) and (b) June 1, 2005 (Varotsos *et al.*, 2008a).

occurred at the altitude region of maximum ozone partial pressure. The meteorological pattern in the latter case is characterized by extension of the polar vortex to mid-latitudes of the Northern Hemisphere, resulting in more intense northwestern circulation above this site. On April 1, 1997 (Figure 4.8a) and February 18, 2005 (Figure 4.8b) a cut-off low system was observed over the region. The ozone profile contains a thick layer in the lower stratosphere. This is in agreement with the observation that cut-off low events seem to be combined with high ozone partial pressure at low levels. At the same time a laminated structure was observed in the region of 8–13 km (Figure 4.8a) or 10–15 km (Figures 4.8b). On February 6, 1997 (Figure 4.9a) the most interesting feature in the vertical ozone structure was the ozone depletion effect that occurred in the maximum ozone partial pressure altitude region. This new finding of strong ozone depletion, which occurred at the level of ozone maximum concentration, should be considered in the context of conventional knowledge (i.e., maximum ozone depletion occurs in the layer just below the layer of ozone maximum partial pressure). In the case of March 15, 2005 (Figure 4.9b) the ozone profile is characterized by scalable ozone increment from the region of characteristic ozone minima at 100 hPa, up to the ozone maximum concentration region. During this day the site was on the southern side of the trough during cyclonic flow.

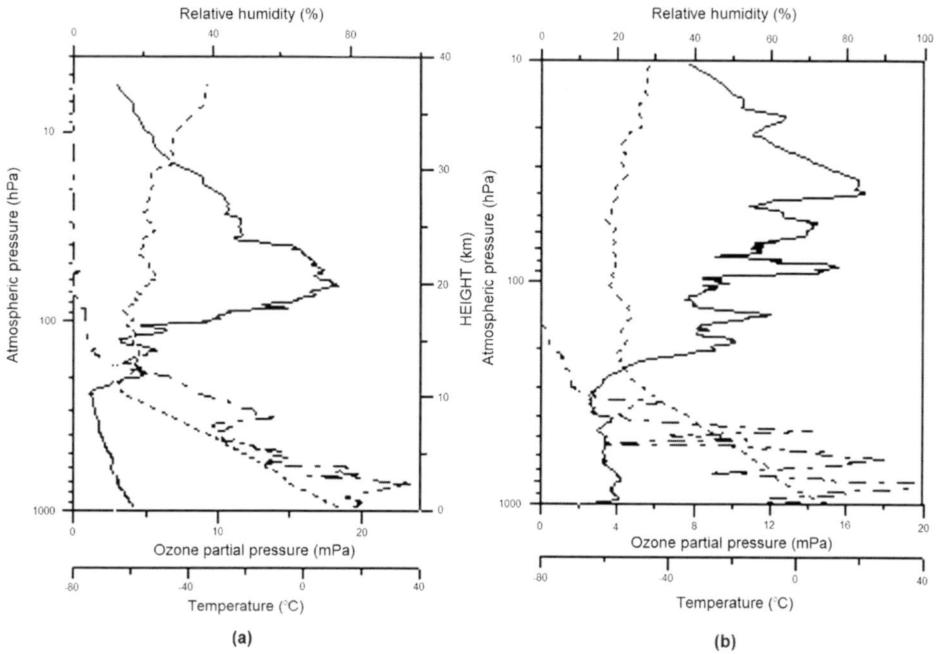

Figure 4.7. Ozone partial pressure (——), relative humidity (– · –), and temperature (– – –) over Athens (a) January 10, 1997 (Varotsos *et al.*, 1999) and (b) April 18, 2005 (Varotsos *et al.*, 2008a).

Kahya *et al.* (2005) discussed the laminae characteristics in the vertical profiles of ozone concentrations at three locations representing the central (Praha, Czech Republic), eastern (Legionowo, Poland) and southeastern (Ankara, Turkey) parts of Europe. A total of 785 soundings between January 1994 and October 2003 with varied periods were analyzed. It was found that lamination is frequent in winter and spring in these locations. The mean magnitude and depth of the laminae in these stations varied depending on the latitudes of the locations. Additionally, in order to understand the influence of tropopause heights on laminae structure, the laminae were grouped according to high and low tropopause heights. The frequency of laminae for both lower and higher tropopause groups for winter and spring seasons are close to each other.

Bojkov (1999) pointed out that, up to 1996 in northern mid-latitudes, combined OVP trends (SAGE I/II, Umkehr, ozonesondes) were negative at all altitudes between 12 and 50 km and were statistically significant at the 2σ level. The downward trend was largest near 40 km and 15 km ($\sim$7.5% per decade) and smallest at $\sim$30 km (2% per decade). Peak ozone loss was observed in the 15–20 km layer. The largest trends were in winter and spring. Annual ozone variation was most clearly pronounced in the subpolar stratosphere of both hemispheres at the 1.5, 8.0 and 40 hPa levels, where the contribution of annual components to total ozone variability

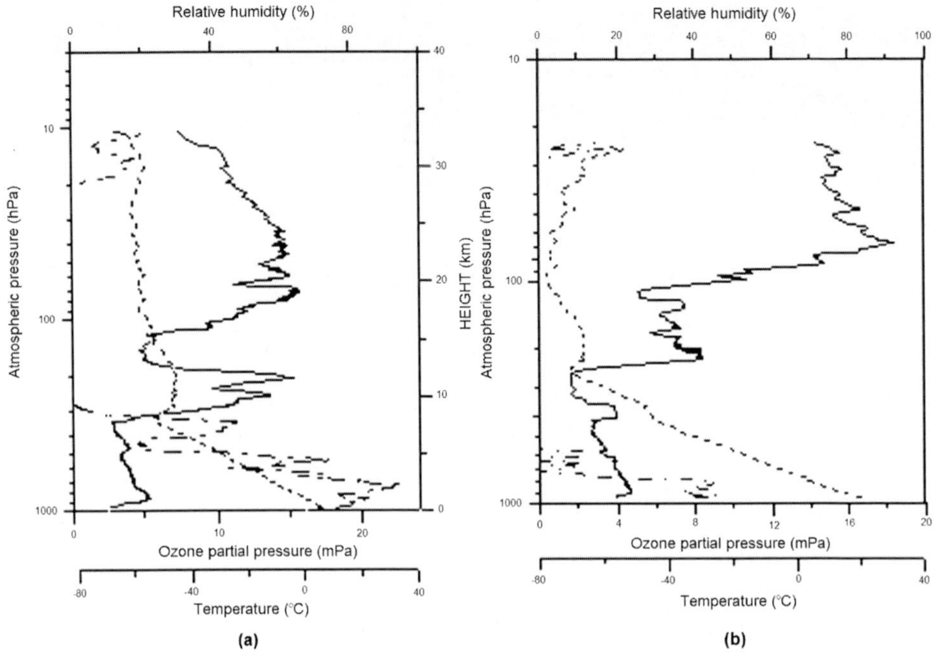

Figure 4.8. Ozone partial pressure (——), relative humidity (– · –), and temperature (– – –) over Athens (a) April 1, 1997 (Varotsos *et al.*, 1999) and (b) February 18, 2005 (Varotsos *et al.*, 2008a).

exceeds 80%. The minimum amplitude of annual variations takes place in the mid and upper stratosphere near 10°N. The maximum amplitude of averaged semi-annual variations was observed near the 2–3 hPa levels in subpolar latitudes, decreasing towards the mid-latitudes. Semi-annual ozone variation in the tropical middle stratosphere is comparable with, or even stronger than, annual variations. The negative trend of annual variation amplitude ($\sim$0.1 ppm yr^{-1}) was observed in the upper stratosphere of southern subpolar latitudes. Interannual changes in the amplitudes of both annual and semi-annual variations are small as a rule (except in the tropical mid-stratosphere, where the influence of the El Chichón eruption was substantial, and the subpolar upper stratosphere in the Southern Hemisphere). The vertical distribution of the ozone trend shows distinct negative trends at about 18 km in the lower stratosphere with the largest declines over the poles, and above 35 km in the upper stratosphere. A narrow band of large negative trends extends into the tropical lower stratosphere (Brunner *et al.*, 2006).

Assessments of the relative contributions of photochemical and dynamical processes to ozone concentration field formation confirmed earlier conclusions. Ozone distribution in the lower stratosphere (below the 30 hPa level) is controlled (especially in winter) mainly by large-scale atmospheric circulation, while in the upper stratosphere (above the 5–10 hPa levels) radiative and photochemical processes dominate (except at subpolar latitudes in winter when atmospheric trans-

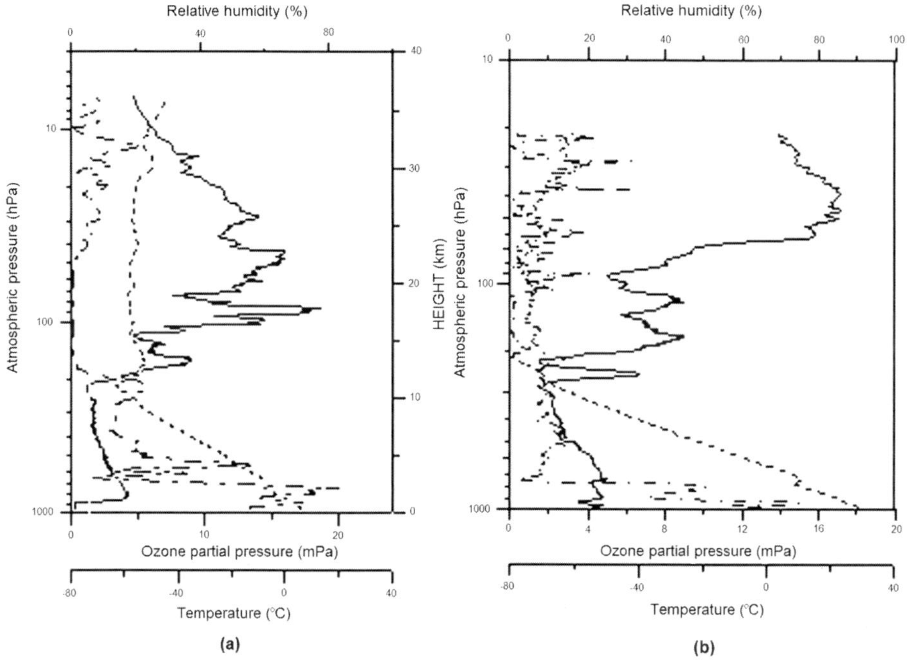

Figure 4.9. Ozone partial pressure (——), relative humidity (– · –), and temperature (– – –) over Athens (a) February 6, 1997 (Varotsos *et al.*, 1999) and (b) March 15, 2005 (Varotsos *et al.*, 2008a).

port is also important). In the tropics this zone is located at a lower level because of the shorter photochemical lifetime in the equatorial belt under conditions of high insolation.

London and Liu (1992) pointed out that the observed ozone vertical gradient is small in the troposphere but increases rapidly in the stratosphere. The seasonal variation at a typical mid-latitude station (Hohenpeissenberg) showed a summer maximum in the low to middle troposphere, shifting to a winter–spring maximum in the upper troposphere and lower stratosphere and a spring–summer maximum at 10 hPa. The amplitude of annual variation increased from a minimum in the tropics to a maximum in polar regions. The amplitude increased with height at all latitudes up to about 30 hPa where the phase of annual variation changes abruptly. Annual long-term ozone trends were significantly positive at about +1.2% per year in the mid-troposphere (500 hPa) and significantly negative at about $-0.6\%\ yr^{-1}$ in the lower stratosphere (50 hPa). The results demonstrated the importance of ozonesonde observations and illustrated the need of combined analysis of both ozonesonde and satellite OVP data.

The very important progress in studying both ozone dynamics near the tropopause and stratosphere–troposphere ozone exchange resulted from the work of the Measurement of Ozone and Water Vapor by the Airbus In-Service Aircraft (MOZAIC) program over a two-year period, from September 1994 to August 1996

(Thouret *et al.*, 1998a, b) (see Section 1.7.4.3). This program was the continuation of a number of dedicated aircraft campaigns: Global Atmospheric Measurements Experiment on Tropospheric Aerosols and Gases (GAMETAG), Stratospheric Ozone Experiment (STRATOC), Tropospheric Ozone Experiment (TROPOZ), Pacific Exploratory Mission-West (PEM-West), Transport and Atmosphere Chemistry Near the Equatorial Atlantic/southern Africa Fine-Atmosphere Research Initiative (TRACE A/SAFARI-92), Troposphärische Ozon (TROZ) project, Global Air Sampling Program (GASP).

From data collected at cruise level by the five Airbus A340s using automatic ozone-measuring devices (dual-beam UV absorption instruments) an accurate ozone climatology at 9–12 km altitude was generated over the Northern Hemisphere (0°–80°N, 130°W–140°E), and down to 30°S over South America and Africa. Between September 1994 and December 1997, 7,500 flights were made over continental landmasses and the Atlantic Ocean. Measurement accuracy was estimated at $\pm(2\,\text{ppbv} + 2\%)$, but much better in-flight levels were in fact observed: the average discrepancy (between different devices) ranging from 1 ppbv at tropospheric concentrations to a few ppbv at stratospheric concentrations. MOZAIC data for the whole troposphere and lower stratosphere (0–12 km) were compared with ground-based observations made at eight stations of the Ozone Sounding Network (OSN): Hohenpeissenberg, Wallops Island, Tateno, Palestine, Pretoria, Goose Bay, Biscarosse, and Poona. This comparison demonstrated a reasonably high level of agreement, despite the different nature of the programs (techniques, platforms, sampling frequencies, spatial distribution, and operation periods). Mean concentrations derived from ozonesondes are about 3 to 13% higher than those obtained by the MOZAIC program in the free troposphere in a similar geographical location (Thouret *et al.*, 1998b). Results of intercomparisons supported the assumption that it is not advantageous to scale ozonesonde data to the overhead ozone column; scaling appears to cause overestimation of tropospheric O_3 concentrations (by about 3–6% at Hohenpeissenberg) and to cause more scatter in ozonesonde–MOZAIC differences. The correspondence between the OSN and MOZAIC ozone climatology confirms their usefulness for validation studies.

Thouret *et al.* (1998a) pointed out that there are still large uncertainties in the budget of atmospheric ozone and especially with regard to the relative importance of photochemical formation and exchanges between the stratosphere and troposphere, and in determining future ozone trends. In this context certain regular features of ozone spatial distribution were analyzed. Thouret *et al.* (1998a) reached the conclusion that north of 35°N the distribution at 9–12 km altitude is dominated by the influence of ozone-rich air of stratospheric origin; farther south, ozone-poor air from the troposphere prevails. To identify the stratospheric and tropospheric components and to help in interpreting the data a classification based on a threshold of 100 ppbv of ozone was used.

Distinct seasonal ozone variability was observed in stratospheric air (maximum in spring, minimum in winter) and in tropospheric air (maximum in summer, minimum in autumn), related to the intensities of dynamic (tropopause variation, stratosphere/troposphere exchange) and chemical (photochemistry) processes. The tropics exhibit variations that are heterogeneous in time and space, and reflect the influences of active photochemical processes, deep convection, and biomass burning emission. Ozone concentrations decrease with latitude in both the stratosphere and troposphere. As far as the ozone vertical profile is concerned, it is more strongly expressed in the stratosphere (with a strong vertical gradient) than in the troposphere, where ozone distribution is much more homogeneous. Smaller ozone concentrations over the Atlantic Ocean than over the continents, because of zonal variation of the polar front and the position of ridge/trough pressure systems, are important features of their geographical distribution.

A substantial contribution to studies of ozone vertical profiles has been made by various ground-based and aircraft remote-sensing observations in which lidar soundings played a special role. For instance, the airborne UV differential absorption lidar (DIAL) system participated in the Tropical Ozone Transport Experiment/Vortex Ozone Transport Experiment (TOTE/VOTE) in late 1995/early 1996 (Grant *et al.*, 1998). This system allowed retrieval of the ozone-vertical profile from approximately 2 km above the aircraft up to about 16 km with a vertical resolution of about 1.3 km (the horizontal resolution was 70 km). An important purpose of ozone soundings were intercomparisons with results from a ground-based DIAL as well as from a number of satellite stratospheric instrumentation: HALOE, MLS, and SAGE II (see Sections 2.5.1.4, 2.5.3.4, 2.5.1.1). Grant *et al.* (1998) pointed out that ozone profiles generally agreed within random error estimates for various instruments in the middle of profiles in the tropics, but regions of significant systematic differences, especially near or below the tropopause or at higher altitudes, were also found. Comparisons strongly suggested that the airborne UV DIAL system can play a valuable role as a mobile lower-stratospheric validation instrument.

Solomonov *et al.* (2003) applied microwave remote-sensing techniques for soundings of the ozone layer using the 142.175 GHz O_3 line ($10_{0.10}$–$10_{1.9}$ rotational transition) located in the 2 mm transparency window. Observations were conducted near Moscow between January 1987 and April 1990. The results indicated the presence of ozone enhancement above 27 km because of the impact of stratospheric warming. Large-scale atmospheric circulation was observed to have a very substantial influence on ozone change.

Kawabata *et al.* (1997) conducted ground-based radiometry observations of strato-mesospheric ozone by employing a superconductive NbSIS receiver cooled by a closed cycle 4 K refrigerator. The $J = 6_{1.5}$–$6_{0.6}$ (110.836 GHz) O_3 line was measured. Sporadic observations have been carried out on the Higashiyama Campus of Nagoya University (35°N, 137°E, 70 m asl) since January 1992. The results obtained indicate that the amplitudes of seasonal variations are small at the 20, 3.2, and 0.056 hPa levels and the phase or type of seasonal variation appears to change at these altitudes. Well-known seasonal variations were observed below the 20 hPa level. The summer O_3 maximum caused by photochemical production can be clearly seen at the 5.6 and 10 hPa levels. The earlier-discovered temperature-dependent ozone distribution cycle with a winter maximum (Froidevaux *et al.*, 1994) was observed in the altitude range between 1.8 and 0.1 hPa. There was a similarity between observed seasonal variation at the 1.8 hPa level and UARS MLS data at the 2.2 hPa level in the mid-latitudes (32°–64°). Strong enhancement of the ozone mixing ratio between 1.0 and 0.1 hPa was observed in the winters of 1992/1993 and 1994/1995 but not in the winter of 1993/1994. It can be assumed that seasonal variation in the layer between 1.0 and 0.1 hPa are of the QBO type, but the period of observations was not long enough for a reliable assessment.

Sandor and Clancy (1998) carried out ground-based microwave emission measurement of HO_2 (at 254.55153 and 268.56586 GHz), O_3 (249.7886 GHz), and $H_2{}^{18}O$ (203.40752 GHz) at the National Radio Astronomy Observatory (Kitt Peak, Arizona, 32°N, 112°W). Microwave ($\lambda \sim 1.2$ mm) emission of the upper stratosphere and mesosphere (45–80 km) was measured in April 1992, January 1993, and April and December 1996 with the purpose of studying HO_2, OH, and H rapid interconversion during HO_x catalytic ozone destruction cycles. Comparison between co-ordinated diurnal (09:00–18:00 local time) measurements of mesospheric O_3 with model calculations revealed a model unpredictability of ozone, which can be resolved with a 40% reduction in the rate coefficient of the reaction $H_2O + O_3 \rightarrow 2OH + O_2$.

Substantial contributions to ozone vertical profile studies were presented at the International Ozone Symposium in 1996 (Bojkov and Visconti, 1998). Grainger and Atkinson (WMO, 1999) carried out the first comprehensive global three-dimensional six-hourly analyses of ozone mixing ratio data from satellite observations (MLS HALOE, SAGE II, SBUV/2, TOVS), which were interpolated onto a 2.5° grid with 19 levels between 1.0 and 0.1 hPa (data for October 1994 were considered). There are still, however, some problems with the harmonization of observational data from various platforms.

Chan *et al.* (1998) discussed OVP observations (ozonesonde data) in Hong Kong, which demonstrates the complicated structure of ozone vertical profiles (basically bimodal with a laminated structure when observed in the lower stratosphere in winter and spring).

Cunnold and Wang (1998) investigated the effects of temperature uncertainty on the interpretation of ozone trends in the upper stratosphere and emphasized that estimates of long-term trends in SAGE ozone measurements and in the solar cycle response of ozone can be different depending on whether they are based on ozone concentrations in altitude levels or ozone mixing ratios on pressure surfaces.

De Luisi *et al.* (1998) discussed aerosol corrections to retrieved Umkehr ozone profiles. A new Brewer Umkehr retrieval algorithm was described by Elansky *et al.* (1998). A number of other papers were also devoted to improving Umkehr retrieval algorithms. A new version of the Umkehr algorithm, UMK04, has recently been developed (Petropavlovskikh *et al.*, 2005a, b).

A new version of SBUV(/2) data (version 8) was described by Bhartia *et al.* (2004).

Corlett and Monks (2001) made a comparison of total column ozone data retrieved from the Global Ozone Monitoring Experiment (GOME), the TIROS Operational Vertical Sounder (TOVS), and the Total Ozone Mapping Spectrometer (TOMS) for 1996, 1997, 1998, and 1999. Statistical and spatial difference analyses were performed on a range of spatiotemporally averaged datasets. An analysis of globally averaged column ozone values showed a consistent offset with TOVS and TOMS values being consistently higher than GOME by ~10–13 DU averaged over the 4-year period. A 4%–5% drift was noted between the years 1996/1997 and 1998/1999 in the magnitude of the offset. The drift was identified as an increased offset of +25–30 DU between TOVS/TOMS and GOME occurring over latitudes above 70°N during 1998/1999, and is a result of TOVS/TOMS ozone columns being higher during 1998/1999 than in 1996/1997. Seasonal and latitudinal trends were noted in global differences. In particular, TOMS and TOVS ozone values are consistently higher than GOME in the Southern Hemisphere from 30° to 90°S. TOVS and GOME ozone columns show good agreement between 20°S and 20°N, with TOMS values approximately 10–15 DU higher than both TOVS and GOME in the same region. All three sensors show reasonable agreement between 20° and 60°N. However, there is no agreement above 60°N, where TOVS columns are higher than TOMS columns which, in turn, are higher than GOME columns. Results from a spatial difference analysis indicated further differences between GOME and TOVS ozone values that were not obvious from global or latitudinal analysis owing to cancellation effects, including an area over Indonesia where GOME columns are higher than TOVS columns.

Fujimoto *et al.* (1998) conducted an intercomparison of lidar, ozonesonde, and SAGE II data using lidar and ozonesonde observations made in Tsukuba (36°N, 140°E). The results indicate that SAGE II and lidar data agree well enough (within 10% in the stratosphere) as also do ozonesonde and lidar data (below 32 km).

The sources of OVPs derived from BUV observations since 1970 were listed in Table 2.4 and are indicated in Figure 4.10 (top).

Long-term observations of the ozone layer with balloonborne instruments allow winter ozone altitude profiles to be compared between the Antarctic and Arctic regions (WMO, 2010a, b). Inspection of Figure 4.10 (bottom) shows that at the South Pole (left panel), a normal ozone layer was observed to be present between 1962 and 1971. In more recent years, as shown here for October 9, 2006, ozone is almost completely destroyed between 14 and 21 km (9 to 13 miles) in the Antarctic in spring. Average October values in the last decades (1990–2009) are 90% lower than pre-1980 values at the peak altitude of the ozone layer (16 kilometers). In contrast, the Arctic ozone layer is still present in spring as shown by the average March profile for 1991–2009 obtained over the Ny-Ålesund site (right panel). No Ny-Ålesund data are available for the 1962–1971 period before significant ODS destruction began. Some March profiles do reveal significant depletion, as shown here for March 29, 1996. In such years, winter minimum temperatures are generally lower than normal, allowing PSC formation for longer periods (it was also the case in 2011—Varotsos *et al.*, 2012). Arctic profiles with depletion similar to that shown for October 9, 2006 at the South Pole have never been observed. The number in parentheses for each profile is the total ozone value in Dobson units (DU). Ozone abundances are shown here as the pressure of ozone at each altitude using the unit "milli-Pascals" (mPa) (100 million mPa = atmospheric sea level pressure).

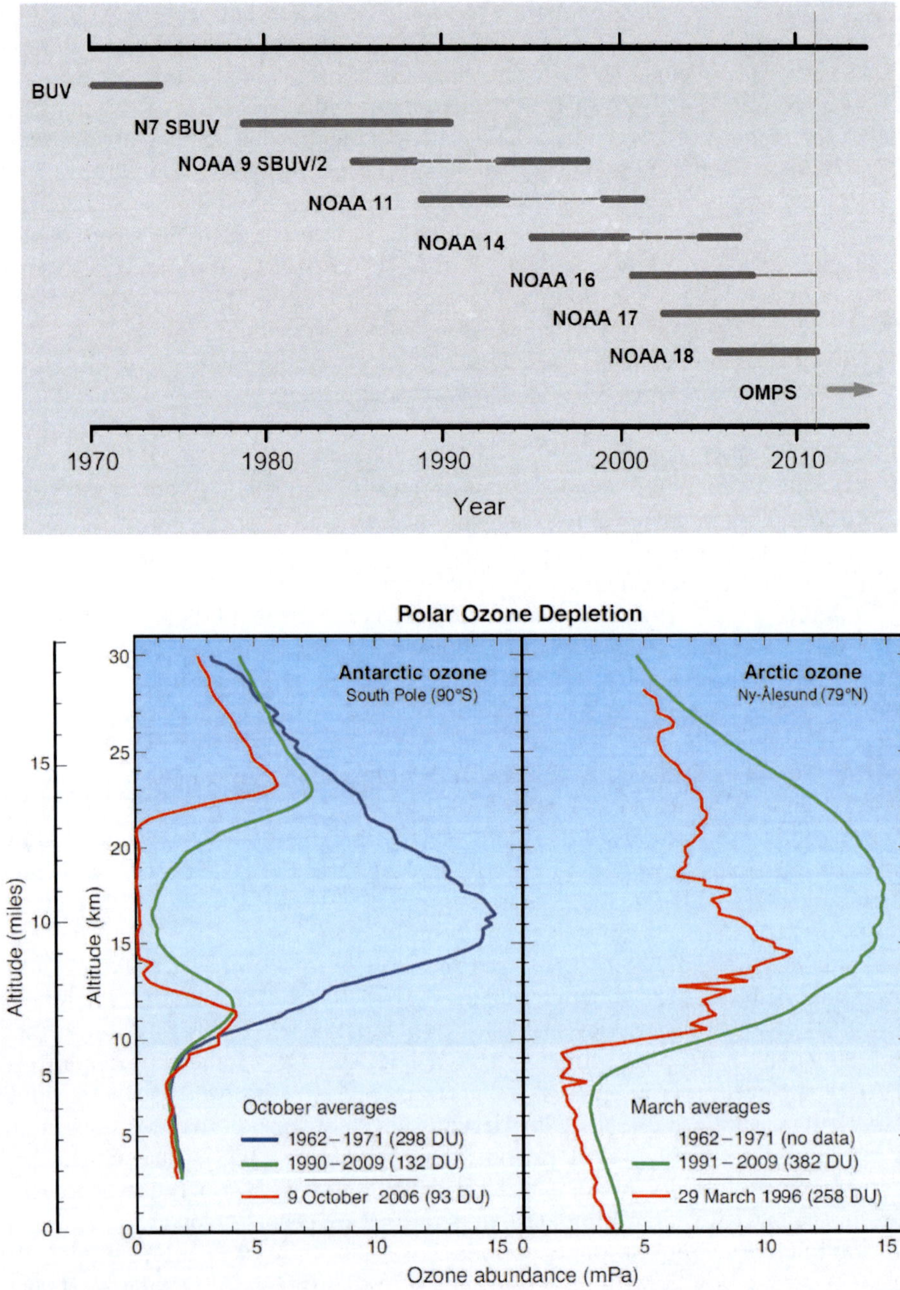

Figure 4.10. (Top) A series of BUV instruments have been measuring ozone vertical distribution since 1970. (The red dashed segments indicate periods when instruments are near the terminator as a result of orbit drift) (Bhartia and McPeters, 2011). (Bottom) Vertical distribution of Arctic and Antarctic ozone (Fahey and Hegglin, 2011).

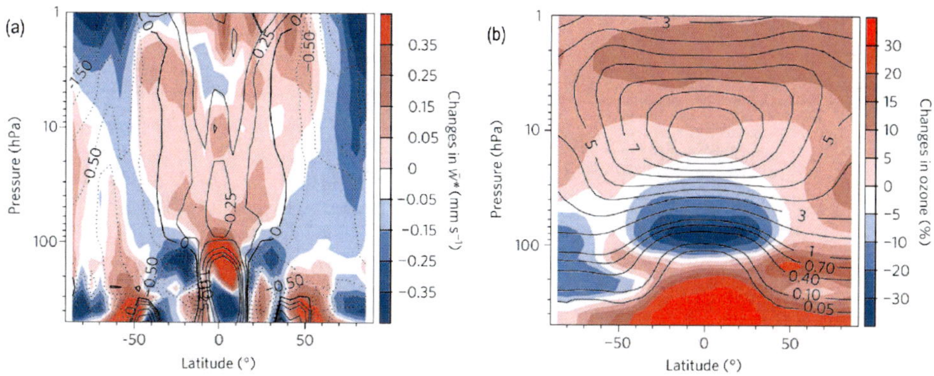

Figure 4.11. Vertical–latitudinal cross-sections of past values (1960–1970, black contours) and long-term changes (differences between 1960–1970 and 2090–2100, color shading) for annual mean vertical residual velocity ($\bar{W}^*$) (a) and ozone (b) using an ensemble mean of three simulations from the CMAM. For ozone, the past values are shown in units of ppmv and long-term changes as relative changes (Hegglin and Shepherd, 2009).

Hegglin and Shepherd (2009) recently addressed the question of how ozone is expected to change in the coming decades. The results obtained are depicted in Figure 4.11.

4.3 GENERAL FEATURES OF OZONE GLOBAL DISTRIBUTION

Since the basic features of TOC global distribution were discussed in Chapter 1, we shall dwell only upon a number of recent results. Note, however, that the first fundamental attempt to substantiate global empirical models of total ozone and ozone vertical profiles (for the 10–32 km layer) was made by Bekoryukov (1991), who used for this purpose all available (at the beginning of the 1990s) ground-based observational data.

A new stage in the study of global ozone characteristics was marked by the publication of Bojkov and Fioletov (1995), who used all available TOC data from over 150 past and present Global Ozone Observing System (GO_3OS) stations (after careful quality re-evaluation) to analyze the dynamics of observed global ozone. The importance of ground-based observations is due to their long time series in comparison with all other data. Bojkov and Fioletov (1995) had the principal aim of deducing basic ozone characteristics both for pre-ozone hole and during "ozone hole" time periods. TOMS data facilitated estimation of the longitudinal inhomogeneity of TOC distribution, which made it possible to use ground-based data for establishing long-term zonal as well as hemispheric and global ozone variation from 1964 to 1994. The difference between estimations of monthly zonal variation from ground-based and TOMS data for the overlapping period of 1979–1993 was less than 1% at latitudes 40°S–60°N. Ozone changes are several times larger than possible errors in estimated values; therefore, the results are highly reliable.

The results obtained by Bojkov and Fioletov (1995) indicate that the Northern Hemisphere average TOC value was ~312 DU and the southern value was ~300 DU in the pre-ozone hole decades (1964–1980), and that the global average from 1984 to 1993 was lower by ~3%: from (306 ± 1) DU down to (298 ± 2) DU. The Southern Hemisphere contributed ~64% of overall ozone decline. Levels of the annual ozone maximum were reduced by 5.8% in the Southern Hemisphere and 3.2% in the Northern Hemisphere, and the levels of ozone minimum by 2.1% and 1.2%, respectively. As far as temporal global mean TOC variation is concerned, three major components of natural ozone variation (annual cycle, solar cycle, and QBO) can be clearly identified. There are two major maxima in the temporal TOC spectrum: 26–28 months (QBO) and 100–130 months (solar cycle). A rather small ENSO signal (about 40 months) has also been identified. A major component of global ozone variation is its annual cycle. The annual TOC course is antisymmetric between the two hemispheres, with large differences both in absolute values and time of appearance of the maximum in polar regions.

Figure 4.12 data from Bojkov and Fioletov (1995) illustrate latitude–time cross-sections of TOC deviations from 1964 to 1980 "norms" (TOMS data) in percent (the calculations were made for 5° latitude zones). These data clearly confirm (the middle and lower part of Figure 4.12 are relevant to overlapping ground-based and TOMS data) the agreement between ground-based and satellite observations. Clearly seen are the appearance of QBO-related change, the very dramatic decline in the southern polar region, well-known TOC anomalies in 1983, 1985, and 1987, and the very strong decline in the 1992–1993 northern winter, etc.

The analysis of long-term TOC variation for the Northern and Southern Hemispheres indicated that a weak decline which started in the 1970s became a major change in hemispheric ozone during the 1980s. This decline was interrupted in the Northern Hemisphere between 1981 and 1982 (solar maximum years). The overall decline in the Southern Hemisphere was so strong between 1984 and 1985 that annual TOC maximum levels were almost always lower than the annual minimum levels of the pre-1980 period.

The mean latitudinal–time cross-sections are shown on Figure 4.13 for two time intervals. Comparison of the data for these time periods demonstrates very substantial change in the southern polar region. Bojkov and Fioletov (1995) pointed out that hemispheric differences in TOC distribution are more pronounced in months of increasing ozone, and in the value of the month in which maximum is achieved, than during the ozone-declining phase. Figure 4.14 illustrates the differences between the two periods in detail. The difference is marked in winter–spring over northern polar and upper mid-latitudes, reaching more than 6%. The difference is lower in spring and autumn. Southward of 60°S from September to November the difference is more than 15% with a maximum of 35% poleward of 80° in October. Table 4.6 shows long-term TOC trends that are characterised by pronounced seasonal dependence varying by ~3% to 8% (and even more for the Southern Hemisphere) on the cumulative decline since 1970 for mid-latitudinal bands (35–65°). The ozone decline

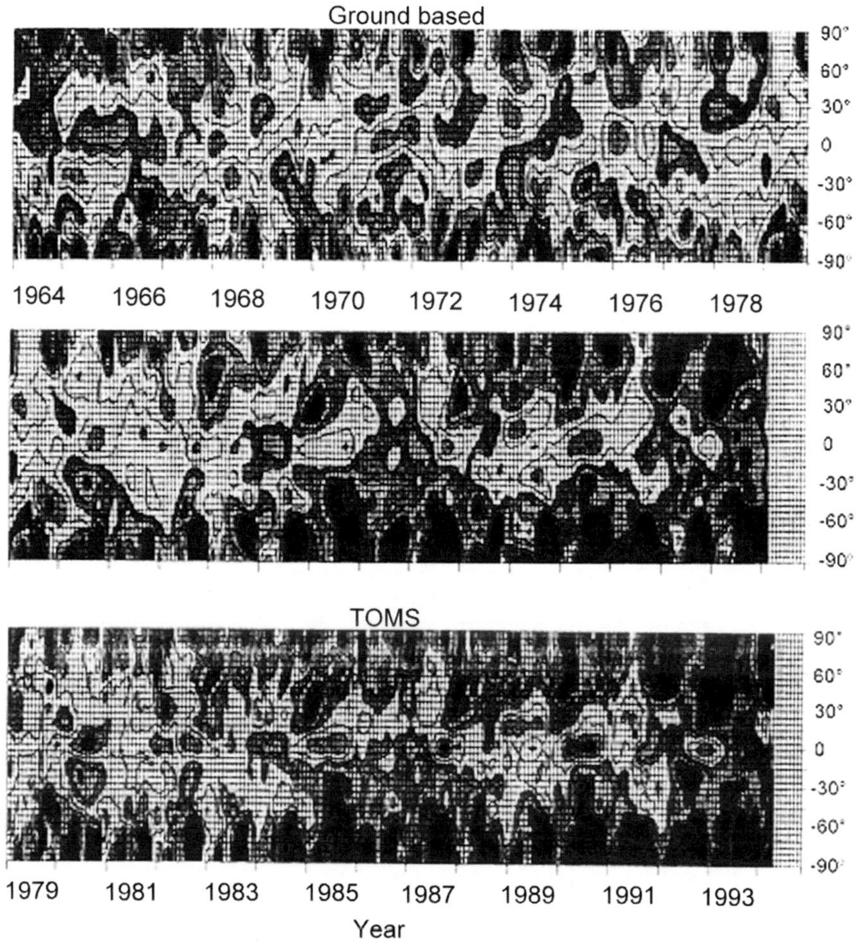

Figure 4.12. Latitude–time cross-sections of total ozone deviations from 1964 to 1980 "norms" in percent: (top) and (middle) ground-based data; (bottom) TOMS. Deviations below −15% are plotted in the same color as for the interval −13 to −15% (Bojkov and Fioletov, 1995).

during the "ozone hole" period is almost twice as large in comparison with 1970. The cumulative year-round global TOC decline is $(4.8 \pm 0.6)\%$, but over mid and polar latitudes it is more than 7%.

A principal conclusion reached by Bojkov and Fioletov (1995) was to reinforce the conclusions from the 1970s and 1980s which led up to the Montreal Protocol (see Chapter 5); this was that natural variation cannot explain the observed ozone decline, indicating the likelihood of anthropogenic influence. Another important conclusion was that the use of complementary ground-based and space-based ozone-observing facilities had great information value from the viewpoint of the analysis of global ozone variability.

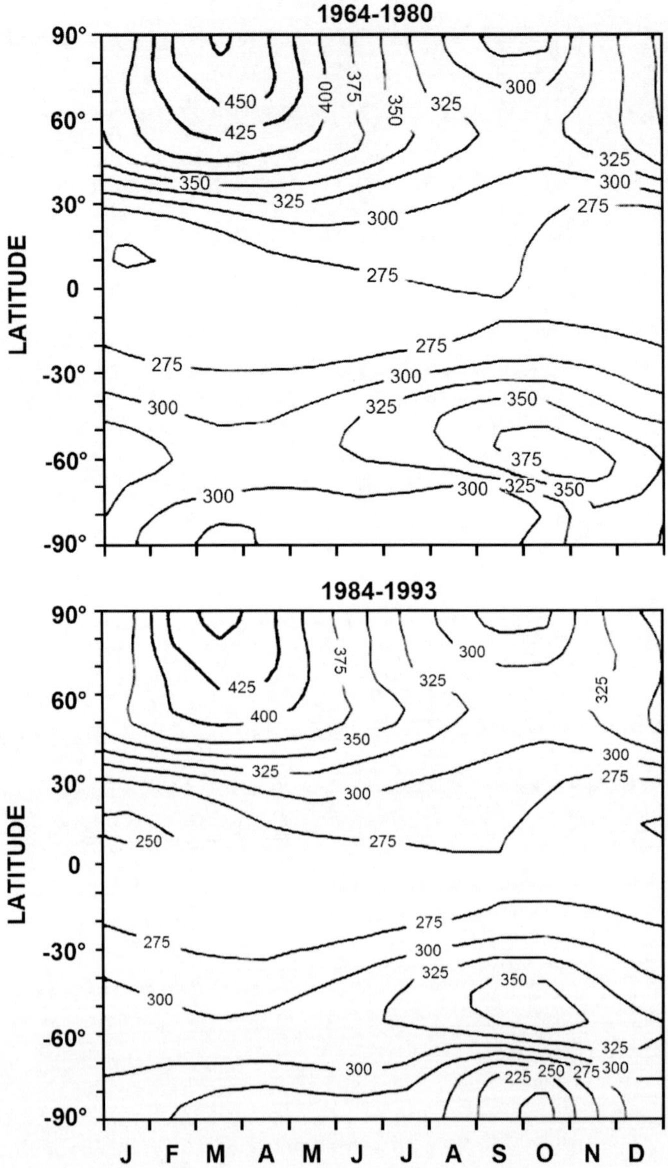

Figure 4.13. Variation in total ozone with latitude and season (in milliatmosphere centimeters) for two time intervals: (top) 1964–1980 and (bottom) 1984–1993 (Bojkov and Fioletov, 1995).

A new very important step forward in understanding the basic features of ozone vertical profiles and their trends was taken by scientific assessments carried out jointly by the WCRP Project on Stratospheric Processes and Their Role in Climate (SPARC) and the International Ozone Commission (IOC), in close

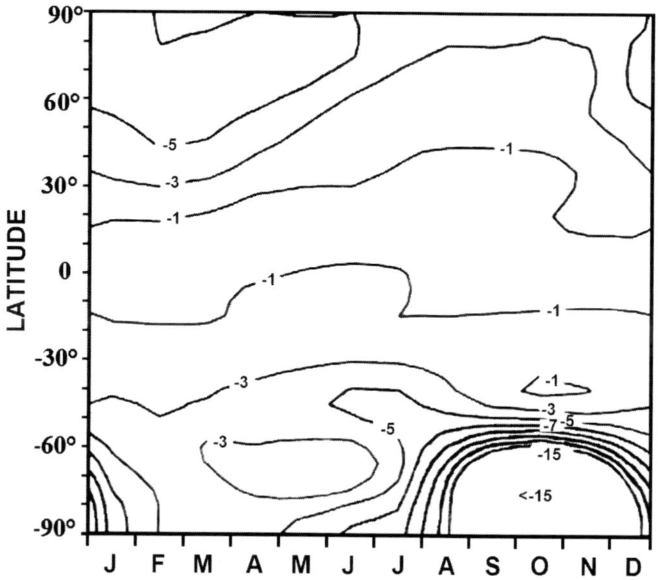

Figure 4.14. Difference (in %) between total ozone values for the two periods (1964–1980 and 1984–1993) reported in Figure 4.12. Southward of 60°S from September to November the difference is more than 15% with a maximum of 35% poleward of 80° in October (Bojkov and Fioletov, 1995).

co-operation with the WMO's Global Atmospheric Watch Program (GAW). The principal aim of SPARC/IOC/GAW "Assessment of Trends in the Vertical Distribution of Ozone" in Harris *et al.* (1998) was the critical review of measurements and trends of the ozone vertical profile (OVP). Analysis of all available information resulted, for the first time, in a single profile of observed ozone depletion at northern mid-latitudes.

Interannual anomalies in the vertical profiles of stratospheric ozone, in both equatorial and extratropical regions, have been shown by Tegtmeier *et al.* (2010) to exhibit a strong seasonal persistence: namely, extended temporal autocorrelations during certain times of the year. They investigated the relationship between this seasonal persistence of equatorial and extratropical ozone anomalies using the SAGE-corrected SBUV dataset. For the regions of the stratosphere where ozone is under purely dynamical or purely photochemical control, the seasonal persistence of equatorial and extratropical ozone anomalies arises from distinct mechanisms but preserves an anticorrelation between tropical and extratropical anomalies established during the winter period. In the 16–10 hPa layer, where ozone is controlled by both dynamical and photochemical processes, equatorial ozone anomalies exhibit a completely different behavior compared to ozone anomalies above and below in terms of variability, seasonal persistence, and especially the relationship between equatorial and extratropical ozone. Cross-latitude–time correlations show that for the 16–10 hPa layer, Northern Hemisphere extratropical ozone anomalies show

Table 4.6. Long-term total ozone trends derived from ground-based data. Estimates were made for "natural" ozone seasons (DJFM, December–March; MJJA, May–August; SON, September–November) using data from January 1964 to February 1994, with a linear trend fit over the 1970–1994 and 1980–1994 periods. Values are given in % per decade $\pm 2\sigma$ (Bojkov and Fioletov, 1995).

Region	*DJFM*	*MJJA*	*SON*	*Year-round*
Slope from 1970 to 1994				
35°N–65°N	-3.2 ± 1.0	-1.1 ± 0.9	-1.1 ± 0.6	-1.9 ± 0.8
10°N–35°N	-1.3 ± 0.8	-0.3 ± 0.7	-0.5 ± 0.7	-0.7 ± 0.7
10°N–10°S	-0.2 ± 0.9	-0.5 ± 0.9	-0.2 ± 0.9	-0.2 ± 0.7
10°S–35°S	-1.2 ± 0.5	-1.1 ± 0.7	-1.4 ± 0.7	-1.2 ± 0.4
35°S–65°S	-1.6 ± 0.8	-2.6 ± 0.8	-3.5 ± 1.3	-2.4 ± 0.7
35°N–90°N	-3.4 ± 1.1	-1.3 ± 1.0	-1.0 ± 0.7	-2.0 ± 0.8
35°S–90°S	-1.8 ± 1.0	-2.7 ± 0.8	-6.0 ± 1.6	-3.2 ± 0.8
Northern Hemisphere	-2.2 ± 0.7	-0.8 ± 0.7	-0.6 ± 0.5	-1.3 ± 0.6
Southern Hemisphere	-1.3 ± 0.7	-1.8 ± 0.5	-3.5 ± 0.8	-2.0 ± 0.5
Global	-1.8 ± 0.5	-1.3 ± 0.4	-2.1 ± 0.5	-1.7 ± 0.4
Slope from 1980 to 1994				
35°N–65°N	-5.8 ± 1.7	-2.6 ± 1.5	-2.5 ± 1.0	-3.8 ± 1.3
10°N–35°N	-2.3 ± 1.4	-0.6 ± 1.3	-1.1 ± 1.1	-1.3 ± 1.1
10°N–10°S	-0.6 ± 1.5	-1.3 ± 1.5	-0.3 ± 1.5	-0.8 ± 1.3
10°S–35°S	-1.8 ± 1.2	-2.0 ± 1.2	-2.3 ± 1.3	-2.0 ± 0.8
35°S–65°S	-3.6 ± 1.2	-4.9 ± 1.3	-7.3 ± 2.0	-5.0 ± 1.0
35°N–90°N	-6.1 ± 1.8	-3.0 ± 1.5	-2.5 ± 1.1	-4.1 ± 1.2
35°S–90°S	-4.2 ± 1.5	-5.1 ± 1.3	-11.9 ± 2.6	-6.4 ± 1.2
Northern Hemisphere	-4.0 ± 1.1	-1.9 ± 1.1	-1.6 ± 0.9	-2.6 ± 0.9
Southern Hemisphere	-2.7 ± 1.0	-3.4 ± 0.8	-6.6 ± 1.5	-3.9 ± 0.8
Global	-3.5 ± 0.5	-2.7 ± 0.5	-4.3 ± 0.7	-3.4 ± 0.4

the same variability as equatorial ozone anomalies but lagged by 3–6 months. High-correlation coefficients were observed during the timeframe of seasonal persistence of ozone anomalies, which is June–December for equatorial ozone and shifts by approximately 3–6 months when going from the equatorial region to Northern Hemisphere extratropics. Thus, in the transition zone between dynamical and photochemical control, equatorial ozone anomalies established in boreal summer/autumn are mirrored by Northern Hemisphere extratropical ozone anomalies with a time lag similar to transport timescales. Equatorial ozone anomalies established in boreal winter/spring are likewise correlated with ozone anomalies in the Southern Hemisphere extratropics with a time lag comparable to transport timescales, similar to what is seen in the Northern Hemisphere. However, correlations between equatorial and Southern Hemisphere extratropical ozone in the 10–16 hPa layer were found to be weak. Sofieva *et al.* (2009) presented spatiotemporal distributions of the tertiary ozone maximum (TOM), based on GOMOS data for 2002–2006. The tertiary ozone maximum was typically observed in the high-latitude winter mesosphere at an altitude of ~72 km. Although the explanation for this phenomenon has been found recently—low concentrations of odd hydrogen cause subsequent decrease in odd oxygen losses—models had previously shown significant deviations from existing observations. Good coverage of polar night regions by GOMOS data made it possible for the first time to obtain spatiotemporal observational distributions of nighttime ozone concentration in the mesosphere. The distributions obtained from GOMOS data exhibit specific features, which are variable from year to year. In particular, due to the long lifetime of ozone under polar night conditions, the downward transport of polar air by meridional circulation is clearly observed in the tertiary ozone maximum time series. Although the maximum tertiary ozone mixing ratio is achieved close to the polar night terminator (as predicted by theory), TOM can be observed also at very high latitudes, not only at the beginning and at the end but also in the middle of winter. We have compared the observational spatiotemporal distributions of the tertiary ozone maximum with that obtained using WACCM (Whole Atmosphere Community Climate Model) and found that the specific features are reproduced satisfactorily by the model.

Many recent studies have been devoted to studies of various aspects of stratospheric ozone dynamics. One of the key aspects is connected with assessment of the role of stratosphere–troposphere exchange in stratospheric ozone dynamics.

4.3.1 Stratosphere–troposphere exchange

Many studies show that there is an increasing trend of background values of ozone in the troposphere (Bojkov, 1986a). This is of serious concern because of the damage caused by high ozone concentrations to human beings, animals, and plants, and also because of possible climatic effects. Stratospheric ozone leaks into the troposphere forming a natural background. In addition, man-made and natural emissions of nitrogen oxides and hydrocarbons lead to the production of ozone in the sunlit troposphere.

Until recently knowledge of the behavior of ozone in the troposphere and transfer across the upper-troposphere/lower-stratosphere boundary was mostly obtained from ground stations operating ozonesonde balloons and from data obtained from aircraft-flown instruments (see Section 1.7.4.3 on the MOZAIC program) (Fischer *et al.*, 2008; Schnadt Poberaj *et al.*, 2009). More recently, information has become available from satellite systems. Changes in the stratosphere–troposphere exchange (STE) of ozone over the last few decades have altered tropospheric ozone abundance and are likely to continue doing so in the coming century as climate changes. An important issue in determining tropospheric ozone concentrations from spaceborne data is to discriminate accurately between the troposphere and the stratosphere. Some success was obtained from indirect methods based on subtracting stratospheric ozone from the value of TOC. But recently direct observation of the tropospheric ozone profile has become possible using data from systems such as GOME (see Section 2.3.5), HIRDLS (see Section 2.5.3.8), MLS (see Section 2.5.3.4), and OMI (see Section 2.3.2). Strategically designed ozonesonde networks have transformed sampling in the upper troposphere and lower stratosphere by using consistent temporal and vertical (100 m) resolution. At the same time they are essential components of satellite validation and monitoring of ozone profiles (Thompson *et al.*, 2011). In particular, two strategic networks—SHADOZ (Southern Hemisphere Additional Ozonesondes) and IONS (INTEX Ozonesonde Network Study)—are described (see Section 6.6.3). All of this has made it possible to study ozone budgets in the upper troposphere/lower stratosphere and the transport of ozone across the tropopause boundary in much more detail than was possible previously (Sørensen and Nielsen, 2001). Hsu and Prather (2009) combined an updated linearized stratospheric ozone chemistry model (Linoz v2) with a parameterized chemistry of polar stratospheric cloud (PSC) model, a 5-year (2001–2005) sequence of the European Centre for Medium-Range Weather Forecasts (ECMWF) meteorology data, and the University of California, Irvine (UCI) chemistry–transport model (CTM), to examine variations in stratosphere–troposphere exchange ozone flux and how it perturbs tropospheric ozone. They estimated stratosphere–troposphere exchange ozone flux in 2001–2005 as $290\,\mathrm{Tg\,yr^{-1}}$ in the Northern Hemisphere and $225\,\mathrm{Tg\,yr^{-1}}$ in the Southern Hemisphere.

In order to assess the anthropogenic component of free atmospheric ozone, the nature of stratosphere–troposphere exchange must be well understood. It was believed until recently that the main transfer of stratospheric ozone occurs in tropopause folding events associated with rapid cyclogenesis. Varotsos and Kondratyev (1995) summarized the level of knowledge in this field and discussed relevant observations made in Athens.

Major tropopause folding events occur at the flanks of large-scale northern flow. A threefold tropopause structure model has been proposed to simulate such events. According to this model, the sounding station in Athens is situated at such a latitudinal band (36°–38°N) that stratospheric air intrusions should occur by means of tropopause foldings between the polar and mid-latitude troposphere. This fact makes the Athens region subject to the influence of the polar front and sometimes to the impact of the subtropical jetstream. The influence of the meteorological situation has been revealed by observations of the laminated structure of the vertical profile of ozone over Athens discussed above.

Observations made near a cut-off low in the upper troposphere suggest that cut-off lows may also contribute to the removal of ozone from the stratosphere. But a limited number of studies have been conducted so far, and there still remain many uncertainties about this type of stratosphere–troposphere exchange.

In the following we examine stratosphere–troposphere ozone exchange over Athens, Greece, during the winter 1991–1992 EASOE campaign, in relation to the presence of a cut-off low in the upper troposphere or a possible tropopause folding, at a time when the station is close to the flank of the upper-level trough. Note that this was the first time such an intensive sounding campaign was performed in Athens. We examined four ozone soundings that correspond to the above meteorological situations. Note also that the nearest ozone-sounding station to Athens (Hohenpeissenberg, Germany, 48°N, 11°E) is too far away to obtain information about the situation over the Athens region, which seems to be quite interesting from the atmospheric dynamics point of view, as mentioned above.

A cut-off low system developed over the Athens region on March 18 and 20, 1992. On the first day (March 18) the lowest height for the pressure level of 200 hPa during our campaign was observed. This resulted in a very low tropopause (approaching the 400 hPa level). Varotsos *et al.* (1994a, b) also observed very high ozone concentration values with a maximum of 190 mb and a strong secondary maximum (about 120 mb) in the lower stratosphere (about 5 km deep). A synoptic map showed that the site was influenced by northern circulation, which brings ozone-rich air to the region. From another point of view, it may be said that a cut-off low event is combined with high ozone partial pressure at low levels. Because stratospheric air was then at higher pressure than normal the ozone partial pressure was similarly increased. What is remarkable about this is the partial pressure minimum, near 15 km, which our ozone data clearly show. The ozone profile on that day did not show any characteristics of detectable cross-tropopause exchange.

On March 20 (the day of the next ozone sounding) Varotsos *et al.* (1994a, b) observed that the cut-off low system was moving eastwards and that the sounding site was influenced by polar air. The intensity of the low weakened and consequently the height of the tropopause increased.

On March 18, the already observed laminated ozone structure in the region of 100–300 hPa became much more pronounced (see also Section 4.2). This is in agreement with the conclusions made earlier, according to which the laminar structure of the ozone profile over Athens is strongly associated with the establishment of northwestern circulation over the region. Comparing the tropospheric temperature profiles of both days it was concluded that on March 20 it was higher by 3–5°C. This agrees fairly well with findings of similar studies of cut-off low systems. This might be associated with cooling of the lower stratosphere and lifting of the tropopause, both of which were in fact observed on that day. The most important feature observed on March 20 was a significant tropopause peak of ozone (about 10 mb) at the 600 hPa level with a sharp decrease in relative humidity suggesting that this air might originate from the stratosphere. One can also see, on the same day, that ozone partial pressure increases sharply before reaching the tropopause, which demonstrated another example of stratospheric ozone intrusion into the troposphere.

On March 27 and April 8 the sounding site was situated on the eastern edge of the upper-level trough. The strong laminae observed during those days were also an indication that it was close to the vortex boundaries. As mentioned above, in this type of situation folding occurs, which creates the potential for troposphere–stratosphere exchange. In fact, Varotsos *et al.* (1994a–c) observed tropospheric ozone peaks of 5–10 mb in the 400–600 hPa level. These peaks were less pronounced than the broad peak observed on March 20 under a cut-off low regime. Also, these peaks were not associated with a sharp decrease of relative humidity. In addition, on April 8 Varotsos *et al.* (1994c) observed a sharp tropospheric ozone increase, before the beginning of cross-tropopause exchange. A slight positive temperature gradient right above the tropopause, as observed on the same day, is another indication of ongoing troposphere–stratosphere exchange.

There is a close relationship between ENSO and stratosphere–troposphere exchange (STE); different patterns of circulation and meteorology in El Niño and La Niña years result in significant differences in the transport of ozone-rich air from the stratosphere to the troposphere. Zeng and Pyle (2005) calculated an anomalously large increase of STE following a typical El Niño year, most evident after the 1997–1998 El Niño event. La Niña events result in a decrease of STE. STE is one factor affecting tropospheric ozone; we find the highest level of ozone after the 1997–1998 El Niño.

Jing *et al.* (2004) investigated isentropic ozone exchange between the extratropical lower stratosphere and the subtropical upper troposphere in the Northern Hemisphere. The quantification method is based on potential vorticity (PV) mapping of Stratospheric Aerosol and Gas Experiment (SAGE II) ozone measurements and contour advection calculations using the NASA Goddard Space Center Data Assimilation Office (DAO) analysis for 1990. The magnitude of annual isentropic stratosphere-to-troposphere ozone flux is calculated to be approximately twice that directed from the troposphere into the stratosphere. The net effect is that some 4.6×10^{10} kg yr^{-1} of ozone are transferred quasi-horizontally from the extratropical lower stratosphere into the subtropical upper troposphere between the isentropic surfaces of 330 and 370 K. Estimated monthly ozone fluxes show that isentropic cross-tropopause ozone transport is stronger in summer and autumn than in winter and spring, and this seasonality is more obvious at the upper three levels (i.e., 345, 355, and 365 K) than at 335 K. The distributions of estimated monthly ozone fluxes indicate that isentropic stratosphere-to-troposphere ozone exchange is associated with wave breaking and occurs preferentially over the eastern Atlantic Ocean and northwest Africa in winter and over the Atlantic and Pacific Oceans in summer.

Collins *et al.* (2003) investigated the impact of circulation changes in a changed climate on the exchange of ozone between the stratosphere and the troposphere. They identified an increase in the net transport of ozone into the troposphere in a future climate of 37%, although a decreased ozone lifetime means that the overall tropospheric burden decreases. There are regions in the mid-latitudes to high latitudes where surface ozone is predicted to increase in the spring. However, these increases are not significant. Significant ozone increases are predicted in regions of the upper troposphere. The general increase in the stratospheric contribution to tropospheric ozone under a climate change scenario indicates that the stratosphere will play an even more significant role in the future.

The downward flux of ozone across the mid-latitude (30°–60°) tropopause is calculated using changes in potential vorticity (PV) and the statistical relationship between PV changes and ozone flux. This approach separates the dynamical aspects of transport from the seasonal cycle of ozone in the lowermost stratosphere. Northern Hemisphere ozone flux is larger than that calculated in the Southern Hemisphere during 2000. The mid-latitude Southern Hemisphere total horizontal area of exchange is equal to or slightly greater than the area of exchange in the Northern Hemisphere throughout an annual cycle. The mean changes in potential vorticity of parcels near the tropopause are also similar or slightly greater in the Southern Hemisphere, suggesting that Northern Hemisphere and Southern Hemisphere downward airmass transport to the troposphere are comparable. These results show that the greater Northern Hemisphere mid-latitude ozone flux is mostly due to the amount of ozone available for exchange rather than dynamical differences near the tropopause level at these latitudes (Olsen *et al.*, 2003).

The present generation of tropospheric chemistry models applies horizontal and vertical model resolutions that are sufficiently fine to represent synoptic-scale processes. Roelofs *et al.* (2003) compared simulations of a tropopause folding event on June 20–21, 2001 from six tropospheric ozone models with tropospheric ozone profiles observed at Garmisch-Partenkirchen (Germany). This event involved airmasses of stratospheric origin and of North Atlantic and North American tropospheric origin. Two coupled chemistry–climate models, three chemistry–transport models, and one chemistry–trajectory model were involved in the intercomparison. The models did not explicitly include stratospheric chemistry, and stratospheric ozone was parameterized instead. The horizontal resolution of the Eulerian models—T42 ($2.8° \times 2.8°$) or finer—appeared to be adequate to represent two prominent features: namely, the stratospheric intrusion descending from the upper troposphere to about 4 km altitude on the first day and an ozone-poor airmass of marine origin in the lower troposphere on the second day. Major discrepancies between model results and observations were the underestimation of ozone levels in the intrusion, too strong downward transport of ozone between the lower stratosphere and the upper troposphere on the first day, and too fast and deep descent of the intrusion. Accurate representation of ozone levels in the intrusion was found to depend directly on the accuracy of simulated ozone in the lower stratosphere. Additionally, for Eulerian models a relatively coarse vertical resolution in the tropopause region may have added to inaccuracies in simulated ozone distributions.

Cristofanelli *et al.* (2009) developed an algorithm to detect airmass transport events from the upper troposphere/lower stratosphere (UT/LS) to the middle troposphere of the southern Himalayas on a daily basis at the Everest-Pyramid

GAW station (EV-PYR, 5,079 m asl, Nepal). In particular, *in situ* surface ozone and atmospheric pressure variations as well as total ozone values from OMI satellite measurements were analyzed. Further insight was gained from three-dimensional backward trajectories and potential vorticity calculated with the LAGRANTO model. According to algorithm outputs, 9.0% of the considered dataset (365 days from March 2006 to February 2007) was influenced by this class of phenomena with a maximum of frequency during dry and pre-monsoon seasons. During 25 days of UT/LS transport events for which any influence of anthropogenic pollution was excluded, the daily ozone mixing ratio increased by 9.3% compared to seasonal values. This indicates that under favorable conditions downward airmass transport from the UT/LS can play a considerable role in determining the concentrations of surface ozone in the southern Himalayas.

Hegglin *et al.* (2010) reported on a multimodel assessment of the performance of chemistry–climate models (CCMs) in the extratropical upper troposphere/lower stratosphere (UT/LS). Process-oriented diagnostics were used to validate the dynamical and transport characteristics of 18 CCMs using meteorological analyses and aircraft and satellite observations. The main dynamical and chemical climatological characteristics of the extratropical UT/LS are generally well represented by the models, despite the limited horizontal and vertical resolution. The seasonal cycle obtained for the lowermost stratospheric mass was realistic, but with a wide spread in its mean value. A tropopause inversion layer is present in most models, although the maximum in static stability is located too high above the tropopause and is somewhat too weak, as can be expected from limited model resolution. Similar comments apply to the extratropical tropopause transition layer. The seasonality in lower-stratospheric chemical tracers is consistent with the seasonality in the Brewer–Dobson circulation. Both vertical and meridional tracer gradients are of similar strength to those found in observarions. Models that perform less well tend to use a semi-Lagrangian transport scheme and/or have a very low resolution. Two models, and the multimodel mean, scored consistently well on all diagnostics, while seven other models scored well on all diagnostics except the seasonal cycle of water vapor. Only four of the models were consistently below average. The lack of tropospheric chemistry in most models limits their evaluation in the upper troposphere. Finally, the UT/LS is relatively sparsely sampled by observations, limiting the ability to quantitatively evaluate many aspects of model performance (for details see Hegglin *et al.*, 2010).

4.3.2 Low-ozone pockets

As pointed out earlier, ozone spatiotemporal variability is characterized by a high level of complexity. One of the features of this complexity is the earlier-discovered low-ozone pocket (Morris *et al.*, 1995). Morris *et al.* (1995, 1998) emphasized that, although the lowest ozone mixing ratios are typically found in the middle stratosphere within the winter circumpolar vortex (according to both chemistry and dynamics), they were also frequently observed—according to satellite remote sounding using Limb Infrared Monitor of the Stratosphere (LIMS) and Microwave Limb

Sounder (MLS) instrumentation—outside the vortex. Pockets of low ozone were registered on numerous occasions in anticyclones at mid to high latitudes in the middle stratosphere, typically in the 20–2 hPa layer (27–43 km or 750–1,250 K) (Manney *et al.*, 1995).

According to analysis of the LIMS data, ozone pockets are formed during mid to late winter in conjunction with stratospheric wave-breaking events. During such events, airmasses with relatively high amounts of ozone move poleward over the polar cap, while a poor or low-ozone region forms in a strong anticyclone. Airmasses within the anticyclones at altitudes near 850 K (~32 km) contain ~25% less ozone than the surrounding air. Trajectory calculations have revealed that much of the air within the pockets originated in the tropics or subtropics at higher altitudes several weeks earlier.

To achieve a better understanding of the nature of low-ozone pockets, Morris *et al.* (1998) examined in detail an event in February–March 1993 using a Lagrangian model with initialization based on observations from the UARS. The results indicated that the model is able to realistically reproduce observational data and explain observed variation in terms of chemistry and dynamics. The primary mechanism responsible for development of these low-ozone regions is the isolation of air at high latitudes for periods of time long enough that significant ozone loss toward the local photochemical equilibrium can occur. Such behavior is different from that of the surrounding airmasses which move from mid to high latitudes and back again over the course of a few days.

A case study of March 9, 1993 considered by Manney *et al.* (1995) showed that the polar vortex was shifted off the pole during a stratospheric warming event, and a strong anticyclone (the Aleutian high) was set up over North America. Each of these dynamical features coincided with a region of low ozone. Relatively high values of ozone (more typical of mid-latitudes) extended over the polar cap, separating the two regions of low ozone. Morris *et al.* (1998) mentioned that the formation of the low-ozone region centered over Europe, inside the polar vortex, developed normally, whereas the one in the anticyclone over Canada evolved quite differently and therefore required special analysis. Comparison of ozone and water vapor distribution showed that this low-ozone pocket did not come from the vortex. The water vapor data strongly suggested tropical origins for the air contained within the pockets.

Manney *et al.* (1995) pointed out a number of issues, related to the formation of low-ozone pockets, that needed to be explained by modeling efforts, including the following:

(1) the pockets appear in the anticyclone several days after the peak of stratospheric warming;
(2) they are confined vertically to the region between 15 and 5 hPa;
(3) they appear downstream of a warm region;
(4) they originate in the tropics;
(5) they contain air parcels that have been confined together for 1–3 weeks;
(6) they contain some amounts of HNO_3.

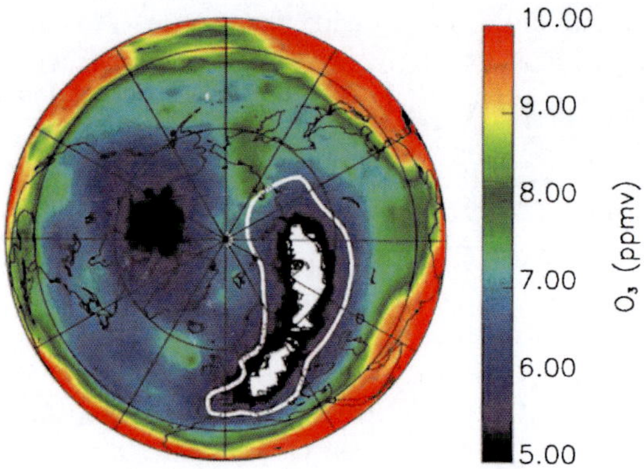

Figure 4.15. Barnes map (Barnes and Simeth, 1986) of gridded output ozone from a Lagrangian chemical model (Kawa *et al.*, 1993) at 00:00 GMT on March 9, 1993 at the 850 K surface. The Lagrangian model reproduced the observed ozone gradients, including low ozone in both the vortex and the anticyclone. White regions indicate data values below 5 ppmv (Morris *et al.*, 1998).

The purpose of the numerical modeling carried out by Morris *et al.* (1998) was to answer these questions by examination of the February–March events at the 850 K surface. The calculated ozone distribution at 00.00 GMT on March 9 (Figure 4.15) compared well with Microwave Limb Sounder (MLS) ozone data. In both cases when low ozone (~5 ppmv) occurred within the vortex boundary, high ozone values were observed in the tropics (>10 ppm) and low values (<6 ppmv) in the region of the anticyclone over northern Canada. Calculations indicated that local values measured by the Microwave Limb Sounder and predicted by the model matched the photochemical equilibrium values for latitudes between 15° and 45°N. North of 45°, however, the observed ozone exceeded the photochemical equilibrium values.

Figure 4.16 illustrates the extent to which air within the anticyclone was isolated at high latitudes. For this purpose, for each air parcel included in the chemical calculations, calculations were made of the equivalent solar latitude (ESL), a quantity related to the dynamic history of an air parcel (Morris *et al.*, 1998). The ESL value characterizes the latitude at which an air parcel, with a given solar exposure, would be found if it had remained at a fixed location in the atmosphere. ESL relates the solar-weighted average latitude at which an air parcel has been found over the previous 2-week period: parcels with a high ESL have spent most of their time at high latitudes, while those with a low ESL have spent most of their time at low latitudes. The ESL field is very highly correlated with the MLS ozone field. High ESL values within the vortex and the anticyclone confirm that air parcels within these regions have been isolated at high latitudes for the previous 2 weeks.

The vertical boundaries of low-ozone pockets (20–2 hPa) were successfully simulated by examination of the photochemical lifetime of ozone as a function of latitude and potential temperature obtained using the 3-D chemistry–transport

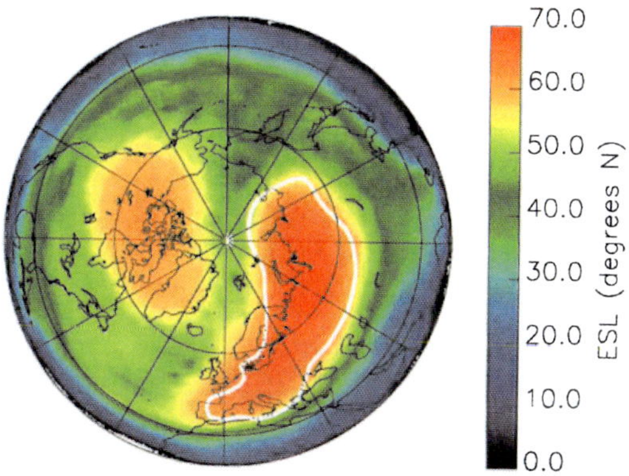

Figure 4.16. Equivalent solar latitude (see text) along 14-day diabatic back trajectories for air parcels used in the chemistry–trajectory model on March 9, 1993 at the 850 K surface. Parcels surrounding the anticyclone have been exposed to significantly more sunlight than those trapped inside it. Solar exposure associated with parcels outside the anticyclone is typically found at 40°–50°N, while solar exposure inside the anticyclone is typically found near 70°N (Morris *et al.*, 1998).

model developed by Douglass *et al.* (1996). Morris *et al.* (1998) showed that the layer in which ozone pockets evolve is one in which the photochemical lifetime of ozone is similar in magnitude to the duration of strong containment provided by dynamical features, such as Aleutian anticyclones. The near balance between these photochemical and dynamical timescales makes development of low-ozone pockets possible at these altitudes (see Sections 6.6.2 and 6.6.3.1).

Sensitivity calculations made by Morris *et al.* (1998) revealed that temperature history and specific chemical initialization (including O_3, overhead column O_3, NO_x, Cl_x, and total inorganic chlorine) are less important for the development of low-ozone pockets than isolation at high latitudes.

Similar results were obtained by Nair *et al.* (1998) who undertook a comprehensive study of December 1992 and February/March 1993 events, using a Lagrangian photochemical model, to simulate the time-dependent ozone loss rate within parcels of air following specific trajectories leading to observed low-ozone pockets and within parcels leading to neighboring regions with higher ozone. These calculations were made for several different altitudes. The Caltech/JPL photochemical model was adapted to operate in a Lagrangian mode. The full model atmosphere consists of 40 levels from the surface to 0.316 hPa. Table 4.7 presents important reactions for odd oxygen—the sum of O, $O(^1D)$, and O_3—balance. The list of species taken into account in the model is presented in Table 4.8. The model successfully simulates the observed formation of low-ozone pockets in the 6–10 hPa altitude range within measurement uncertainties but overestimates the ozone loss rate at higher altitudes. Under the conditions considered, the low-ozone levels are due to a decrease in the odd oxygen production rate, and not to an increase in the loss rate by reaction with halogen species, as in the "classical" ozone hole. Trajectory

Table 4.7. Important reactions for odd oxygen balance (Nair *et al.*, 1998).

Number	*Reaction*	ΔO_x	*Number*	*Reaction*	ΔO_x
(R1)	$O_2 + h\nu \rightarrow 2O$	+2	(R147)	$N + NO \rightarrow N_2 + O$	+1
(R2)	$O_2 + h\nu \rightarrow O + O(^1D)$	+2	(R148)	$N + NO_2 \rightarrow N_2O + O$	+1
(R5)	$HO_2 + h\nu \rightarrow OH + O$	+1	(R149)	$NO + O + M \rightarrow NO_2 + M$	−1
(R8)	$H_2O + h\nu \rightarrow 2H + O$	+1	(R150)	$NO + O_3 \rightarrow NO_2 + O_2$	−1
(R10)	$NO + h\nu \rightarrow N + O$	+1	(R158)	$NO_2 + O_3 \rightarrow NO_3 + O_2$	−1
(R11)	$NO_2 + h\nu \rightarrow NO + O$	+1	(R166)	$NO_3 + O \rightarrow O_2 + NO_2$	−1
(R12)	$NO_3 + h\nu \rightarrow NO_2 + O$	+1	(R193)	$O + ClO_2 \rightarrow ClO + O_2$	−1
(R16)	$N_2O_5 + h\nu \rightarrow NO + NO_3 + O$	+1	(R194)	$O + ClO_2 + M \rightarrow ClO_3 + M$	−1
(R24)	$ClO + h\nu \rightarrow Cl + O$	+1	(R195)	$O + Cl_2O \rightarrow 2ClO$	−1
(R25)	$ClOO + h\nu \rightarrow ClO + O$	+1	(R196)	$O + ClNO_3 \rightarrow O_2 + ClONO$	−1
(R26)	$ClO_2 + h\nu \rightarrow ClO + O$	+1	(R197)	$O + HCl \rightarrow OH + Cl$	−1
(R36)	$ClNO_3 + h\nu \rightarrow O + ClONO$	+1	(R198)	$O + HOCl \rightarrow OH + ClO$	−1
(R83)	$CO_3 + h\nu \rightarrow CO + O$	+1	(R199)	$O(^1D) + Cl_2 \rightarrow Cl + ClO$	−1
(R95)	$2O + M \rightarrow O_2 + M$	−2	(R200)	$O(^1D) + HCl \rightarrow Cl + OH$	−1
(R99)	$O + O_3 \rightarrow 2O_2$	−2	(R201)	$O(^1D) + HCl \rightarrow ClO + H$	−1
(R102)	$O(1D) + O_3 \rightarrow 2O_2$	−2	(R202)	$O(^1D) + CHFCl_2 \rightarrow CFCl_2 + OH$	−1
(R104)	$O + H_2 \rightarrow OH + H$	−1	(R203)	$O(^1D) + CHF_2Cl \rightarrow CF_2Cl + OH$	−1
(R105)	$O + OH \rightarrow O_2 + H$	−1	(R204)	$O(^1D) + CCl_4 \rightarrow CCl_3 + ClO$	−1
(R106)	$O + HO_2 \rightarrow OH + O_2$	−1	(R205)	$O(^1D) + CFCl_3 \rightarrow CFCl_2 + ClO$	−1

(R107)	$O + H_2O_2 \rightarrow OH + HO_2$	−1	(R206)	$O(^1D) + CF_2Cl_2 \rightarrow CF_2Cl + ClO$	−1
(R108)	$O(^1D) + H_2 \rightarrow H + HO$	−1	(R207)	$O(^1D) + CF_3Cl \rightarrow CF_3 + ClO$	−1
(R109)	$O(^1D) + H_2O \rightarrow 2OH$	−1	(R217)	$O(^1D) + CF_3CFCl_2 \rightarrow Cl + 4F + ClO$	−1
(R112)	$H + O_3 \rightarrow OH + O_2$	−1	(R219)	$O(^1D) + COCl_2 \rightarrow 2ClO + CO$	−1
(R115)	$H + HO_2 \rightarrow H_2O + O$	+1	(R220)	$O(^1D) + COFCl \rightarrow ClO + FCO$	−1
(R116)	$2OH \rightarrow H_2O + O$	+1	(R259)	$Cl + O_3 \rightarrow ClO + O_2$	−1
(R118)	$OH + O_3 \rightarrow HO_2 + O_2$	−1	(R313)	$ClO + O \rightarrow Cl + O_2$	−1
(R122)	$HO_2 + O_3 \rightarrow OH + 2O_2$	−1	(R315)	$ClO + O_3 \rightarrow ClO_2 + O_2$	−1
(R125)	$O + NO_2 \rightarrow NO + O_2$	−1	(R316)	$ClO + O_3 \rightarrow ClOO + O_2$	−1
(R126)	$O + NO_2 + M \rightarrow NO + M$	−1	(R331)	$ClO_2 + O_3 \rightarrow ClO + 2O_2$	−1
(R127)	$O + HNO_3 \rightarrow OH + NO_3$	−1	(R353)	$O + ClCO \rightarrow Cl + CO_2$	−1
(R128)	$O + HO_2NO_2 \rightarrow OH + NO_2 + O_2$	−1	(R358)	$O + ClCO_3 \rightarrow Cl + O_2 + CO_2$	−1
(R129)	$O + N_2O_5 \rightarrow 2NO_2 + O_2$	−1	(R491)	$O + CH_4 \rightarrow CH_3 + OH$	−1
(R130)	$O(^1D) + N_2 + M \rightarrow N_2O + M$	−1	(R494)	$O + H_2CO \rightarrow OH + HCO$	−1
(R131)	$O(^1D) + N_2O \rightarrow 2NO$	−1	(R496)	$O(^1D) + CH_4 \rightarrow CH_3 + OH$	−1
(R132)	$O(^1D) + N_2O \rightarrow N_2 + O_2$	−1	(R497)	$O(^1D) + CH_4 \rightarrow CH_3O + H$	−1
(R133)	$O_3 + CH_3ONO \rightarrow CH_3ONO_2 + O_2$	−1	(R498)	$O(^1D) + CH_4 \rightarrow H_2CO + H_2$	−1
(R143)	$N + O_2 \rightarrow NO + O$	+1	(R532)	$CH_3 + O \rightarrow H_2CO + H$	−1
(R144)	$N + O_3 \rightarrow NO + O_2$	−1	(R535)	$CH_3 + O_3 \rightarrow CH_3O + O_2$	−1

Where $O(^1D)$ denotes the excited oxygen atom.

Table 4.8. Species in photochemical models (Nair *et al.* 1998).

O	$O(^1D)$	H	H_2	OH	HO_2
H_2O_2	N	NO	NO_2	NO_3	N_2O_3
N_2O_4	N_2O_5	HNO	HNO_2	HO_2NO_2	CH_3ONO
CH_3ONO_2	$CH_3O_2NO_2$	Cl	Cl_2	ClO	ClOO
ClO_2	ClO_3	Cl_2	Cl_2O_2	Cl_2O_3	ClNO
$ClNO_2$	ClONO	$ClNO_3$	HCl	HOCl	CH_2Cl
CCl_3	$CFCl_2$	CF_2Cl	CH_3Cl	CH_2FCl	$CHFCl_2$
CHF_2Cl	CCl_4	$CFCl_3$	CF_2Cl_2	CH_3CCl_3	CF_3CCl_3
CF_3CFCl_2	ClCO	CHClO	$COCl_2$	COFCl	CH_2ClO
CH_3OCl	CH_2ClO_2	CCl_3O_2	$CFCl_2O_2$	CH_2ClO_2H	O_2ClNO_3
CCl_3NO_4	$CFCl_2NO_4$	CH_3	CO	CO_2	HCO
H_2CO	CH_2	CH_3O	CH_2O_2	CH_2OOH	CH_3OH
CH_3O_2	CH_3OOH	$HOCH_2OO$			

Where $O(^1D)$ is an excited oxygen atom.

calculations for both events considered indicate that when air is held at high solar zenith angles for periods of time longer than the photochemical lifetime of ozone, a low-ozone pocket forms. Air outside these pockets and outside the polar vortex typically swings back and forth between mid and high latitudes, such that the average production rate of odd oxygen is high relative to that of air in the low-ozone pocket. Nair *et al.* (1998) underlined the fact that significant uncertainties remain in our understanding of photochemical processes controlling ozone in the upper stratosphere; this is still the case.

In the context of the stratosphere–troposphere exchange problem (see also Section 4.3.1), Pan *et al.* (1997) analyzed hemispheric asymmetries and seasonal variations of lowermost stratospheric water vapor and ozone, derived from SAGE II data (the lowermost stratosphere is defined as that portion of the stratosphere between the tropopause and the approximately 380 K potential temperature surface). Pan *et al.* (1997) emphasized that the lowermost stratosphere makes an interesting case for the study of transport, since the dynamics of this region are influenced by both downward circulation from the so-called overworld (where the extratropical tropopause intersects potential temperature surfaces from about 300 to 380 K) and stratosphere–troposphere exchange across the extratropical tropopause. Trace gases of tropospheric origin can enter this region through the tropical tropopause into the overworld and then be transported into this region via downward circulation, or they can enter directly from the troposphere as part of extratropical stratosphere–troposphere exchange. According to Pan *et al.* (1997), both water vapor and ozone data suggest that troposphere–stratosphere exchange across the extratropical tropopause has a significant influence on the lowermost stratosphere, especially for the Northern Hemisphere in the summer season.

As discussed earlier, the existence and seasonal variability of the Atlantic subtropical potential vorticity barrier, controlling stratosphere–troposphere exchanges between the lowermost extratropical stratosphere and the upper equatorial troposphere, was studied within the Measurements of Ozone by Airbus In-Service Aircraft (MOZAIC) project. In this context Cammas *et al.* (1998) investigated episodes of high-ozone mixing ratios sampled along MOZAIC flight tracks over the subtropical Atlantic with mixing ratios exceeding 100 ppbv on length scales larger than 200 km on flights between 11 and 12 km (a total of 154 high-ozone episodes were extracted from the MOZAIC database from August 1994 to April 1997). All these high-ozone episodes were observed north of 15°N and had lengths ranging from subsynoptic to synoptic scales (see also Section 4.2).

Cammas *et al.* (1998) demonstrated that the barrier effect, located at 15°N over the central/eastern Atlantic, fits with the southernmost latitude of the subtropical jetstream during the period of interest. South of the jetstream, within the latitude band where the ITCZ oscillates, tens of ozone-rich transients (high-ozone episodes with length scales smaller than 80 km) were sampled within the upper equatorial Atlantic troposphere (9–12 km). The origin of these tropical ozone-rich transients is not clear as yet. Cammas *et al.* (1998) pointed out that some of the ozone-rich transients may be interpreted as the final result of tropopause foldings and small-scale mixing processes. Analysis of the seasonal variability of subtropical barriers indicates that this variability clearly agrees with that of the position of the subtropical jetstream as derived from mean isotach data.

Liu, Y. *et al.* (2009) studied the distribution and transport of stratospheric O_3 during the remarkable stratospheric sudden warming event observed in January 2004 in the northern polar region. Comparison between observations by the MIPAS instrument on board the ENVISAT spacecraft and model simulations shows that the evolution of the polar vortex and of planetary waves during the warming event played an important role in controlling the spatial distribution of stratospheric ozone and downward ozone flux in the lower stratosphere and upper troposphere (UT/LS) region. Compared to the situation during the winter of 2002–2003, lower ozone concentrations were transported from the polar region to mid-latitudes, leading to exceptionally large areas of low ozone concentrations outside the polar vortex and low-ozone pockets in the mid-stratosphere. Unusually long-lasting stratospheric westward winds (easterlies) during the 2003–2004 event greatly restricted the upward propagation of planetary waves, causing the weak transport of ozone-rich air originating from low latitudes to the middle polar stratosphere (30 km).

Earth Observing System (EOS) MLS observations of ozone have been analyzed along with meteorological data from the U.K. Meteorological Office using a new automated algorithm to detect low-ozone pockets (LOPs) in stratospheric anticyclones. The algorithm is illustrated with a case study, and all LOPs identified in over 3 years of MLS data are shown. Daily averaged LOP ozone mixing ratios are 5–30% lower than ozone mixing ratios located directly outside an LOP. In the middle stratosphere, near 50° latitude and the International Date Line, large LOPs are identified on over 60% of Northern Hemisphere winter days and 50% of Southern Hemisphere spring days. This suggests that an LOP is present in the Aleutian and Australian anticyclones nearly all the time during these seasons. Ozone reductions in the heart of LOPs are typically about 20% (10%) in the Northern Hemisphere (Southern Hemishere). Thus, LOP contribution to seasonal ozone loss in anticyclones is estimated at up to about 12% in Northern Hemisphere winter and about 5% in Southern Hemisphere spring. The average total column ozone loss from individual LOPs is estimated at about 3–6% (Harvey *et al.*, 2008).

A record minimum in total column ozone over the Tibetan Plateau in December 2003 was analyzed using both satellite observations and a chemistry–transport model (MOZART-3). Ozone profiles observed by MIPAS and GOMOS over the ozone minimum region (25°–35°N, 85°–110°E) showed pronounced ozone reduction in both the upper troposphere and lower stratosphere (UT/LS) and mid-stratosphere regions. MOZART-3 simulation indicated that ozone reduction in the UT/LS region is generally caused by uplift of the local tropopause and northward transport of tropical ozone-poor air associated with an anomalous anticyclone in the upper troposphere. MIPAS observations suggested that the displacement of low-ozone pockets is responsible for ozone decline in the mid-stratospheric region and contributes to 29–46% of total column ozone reduction (Liu, X. *et al.*, 2010a, b).

4.4 THE NON-LINEAR NATURE OF OZONE VARIABILITY; DETRENDED FLUCTUATION ANALYSIS (DFA)

The atmosphere is a complex system and time series of data emerging from complex systems are typically governed by the interplay of random and deterministic mechanisms. Consequently, the laws of physics that govern atmospheric phenomena are mostly non-linear. Any attempt to predict change in the atmosphere's ozone distribution is impossible without consideration of the complexity of all interactive processes including the chemistry and dynamics of the atmosphere (Brandt *et al.*, 2003; Crutzen *et al.*, 1999; Dameris *et al.*, 2005; Ebel, 2001; Grytsai *et al.*, 2005; Kirk-Davidoff *et al.*, 1999; Kondratyev and Varotsos, 2000; Lawrence *et al.*, 1999; Sander, 1999; Stohl *et al.*, 2004). The application of conventional Fourier spectral analysis to the time series of atmospheric quantities reveals that these are usually non-stationary (correlation functions are not invariant under time translation) (Chen *et al.*, 2002 and references therein). Quite often these non-stationarities (e.g., trends and cycles) conceal existing correlations (or other intrinsic properties) and, therefore, new analytical techniques capable of eliminating non-stationarities in the data should be employed (Hu *et al.*, 2001; Varotsos and Kirk-Davidoff, 2006; Varotsos, C.A. *et al.*, 2005, 2006; Varotsos, P.A. *et al.*, 2003a–c).

Characteristic times of such systems may vary over a large range. As a consequence, one often observes a non-exponential decay of correlations, of which stretched exponential and algebraic decay are two examples. Precise classification of the decay of correlations is of major importance for the analysis of various natural, technical, and economic systems. Direct estimate of the correlation function of a time series is known to be limited to rather small time lags, and also the determination of the power spectrum is hampered by large statistical uncertainties if one goes to those low frequencies that reflect the long-time behavior of the system. More recently, alternative methods have been suggested, particularly wavelet techniques (e.g., Koscielny-Bunde *et al.*, 1998), fluctuation analysis and detrended fluctuation analysis (DFA) introduced by Peng *et al.* (1994) (e.g., Ausloos and Ivanova, 2001; Chen *et al.*, 2002; Efstathiou and Varotsos, 2010; Varotsos, C.A. *et al.*, 2003, 2004, 2005, 2006, 2007, 2008b; Weber and Talkner, 2001).

Consideration of the decay of correlations in time can yield insight into the dynamics of a complex system. A randomly forced first-order linear system with memory should have fluctuations whose autocorrelation decays exponentially with lag time, but a higher order system will tend to have a different decay pattern. However, direct calculation of the autocorrelation function is usually not the best way to distinguish among different decay patterns at long lags, due to noise superimposed on the data (Galmarini *et al.*, 2004; Kantelhardt *et al.*, 2002; Stanley, 1999). The pattern of the decay of autocorrelation with increasing time lag can be obtained by using DFA, which transforms the decay with increasing lag into an increase in noise amplitude as the timescale increases; this is less sensitive to statistical errors (Talkner and Weber, 2000). DFA can be applied to time series of a quantity that fluctuates in a complex apparently erratic manner; it is based on methods of time series analysis that were originally used in statistical physics for the study of critical phenomena where fluctutions at all timescales occur. Some of the first examples of the application of DFA were to the analysis of human heartbeat time series (Peng *et al.*, 1995) and daily temperature records from central Europe (Talkner and Weber, 2000).

4.4.1 Long-memory processes in global ozone and temperature variations

Varotsos and Kirk-Davidoff (2006) used DFA to compare the long-range correlations of fluctuations of total ozone (TOC) and tropospheric brightness temperature (TRT) and to determine if they exhibit persistent long-range correlations. This means that TOC or TRT fluctuations at different times are positively correlated, and the corresponding autocorrelation function decays more slowly than exponential decay with increasing lag, possibly by power law decay. In other words, "persistence" refers to the "long-term memory" within the TOC (or TRT) time series (Collette and Ausloos, 2004; Hu *et al.*, 2001; Stanley, 1999; Varotsos, 2005a). Previous workers (e.g., Talkner and Weber, 2000) investigated the decay of autocorrelation in surface temperature records at individual stations, but not in mid-tropospheric temperatures averaged over large regions. Camp *et al.* (2003) carried out an empirical orthogonal function study of the spatiotemporal patterns of tropical ozone variability and calculated the quantitative contribution of low-frequency oscillations (e.g., Quasi-Biennial Oscillation, El Niño-Southern Oscillation, decadal oscillation) to variation in ozone deseasonalized data, but did not discuss the decay of autocorrelation.

DFA which has proved to be useful in a large variety of complex systems with self-organizing behavior (e.g., Chen *et al.*, 2002; Peng *et al.*, 1994; Varotsos, 2005a; Varotsos, P.A. *et al.*, 2003a, b; Weber and Talkner, 2001) can be broken down into steps as follows:

(1) We consider the time series $Z(i)$ $(i = 1, \dots, N)$, describing N observations of some quantity at equally spaced intervals in time. This is a bounded series (bounded above and below) in time (i.e., in bounded time). In the present context this is a series of deseasonalized observations of TOC or TRT. Fluctuations of the N observations $Z(i)$ from their mean value Z_{ave} are given by:

$$y(i) = Z(i) - Z_{\text{ave}} \tag{4.1}$$

The time series $Z(i)$ is first converted into an unbounded $y(k)$ where:

$$y(1) = [Z(1) - Z_{\text{ave}}], \quad y(2) = [Z(1) - Z_{\text{ave}}] + [Z(2) - Z_{\text{ave}}], \quad \dots,$$

$$y(k) = \sum_{i=1}^{k} [Z(i) - Z_{\text{ave}}] \tag{4.2}$$

$y(k)$ is called a cumulative sum or profile. This summation process converts, for example, a white noise process (i.e., with a flat power spectral density) into a random walk (consisting of successive steps with equal probability).

The summation exaggerates the non-stationarity of the original data, reduces the noise level, and generates a time series corresponding to the construction of a random walk that has the values of the original time series as increments. The new time series, however, still preserves information about the variability of the original time series (Chen *et al.*, 2005; Kantelhardt *et al.*, 2002).

(2) The new time series is divided into non-overlapping boxes (segments) of equal length, n. In each box, a least squares line is fitted to the data; this represents the trend in that box. The y value of the relevant least squares fit k is denoted by $y_n(k)$.

It is worth noting that subtraction of the mean is not compulsory, since it would be eliminated by later detrending in the third step anyway (Kantelhardt *et al.*, 2002). The least squares line in each box may be replaced with a polynomial curve of order l, in which case the method is referred to as DFA-l. Using the best least squares straight line is therefore sometimes called the DFA-1 method.

(3) $[y(k) - y_n(k)]$ represents the departure from the local trend (i.e., the least squares fit value) within the box in which k is situated (i.e., fluctuation from the local trend). The root-mean-square fluctuation, $F(n)$, of the whole series, is computed where

$$F(n) = \sqrt{\frac{1}{N} \sum_{k=1}^{N} [y(k) - y_n(k)]^2} \tag{4.3}$$

(4) The computation of $F(n)$ is then repeated for all timescales n.

Typically, $F(n)$ increases with n, the box size. The existence of a power law relation $F(n) \propto n^{\alpha}$ indicates the presence of scaling (self-similarity) (i.e., fluctuations in small boxes are related to fluctuations in larger boxes in a power law fashion). We can determine the value of α from the slope of a plot of $\log F(n)$ against $\log n$.

If there is no long-range correlation (classical random walk problem, white noise) then $\alpha = 0.5$. Other special values of α are $\alpha = 1$ for the so-called $1/f$ noise (f = frequency) and $\alpha = 1.5$ for Brownian noise (the integration of white noise; Ausloos and Ivanova, 2001). What is of interest to us in studyng long-range correlations, or possible autocorrelations, amongst atmospheric parameters are the following ranges of α:

$0 < \alpha < 0.5$ anticorrelations

$0.5 < \alpha < 1.0$ correlations

$1.0 < \alpha < 1.5$ stronger correlations.

A detailed discussion of the the relation between the fluctuation function $F(n)$, the power spectrum, and the autocorrelation function is given by Talkner and Weber (2000). The power spectrum $S(\omega)$, or power spectral density, can be determined from a Fourier transform of the data ($\omega = 2\pi f$). The autocovariance of the data is given by

$$C(\tau) = \langle [Z(i+\tau) - Z_{\text{ave}}][Z(i) - Z_{\text{ave}}] \rangle \tag{4.4}$$

where τ is some time interval (an integer in the case where we have a discrete dataset and, essentially, the same as the box length n we have used above). $S(\omega)$ and $C(\tau)$ are related by

$$C(\tau) = (1/(2\pi)) \int \mathrm{d}\omega \, S(\omega) \cos(\omega\tau) \tag{4.5}$$

(i.e., the power spectral density, or power density spectrum, is the Fourier transform of the autocovariance $C(\tau)$). Under certain conditions $C(\tau)$ and $S(\omega)$ can be expressed as power laws, $S(\omega) \propto \omega^{-\beta}$ and $C(\tau) \propto \tau^{-\gamma}$, and it is then possible to derive relations between β, γ, and α where $F(n) \propto n^{\alpha}$ (for details see Talkner and Weber, 2000).

Varotsos and Kirk-Davidoff (2006) considered TOC over the tropical and mid-latitudinal zones of both hemispheres from 1964 to 2004 (see Figure 4.17). Inspection of Figure 4.17 shows that this time series is apparently non-stationary, and includes both periodic and aperiodic fluctuations. To avoid obscuring possible scaling behavior in the long-term trend and various frequency peaks induced from well-known cycles the deseasonalized data shown in Figure 4.17 were used and analyzed using the DFA method outlined above. $F(n)$ is now denoted by $F_d(\Delta t)$ where Δt is the number of years included in each box and F_d denotes the root-mean-square fluctuation function.

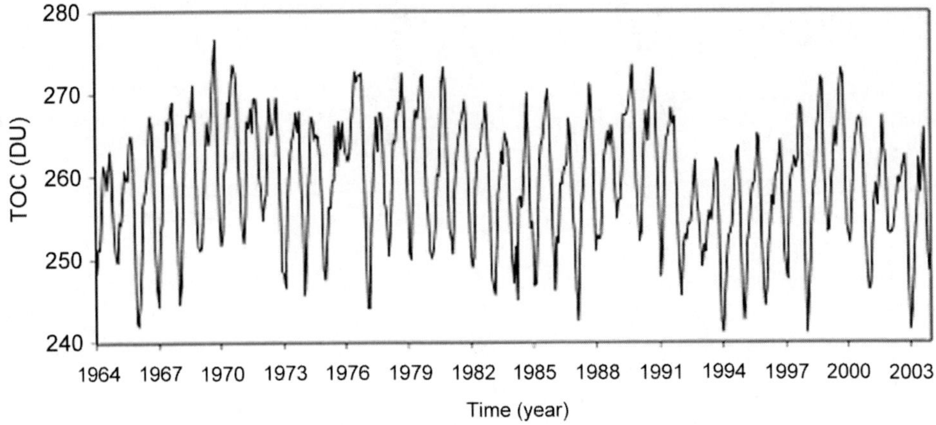

Figure 4.17. TOC mean monthly values in Dobson units (DU) during 1964–2004, over the belt 25°S–25°N derived from the WMO Dobson Network (100 DU = 1 mm thickness of pure ozone on the Earth's surface) (Varotsos and Kirk-Davidoff, 2006).

The results for the tropics and subtropics are shown as log-log plots in Figure 4.18. For the tropics, more strictly for 25°S to 25°N, the log-log plot is a straight line with slope $\alpha = 1.1 \pm 0.04$ (Figure 4.18b). We conclude that TOC fluctuations over the tropics exhibit persistent long-range correlations ($1/f$ noise-like) for the time interval ranging from about 4 months to 11 years. The long-range correlations obtained do not signify the presence of cycles with definite periodicities (i.e., as described by Camp *et al.*, 2003), but rather the existence of dynamical links between long and short timescale behavior.

Results for the subtropics and the mid-latitude zone, specifically for the latitude belts 25°N–60°N and 25°S–60°S during 1964–2004, are shown in Figure 4.18a, c. It will be seen here too that persistent long-range correlations exist for shorter timescales. According to Figure 4.18a for 25°N–60°N, since $\alpha_1 = 1.22 \pm 0.04$ (for timescales shorter than about 2 years), the correlations in TOC fluctuations exhibit stronger memory than those of $\alpha_2 = 0.63 \pm 0.04$ (for timescales from about 2 to 11 years). From Figure 4.18c for 25°S–60°S the persistence of TOC fluctuations is also observed. In particular, for timescales shorter than about 2 years, $\alpha_1 = 1.11 \pm 0.02$, while for longer timescales, $\alpha_2 = 0.64 \pm 0.06$. Thus, the tropics exhibit stronger persistence at long timescales than the extratropics, but approximately equivalent persistence at short timescales (Varotsos, 2005b). For a further discussion of these results see Varotsos and Kirk-Davidoff (2006).

The fact that the persistence of ozone fluctuations changes character at about 2 years in both extratropical bands suggests some connection with the QBO. The result suggests that for timescales longer than 2 years, there is little dynamical memory for ozone fluctuations in the extratropics. It is therefore interesting and puzzling that this change in the character of persistence does not appear in the tropics, where QBO dynamics originate, and where positive TOC deviations are

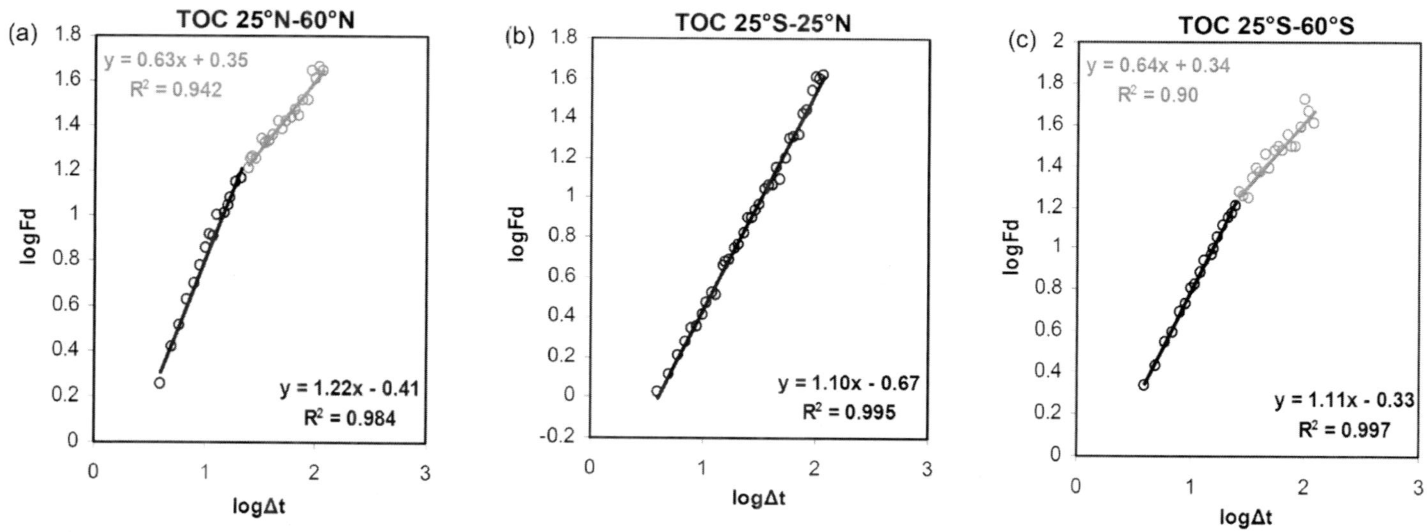

Figure 4.18. Log-log plot of the TOC root-mean-square fluctuation function (F_d) versus temporal interval Δt (in months) for deseasonalized TOC values, observed by the WMO Dobson Network over (a) Northern Hemisphere mid-latitudes, (b) the tropics, and (c) Southern Hemisphere mid-latitudes during 1964–2004 (with the crossover at $\Delta t \approx 28$ months over the mid-latitudes) (Varotsos and Kirk-Davidoff, 2006).

observed to occur a few months before maximum westerlies at 50 hPa (WMO, 2003). The use of DFA with TOC data from the TOMS satellite-flown system yielded results similar to those just described using data from the WMO Dobson Network (Kiss *et al.*, 2007). After the removal of semi-annual, annual, and quasi-biennial background oscillations, the residuals were evaluated by the DFA method and long-range correlations were detected everywhere. Surprisingly, the latitude dependence of zonally averaged correlation exponents exhibits the same behavior as exponents for daily surface temperature records. Additionally, Vyushin *et al.* (2007), analyzing total ozone time series over 60°S–60°N from 1979 to 2005, showed that most latitude bands exhibit long-range correlated behavior, meaning that ozone autocorrelation functions decay by a power law rather than exponentially, thus verifying the earlier results by Varotsos and Kirk-Davidoff (2006).

Furthermore, the log-log plot derived from application of DFA-1 to global D-TOC data reveals that $\alpha = 1.1 \pm 0.02$, suggesting that the strong persistence in the tropics and mid-latitudes discussed above dominates the variability of the global ozone layer. In summary, TOC fluctuations over the tropics, extratropics, and mid-latitudes of both hemispheres, as well as globally, exhibit persistent long-range correlations for all time lags between about 4 months to 11 years. Over the extratropics, this persistence becomes weaker for time lags between about 2–11 years. These findings are consistent with the preliminary results presented by Varotsos (2005b, c).

Results of the application of DFA to TRT by Varotsos and Kirk-Davidoff (2006) are shown in Figure 4.19. The data used were passive microwave temperature soundings from Microwave Sounding Units (MSU channel 2 from the satellites TIROS-N, NOAA-6 to NOAA-12, and NOAA-14) and Advanced Microwave Sounding Units (AMSU channel 5 from the satellites NOAA-15 to NOAA-17 and Aqua) for the period 1978–2004. A detailed description of the multisatellite data analysis and the development technique of the time series of bias-corrected globally averaged daily mean (pentad-averaged) TRT is given by Vinnikov and Grody (2003). Furthermore, to account for the effects of dynamical perturbations on TOC, TRT is often used as an index of dynamical variability (Chandra *et al.*, 1996). Results for the zones 25°N–60°N and 25°S–60°S are shown in Figure 4.19a, c. These results show that long-range power law correlations exist in both hemispheres for a time interval ranging from about 20 days to 7 years, with $\alpha = 0.80 \pm 0.01$. Thus, global TRT fluctuations exhibit long-range persistence. These results were confirmed by DFA-2 to DFA-7 analysis yielding α-values ranging from 0.78 to 0.86 and the persistence found above was confirmed by applying DFA-1 to shuffled TRT anomalies, which again showed no persistent fluctuations. Figure 4.19b shows the results of applying DFA to the TRT time series averaged in pentads of days over the belt 25°S–25°N. It can be seen that there is a crossover between two linear regions at a Δt value of approximately 28 months. Since $\alpha_1 = 1.13 \pm 0.04$ (for timescales shorter than about 2 years) the fluctuations in TRT exhibit long-range correlations, while for timescales from about 2 to 7 years (e.g., El Niño-Southern Oscillation) they follow a random walk ($\alpha_2 = 0.50 \pm 0.04$).

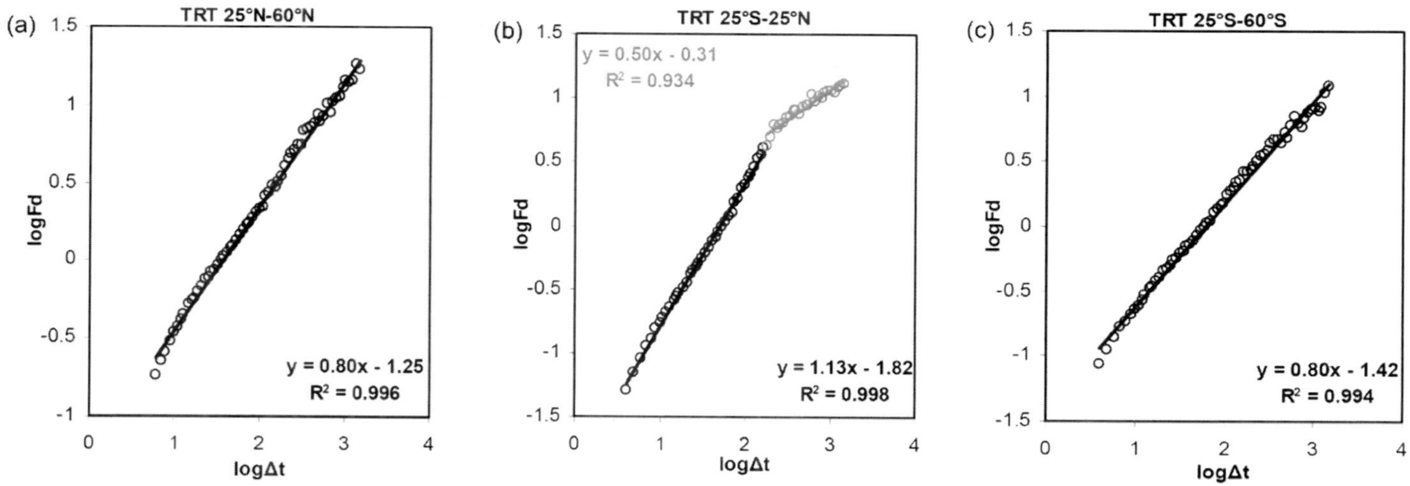

Figure 4.19. Log-log plot of the TRT root-mean-square fluctuation function (F_d) versus time interval Δt (averaged in pentads of days) for deseasonalized TRT (mid-tropospheric temperature) values, observed by the multisatellite instrumentation over the tropics (b), and mid-latitudes of both hemispheres (a) and (c) during 1978–2004 (Varotsos and Kirk-Davidoff, 2006).

Taking these results for TOC and TRT behavior together shows the opposite behavior for mid-tropospheric temperature and ozone fluctuations. For ozone fluctuations, persistence at long timescales is strongest in tropics, but weaker in mid-latitudes, while temperature fluctuations show strong persistence in the mid-latitudes and a random walk in the tropics. Various mechanisms might account for these differences. Greater persistence is in general a result of either stronger positive feedbacks or larger inertia. Thus, the reduced slope of the power distribution of temperature in the tropics at long timescales, compared to the slope in the mid-latitudes could be connected to the poleward increase in climate sensitivity (due to latitude-dependent climate feedbacks) predicted by global climate models. However, this argument applies to surface temperature rather than to mid-tropospheric temperatures, which are expected to increase more or less uniformly with latitude.

The latitude dependence of the persistence in ozone fluctuations can be rationalized as follows. Zonal mean TOC fluctuations in the mid-latitudes are governed largely by motions of the jetstream, which marks the boundary between low tropopause heights on the poleward side (and more TOC) and higher tropopause heights (and less TOC) on the tropical side (Hudson *et al.*, 2003). Such variations might be expected to show relatively low persistence beyond time-scales of a few months (otherwise, seasonal weather prediction would be easier). Thus, the difference in persistence patterns between ozone and temperature could arise because the TOC distribution is more closely tied to gradients in temperature (which are associated with the jet position) than to the temperature itself. Column ozone fluctuations in the tropics are more closely tied to the QBO and ENSO (Camp *et al.*, 2003), and so would be expected to exhibit some persistence at timescales of more than two years.

Although many coupling mechanisms between ozone and temperature are known, the net effect of interactions and feedbacks is only poorly understood and quantified. The dynamical features of the atmosphere are not simple to explain. Such features present appealing targets for model–data intercomparison, since models will not have been tuned to present them. This means that the intrinsic properties of input model data must be retained and compared with those of model output data. Comparison of DFA results with the results of a similar analysis of model output should provide a robust test of the fidelity of the model dynamics to atmospheric dynamics, and could improve the prediction of global warming and future ozone depletion, or recovery, under different climate change and halogen-loading scenarios.

4.4.2 Long-term memory dynamics of total ozone content

It has long been recognized that the significant correlation, observed between long-period variations in total ozone content (TOC) and the annual ring density chronologies of coniferous trees (especially spruce), is derived from the response of the tree ecosystem to systematic deviations in UV-B solar radiation from the climatic norm (Cracknell and Varotsos, 2007b; Efstathiou *et al.*, 1998; Kondratyev *et al.*, 1995; Varotsos, 1994, 2005a; Varotsos, C.A. *et al.*, 1995, 2001a, 2003).

In this context, Zuev (2005) used dendrochronological (tree ring dating) data to reconstruct the behavior of the ozonosphere and proposed TOC indices which have been reconstructed to look back several centuries for some Eurasian regions and denote TOC temporal variability at times where no TOC observations exist. Results show the behavior of TOC indices reconstructed for all presented Eurasian regions delving many centuries into the past is characterized by pronounced long-period variations, with periods coinciding with the cycles of solar activity. These variations for geographically widely spaced regions are out of phase due to spatio-temporal inhomogeneity of the TOC field at the middle and high latitudes, where ozone in the lower stratosphere plays the role of passive tracer of atmospheric circulation (Chandra and Varotsos, 1995; Chandra *et al.*, 1996; Cortesi *et al.*, 2007; Gernandt *et al.*, 1995; Kondratyev and Varotsos, 1995a, b; Kondratyev *et al.*, 1994; Schulz *et al.*, 2001; Varotsos and Cracknell, 1993, 1994, 2004; Varotsos, P.A. *et al.*, 1982).

This retrospective information on the behavior of the ozonosphere is very important because some unprecedented events reveal that the physics of atmospheric ozone are not understood well (e.g., Cartalis and Varotsos, 1994). For instance, until September 2002, major sudden stratospheric warming was considered to be a "forbidden" event in Antarctica. However, such an event has actually occurred due to very strong planetary waves that caused the southern polar vortex (and the ozone hole) to split into two pieces (Varotsos, 2002a, b, 2003a, b, 2004a, b).

Zuev and Efstathiou (2006) confirmed the long-range power law persistence of TOC, by applying the DFA method (of order 1) to the retrospective time series of TOC constructed by Zuev (2005), during the period 1626–2002. Application of the DFA method to the detrended time series of reconstructed TOC values revealed persistent long-range power law correlations with characteristics of $1/f$ noise, for two areas (Arosa in Switzerland and the Poluy River in Russia) and for all time lags between 4 years and 90 years. Furthermore, an attempt was made to shuffle de-trended reconstructed TOC time series; the DFA method applied to that shuffled data revealed white noise and proved that long-range persistence was derived from intrinsic self-similarity and not from the time sequence of TOC values.

The DFA method that was used in deriving the results of Varotsos and Kirk-Davidoff (2006), which were described in the previous section, was the linear DFA (i.e., in each box a linear least squares fit to the data was used). In outlining the method it was mentioned that instead of a linear fit one could use a least squares fit with a polynomial of order l in each box and the method is then described as DFA-l. Efstathiou *et al.* (2009) applied the multifractal DFA-l method (of order $l = 1$–5) to a detrended long time series of reconstructed TOC values, derived from the dendro-chronological data at Arosa (46.47°N, 9.40°E) and Poluy River (66.1°N, 68.3°E) for the period 1626–2002.

The DFA exponents revealed persistent long-range power law correlations for all time lags between 4 years and 90 years and a positive trend with increasing order of the multifractal DFA method. Table 4.9 presents the DFA exponent α, for the orders 1–5, which varies from 1.15 to 1.39 (for Arosa) and from 0.98 to 1.40 (for Poluy). Results show that the polynomial least squares curve, fitted to the τ data points of each class, gives a better local coefficient of determination and exhibits strong TOC fluctuation for all time lags between 4 years and 90 years. Whereas, in the case of the DFA-3 for Arosa, subtraction of the locally polynomial fitted trend (of order 3) in each class seems not to be efficiently fitted to the examined TOC time series, exhibiting the lower α-value.

Table 4.9. Multifractal DFA exponents α for reconstructed TOC data at Arosa and Poluy ($R^2 \geq 0.96$ in each case) (Efstathiou *et al.*, 2009).

	Arosa	*Poluy*
DFA-1	1.15	0.98
DFA-2	1.23	1.12
DFA-3	1.21	1.22
DFA-4	1.27	1.29
DFA-5	1.39	1.40

Efstathiou *et al.* (2009) also attempted to smooth a detrended reconstructed TOC time series by employing a 12-year moving average (MA), in order to eliminate any cycle with a period less than or equal to 11 years (Varotsos and Cracknell, 2004). The simple DFA method (of order 1) applied to the detrended and smoothed reconstructed TOC data revealed that the α-exponent equals 1.6 for Arosa and 1.45 for Poluy. That α-value, however, is biased by application of moving average filtering, which enhances correlations among the data and increases the DFA exponent. In order to estimate the magnitude of that increase, a long time series of fractional Gaussian noise (fGn) and fractional Brownian motion (fBm) was produced and filtered by using the 12-point MA technique. In the following, the simple DFA method was applied to the above-described time series and the α-exponent of the smoothed time series was plotted versus the α-exponent of the original fGn or fBm time series. The results indicated the linear relation $\alpha_{\text{smoothed}} = 0.7 + 0.7\alpha_{\text{original}}$ (N. Sarlis, pers. commun.). After that, an attempt was made to reduce the derived DFA exponents for the detrended and smoothed reconstructed TOC data to the original values, which equalled 1.28 and 1.1 for Arosa and Poluy, respectively, confirming the existence of strong long-range power law correlations (see Figure 4.20).

On the other hand, an attempt was made to smooth the initial detrended reconstructed TOC time series by using the classical Wiener method, removing Quasi-Biennial Oscillation, Southern Oscillation, and solar cycle components from the respective periodogram. The derived DFA exponent was again close to 1.2 for both areas.

Finally, application of the simple DFA method to yearly increments of reconstructed TOC data at Arosa and Poluy exhibited a crossover at approximately 11 years and the values of the slope before (after) the crossover point revealed persistent (antipersistent) long-range power law correlations in the fluctuations of reconstructed TOC increments, for time scales between about 4 years and 11 years (11–90 years) (see Figure 4.20b) (for details see Efstathiou *et al.*, 2009). The crossover point obviously coincides with a solar activity cycle and it is worth noting that TOC increments as a result of any external trend are dismissed (Varotsos, 1989).

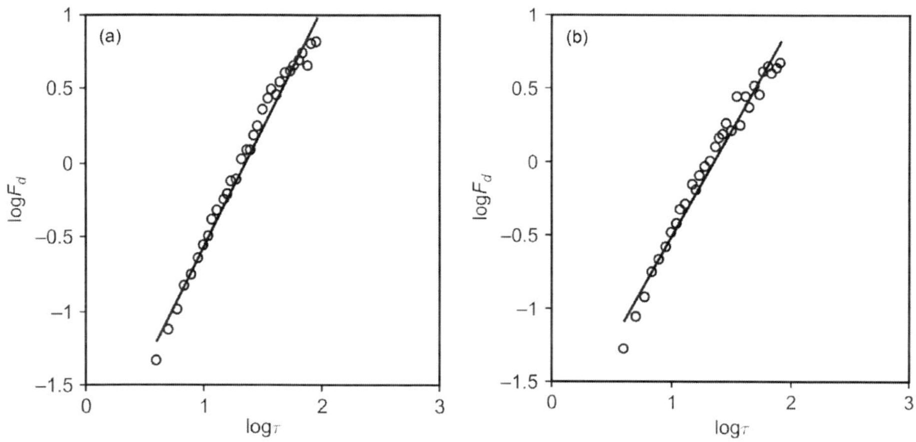

Figure 4.20. Log-log plot of the RMS fluctuation function, F_d, for the smoothed (12-year moving average) and detrended reconstructed TOC time series, versus time interval τ (in years) and the respective best fit equation and correlation coefficient at (a) Arosa ($y = 1.60x - 2.16$ and $R^2 = 0.98$) and (b) Poluy River ($y = 1.45x - 1.96$ and $R^2 = 0.97$). Note that $\alpha_{\text{smoothed}} = 0.7 + 0.7\alpha_{\text{original}}$ (Efstathiou *et al.*, 2009).

4.4.3 Scaling behavior of the global tropopause

Having considered the application of DFA to ozone and troposphere brightness temperature, we now consider its application to the study of time variations of the height of the tropopause (Varotsos *et al.*, 2009). The location of the tropopause—the boundary between the troposphere and stratosphere—can be defined in a variety of ways, according to the dynamic, thermal, and physico-chemical properties of the atmosphere (Hoinka, 1998, 1999; Reichler *et al.*, 2003).

Conventionally, the tropopause is a thin layer separating the turbulently mixed troposphere (where temperature decreases with height) from the stable stratified stratosphere (where temperature is constant or increases with height). The thermal tropopause is operationally defined as the lowest level at which the lapse rate decreases to 2°C km^{-1} or less, provided that the average lapse rate between this level and all higher levels at intervals of 2 km does not exceed 2°C km^{-1} (WMO, 1957). However, a number of observational studies suggest that the tropopause is a transition zone, which is also marked by changes in the chemical composition of the atmosphere.

Precise knowledge of the spatiotemporal structure of this transition zone is necessary to evaluate the exchange of mass, water, and chemical constituents between the troposphere and the stratosphere (Allen *et al.*, 2008; Varotsos, C.A. *et al.*, 1994a–c, 2004). Furthermore, under the physicochemical conditions that prevail in the upper troposphere–lower stratosphere, several gas–solid reactions involving chlorine and bromine can directly or indirectly cause ozone destruction, depending on the variability of solid particle properties (e.g., conductivity, dielectric constant) with air temperature and pressure (Kostopoulos *et al.*, 1975; Varotsos, P.A. 1978, 1981).

Changes in tropopause height may be a useful fingerprint of human effects on climate. That is because the tropopause height should do the following: (1) increase in response to warming of the troposphere and cooling of the stratosphere caused by increases in well-mixed greenhouse gases; (2) increase due to cooling of the stratosphere induced by stratospheric ozone depletion (Chandra and Varotsos, 1995; Varotsos, 1989, 2002a, b, 2004a, b; Varotsos and Cracknell, 1993, 1994); (3) decrease after stratospheric warming and tropospheric cooling resulting from massive volcanic eruptions (Kondratyev and Varotsos, 1995a–c). The height of the 100 hPa level is a good estimate for the tropopause height in the tropics, the 200 hPa level is appropriate in the mid-latitudes, and the 300 hPa level is a good estimate for both polar regions. The temperature at 100 hPa represents a good approximation to global tropopause temperature (Hoinka, 1998).

Of particular interest is the tropical tropopause, since it is widely accepted that most water vapor—a primary greenhouse gas (GHG)—enters the stratosphere here; hence, tropical tropopause conditions have a strong influence on stratospheric water vapor distribution. The large vertical temperature gradients both above and below the tropical tropopause act as a strong constraint on the size of tropopause height changes that can be caused by predominantly tropical modes of variability—such as the El Niño/Southern Oscillation (ENSO). By contrast, ozone-induced and GHG-induced cooling of the stratosphere tend to be largest in high latitudes in both hemispheres (Cortesi *et al.*, 2007; Efstathiou *et al.*, 2003; Gernandt *et al.*, 1995; Kondratyev *et al.*, 1994; Sausen and Santer, 2003).

Seidel and Randel (2006) examined global tropopause variability on synoptic, monthly, seasonal, and multidecadal timescales by applying conventional statistical tools to 1980–2004 radiosonde data. They found that the association of tropopause height and stratospheric temperature trends, together with the presence of a significant Quasi-Biennial Oscillation signal in tropopause height, suggests that at these lowest frequencies the tropopause is primarily coupled with stratospheric temperatures. They came to the conclusion that long-term changes in the tropopause (considered as an indicator of climate change) may carry less information about changes throughout the vertical temperature profile than had been suggested by previous studies using global climate models.

Varotsos *et al.* (2009) applied the DFA method to time fluctuations of the global tropopause height using the data of Seidel and Randel (2006) and discussed in detail whether tropopause height fluctuations at different times are positively correlated, thus exhibiting persistent long-range correlations. These radiosonde observations from 100 stations (listed in Table 1 of Seidel and Randel, 2006) came from the Integrated Global Radiosonde Archive at the NOAA National Climatic Data Center (NCDC). These stations were selected on the basis of the length and completeness of their archived daily data record, with the aim of covering the tropics (centered on the equator), the northern and southern subtropics (centered on 27.5° latitude), the mid-latitudes (centered on 51.4° latitude), and the polar zones (centered on 77.1° latitude).

Results of application of the DFA method by Varotsos *et al.* (2009) to monthly anomalies in tropopause height (Z) for the tropics are shown in Figure 4.21a, where

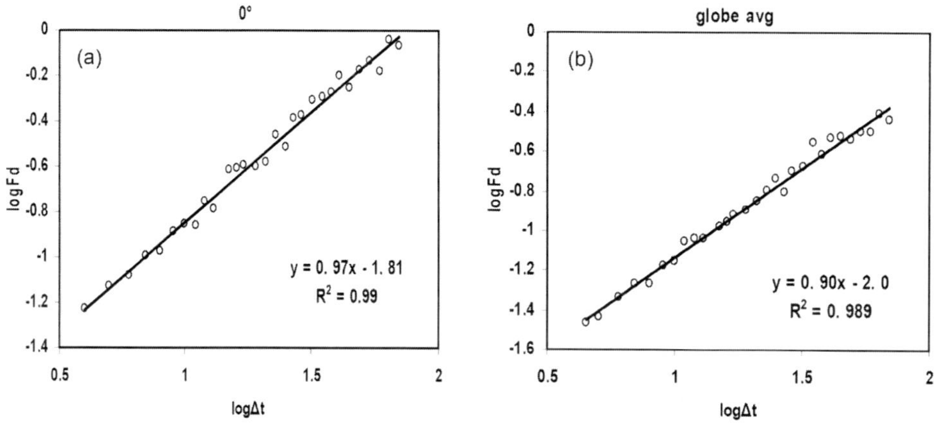

Figure 4.21. Log-log plot of tropopause height (Z) root-mean-square fluctuation function ($F_d(\Delta t)$) versus time interval Δt (in months) for the detrended and deseasonalized monthly Z values observed by the radiosonde network over the equator and the globe during 1980–2004 (Varotsos *et al.*, 2009).

a log-log plot of the root-mean-square fluctuation function $F_d(\Delta t) = F(n)$ is shown. Since $\alpha = 0.97 \pm 0.04$ for latitude $0°$, it was concluded that Z fluctuations over the equator exhibit persistent long-range correlations ($1/f$ noise-like) for time intervals ranging from about 4 months to about 6 years (statistically significant). Note that the derived $1/f$ noise-like behavior of the tropical tropopause is of particular interest, because most water vapor enters the stratosphere here. It is also worth noting that the long-range correlations obtained do not signify the presence of cycles with definite periodicities (i.e., as described by Camp *et al.*, 2003), but rather the existence of dynamical links between long and short timescale behavior. Corresponding results for global average Z are shown in Figure 4.21b, where—since $\alpha = 0.90 \pm 0.04$—strong persistence of global average Z fluctuations is observed and the time interval ranges from 4 months to 6 years (ENSO), approximately.

Corresponding results for the subtropics, the mid-latitudes, and high latitudes of both hemispheres using deseasonalized and detrended Z data for 1980–2004 showed that, once again, persistent long-range correlations exist in all latitude bands of both hemispheres. The correlations in Z fluctuations exhibit increasing memory going from the high to low latitudes and show that memory is stronger in the Northern Hemisphere than in the Southern Hemisphere.

As was done with the results for TOC and TRT, Varotsos *et al.* (2009) tested these results for the height of the tropopause in two ways. The first was to apply the DFA-l method to the same dataset. Results obtained from the DFA-l analysis did not show any substantial deviations from those obtained with DFA-1 shown in Figure 4.21 and in the corresponding results for other latitudes. Second, the linear DFA method (DFA-1) was applied to randomly shuffled Z data over the tropics. The result obtained from this analysis was $\alpha = 0.51 \pm 0.02$ and, thus, it was concluded that this persistence stems exclusively from the sequential ordering of Z values and is not a result of the distribution of Z values. Similar results were also obtained for the Z time series over the subtropics, mid-latitudes, and polar latitudes in both hemispheres.

We therefore see that, in addition to persistence in TOC and TRT which we discussed earlier, DFA demonstrates the existence of persistent long-range correlations in Z for all time lags between about 4 months and 6 years. Over the tropics, this persistence becomes stronger approaching the $1/f$ mode. Regarding a plausible explanation for these findings it is worth recalling that greater persistence is, in general, a result of either stronger positive feedbacks or larger inertia. Bearing in mind that long-range dependence and long memory are synonymous notions, this is equivalent to saying that the global tropopause height exhibits long memory that is associated with fractal behavior.

One of the difficulties in explaining this situation is that it is desirable to explain the cause that generates the long memory effect in column ozone, tropopause properties, etc. and then to proceed to estimation of the consequences. Unfortunately, the present stage of relevant studies can be considered little more than initial and preliminary. What is missing is some kind of theory as to the cause of long-range correlations in tropopause height.

The interesting question is to try to establish the cause of this persistence in ozone concentration and in the properties of the troposphere, but attempts to address that question are in their infancy. In the meantime. Varotsos *et al.* (2009) proposed that a useful step towards achieving an understanding of the cause of the long-term persistence in these properties would be to repeat the DFA studies on some other climate-related properties. They considered the mean monthly land and sea surface temperature (LSST) anomalies derived from the datasets of the Climatic Research Unit (*http://www.cru.uea.ac.uk/cru/data/temperature/*) from January 1850 to August 2008. To make a dataset with global coverage, which includes coastal and island grid boxes, land and marine data were blended into a combined average (see Jones *et al.*, 2001; Rayner *et al.*, 2006). Their DFA results revealed strongly persistent long-range power law correlations, with a scaling exponent of almost 0.86, for all time lags between 4 months and 39 years. The strong persistence found signifies that fluctuations in LSST anomalies, from small time intervals to larger ones (up to 39 years), are positively correlated in a power law fashion. This result suggests that the correlations between fluctuations in LSST anomalies do not obey classical Markov-type stochastic behavior (exponential decrease with time), but display more slowly decaying correlations. The latter conclusion is consistent with recent predictions of climate models and theoretical expectations, according to which the upper troposphere should be warming faster than the surface. In tropical regions, in particular, cooling of the tropical tropopause is a dynamical result of tropospheric convection, which in turn partially depends upon sea surface temperatures. As with their previous DFA studies, Varotsos *et al.* (2009) applied the DFA method to randomly shuffled data of LSST anomalies and the exponent obtained was $\alpha = 0.53 \pm 0.01$. This demonstrates the existence of long-term memory in the time series of LSST anomalies (Varotsos, 2005b; Varotsos *et al.*, 2009).

4.4.4 Scaling properties of air pollution at the surface; surface ozone (SOZ)

Given that air and particulate matter pollution may produce adverse effects in human health, it is urgent to develop reliable tools and models to predict pollutant concentrations. Many attempts have already been made in this direction, which are mainly based on classical statistical methods and neural network techniques (Gardner and Dorling, 1998; Perez and Trier, 2001; Windsor and Toumi, 2001). There is convincing evidence that over the last two decades the anthropogenic emissions of air pollutants (e.g., NO_x) in metropolitan areas are decreasing. Also, long-term measurements show that peak values of surface ozone (SOZ) concentration have reduced, since the late 1980s, while there is an upward trend of low SOZ values. In Europe, the summer mean of daily maximum SOZ concentration is of the order of 40–60 ppb over continental regions and lower (20–40 ppb) at the boundaries. In a few metropolitan areas in Europe, the gradients of NO_x are large with concentrations at the boundaries of the order of 0.1 ppb to concentrations over 50 ppb (Kondratyev and Varotsos, 2000).

As a final example of the application of DFA we consider the case of surface ozone (SOZ); this is a secondary pollutant, which has traditionally been used as the severity measure of photochemical smog episodes, because it is formed by photochemical reactions involving oxides of nitrogen ($NO + NO_2 \rightarrow NO_x$) and volatile organic compounds. Primary or secondary pollutant species are also PM10, which is a part of particulate matter and refers to coarse particles with aerodynamic diameters of less than or equal to 10 μm.

Varotsos *et al.* (2005a) applied DFA to the study of SOZ over the greater Athens area, Greece. This area was chosen because it has significant air pollution problems, due to high population density (considerable emissions of air pollutants), intense sunshine, and characteristic features of the topography (a basin surrounded by mountains). The SOZ data for Athens were obtained from several of the stations of the National Air Pollution Monitoring Network (NAPMN), which is run under the auspices of the Department for Air Quality of the Greek Ministry for the Environment, Physical Planning, & Public Works.

To overcome the natural non-stationarity of the data due to long-term and seasonal trends, all these data were deseasonalized and detrended (Chen *et al.*, 2005; Varotsos *et al.*, 2001b). Daytime and nighttime values were considered separately. In both cases for time lags between 4 days and 5 years, SOZ fluctuations behave persistently with the following feature: during daytime these correlations are more intense (i.e., have stronger memory) compared to those during nighttime. Similar results were also obtained for NO_x and PM10–PM2.5 (Varotsos, P.A. *et al.* 2005a, b).

For the time being it is a difficult and challenging task to give a reliable explanation about the origin of long-range persistence not only of air pollution, but in many physical systems, (e.g., Stanley, 1999; Varotsos, 2005c; Windsor and Toumi, 2001; Zhu and Liu, 2003). According to current understanding, to ascertain the physics behind the aforementioned findings, there needs first to be clarification that the research interest is focused on the intrinsic dynamics of the time evolution of basic air pollutants. Air pollutants must be definitely considered as a complex system, since these are subject to multicomponent control mechanisms (e.g., atmospheric dynamics and chemistry). To this end, it is necessary to distinguish trends (caused by external effects) from long-range fluctuations intrinsic in the data. The main finding of the analysis presented here is that persistent long-range correlations exist with the time evolution of basic air pollutants at the Athens basin. From the physical point of view various mechanisms exist for generating long-range dependence such as aggregation, non-linear interactions, and self-criticality. Experiments confirm that correlations do indeed decay exponentially, but if the system is at its critical point, then rapid exponential decay turns into long-range power law decay (Zhu and Liu, 2003). The physical meaning of this turning point (from exponential decay to power law decay) is that long-range correlations in the time series of air pollutants exhibit "stronger memory" (the processes involved "forget" their past behavior more slowly).

The main findings that can be deduced from the above-mentioned investigation are the following:

(1) Persistent long-range power law correlations in the fluctuations of daytime and nighttime ozone on the timescale of 4 days to 5 years were detected with more intense correlations ("stronger memory") during daytime. Fluctuations in nitrogen oxides exhibit similar behavior.
(2) For the time interval ranging from about 4 hours to 9 months persistent long-range power law correlations in particulate matter fluctuations prevail over Athens. Presumably, these reflect short and longer term source/meteorological features. Persistency from about 4 h to 15 days is also exhibited by fluctuations in PM2.5 concentrations measured in Baltimore (Maryland) reflecting remarkably similar processes driving time lags of about 4 h .

The aforementioned long-range correlations to the time evolution of basic air pollutants may be considered that are related to the self-organized critical behavior of the atmosphere. The source of long-range persistence of air pollution merits further investigation.

Although the above-mentioned findings provide power law relationships that are practically simple simulating models for O_3, NO_x, PM10–PM2.5 fluctuations, they might also be helpful for the development of new models for the reliable prediction of future air pollutant concentrations under different scenarios. Furthermore, the results of already existing operational prediction models should be checked as to whether they demonstrate scaling behavior in the time evolution of the air pollutants discussed above.

4.5 IMPACTS OF THE SOLAR ECLIPSE OF MARCH 29, 2006 ON SURFACE OZONE AND RELATED AIR POLLUTANTS

Solar eclipses, despite being a rare natural phenomenon, provide an opportunity to investigate how photochemical processes in the atmosphere react to comparatively fast solar radiation changes. Variations in stratospheric composition caused by natural processes like a solar eclipse have been studied by many workers (Bojkov, 1968; Burnett and Burnett, 1985; Elansky *et al.*, 1983; Gogosheva *et al.*, 2002; Mims and Mims, 1993; Tzanis, 2005; Wuebbles and Chang, 1979; Zerefos *et al.*, 2000). In particular, variation in total ozone (TOC) content during a solar eclipse event has been attributed mainly to cooling of the atmosphere and, therefore, the generation of internal gravity waves (Chimonas and Hines, 1970, 1971; Zerefos *et al.*, 2007) as well as alteration of the photochemical balance. However, the number of studies exploring the effects of a solar eclipse on surface temperature and surface winds as well as on surface ozone (SOZ) and its precursors is relatively small (Fernandez *et al.*, 1993; Kolev *et al.*, 2005; Srivastava *et al.*, 1982; Tzanis, 2005; Zerefos *et al.*, 2001b).

Experimental data obtained from different observational sites in Bulgaria during the solar eclipse of August 11, 1999 demonstrated that the influence of the phenomenon was manifested with a certain delay and maximal impact of the eclipse on meteorological parameters (temperature, humidity, and wind speed and direction) and surface ozone concentration was revealed to be 7 and 10 minutes after maximal eclipse, respectively (Kolev *et al.*, 2005). A decrease of around 10–15 ppbv in SOZ concentration was observed at Thessaloniki, Greece, during the solar eclipse of August 11, 1999 (Zerefos *et al.*, 2001b), while the percentage change of SOZ concentration at Athens, Greece, was maximized one hour after solar eclipse maximum and the largest values of SOZ percentage change were observed at the Patision station, an urban station located in the central part of the Athens basin (Tzanis, 2005; Tzanis *et al.*, 2008). Ozone profile measurements over Thessaloniki during the solar eclipse of August 11, 1999 also indicated an ozone decrease up to 2 km with a lag time between eclipse maximum and the maximum of induced ozone decrease (Tzanis *et al.*, 2008; Zerefos *et al.*, 2001b).

Surface ozone and its precursors as well as its relationship with meteorological parameters in the Greater Athens area, which is one site in the Mediterranean region where very frequent photochemical pollution episodes occur, have been discussed in a number of publications (Cartalis and Varotsos, 1994; Kondratyev and Varotsos, 2001a, b; Varotsos and Kondratyev, 1995; Varotsos *et al.*, 2001a, b; Ziomas *et al.*, 1998). Although the most important chemical mechanisms involved in photochemical pollution have already been identified and studied, further investigation is necessary because this atmospheric phenomenon is a very complex process involving meteorological, topographic, emission, and chemical parameters (Varotsos, C.A. *et al.*, 2004).

Tzanis *et al.* (2008) studied the behavior of surface ozone and nitrogen dioxide concentrations, solar ultraviolet radiation, and variations in various meteorological parameters (temperature, relative humidity, and wind speed and direction) at the ground before, during, and after the total solar eclipse that took place on March 29, 2006 over Athens, Greece (38°N, 23°E). This eclipse started at 12:31 LT, reached maximum solar coverage (84%) at 13:48, and ended at 15:04. Tzanis *et al.* used measurements of SOZ and NO_2 concentrations along with meteorological measurements, taken at four monitoring stations (Patision, N. Smirni, Geoponiki, and

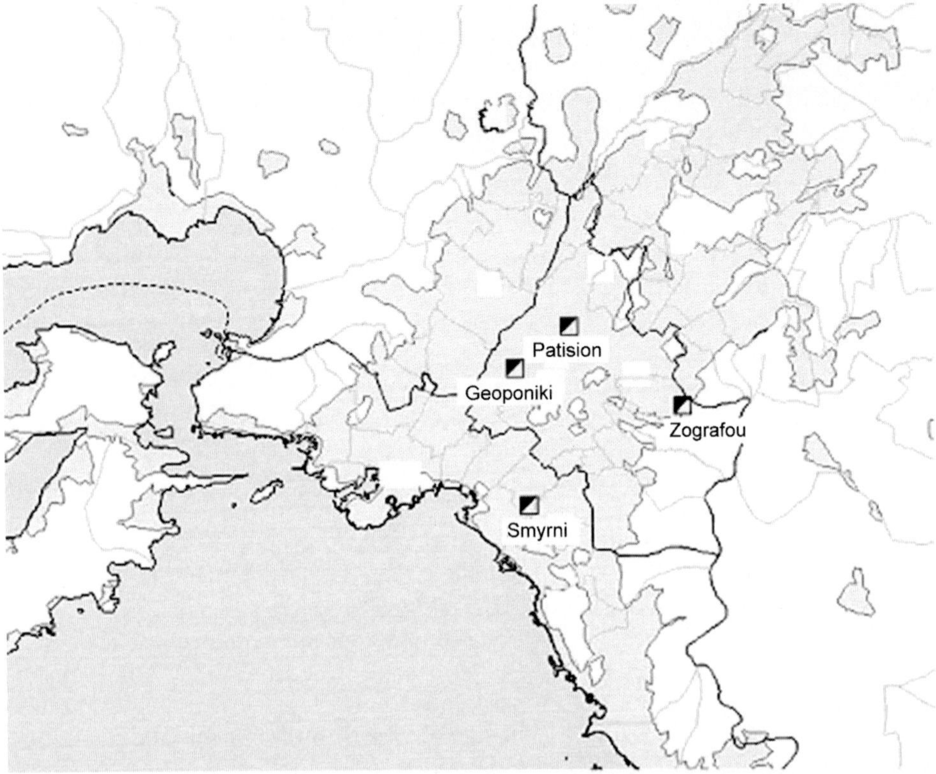

Figure 4.22. Map with the locations of the monitoring stations.

Zografou) of the local air pollution monitoring network (National Service for Air Pollution Monitoring) in the Athens basin (see Figure 4.22).

Figure 4.23 presents SOZ measurements and expected SOZ values as derived from the fitted curve of measurements before, during, and after the eclipse. The percentage changes of SOZ at these stations during the eclipse are shown in Table 4.10. According to Figure 4.23 and Table 4.10, the percentage change of SOZ is greatest one hour after solar eclipse maximum at each of the stations. The percentage change of SOZ was largest at the Patision station, which is an urban station located in the central part of the Athens basin. This is very similar to the behavior that was observed before, during, and after the solar eclipse of August 11, 1999.

The observed behavior of SOZ during the solar eclipse is related to the effect caused by gradual decrease of solar radiation on photochemical reactions within the planetary boundary layer. The decrease in SOZ concentration started a few minutes after the beginning of the eclipse and reached its maximum after solar eclipse maximum as a consequence of the further fall in sunlight intensity that decreased the efficiency of photochemical ozone formation. Solar radiation started to increase after eclipse totality while the SOZ concentration started to increase about one hour later and returned to its ordinary behavior several minutes after the end of the solar

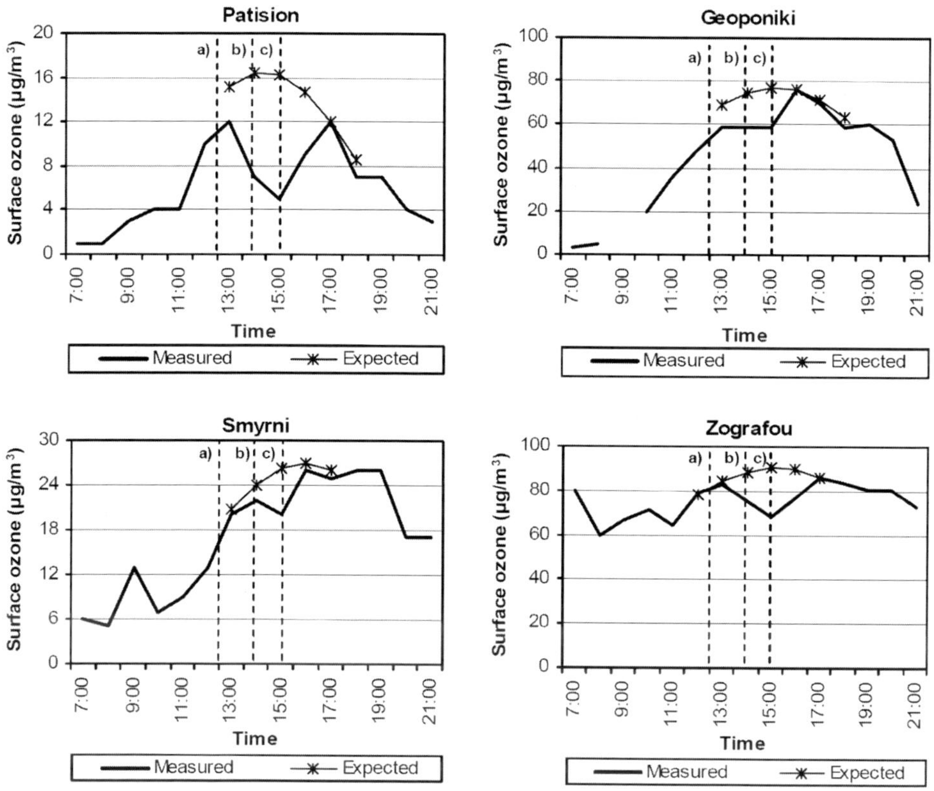

Figure 4.23. Surface ozone measurements and the expected surface ozone values as derived from the fitted curve of measurements before, during, and after the eclipse event of March 29, 2006 at four stations (Patision, Smyrni, Geoponiki, and Zografou) in the Athens basin: (a) beginning of the solar eclipse, (b) solar eclipse maximum, (c) end of the solar eclipse (Tzanis *et al.*, 2008).

Table 4.10. Calculated percentage change of surface ozone (at four stations) throughout the eclipse event of March 29, 2006 at Athens, Greece (Tzanis *et al.* 2008).

	Surface ozone			
Time	*Patision*	*Geoponiki*	*Smyrni*	*Zografou*
13:00	−21.04	−14.41	−3.45	−1.38
14:00	−57.39	−20.96	−8.44	−15.24
15:00	−69.26	−23.51	−23.84	−24.82
16:00	−38.82	−0.001	−3.85	−14.11

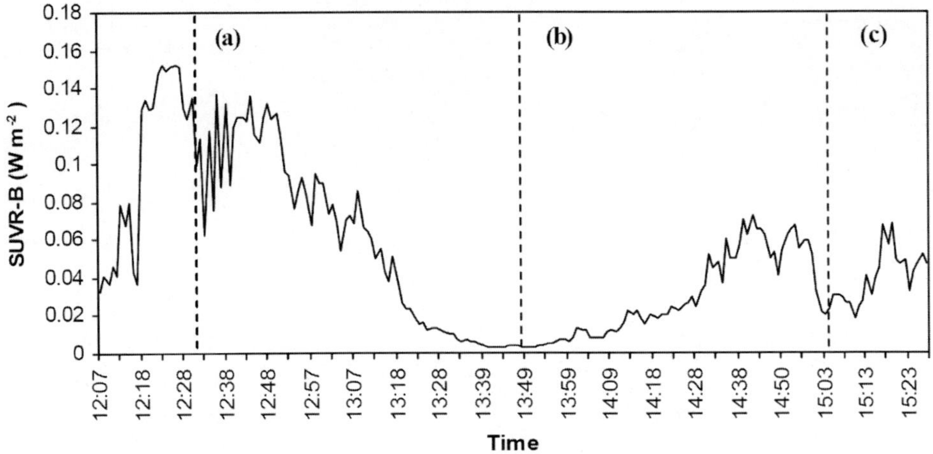

Figure 4.24. SUVR-B (312 nm) measurements derived from the MICROTOPS II sun-photometer (Solar Light Co., Inc.) before, during, and after the solar eclipse of March 29, 2006, at Athens: (a) beginning of the solar eclipse, (b) solar eclipse maximum, (c) end of the solar eclipse (Tzanis *et al.*, 2008).

eclipse. SOZ variations detected at Athens before, during, and after the solar eclipse of March 29, 2006 demonstrated that the influence of the phenomenon was manifested with a certain delay.

The observed solar UV radiation measurements at 312 and 365 nm obtained by two different instruments before, during, and after the solar eclipse of March 29, 2006 at Athens are shown in Figures 4.24 and 4.25. The solar UV radiation at 312 nm measured by the two instruments reached the reduced values of 97% and 93% at the eclipse maximum while the change in solar UV radiation at 365 nm was reduced to about 93%. This is in close agreement with the observed reduction of incoming solar radiation as derived from measurements conducted at Thission station (in the center of Athens) by Founda *et al.* (2007).

Figure 4.26 shows the variation of surface NO_2 concentration before, during and after the solar eclipse as well as the nitrogen dioxide variations observed on March 28, 2006, a day with similar meteorological conditions for the Patision, Smyrni, Geoponiki, and Zografou stations.

Comparison of Figures 4.23 and 4.26 shows that the observed reduction of SOZ concentration after the beginning of the solar eclipse is accompanied by an increase of NO_2 concentration at each station. Furthermore, the low photolysis rates of NO_2 may be the most likely reason for the reduced surface ozone values during the eclipse event due to the fact that the primary photochemical source of ozone in the troposphere is photolysis of NO_2 in the presence of molecular oxygen (Kondratyev and Varotsos, 2000). For a further discussion and for the details of the observed behavior of various surface meteorological parameters during the eclipse see Tzanis *et al.* (2008).

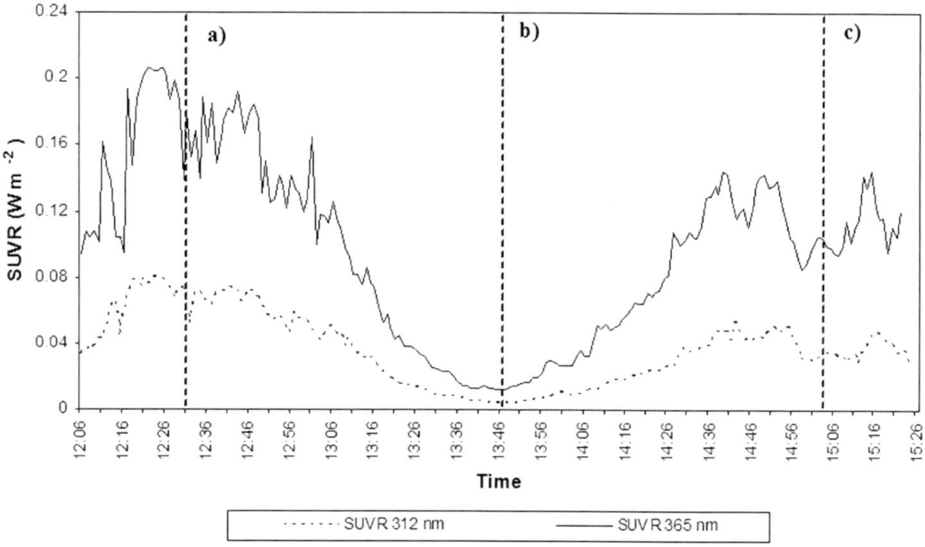

Figure 4.25. SUVR (312 and 365 nm) measurements derived from the VLX-3W radiometer (Vilber-Lourmat, France) before, during, and after the solar eclipse of March 29, 2006, at Athens: (a) beginning of the solar eclipse, (b) solar eclipse maximum, (c) end of the solar eclipse (Tzanis *et al.*, 2008).

4.6 LONG-TERM COUPLING BETWEEN TOC AND TROPOPAUSE PROPERTIES

It has long been known that short-term fluctuations in TOC originate from synoptic-scale baroclinic disturbances in the upper troposphere and lowermost stratosphere (below about 20 km), which affect ozone through the depth of the lowermost stratosphere (Chandra *et al.*, 2002; Ebel *et al.*, 1997; Singh *et al.*, 2002; Spänkuch and Schulz, 1995). Therefore, TOC is to some extent a tracer for large-scale meteorological processes in the upper troposphere and lowermost stratosphere (Holton *et al.*, 1995). In particular, TOC tends to be highest on the cyclonic side of upper-level jetstreams and in the region of isolated cyclonic vortices (cut-off lows). Tzanis (2009a) investigated vertical temperature profiles and column ozone data to explore the relationship between these two variables at various levels in the troposphere and lower stratosphere over Athens, Greece, during the period 1992–2004. The results showed an out-of-phase relationship between total ozone and temperature in the troposphere and an in-phase relationship in the lower stratosphere. Such a phase change is characteristic of all seasons. A review of temperature and ozone variations in the stratosphere is given by Angell (1980).

The observed connection between TOC and tropopause height (Z) is valid not only on short timescales but also on long timescales. Longer term changes are associated with planetary-scale circulation anomalies, such as the Northern Hemisphere annular mode (NAM), which elevate the tropopause at certain long-

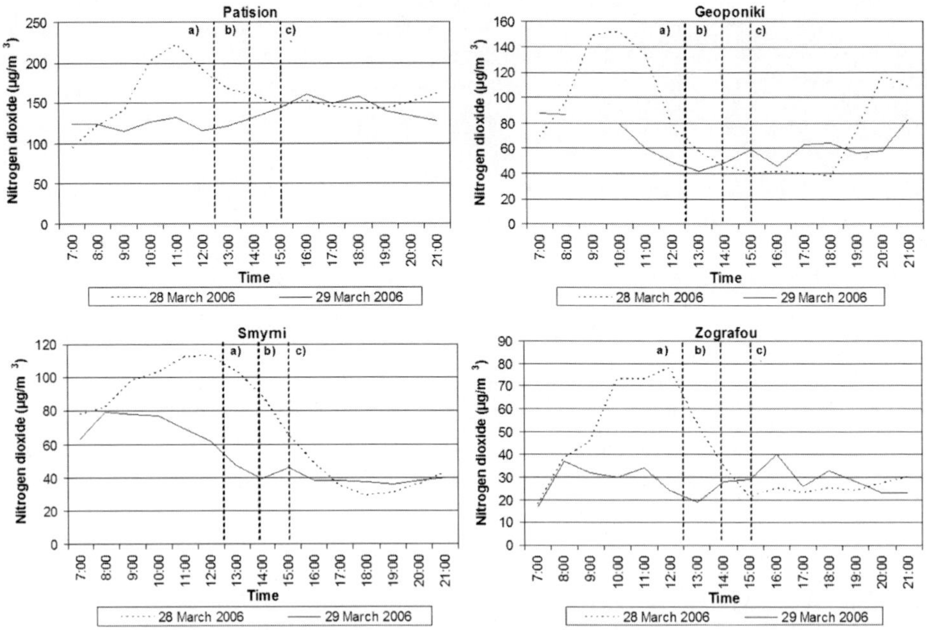

Figure 4.26. Nitrogen dioxide measurements before, during, and after the solar eclipse of March 29, 2006 at four stations (Patision, Smyrni, Geoponiki, and Zografou) in the Athens basin: (a) beginning of the solar eclipse, (b) solar eclipse maximum, (c) end of the solar eclipse. The dashed line presents the nitrogen dioxide variations observed on March 28, 2006 (a day with similar meteorological conditions) (Tzanis *et al.*, 2008).

itudes and lower it at others. However, the effect of tropospheric circulation on TOC, and its relation to Z, is not well understood at present (Krzyścin *et al.*, 2001; Thuburn and Craigh, 1997; WMO, 2003).

Changes in Z are related to changes in tropospheric and lower stratospheric temperatures. An increase in the former and a decrease in the latter—as expected, for example, with increases in CO_2—both lead to a higher tropopause. In each case, a 1°C change leads to an increase in Z of about 160 m. The observed lower-stratospheric cooling attributable to ozone decreases (see Chapters 5 and 6) would be expected to raise Z on radiative grounds alone (an effect that has not been quantified yet). By contrast, sulfate aerosols are calculated to cool the troposphere and therefore lower the tropopause. Other natural and anthropogenic forcings will also act to change Z (Hoskins, 2003). For instance, a recent analysis by Santer *et al.* (2003) suggests that, over the past 20 years, the observed globally averaged Z has increased by ~200 m.

Notwithstanding these concerns, several studies have attempted to estimate the dynamical contribution to TOC trends by statistical regression against Z or the tropospheric circulation indices that are correlated with it. There appears to be a consensus between different statistical studies that, on decadal timescales, a significant fraction (20–40%) of observed Northern Hemisphere mid-latitude TOC change is associated with changes in tropospheric circulation. Thus, it appears that, although ozone decreases may have led to changes in tropospheric circulation (including *Z*) that are qualitatively consistent with observed changes, they are of insufficient magnitude, implying that other factors must contribute. So, while there is a certain self-consistency between the various dynamical trends—without a clear mechanism connecting the NAM, TOC, and *Z*—any such connection in the context of long-term trends remains speculative. This question will probably not be able to be answered until the origin of changes in individual dynamical quantities has been identified (WMO, 2003). Furthermore, the interrelationships between ozone and circulation parameters in the lowermost stratosphere are poorly quantified at present, and estimates of circulation effects on decadal changes in ozone are highly uncertain.

Varotsos, C.A. *et al.* (2004) attempted to investigate these problems by analyzing data on TOC and *Z* from 10 stations distributed around the Northern Hemisphere (see Table 4.11). Special attention was given to the area of Greece and, in particular, to Athens (37.9°N, 23.7°E). The possibility of using *Z* variations as an indicator of dynamical variability in TOC is very important because of the widely discussed problem regarding the time of ozone recovery (Hoskins, 2003; Santer *et al.*, 2003; Varotsos, 2002a, b, 2003a, b) (see Section 6.5). Initially, the characteristic features of the first two thermal tropopauses were examined alongside the column ozone over the area of Greece. Then this examination was expanded to the remaining geographical sites in order to explore the spatial distribution of the relationship between the column ozone and tropopause properties.

The data used in this study were daily vertical temperature profiles and daily TOC observations obtained at Athens during the period January 1984–March 2002. The temperature profiles were obtained from radiosondes launched by Athens University (37.9°N, 23.7°E) and by the Hellenic Meteorological Service (from sites around 10 km apart). In addition, the available archives of radiosoundings performed at Thessaloniki (40.4°N, 23.0°E) and Heraklion (35.2°N, 25.7°E) were also used occasionally in order to confirm the results obtained from the Athens observations. Daily TOC observations for the period from January 1984 to September 1991 were obtained from TOMS (see Section 2.3.1) and from the Dobson spectrophotometer 118 at Athens University. Furthermore, the ozone vertical profiles were measured by ozonesondes at Athens University during the period from December 1991 to April 2001 (Varotsos *et al.*, 1994a, b, 2001a). Similar observations of TOC and tropopause properties obtained at the other sites are listed in Table 4.11.

4.6.1 Occurrence frequency of tropopause height

Data for the two seasons—May/June/July and November/December/January—were selected for consideration separately. The histograms in Figure 4.27 illustrate the frequency of occurrence of various tropopause heights for these two periods. It can be seen from Figure 4.27 that the November/December/January data have a single maximum, while the May/June/July data have a principal maximum and also a

Table 4.11. Sites used for detection of the relationship between TOC and tropopause properties (Varotsos, C.A. *et al.*, 2004).

Station name	*Latitude* (°)	*Longitude* (°)
Thiruvananthapuram, India	80N	76E
New Delhi, India	28N	77E
Cairo, Egypt	30.0N	31.3E
Heraklion, Greece	35.2N	25.7E
Athens, Greece	37.9N	23.7E
Thessaloniki, Greece	40.4N	23.0E
Sapporo, Japan	43.0N	141.3E
Bismarck, North Dakota	46.7N	100.8W
Hohenpeissenberg, Germany	47.8N	11.0E
Belsk, Poland	51.8N	20.8E

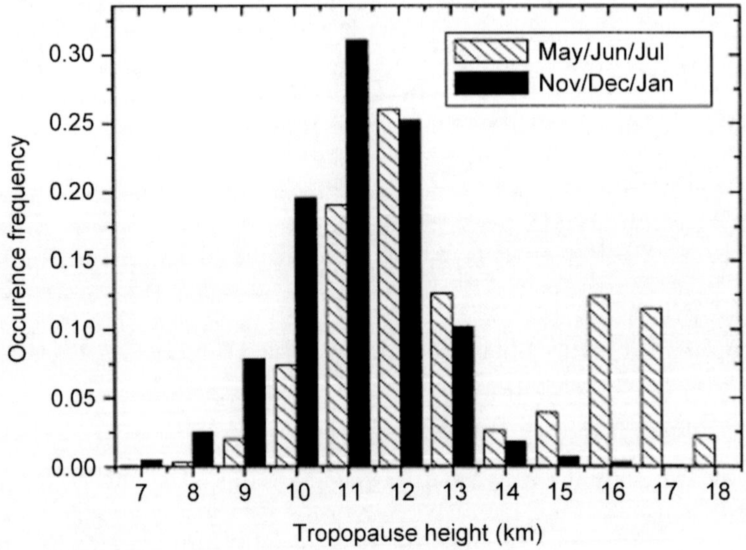

Figure 4.27. Occurrence frequency for the first tropopause height classes at Athens during 1984–2002 (deduced from 12,728 balloon ascents) (Varotsos, C.A. *et al.*, 2004).

secondary one at a higher altitude. It was noted that no secondary maximum was found by applying a similar analysis to May/June/July radiosonde data from Hohenpeissenberg (47.8°N, 11.0°E) (Steinbrecht *et al.*, 1998). The distributions shown in Figure 4.27 do not change if the radiosoundings performed at 00:00 and 12:00 are plotted separately.

Varotsos, C.A. *et al.* (2004) further explored the secondary maximum in the May/June/July distribution by considering the tropopause data stored in the NCEP/NCAR Reanalysis database and provided by the NOAA-CIRES Climate Diagnostics Center, Boulder, Colorado from their website at *http://www.cdc.noaa.gov/* Very close agreement was found between the temperature deviations and trends of the tropopause above Athens, as derived from *in situ* and reanalysis data. However, no evidence of a secondary maximum in the tropopause height was found in the reanalysis data. This was thought to be due to the fact that the secondary maximum in May/June/July distribution reflects day-to-day variations of the local weather conditions prevailing at Athens, Greece, which cannot be reproduced by the NCEP/NCAR Reanalysis database (for details see Varotsos, C.A. *et al.*, 2004).

Varotsos, C.A. *et al.* (2004) gave some further consideration to the second thermal tropopause. Figure 4.28 illustrates the seasonal variability of the first and second thermal tropopause above Athens. The interesting point in Figure 4.28 is that

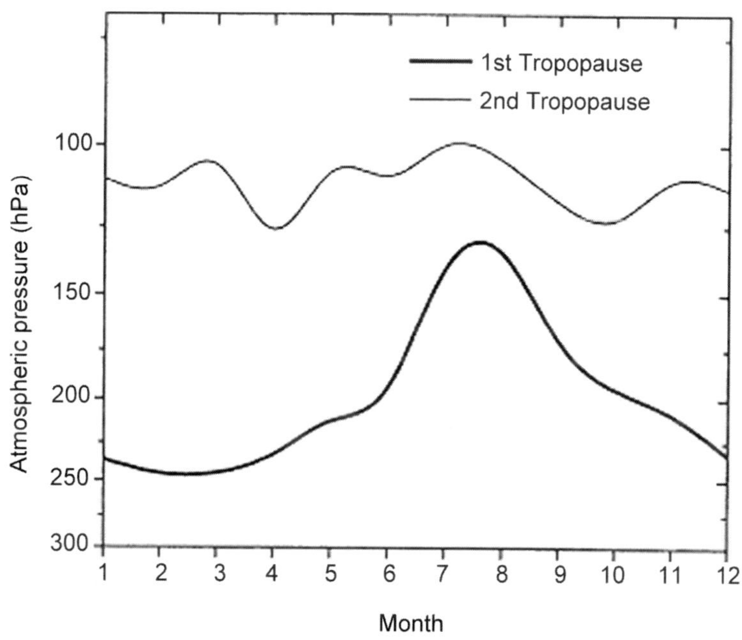

Figure 4.28. Seasonal variability of the first and second thermal tropopause over Athens, Greece (Varotsos, C.A. *et al.*, 2004).

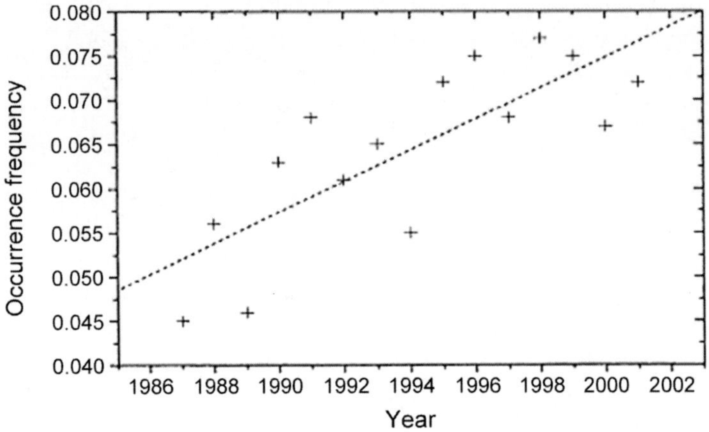

Figure 4.29. Relative frequency of occurrence of the second tropopause over Athens, Greece (Varotsos, C.A. *et al.*, 2004).

the seasonal variability of the second thermal tropopause does not follow the well-known pattern of the annual cycle of the first thermal tropopause, denoting that their origin is probably different. Figure 4.29 depicts the frequency of occurrence of the second thermal tropopause, demonstrating that the second thermal tropopause has been more frequent in recent years.

Figure 4.30 illustrates the temperature evolution of both thermal tropopauses observed at Athens, Greece. The most interesting feature in Figure 4.30 is that the temperatures of both tropopauses change approximately in phase, following a statistically significant negative trend (2σ), which means that the height of both tropopauses is increasing (at a higher rate for the second tropopause). This trend was found to be present in all seasons. It is worth noting that the contribution to TOC of ozone content in the layer at 12–16 km is quite substantial.

It can be seen from Figure 4.29 that in recent years the frequency of occurrence of the second thermal tropopause over Greece has been increasing, while Figure 4.30 shows that it has been reaching progressively higher altitudes and hence lower temperatures. This increase in Z may primarily be a result of both stratospheric cooling (due to stratospheric ozone depletion) and tropospheric warming (due to enhancement of the atmospheric greenhouse effect). However, the relative importance of these two factors is still uncertain. Santer *et al.* (2003) emphasized that anthropogenically driven tropospheric warming is an important factor in explaining modeled changes in Z. In particular, they suggested that, over the period 1979–1999, roughly 30% of the increase in Z is explained by warming of the troposphere, which is induced by greenhouse gases (GHGs). In parallel, some natural forcings, such as large explosive volcanic eruptions, have the opposite effect of warming the lower stratosphere and cooling the troposphere leading to a decrease of Z (Hoskins, 2003; Kondratyev and Varotsos, 2000).

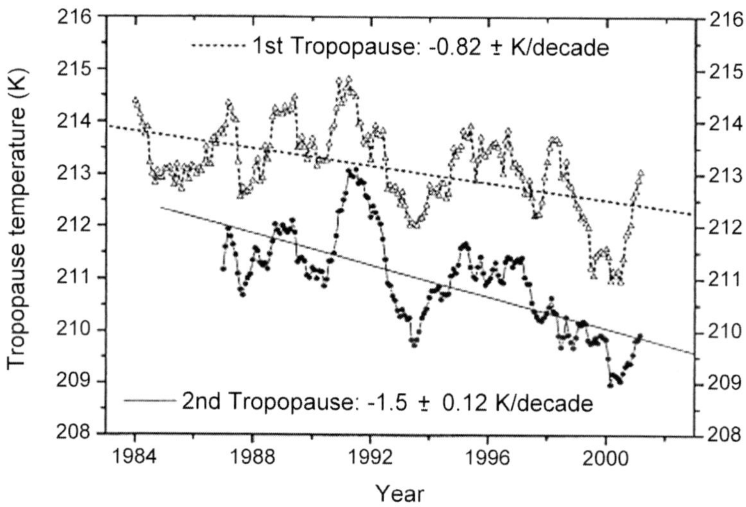

Figure 4.30. Temperature trends (2σ) of the first and second tropopause over Athens, Greece (Varotsos, C.A. *et al.*, 2004).

4.6.2 Association between tropopause properties and TOC

Varotsos, C.A. *et al.* (2004) established an inverse relationship between TOC and Z which is presented in Figure 4.31. As can be seen, any decrease in TOC is always accompanied by an increase in Z, at a rate of $-8.5\,\mathrm{DU\,km^{-1}}$ in summer and $-11.2\,\mathrm{DU\,km^{-1}}$ in winter. As expected, TOC is higher in summer than in winter. Similar results were obtained for Thessaloniki, Heraklion, and in Germany, where Hoinka *et al.* (1996) reported that the vertical displacement of the tropopause over Germany by one kilometer was associated with an ozone difference of 13 DU.

For the remainder of this chapter let us keep in mind that the annual variation of ozone in the troposphere and lower stratosphere region has been studied using all the available ozonesonde data collected between 1991 and 2001 at the Athens station.

The climatology of ozone deduced from the Athens ozonesonde data exhibits a prominent annual cycle in the upper-troposphere and lower-stratosphere region (Figure 4.32). The observed change in the phase of the annual cycle from late spring–summer at 500 hPa to spring at 200 hPa and to late winter–spring at 100 hPa shows a change in ozone control from photochemical to dynamical. The latter is in close agreement with the relevant findings of Rao *et al.* (2003) deduced from ozonesonde data collected between 1994 and 2001 at several northern European stations. In addition, Figure 4.32 shows that upward movement of the ozone profile is accompanied by an increase in annually averaged Z. This observed shift of the ozone profile in the lower stratosphere leads to excessive ozone depletion. This part of the TOC trend (0.5–1% per decade) is clearly associated with upward movement of the ozone profile and, thus, should not be considered as a direct result of anthropogenic chemical ozone depletion. Krzyścin *et al.* (1998) and Forster and

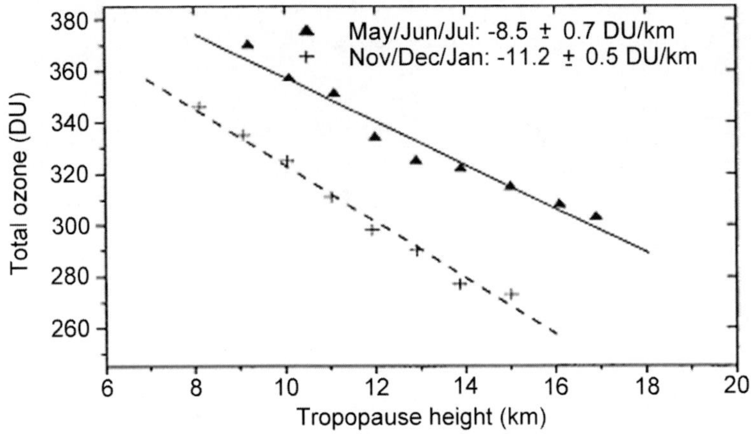

Figure 4.31. Relation between TOC and tropopause height for average TOC values of different tropopause height classes, as deduced from Athens observations (Varotsos, C.A. *et al.*, 2004).

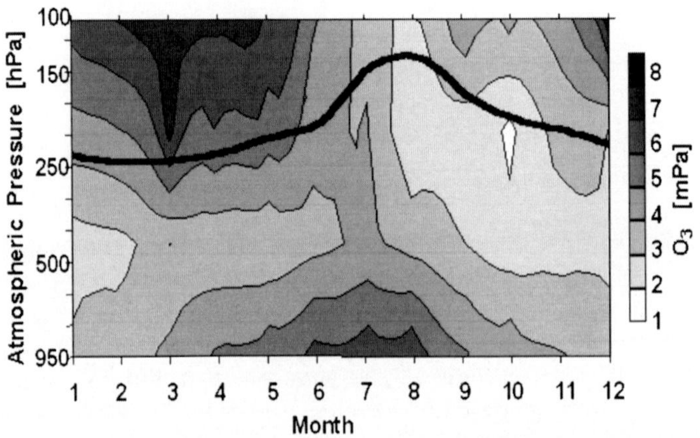

Figure 4.32. Seasonal variability in ozone partial pressure in the troposphere and the lower stratosphere at Athens (Greece) as deduced from ozonesonde ascents throughout the period 1991–2001. The bold curve denotes the first thermal tropopause (Varotsos, C.A. *et al.*, 2004).

Tourpali (2001) reached similar conclusions by performing observation analysis at locations in the Northern Hemisphere since 1979.

Furthermore, an alternative way to understand the dynamics of ozone in the upper-troposphere and lower-stratosphere region is to investigate the relative position of the first tropopause and ozonepause (i.e., the height where ozone partial pressure reaches its minimum value). In this respect, Figure 4.32 shows that in winter the ozonepause lies below the first thermal tropopause (ozone as it penetrates the troposphere) and the second thermal tropopause (shown in Figure 4.28, near 100 hPa). The latter is consistent with the generally approved ozonepause pattern, according to which—in winter in polar and moderate latitudes—the ozonepause most frequently lies below the tropopause and—in equatorial latitudes in both the summer and winter—the ozonepause generally lies above the tropopause.

From the observation results mentioned above it appears that the most efficient mechanism modifying TOC amounts is vertical displacement of the tropopause. This confirms that tropospheric circulation affects ozone distribution, with the direct effects confined to the lowermost stratosphere (Canziani *et al.*, 2002; WMO, 2003). It is actually an influence to the lowermost stratosphere from below, which is also sensitive to the Brewer–Dobson circulation from above. In this respect, Krzyscin (2002) suggested a 60% ozone reduction in the lower stratosphere and an approximately 50 hPa upward shift of the thermal tropopause is observed during mini-hole events (James *et al.*, 1997). Therefore, increases in both Z and the frequency of mini-hole events induce decreases in ozone over northern mid-latitudes. However, these induced decreases in ozone must not be regarded as added, because the two causal events are not independent of each other (WMO, 2003).

The same data analysis that was performed at the Athens station was repeated for the Hohenpeissenberg station during 1967–2002 and showed that any TOC decrease is always accompanied by an increase in Z, at a rate of $-11.6\,\mathrm{DU\,km^{-1}}$ in summer and $-18.9\,\mathrm{DU\,km^{-1}}$ in winter. The correlation coefficient between TOC and Z is -0.72 for the summer period and -0.63 for the winter period (95% confidence level). The analysis applied to radiosonde data obtained at Belsk (51.8°N, 20.8°E) showed a correlation coefficient between TOC and Z of -0.67 for the summer period and -0.60 for the winter period (both significant at the 95% confidence level).

An analysis similar to that described above was also carried out for Cairo (30.0°N, 31.3°E) for 1968–2002, where the correlation coefficient of -0.43 between TOC and the tropopause temperature was found (statistically significant at the 95% confidence level). Applying the same analysis for Sapporo (43.0°N, 141.3°E) for 1976–2002, it was found that this correlation coefficient becomes -0.70 (statistically significant at the 99% confidence level). Finally, for Bismarck (46.7°N, 100.8°W) during 1962–2002 there was a correlation coefficient of -0.3. It should be noted that some other stations in the U.S.A. (Boulder, Colorado and Wallops Island, Virginia) also gave statistically non-significant correlation coefficients like that of Bismarck.

From the puzzle of the aforementioned correlation coefficients between TOC and Z at various geographical sites, it is clearly evident that there exists longitudinal and latitudinal variability. The most characteristic feature is that the strongest signal in the inverse relationship between TOC and Z appears to be present in eastern mid-latitudes of the Northern Hemisphere.

However, in an attempt to verify whether there really are such large longitudinal and latitudinal differences in the ozone and temperature fields in the upper-troposphere and lower-stratosphere region, the NCEP-NCAR reanalysis tropopause and 50 hPa data for the aforementioned locations were examined along with the corresponding TOC time series deduced from TOMS observations. This analysis thus confirms the above-mentioned observation result that the strongest anticorrelation between TOC and Z occurs in eastern mid-latitudes of the Northern Hemisphere. However, the analysis based on TOC and 50 hPa temperature data at the aforementioned locations showed a more precise relationship between these two variables: a 10 DU change in TOC corresponds to roughly a 1 K change of 50 hPa temperature. This relationship is very important for quantification of natural variations in total ozone. We shall return to the question of the relation between TOC and tropopause properties in Chapter 6 (Section 6.6.4) when we discuss trends in ozone concentration.

4.6.3 The tropopause; summary

Let us summarize the main points discussed in this section.

(1) The long-term variability of Z over Athens shows that the winter distribution of the frequency of occurrence of the tropopause exhibits a single maximum around 11.3 km, while the corresponding summer distribution exhibits a main maximum around 11.8 km and also a subsidiary one around 16.3 km. It is shown that the secondary (elevated) maximum originates by the following mechanism: when the Athens site is located below the cyclonic shear side of the subtropical jet then Z is at approximately 12 km, whilst it reaches around 16 km when the Athens site is located below the anticyclonic shear side of the subtropical jet.

(2) The observed second thermal tropopause over Greece is characterized by a temporal increase in the frequency of occurrence in all seasons. It seems that transport in the upper troposphere and lower stratosphere originating in the equatorial region forces the appearance of a multiple tropopause.

(3) Temporal evolution of the temperatures of both thermal tropopauses in all seasons shows that they change approximately in phase, following a statistically significant negative trend. This equivalently means that the heights of both thermal tropopauses in all seasons are significantly increasing (at a higher rate for the second tropopause). This increase in Z may primarily be attributed to natural and anthropogenic forcings that result in both stratospheric cooling (due to stratospheric ozone depletion) and tropospheric warming (due to enhancement of the atmospheric greenhouse effect).

(4) The vertical displacement of the tropopause over Greece by one kilometer is associated with an ozone anomaly of 10 Dobson units. Observational analysis of the vertical ozone distribution as it is derived from ozonesonde ascents performed at Athens during 1991–2001 shows that there has been an upward movement of the ozone profile accompanied by an increase in annually averaged Z. This observed shift of the ozone profile in the lower stratosphere has led to excessive ozone depletion resulting in a TOC trend around 0.5–1% per decade, which should not be considered as a direct result of anthropogenic chemical ozone depletion.

(5) The inverse relationship between TOC and Z at various geographical sites shows both longitudinal and latitudinal variability with the strongest signal in eastern

mid-latitudes of the Northern Hemisphere. However, a more precise relationship between TOC and 50 hPa temperature holds: a 10 DU change in TOC corresponds to a roughly 1 K change in 50 hPa temperature. This relationship is very important for quantification of natural variations in total ozone.

5

The Montreal Protocol

5.1 INTRODUCTION

In this chapter we discuss the effects of human activities on atmospheric ozone and, in particular, ozone depletion. We divide ozone depletion into two distinct parts; one is the steady depletion that occurs to a lesser or greater extent over the whole of the globe and is an underlying trend beneath all the various large daily, seasonal, and regional fluctuations; and the other is the dramatic reduction of stratospheric ozone in the Antarctic spring, usually referred to as the (Antarctic) ozone hole. A similar, but smaller effect, to the Antarctic ozone hole has also been observed more recently in the Arctic. We shall describe the discoveries and events that led to the adoption of the Montreal Protocol which was drawn up in 1987 and which came into effect in 1989. However, when the Montreal Protocol was drawn up the relevant scientific evidence was far from complete. Very readable accounts of the steps involved in this process are given in the book by Sharon Roan (1989), albeit in a slightly journalistic format and with a U.S. bias, and in the book by Richard Benedick (1991) which concentrates on the diplomatic aspects of the negotiations leading to the Montreal Protocol, rather than on the science. The contribution of remote sensing in the scientific work leading up to the establishment of the Montreal Protocol and the monitoring of its success is discussed by Cracknell and Varotsos (2009), part of a special issue of the *International Journal of Remote Sensing* (Volume 30, Nos. 15–16, August 2009); this chapter draws substantially on that article.

Unlike many other international treaties, the Montreal Protocol was not set in stone when it was originally formulated but allowance was made for subsequent modifications to be made to the Protocol in the light of later scientific and technological advances. This is achieved by annual Meetings of the Parties (i.e., the countries that signed/ratified the Protocol) and from time to time the Meetings of the Parties have revised or extended the Protocol. Thus, we shall also discuss the developments that have occurred subsequent to the Montreal Protocol.

It is now generally accepted that ozone depletion is mainly caused by CFCs (chlorofluorocarbons), which are entirely man-made chemicals that do not occur in nature (the first was synthesized by Thomas Midgley in 1928), and by oxides of nitrogen, NO_x, which are produced both by natural processes (e.g., lightning strikes) and by human activities (combustion processes). It took quite a while for this to be accepted scientifically and even longer for it to be accepted politically at the international level, leading to the international agreement which was made to phase out the production and use of CFCs (namely, the Montreal Protocol).

The Montreal Protocol is an international treaty which was opened for signature in 1987 and which came into force in January 1989. It is concerned with attempting to protect the ozone layer by phasing out the production and use of various chemicals, principally CFCs (chlorofluorcarbons), which are held to be responsible for the destruction of atmospheric ozone. It is worthwhile to point out that very nearly everything which is known about the ozone in the atmosphere has been established as a result of remote-sensing observations (Cortesi *et al.*, 2007; Kondratyev *et al.*, 1994). The only direct observations of atmospheric ozone are obtained from a limited number of ozone-sounding stations and from a very small number of instruments developed for launching with balloons and rockets (see Section 1.7). Other than this, and a few other experiments using instruments flown in aircraft (see Section 1.7.4), virtually all studies of atmospheric ozone involve remote sensing. In this context we include the extensive series of ground-based ultraviolet spectrophotometers and a whole variety of (airborne and) spaceborne instruments (see Section 1.7.3).

5.2 THE PROPOSITION BY MOLINA AND ROWLAND OF HUMAN RELEASES OF CFCS BEING RESPONSIBLE FOR OZONE DEPLETION

The story of developments leading up to the Montreal Protocol can reasonably be taken to have started with some calculations carried out late in 1973 and published the following year which suggested that man-made CFCs posed a major threat to the ozone layer (Molina and Rowland, 1974). Like all great discoveries this was made against a background of clues that had not been followed up or resolved before. These included the following:

- It had long been known that ozone molecules could easily be decomposed to molecular oxygen by chlorine in the presence of (sun) light.
- Lovelock in 1970–1972 had discovered significant concentrations of CFCs in the atmosphere indicating that these chemicals were very stable and very long-lived in the atmosphere, unlike most other chemicals that are released into the atmosphere.
- There had been extensive "discussions" of a possible threat to the ozone layer from NO_x (various oxides of nitrogen) that would be released if the proposed development of a large fleet of supersonic transport aircraft were to go ahead.

- In some work by Rich Stolarski and Ralph Cicerone, the publication of which was delayed to 1974, there were suggestions that the exhaust of the Space Shuttle (which was then on the drawing board) would provide a source of chlorine that would pose a threat to the ozone layer. In other words, they had shown that Cl is even more efficient than NO at catalyzing the destruction of ozone.

We shall consider these points in a little more detail.

The first serious discussion of ozone in the atmosphere was given by Chapman (1930, 1932) in what came to be known as the Chapman relations:

$$O_2 + h\nu \rightarrow O + O \quad \text{(wavelengths shorter than 240 nm)} \tag{5.1}$$

$$O + O_2 + M \rightarrow O_3 + M \tag{5.2}$$

$$O_3 + h\nu \rightarrow O_2 + O \quad \text{(wavelengths shorter than 900 nm)} \tag{5.3}$$

$$O + O_3 \rightarrow 2O_2 \tag{5.4}$$

where $h\nu$ is the energy of an ultraviolet photon of frequency ν and h is Planck's constant. The first two of these reactions involve the splitting of an ordinary oxygen molecule, O_2, by an ultraviolet (UV) photon into two free oxygen atoms. A free oxygen atom combines with an O_2 molecule, in the presence of some other (catalytic) molecule, M, to form an ozone molecule O_3. Equations (5.3) and (5.4) correspond to the destruction of an ozone molecule by a UV photon. Equation (5.4) could in theory be replaced by $O + O \rightarrow O_2$ but the concentration of O atoms would be so small that (5.4) is much more likely.

Qualitatively, Chapman's theory explained how the concentration of natural ozone is maintained by a balance between reactions creating and destroying ozone. However, from the 1950s, when more quantitative studies were undertaken, rocket experiments were well under way, and extensive data became available for the upper reaches of the atmosphere and for outer space. It began to be concluded that there was less ozone, maybe by a factor of 2, in the atmosphere than would be expected if the Chapman reactions were the only significant reactions. Thus, it became apparent that there must be some additional mechanism involved in the ozone destruction process, even in the natural situation. First, it was suggested that reactive species like oxides of hydrogen (i.e., free radicals such as OH) could be created from water vapor brought into the stratosphere and catalyzing ozone destruction. In general, it could be schematically shown as follows:

$$\begin{array}{llcl}
 & X + O_3 & \rightarrow & XO + O_2 \\
 & O_3 + \text{light} & \rightarrow & O + O_2 \\
 & O + XO & \rightarrow & X + O_2 \\
\hline
\text{net:} & 2O_3 & \rightarrow & 3O_3
\end{array}$$

$$X = NO, OH, Cl, Br$$

$$XO = NO_2, HO_2, ClO, BrO$$

where X represents a compound such as NO, OH, Cl, or Br which reacts with ozone. In the case where X stands for OH, then:

$$HO + O_3 \rightarrow HO_2 + O_2 \tag{5.5}$$

$$HO_2 + O \rightarrow HO + O_2 \tag{5.6}$$

These two reactions together are equivalent to

$$O + O_3 \rightarrow 2O_2 \tag{5.4}$$

and are frequently designated as the HO_x catalytic chain. The chain is catalytic because the HO radical is not consumed but can be used over and over again to destroy ozone molecules (Bates and Nicolet, 1950; Dütsch, 1970; Hampson, 1964). However, when further measurements of the rates of these reactions were made, it became clear that this proposed mechanism by itself was not sufficient to close the gap between calculated production and destruction rates for ozone. As explained by Rowland (2006, 2009) this is largely because the HO_x reaction sequence is most significant at the top of the stratosphere, above the altitudes where most of the ozone is found. However, this suggestion of the possible role of the HO radical caused people to think beyond the Chapman reactions and brought free radical reactions into the picture.

Crutzen (1969) suggested that it was not oxides of hydrogen but oxides of nitrogen from natural sources which were responsible for ozone destruction. A catalytic reaction involving NO and NO_2 was suggested:

$$NO + O_3 \rightarrow NO_2 + O_2 \tag{5.7}$$

$$NO_2 + O \rightarrow NO + O_2 \tag{5.8}$$

Again these two reactions together are equivalent to

$$O + O_3 \rightarrow 2O_2 \tag{5.4}$$

and the NO, like the HO above, is behaving as a catalyst. Both the HO and NO can be generated by natural causes. Crutzen (1970) went on to suggest that oxides of nitrogen entering the atmosphere as a result of the use of artificial fertilizers could lead to human-induced destruction of atmospheric ozone, but this effect was dismissed as being very small. At about that time there was a proposal for the construction in the U.S.A. of a fleet of SSTs (supersonic jet transport aircraft) which would fly in the lower stratosphere. There was considerable controversy regarding this proposal and one aspect of the controversy was related to the question of whether the water vapor and/or oxides of nitrogen emitted by these aircraft would damage the ozone layer (Crutzen, 1971; Johnston, 1971). In the event, it was concluded that the threat to the ozone layer from oxides of nitrogen from supersonic jet aircraft was less than had been suggested. However, the fleet of supersonic jets was never built anyway; with the subsequent large rises in the price of oil since then they would probably have been grounded anyway—as Concorde had been—for reasons of fuel efficiency.

Another catalytic reaction proposed by Stolarski and Cicerone (1974) leading to the destruction of atmospheric ozone involves chlorine oxide, ClO, in reactions similar to those involving HO and NO:

$$ClO + O_3 \rightarrow ClO_2 + O_2 \tag{5.9}$$

$$ClO_2 + O \rightarrow ClO + O_2 \tag{5.10}$$

Once more, these two reactions together are equivalent to

$$O + O_3 \rightarrow 2O_2 \tag{5.4}$$

and the ClO, like the HO and NO, is a catalyst. This mechanism is not unreasonable. CH_3Cl occurs naturally and acts as a source of stratospheric ClO_x, whereas other partially halogenated alkanes are anthropogenic in origin (see below and Section 5.4). It should be noticed in passing that all the radicals involved in these three catalytic mechanisms are electrically neutral species which contain an odd number of electrons, HO (9), NO (15), ClO (17), HO_2 (17), NO_2 (23), ClO_2 (25). By contrast, the stable molecules H_2O, O_2, N_2, CO_2, etc. contain an even number of electrons.

As mentioned above the suggestion that ClO might be involved in the destruction of (stratospheric) ozone led Stolarski and Cicerone in 1973 to study the threat to the ozone layer from chlorine in the exhaust of the Space Shuttle, but it was concluded that the effect would be small. However (perhaps under pressure from NASA, their sponsors), rather than dwell on the Shuttle as a source of chlorine the paper by Stolarski and Cicerone (1974) concentrated on natural sources of chlorine (e.g., from volcanoes), leaving it to Wofsky and McElroy (1974) to publicize the Shuttle threat.

In 1970 James Lovelock had constructed an electron capture gas chromatograph (Electron Capture Detector (ECD)) which was capable of measuring the concentration of CFCs the air. He made some preliminary measurements of the concentration of CFCs in southern England and western Ireland in 1970 and found the concentration of CCl_3F to be $>10^{-11}$ by volume. He followed this by taking measurements on the voyage of RRS *Shackleton* from the U.K. to Australia in 1971–1972 (Lovelock, 1971, 1972; Lovelock *et al.*, 1973). He studied the observed distribution of CCl_3F in the atmosphere from about 50°N to 70°S and found that it agreed with the predicted distribution from models of the behavior of an ideal inert gas released at an exponentially increasing rate, doubling every two years, in the band between 35°N and 55°N. He concluded that the lifetime of CFCs in the atmosphere was extremely long. A thoroughly readable account of James Lovelock's work on the ECD and the detection of CFCs in the atmosphere can be found in his autobiography (Lovelock, 2000).

The scientific achievement of Molina and Rowland in late 1973 was to build on some of these results and to realize that since the CFCs released into the atmosphere remained there for a very long time, they would eventually reach the stratosphere, where they would ultimately decompose to give free chlorine that could severely deplete the ozone layer and pose a serious threat to human, and other, life (Molina and Rowland, 1974). At that time it was well known that atmospheric ozone concentrations fluctuated temporally and spatially, but there was no clear evidence of systematic depletion of atmospheric ozone and its attribution to man-made chemicals, such as CFCs, etc. That conclusion was then reached rather rapidly. There was a long way to go for this scientific discovery to form the basis of an international agreement and we shall follow that development in the next section.

5.3 THE SCIENCE FROM 1974 TO 1985

In this section we rely quite heavily on the account given in the book by Roan (1989) which, as we have already pointed out, is a bit journalistic—see *http://www.albany.edu/faculty/rgk/atm101/o3histor.htm* (accessed November 21, 2008). The story covers 1974 to 1987, when the Montreal Protocol was formulated, with the key point along the way being the "discovery" of the Antarctic ozone hole in 1983–1984. The chronology is summarized in Table 5.1.

By 1974 the production of CFCs and their use in refrigeration units and aerosol spray cans had become a major industry. It had started in 1928 with the search for a replacement of ammonia and sulfur dioxide, which are both highly toxic, as the working gases in refrigerators. The replacement had to be non-toxic, non-flammable, and stable and Thomas Midgley produced and proposed the compounds which came to be known as chlorofluorocarbons, or CFCs for short, which seemed to be suitable for the purpose. Formally, one can think of a hydrocarbon, most commonly methane CH_4, in which the hydrogen atoms are all replaced by halogen atoms (fluorine, chlorine, bromine, or iodine), most commonly fluorine and chlorine. The chemical industry went into production and CFCs quickly replaced ammonia and sulfur dioxide as the working substance in refrigeration units. Subsequently, CFCs came to be used as the propellant in aerosol spray cans. This had all come about in the belief that CFCs were safe. Figures quoted by Roan (1989) showed that CFC production and use in 1974 generated U.S.$8 billion in business and employed 200,000 people. This would not easily be sacrificed for a scientific hypothesis that was not proven.

The paper suggesting CFCs as a source of chlorine in the stratosphere and its possible role in the destruction of the ozone layer was published in June 1974 (Rowland and Molina, 1974). Their theory was discussed publicly for the first time at the American Chemical Society meeting in Atlantic City in September 1974 and received sufficient publicity that in October 1974 the National Academy of Sciences, a U.S. government-sponsored organization, began to conduct a study on the validity of the CFC–ozone theory (Figure 5.1). By early 1975 the theory had attracted widespread attention and it was hotly debated at the American Chemical Society Meeting in Philadelphia in May 1975. A Committee on the Inadvertent Modification of the Stratosphere (IMOS) was set up early in 1975 as a Presidential task force to advise the U.S. government on what action should be taken on CFCs and aerosol sprays. In June 1975 IMOS reported that there was "legitimate cause for concern" and that restrictions on CFCs might be necessary and proposed that if the National Academy of Sciences found that CFCs were hazardous then the federal government should legislate to restrict or ban their use. It also recommended that the U.S. should alert other countries to the possible dangers of using CFCs. However, at

Table 5.1. Chronology of events leading up to the Montreal Protocol (compiled by Cracknell and Varotsos, 2009 from Roan, 1989 and Benedick, 1991).

December 1973	Rowland and Molina discover that CFCs can destroy ozone in the stratosphere.
June 1974	Rowland and Molina's paper on their discovery is published in *Nature*.
September 1974	Rowland and Molina discuss their theory publicly for the first time at the American Chemical Society meeting in Atlantic City.
October 1974	A U.S. government committee recommends that the National Academy of Sciences conduct a study on the validity of the CFC–ozone theory.
December 1974	First U.S. government hearings are held on the CFC–ozone theory.
May 1975	The CFC–ozone theory is hotly debated at the American Chemical Society Meeting in Philadelphia.
June 1975	Johnson Wax, the fifth largest manufacturer of aerosol sprays in the U.S.A., announces it will stop using CFCs in its products. Oregon becomes the first U.S. state to ban CFCs in aerosol sprays.
April 1976	Nairobi: UNEP Governing Council calls for international meeting.
March 1977	Washington: UNEP meeting recommends "World Plan of Action on the Ozone Layer" and establishes Coordinating Committee on the Ozone Layer (CCOL) to produce annual science reviews.
April 1977	Washington: U.S. hosts first intergovernmental meeting to discuss international regulation of CFCs.
1978	U.S.A. bans use of CFCs in non-essential aerosol sprays (followed by Canada, Norway, and Sweden).
December 1978	Munich: Federal Republic of Germany hosts second international conference on regulating CFCs.
1980	European Community reduces aerosol use by 30% and enacts a production capacity cap.
April 1980	Nairobi: UNEP Governing Council calls for reductions in CFC-11 and CFC-12 plus a production capacity cap. The EPA announces the U.S. intention to freeze all CFC production at 1979 levels.
August 1981	NASA scientist Donald Heath announces that satellite records show ozone has declined 1%.

(*continued*)

Table 5.1. Chronology of events leading up to the Montreal Protocol (*cont.*)

May 1981	Nairobi: UNEP Governing Council recommends a convention for protection of ozone layer.
January 1982	Stockholm: UNEP convenes Ad Hoc Working Group of Legal and Technical Experts for the Preparation of a Global Framework Convention for the Protection of the Ozone Layer.
1982–1985	Geneva: UNEP Working Group negotiates a convention on research, monitoring, and data exchange but fails to get an agreement.
April 1983	During international talks, Norway, Sweden, and Finland submit a plan for a worldwide ban of CFCs in aerosols and limitations on all uses of CFCs.
June 1984	At a scientific meeting in Germany, Rowland reports his calculations on heterogeneous reactions involving hydrogen chloride and chlorine nitrate reactions that could significantly speed up ozone depletion.
March 1985	Vienna Convention for the Protection of the Ozone Layer adopted.
May 1985	A British research group led by Joe Farman publishes data in *Nature* showing a 40% ozone loss over Antarctica during the austral spring (i.e., the Antarctic ozone hole).
August 1985	NASA's Donald Heath shows satellite photos confirming the existence of the ozone hole over Antarctica.
January 1986	EPA releases its Stratospheric Ozone Protection Plan which calls for new studies to determine whether additional CFC regulations are needed.
March 1986	Atmospheric scientists meeting in Boulder (Colorado) discuss plans for an expedition to Antarctica (NOZE) to study ozone depletion there.
May 1986	A NASA-UNEP report states that damage to the atmosphere has occurred.
June 1986	Papers are published by two research groups indicating chemicals and polar stratospheric clouds are responsible for ozone loss over Antarctica. Hearings on ozone depletion and greenhouse warming are held in Washington. Thomas announces that some government intervention may be needed to halt emissions of gases that could harm the atmosphere. Scientists testify that greenhouse warming has begun due to emissions of gases such as CFCs. CFC manufacturers suggest that safe substitutes for the chemicals might be possible but the price will be high.

August 1986	Thirteen U.S. scientists depart for Antarctica on the National Ozone Expedition (NOZE).
September 1986	Leesburg (Virginia): UNEP workshop on control strategies.
November 1986	Netherlands: CCOL meeting on effects of ozone depletion. U.S. negotiating position ("Circular 175") approved.
December 1986	Geneva: first round of protocol negotiations. United States and Soviet Union agree on joint ozone research.
February 1987	Vienna: second round of protocol negotiations.
February–June 1987	Washington: U.S. Domestic Policy Council reconsiders U.S. negotiating position.
April 1987	Würzburg: UNEP meeting on science models. Geneva: third round of protocol negotiations.
June 1987	Venice: economic summit declaration lists stratospheric ozone depletion first among environmental concerns. President Reagan formally approves final U.S. position. Brussels: UNEP informal negotiations among key delegation heads.
July 1987	The Hague: UNEP protocol legal drafting group convene.
September 1987	Montreal: final round of negotiations. Protocol on Substances that Deplete the Ozone Layer adopted.

a lower level, two other things happened in June 1975. One was that Oregon became the first State in the U.S.A. to ban the sale of CFCs in aerosols, as a precautionary measure even though at that stage there might still be doubt about the science. The second was that Johnson Wax, the fifth largest manufacturer of aerosol sprays in the U.S., announced it would stop using CFCs in its aerosol products "in the interests of our customers during a period of uncertainty and scientific enquiry." The next few years saw hardening of the scientific evidence. There were direct measurements detecting CFCs in the stratosphere (note that these measurements were made by instruments flown in aircraft and carried by balloons, as this was before appropriate satellite remote-sensing instruments were available). Experiments were also conducted demonstrating that CFCs were dissociated by UV, producing chlorine. The National Academy of Sciences released its report in September 1976 supporting the Rowland–Molina hypothesis, but saying that government action on CFC regulations should be postponed (the story behind the weakness of this report, especially regarding the question of chlorine nitrate, is told in chapter 5 of Roan, 1989). Although some other States introduced anti-CFC legislation, there was no rapid progress towards a national U.S. ban on CFCs. However, in May 1977 several U.S. government agencies announced joint plans for a timetable to ban the

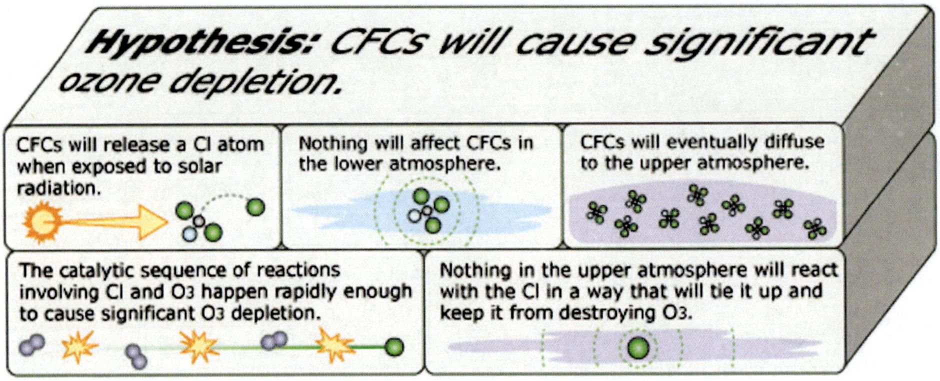

Figure 5.1. Schematic diagram of the CFC impact on the ozone layer (*undsci.berkeley.edu*).

"non-essential" uses of CFCs (i.e., principally in aerosol sprays) by October 1978. Companies were required to stop manufacturing CFCs for non-essential purposes by October 15, 1978 and to stop using them in non-essential products by December 15, 1978. Non-essential purposes meant aerosol sprays; alternatives to the use of CFCs in aerosol sprays were also beginnng to appear. Other uses of CFCs were not proposed to be affected at that stage.

> "For an environmental movement that had started out so convincingly, the next eight years [from 1977] proved astonishing as the momentum to protect the earth's ozone layer from manmade pollutants crumbled" (Roan, 1989).

Following the ban on CFCs in aerosol sprays in the U.S.A. there was talk of introducing restrictions on the production of CFCs for non-aerosol spray purposes, but nothing came of it. The reasons cited include politics, economic pressures, and scientific confusion. Until 1977 the U.S. had taken the lead, but after that the initiative began to move into the international arena, although international activities consisted mostly of talk and only a very small number of countries (Canada, Norway, and Sweden) followed the U.S.A. in introducing a ban on the use of CFCs in aerosol sprays. In March 1977 the United Nations Environmental Program (UNEP) held the first international meeting to discuss ozone depletion. In 1983 Canada, Finland, Sweden, Norway, Switzerland, and the U.S.A. (the so-called "Toronto Group") produced a draft proposal for a worldwide ban on CFCs in aerosol sprays and restrictions on all uses of CFCs (Barratt-Brown, 1991). In March 1985 an international agreement, the Vienna Convention, was signed. This imposed no restrictions on anyone, but it set out a plan for research, exchange of information, and negotiations with the intention of leading to a protocol to protect the ozone layer, which was to be negotiated through UNEP.

Scientific problems arose from the fact that the detection of ozone depletion during this period was difficult, stratospheric chemistry was in its infancy, and direct observation of the chemical species involved in the various models of ozone destruction in the stratosphere was very difficult (see Roan, 1989, pp. 134–136). Ozonesonde data were far too sparse to be of any use. Although there were several dozen Dobson stations, mostly in the Northern Hemisphere, and some of them had very long records of TOC, it was not easy to search for evidence of ozone depletion from these data. Of course, if there were any depletion due to CFCs it would only have started in the mid-1970s.

Detection of a depletion of a few percent over, say, the last 10 years involved looking for a very small trend in data that fluctuate widely from day to day, from one season to another, from place to place, and throughout the 11-year sunspot cycle. Detecting global ozone changes from Dobson data was also difficult because of variations in the maintenance and calibration procedures in different stations. For several years, studies of data from Dobson stations (many sponsored by the CFC industry or the Chemical Manufacturers Association) had shown no change in total ozone, and industry officials gave great credence to the long history of the Dobson Network.

The situation in the early 1980s was summarized by Watson *et al.* (1986). As well as demonstrating the ozone hole, the TOMS data quoted by Heath in 1988 were beginning to show evidence of a general decline in TOC over the whole globe (see WMO, 2007 and references therein). Statistical analyses for a global trend in total column ozone using Dobson spectrophotometer data had not shown a statistically significant trend during the 11-year period 1970–1980. However, different groups using Dobson Network data for the period 1970–1983 reported different results. One study reported that there was no statistically significant trend, namely $(-0.17 \pm 0.55)\%/\text{decade}$, while another study reported that the trend was statistically significant, namely $(-1.10 \pm 0.47)\%/\text{decade}$. In the second study it was reported that negative trends were primarily the result of abnormally low total ozone values in the Northern Hemisphere during early 1983, and that the reason for the disagreement between the different results may be due to the different methods of area weighting of the data. There was, however, the possibility that ozone loss in the Northern Hemisphere at that time could be due to the eruption of the El Chichón volcano in April 1982 (Mantis *et al.*, 1986). Harris and Rowland (1986) examined the historical record of Dobson spectrophotometer values of TOC from Arosa in the Swiss Alps. This station has been obtaining TOC data since August 1931. Although the complete dataset contains individual TOC measurements for most days, statistical treatments usually concentrate on monthly or annual average ozone levels. Rather than considering mean annual values of TOC (i.e., neglecting possible differences in seasonal variations) Harris and Rowland (1986) considered the monthly averages as 12 distinct datasets, one for each month of the year. They separated the 56-year data series from Arosa into two datasets: one for the 39-year period 1931–1969 and the other for the 17-year period 1970–1986. The result is shown in Figure 5.2 and this figure shows that there was little change in TOC in the summer but significant average depletion in the winter months of 1970–1986 relative to 1931–1969. Subsequent analysis of monthly average data from two U.S. stations, which are at the same latitude as Arosa but which only began taking data in 1963, also exhibited wintertime losses when TOC data from 1976–1986 and 1965–1975 were compared.

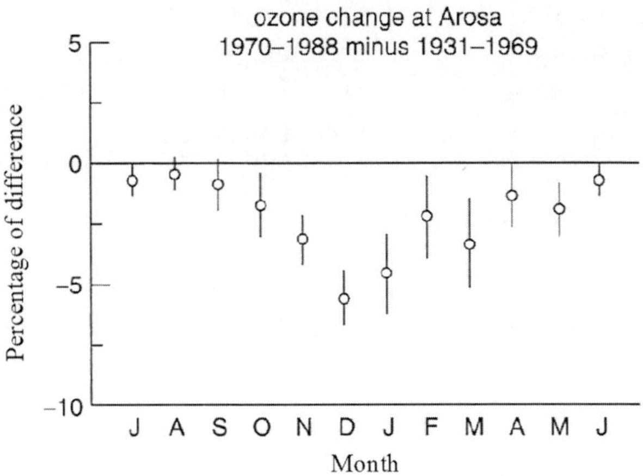

Figure 5.2. Differences in average ozone concentrations 1970–1988 versus 1931–1969 (Harris and Rowland, 1986).

This comparison of consecutive 11-year periods minimizes any influence of 11-year solar cycle variations in UV emission on the evaluation of longer term trends. Subsequently, the WMO/NASA Ozone Trends Panel examined all 18 of the ozone stations between 35°N and 60°N that had at least 22 years of data (again, enough for two complete 11-year solar cycles), and found that all of them had less wintertime ozone in 1976–1986 than in the earlier period (as shown in Figure 5.3) (Rowland, 2006, 2009; Rowland *et al.*, 1989; WMO, 1990). Previous statistical analyses had not detected ozone loss because they had not included the most recent years of new data and had generally only considered annual averages and not studied the individual months or seasons of the years separately (Rowland 2006, 2009; WMO, 1986).

It was also in this period that satellite data related to atmospheric ozone were

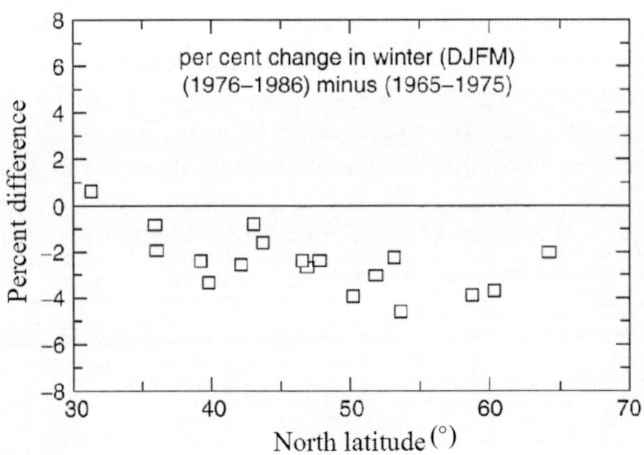

Figure 5.3. Percentage change in ozone 1976–1986 versus 1965–1975, Northern Hemisphere Dobson stations (Rowland *et al.*, 1989).

just beginning to become available. In 1981 Heath (see WMO, 1999) produced evidence from satellite data of a 1% loss in ozone at a height of 40 km from 1970 to 1979. These Dobson and satellite results were not of a nature to provide evidence of ozone loss due to CFCs that would convince scientific skeptics, politicians, or industrialists with a vested interest in CFC manufacturing. That evidence came dramatically with the discovery of the Antarctic ozone hole.

In late 1984 a major cooperative international scientific research program was launched to summarize for governments the best scientific information available at that time on whether human activities represent a substantial threat to the ozone layer. This research was sponsored by NASA, NOAA, the U.S. Federal Aviation Administration, UNEP, the World Meteorological Organization (WMO), the West German Ministry for Research and Technology, and the Commission of the (then) European Communities. The published result (WMO, 1986) was the most comprehensive study of the stratosphere undertaken at that time; it ran to three volumes containing nearly 1,100 pages of text plus 86 reference pages listing hundreds of articles. It covered all aspects of atmospheric ozone, including separating the effects of anthropogenically produced trace gases on the atmosphere from such naturally occurring phenomena as solar activity and volcanic eruptions (see also Kondratyev and Varotsos, 1995a, b). The research found that the concentrations of CFC-11 ($CFCl_3$) and CFC-12 (CF_2Cl_2) in the atmosphere had nearly doubled between 1975 and 1985. The research also found that the ozone layer was threatened not only by CFC-11 and CFC-12, which had been the original focus of international scientific concern, but also by other fully halogenated alkanes, which included the related CFC-113 ($C_2F_3Cl_3$), CFC-114 ($C_2F_4Cl_2$), and CFC-115 (C_2F_5Cl) and two bromine compounds, halon-1211 (CF_2ClBr) and halon-1301 (CF_3Br). It was estimated that continued emissions of CFC-11 and CFC-12 at the 1980 rate could reduce the ozone layer by about 9% on a global average by the last half of the 21st century, with even greater seasonal and latitudinal declines. Several different researchers made predictions of the anticipated loss of ozone and there were some differences between the various predictions, but the general message was clear. It was also noted that, molecule for molecule, CFCs are far more powerful than CO_2 as greenhouse gases.

It would not be fair to conclude this section without making the point that there were serious scientists who doubted the evidence for ozone depletion and who disputed the connection between CFCs and this depletion. While these matters are now only of historical interest, one useful account of many of these objections is given by Maduro and Schauerhammer (1992). One example of these arguments is the claim that the existence of the Antarctic ozone hole (see the next section) had been known in 1956 and that it could not, therefore, be due to CFCs.

5.4 THE OZONE HOLE

A major step forward in reducing opposition to the idea that CFCs, etc. posed a threat to the ozone layer came with the publication in *Nature* of the results of

Farman *et al.* (1985) from their long records of Dobson UV spectrophotometer measurements at two British Antarctic Survey stations: Halley Bay (now known simply as Halley after 1977; 75°35′S, 26°34′W) and Faraday Base on Argentine Island (65°15′S, 64°16′W), now known as Vernadsky and operated by the Ukraine since February 1996, about 1,000 miles (1,600 km) northwest of Halley (see also Chubachi, 1984, 1985). These results, which are now very well known, showed that in the Antarctic spring in October 1984 the total ozone content (TOC) of the air column dropped from its normal value of around 300 DU to about 200 DU and that over the previous 10 years the TOC in October had been declining steadily year on year. These were, of course, only measurements at two points, albeit far apart from one another. However, TOMS, the Total Ozone Monitoring Spectrometer, had been flown on the NASA experimental spacecraft Nimbus-7 since 1978. Low TOC values of 200 DU were indeed deduced from the TOMS data but, prior to the release of the results of Farman *et al.* (1985), NASA scientists had no way of telling whether the results were genuine or due to some error in the instrument or the algorithm used for processing the data. After the results of ground measurements were released the satellite data were re-examined and it was found that each spring there was a massive drop in TOC, not just at Halley Station but over a huge area of Antarctica equal to about the size of the continental U.S. Subsequent data from ozonesondes showed that when the ozone hole is at its deepest the ozone concentration between about 14 km and 21 km altitude is virtually zero; this is illustrated very clearly in Figure 5.4 from 2006 (Hofmann *et al.*, 2009). Armed with data from 1978 up to the present time gathered by TOMS and its various successors flown in space, the scientific community have been closely monitoring the Antarctic ozone hole and its evolution each year.

Bhartia (2009) discussed the role played by satellite total ozone-measuring instruments in the study of stratospheric ozone depletion, in general, and in the discovery of the ozone hole, in particular. The first Total Ozone Monitoring Spectrometer (TOMS) was flown in 1978 on the Nimbus-7 satellite (see Section 2.3.1) and the dramatic decreases in ozone concentration in the Antarctic spring, which came to be known as the ozone hole, can be seen clearly in TOMS data. The question therefore arises as to why NASA did not report the discovery of the Antarctic ozone depletion until late 1985 and, thereby, claim the credit for the discovery of the ozone hole. This was initially described as a computer error and is still often described in that way in both scientific and popular media.

Bhartia argued that it is not that simple. He pointed to the fact that "the systematic measurement of stratospheric ozone from satellites started in April 1970 with the launch of the Nimbus-4 satellite ..." (see Table 2.4) and by 1978 was well established with the launch of TOMS and SBUV on Nimbus-7 in October 1978. TOC can also be determined from one channel of the TOVS (TIROS Operational Vertical Sounder) which has also been flown in space from 1978 (on TIROS-N and its successors) but the values are less accurate than the TOMS values (see Sections 2.2.1 and 2.3.1). However, atmospheric sounding, including the retrieval of total ozone, from satellite data is a mathematically ill-posed problem and so one needs external, or *a priori*, information (see, e.g., Rodgers,

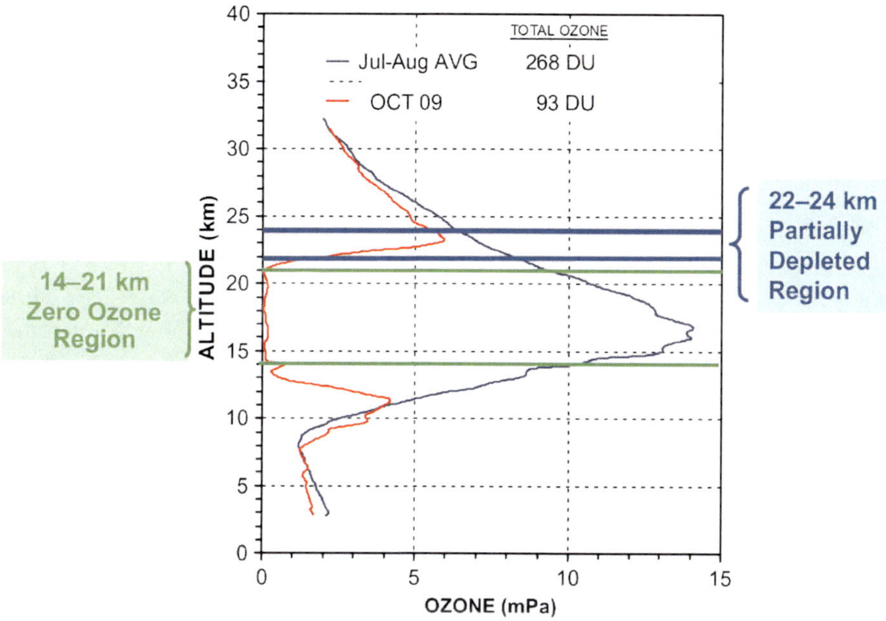

Figure 5.4. South Pole ozone profiles in 2006 before the ozone hole developed (blue) and for the sounding that displayed the minimum ozone (red). The regions of interest at 14–21 km and 22–24 km are delineated (Hofmann *et al.*, 2009).

2000). The approach proposed by Mateer (1971) involved using a set of standard ozone profiles obtained from ozonesondes to estimate total column ozone from satellite-measured radiances. This is based on the fact that variations in the lower ozone column reflect large-scale variations in the lower-stratospheric ozone profile. This innovative idea works extremely well and continues to be used by modern satellite algorithms. By the early 1980s total ozone measurements from space were becoming comparable in quality with measurements from ground-based stations (Bhartia *et al.*, 1984) and profiles obtained from ozone soundings played a key role in these retrievals.

To construct the standard profiles for use in the retrievals, Mateer used ozonesonde data below 25 km and rocketsonde data above that height.

> "Before the discovery of the ozone hole, there were virtually no ozone profiles from sondes that contained less than 180 DU of total ozone; hence the lower range of validity of satellite retrievals was set at 180 DU. Lower values were flagged as invalid since they could not be correctly retrieved using the standard profiles" (Bhartia, 2009, p. 185).

The SBUV and TOMS data from October 1983 were processed at NASA in August 1984 and it

> "became clear that both satellite instruments were measuring far more UV radiation than that expected from a 180 DU total ozone standard profile. To analyse these data the standard ozone profiles were arbitrarily extrapolated to lower ozone values" (Bhartia, 2009, p. 185).

The first complete map of the Antarctic ozone hole was made in late 1984 using Nimbus-7 TOMS data (see Figure 1 of Bhartia, 2009). This map and the ozone profiles derived from the SBUV instruments were presented at the IAGA/IAMA symposium in Prague, Czechoslovakia, 1985 (Bhartia *et al.*, 1985). Several figures from this presentation were later reproduced by Callis and Natarajan (1986). Stolarski *et al.* (1986) used the Nimbus-7 Total Ozone Mapping Spectrometer (TOMS) measurements to show that the phenomenon was widespread, covering nearly all of Antarctica. TOMS makes measurements of column ozone amounts but gives no data about the vertical structure of the ozone layer. Had NASA in 1984 rushed out into the scientific world with their map and challenged experimental workers to verify whether or not these very low values were real they might have been credited with the discovery of the ozone hole—but they didn't. Hindsight is a great thing. Nevertheless, credit should be given where credit is due. The measurements of Chubachi (1984, 1985) and of Farman *et al.* (1985) were point measurements and could give no idea of the spatial extent of the hole, while the satellite data were quickly able to delineate the spatial extent of the hole. The strength of satellite instruments lies in their capability to expand the limited view provided by ground-based and, later, aircraft instruments to provide a global view. This enabled satellite data to play a key role in making the public and policy makers aware of the seriousness of the ozone depletion problem, which led to the phase-out of ozone-destroying chemicals and eventually to the international agreement of the Montreal Protocol.

There is no doubt that the effect that publication of the work of Farman *et al.* (1985) had on the "discovery" of the ozone hole in the Antarctic was enormous, one might even say electrifying, but it is misleading to credit them with the "discovery" of the hole. We should make two points for the record. First, the existence of anomalously low values of TOC at Halley Station during the Antarctic spring and their persistence until November were described by the great Oxford ozone scientist G.M.B. Dobson in his book *Exploring the Atmosphere* (Dobson, 1963b). This, of course, was a natural situation and the term "ozone hole" was not used then. Moreover, Dobson had no way of knowing about its areal extent. Although the present hole is much deeper, with ozone levels of only around 100–150 DU, we have no information as to whether its area is larger than that of the original natural hole. Second, almost identical measurements to those of Farman *et al.* (1985) at the British Antarctic Survey (BAS) were published a year or more earlier. Measurements that were made in 1982–1983 at the Japanese Antarctic station showed a decrease in TOC and were reported in a poster at a meeting in Thessaloniki in September 1984 (and subsequently published in the proceedings of that meeting; Chubachi, 1985). It is claimed that when Farman saw this poster at that meeting he was stimulated to publish the BAS work. The success of Farman *et al.* (1985) while Chubachi's work

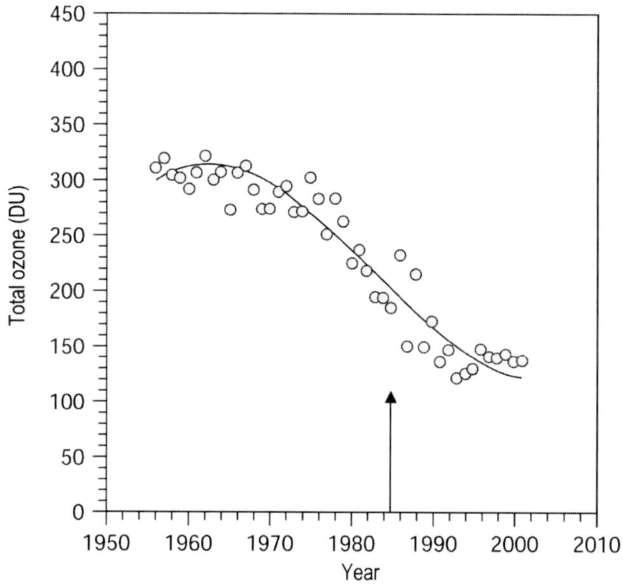

Figure 5.5. Total ozone over Halley Bay (75.5°S, 26.65°W), Antarctica, in October (southern spring) (*http://www.atmos.washington.edu/~davidc/ATMS211/Lecture7-slides-PDF.pdf*).

was largely ignored, came from the fact that (a) they published their work in *Nature* and not in the proceedings of a specialized conference, (b) they were able to compare the 1984 data with the long historical dataset for Halley Station (see Figure 5.5) which showed the increasing trend in the depth of the hole through the late 1970s and early 1980s, (c) they made some initial attempts to relate the ozone hole, in general terms, with ozone-destroying chemicals/radicals possibly produced by human activities, (d) the Halley Station is farther south (75°S) than the Japanese Syowa Station (69°S) and closer to the center of the hole, and (e) the discussion of Chubachi's paper at the Thessaloniki conference in 1984 was inadequate.

In the previous section, we mentioned that the existence of the Antarctic ozone hole had been claimed to be known in 1956 and that it could not, therefore, be due to CFCs. Maduro and Schauerhammer (1992) quote a paragraph from G.M.B. Dobson's classic book *Exploring the Atmosphere* (Dobson, 1963b, pp 128-129, see also Dobson, 1968); this related to ozone data from Halley Station collected during the International Geophysical Year (1956) (see Section 1.1). Comparisons had been made between TOC at Halley and at Spitsbergen in the Northern Hemisphere (latitude 78°54′N) and the values of TOC at Halley in September and October 1956 were about 150 DU lower than the values at Spitsbergen 6 months earlier. Dobson (1968) discussed this:

> "One of the more interesting results on atmospheric ozone which came out of the IGY (International Geophysical Year) was the discovery of the peculiar annual

variation of ozone at Halley Bay (76°S, 26°W). The annual variation of ozone at Spitzbergen was fairly well known at that time, so, assuming a six months difference, we knew what to expect. However, when the monthly telegrams from Halley Bay began to arrive and were plotted alongside the at Spitzbergen curve, the values for September and October 1956 were about 150 units lower than was expected. We naturally thought that Evans has made some large mistake or that, in spite of checking just before leaving England, the instrument had developed some fault. In November the ozone values suddenly jumped up to those expected from the Spitzbergen results. It was not until a year later, when the same type of annual variation was repeated, that we realized that the early results were indeed correct and that Halley Bay showed a most interesting difference from other parts of the world. It was clear that the winter vortex over the South Pole was maintained late into the spring and that this kept the ozone values low. When it suddenly broke up in November both the ozone values and the stratosphere temperatures suddenly rose."

There are suggestions that observations made in 1958 at the French Antarctic station of Dumont d'Urville show significant ozone depletion. These measurements were made using a photographic spectrometer and are subject to large errors, in the region of 20–50%. The observations disagree with measurements made by Dobson spectrophotometers at the same time. There was also no correlation between photographic measurements and stratospheric conditions and all they show is random scatter.

According to Taylor (2011), in a review of 40 years of satellite meteorology at Oxford University, Dobson's achievements in using ground-based remote sensing to understand stratospheric ozone have many resonances with those of a later generation of Oxford scientists using satellite-based techniques to study not just ozone but its chemical partners along with the temperature and dynamics of the entire atmosphere and its analogues on other planets. Taylor argues that 40 years seems to be a natural cycle time for major ideas and approaches to fulfill themselves and then subside. Of course, space observations continue to be a key part of meteorological research and operational data gathering, as indeed does the global network of Dobson spectrophotometers.

Dobson's conclusion (1963, p. 129) was that "much further work will be necessary before these conditions are fully understood." The 1956 values of TOC are given along with the 1996 values in Table 5.2, in which it will be noticed that the 1996 TOC values in the Antarctic spring are much lower than the 1956 values. While it is interesting and important to know that there appears to have been a natural ozone hole before there was any anthropogenic cause, it is regrettable that we have very little quantitative information on the depth and extent of the natural ozone hole. Whatever natural hole there may have been was clearly very much smaller than the present hole.

The satellite data from TOMS played an enormously important role in confirming that the measurements of Chubachi and Farman *et al.* (1985) were not just isolated freak measurements at three points, but that the dip in ozone occupied

Table 5.2. Values of TOC (DU) from Spitsbergen and Halley in 1956 and Halley in 1996.

Spitsbergen									
	Feb	*Mar*	*Apr*	*May*	*Jun*	*Jul*	*Aug*	*Sep*	*Oct*
1956	440	470	450	400	350	320	300	280	280
Halley									
	Aug	*Sep*	*Oct*	*Nov*	*Dec*	*Jan*	*Feb*	*Mar*	*Apr*
1956	300	300	300	330	350	320	300	280	280
1996	172	155	149	181	260	278	265	245	242

Available at *http://www.antarctica.ac.uk/met/jds/ozone/history.html*

an area larger than the continental U.S. It was possible to make a movie showing the appearance of the hole and its final disappearance in November or thereabouts. This was enormously powerful in terms of convincing the public and politicians that there was a grave environmental problem that needed early and urgent attention. Discovering the Antarctic ozone hole and using satellite imagery to generate spectacular moving pictures of its appearance, evolution, and disappearance played an enormous role in contributing to ratification of the Montreal Protocol. But the mechanism involved in the formation of the hole was initially far from obvious and various theories were put forward, some of them not involving the degradation of CFCs in the stratosphere at all. There were three basic theories: (1) the dynamical theory, which proposed that the ozone loss could be produced by reversal of the normal downwelling circulation over Antarctica in which low-ozone air was lifted up from the troposphere into the ozone hole region (Tung, 1986; Tung *et al.*, 1986); (2) the solar theory, in which solar production of odd nitrogen species initiated the NO_x catalytic cycle for ozone destruction; and (3) halogen chemistry due to CFCs in accordance with the theory of Molina and Rowland. In the original paper by Farman *et al.* (1985) it was suggested that the ozone hole was related to the increase in chlorofluorocarbons in the stratosphere. However, it was quickly realized that the mechanisms then known could not explain the massive loss of ozone occurring at that time.

If the decomposition products of CFCs were involved then they must be acting catalytically for the hole to develop so suddenly and it was proposed that the chemicals were being stored in polar stratospheric clouds in the polar vortex during the winter and suddenly released when the Sun returns to the Antarctic in the spring. Various catalytic reactions were proposed and it was also shown that homogeneous (gas-phase) reactions were not adequate to provide an explanation and that heterogeneous reactions on the surfaces of particles in polar stratospheric clouds must be involved (McElroy *et al.*, 1986; McKenzie *et al.*, 1999; Molina and Molina, 1987; Santee *et al.*, 2005; Solomon *et al.*, 1986). These reactions are strongly temperature dependent, a fact that had led to the small-sized Antarctic ozone hole and its subsequent split in September 2002, due to the unprecedented occurrence of major sudden stratospheric warming (Varotsos, 2002a, b). For each one of these three hypotheses some key measurements were identified. In the case of dynamics, it was very important to measure tracers of atmospheric motion (such as N_2O) to be able to relate ozone to its source region. For the solar theory, examining the possible role of NO_x catalytic cycles and measurements of NO and other NO_y species were critical. Finally, for the halogen cycles, measurements of active chlorine were most important. It was also essential to obtain measurements of ozone itself as well as of various meteorological parameters. The Antarctic ozone hole is the most recognizable part of the ozone depletion problem, and is a classic example of the effects of human-produced compounds on our atmosphere on a very large scale. The Antarctic ozone hole develops during the austral winter and reaches its deepest levels by early spring (late September to early October). The extent of the hole has been assessed from satellites by calculating the average aerial coverage of depletion during the September–October period. Subsequent work has shown that ozone is completely destroyed in the 14–21 km layer by early October (see Figure 5.4).

5.5 THE ROLE OF REMOTE SENSING IN THE LEAD-UP TO THE MONTREAL PROTOCOL

The term "remote sensing" is not restricted to the use of data from instruments flown on aircraft or satellites. The Dobson (or Brewer) spectrophotometer can legitimately be regarded as a remote-sensing instrument because it gathers information about an object or system (i.e., the ozone in the atmosphere) using radiation and not by direct contact with the ozone. Thus, detection of the Antarctic ozone hole using Dobson spectrophotometers was performed by a remote-sensing technique. Moreover, the spectacular demonstration of the extent of the hole and the monitoring of its development, evolution, and decay each year was performed using satellite data from TOMS which was flown on satellites from 1978. This had an enormous impact on the public and on politicians. What was not available before the Montreal Protocol was signed were satellite-flown instruments for detecting and monitoring the other trace gases in the stratosphere that were crucial to testing theories of the mechanism behind formation of the ozone hole. At that stage information had to come from aircraft and balloons. A series of balloon experiments was performed in Texas and they were able to detect chlorine oxide, one of the decomposition products of CFCs and thought to be capable of playing a key role in the destruction of stratospheric ozone (Anderson *et al.*, 1977, 1980; Weinstock *et al.*, 1981).

5.6 THE NOZE AND AAOE EXPEDITIONS

Following the publication of Chubachi (1984, 1985) and Farman *et al.* (1985) and evidence from the satellite data from TOMS, in March 1986 a group of scientists from NOAA and the University of Wyoming met in Boulder (Colorado) to discuss the evidence for ozone depletion. The Wyoming Group had already arranged, through the Office of Polar Programs of the U.S. National Science Foundation, to conduct extensive ozone balloon soundings during the 1986 Antarctic spring of 1986 (August–September). It was decided to extend this to an expedition, the U.S. National Ozone Expedition (NOZE), involving researchers from both NOAA and NASA as well as universities. The main objective was to gain data about what was happening in the ozone hole and decide among the various rival theories as to why and how the hole occurred.

In 1985 there were virtually no observations of the trace gases involved in Antarctic stratospheric ozone chemistry—although some ground-based column measurements of trace gases were being made at Palmer Station (64°46.45′S, 64°03.27′W) on the Antarctic Peninsula (Cronn *et al.*, 1986) by using a scanning UV spectroradiometer. Column ozone abundance was derived from the ratio between measured irradiances at 300 and 313.5 nm by means of theoretical calculation of this ratio as a function of total ozone amount (Cronn *et al.*, 1986; Lubin and Frederick, 1990). It was also known that the lower stratosphere was extremely cold and that polar stratospheric clouds (PSCs) were persistent Antarctic winter features (Douglass and Stanford, 1982; McCormick *et al.*, 1982). Finally, meteorologists knew that the Antarctic stratosphere was contained within an enormous vortex. This vortex forms in the autumn as the polar stratosphere radiatively cools creating a thermal contrast with mid-latitude air. As the cooler polar air sinks, warmer air moves northward to replace the sinking cool air and a strong wind jet forms when the Coriolis force deflects the air eastward. This vortex is sometimes called the polar night jet. Aerosol/ozonesonde measurements by Hofmann *et al.* (1986) showed that air on the cyclonic (poleward) side of the jet had very different characteristics from air equatorward of the jet. It was also apparent from TOMS ozone images that ozone depletion was roughly confined to the interior of the vortex. Observations also showed that there appeared to be a springtime cooling trend in the lower stratosphere that was highly correlated with ozone decrease (Angell, 1986; Chubachi, 1986; Newman and Schoeberl, 1986; Sekiguchi, 1986). These papers summarized the data available to the community prior to the NOZE ground-based expedition in the summer of 1986.

To commence taking measurements before the ozone hole started to develop, the scientists needed to go to McMurdo Station in Antarctica before the end of the winter (i.e., in late August, which was earlier than scientists would usually go there). This involved U.S. Navy flights from Christchurch, New Zealand, which generally were used for operational rather than scientific purposes. The scientific team arrived at McMurdo Station on Ross Island on August 22, 1986 and ozonesonde measurements of the atmospheric profiles of ozone were conducted from August 24 until mid-November using balloons that were specially designed to withstand the very low stratospheric temperatures (down to −90°C). The ozone profiles determined before the ozone hole developed and when the depletion had reached its maximum showed that depletion was almost complete at altitudes between 12 and 20 km; measurements in later years showed that the virtually complete depletion

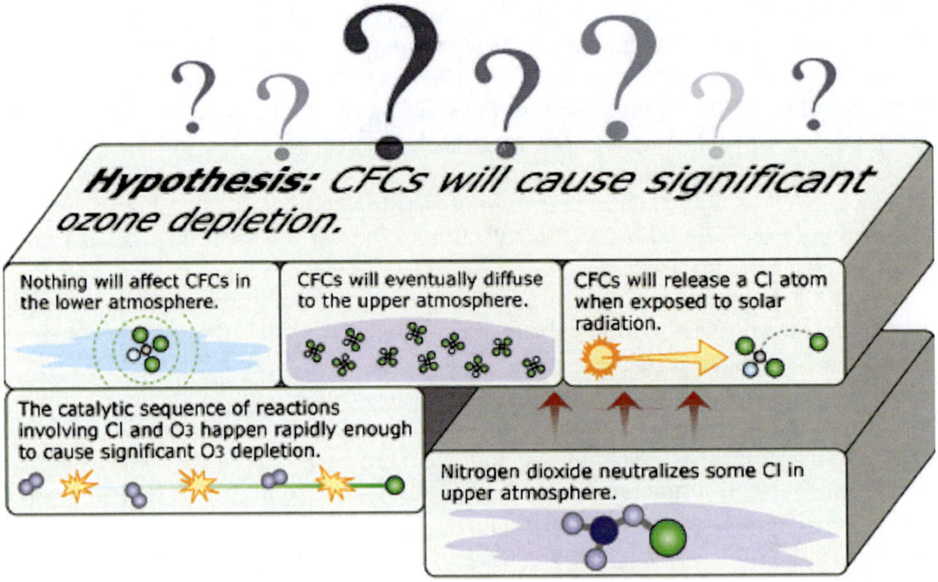

Figure 5.6. Schematic diagram of the role of NO_2 in ozone depletion (*undsci.berkeley.edu*).

extended to an altitude of about 24 km (Figure 5.4) A low concentration of NO_2 would argue against the solar theory, which would need NO_2 molecules to act as a catalyst in ozone desctruction; ClO_2 molecules would be strong evidence in favor of a chemical theory of ozone destruction (Figure 5.6). Ground-based measurements using ultraviolet and visible light spectrometers could detect molecules such as NO_2 and ClO_2. It was shown from balloonborne measurements of stratospheric aerosol particles that the effects of the El Chichón volcanic eruption in 1982 had almost disappeared and could not be related to ozone loss. It was also shown that there was no evidence in particle measurements for the dynamical theory which postulated vertical motions in the polar vortex as being responsible for removing ozone. The concentration of NO_2, which of course can arise from natural or anthropogenic causes, was not found to be sufficient to support the solar theory. The discovery of ClO_2 was strong evidence for a chemical explanation, rather than a dynamical explanation, of the origin of the Antarctic ozone hole and so the initial evidence relating Antarctic ozone depletion to chlorine chemistry was provided by the results from NOZE. It was clear that more data were needed and so NOZE was followed in 1987 by another expedition, the Airborne Antarctic Ozone Expedition (AAOE), which included airborne experiments (see Kurylo, 2009). Preliminary results of the NOZE expedition seemed clearly to favor a chemical explanation, but the final death blow to the dynamical theory only came with the results of the AAOE one year later. Schoeberl and Rodriguez (2009) recalled that in 1985 it was well known that column ozone data could be strongly influenced by the dynamics of the lower stratosphere. They also pointed out that TOMS was originally developed for studying upper-tropospheric motion. Schoeberl and Krueger (1983) showed that Southern Hemisphere large-scale waves were clearly visible in the TOMS data and these ozone waves were caused by the upward and downward motion of the tropopause, which marks the boundary between the stratosphere with its high concentration of ozone and the troposphere with its low concentration of ozone. Low ozone concentration was correlated with lower temperatures and tropopause uplift while higher ozone concentration was correlated with higher temperatures and downwelling.

The AAOE benefited significantly from the first National Ozone Expedition (NOZE). During the AAOE, high-flying aircraft were also used: a DC-8 flying at around 42,000 ft (=12,800 m) and an ER-2 flying at around 60,000 ft (=18,300 m) under very hostile conditions right inside the ozone hole. NOZE had already provided initial evidence relating Antarctic ozone depletion to chlorine chemistry. It was fortunate that NASA had been preparing to conduct the Stratosphere and Troposphere Exchange Project (STEP) in Darwin, Australia, early in 1987. A lot of instrumentation, predominantly for studying ozone and dynamics, had already been assembled and tested for the STEP campaign when the AAOE campaign was formulated. However, it was decided that some additional instrumentation would be needed, which included the Harvard ClO/BrO instrument which had been flown in high-altitude balloons but had never been flown on an aircraft. It was a major achievement to get this instrument working on a high-altitude aircraft in such a short period of time (see Roan, 1989, pp. 185–188).

In the austral spring of 1987, two major field missions were mounted to explore the ozone hole phenomenon: the AAOE and NOZE II (the second NOZE campaign) based, respectively, at Punta Arenas Airfield, Chile, on the Straits of Magellan, and at McMurdo Base, Antarctica. In addition, scientists from the Geophysical Monitoring for Climatic Change program of the National Oceanic and Atmospheric Administration supplied their regular ozonesonde data from the South Pole. The missions generated papers which were accepted for a *Journal of Geophysical Research* Special Issue as follows: 52 from AAOE, eight from NOZE II, one from NOZE I, and one from GMCC (NOAA) (Tuck *et al.*, 1989).

The success of the AAOE campaign lay in the fact that it provided important corroboration of the results of NOZE. From a series of flights from mid-August to mid-September it was possible to see the build-up of ClO without significant change in ozone until the return of sunlight to the polar region, when the catalytic cycle could commence and the destruction of ozone could begin.

By the time of the AAOE and NOZE II the Montreal Protocol had already been signed. At the time the Protocol was signed "there was no observed ozone loss directly linked to CFCs that could be put on the table" and it was signed "on the belief—on the theoretical prediction originated by Rowland and Molina—that if chlorine continued to increase then, ultimately, there would be a substantial loss of ozone particularly in the winter and in the high latitude" (Roan, 1989, p. 207).

Following the AAOE there were several further airborne campaigns, not just in the Antarctic but in the Arctic as well. These included (Pommereau, 2009):

- The first Airborne Arctic Stratosphere Expedition (AASE 1) from Stavanger (Norway) conducted in 1989 by NASA when it was found that active chlorine abundances in the Arctic approached those measured in the Antarctic, followed two years later by AASE II based in Bangor (Maine).
- The Franco-German CHEOPS (CHEmistry of Ozone in the Polar Stratosphere) campaign in Kiruna in northern Sweden in early 1987, followed by CHEOPS II and CHEOPS III (1989–1990) (Pommereau and Schmidt, 1991).
- The European Arctic Stratospheric Ozone Experiment (EASOE) (see Section 2.6.1) was carried out in 1991–1992 (Pyle *et al.*, 1994), followed by the Second European Stratospheric Arctic and Mid-latitude Experiment (SESAME) (see Section 2.6.2), and the Third European Stratospheric Experiment on Ozone) (THESEO) (see Section 2.6.3) in 1998–1999.
- The POLARIS (Photochemistry of Ozone Loss in the Arctic Region In Summer) campaign in the summer of 1997 was designed to understand the seasonal behavior of polar ozone as it changed from high concentrations in the spring to very low concentrations in the autumn, attributed to the increased role of NO_x catalytic cycles during periods of prolonged solar illumination in the summer period (Kurylo, 2009).
- SOLVE-THESEO (see Section 2.6.4), the first joint U.S.–Europe effort to study Arctic ozone loss, was an effort to extend THESEO beyond the winter of 2000, after the offer from NASA to participate in validating the SAGE II instrument which was launched in late 1999 on the Russian METEOR satellite (see Section 3.5.1) as well as the measurements of all satellites in operation at that time: in fact, SAGE-METEOR was not launched in time for the campaign, but the decision to carry on was made since it offered a unique opportunity for joint measurements from a variety of platforms using the best instruments developed on both sides of the Atlantic during the previous 15 years. With the improved and more extensive instrumentation of SOLVE/THESEO it was possible to map ozone loss rates as a function of NO_2. Thus, it was possible to characterize the interplay between the NO_x and ClO_x catalytic cycles in the depletion of ozone. One of the additional key findings was the discovery of some rather large ice particles, now commonly referred to as nitric acid trihydrate (NAT) rocks, which may play a significant role in denitrification processes and (depending on seasonal conditions) could actually lead to enhanced ozone loss in the Arctic (Kurylo, 2009).
- The SOLVE II-VINTERSOL campaign in January–February 2003. This joint U.S./European effort further refined our understanding of the role of meteorological variability on the susceptibility of the Arctic region to significant seasonal ozone depletion. The campaign also marked the use (from a U.S. perspective) of heavy-lift high-altitude balloon payloads in conjunction with research aircraft to study polar ozone. Lidar measurements from the NASA DC-8 played an

important role in quantifying the amount of ozone loss during that Arctic winter (Kurylo, 2009).

The results of these various European campaigns were presented in a number of workshops/symposia (WMO, 1990, 1991, 1995, 1999, 2003).

At the time of the original NOZE and AAOE expeditions, identifying the chemicals present in the polar vortex could only be done with ground-based and airborne experiments. Satellite data about the various trace gases only became available later on, particularly as a result of the development and launch of the Upper Atmosphere Research Satellite (UARS), which was launched in 1991. This satellite was constructed very largely to help in the understanding and monitoring of the ozone problem. As we now know, to understand the ozone hole required not only confirmation of the presence of the decomposition products of CFCs in the stratosphere but also an extensive study of heterogeneous chemical reactions, which are now believed to be occurring on the surfaces of particles in polar stratospheric clouds.

5.7 THEORIES OF THE OZONE HOLE

Under "normal" mid-latitude conditions, the primary catalytic cycle by which chlorine destroys ozone is

$$Cl + O_3 \rightarrow ClO + O_2 \tag{5.11}$$

$$ClO + O \rightarrow Cl + O_2 \tag{5.12}$$

$$O + O_3 \rightarrow 2O_2 \tag{5.4}$$

In the above reactions, (5.12) is the "rate-limiting step" (i.e., the reaction that determines the overall rate of ozone removal).

If chlorine released by chlorofluorocarbons (CFCs) is the mechanism to explain the rapid Antarctic ozone decrease, then there are two problems to resolve. In the first place, over 98% of the chlorine released from CFCs is converted to the non-reactive reservoir species HCl and $ClONO_2$. Thus, a mechanism must be found to dramatically alter the partitioning of chlorine in the Antarctic stratosphere in favor of the reactive radical species ClO. In addition, the above catalytic cycle requires sufficient concentrations of atomic oxygen, but the required concentrations to complete (5.12) are not present under the low solar zenith angle conditions of the Antarctic spring. So, the second problem to be solved is to find a catalytic cycle that does not require substantial amounts of atomic oxygen.

Following the publication of Farman *et al.* (1985), *Nature* published two papers that hypothesized that the ozone hole had a chemical cause. To solve the first problem, Solomon *et al.* (1986) and McElroy *et al.* (1986) suggested that more radical chlorine could be made available for ozone loss by moving it from the reservoir gases, HCl and $ClONO_2$, through heterogeneous reactions on the surfaces

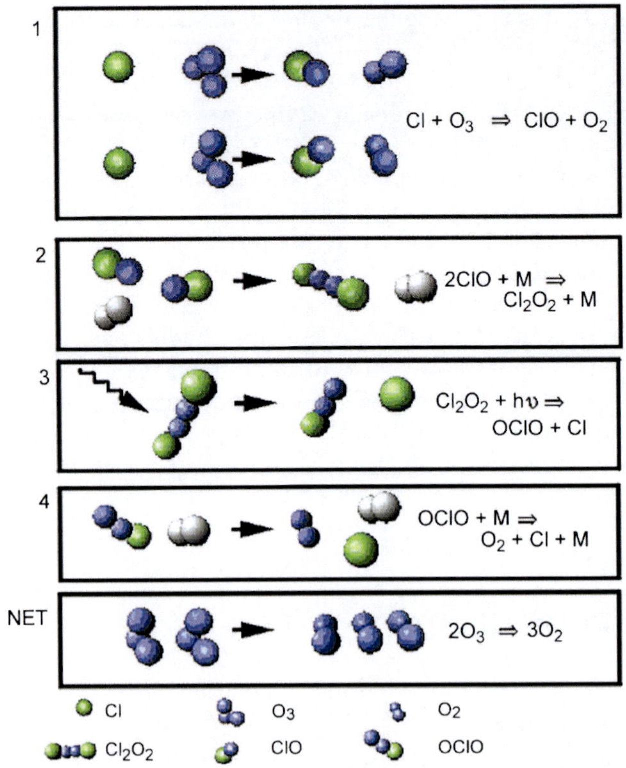

Figure 5.7. Schematic diagram of polar ozone destruction (*ccpo.odu.edu*).

of polar stratospheric clouds (PSCs). Normally, these two gases react slowly, if at all, but the following heterogeneous reaction could speed up the process:

$$HCl\ (particle) + ClONO_2\ (particle) \rightarrow Cl_2 + HNO_3 \qquad (5.13)$$

Then, upon the return of sunlight at the beginning of Antarctic spring, Cl_2 photolyzes to yield reactive chlorine (Figure 5.7). Of course, a problem with this mechanism is how to prevent reformation of the reservoir gas $ClONO_2$. The HNO_3 formed in (5.13) will also photolyze in weak sunlight forming NO_2, which can then react with ClO reforming $ClONO_2$. The net result is that the reservoir shifts from HCl and $ClONO_2$ to mostly $ClONO_2$, which would reduce the reactive chlorine. The heterogeneous reaction could partly solve this problem if the HNO_3 remains on the ice, thus preventing it from competing in gas-phase reactions. In addition, both of the above papers suggested the heterogeneous reaction

$$ClONO_2 + H_2O\ (particle) \rightarrow HOCl + HNO_3 \qquad (5.14)$$

as a way to recycle $ClONO_2$ back to nitric acid and, in addition, producing reactive chlorine by photolysis of HOCl. Laboratory studies suggested that both (5.13) and (5.14) occur rapidly on the surfaces of reacting vessels (Molina *et al.*, 1985; Rowland *et al.*, 1986), but no information was available on the rates of these reactions on typical PSCs, but both McElroy *et al.* (1986) and Solomon *et al.* (1986) assumed rates sufficiently rapid to effect the necessary chlorine repartioning.

The second problem is finding a catalytic cycle that does not need significant atomic oxygen. To solve this second problem, Solomon *et al.* (1986) suggested the following cycle:

$$HO + O_3 \rightarrow HO_2 + O_2 \tag{5.5}$$

$$Cl + O_3 \rightarrow ClO + O_2 \tag{5.11}$$

$$HO_2 + ClO \rightarrow HOCl + O_2 \tag{5.15}$$

$$HOCl + h\nu \rightarrow OH + Cl \tag{5.16}$$

of which the net result is

$$2O_3 \rightarrow 3O_2 \tag{5.17}$$

with photolysis of HOCl providing the rate-limiting step. Even though there was some uncertainty in the kinetic data available at the time, the observed ozone loss rate could not be explained through the above cycle without modification of the reaction rates. Nonetheless, it should be noted that the proposed cycle does not require atomic oxygen.

McElroy *et al.* (1986) suggested the cycle

$$Br + O_3 \rightarrow BrO + O_2 \tag{5.18}$$

$$Cl + O_3 \rightarrow ClO + O_2 \tag{5.11}$$

$$BrO + ClO \rightarrow BrCl + O_2 \tag{5.19}$$

$$BrCl + h\nu \rightarrow Br + Cl \tag{5.20}$$

which also has the net result

$$2O_3 \rightarrow 3O_2 \tag{5.17}$$

This mechanism would require BrO concentrations of about 25 pptv. Later stratospheric measurements showed that concentrations of BrO are closer to 5 pptv (Brune *et al.*, 1988), such that this mechanism actually ends up accounting for about 20% of the ozone loss, but not all the loss that is observed. The reaction of BrO + ClO can also occur through other channels, one of which produces ClO_2, which, when photolyzed, would interrupt the cycle. On the other hand, since ClO_2 absorption lines are detectable in the visible, a sensitive spectrometer should be able to detect anomalous amounts of ClO_2 in the stratosphere. In 1986 NOZE included instruments to measure column ClO_2 (Solomon *et al.*, 1987) and ClO amounts (de Zafra *et al.*, 1987). In the absence of any other mechanisms, detection of ClO_2 and high levels of ClO would probably support the chemical theories or some variant.

5.8 DIPLOMACY, 1974–1989; FORMULATION AND RATIFICATION OF THE MONTREAL PROTOCOL

Finding a solution to the ozone problem, which affects the whole world, requires being done in a global context. This task was undertaken by the United Nations Environment Program (UNEP), with its headquarters in Nairobi and led at that time by Mostafa Tolba, an Egyptian scientist. The chronology of events from the initial work of Molina and Rowland in late 1973 to January 1989 when the Montreal Protocol came into force has already been summarized in Table 5.1 which was compiled from chronologies in the book by Roan (1989) and the book by Benedick (1991). Richard Benedick was a U.S. diplomat and headed U.S. preparations that led to the Montreal Protocol. His book gives a detailed account of the various stages in the process from a postulate by two scientists that the world's ozone layer was in danger from man-made CFCs to an international agreement to protect the ozone layer. We shall not go through all the stages in detail but merely highlight one or two points from that story. The book by Roan (1989) is more concerned with the science and restricts the political discussion to events in the U.S.A.

The first expression of intergovernmental concern about "modification of the ozone layer due to human activities and some possible geophysical consequences" was issued as a formal WMO Statement in 1975. The WMO Seventh World Meteorological Congress, which was attended by representatives from over 130 governments, stated in May 1975: "... there is an urgent need for more studies (and for a definite review of these studies) to determine the extent to which man-made pollutants might be responsible for reducing the quantity of ozone in the stratosphere." The Congress also "stressed the need to determine the role played by chlorofluoromethanes (used in refrigerants and aerosol cans) in destroying ozone."

Furthermore, the WMO decided to initiate the WMO Global Ozone Research and Monitoring Project with the main thrust of encouraging the study of ozone-destroying compounds, the calibration of instruments, data re-evaluation, the assessment of ozone changes, and advising governments on the state of the ozone layer (Bojkov and Balis, 2009). In 1976, the UNEP Governing Council called for action. In March 1977, the UNEP Executive Director Dr. Mostafa Tolba called a meeting of experts designated by governments and intergovernmental organizations in Washington, D.C. An overview of the state of the ozone layer was presented by the WMO. The meeting recommended a World Plan of Action (in which the WMO would be responsible for monitoring, research, and assessments and UNEP for a broad catalytic role and attending to legal aspects). The UNEP Governing Council approved the proposed Plan of Action and the establishment of a Coordination Committee on the Ozone Layer (CCO3L); annual meetings started in 1977 and continued until 1986. Based on the contributions of hundreds of scientists and reviews by the WMO, negotiations for the Global Framework Convention, initiated by the Governing Council (GC) of UNEP, started in 1981 with the first meeting of the Working Group (WG) of Legal and Technical Experts for the Elaboration of the

Framework Convention held in Stockholm in January 1982, guided by the former Executive Director of UNEP, Dr. Mostafa Tolba.

UNEP began by funding a WMO technical conference on research on CFCs and stratospheric ozone carried out in the U.S.A. This meeting produced the first official statement of international scientific concern about CFCs and atmospheric ozone (WMO, 1975). In 1977 UNEP sponsored a meeting of governments and international agencies in Washington, D.C., which drafted a "World Plan of Action on the Ozone Layer" that recommended intensive international research and monitoring of the ozone layer and gave UNEP a central co-ordinating responsibility for promoting research and gathering relevant economic and scientific data (UNEP, 1977). Moreover, UNEP established the Coordinating Committee on the Ozone Layer (CCOL) which produced periodic reports. The question of introducing international controls on CFC production was formally raised at an intergovernmental meeting hosted by the U.S.A. in Washington, D.C., in April 1977. Various subsequent meetings were convened to discuss the science, to discuss possible restrictions on the manufacture of CFCs, and to discuss the legal framework for a convention to protect the ozone layer. A significant stage came with a meeting in Vienna of representatives of 43 nations in March 1985 at which a convention, the Vienna Convention, was signed which placed an obligation on nations to take appropriate measures to protect the ozone layer, but did not define what these measures might be. In the words of Benedick (1991) "The Vienna Convention was itself a considerable accomplishment. It represented the first effort of the international community formally to deal with an environmental danger before it erupted." The text of the Vienna Convention is reproduced in Appendix A of the book by Benedick (1991). At this meeting there was a resolution, which was not part of the Convention, that authorized UNEP to convene a workshop to pave the way for achieving a legally binding protocol by 1987. There are a lot of interesting historical details associated with the final stages leading up to the signing of the Montreal Protocol and they are described in the book by Benedick (1991); the text of the Protocol itself will be found in Appendix B of that book.

The point was made by Benedick (1991) that "At the time of the negotiations and signing, no measurable evidence of damage existed. Thus, unlike environmental agreements of the past, the treaty was not a response to harmful developments or events but rather a preventive action on a global scale." Later, Benedick (2009) elaborated on this:

> "Unlike prior environmental agreements, the Montreal Protocol was not a response to existing conditions, but rather preventive action on a global scale. By the time the evidence on issues such as ozone layer depletion and climate change [was] beyond dispute, the damage could be irreversible. It would be too late to avoid serious harm to human life and the environment as well as draconian future costs to society. Thus, political leaders need[ed] to act even while there [were] still scientific ambiguities and uncertainties, based on a responsible balancing of the risks and costs of delay."

Substantial evidence of the damage did, of course, emerge later on and the particular role played by remote sensing in monitoring ozone depletion, studying the effectiveness of the Montreal Protocol in preventing ozone depletion, and in studying the chemical processes involved in the destruction of stratospheric ozone was presented in a special issue of the *International Journal of Remote Sensing* (Volume 30, Issues 15 and 16, 2009).

5.9 REASONS FOR THE SUCCESS IN REACHING INTERNATIONAL AGREEMENT IN MONTREAL

Producing a document, the Montreal Protocol, which was eventually ratified by all nations, was a major achievement. Ever since its ratification, it has proved in the following 20 years to be very successful in terms of moving towards its objective: namely, protecting and restoring the ozone layer.

Benedick (1991, chapter 1, pp. 5–7) analyzes the reasons for the success in reaching international agreement in Montreal. He cites several key factors, which we summarize below, and finds some pointers in them for people who are trying to find solutions to other major environmental problems, particularly global warming. The factors he identifies are:

- The *indispensable role of science* in ozone negotiations. Scientists were drawn out of their laboratories and into the political arena. Political and economic decision makers needed to understand the scientists, to fund the necessary research, and to be prepared to undertake internationally co-ordinated actions based on realistic and responsible assessments of risk.
- The *power of knowledge and public opinion* was a formidable factor in the achievement at Montreal. A well-informed public was the prerequisite to mobilizing the political will of governments and to weakening industry's resolve to defend the continued manufacture of CFCs.
- Because of the global scope of the issues, the *activities of a multinational institution* were critical to the success of negotiations. This role was played by UNEP, and Dr. Mostafa Tolba, the executive director of UNEP, was the driving force behind the success of the negotiating process.
- An *individual nation's policies and leadership* made a major difference. This leadership was provided by the United States. The U.S. government was the first to take action against CFCs and played a major role in developing the Montreal Protocol and campaigning for its acceptance by the international community. Other countries that played leading roles were Canada, Germany, the Nordic countries, Australia, and New Zealand.
- *Private sector organizations* (i.e., environmental groups and industry) participated actively, though with different objectives. Environmental groups warned of the risks, lobbied governments to act, and promoted research and legislation. Industrial organizations initially sought to oppose restrictions on the

manufacture and use of CFCs, but later co-operated in finding substitutes that did not damage the ozone layer.

- The *process* that was adopted in reaching the ozone accord was itself a determining factor. This involved subdividing the problem into manageable components during a pre-negotiating phase. Extensive preliminary scientific and diplomatic groundwork enabled subsequent negotiations to move forward relatively rapidly.
- Once the serious nature of ozone depletion was realized, the crucial and successful discussions of 1986–1987 were both small in attendance and short in duration by later standards. For example, the first round of one-week negotiations in 1986 was attended by only 20 nations and three non-governmental organizations (NGOs), while at the decisive 1987 Montreal Conference itself, there were only about 60 national delegations (see Benedick, 1991 for details).

Benedick makes another important point. This is that the Montreal Protocol is not like traditional international treaties which seek to cement the *status quo*. It was deliberately designed as a flexible and dynamic instrument which can readily be adapted to evolving conditions and developing scientific knowledge (Benedick, 1991; Gonzalez, 2009; Sarma and Taddonio, 2009). Every year there are Meetings of the Parties (i.e., the countries that have ratified the Protocol), and over the years these Meetings have made substantial modifications to the Protocol based on scientific, economic, environmental, and technological assessments of changing situations. So far there have been four amendments, the London Amendment (in 1990, in force in 1992), the Copenhagen Amendment (in 1992, in force in 1994), the Montreal Amendment (in 1997, in force in 1999), and the Beijing Amendment (in 1999, in force in 2002). By the end of 2011, the dedicated Multilateral Fund (MLF) will have provided nearly U.S.$3 billion in "agreed incremental costs" in 3-year cycles based on an independent calculation by the Protocol's Technology and Economic Assessment Panel (TEAP) of the funding needed to ensure that developing country Parties can meet their mandatory obligations for ODS control measures (Molina *et al.*, 2009). Facing the threat of sea level rise (Velders *et al.*, 2007), a group of island nations led by the Federated States of Micronesia and Mauritius proposed an amendment to the Montreal Protocol that would provide jurisdiction over the production and consumption of HFCs and that would use the technical expertise and administrative structure of that treaty to start quickly phasing out HFCs with high GWP (Molina *et al.*, 2009). The United States, Canada, and Mexico supported them by making a similar joint proposal (Molina *et al.*, 2009).

Since then several Meetings of Parties (MOPs) have taken place:

— *MOP-15 and the first Extraordinary MOP*: MOP-15, held in Nairobi (Kenya), in 2003, resulted in decisions on issues including the implications of the entry into force of the Beijing Amendment. Delegates could not reach agreement and took the unprecedented step of calling for an "extraordinary" MOP. The first Extraordinary Meeting of Parties to the Montreal Protocol (ExMOP-1) took place in March 2004, in Montreal (Canada). Parties agreed to critical use exemptions

(CUEs) for methyl bromide for 2005 only. The introduction of a "double-cap" concept distinguishing between old and new production of methyl bromide was central to this compromise. Parties agreed to a cap on new production of 30% of parties' 1991 baseline levels, meaning that where the capped amount was insufficient for approved critical uses in 2005, parties were required to use existing stockpiles.

— *MOP-16 and Ex-MOP-2*: MOP-16 took place in Prague (Czech Republic), in 2004. Work on methyl bromide exemptions for 2006 was not completed and parties decided to hold a second ExMOP. ExMOP-2 was held in July 2005, in Montreal (Canada). Parties agreed to supplementary levels of critical use exemptions (CUEs) for 2006. Under this decision, parties also agreed that: CUEs allocated domestically that exceed levels permitted by the MOP must be drawn from existing stocks; methyl bromide stocks must be reported; and parties must "endeavor" to allocate CUEs to the particular use categories specified in the decision.

— *COP-7/MOP-17*: MOP-17 was held jointly with the seventh Conference of Parties to the Vienna Convention (COP-7) in Dakar (Senegal) in December 2005. Parties approved essential use exemptions for 2006 and 2007, supplemental CUEs for 2006 and CUEs for 2007, and production and consumption of methyl bromide in non-Article 5 parties (i.e., developed/industrialized countries) for laboratory and analytical critical uses. Other decisions included a U.S.$470.4 million replenishment of the MLF for 2006–2008, and agreement on terms of reference (ToR) for a feasibility study on developing a monitoring system for the trans-boundary movement of controlled ODS.

— *MOP-18*: MOP-18 took place in New Delhi (India) from October 30–November 3, 2006. Parties adopted decisions on, *inter alia*: future work following the Ozone Secretariat's workshop on the Special Report of the Intergovernmental Panel on Climate Change and the Technical and Economic Assessment Panel (TEAP); difficulties faced by some Article 5 parties manufacturing CFC-based metered dose inhalers (MDIs); treatment of stockpiled ODS relative to compliance; and a feasibility study on developing a system for monitoring the trans-boundary movement of ODS.

— *MOP-19*: MOP-19 took place in Montreal (Canada) in September 2007. Delegates adopted 29 decisions, including: accelerated phase-out of HCFCs; essential use nominations and other issues arising out of the 2006 reports of the TEAP; critical use nominations for methyl bromide; and monitoring trans-boundary movements and illegal trade in ODS.

— *COP-8/MOP-20*: MOP-20 was held jointly with COP-8 of the Vienna Convention in Doha (Qatar) in November 2008. Parties agreed to replenish the MLF with U.S.$490 million for 2009–2011 and adopted other decisions concerning, *inter alia*: the environmentally sound disposal of ODS; approval of 2009 and 2010 CUEs for methyl bromide; and compliance and reporting issues. This meeting was also the Protocol's first paperless meeting.

— *MOP 21*: MOP-21 took place in Port Ghalib (Egypt) from November 4–8, 2009 and adopted decisions on: alternatives to HCFCs; institutional strengthening;

essential uses; environmentally sound management of banks of ODS; methyl bromide; budget; and data and compliance issues. Delegates considered a proposal to amend the Montreal Protocol to include HFCs, but this was not agreed.

— *MOP 22*: MOP-22 took place in Bangkok (Thailand) from November 8–12, 2010. MOP-22 adopted 16 substantive and several procedural decisions, including: the terms of reference (ToR) for the Technical and Economic Assessment Panel study on the replenishment of the Multilateral Fund; the ToR for the evaluation of the financial mechanism; assessment of technologies for ozone-depleting substance (ODS) destruction; budget; and data and compliance issues. MOP-22 was not able to make progress on low global warming potential alternatives, or ODS destruction, which many delegates said were issues key to the long-term future of the Protocol. Although the draft decisions to amend the Montreal Protocol to include hydrofluorocarbons (HFCs) were not successful, their proponents remained committed to addressing HFCs through the Montreal Protocol in the future.

Benedick quotes Dr. Mostafa Tolba of UNEP as saying (rather optimistically) "The mechanisms we design[ed] for the [Montreal] Protocol will—very likely—become the blueprint for the institutional apparatus designed to control greenhouse gases and adaptation to climate change." However, such early optimism appears to have been unfounded. In more recent papers Benedick (2007, 2009) pointed out that the success of the negotiators of the Montreal Protocol has not, in fact, been followed up in more recent climate negotiations aimed at finding a successor to the Kyoto Protocol. Many ODSs are also greenhouse gases (see also Section 5.3) and it has been argued (Velders *et al.*, 2007) that the climate protection already achieved by the Montreal Protocol alone is far larger than the reduction target of the first commitment period of the Kyoto Protocol. Moreover, additional climate benefits that are significant compared with the Kyoto Protocol reduction target could be achieved by actions under the Montreal Protocol, by managing the emissions of substitute fluorocarbon gases and/or implementing alternative gases with lower global warming potentials (Velders *et al.*, 2007). There has been a long-running succession of United Nations global environmental mega-conferences and negotiations on climate change which have been held every year since 1995 and which involve several thousand official delegates from over 190 countries, together with hundreds of NGO representatives—and all in the full glare of the world's media. These conferences achieve very little and Benedick (2007, 2009) has argued that there are several useful lessons which could have been learned from the history of the Montreal Protocol. In particular, he argues that small more focused meetings taking place outside the glare of media publicity could achieve more success. He has suggested that climate negotiations should be disaggregated, or separated, into a number of components such as: energy technology research and development; emission reduction policies for particular sectors (e.g., transportation, agriculture and forestry) and their adaptation to changing climatic conditions; regional co-operation to combat or accommodate climate change. It can be argued that there is no reason why every aspect of a complicated scientific and environmental problem must be addressed by every nation at the same time and in the same place. This is particularly true for climate change negotiations. In reality, only 24 nations (half of them "developing" countries) together account for about 80% of global greenhouse gas emissions, while the remaining 170 nations each contribute less than a fraction of 1% (Baumert *et al.*, 2005).

5.10 RATIFICATION OF THE MONTREAL PROTOCOL

The Montreal Protocol was eventually ratified by virtually all countries; the status of ratification as at November 25, 2009 is shown in Table 5.3. It is widely held (see, e.g., Sarma and Taddonio, 2009) that it is one of the most successful international treaties for three clear reasons:

(1) it is based on science that is constantly updated and taken on board;
(2) it applies the "precautionary principle" in setting technically and administratively challenging but feasible goals for phasing out ozone-destroying substances (ODSs) *before* the adverse impacts of ozone become catastrophic; and
(3) it is successful at motivating the development, commercialization, and transfer of technology to developing countries, which largely have not been responsible for causing the problem of ozone depletion.

A very good example of the updating of the Protocol is that it started in 1987 with only a freeze on halon production and a 50% reduction in CFCs. "Halon" is a general term for a derivative of methane CH_4, or ethane C_2H_6, where one or more of the hydrogen atoms have been replaced by halogen atoms. CFCs are particular examples where all the H atoms have been replaced by fluorine or chlorine atoms. Wikipedia lists a number of halons that have been used in agriculture, dry cleaning,

Table 5.3. Ratification of the Montreal Protocol.

Instrument	*Number of countries having ratified*
Vienna Convention	196
Montreal Protocol	196
London Amendment	194
Copenhagen Amendment	191
Montreal Amendment	179
Beijing Amendment	161

As at November 25, 2009 (UNEP Ozone Secretariat).

Table 5.4. Halons used in various applications.

Halon 10001	Iodomethane (methyl iodide)	CH_3I
Halon 1001	Bromomethane	CH_3Br
Halon 1011	Bromochloromethane	CH_2ClBr
Halon 104	Carbon tetrachloride	CCl_4
Halon 1103	Tribromofluoromethane	$CFBr_3$
Halon 112	Dichlorofluoromethane	$CHFCl_2$
Halon 1201	Bromodifluoromethane	CHF_2Br
Halon 1202	Dibromodifluoromethane	CF_2Br_2
Halon 1211	Bromochlorodifluoromethane	CF_2ClBr
Halon 122	Dichlorodifluoromethane	CF_2Cl_2
Halon 1301	Bromotrifluoromethane	CF_3Br
Halon 14	Tetrafluoromethane	CF_4
Halon 242	1,2-Dichlorotetrafluoroethane	$C_2F_4Cl_2$
Halon 2402	Dibromotetrafluoroethane	$C_2F_4Br_2$
Halon 2600	Hexafluoroethane	C_2F_5

fire suppression, and other applications (see Table 5.4). The substances initially included in the Montreal Protocol in 1987 are listed in Table 5.5. A useful summary of the chemical formulae and nomenclature of CFCs and related compounds is given in Annex V of an IPCC Special Report (IPCC, 2005). New substances have been added to the Montreal Protocol's list from time to time, and at present the Protocol aims to phase out 96 ozone-depleting substances (ODSs) that are used, or have been used, in thousands of products; it provides for this to be done by changing to ozone-safe technologies throughout the world in a specified time frame. In particular, hydrochlorofluorocarbons (HCFCs), which were originally seen as substitutes for CFCs, were later added to the Montreal Protocol's list of substances; they are being phased out more because they are powerful greenhouse gases rather than because of damage they might cause to the ozone layer. HCFCs were scheduled to be phased out by non-Article 5 parties (i.e., developed/industrialized countries) by 2030 and 10 years later by Article 5 parties (i.e., developing countries and countries with emerging economies, like the post-Soviet Union collapse countries). However, at the Meeting of Parties in Montreal in September 2007, the Parties approved an adjustment to accelerate the phasing out of HCFCs.

Table 5.5. Substances originally included in the Montreal Protocol.

Group	*Substance*	*Ozone-depleting potential*[a]
	Group I	
CFC-11	$CFCl_3$	1.0
CFC-12	CF_2Cl_2	1.0
CFC-113	$C_2F_3Cl_3$	0.8
CFC-114	$C_2F_4Cl_2$	1.0
CFC-115	C_2F_5Cl	0.6
	Group II	
Halon-1211	CF_2ClBr	3.0
Halon-1301	CF_3Br	10.0
Halon-2402	$C_2F_4Br_2$	6.0

[a] These are the values of the ozone-depleting potentials estimated in 1987; they are subject to revision in the light of subsequent knowledge (Montreal Protocol, Appendix A).

According to Sarma and Taddonio (2009), more than 240 industrial sectors that used ozone-depleting substances in their products had replaced most of them within 10 years. Sarma and Taddonio (2009) said:

> "This was a no-compromise market transformation: in addition to being ozone-safe, alternatives and substitutes are equally safe or safer for the Earth's climate, more energy-efficient, lower in toxicity, superior in safety, and more reliable and durable. Often, alternatives have reduced costs to businesses and increased employment. Corporate, military, environmental, and citizen stakeholders are proud of what they have accomplished through consensus, cooperation, and regulation.
>
> Generally speaking the replacement of ozone depleting substances was accomplished with no compromise in environmental health and safety, and in a majority of cases all aspects of environmental performance were improved."

Complying with the demands of the Montreal Protocol was a major task even in the developed countries. The governments of developed countries took various steps to comply with the Montreal Protocol; these included the adoption of policies, making regulations, conducting awareness and education campaigns, and introducing financial incentives and disincentives. Not only were ozone-depleting substances used by major sectors of industry, they were also used by numerous smaller sectors. A further complication was that there were various laws and policies on safety requirements that were formulated in terms of using substances that were later found to be ozone-depleting substances:

> "Fire safety laws for weapons systems, aircraft, ships, and racing cars, for example, often mandated the use of halon fire suppression systems. Quarantine standards often encouraged or required methyl bromide for certain pests and products. CFC refrigerants were required by law in some areas, and both industrial and military cleaning standards required cleaning with CFC-113. In these cases, enterprises and governments seeking change organized strategic projects to remove specific barriers" (Sarma and Taddonio, 2009).

Complying with the Montreal Protocol was rather different for developing countries. First, their production and use of ozone-depleting substances was relatively small. Second, they were able to argue—not unreasonably—that it was the developed/industrialized nations that were largely responsible for having damaged the ozone layer. Compliance was only achieved because money was made available to developing countries to help them to comply with the Protocol. The Montreal Protocol provided for technology transfer to developing countries. According to Article 10A of the Protocol,

> "each Party shall take every practicable step, consistent with the programs supported by the financial mechanism, to ensure that the best available, environmentally safe substitutes and related technologies are expeditiously transferred to Parties operating under paragraph 1 of Article 5 (i.e. developing countries) [and that] transfers ... occur under fair and most favourable conditions."

Financial assistance to developing countries was therefore a critical part of institutional arrangements; the Protocol created a financial mechanism to assist developing countries to implement the control measures. This was done by the Multilateral Fund (MLF) for the implementation of the Montreal Protocol on substances that Deplete the Ozone Layer. This was established under Article 10 of the Montreal Protocol in 1990 to help developing countries to phase out the production and consumption of ozone-depleting substances. The Fund started out with U.S.$240 million for the period 1991–1993. The Fund has been replenished every 3 years after needs assessments made by MOPs. The three-year replenishments rose to U.S.$400 million by 2009–2011. By November 7, 2008 the Executive Committee of the MLF had approved the expenditure of U.S.$2.4 billion to support about 5,789 projects and activities in 144 developing countries, planned to result in phasing out consumption of more than 427,000 ODP t (see Section 6.2.1) and the production of about 173,616 t of ozone-depleting substances.

Sarma and Taddonio (2009) cited a number of reasons for the success of the Montreal Protocol:

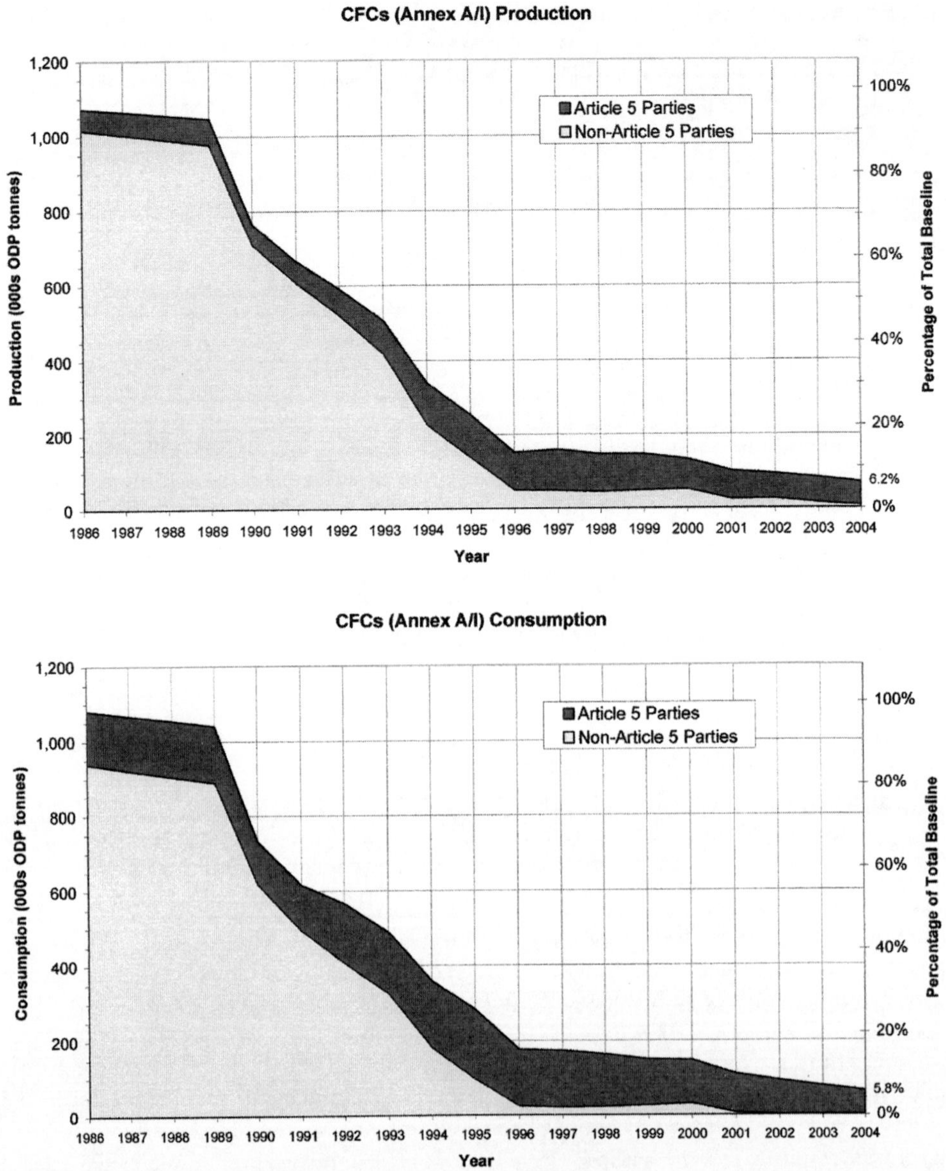

Figure 5.8. (a) Production and (b) consumption trends for (Annex A/I) CFCs (UNEP, 2005).

- One very important factor was leadership from a wide variety of organizations and people from governments, international organizations, non-governmental organizations, industry associations, scientists, engineers, and many others who took early action to tackle the problem of ozone depletion and inspired others to follow their example.

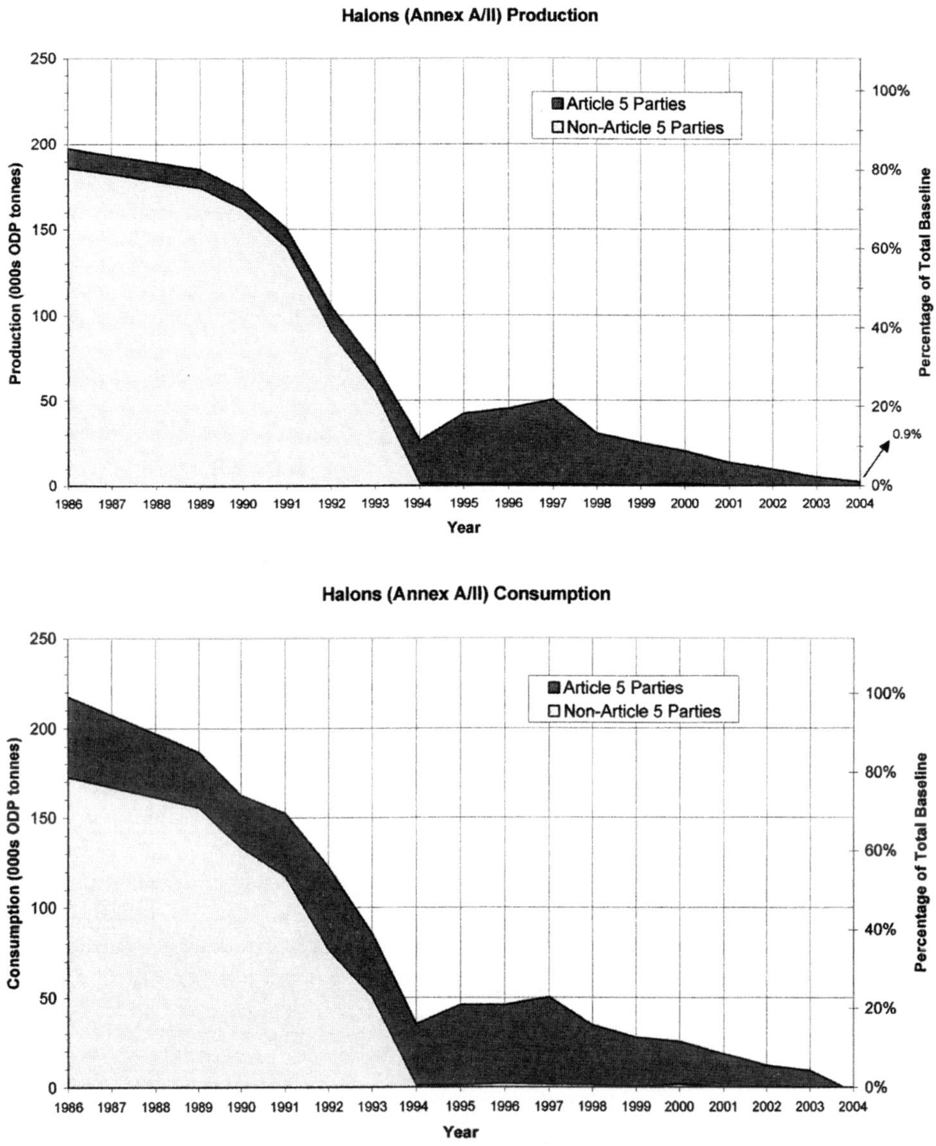

Figure 5.9. (a) Production and (b) consumption trends for halons (UNEP, 2005).

- Institutional arrangements were also important, including the Ozone Secretariat at UNEP in Nairobi and the individual Ozone Units—focal points for action on the Montreal Protocol in each of its Parties. Ozone Unit focal points are a dedicated group of professionals who understand their mission and have formed both formal and informal networks with the broader ozone community to carry it out.

- Another very important factor was the acceptance of the "principle of common but differential responsibility". Thus, it was acknowledged that it was the developed countries which had produced and used most of the CFCs and therefore were largely responsible for the problem of ozone depletion and that the developing countries were much less responsible for the problem. It was recognized that the developed countries possessed greater resources, in terms of technological and financial resources, to tackle the problems involved in complying with the Protocol. Thus, developing countries were given longer than developed countries to phase out the manufacture and use of ozone-depleting substances and they were provided with financial assistance via the MLF (see above).
- Article 6 of the Montreal Protocol states "... the Parties shall assess the control measures ... on the basis of available scientific, environmental, technical and economic information." Thus, the Scientific Assessment Panel, the Environmental Effects Assessment Panel, and the Technology and Economic Assessment Panel were set up. From time to time these Panels produce reports which are made available to the Parties in readiness for their annual meetings, at which they consider possible amendments to the Protocol. The latest of these was produced in 2010 (see Section 5.9).
- The ability of those involved to identify and remove barriers to technology transfer, including through changes in national laws and voluntary codes.

An extensive set of tables and graphs showing the production and consumption trends of CFCs, halons, HCFCs, etc. is given in an Ozone Secretariat report (UNEP, 2005) (see Figures 5.8 and 5.9). This shows that compliance with the Montreal Protocol has been excellent so far. However, compliance with the Protocol in terms of restricting the emission of ozone depletions substances into the atmosphere is only the beginning. Ultimately, the success, or otherwise, of the Montreal Protocol will have to be judged by whether (a) first, the decline in stratospheric ozone concentration is halted and (b) second, if that is achieved, whether a return to the pre-CFC levels of ozone can be achieved. We shall turn to the consideration of this in the next chapter.

6

The study of atmospheric ozone since 1987

6.1 INTRODUCTION

Following the agreement of the Montreal Protocol in 1987 a great deal of research work has been done on atmospheric ozone and a considerable amount of that research was directed towards monitoring the effectiveness of the Montreal Protocol. As we said at the end of Chapter 5, there are various stages involved: (a) monitoring the production and use of ozone-depleting substances; (b) monitoring the concentration of these substances in the atmosphere, both in the troposphere and in the stratosphere; (c) monitoring the trends in ozone concentration in the troposphere; and (d) monitoring the ozone concentration in the stratosphere both in equatorial areas and mid-latitudes and also in the polar regions. The main datasets of TOC and ozone profiles are listed and described briefly in Appendix 3A of WMO (2007). Apart from monitoring the success of the Montreal Protocol so far, the data collected are also important in contributing to models that can be used to make predictions about the rate of recovery of the ozone layer in the future. This work has been reported, in the usual way, in scientific journals and at the Quadrennial Ozone Symposia. However, in discussing the annual Meetings of the Parties, Article 6 of the Montreal Protocol states "... the Parties shall convene appropriate panels of experts (and) the panels will report their conclusions ... to the Parties." Thus, from time to time, the Scientific Assessment Panel produces reports that provide convenient summaries and analyses of recent relevant research. These reports and their timings are listed in Table 6.1.

6.2 THE REDUCTION OF OZONE-DESTROYING CHEMICALS IN THE ATMOSPHERE

The Montreal Protocol was implemented as a result of the careful science that identified human-produced ozone-depleting substances (ODSs) as threats to the

Table 6.1. Scientific assessment reports.

Year	*Policy process*	*Scientific assessment*
1981		*The Stratosphere 1981: Theory and Measurements.* WMO No. 11.
1985	Vienna Convention	*Atmospheric Ozone 1985.* Three volumes. WMO No. 16.
1987	Montreal Protocol	
1988		*International Ozone Trends Panel Report 1988.* Two volumes. WMO No. 18.
1989		*Scientific Assessment of Stratospheric Ozone: 1989.* Two volumes. WMO No. 20.
1990	London Adjustment & Amendment	
1991		*Scientific Assessment of Ozone Depletion: 1991.* WMO No. 25.
1992		*Methyl Bromide: Its Atmospheric Science, Technology, and Economics* (Montreal Protocol Assessment supplement). UNEP (1992).
1992	Copenhagen Adjustment & Amendment	
1994		*Scientific Assessment of Ozone Depletion: 1994.* WMO No. 37.
1995	Vienna Adjustment	
1997	Montreal Adjustment & Amendment	
1998		*Scientific Assessment of Ozone Depletion: 1998.* WMO No. 44.
1999	Beijing Adjustment & Amendment	
2002		*Scientific Assessment of Ozone Depletion: 2002.* WMO No. 47.
2006		*Scientific Assessment of Ozone Depletion: 2006.* WMO No. 50.
2007	19th Meeting of Parties	
2010		*Scientific Assessment of Ozone Depletion: 2010.* WMO No. 52.

ozone layer. The discovery of the Antarctic ozone hole brought this threat to the immediate attention of both the science and policy maker communities. Because of the agreement, the levels of ODSs are declining in ground measurements (WMO, 2007), while satellite observations of HCl show that stratospheric chlorine levels are also declining (Anderson *et al.*, 2000; Froidevaux *et al.*, 2006).

Nearly two decades after the signing of the Montreal Protocol, as stated in the 2006 *Scientific Assessment of Ozone Depletion* (WMO, 2007): "the total combined abundances of anthropogenic ozone-depleting gases in the troposphere continue to decline from the peak values reached in the 1992–1994 time period and the combined stratospheric abundances of the ozone-depleting gases show a downward trend from their peak values of the late 1990s." Concerning changes in the ozone field, WMO (2007) also stated that: "in the northern hemisphere the midlatitude column ozone values for the 2002–2005 period were approximately 3% below pre-1980 values, essentially the same as in the 1998–2001 period, the ozone abundances in the midlatitude lowermost stratosphere (12–15 km) show a strong decrease between 1979 and the mid-1990s, followed by an overall increase from 1996 to 2004, giving no net long-term decrease at this level and the slowing of the decline and leveling off of midlatitude upper stratospheric (35–45 km) ozone over the past decade have very likely been dominated by the corresponding changes in stratospheric ozone-depleting gases." In addition, among the basic conclusions of WMO (2007) are the following: "The Montreal Protocol is working: There is clear evidence of a decrease in the atmospheric burden of ozone-depleting substances and some early signs of stratospheric ozone recovery [while] long-term recovery of the ozone layer from the effects of ozone-depleting substances is expected to span much of the 21st century and is estimated to occur later than projected in the previous Assessment (2002)." These statements rely on continuous and long-term collection of reliable ozone data (Cracknell and Varotsos, 2007b).

By the very nature of the problem of ozone depletion by man-made CFCs, these compounds are long-lived and so one should not expect any rapid reduction in their concentration in the atmosphere following the ban on their manufacture and use. An exception is CH_3CCl_3 which has a short lifetime in the atmosphere of about 5 years because of its reactivity towards OH radicals, and major releases to the atmosphere stopped early in the 1990s in response to the scheduled ban. CH_3CCl_3 concentrations had reached about 140 ppt by 1992, and had decreased to about 22 ppt by mid-2004. With three Cl atoms per molecule, this change reduced the total organochlorine content in the troposphere by about 350 ppt. However, with so much of the CH_3CCl_3 already gone and no other important short-lived contributor to the total organochlorine burden left in the atmosphere, this initial rapid reduction in total chlorine will not continue. Instead, the reduction will proceed at the average of the 2 and 1% per year rates, respectively, calculated for CCl_3F and CCl_2F_2. For example, the total ban on CCl_2F_2 is expected to reduce its atmospheric burden from 540 ppt by only about 50 ppt in the next decade (Rowland, 2006, 2009). According to Roots *et al.* (2009), the use of CFCs in Estonia decreased from 199 t in 1997 to zero in 2002.

Observations from surface stations have been combined with ODS emission estimates to predict future ground levels of ODSs such as given in scenario Al

(Daniel *et al.*, 2007). This scenario assumes that the Montreal Protocol will be adhered to and projects a steady decline of most ODSs over the coming decades. Future EESC values depend on current values, estimates of future emissions, and the lifetimes of a number of species in the atmosphere. Between 1950 and 1980, Antarctic EESC levels more than doubled. However, ozone levels showed only modest changes from 1957 to 1980. After 1980, Antarctic ozone levels decreased radically. Because only small losses were observed prior to 1980, this year is conventionally chosen as a reasonable level of normal conditions. Based upon the scenario Al estimate, Antarctic EESC is projected to return to the 1980 level in 2067. This return to 1980 levels is dominated by decreases of CFC-11, carbon tetrachloride (CCl_4), halon-1211, and methyl chloroform (CH_3CCl_3).

6.2.1 Ozone depletion potential (ODP)

Ozone depletion potentials are indices that provide a simple way to compare the relative ability of various ODSs to destroy stratospheric ozone (Fisher *et al.*, 1990a–c; Solomon *et al.*, 1992; Wuebbles, 1983). The concept of the ODP has been discussed extensively in previous WMO reports (WMO, 1995, 1999, 2003). ODPs are often calculated assuming steady-state conditions with constant emissions; for compounds that are removed by linear processes, this is equivalent to assuming an emission pulse and integrating over the entire decay of the compound (Prather, 2002; Prather *et al.*, 1996). Steady-state ODPs are integrated quantities that are independent of time and are distinct for each halocarbon species.

The ODP of a well-mixed ozone-destroying species i is given by:

$$\mathrm{ODP}_i = \frac{\text{global ozone loss due to unit mass emission of } i}{\text{global ozone loss due to unit mass emission of CFC-11}} \tag{6.1}$$

This can be calculated using computer models, with accuracy depending on the model's ability to simulate the distribution of the halocarbon under consideration and the ozone loss associated with it. Because ODPs are defined relative to the ozone loss caused by CFC-11, ODP values demonstrate less sensitivity to photochemical modeling errors than do absolute ozone loss calculations. Time-dependent ODPs can also be calculated (Solomon and Albritton, 1992), which reflect the different time-scales over which the compound and reference gas (the chlorofluorocarbon CFC-11) liberate chlorine and bromine into the stratosphere. Compounds that have shorter atmospheric lifetimes than CFC-11 have ODPs that decrease with increasing integration time and compounds that have longer atmospheric lifetimes than CFC-11 have ODPs that increase with increasing integration time. Solomon *et al.* (1992) proposed a semi-empirical approach to estimate ODPs which approximates and simplifies the accurate representation of ozone photochemistry and provides an observational constraint to model-based results. In this approach, measurements of correlations between halocarbons are used to evaluate chlorine and bromine relative stratospheric release values. These release rates are used with lifetimes, molecular weights, and the number and type of halogen atoms per molecule to estimate the effect of a small source gas pulse emission on stratospheric ozone depletion. In the case of bromine and iodine, the catalytic efficiency for ozone destruction relative to chlorine is needed and differs from unity partly due to the different partitioning of halogen chemical families. For long-lived chlorocarbons and bromocarbons that are well mixed in the troposphere, the semi-empirical ODP definition is expressed by:

$$\mathrm{ODP}_i = (\alpha n_{\mathrm{Br},i} + n_{\mathrm{Cl},i}) \frac{f_i}{f_{\mathrm{CFC\text{-}11}}} \frac{\tau_i}{\tau_{\mathrm{CFC\text{-}11}}} \frac{M_{\mathrm{CFC\text{-}11}}}{M_i} \frac{1}{3} \tag{6.2}$$

where f is the fractional halogen release factor; α is the relative effectiveness of bromine compared with chlorine for ozone destruction; τ is the global lifetime; M is the molecular weight; and n_{Cl} (n_{Br}) is the number of chlorine (bromine) atoms contained in the compound. CFC-11 subscripts indicate quantities for CFC-11 and i subscripts denote quantities pertaining to the compound, i, for which the ODP is being calculated (WMO, 2007, chapter 8). For very short-lived substances (VSLSs), the location and season of emission affect the amount of halogen that can reach the stratosphere; equation (6.2) is not appropriate for these gases.

A detailed explanation of global lifetimes, τ_i, which appear in equation (6.2) can be found in the fifth WMO Scientific Assessment (WMO, 2003) and tables of the lifetimes of various halocarbons are given in Tables 8-1 and 8-2 of the sixth Scientific Assessment (WMO, 2007); Table 8-1 is reproduced as Table 6.2. As to the fractional release factor f_i that appears in equation (6.2), $f_i(x, y, z, t)$ at a location (x, y, z) at time t is defined by:

$$f_i(x, y, z, t) = \frac{\rho_{i,\mathrm{entry}} - \rho_i(x, y, z, t)}{\rho_{i,\mathrm{entry}}} \tag{6.3}$$

where $\rho_i(x, y, z, t)$ denotes the mixing ratio of the halocarbon at position (x, y, z) and time t; and $\rho_{i,\mathrm{entry}}$ is the mixing ratio of species i in the air parcel when it entered the stratosphere (WMO, 2003).

Values of the fractional release factors of various halocarbons are given in Table 6.2. Finally, with regard to equation (6.2), it is an oversimplification to regard α, the relative effectiveness of Br compared to Cl for ozone destruction, as a constant. In fact, it represents a globally integrated result with sensitivity to many factors, including the kinetic parameters for chlorine and bromine species, the amount of inorganic bromine and inorganic chlorine in the background atmosphere, and atmospheric transport. The value of α was assumed to be 45 in the fifth Scientific Assessment (WMO, 2003), based partly on the results from Daniel *et al.* (1999) and Ko *et al.* (1998). In chapter 8 of the sixth Scientific Assessment (WMO, 2007) a number of values of α were quoted—52, 57, 61, 64, and 71—resulting from several different models and assumptions. The conclusion in that Assessment was to recommend a value of $\alpha = 60$ for global ozone destruction and a value of $\alpha = 65$ for the Antarctic. The effectiveness of fluorine in ozone destruction was regarded as negligibly small (Ravishankara *et al.*, 1994; Wallington *et al.*, 1995).

Detection of the first two stages of ozone recovery (reductions in the rate of decline and then increasing ozone attributable to decreases in ODSs) has been

Table 6.2. Lifetimes, relative fractional halogen release factors, and ozone depletion potentials for halocarbons. ODPs recommended and ODPs adopted in the Montreal Protocol are included (WMO, 2007).

*Halocarbon**	*Lifetime* (years)	*Relative fractional release factor*[1]	*Semi-empirical ODP*	*ODP in Montreal Protocol*
		Annex A-I		
CFC-11	45	1	1.0	1.0
CFC-12	100	0.6	1.0	1.0
CFC-113	85	0.75	1.0	0.8
CFC-114	300	0.28 ± 0.02[2]	1.0	1.0
CFC-115	1,700		0.44[†]	0.6
		Annex A-II		
Halon-1301	65	0.62	16	10.0
Halon-1211	16	1.18	7.13	3.0
Halon-2402	20	1.22	11.5	6.0
		Annex B-II		
Carbon tetrachloride	26	1.06	0.73	1.1
		Annex B-III		
Methyl chloroform	5	1.08	0.12	0.1
		Annex C-I		
HCFC-22	12	0.35	0.05	0.055
HCFC-123	1.3	1.11	0.02	0.02
HCFC-124	5.8	0.52	0.02	0.022
HCFC-141b	9.3	0.72	0.12	0.11
HCFC-142b	17.9	0.36	0.07	0.065
HCFC-225ca	1.9	1.1	0.02	0.025
HCFC-225cb	5.8	0.5	0.03	0.033

Annex E				
Methyl bromide	0.7	1.12	0.51	0.6
Others				
Halon-1202	2.9		1.74	
Methyl iodide	See Chapter 2 of WMO (2007)		See Chapter 2 of WMO (2007	
Methyl chloride	1	0.8	0.02	

* Chemical formulae for the halocarbons are listed in Tables 5.4 and 5.5.
† Model-derived value (WMO, 2003).
[1] From WMO (2003), Table 1-4, except for the value for CFC-114. For the EESC calculations in Section 1.8 of WMO (2003), slightly different relative fractional release factors were used by mistake for the halons.
[2] From Schauffler *et al.* (2003).

demonstrated in many locations in the atmosphere, and the emphasis has now shifted to better understand what processes will affect the long-term full recovery of the ozone layer, including the future evolution of equivalent effective stratospheric chlorine (EESC) and future trends in stratospheric temperatures. Because of the strong dependence of ozone recovery on changes in EESC, understanding and reducing the uncertainties in projections of stratospheric halogen loading was a key topic (Figure 6.1).

An interesting estimate of the extent of ozone depletion in the Antarctic ozone hole had there been no Montreal Protocol was made by Newman *et al.* (2009) and assuming the concentration of CFCs increased at the rate of 3% per annum (see Figure 6.1). The TOC in this scenario in October for 70°S to 90°S is depicted by the black line in this figure and the three inset images show ozone concentrations in the ozone hole in this scenario for 1980, 2020, and 2060. The green line represents a simulation where CFCs were supposed fixed at 1960 levels (i.e., before serious depletion of ozone was detected). The future (past) reference simulation is shown in red (dark blue).

6.2.2 Equivalent Effective Stratospheric Chlorine (EESC)

Stratospheric ozone destruction is related more directly to the actual amount of chlorine and bromine in the stratosphere than to the amount of halogenated hydrocarbons emitted into the atmosphere. Due to the established relationship between

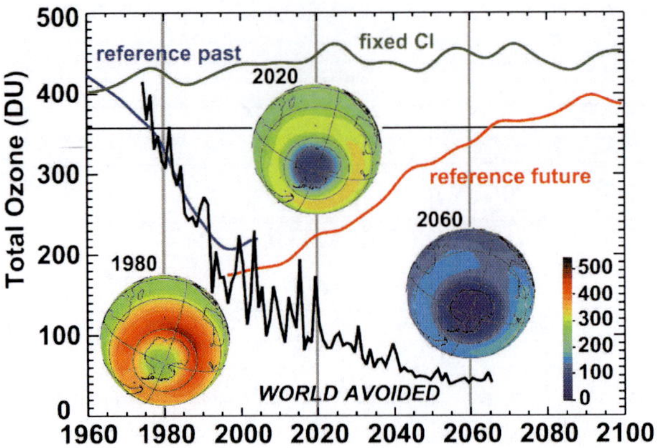

Figure 6.1. Antarctic ozone in October (averaged poleward of 70°S) for four different simulations from the NASA/GSFC coupled chemistry–climate model. The thick black line is an ozone simulation for the "world avoided" (i.e., assuming there was no Montreal Protocol and CFCs increased at 3% per year. The future (past) reference simulation is shown in red (dark blue). A simulation with CFCs fixed at 1960 levels is shown in green. The latter curves are smoothed with a Gaussian filter with a half-amplitude response of 20 years, except for the "world avoided", which is unsmoothed. The inset false-color images show 1980, 2020, and 2060 with 20 DU color increments (see inset scale) for the "world avoided" simulation (Newman *et al.*, 2008).

stratospheric ozone depletion and inorganic chlorine and bromine abundances, the temporal evolution of chlorine and bromine-containing source gases is an important indicator of the potential effects of anthropogenic activity on the health of stratospheric ozone. Indices have been developed to demonstrate this halogen evolution in a simple manner. They account for the greater per-atom potency of stratospheric bromine (Br) compared with chlorine (Cl) in its ozone destructiveness with a constant factor, α, and include the varying rates at which Cl and Br will be released in the stratosphere from different source gases. The EESC (equivalent effective stratospheric chlorine) is an index reflecting the amount of ozone-depleting chlorine and bromine in the stratosphere. It was introduced as a simple combination of ODSsbwhich was developed to relate halogen evolution to tropospheric source gases in a simple manner (Daniel *et al.*, 1995). EESC is a sum of ODSs which accounts for: (i) transit time to the stratosphere, (ii) the greater potency of stratospheric bromine (Br) compared to chlorine (Cl), and (iii) the varying fractions with which Cl and Br are released in the stratosphere from different source gases (i.e., fractional release).

EESC provides a simple index that relates the time evolution of long-lived surface abundances of ODSs with the ozone-destructive ability of stratospheric halogens that come from these long-lived source gases (WMO, 1995, 1999, 2003). The contributions of very short-lived chlorine and bromine-containing sources gases and of tropospheric inorganic halogens generally have been neglected. EESC is defined as

$$\mathrm{EESC} = f_{\mathrm{CFC\text{-}11}} \left[\sum_{\substack{\text{Cl-containing}\\ \text{halocarbons}}} n_i \frac{f_i}{f_{\mathrm{CFC\text{-}11}}} \rho_{i,\mathrm{entry}} + \alpha \sum_{\substack{\text{Br-containing}\\ \text{halocarbons}}} n_i \frac{f_i}{f_{\mathrm{CFC\text{-}11}}} \rho_{i,\mathrm{entry}} \right] \tag{6.4}$$

(Daniel *et al.*, 1995), where n_i is the number of chlorine or bromine atoms in the source gas, i; $f_i/f_{\mathrm{CFC\text{-}11}}$ represents the efficiency of stratospheric halogen release relative to that of CFC-11, denoted by $f_{\mathrm{CFC\text{-}11}}$; and $\rho_{i,\mathrm{entry}}$ is the tropospheric mixing ratio of source gas i when it entered the stratosphere. Traditionally, $\rho_{i,\mathrm{entry}}$ is calculated assuming a simple time lag Γ from the surface observations, i.e.,

$$\rho_{i,\mathrm{entry}}(t) = \rho_i(t) - \Gamma \tag{6.5}$$

where $\rho_i(t)$ is the surface mixing ratio at time t. In WMO (2003), EESC was estimated assuming $\Gamma = 3$ years (typical of the lower mid-latitude stratosphere) to obtain a value appropriate for relating to mid-latitude-averaged ozone loss; however, Γ is longer, about 6 years, over the poles and in the upper stratosphere. Effective equivalent chlorine (EECl) (Montzka *et al.*, 1996) is another index, which is similar to EESC but which includes no consideration of the transport lag time.

To retain the simplicity of the EESC index, several assumptions are generally made in its calculation. One such assumption is that the stratospheric entry mixing ratio for a given time is taken to be the surface mixing ratio at that time. This could be an overestimate of the EESC contribution for shorter lived gases, like CH_3Br, whose abundance may be reduced before reaching the tropopause. Montzka *et al.* (2003a) accounted for this by reducing the surface mixing ratio by 7% in the calculation of EESC. A second assumption generally made in EESC calculations is the neglect of the transport time from the surface to the tropopause. Although the particular location of the surface emission affects the transport time to the tropopause, this is generally not considered.

Newman *et al.* (2006) extended the method of calculating EESC to account for the lack of a single transit time from the surface to a stratospheric location (i.e., stratospheric air is composed of air characterized by a range, or spectrum, of transit times or ages) and that the fractional release values depend on the age of air. Newman *et al.* calculated $\rho_{i,\mathrm{entry}}$ using

$$\rho_{i,\mathrm{entry}}(t) = \int_{-\infty}^{t} \rho_i(t) G(t - t')\, \mathrm{d}t' \tag{6.6}$$

where $G(t)$ is the age spectrum, assumed to be an inverse Gaussian function with mean Γ and width Δ (see Equation 9 of Waugh and Hall, 2002); and $f_i = f_i(\Gamma)$. Including an age spectrum does not affect EESC evolution when the temporal trend is constant, but it acts to smooth changes over time when the trend is varying. Specifically, when an age spectrum is included, the period close to when maximum EESC occurs is characterized by a "flatter" or less peaked EESC time series.

Equivalent effective stratospheric chlorine (EESC) peaked in the late 1990s to early 2000s, and so there are now several years of observations since that peak. Fractional release accounts for ODS disassociation in the stratosphere relative to the amount that passed through the tropopause. The EESC trend model has been used in several recent statistical analyses of ozone (e.g., Dhomse *et al.*, 2006; Fioletov and Shepherd, 2005; Guillas *et al.*, 2004; Newman *et al.*, 2004; Stolarski *et al.*, 2006b; Yang *et al.*, 2005).

The sixth Scientific Assessment (WMO, 2007) notes one or two issues with the use of EESC. The calculation of EESC depends on the age of the air, which varies with altitude and latitude (see Section 6.2.1). The age of air is about 3 years in the mid-latitude lower stratosphere, but it is longer, about 6 years, over the poles and in the upper stratosphere. Ozone dilution from the Antarctic, and to a lesser extent from the Arctic, affects lower-stratospheric ozone over the mid-latitudes and therefore, strictly speaking, the effect of polar EESC should be taken into account when mid-latitude trends are considered. However, in practice, this 3-year difference in the age of air only has a limited effect on the estimated EESC-related regression coefficient. For example, the annual trend for 60°S–60°N is $(-1.6 \pm 0.17)\%$ and $(-1.3 \pm 0.15)\%$ per decade (2σ uncertainties) for EESC calculated with an age of air of 3 and 6 years, respectively. However, the age of air plays a major role when ozone recovery is considered (see Sections 7.2.1, 7.2.2, 7.6).

6.3 GROUND-BASED AND OZONESONDE DATA ON OZONE DEPLETION

Generally speaking, before the advent of satellite data, the records of traditional atmospheric ozone measurements (from ground-based spectrophotometers and ozonesondes) were not used to investigate possible ozone depletion. This was because, for a long time, no one suspected that the ozone concentration might be decreasing. Moreover, there were also problems over standardization of instruments and techniques (as we have noted in Section 1.2.2).

In any attempt to detect the recovery of the ozone layer (which is the main purpose of the implementation of the Montreal Protocol), the trend in total ozone data during the past decade (or so) must be examined. However, short-term trends in total ozone cannot be inferred from long-term trends and *vice versa*, if data are not statistically reliable. Slowing of long-term global ozone losses has been observed during the past decade and a recovery of the ozone layer is forecast for the coming decades. Therefore, the accuracy requirements for ozone measurements are increasing, which, in turn, requires corresponding analysis of measurement errors. This means that it is necessary to maintain all Dobson instruments well calibrated. It should be kept in mind that the recovery expected after implementation of the Montreal Protocol is likely to be detected in some areas sooner than others because of natural variability in the ozone field. Systematic errors in Dobson total ozone measured at the South Pole have been studied by Bernhard *et al.* (2005), who showed that Dobson underestimation of total ozone under ozone hole conditions reached up to 4% but could be corrected for by accounting for the temperature dependence of the ozone cross-section and accurate airmass calculation. Corrected Dobson measurements compared with a collocated SUV-100 spectroradiometer and version 8 overpass data from TOMS show no clear dependence on total ozone. Probably, similar errors can be deduced from the results for the three Antarctic stations. Note that the EP-TOMS data have been compared with preliminary SAOZ measurements in Antarctica for Rothera (U.K.), and significant negative dependence of ΔTOC on low TOC also exists.

On the basis of both the magnitude and autocorrelation of the noise from Dobson measurements, estimates of the time required to detect a fixed trend in ozone at various locations around the world can be made by systematically calibrating all the Dobson spectrophotometers that participate in the Dobson WMO Network (see Section 1.4). These predictions require the development of certain models that can be used to estimate the time required to detect predicted trends in different areas of the world. The analysis should be based on our current understanding of ozone chemistry, full compliance with the Montreal Protocol and its amendments, and no intervening factors (such as major volcanic eruptions or enhanced stratospheric cooling). Should the recovery be slower than predicted in various forums by several models (e.g., owing to the effect of greenhouse gas emissions) or should measurement sites be perturbed, even longer times would be needed for detection.

After filtering out known natural variability (interannual and intra-annual cycles) TOC decline in both hemispheres is strongest during the winter and spring (higher than 6–7% per decade) and is only half that during the summer and autumn. According to WMO (1999), TOC declined during all months of the year in the mid and high latitudes. In the Northern Hemisphere, trends are much larger in winter and spring (about −3 to −6% per decade), than in summer and autumn (about −1 to −3% per decade).

Due to new evidence, ozone changes in the 21st century are expected to encompass the period of "recovery" of ozone from the influence of ODSs that have been released in anthropogenic activities. ODSs have led to a decline in global ozone amounts, with measurable changes beginning in the 1980s and the largest changes found in winter/spring polar stratospheres. The total abundance of ODSs is in decline in the troposphere and stratosphere as a result of effective actions undertaken as part of the Montreal Protocol. The start of the decline in EESC is a metric used to gauge the overall burden on ozone-depleting halogens in the stratosphere from ODSs and marks the conceptual start of the ozone recovery process. Continued analysis of ozone measurements during the 21st century is expected to reveal and confirm three key stages in the recovery process:

(i) The slowing of ozone decline, identified as the occurrence of a statistically significant reduction in the rate of decline in ozone due to changing EESC.
(ii) The onset of ozone increases (turnaround), identified as the occurrence of statistically significant increases in ozone above previous minimum values due to declining EESC.

(iii) The full recovery of ozone from ODSs, identified as when ozone is no longer affected by ODSs. In the absence of changes in the sensitivity of ozone to ODSs, this is likely to occur when EESC returns to pre-1980 levels (WMO, 2007).

London and Liu (1992) used ozonesonde observations to analyze long-term tropospheric and lower-stratospheric ozone variation over 20 years for the four seasons. Their results indicated the presence of minimum ozone-mixing ratios in the equatorial and tropical troposphere except in regions where net photochemical production is dominant. In the mid and upper troposphere, and lower stratosphere to the 50 hPa level, ozone concentration increases from the subtropics to subpolar latitudes in both hemispheres. In the middle stratosphere the ozone-mixing ratio is maximum over the tropics.

Spänkuch *et al.* (1998) suggested a multiple regression approach to assess the amount of layer ozone ΔO_3 from predictors accessible from model outputs of meteorological analysis, in addition to TOC (which is the most important predictor). This approach was verified using Hohenpeissenberg data. Claude *et al.* (1998) analyzed TOC and OVP trends obtained by Dobson and ozonesonde observations in Hohenpeissenberg (both since 1967). The trend for lower-stratospheric ozone (12–17 km) was −5% per decade, obtained from OVP data for the Jungfraujoch site in the Swiss Alps. De Mazière *et al.* (1998) suggested a climatological ozone/vertical ozone distribution model. Mlch and Halenka (1998) studied the fine structure (lamination) of OVP using ozonesonde data. The validation and results of 9 years of lidar (DIAL) sounding at Hohenpeissenberg were considered by Steinbrecht *et al.* (1998). Since 1987, lidar, in agreement with SAGE, has shown significant ozone depletion by 0.5 to 1% per year between 35 and 40 km. Stübi *et al.* (1998) discussed harmonization of the Payerne Brewer/Mast ozone-sounding series, which started in 1966.

Furrer *et al.* (1992) discussed the results of long-term Dobson and balloon ozone measurements in the vicinity of Berlin. The results of observations over 24 years (1964–1988) indicated an increase in ozone concentration within the middle stratosphere (below 31 km height) as well as in the troposphere but, on the other hand, ongoing ozone depletion in the lower stratosphere became evident. The large-scale vertical redistribution of atmospheric ozone in the troposphere and in the lower stratosphere is in agreement with model calculations and trend predictions, because of change in the chemical composition and ozone photochemistry induced by anthropogenic emissions of ozone-depleting substances.

6.4 PIECEWISE LINEAR TRENDS IN OZONE DEPLETION

Many of the earlier studies of ozone depletion involved description of long-term changes in ozone concentration due to ODSs in terms of linear trends, estimated using regression analysis (WMO, 1995, 1999). The decision to fit a linear trend was based on the expected response to the approximately linear increase in ozone-depleting substances (ODSs). However, it became clear that the trend of ODSs from

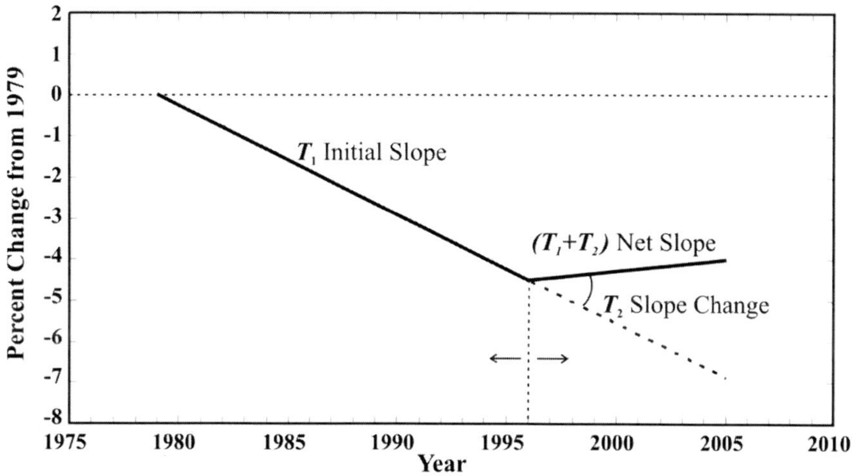

Figure 6.2. Hockey stick statistical trend model in the style of Reinsel *et al.* (2002, 2005) (Yang *et al.*, 2009).

the late 1970s to the 1990s and beyond was no longer linear with time (see Section 6.3). There is some considerable evidence that, whereas there was a steady decline in TOC until the early 1990s, this decline was arrested and even possibly reversed round about that time. The piecewise linear trend concept, in which different linear fits are used before and after a turning point, was then used in several trend analyses; this is sometimes referred to as the hockey stick algorithm (Miller, A.J. *et al.*, 2006; Newchurch *et al.*, 2003; Reinsel *et al.*, 2002, 2005). Figure 6.2 illustrates the fundamental concept of the algorithm. Anticipating a trend change of TOC from implementation of the Montreal Protocol, the algorithm is designed with an inflection point at a specific time. This results in an initial slope before the inflection (T_1), a slope change at the inflection point (T_2). and a subsequent net slope ($T_1 + T_2$) after the inflection. If the net slope is positive and the coefficient is statistically significant then it can be argued that total ozone has begun to recover. The time chosen for the inflection point needs to be determined, and from the results of Miller, A.J. *et al.* (2006) January 1996 was selected (see Figure 6.2) (Yang *et al.*, 2009).

Various other dates for the inflection point, scattered around 1996, have been obtained in different pieces of work. A reversal of the decline in TOC from 1991 was found by Varotsos *et al.* (2000) using TOMS data and ground-based Dobson spectrophotometer data for Athens, Greece (38°N, 24°E) (see Figure 6.3). Krzyścin *et al.* (2005), using ground-based observations and merged TOMS/SBUV data, found a breakpoint at 1994, while Reinsel *et al.* (2002, 2005), using monthly zonal average total ozone data from both TOMS and SBUV satellite sources and ground-based instruments over 1978–2002 data, obtained a breakpoint at 1996. Reinsel *et al.* (2005) went into greater detail and found that, for latitude zones of 40° and above in both hemispheres, large positive and significant estimates of a change in trend (since 1996) were obtained (of the order of 1.5 to 3 DU per year).

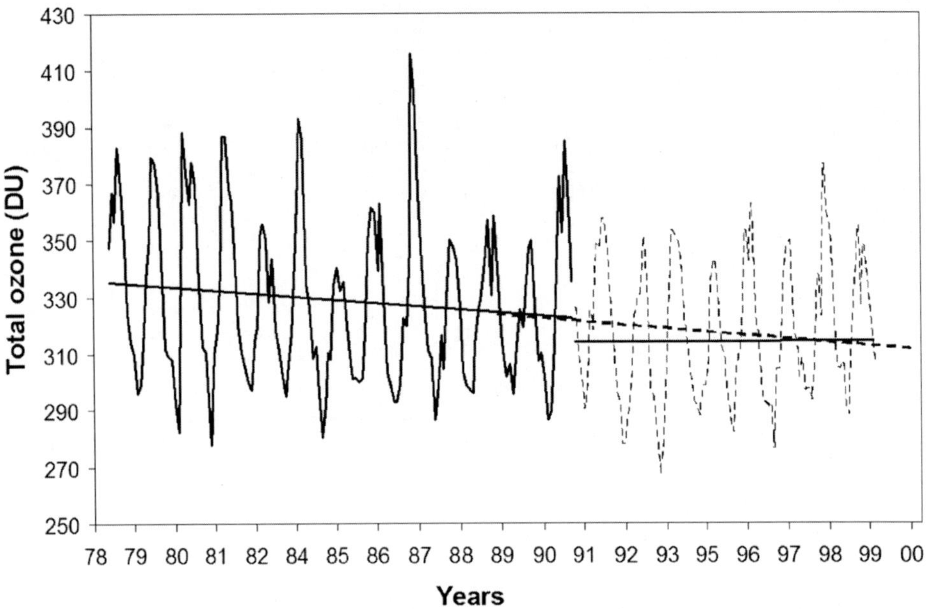

Figure 6.3. Monthly TOC over Athens, Greece, from 1979 to 1999. The solid thick line represents the least squares fit to TOC up to May 1991 and is extended as a dashed line to July 1999. The solid thin line represents the least squares fit to TOC from June 1991 to July 1999 (Varotsos *et al.*, 2000).

SAGE I and SAGE II satellite data provided an important source of information concerning ozone/vertical ozone profiles (see Section 2.5.1.1). McCormick *et al.* (1982, 1992) discussed SAGE data with the aim of revealing the basic features of annual, semi-annual, and quasi-biennial variation as well as long-term trends that depend on altitude and latitude. The results indicate, for instance, the existence of an ozone depletion trend ($\sim$ few percent) in mid-latitudes near 40 km altitude between 1980 and 1986, whereas ozone changes above and below this level were much smaller. During the whole period considered, ozone depletion at the height of 25 km was about 3%. Total ozone content above 17 km averaged for the latitude interval 65°S–65°N was characterized by a linear trend of about $(-0.30 \pm 0.19)\%$ per year (or -3.6% for the whole period of observations from February 1979 to April 1991). Statistically significant ozone trends were concentrated near the 23 km altitude (negative trends at all latitudes) and in the 38–50 km layer (10°N–35°S), where a negative ozone trend was observed. Column ozone losses occurred below 25 km, with most of the loss coming from the region between 17 and 20 km. McCormick *et al.* (1992) emphasized that the concentration of mid-latitude ozone depletion in the lower stratosphere (17–20 km) imposes certain constraints on the choice of potential mechanisms that were responsible for this depletion. Fussen (1998) assessed the biases in long-term ozone trends from SAGE II data because of the impact of volcanic aerosol.

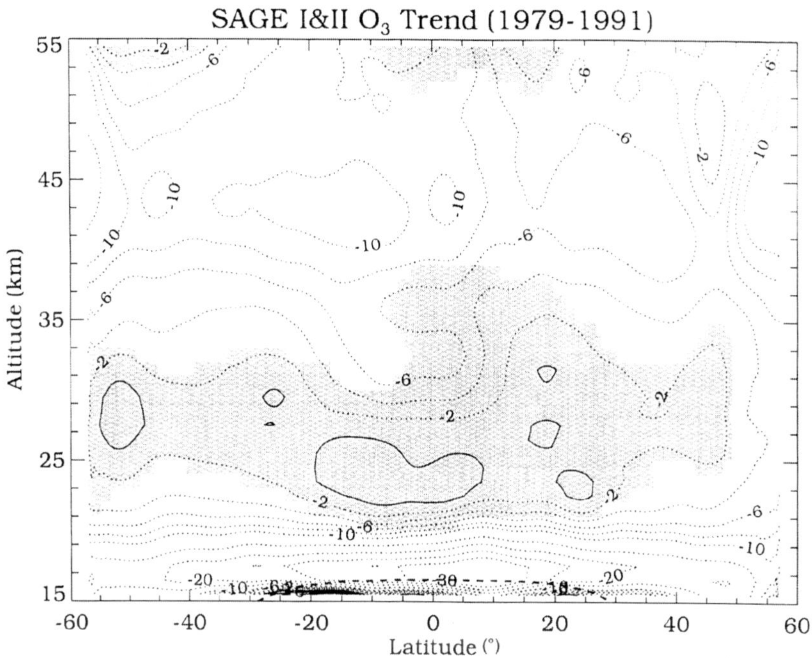

Figure 6.4. Trends calculated for SAGE I/II for 1979–1991. Hatched areas indicate trends calculated to be insignificant at the 95% confidence level. The dashed line indicates the tropopause. The altitudes of SAGE I measurements have been adjusted by 300 m (WMO, 1995).

Simultaneous observations of O_3, H_2O, and aerosols made it possible to analyze interrelationships between mixing ratios of ozone and aerosol within the circumpolar vortex in Antarctica during winter and spring (when the ozone hole is formed) and also the synchronous variability of ozone and water vapor.

The most representative statistical information on ozone vertical profiles was obtained on the basis of satellite and ozonesonde observations. These results were considered in detail in the UNEP Scientific Assessment (WMO, 1995, 1999). Only some of the results are mentioned here.

The results presented in Figure 6.4 show OVP trends derived from SAGE I/II data for 1979–1991. Figure 6.5 illustrates an intercomparison of SBUV, SAGE I, SAGE II, Umkehr, and ozonesondes where these data are in reasonable agreement between them. In addition, Miller *et al.* (1995) demonstrated a comparison of trends in the vertical distribution of ozone between 30°N and 50°N during the 1980s. They showed that there is no significant ozone depletion in the 25–30 km layer. The agreement continues down to about 20 km, where statistically significant reductions of $(7 \pm 4)\%$ per decade were observed between 30°N and 50°N by both ozonesonde and SAGE I/II. According to SAGE data there was a decrease of more than 20% per decade in the equatorial region just above the tropopause between about 30°N and 30°S. In absolute terms this loss is small because of a rather low ozone concentration

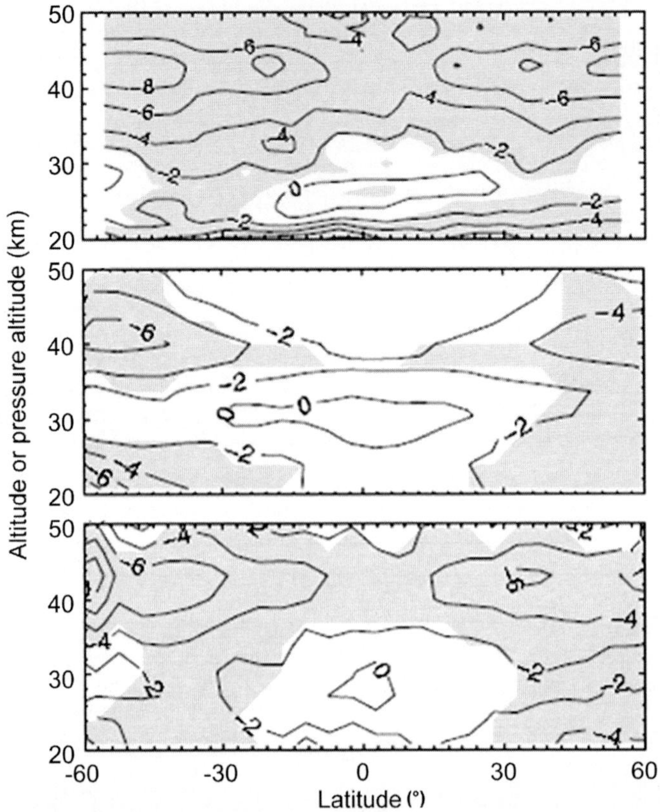

Figure 6.5. Latitude–altitude cross-section of stratospheric ozone trends from various data sources. The plots display trend estimates from various published datasets: (a) SAGE I + II dataset (adapted from Randel and Wu, 2007); (b) the NASA merged SBUV(/2) dataset; (c) SAGE-corrected SBUV(/2) dataset (McLinden *et al.*, 2009). For panel (c) SAGE I + II data are converted from Randel *et al.* (2009). Shading indicates trends are significant at the 2σ level. Trend contours are every 2% per decade (*source*: WMO, 2010a).

at these altitudes. The height of maximum ozone decrease is about 16 km and the region of decrease becomes broader away from the equator. At northern mid-latitudes the SAGE trend at 16–17 km is (-20 ± 8)% per decade, compared with an average trend from the ozonesondes of (-7 ± 3)% per decade. Over the longer period (1968–1991) the ozonesonde record indicates a trend of (-4 ± 2)% per decade at 20 km at northern mid-latitudes. At 35–45 km, ozone declined (5–10)% per decade at 30–50°N and slightly more at southern mid-latitudes.

Terao and Logan (2007) showed that there are differences up to 4%/decade between SAGE trends in altitude and pressure co-ordinates if National Center for Environmental Prediction (NCEP) temperature reanalysis data are employed for the conversion. In this context, Figure 6.5 illustrates the results from an analysis of SBUV(/2) and SAGE ozone time series, which reveal that this difference can be as much as 4% in the upper stratosphere if SAGE trends calculated in number density versus altitude are compared to SBUV(/2) partial pressure versus pressure trends (Figure 6.5). However, if SAGE data are converted to the same units as SBUV(/2) utilizing temperature data with proper temperature trends (Randel *et al.*, 2009) and are adjusted to match SBUV(/2) vertical resolution as was done in the SAGE-corrected SBUV dataset (McLinden *et al.*, 2009), the ozone trends derived from SAGE and SBUV(/2) are consistent at the 1–2%/decade level, roughly that of trend uncertainties (WMO, 2010a).

The vertical profiles of trends in ozone over northern mid-latitudes (35°–60°N) derived from SAGE, SBUV, and Umkehr instruments are very similar for 30–38 km. SAGE trends are larger than SBUV and Umkehr trends above 40 km (Figure 6.6). Similarly, over 35–60°S, SAGE and SBUV trends are essentially the same as those for 30–38 km and SAGE trends are larger above 40 km. Despite the difference in the magnitude of the trend, both SAGE and SBUV show that the decline at 40 km over Southern Hemisphere mid-latitudes is slightly larger than that over Northern Hemisphere mid-latitudes, but the difference is within the error bars. A slightly stronger decline over the Southern Hemisphere was previously reported from SAGE data in WMO (2003).

Analysis of trends revealed the presence of statistically significant trends for all altitudes within the 12–50 km layer with two distinct maxima near 15 km and

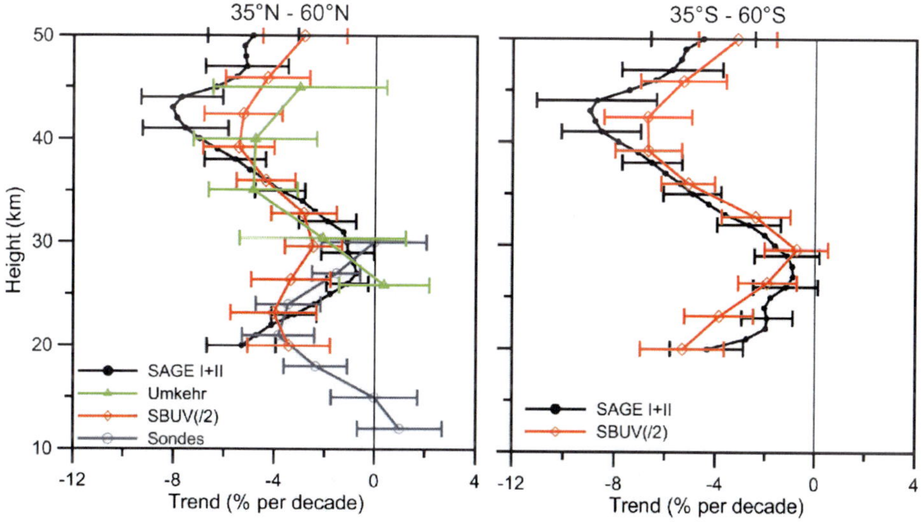

Figure 6.6. Vertical profile of ozone trends over northern and southern mid-latitudes estimated from ozonesondes, Umkehr, SAGE I, SAGE II, and SBUV(/2) for the period 1979–2004. The trends were estimated using regression to an EESC curve and converted to %/decade using the variation of EESC with time in the 1980s. The trends were calculated in geometric altitude co-ordinates for SAGE and in pressure co-ordinates for SBUV(/2), sondes, and Umkehr data, and then converted to altitude co-ordinates using the standard atmosphere. The 2σ error bars are shown (WMO, 2007).

40 km. According to new observation data an ozone concentration decrease from −6% to −8% per decade takes place at 40–45 km but is characterized by substantial annual variations in the winter maximum (noticeable hemispheric differences are absent). In the lower stratosphere (at altitudes below 30 km), the principal information comes from SAGE (20–30 km) and ozonesonde (up to 27 km) observations. At 20 km the ozone concentration trend varies (due to ozonesonde data for different points) from −3% to −11% per decade. Annual variations (which are absent above 20 km) are noticeable mainly in the 10–20 km layer, but are characterized by substantial geographic variability. According to data from European stations, maximum ozone drop is observed in winter and spring, whereas Canadian data revealed the maximum in spring and summer (Harris *et al.*, 1998). There is agreement between satellite and ozonesonde observations with regard to annual variation in the 10–20 km layer in the mid-latitudes with a distinct summer/spring maximum in the Northern Hemisphere and much weaker variation in the Southern Hemisphere.

Steinbrecht *et al.* (2009) used data from a variety of satellite-flown instruments along with data from five ground stations from 45°S to 48°N to study the ozone concentration in the 35–45 km range in the stratosphere. From 1979 until the late 1990s, all the available data showed a clear decline of ozone near 40 km, by 10% to 15%. This decline did not continue in the following 10 years. At some sites, ozone at 40 km appears to have increased since 2000, consistent with the beginning of a decline of stratospheric chlorine.

Tatarov *et al.* (2009) used differential absorption lidar (DIAL) (see Section 1.6.5 for a description of the DIAL technique) at Tsukuba (36°N, 140°E), Japan, along with data from SAGE II (Stratospheric Aerosol and Gas Experiment) and NCEP (National Center for Environmental Prediction) data to study the long-term trends of ozone concentration after subtraction of all natural variations.

Figure 6.7 shows the ozone trend after subtraction of QBO and solar cycle effects for the altitude range from 30 to 40 km (upper stratosphere). This figure clearly shows the different behavior of the ozone trend for the periods before and after about 1998. One can see the strong steady decrease of ozone concentration at 30–40 km altitude from 1988 to the end of 1997. The trend from the lidar data was $(-6.0 \pm 0.5)\%$ per decade and the trend from the SAGE II data was $(-5.2 \pm 0.4)\%$ per decade. After 1998 there was only a negligible decrease of ozone concentration obtained by both instruments, namely $(-0.8 \pm 1.1)\%$ per decade from the lidar data and $(-0.9 \pm 1.0)\%$ per decade according to SAGE II data. It was not possible to draw any strong conclusion about decreasing ozone concentration after 1998, because of the high standard deviations for both datasets.

Sahoo *et al.* (2005) studied the variability in TOC over the Indian subcontinent, which, of course, includes tropical areas in the south and mid-latitude areas in the north; their work demonstrates clearly the difference between these two areas. They used satellite and ground-based observations from 1979 to 2003. They observed that the declining trend of TOC over the northern parts of India was higher compared with other parts of India, where the TOC was almost stable. Ganguly and Iyer (2009) used total ozone data obtained from TOMS and stratospheric column ozone (SCO)

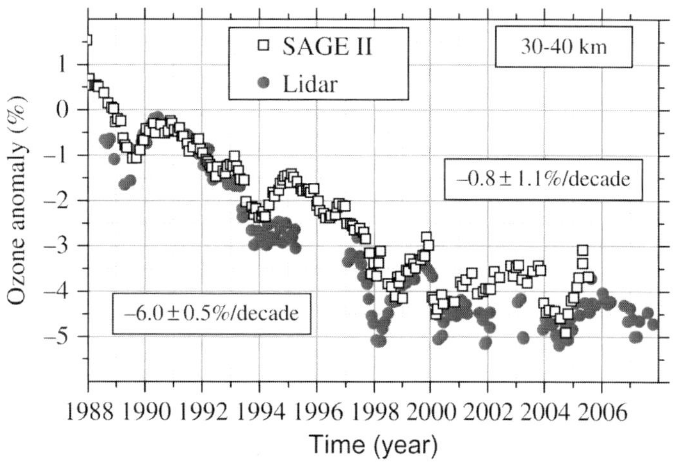

Figure 6.7. Ozone anomalies after subtraction of annual cycle, QBO, and solar cycle effects at altitude range from 30 to 40 km. Monthly means values are presented. Data are smoothed by a 6-month running mean (from Tatarov *et al.*, 2009).

obtained using the Convective Cloud Differential (CCD) method of Ziemke *et al.* (2005) to study the SCO over 5° wide latitude belts of the Indian subcontinent from 10°N to 35°N (five belts). This was done for three periods, 1979–1988, 1989–1992, and 1997–2001. Unlike places located at higher and middle latitudes, where severe ozone depletion has been reported (Chandra *et al.*, 1996; Kondratyev *et al.*, 1994; Varotsos and Cracknell, 1994; Varotsos and Kirk-Davidoff, 2006; WMO, 1989) the SCO and TOC at Indian latitude belts exhibited a small decreasing trend, which is larger for higher latitudes, for the period 1979–1988. This decreasing trend was found to increase still further at all latitude belts during the period 1989–1992. In the third period studied Ganguly and Iyer observed a reduction in the decreasing trend in SCO and TOC in the higher latitudes (25°N–35°N) while places located within the range 10°N–25°N showed a small increasing trend in SCO and TOC. The trends for SCO and TOC (see Table 6.3) show that the decreasing trend in TOC is slightly less than that of SCO; however, it should be noted that the periods over which the trends were calculated are relatively short and the significance of the trends cannot be determined.

Krzyścin and Rajewska-Więch (2009) studied Umkehr data (see Sections 1.6.2, 1.6.3 for a description of the Umkehr technique) from the Dobson spectrophotometer at Belsk (52°N, 21°E) for 1963 to 2007. Trends in ozone content in Umkehr layers were calculated in time series using natural ozone fluctuations due to 11-year solar activity, but with QBO removed. They found a negative trend in TOC of about ~3.5% per decade for the period 1980–1995. After that, from 1996–2007, the trend in TOC was not statically significant. The trends in individual Umkehr layers are shown in Table 6.4. Ozone depletion in the period 1980–1995 is mostly due to ozone depletion in the lower and middle stratosphere (up to 23.5 km). After that the trends in TOC and in lower and mid-stratospheric ozone

Table 6.3a. Comparison of annual changes in stratospheric column ozone (SCO) ($DU\,y^{-1}$) from 1979–1988, 1989–1992, and 1997–2001 in different Indian latitude belts before and after the implementation of the Montreal Protocol in India (from Ganguly and Iyer, 2009).

Latitude belt (°N)	*Annual trend of SCO* *1979–1988*	*Annual trend of SCO* *1989–1992*	*Annual trend of SCO* *1997–2001*	*SCO recovery between 1989–1992 and 1997–2001*
10–15	−0.036	−0.212	+0.031	+0.243
15–20	−0.066	−0.268	+0.017	+0.285
20–25	−0.134	−0.371	+0.013	+0.384
25–30	−0.158	−0.309	−0.014	+0.295
30–35	+0.177	−0.333	−0.067	+0.266

Table 6.3b. Comparison of annual changes in total ozone content (TOC) ($DU\,y^{-1}$) from 1979–1988, 1989–1992, and 1997–2001 in different Indian latitude belts before and after the implementation of the Montreal Protocol in India (from Ganguly and Iyer, 2009).

Latitude belt (°N)	*Annual trend of TOC* *1979–1988*	*Annual trend of TOC* *1989–1992*	*Annual trend of TOC* *1997–2001*	*TOC recovery between 1989–1992 and 1997–2001*
10–15	−0.032	−0.225	+0.122	+0.347
15–20	−0.053	−0.293	+0.115	+0.408
20–25	−0.088	−0.298	+0.093	+0.391
25–30	−0.132	−0.274	−0.006	+0.268
30–35	−0.163	−0.323	−0.071	+0.252

are not statistically significant. In the upper stratosphere—above ~33 km (i.e., layer 7 and combined layers 8, 9, and 10)—there are positive trends of ~3–5% per decade (statistically significant) in the period 1996–2007. Thus, the trend analyses presented by Krzyścin and Rajewska-Więch (2009) suggest a positive change in trend after 1995 but no turnaround in TOC or in lower-stratosphere ozone. According to Krzyścin and Rajewska-Więch (2009), in the post-Pinatubo period, large negative trends in TOC during summertime combined with large positive trends in TOC during the autumn–winter period resulted in a decrease in annual peak-to-trough differences from those of the pre-Pinatubo period.

Yang *et al.* (2009) used the solar backscatter ultraviolet TOC measurements from Nimbus-7 SBUV and operational NOAA-9, -11, -14, -16, and -17 SBUV/2

Table 6.4. Linear trends (% per decade) at Belsk in selected Umkehr layers and corresponding TOC for the periods 1980–1995 and 1996–2007 (from Krzyścin and Rajewska-Więch, 2009).

Layer	*Altitude range* (km)	*Trend 1980–1995*	*Trend 1996–2007*
2 + 3	10.3–19.1	−9.9 ± 3.2	−0.7 ± 3.5
4	19.1–23.5	−2.0 ± 1.3	1.0 ± 1.7
5	23.5–28.0	0.6 ± 1.3	0.4 ± 1.5
6	28.0–32.6	0.2 ± 1.4	0.1 ± 1.6
7	32.6–37.5	−0.9 ± 1.8	3.2 ± 1.5
8 + 9 + 10	>37.5	0.5 ± 1.6	4.5 ± 1.3
1–10	TOC	−3.5 ± 1.6	0.4 ± 1.5

data to compile a long-term SBUV/2 total ozone dataset. This compilation, which aimed at achieving the level of quality suitable for trend analysis, followed the methodology of Miller *et al.* (2002), but with several improvements. The dataset started from 1979 and went through to 2007, but the two years of data following the Mt. Pinatubo eruption in June 1991—from July 1991 to June 1993—were excluded from their trend analysis of the data. Using the CUSUM—the cumulative sum (CUSUM) is an analysis in which the cumulative sum of the differences in time between the data and an assumed model is used to characterize the data relative to the model—algorithm (Newchurch *et al.*, 2003), for assessing the recovery of lower-stratospheric ozone, Yang *et al.* (2009) chose the inflection at the start of 1997. Based on the data up to 2005, Weatherhead and Anderson (2006) had used 1996 as the inflection and reported a plausible ozone turnaround.

Since the SBUV series instrument requires backscattered UV radiation to determine how much ozone is in the atmosphere, the range of latitudes able to provide measurements varies with season as the polar night region waxes and wanes. In addition, the NOAA operational polar-orbiting satellites used after 1989 were not in a stable equator-crossing orbit. As a result, the latitudinal range of observations decreased, especially near the solstices. The net result is that use of latitudes beyond 60°N and 60°S would misrepresent the true annual cycle. Consequently, although Yang *et al.* (2009) analyzed the SBUV/2 data in 5° latitude zones centered from 80°N to 80°S, they only used data from 50°N to 50°S in their trend analysis. To eliminate inter-satellite biases, an overlap of at least one year was achieved, except for NOAA-11 and NOAA-16 when there was a 6-month overlap. The mean difference between current and succeeding satellites for each zone was then computed and all the satellite data were then adjusted to Nimbus-7 data.

Yang *et al.* (2009) computed monthly average total ozone in each latitude band for the period January 1979 to December 2007. Each month's individual value was

Table 6.5. Trend results for TOC computed by Yang *et al.* (2009) for 45°N, 45°S, and integrated 50°N–50°S.

Trend term	45°N	45°S	50°N–50°S
Trend (T_1) (%/decade)	−3.69(1.16)	−3.30(1.13)	−2.43(0.40)
Trend change (T_2) (%/decade)	5.02(2.65)	2.84(2.60)	3.67(0.89)
Net slope ($T_1 + T_2$) (%/decade)	1.33(2.49)	−0.46(2.12)	1.24(0.83)
Solar (%/100f10.7)	2.74(3.78)	5.30(7.00)	8.90(2.76)
AO (%/AO)	−5.29(1.56)	0.41(1.42)	0.07(0.60)
AAO (%/AAO)	−0.89(1.50)	−4.99(1.40)	−0.62(0.58)
QBO1 (%/QBO1)	−0.23(0.17)	−0.43(0.16)	−0.18(0.06)
QBO2 (%/QBO2)	−0.16(0.12)	−0.11(0.12)	−0.02(0.04)
R^2	0.58	0.69	0.83

then subtracted from that month's 29-year average to compute the monthly anomaly. Percentage anomalies are determined by dividing by each month's long-term average. The algorithm of linear regression was applied to the three ozone time series of zonal averages: 45°N (45N), 45°S (45S), and integrated from 50°N to 50°S (50NS). Allowance was made for the effects of variations due to the solar cycle, the Arctic Oscillation (AO), the Antarctic Oscillation (AAO), and the Quasi-Biennial Oscillation (QBO) (see Sections 4.2, 4.4). Randel and Wu (1996) separated the global QBO signal into a set of two orthogonal modes, QBO1 and QBO2, which can account for latitudinal variations.

The results obtained in the trend analysis by Yang *et al.* (2009) are presented in Table 6.5; they are briefly described as follows:

1. For 45°N both the trend and trend change are statistically significant at better than the 95% level of confidence and the trend change term is positive so that the net slope is positive. However, the net slope value of 1.33 is far from being statistically significant so that it could not be stated with confidence that, for this latitude, TOC has been increasing with time since 1996.
2. For 45°S the results were quite different. While the trend and trend change terms are each statistically significant, the trend change coefficient is smaller than the trend term so that the net slope remains negative though not statistically significant from zero. For this region, it could not be argued that TOC is increasing, but perhaps is decreasing at a slower rate than prior to 1996.
3. For the integrated domain of 50°N–50°S, the averaging process diminishes greatly the standard errors of the terms. The result is that the net slope for the integrated domain is +1.24% per decade and is statistically significant at

the 95% confidence level. Also, the solar coefficient is statistically significant, indicating the tropical dominance of this effect.

Instead of fitting a linear (or piecewise linear) trend to the ozone time series, the changes in ozone can be analyzed by including a term in the statistical significance that depends on the stratospheric burden of ODSs, such as the EESC or a similar function, rather than just the concentration of ODSs.

The contributions of the various explanatory variables to ozone fluctuations are illustrated in Figure 6.8(a) (WMO, 2007). The figure shows area-weighted seasonal mean total ozone values for the region 60°S–60°N estimated from ground-based and satellite data, and the contribution (in Dobson units, DU) of the natural components in ozone variability and the EESC contribution to TOC from 1964 to 2006 (WMO, 2007). The EESC-related term included in the regression model allows for possible seasonal variations. TOC for the region 60°S–60°N, with the natural components removed, is shown in Figure 6.8b. The plot illustrates the good agreement between long-term changes in ozone and EESC (WMO, 2007). It can be seen from Figure 6.8b that the trends in EESC and in ozone are very nearly linear for 1979–1989; thus, ozone depletion can be expressed in terms of linear changes for this time period, with results reported in ozone changes (% or DU) per decade (as in Stolarski *et al.*, 2006a, b); this makes it easier to compare the results with linear trends reported earlier. For a period that includes the non-linear region, ozone depletion can be expressed as simple differences between two dates.

It should also be noted that care needs to be taken when comparing trends derived from the vertical distributions of ozone concentration expressed in different vertical co-ordinate systems (i.e., heights in km, pressure in hPa, or temperature in degrees K). Rosenfield *et al.* (2005) demonstrated that, between 1979 and 1997, ozone trend differences at 3 hPa are as large as 1–2%/decade depending on how ozone changes are reported, with the largest differences being in southern high latitudes. The most negative trend is computed when ozone profiles are given in terms of heights expressed in km, while the least negative trend is obtained when ozone profiles are expressed in terms of pressure levels in hPa. As a practical example, ozone profiles obtained from the SAGE instrument are expressed in terms of altitude in km, while ozone profiles obtained from Solar Backscatter Ultraviolet (SBUV) data are expressed in terms of pressure in hPa. Thus, upper-stratospheric ozone trends computed from SAGE data will be larger than those computed from SBUV data when the trends are computed in the natural vertical co-ordinate for each instrument. For comparison with SBUV, SAGE ozone profiles are typically converted from concentrations for altitudes in km to mixing ratios for pressure levels in hPa (e.g., McPeters *et al.*, 1994; Wang *et al.*, 1996). However, errors in the temperature data used to carry out this conversion can then affect computed trends. Li *et al.* (2002) estimated that adjustment for temperature trends would reduce differences in SBUV trends relative to SAGE by about 1%/decade, in agreement with Rosenfield *et al.* (2005).

When long-term changes in global ozone were reviewed in the 2002 WMO report (WMO, 2003), global total ozone values were about 3% below the pre-

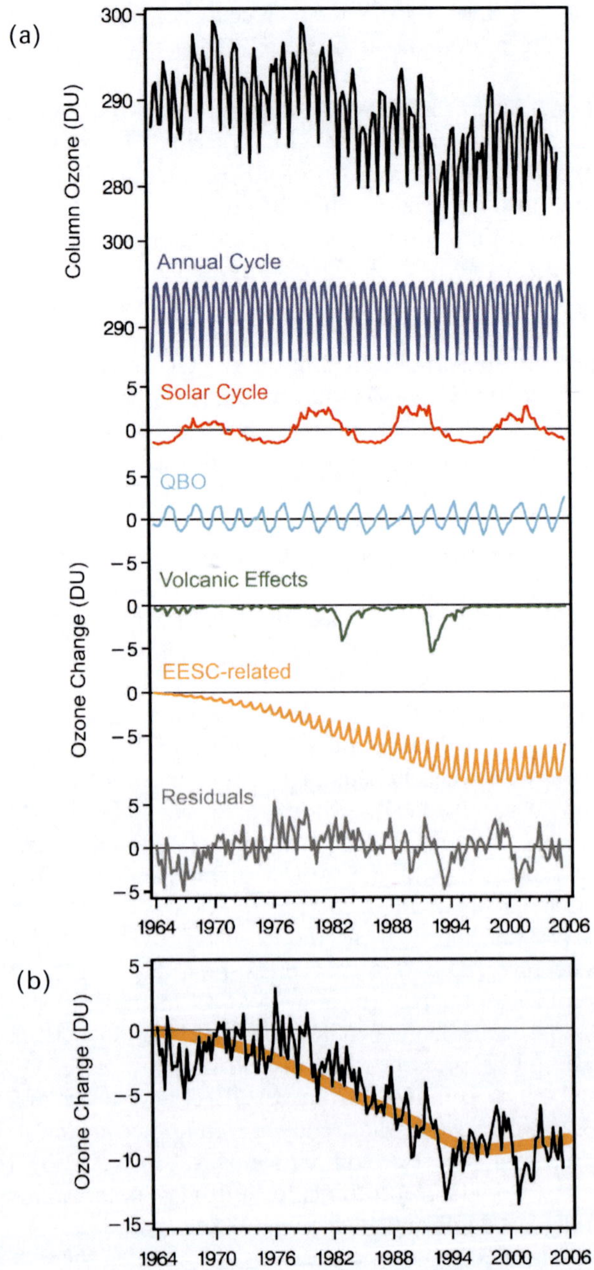

Figure 6.8. (a) Ozone variations for 60°S–60°N estimated from ground-based data and individual components that comprise ozone variations. (b) Deseasonalized area-weighted total ozone deviations estimated from ground-based data adjusted for solar, volcanic, and QBO effects, for 60°S–60°N. The thick yellow line represents the EESC curve scaled to fit the data from 1964–2005 (WMO, 2007).

1980 levels. This reduction had occurred primarily in the mid- to high latitudes of both hemispheres, with no significant trends reported in the tropics. Clear differences between the two hemispheres were noted. Annually averaged total ozone over the northern mid-latitudes was 3% lower than pre-1980 levels, with twice as much decline in winter than in summer. Over the southern mid-latitudes, by contrast, a long-term decrease of 6% had occurred, roughly constant through the year. WMO (2003) concluded that models including observed changes in halocarbons, other source gases, and aerosols captured the long-term behavior in the mid-latitudes. At that time, the differences between the Northern and Southern Hemispheres were not explained. However, the report (WMO, 2003) did conclude that there was increased evidence that the observed changes in atmospheric dynamics had a significant influence on ozone over northern mid-latitudes on decadal timescales. It also noted that the magnitude of these influences, largest in the lower stratosphere, had not been quantified unambiguously. At higher altitudes (35–50 km), ozone declining trends were up to 8%/decade over the mid-latitudes in both hemispheres, consistent with the expectation due to observed changes in chlorine concentrations.

Karol *et al.* (2009) made the point that studies of TOC content and trends in extratropical northern latitudes as measured at ground-based stations and from satellites have generally been carried out for zonally averaged latitudinal bands, but by contrast there have been very few studies of longitudinal variations of TOC. Karol *et al.* (2009) conducted a statistical analysis of TOC data from Russian and former U.S.S.R. territory for 1973–2005 from the ground-based network (32 stations), which is spread between 50°N and 70°N and extends from 30°E to 160°E. These stations measure TOC and are distributed almost uniformly over the territory of the former U.S.S.R.; they began functioning in the 1960s. Filter ozonometers (M-83) were initially used at these stations and, after the installation of new filters in 1973, TOC measurement results have fitted well with averaged TOC data of other ground-based stations in the Northern Hemisphere, which used Brewer and Dobson spectrophotometers (Bojkov *et al.*, 1994; Fioletov *et al.*, 1999, 2002; Ionov *et al.*, 2002; Shalamyansky, 1993). M-83 filter ozonometers were replaced by M-124 ozonometers (see Section 1.4) in the late 1980s; these instruments are regularly calibrated against the standard Dobson spectrophotometer in St. Petersburg.

Russia is usually divided into five climatic regions: north (NWR) and south (SWR) European Russia, western (WS) and eastern (ES) Siberia, and the Far East and Chukotka (FE). Each of these regions contains five to eight stations providing a continuous TOC record for around 30 years. Figure 6.9 presents seasonal variations of mean monthly TOC averaged over the station data for each of these regions, for 1973–2002. The TOC data for the Transcaucasus and Central Asia region are also shown in Figure 6.9, but were not included in the analysis of Karol *et al.* (2009) because measurements were discontinued in the early 1990s. This figure reveals significant regional TOC differences in the winter–spring seasons, but very little apparent TOC difference between the regions in the summer period June–August (the details are discussed at length by Karol *et al.*, 2009).

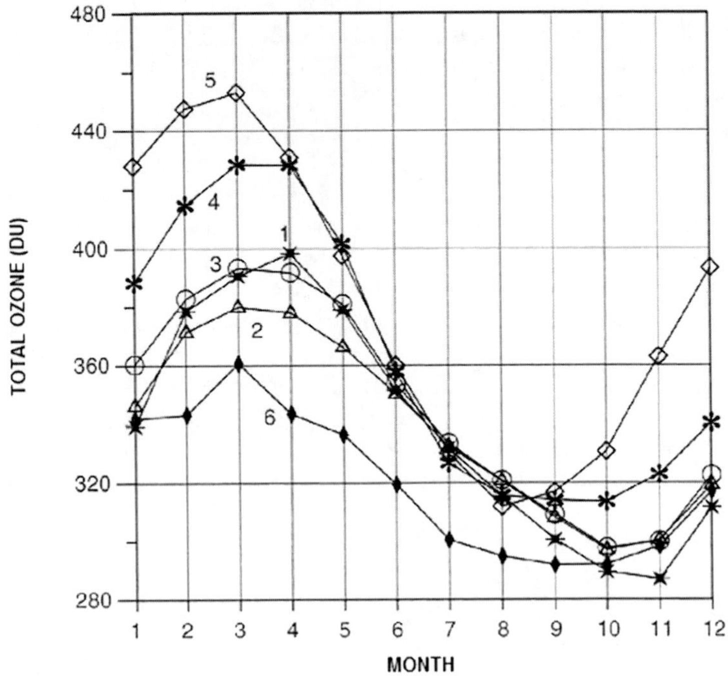

Figure 6.9. Mean seasonal variations of monthly TOC averaged over the NWR (1), SWR (2), WS (3), ES (4), FE (5), and Transcaucasus and Central Asia (6) regions over the period 1973–2002 (Karol *et al.*, 2009).

Karol *et al.* (2009) studied three sets of TOC data for each of the five regions—A (annual data), B (December–March), and S (June–August)—and analyzed the temporal ozone variations for the residual A′, B′, and S′ detrended TOC series after subtracting their linear trends from the initial A, B, and S. This was done for the period 1973–1992 for TOC decrease and also for the period 1993–2005 for ozone partial recovery. The 1992/1993 trend sign change in the above analysis reflects the joint influence of the 1991 eruption of the Pinatubo volcano and the effects of the Montreal Protocol coming into action.

Table 6.6 presents these trends and the significance of their deviations from zero (in the denominators). The maximal trends for both signs are for the winter B series, but with minimal confidence levels. The weakest trends are for the S (summer) TOC series. From this table we see that the TOC decrease rate in 1973–1992 was slightly higher in European Russia. The TOC recovery in 1993–2005 was a little faster in the Asian part of the country, but this distinction and all trends have very low significance due to large interannual variations of TOC.

These results demonstrate significant longitudinal differences in the structure of TOC fields and their evolution in time over Russia from west to east. These differences are statistically significant and almost permanent during the long period of

Table 6.6. Total ozone linear trends (DU/year, in numerators) and their deviation from zero significance (%, in denominators) for periods I, 1973–1992 and II, 1993–2005 in the A (annual), B (Dec.–Mar.), S (Jun.–Jul.) total ozone time series (from Karol *et al.*, 2009).

Period	*Time series*	*Regions*				
		1. NWR	2. SWR	3. WS	4. ES	5. FE
I. 1973–1992	A	−1.65/82	−1.21/78	−1.01/70	−1.62/76	−1.41/80
	B	−2.53/77	−1.55/74	−1.33/65	−1.36/65	−1.50/73
	S	−0.76/65	−0.99/70	−1.20/75	−1.10/73	−0.90/70
II. 1993–2005	A	1.06/*	1.71/79	1.1/66	0.08/76	1.60/72
	B	1.57/*	1.88/71	2.05/73	3.43/77	2.34/70
	S	0.19/*	1.41/79	0.0/*	0.32/*	0.27/*

* Significance is less than 65%.

ozone depletion in the 1980s to the first years of the 1990s and also in the current period of ozone partial recovery. The two phases of QBO west (W) and east (E) at the 30–50 hPa levels for January–March of each year are considered. The years of high winter and annual ozone and the warm Arctic lower stratosphere begin with the E QBO phase, when meridional stratospheric air transport in winter was more intense compared to the years with the W QBO phase (see, e.g., Labitzke and van Loon, 1997b; WMO, 2003). During the years of low winter–annual ozone in the cold Arctic stratosphere, the QBO had a W phase, and in some years the E phase changed to a W phase at upper levels.

On the basis of these results, Karol *et al.* (2009) proposed a working hypothesis of ozone dynamics over the subarctic zonal belt and northern Eurasia: stratospheric ozone is transported by planetary waves from its maximal content in the tropical lower stratosphere to high northern latitudes and to the Arctic stratosphere, predominantly during the QBO E phase and in the longitudes of the Far East (5) region and Western Pacific Ocean along the western periphery of the known North Pacific winter stratospheric anticyclone. Then, zonal flow distributes this ozone over the relevant zonal belt to the east and to the west and ozone comes to the stratosphere of European Russia, having a short "memory" of QBO and being influenced by the 11-year solar cycle. Further investigations are needed to examine this hypothesis and find the influence of other factors (e.g., the Arctic and North Atlantic Oscillations) on such ozone transport over Russia.

Whereas most investigations of ozone depletion consider average values of ozone concentration (e.g., monthly averages or yearly averages), Patil and Revadekar (2008) have adopted a rather different approach to studying trends in ozone; this makes it difficult to compare their results with the results of other trend analyses. Revadekar and Patil (2009) considered the frequency of low ozone days and high ozone days and the lowest ozone concentration and the highest ozone concentration over northern India in winter for the periods 1979–1988 and 1997–2005. They used daily values of TOC from TOMS on Nimbus-7 (1979–1988) and the Earth Probe (1997–2005) satellites and obtained TOC values for eight locations in India north of 20°N. To compare recent changes with the extremes prior to the Montreal Protocol, the 10th and 90th percentile values were computed for each of these stations based on daily ozone data during the winter months (January and February) for the period 1979–1988. These were then used as thresholds to identify, at each station, whether each day was a low, high, or normal ozone day. A best fit linear trend was computed for each of the selected stations for frequencies of low/high ozone days and also for the lowest/highest ozone values for both the periods 1977–1988 and 1997–2005.

6.5 RECOVERY OF THE OZONE LAYER; THE POLAR REGIONS

As mentioned in Chapter 1, strong ozone depletion has been recorded in the lower polar stratosphere in both hemispheres. Nowadays, there is a consensus in the scientific community that anthropogenic activities are resulting in an increase in tropospheric ozone (e.g., Cartalis and Varotsos, 1994; Jacovides *et al.*, 1994; Kondratyev and Varotsos, 2001a, b; Varotsos *et al.*, 2001a) and a decrease in the stratospheric ozone (e.g., Varotsos *et al.*, 2000). In particular, stratospheric ozone depletion is considered as one of the strongest anthropogenic signals in the Earth system. Nearly two decades after the signing of the Montreal Protocol, stratospheric ozone loss is as severe as ever over the Arctic, and the timing of ozone recovery is uncertain. In this context, the "Ozone layer and UV radiation in a changing climate" (ORACLE-O3) project running during the International Polar Year (IPY) 2007–2008 implies quantification of polar ozone losses in both hemispheres by employing ground-based and satellite observations.

In the Antarctic, ozone depletion became evident in the 1980s, with most of the ozone loss occurring in the lower stratosphere and attributed to chemical ozone destruction mechanisms (WMO, 1999). We discussed the Antarctic ozone hole in some detail in Section 5.4. In the early days following the discovery of the ozone hole in the Antarctic spring relatively little attention was paid to the question of whether a similar phenomenon might exist in the Arctic. We noted that the ozone hole is not actually a hole but a region of heavily depleted ozone in the atmosphere that is defined, slightly arbitrarily, as a region where the total ozone column (TOC) is less than 220 DU; the hole only survives for a few months each year. The Antarctic ozone hole was first discovered from Dobson spectrophotometer data (Chubachi, 1984, 1985; Farman *et al.*, 1985), but its extent and temporal evolution are continuously monitored from satellite data obtained from the Total Ozone Monitoring Spectrometer (TOMS) and its successors, such as Microwave Limb Sounder (MLS) and ozonesonde ascents (Antón *et al.*, 2009; Manney *et al.*, 1997). After a while, serious studies of TOC in the Arctic spring began to be undertaken and a similar, but smaller, ozone hole was found there too.

Chemical ozone losses in the lower stratosphere in the Arctic have been observed from the beginning of the 1990s (von der Gathen *et al.*, 1995). The strongest ozone loss in the lower stratosphere occurs in the Antarctic in October and in the Arctic in March. It is commonly accepted that these ozone losses are mainly caused by polar stratospheric clouds (PSCs) and related chemical destruction of ozone inside the circumpolar vortex (see Sections 5.6 and 6.5.4). These clouds provoke further decrease in the temperature of captured circumpolar air. The heterogeneous chemical reactions that deplete ozone take place at the surfaces of the crystals in these clouds, leading to depletion of the ozone and production of the ozone hole. Ozone-destroying chemical processes are more effective under cold conditions (below -78°C). Consequently, the ozone hole appeared first over the Antarctic because the stratosphere in the Southern Hemisphere is about 5°C colder than in the Northern Hemisphere and PSCs persist there for longer periods. In addition, natural ozone levels in the Arctic spring are much higher than in the Antarctic spring. Therefore, ozone depletion over Antarctica is much more pronounced than over the Arctic. While some considerable attention has been paid in recent years to studying the rate of recovery of the Antarctic ozone hole, very little work has been done on the Arctic ozone hole; this is because its discovery was more recent and it is smaller than the Antarctic ozone hole (see also Section 6.5.4). It could be argued, however, that the formation of even a moderate hole above the Arctic region in the ozone layer, which protects the biosphere of our planet from harmful solar ultraviolet radiation (SUVR) (Alexandris *et al.*, 1999; Katsambas *et al.*, 1997; Kondratyev and Varotsos, 1996a, b; Varotsos *et al.*, 1995a, c), could be a serious problem for the large population in the middle and high latitudes of the Northern Hemisphere.

Kiesewetter *et al.* (2010) described a 29-year dataset of stratospheric ozone from sequential assimilation of solar backscatter UV (SBUV) satellite ozone profile observations into a chemistry–transport model and validated it against independent observations (satellite instruments and ozonesondes). The assimilated dataset showed excellent agreement with ozone profile data obtained from ozonesonde measurements from high-latitude observation sites in both hemispheres and MIPAS satellite observations, including during polar night when no SBUV observations are available. Although they only assimilated ozone profiles, the values of TOC in the assimilated dataset were in good agreement with independent satellite observations from GOME, SCIAMACHY, and TOMS. The dataset can thus be viewed as a consistent long-term dataset closing the gaps in satellite observations in order to investigate high-latitude ozone variability. Kiesewetter *et al.* (2010) then used the assimilated dataset to analyze the development and persistence of both high and low ozone anomalies in the Arctic stratosphere. Ozone anomalies typically develop in the 1,000 K potential temperature ($\sim$35 km) region and slowly descend from there, remaining visible for around 7 months. Anomalies in stratospheric circulation, expressed by the Northern Hemisphere annular mode (NAM) index, show a large influence on ozone anomalies. Extreme phases of the NAM index (strong and weak vortex events) lead to the creation of distinctively shaped ozone anomalies, which first appear in the uppermost stratosphere and then rapidly cover the upper and middle stratosphere, from where they then slowly descend into the lowermost stratosphere within 5 months.

The true long-term decline in winter and spring above the Arctic may be underestimated by a lack of TOMS/SBUV observations in winter, particularly in the cold 1990s. The main contributions to observed interannual variations of total ozone and lower-stratospheric temperature at 50 hPa come from a linear trend (up to −10 DU/decade at high northern latitudes, up to −40 DU/decade at high southern latitudes, and around −0.7 K/decade over much of the globe), from the intensity of polar vortices (more than 40 DU, or 8 K peak to peak), the QBO (up to 20 DU, or 2 K peak to peak), and from tropospheric weather (up to 20 DU, or 2 K peak to peak). Smaller variations are related to the 11-year solar cycle (generally less than 15 DU, or 1 K), or to ENSO (up to 10 DU, or 1 K). These observed variations are replicated well in the simulations. Volcanic eruptions have resulted in sporadic changes (up to −30 DU, or +3 K). At low latitudes, patterns are zonally symmetric. At higher latitudes, however, strong zonally non-symmetric signals are found close to the Aleutian Islands or south of Australia. Such asymmetric features appear in model runs as well, but often at different longitudes than in observations. The results point to the key role of zonally asymmetric Aleutian (or Australian) stratospheric anti-cyclones for interannual variations at high latitudes, and for coupling between the polar vortex strength, the QBO, the 11-year solar cycle, and ENSO.

Ozone observations by GOME-2 on ERS-2 during the Arctic springs of 1997 and 1998 showed that extensive regions of low TOC were observed. In particular, the major decrease is dominant in the lower and mid-stratosphere inside the polar vortex. Also, the winter 1997/1998 was warmer than the year before and less ozone depletion was observed. In spring 1998 an ozone mini-hole event was observed and ozone profiles under ozone mini-hole conditions were derived for the first time.

Krasouski *et al.* (2010) performed extensive studies of the statistics for negative (mini-holes) and positive (mini-highs) anomalies over Europe for the period of 1957–2006 using data derived from satellites and ground-based ozone stations. Their results were exclusively derived from the data of the total ozone fields for the period without employing any additional sources of information and hypothesis. The "strength" of an ozone anomaly is estimated from the maximum total ozone deviation from the long-term monthly average value of ozone for each ozone station registering the anomaly. Analysis of the temporal sequence of an anomaly passing through a set of ozone stations gives information on the trajectory of the anomaly. They selected two historical periods (1957–1979 and 1980–2006) for comparison. Some conclusions drawn from other studies are confirmed by statistical analysis of the frequency and seasonal distribution of the ozone anomalies that are presented by Krasouski *et al.* (2010), whereas some others are not well confirmed with their findings. Ozone anomalies may be classified into two types according to their movement trajectory towards Europe: Atlantic-type and Arctic-type anomalies (see Figure 6.10):

- anomalies of the Atlantic type moving to Europe from the equatorial (path 2 in Figure 6.10) and western (path 1 in Figure 6.10) parts of the Atlantic Ocean; and
- anomalies of the Arctic type moving to Europe from the North Pole (path 3 in Figure 6.10).

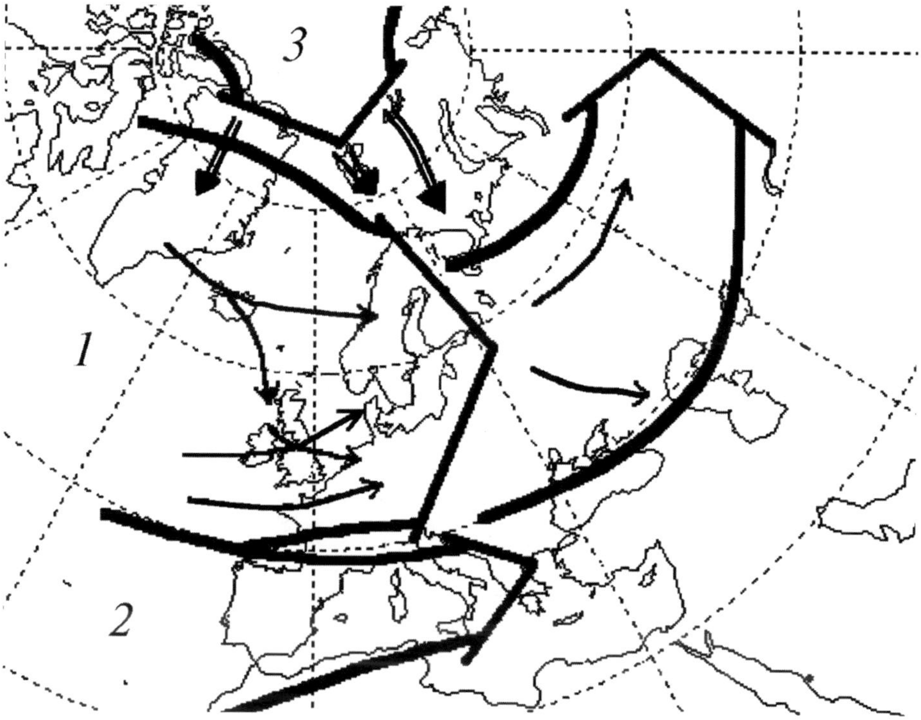

Figure 6.10. Propagation paths of ozone anomalies to Europe (Krasouski *et al.*, 2010).

The statistics confirmed an increase in the number of the Atlantic-type negative ozone anomalies in the annual cycles of the last few decades. The frequency of Arctic-type negative and positive anomalies had practically not changed. The seasonal distributions of the relative frequency for anomalies of all types had not essentially altered. The seasonal distribution of the relative frequency of the deepest negative Atlantic anomalies is narrower than the distribution of average-type anomalies. There are also essential distinctions in the seasonal distribution of the frequency of the Arctic-type "average" and "high" negative anomalies (Krasouski *et al.*, 2010).

To study seasonal ozone variation at high northern latitudes, ozone vertical distributions collected from the Arctic Lidar Observatory for Middle Atmosphere Research (ALOMAR) (69.3°N, 16.0°E) station at Andenes (Norway) and from the stations at Sodankylä (Finland) (67.4°N, 26.6°E) and at Ny-Ålesund (Svalbard) (78.9°N, 11.9°E), were examined by Werner *et al.* (2011). The datasets cover the period from 1994 to 2004. They found a second ozone maximum near 13–15 km, between the tropopause and the absolute ozone maximum near 17–20 km. This maximum is built up by the combination of airmass transport and chemical ozone destruction, mainly caused by the NO_x catalytic cycle, which begins after the polar night and intensifies with increasing day length. Formation of a troposphere inversion layer was observed. This inversion layer is thicker and reaches higher altitudes in winter than in summer. However, temperature inversion during summer is stronger. The formation of enhanced ozone density was observed during the spring–summer period. Ozone is accumulated or becomes poor by synoptic weather patterns just above the tropopause from spring to summer. Seasonal averages show ozone enhancement above the tropopause. The stronger temperature inversion during the summer period inhibits vertical stratosphere–troposphere exchange. Horizontal advection in the upper troposphere and lower stratosphere is enforced during summer. The combination of these mechanisms generates a layer with very low ozone density above the troposphere inversion layer subsisting from June/July until the late autumn.

Syrovatkina *et al.* (2008) performed a synoptic analysis of monthly and daily mean Northern Hemisphere total ozone fields for 1998–2005 using ground-based (Roshydromet) and TOMS measurements. Large interannual changes in the evolution of the stratospheric polar vortex and the North Pacific anticyclone were shown to influence the formation and dynamics of winter–spring ozone fields in the stratosphere of high northern latitudes. The analysis showed considerable variations in the direction of zonal ozone transport from the sector of ozone inflow from low latitudes and accumulation in the Far East depending on winter polar stratosphere temperature and Quasi-Biennial Oscillation (QBO) phase. In years with an easterly QBO phase and a warm polar stratosphere, ozone at the end of winter is transported to northeastern Canada and the Atlantic. In years with an easterly phase and a cold polar stratosphere, ozone transport is directed to northern Eurasia.

Rafanelli *et al.* (2009) reported the beginning of a weak recovery of the ozone layer in the Arctic. This was based on the results of a winter campaign which was initiated in 2007 at Ny-Ålesund (Svalbard) (78.9°N, 11.9°E) with a Brewer spectrophotometer. Because of the lack of sunlight the instrument is operated by UV moonlight. TOMS, MLS, and SCIAMACHY data were also reported.

Negative ozone trends outside the polar regions are not well understood (WMO, 2010a). In order to discuss the relation between strong negative ozone trends in the polar regions and those in lower latitudes, with respect to mean global circulation and transport, data on altitude-dependent ozone trends are useful. Sutton (1994) found, using Lagrangian calculations of transport phenomena in the middle atmosphere, that air flow on seasonal timescales begins in the mesosphere in early winter and is directed towards the upper polar stratosphere. On the other hand, mean hemispheric meridional circulation in the lower stratosphere is also directed from the tropics towards polar altitudes.

In order to study potential connections between globally observed ozone trends and those in polar latitudes, altitude-dependent meridional ozone trends have been studied for the spring season (i.e., October and March monthly zonal means), when the strongest ozone depletion occurs in the lower polar stratosphere. During the period 1985–1995, vertical ozone distributions in the polar stratosphere were mainly determined by ozone soundings. These observations give a detailed description of the variability in ozone up to altitudes of about 30 km (i.e., 10 hPa). Data from higher altitudes can be obtained by using satellite observations such as those from SBUV and SAGE (McCormick *et al.*, 1992).

Altitude-dependent ozone trends were retrieved from satellite SBUV measurements between 1979 and 1993 (NASA/GSFC and NOAA; Hollandsworth *et al.*, 1995; Hood *et al.*, 1993). Over this period two SBUV instruments were in use. The first instrument (SBUV) was on board the Nimbus-7 satellite from 1979 to 1990 and the second (SBUV/2) was on board the NOAA-11 satellite from 1989 onwards. NOAA-11 SBUV/2 data were processed using the version 6 algorithm, in almost the same way as for the SBUV data. The ozone data used here consisted of two separate time series derived from SBUV and SBUV/2 observations, taking into consideration the deviation between observations during normal operation in the 18-month overlap period (Hollandsworth *et al.*, 1995). It is well documented that the retrieval of ozone trends from SBUV data for the lower stratosphere is uncertain, since data for the four lowest layers (i.e., below 32 hPa) primarily reflect the differences between column ozone and the profile amounts above 32 hPa. So, we have compared these SBUV trends with those obtained from long-term ozonesonde observations made at the Syowa Antarctic Station (69°S; 39°E) (Gernandt *et al.*, 1995). The coincidence of both trends—the zonal mean trend from SBUV data and the local trend from ozonesonde data—is fairly good up to altitudes of about 20 hPa (Figures 2.6, 2.14). By considering the result of this intercomparison we conclude that trends calculated from SBUV data are qualitatively reliable for this southern latitude.

For this analysis, zonal means of ozone content for Umkehr layers 1 to 11 (i.e., up to 0.25 hPa) were calculated for 10° latitude bands. The volume mixing ratio for the mean pressure altitude at each Umkehr layer was calculated from zonal mean ozone content. Applying the linear regression technique, the trend of monthly zonal means for October and March for each Umkehr layer was obtained expressed in ppmv per year. The statistical significance of these trends was assessed by means of regression coefficients. By this method a rough profile of altitude-dependent trends was obtained for October and March.

The vertical profile of ozone trends computed from SBUV(/2), Umkehr, and ozonesonde data over northern mid-latitude stations (35°N–60°N) showed that the ozone between 12 and 15 km decreased by about 9% between 1979 and 1995, and increased by about 6% between 1996 and 2009 (Figure 6.11). The increase since the mid-1990s is larger than the change expected from the decline in ODS abundance.

It is evident from Figure 6.11(a) that for 1979–1995 during the winter and spring, there are two significant trend maxima in the lower and upper stratosphere over high latitudes. The analysis for October, when the maximum ozone loss occurs, shows clearly that the zonal mean total ozone loss accounts in the same order of about −0.08 ppmv per year in both the upper stratosphere and the lower stratosphere.

An inspection of Figure 6.11(a) also shows that both maxima of negative trends are clearly separated by an altitude region from about 10 to 20 hPa, where no significant trend is seen except in polar latitudes. Inside the polar vortex, air is being transported downward due to diabatic cooling. Therefore, this ozone-poor air comes to the lower stratosphere, where by PSC-related chemistry, ozone is further depleted in these airmasses. This downward displacement of air might

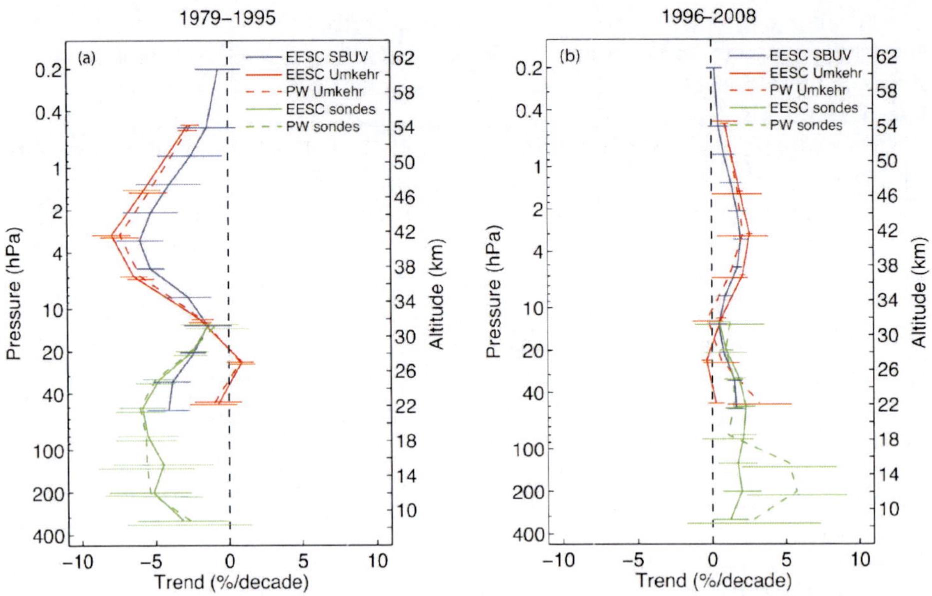

Figure 6.11. Vertical profile of ozone trends over northern mid-latitudes (35°N–60°N) estimated from ozonesonde, Umkehr, and SBUV(/2) measurements for the period 1979–2008: (a) from 1979 to 1995 and (b) from 1996 to 2008. The error bars correspond to the 95% confidence interval (WMO, 2010a).

explain the downward extension of air with decreased ozone from the mid-stratosphere.

The chemical ozone loss in the lower polar stratosphere amplifies the negative trend. So it is assumed that the negative trend in the lower polar stratosphere is caused by both downward transport of air with low ozone concentration and additional chemical loss in the lower polar stratosphere.

The global distribution of ozone trends for October and March shows a clear separation between the negative trend in the upper stratosphere and in the lower stratosphere, except for the spring polar stratosphere (i.e., October in the southern and March in the northern hemisphere. Inside the polar vortex the negative trend is significant (within 95%) at all stratospheric altitudes during spring. While in the upper stratosphere significant negative trends are obtained at all latitudes, in the lower stratosphere close to 30° to 40° latitude in both hemispheres for the months October and March, the trends do not continue to be significant. The trends observed at the upper stratosphere coincide very well with the results of SBUV data analysis (Hollandsworth *et al.*, 1995; Hood *et al.*, 1993), where a negative ozone trend of about 6–8% per decade for the 2–4 hPa layer for October at 60°S is observed. From the same analysis it was found that the upper trend maximum is displaced upwards from the equator to polar latitudes in the summer hemispheres (i.e., in spring and autumn). In the lower stratosphere the trend maximum turns back to mid-latitudes. Along these trajectories the negative trend is increasing with latitude towards both polar regions in the upper stratosphere. Within the polar region the negative trend is amplified downward during spring, and it decreases toward mid-latitudes in the lower stratosphere and upper troposphere. It can also be concluded that the trajectories of the maximum negative trend coincide very well with mean circulation patterns. From the tropical upper stratosphere and mesosphere, air is transported downward and towards polar latitudes by the meridional circulation. The upper trend maximum follows this pattern. On the other hand, a decrease in meridional transport may also cause decrease of ozone towards high latitudes. Inside the vortex, air is transported downward to regions where chemical ozone loss occurs in spring; that is, mainly in September and October in the Antarctic and from January to March in the Arctic (von der Gathen *et al.*, 1995). During spring in the lower stratosphere the increasing chemical ozone loss amplifies the negative trend. The altitude range of significant negative ozone trends is extended towards mid-latitudes in the spring hemisphere.

Figure 6.12 displays the deseasonalized ozone monthly means in three pressure layers (upper, lower, and lowermost stratosphere) based on ozonesonde, Umkehr, and SBUV(/2) observations over Europe (Terao and Logan, 2007; WMO, 2010a). These time series show that ozone levels have stabilized in the upper and lower stratosphere after a decrease from the early 1980s to the mid-1990s.

By considering these results and global meridional circulation a connection might be found between atmospheric regions by transport from the tropical upper stratosphere via the winter polar vortex to the upper troposphere in the mid-latitudes where some portions of the airmass finally arrive with decreased ozone. Along this path the trend increases from tropical to polar latitudes, which might be caused by chemical or dynamic reasons. The trend is significantly amplified by PSC-related chemical ozone loss inside the polar vortex in spring. Because of the convection of air the net ozone loss can be greater when ozone-depleted air is transported downward from the middle stratosphere. Mesoscale exchange processes between polar and middle latitudes may finally transport stratospheric air with depleted ozone into the upper troposphere. Therefore, some impact can appear in the upper troposphere due to stratosphere–troposphere exchange processes in the mid-latitudes.

The area of the Antarctic ozone hole grew dramatically in the period after 1980. The area of the ozone hole is primarily related to the levels of chlorine and bromine and the year-to-year variability of temperature of the Antarctic vortex collar region (Newman *et al.*, 2004). Year-to-year variability of the area is related to temperature, such that a 1 K colder-than-average temperature results in an ozone hole area that is approximately 1,000,000 km^2 larger than average (for peak chlorine loadings). For example, 1988 and 2002 were warmer than average in the polar vortex collar, so the ozone hole area was much smaller than average in those years.

The Montreal Protocol has been successful in that the production and consumption of CFCs declined sharply and, consequently, the concentration of CFCs in the troposphere has been declining for over a decade (WMO, 2007). Around 2000, chlorine levels reached their maximum in the stratosphere as well. Since then, stratospheric chlorine has started to decline. While this has not yet

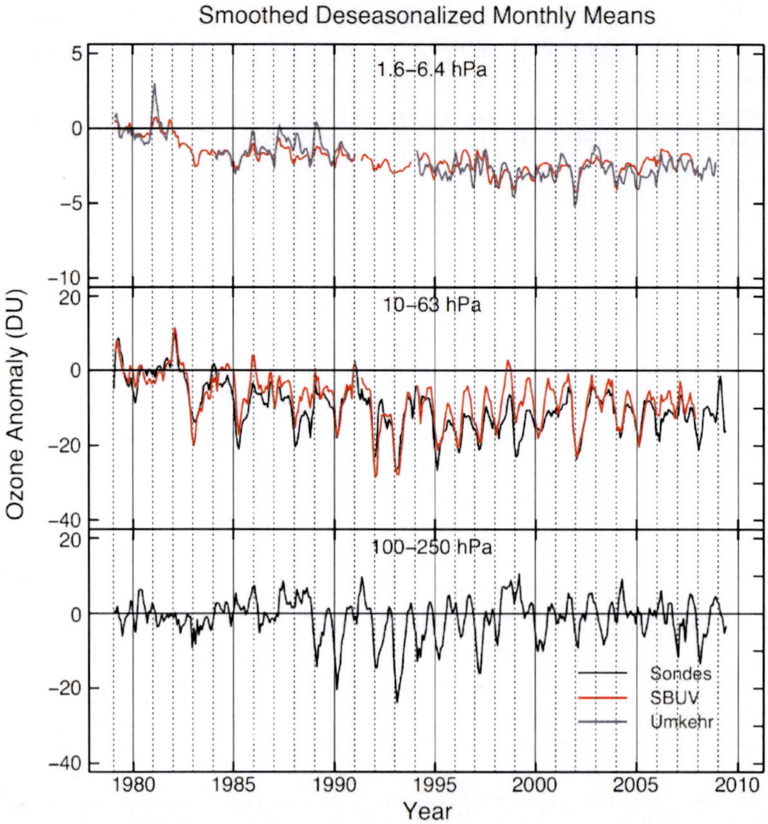

Figure 6.12. Monthly ozone anomalies for Europe as measured by ozonesondes (black line), SBUV(/2) (red line), and Umkehr (blue line) at three pressure layers (WMO, 2010a).

resulted in a substantial reduction in the size of the Antarctic ozone hole (Newman *et al.*, 2006), or of the more variable Arctic ozone depletion (e.g., von der Gathen *et al.*, 1995), possible signs of a beginning recovery have been seen in the 40 km altitude range (more than enough chlorine/bromine and complete destruction of ozone in a certain height range) (Newchurch *et al.*, 2003; Steinbrecht *et al.*, 2006a, 2009; Zanis *et al.*, 2006). One should not expect an immediate recovery of ozone in the lower stratosphere because that layer is saturated with ozone-destroying halogen atoms (i.e., there is more than enough of them to destroy all the ozone in that layer). Steinbrecht *et al.* (2009) found that—particularly for the northern stations they studied—there were substantial super-imposed short-term ozone variations, up to ±5%, many of which were related to the QBO. Finally, slightly elevated ozone levels, a few percent at most, tend to appear during solar maxima, but this solar-cycle-related ozone variation is quite small, between 1% and 4% peak to peak (Steinbrecht *et al.*, 2006b).

6.5.1 Sudden stratospheric warmings

We discussed the discovery of the Antarctic ozone hole in Section 5.3 and considered the explanation of its formation in Section 5.4. An important component in explaining how the ozone hole forms is the presence of a circumpolar vortex in winter and spring seasons, which captures circumpolar air; the air temperature then drops to around −90°C, and this leads to the formation of polar stratospheric clouds (PSCs). Heterogeneous chemical reactions on the surfaces of the particles of these clouds transform comparatively inert $ClONO_2$ and HCl into active forms of chlorine compounds, catalyzing ozone depletion and at the same time combining nitrogen compounds.

Two types of PSCs have been discovered; they consist either of particles of nitric acid trihydrate, concentrated on the nuclei of sulfate aerosols (the prevailing type), or of large particles of water ice. Both types of PSC fulfill the function of stratosphere denitrification (removal of gaseous nitrogen compounds), which is a condition necessary for total ozone decrease. Laboratory investigations confirm that particles of nitric acid trihydrate play a key role in the formation of active chlorine compounds in the Antarctic stratosphere.

According to a September 30, 2002 press release from NASA and NOAA (NASA/NOAA, 2002), the size of the Antarctic ozone hole was around 15 M km^2 during the last two weeks of September 2002, which is well below the more than 24 M km^2 seen in this season for the previous six years. Analysis of the ozone and temperature observations showed that the most important factor for the smaller size of the Antarctic ozone hole in September 2002 was the prevalence of much higher than normal temperatures in the Antarctic stratosphere, which resulted in the limited activity of ozone-destroying processes (Figure 6.13) (Varotsos, 2002a, b).

The reason for the high temperatures in Antarctica is mainly the presence of strong planetary waves (long waves that circle the globe) or planetary-scale weather systems. In addition, the basic polar vortex in the stratosphere of the Southern Hemisphere during September 21–26, 2002 became elongated step by step and distorted until it split into two cyclonic vortices, leading therefore to a corresponding split of the ozone hole (Figure 6.14). This event was attributed to the unprecedented occurrence of major sudden stratospheric warming over Antarctica (induced by strong planetary waves) (NASA/NOAA, 2002; Varotsos, 2002a).

Conventionally, an SSW (an increase in stratospheric temperature by as much as 70°C for a period of a few weeks during midwinter) is accompanied by a situation in which the polar vortex slows down (minor SSW) or even reverses direction (major SSW) in a few days because of the propagation of planetary waves 1 and 2. The WMO's Commission for Atmospheric Sciences has adopted the following definitions for the terms "minor" and "major" SSWs (as stated in the work of McInturff, 1978):

1. A stratospheric warming is called "minor" if a significant temperature increase is observed (i.e., at least 25 degrees in a period of a week or less) at any stratospheric level in any area of the wintertime hemisphere, measured by radiosonde or rocketsonde data and/or indicated by satellite data; and if criteria for major warmings are not met.

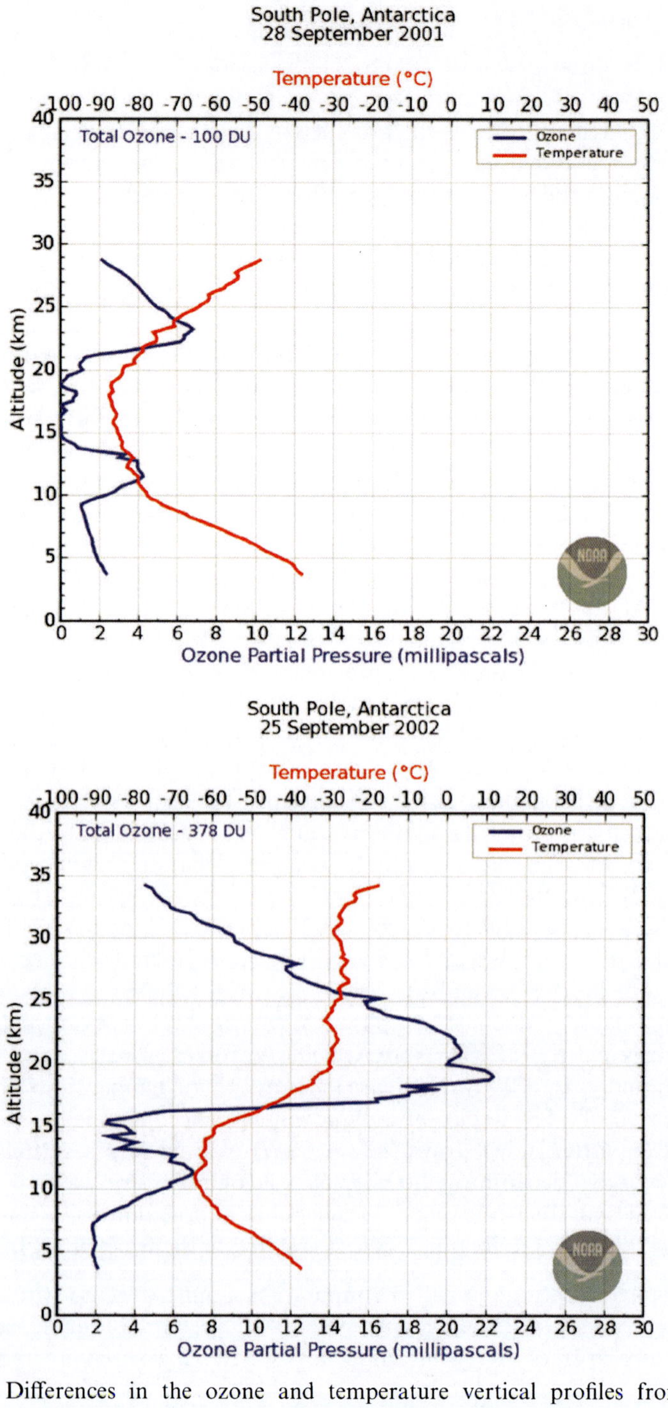

Figure 6.13. Differences in the ozone and temperature vertical profiles from 2001 (cold stratosphere, low ozone) to 2002 (warm stratosphere, high ozone) (Varotsos, 2003b).

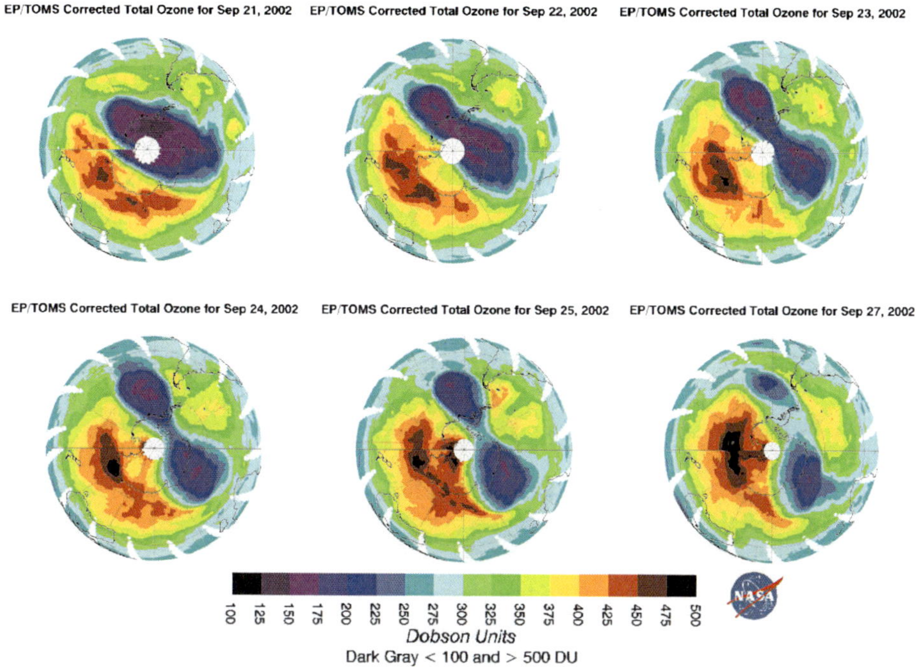

Figure 6.14. Southern hemisphere total ozone content during September 21–27, 2002 obtained from Earth Probe (EP)/Total Ozone Mapping Spectrometer (TOMS) observations (*source*: NASA).

2. A stratospheric warming can be said to be "major" if at 10 hPa or below the latitudinal mean temperature increases poleward from 60 degrees latitude and an associated circulation reversal is observed (i.e., mean westerly winds poleward of 60° latitude are succeeded by mean easterlies in the same area).

SSWs can occur in both the Northern Hemisphere (where they occur about every other winter) and the Southern Hemisphere. However, they are less common in the Southern Hemisphere because of the lower planetary wave activity and the resulting stronger polar vortex in that hemisphere; recently there has only been one major SSW in the Southern Hemisphere and that was in 2002 (Varotsos, 2002a, b, 2003a, b, 2004a, b).

It should be emphasized, however, that the small Antarctica ozone hole in 2002 is not evidence for recovery of the ozone layer, since reduced ozone loss was mainly induced from the unexpected major stratospheric warming that occurred there.

6.5.2 Observation of sudden stratospheric warmings detected in deep underground muon data

Sudden stratospheric warmings (SSWs) can obviously be detected by any one of the various systems available for measuring the temperature of the stratosphere:

radiosondes, rocketsondes, and satellites. However, they can also manifest themselves in a rather unusual way which involves deep underground muon detectors, though such systems are rather unlikely to be useful for routine detection of SSWs (Varotsos *et al.*, 2010).

Osprey *et al.* (2009) reported on intermittent and sudden increases in the levels of muons derived from cosmic rays during winter months observed at the Main Injector Neutrino Oscillation Search (MINOS) underground detector (located in the Soudan Underground Mine State Park, at a depth of 716 m, in Minnesota, at 47.8°N, 92.2°W). They were able to correlate these increases with sudden stratospheric warmings (SSWs).

Figure 6.15 (Figure 3 of Osprey *et al.*, 2009) shows the high correlation between the muon rate in MINOS and effective temperature. This effective temperature was defined by an altitude-weighted average air temperature at 21 pressure levels making use of a weighting function that is related to the relative importance of each level in the production of the muons observed underground; it peaks at about 15 km (see Figure 1 in Osprey *et al.*, 2009; see also Ambrosio *et al.*, 1997; Barrett *et al.*, 1952). This high correlation between the muon rate and effective temperature appears to be present during all the winters presented in Figure 6.15, where an abrupt increase (decrease) in effective temperature is associated with a rapid increase (decrease) in the

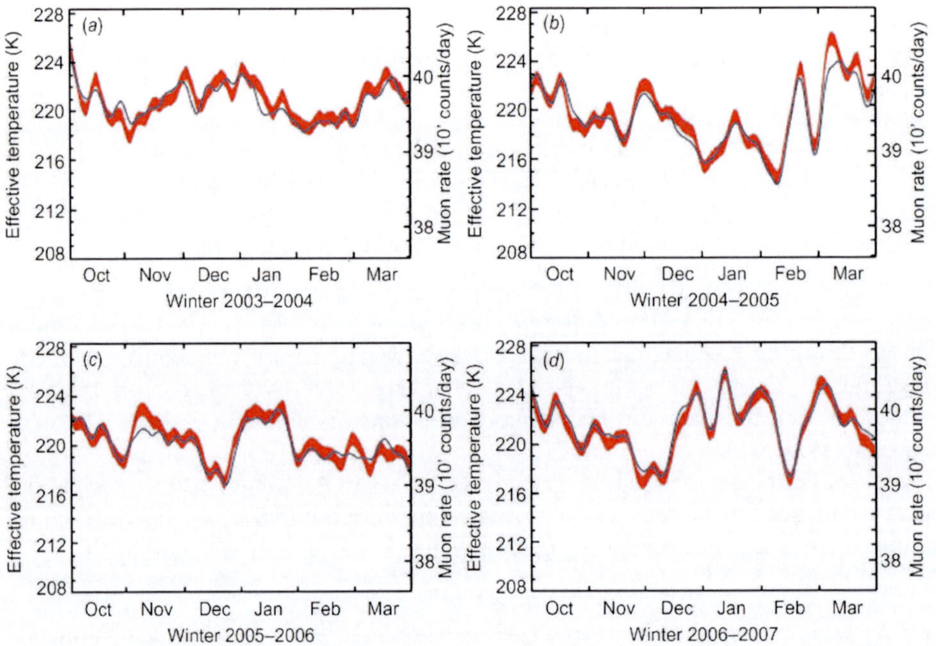

Figure 6.15. Time series of effective temperature for winters 2003–2007 (blue) from ECMWF and daily muon rate ±1 standard deviation registered at the MINOS Far Detector (red). Five-day smoothing has twice been applied to both data (Osprey *et al.*, 2009).

muon rate. Osprey *et al.* (2009) focused especially on the winter of 2004–2005 when an SSW event occurred in February. As can be clearly seen in Figure 6.15b, there was an abrupt increase followed by a sharp decrease in effective temperature and the muon rate, which took place over a period of 2–3 weeks in February during which the muon rate changed by $(4 \pm 1)\%$. The explanation given by Osprey *et al.* (2009) for the increased muon rate during an SSW is *verbatim*:

> "There are two mechanisms by which the condition of the atmosphere affects the muon rate. Firstly, an increase in temperature causes the atmosphere to expand so muons are produced higher up and therefore have a larger probability to decay before being detected. Secondly, the mesons may interact (and thereby be lost) as well as decay. As the temperature increases, the probability of interaction becomes smaller because the local atmospheric density decreases, so more mesons decay, causing an increase in the muon rate. In deep underground detectors where muons with a high surface energy are measured, the second effect dominates and this causes a positive correlation between temperature and muon rate."

The observations of Osprey *et al.* (2009) led Varotsos *et al.* (2010) to examine changes in the vertical profile of the temperature that occur during an SSW, with a view to amplifying this explanation. In all, six temperature profiles, corresponding to five SSWs, were studied: one over Lerwick, Scotland (60°8′N, 1°11′W) in February 2001 (Charyulu *et al.*, 2007; Jacobi *et al.*, 2003); one over Scoresbysund, Greenland (70°29′N, 21°58′W) in February 2002 (Charlton and Polvani, 2007); one over Bern, Switzerland (46°57′N, 7°26′E) in February 2008 (Flury *et al.*, 2009); the one whose effect was observed in the data of Osprey *et al.* (2009); and, finally, two profiles at different latitudes for an SSW that occurred at around 31–33°E in late January/early February 2006. An important observation was that an SSW does not necessarily exhibit an increase in temperature at all heights, but is quite complicated (see Figure 6.16 which shows just one example).

A summary of the behavior of the six profiles (five SSWs) studied is given in Table 6.7. This table shows that in an SSW event: (a) warming only occurs at some heights and that there is cooling, or possibly no change, at some other heights; and (b) there is no common pattern and warming may occur in the upper or lower stratosphere. In addition, an SSW affects all atmospheric layers from the troposphere to the thermosphere (Flury *et al.*, 2009) and is usually accompanied by cooling at other altitudes; moreover, each SSW is different and has latitudinal and longitudinal variations (Hoffmann *et al.*, 2007). Varotsos *et al.* (2010) identified it as a topic for further work to determine whether an SSW always leads to a rise in effective temperature (as defined by Osprey *et al.*, 2009).

Another issue relates to the explanation by Osprey *et al.* (2009) that we quoted *verbatim* above. This concludes by saying "... the second effect dominates and this causes a positive correlation between temperature and muon rate." However, it would also seem to be necessary to take into account the effects of these two mechanisms in the regions above the 37 km altitude in the stratosphere where the

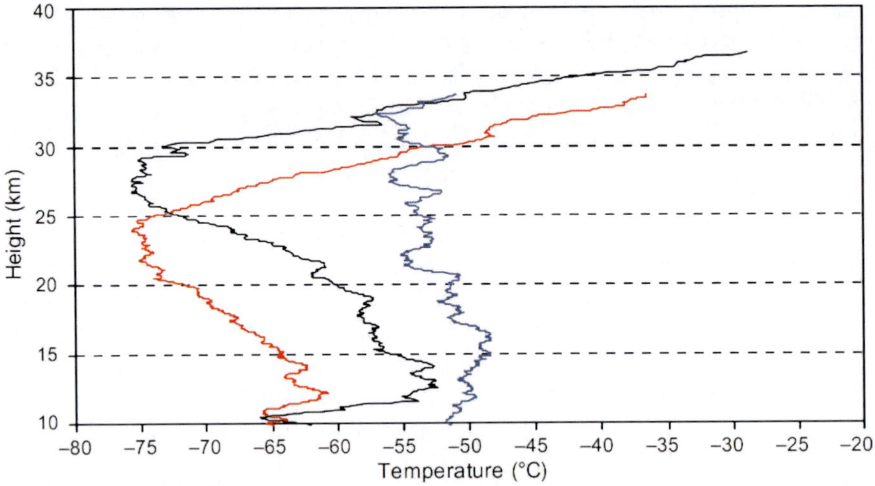

Figure 6.16. Temperature profiles observed on February 16, 2001 (black line) during a major SSW event (February 12–23, 2001) as well as data from before (February 9, 2001, red line) and after (February 28, 2001, blue line) the event over Lerwick, Scotland (Varotsos *et al.*, 2010).

Table 6.7. Warming and cooling at different heights (in km) in some SSWs (Varotsos *et al.*, 2010).

		Atmospheric elevation (km)		
Location	*Date*	*Warming*	*Cooling*	*No change*
Bern	February 2008	35–50	20–25	—
Lerwick	February 2001	<25	>25	
Scoresbysund	February 2002	<30	—	>30
51°N, 33°E	February 2006	22–33; >47	12–22; 33–47	—
80°N, 31°E	February 2006	<27	>27	—
MINOS	February 2005	<37	>37	—

temperature decreases. It was concluded that it was an open question whether effective temperature and the muon flux rate are always positively correlated in an SSW.

6.5.3 The role of the diffusion of gases in ice or an amorphous binary mixture in the polar stratosphere and the upper troposphere

We have seen that heterogeneous chemical reactions involving frozen particles in PSCs play a key role in the depletion of ozone leading to the formation of the

Antarctic ozone hole. Since the rates of heterogeneous chemical reactions depend on both surface and bulk concentrations, it is important to study the diffusion of gases in and on such particulate material.

When a single diffusion mechanism is operative in a solid, the diffusion coefficient, D, is often found to obey an Arrhenius-type behavior, i.e.

$$D = D_o \exp[-E/(k_B T)] \tag{6.7}$$

where the activation energy E and the factor D_0 are essentially temperature independent; and k_B is the Boltzmann constant. In this case, the slope of a $\ln D$ against $1/T$ plot leads to the value of E which, as shown by thermodynamical studies (Varotsos and Alexopoulos, 1979, 1986), coincides with the enthalpy of activation ΔH_{act} of the diffusion process. However, while such findings apply to the rate of diffusion, they do not provide any information on the diffusion mechanism itself.

Early studies of self-diffusion in ice were performed using microtome sectioning and scintillation or mass spectrometric probes ($H_2{}^{18}O$, D_2O, T_2O) near the ice melting point (see Petrenko and Whitworth, 1999 and references therein). These studies led to a diversity of views concerning the mechanism of H_2O isotopic diffusion in ice (Petrenko and Whitworth, 1999). Opinions have been expressed, for instance, that self-diffusion occurs by an interstitial-mediated mechanism for $T > 223$ K, while a vacancy-mediated process is operative for $T < 223$ K (Goto *et al.*, 1986; Hondoh *et al.*, 1989).

More recently new infrared laser resonant desorption (LRD) techniques have been employed to derive diffusion coefficients by depth-profiling of ice films (Livingston and George, 1999, 2001; Livingston *et al.*, 1997, 1998, 2000, 2002; Krasnopoler and George, 1998). These measurements revealed two different rate categories for bulk diffusion in ice. While compounds such as HCl, CH_3OH, HCOOH, and CH_3COOH display diffusion kinetics similar to the kinetics of H_2O self-diffusion, species such as Na and NH_3 exhibit diffusion kinetics appreciably slower than for H_2O self-diffusion. Livingston *et al.* (2002) argued that the diffusion process of compounds of the first category occurs via a vacancy-mediated mechanism in which H_2O vacancies may be required to diffuse to lattice sites adjacent to the impurity compounds prior to the diffusion of this compound into the vacancy sites. Varotsos and Zellner (2010) studied the hetero-diffusion data of compounds of the first category.

Figure 6.17 shows the temperature dependence of kinetic data for bulk diffusion in ice for selected gas phase compounds as measured by infrared LRD depth profiling (Livingston *et al.*, 2002) as well as by chemical titration (for CH_3COOH only; Nehme, 2006) in the form of a $\ln D$ against $1/T$ plot. The original data have been omitted and only the resulting Arrhenius forms are shown. The corresponding Arrhenius parameters are summarized in Table 6.8.

Table 6.8 also includes the temperature range over which these data have been obtained as well as the absolute values of D for a temperature of 180 K. As can be seen from this compilation, there is a very wide range of values of D_0; small values of D_0 correspond to smaller activation energies E and *vice versa*. On the other hand, differences in the absolute values of D at 180 K are much less pronounced.

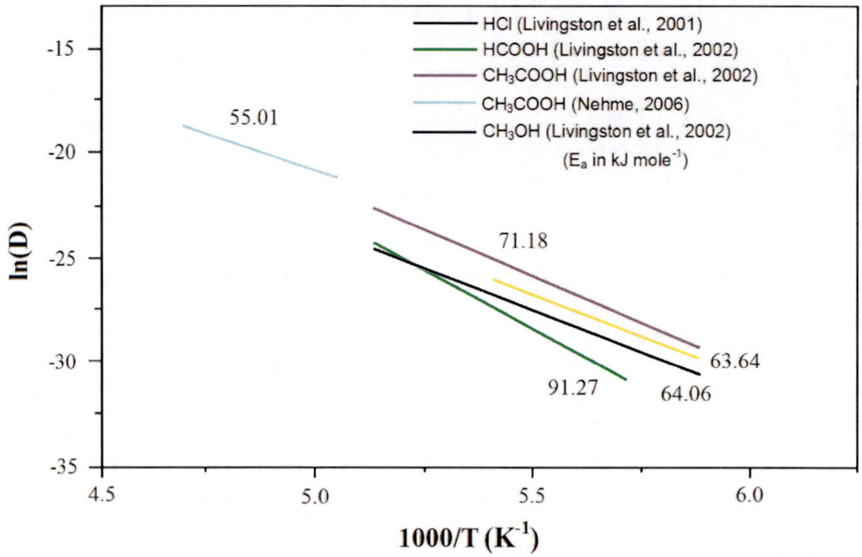

Figure 6.17. Arrhenius plots for temperature-dependent diffusion coefficients for various species in ice as obtained by the LRD depth-profiling technique (Livingston *et al.* 2001, 2002) and chemical titration (Nehme, 2006). Figures on individual lines are activation energies (in kJ $mole^{-1}$) as derived by the relevant authors (see also Table 6.8) (Varotsos and Zellner, 2010).

Table 6.8. Summary of rate parameters for the diffusion coefficients of selected compounds in ice (Varotsos and Zellner, 2010).

	E (kJ mol^{-1})	D_0 ($cm^2\ s^{-1}$)	D (180 K) ($cm^2\ s^{-1}$)	T (K)	*Reference*
HCl	64.06 ± 4.19	$1.5 \times 10^{7\pm0.2}$	4.83×10^{-12}	169.0–194.9	Livingston *et al.* (2001)
HCOOH	91.27 ± 3.77	$8 \times 10^{14\pm0.1}$	2.42×10^{-12}	175.1–194.9	Livingston *et al.* (2002)
CH_3COOH	71.18 ± 2.93 55.01 ± 12.56	$1.0 \times 10^{10\pm0.1}$ $0.813 \times 10^{6\pm3.2}$	2.08×10^{-11} 8.6×10^{-11}	169.9–194.9 198–213	Livingston *et al.* (2002) Nehme (2006)
CH_3OH	63.64 ± 2.93	$2.4 \times 10^{7\pm0.3}$	7.8×10^{-12}	169.4–185.4	Livingston *et al.* (2002)

It is interesting to consider a plot of ln D_0 against the activation energy, E (see Figure 6.18). Inspection of this figure reveals that, for the diffusion of the species selected here, a near linear relation holds. A least squares fit to a straight line leads to a slope of 2.502 and an intercept of −20.365 (when D_0 is measured in cm^2/s and E in kcal/mol) with a correlation coefficient of 0.98. Maximum uncertainty in this slope is at the most around 10%, if one considers experimental errors in the values of E and D_0 extracted from the diffusion data.

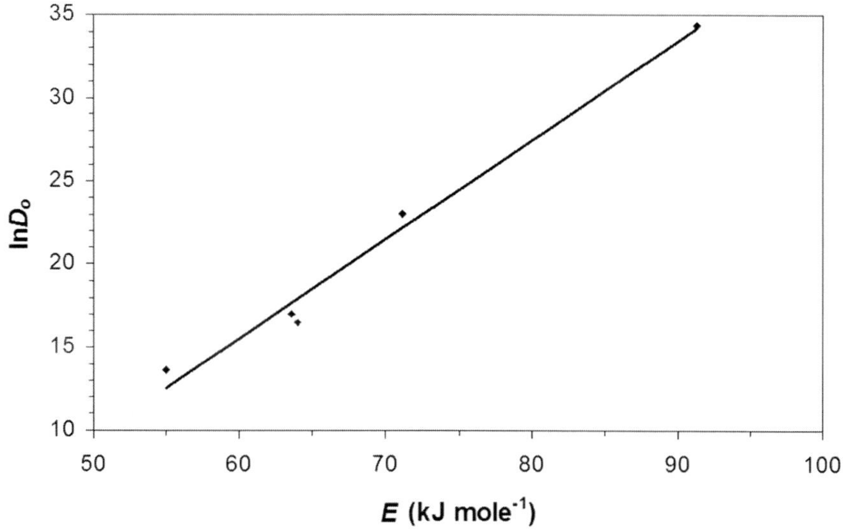

Figure 6.18. Relationship between the factor D_0 of the diffusion coefficient and the activation energy E ($y = 0.5977x - 20.365$, $R^2 = 0.984$) (Varotsos and Zellner, 2010).

According to the theory of diffusion in solids, the temperature dependence of the diffusion coefficient D for a single diffusion mechanism in a given crystal is given by (Varotsos and Alexopoulos, 1986):

$$D = f\alpha^2 \nu \exp[-\Delta G_{\text{act}}/(k_B T)] \tag{6.8}$$

where ΔG_{act} is the Gibbs activation energy for the diffusion process; f is a numerical constant that depends on the mechanism; α is the lattice constant; and ν is the attempt frequency. To a first approximation the attempt frequency depends on the mass of the diffusing species and according to Varotsos and Alexopoulos (1986):

$$\nu^j/\nu^m = (M^m/M^j)^{1/2} \tag{6.9}$$

where M^m, M^j denote the mass of the matrix m and the diffusing species j, respectively; and ν^m, ν^j denote the corresponding attempt frequencies.

The Gibbs energy ΔG_{act} can be decomposed into two terms, an entropy term $\Delta S_{\text{act}} = -\mathrm{d}(\Delta G_{\text{act}})/\mathrm{d}T|_p)$ and an enthalpy term $\Delta H_{\text{act}} = \Delta G_{\text{act}} + T\,\mathrm{d}(\Delta G_{\text{act}})/\mathrm{d}T|_p)$ term, where p is the pressure, whereupon equation (6.8) becomes

$$D = f\alpha^2 \nu \exp(\Delta S_{\text{act}}/k_B) \exp[-\Delta H_{\text{act}}/(k_B T)] \tag{6.10}$$

which can be written alternatively as:

$$D = D_0 \exp[-\Delta H_{\text{act}}/(k_B T)] \tag{6.11}$$

with

$$D_0 = f\alpha^2 \nu \exp(\Delta S_{\text{act}}/k_B) \tag{6.12}$$

As a consequence the temperature dependence of the diffusion coefficient follows a simple exponential behavior, for which the pre-exponential factor D_0 is associated with a dynamic (frequency) term as well as a thermodynamic (entropy) term. For a given matrix, therefore, low values of D_0 reflect low attempt frequencies or low entropies of activation. It should be noted that ΔS_{act} is always positive because the diffusing species is more loosely bound or less ordered at the transition state than at individual lattice sites.

The Gibbs free energy of activation can be further decomposed into the product of a bulk mechanical property and a molecular property (Varotsos, 1976; Varotsos and Alexopoulos, 1978, 1980, 1981, 1984a, 1986; Varotsos *et al.*, 1982, 1985, 1986), i.e.

$$\Delta G_{\text{act},j} = c_{\text{act},j} B\Omega \tag{6.13}$$

where B is the isothermal bulk modulus; Ω is mean volume per atom; and $c_{\text{act},j}$ is a constant independent of temperature and pressure. The additional subscript j indicates that the value of this constant may be different for different diffusing species j. This is called the $cB\Omega$ model. The validity of this model has been tested for various defect processes and different categories of solids (see Varotsos and Alexopoulos, 1986 for the results).

On the basis of equation (6.8), the corresponding entropy, $\Delta S_{\text{act},j}$, and enthalpy, $\Delta H_{\text{act},j}$, terms are found to be given by

$$\Delta S_{\text{act},j} = -c_{\text{act},j}\Omega\left(\beta B + \left.\frac{\mathrm{d}B}{\mathrm{d}T}\right|_p\right) \tag{6.14}$$

$$\Delta H_{\text{act},j} = c_{\text{act},j}\Omega\left(B - T\beta B - T\left.\frac{\mathrm{d}B}{\mathrm{d}T}\right|_p\right) \tag{6.15}$$

where β is the thermal volume expansion coefficient. Thus, by inserting equation (6.14) into equation (6.12) and by also considering equation (6.9) the value of D_0 for the diffusion of species j (labeled hereafter D_{0j}) is given by

$$D_{0j} = f\alpha^2\nu^m(M^m/M^j)^{1/2}\exp\left[(-c_{\text{act},j}/k_{\text{B}})\Omega\left(\beta B + T\left.\frac{\mathrm{d}B}{\mathrm{d}T}\right|_p\right)\right] \tag{6.16}$$

or

$$\ln\{D_{0j}/[f\alpha^2\nu^m(M^m/M^j)^{1/2}]\} = \left[(-c_{\text{act},j}/k_{\text{B}})\Omega\left(\beta B + T\left.\frac{\mathrm{d}B}{\mathrm{d}T}\right|_p\right)\right] \tag{6.17}$$

Together with equation (6.15), this relation shows that—for the diffusion of various species j into the same matrix material, m—when plotting $\ln\{D_{0j}/[f\alpha^2\nu^m(M^m/M^j)^{1/2}]\}$ against $\Delta H_{\text{act},j}$, a straight line should result, the slope of which is solely governed by the elasticity and thermal expansivity properties of the host material. Since in the present case we only consider water ice as a matrix and data for other hosts are not easily available, the variation of diffusion data with these properties has not yet been tested.

In the present case the D_{0j} values vary by several orders of magnitude (i.e., the D_0 value for HCOOH exceeds that of HCl by as much as seven orders of magnitude, see Table 6.7); the error for D (e.g., for the case of HCl) is around 30% as given in Table 2 of Huthwelker *et al.*, 2006). In view of this large variability, the mass effect associated with the frequency term $\nu^m(M^m/M^j)^{1/2}$ which emerged from equation (6.9) can, to a first approximation, be discarded. This explains why the plot of the ln D_{0j} versus $\Delta H_{\text{act},j}$ or E is almost a straight line.

The apparent success of this analysis is, except for other potential sources, mainly due to the invariance of the diffusion mechanism. In all cases considered here this mechanism is identical (i.e., diffusion is governed by a vacancy-controlled mechanism). This in turn leads to the conclusion that diffusion data that do not fall onto the line delineated here correspond to different diffusion mechanisms. This, for instance, applies to NH_3 and Na.

It should be noted that measurements of diffusion in ice are difficult to perform and many measurements have been debated in the past. Thus, for example, it has been argued by Domine and Xueref (2001) that the diffusion coefficients reported by Livingston *et al.* (2002) were too large to be interpreted as molecular diffusion in crystalline ice because they were so far below its melting point and that the ice samples used by Livingston *et al.* (2002) were in fact an amorphous binary mixture with such a degree of disorder that diffusion could be expected to be fast. In the case of the diffusion of HCl in ice the data available until 2006 were reviewed by Huthwelker *et al.* (2006) and were shown to exhibit a large scatter. After careful inspection and study of this scatter, Huthwelker *et al.* (2006) concluded that the value of the diffusion constant can be affected by various factors including the concentration of HCl, hydrate formation, etc. (for further discussion see Varotsos and Zellner, 2010). While it is well recognized that H_2O ice particles play a key role in ozone depletion in the polar stratosphere and contribute to radiative forcing (e.g. Chandra and Varotsos, 1995; Gernandt *et al.*, 1995), it is clear that the use of measurements of diffusion constants in the laboratory need to be treated with some caution before they are applied to the particles in PSCs. If values for the D_{0j} could be estimated (i.e., by considering entropy change during the diffusion process) then the correlation illustrated in Figure 6.18 could be of substantial predictive power for the diffusion coefficients of other species in ice.

6.5.4 Experimental studies of the Antarctic ozone hole and ozone loss in the Arctic

As we have mentioned above, what little evidence there is suggests that there was a small Antarctic ozone hole in 1956, long before the effect of CFCs came to be felt. But how deep it was and how large an area it covered we have no way of knowing. The Antarctic ozone hole has been studied extensively since the mid-1970s. Balis *et al.* (2009) described the use of data from the TOMS, SBUV/2, and OMI satellite-flown instruments to study the area and depth of the Antarctic ozone hole (see Figure 6.19).

The area of the ozone hole increased dramatically during the 1980s and since then it has more or less stabilized, but, so far it shows no sign of reducing. The lowest

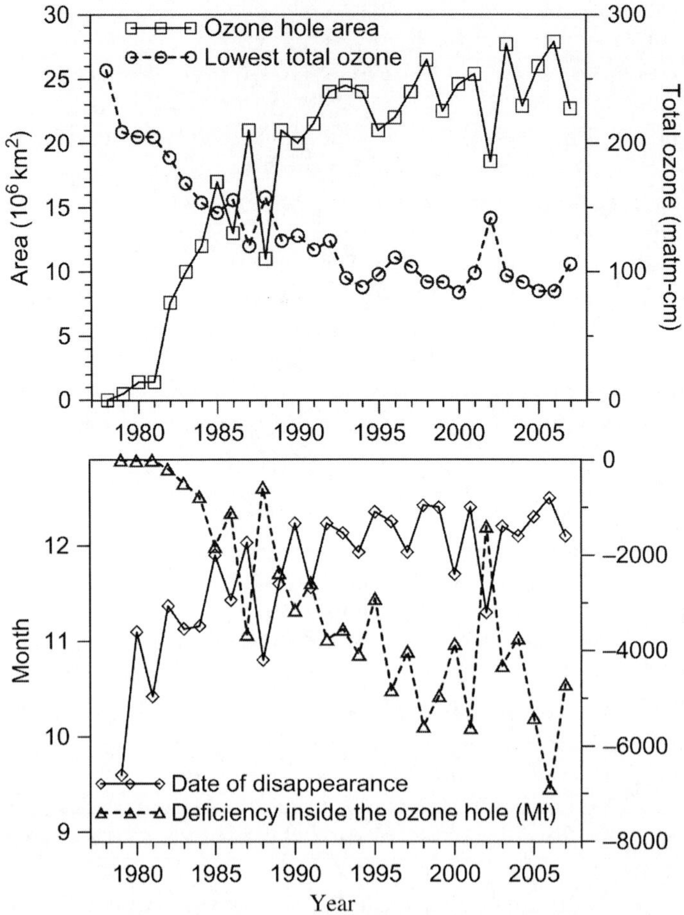

Figure 6.19. (Upper panel) Maximum day area covered by the ozone hole (squares), and lowest total column ozone measured each year (circles). (Lower panel) Ozone mass deficiency (O_3MD) as the difference from pre-1976 values integrated for 105 days (September 1–December 15) within the ozone hole area (triangles), and the date on which values <220 DU disappeared (diamonds) (Balis *et al.*, 2009).

total ozone values within the ozone hole usually occur in late September and early October. Minimum total ozone values had fallen from around 200 DU in the early 1980s to around 150 DU in the mid-1980s and since then to values slightly less than 100 DU in many subsequent years. Balis *et al.* (2009) also calculated the date of disappearance of the ozone hole, taken to be the date when no TOC values less than 220 DU remained (see Figure 6.19, lower panel). The anomalously early disappearance of the ozone hole in October of 1988 and 2002, with relatively smaller ozone deficiency, was due to an early breakdown of a warmer stratospheric polar vortex (e.g. Varotsos, 2003a, b; WMO, 2007). Balis *et al.* (2009) also calculated the ozone mass deficiency (O_3MD) within the ozone hole, including only those values where the magnitude of the deficiency exceeded 10% (see Figure 6.19, lower panel). Total ozone content (TOC) in Dobson units can easily be expressed in $kg\,m^{-2}$ and the O_3MD was taken to be the departure of the mass of ozone from its pre-1976 value (see also Figure 6.19, lower panel). Balis *et al.* (2009) also studied the duration of the ozone hole in relation to its size; see Table 6.9 which gives the number of days in each season for which the area of the ozone hole exceeded 10, 15, 20, and 25 M km^2.

Balis *et al.* (2009) also calculated the zonal average O_3MD over two ranges of latitude in the Southern Hemisphere for the four months from September to December (see Figure 6.20, upper panel). From this figure it can be seen that the O_3MD below 60°S increased rapidly during the early 1980s and since then has been increasing less rapidly, with the exception of 2002, reaching a record high value in 2006. Year-to-year variability of the O_3MD in this polar region is highly correlated with the corresponding year-to-year variability of seasonal averaged temperatures in the lower stratosphere (see Figure 6.20, lower panel). By contrast in the 35°–55°S zone the O_3MD has not shown any major decline since the early 1980s.

Table 6.9. Number of days when the Antarctic ozone hole area (total ozone < 220 DU) covered a sunlit surface greater than 10, 15, 20, and 25 M km^2 from 1995 to 2006 (from Balis *et al.*, 2009).

Year	>10	>15	>20	>25
1995	75	68	36	—
1996	88	70	35	
1997	7	56	25	
1998	101	70	48	20
1999	98	79	50	
2000	75	54	41	7
2001	97	82	53	17
2002	18	5		
2003	68	54	38	14
2004	75	54	21	
2005	81	64	50	6
2006	97	89	61	21

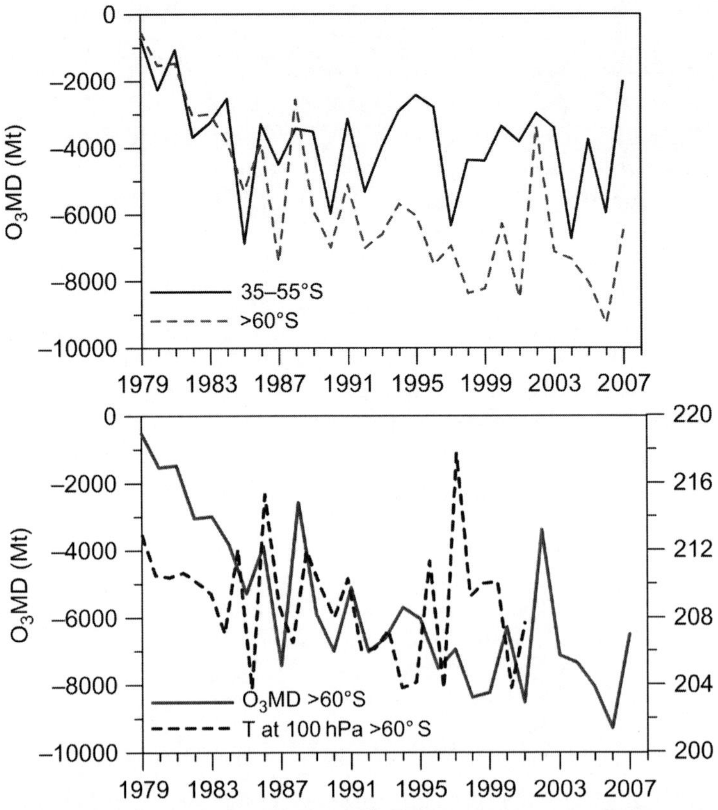

Figure 6.20. (Upper panel) O_3MD from pre-1976 averages, inside the −10% deviation contours, integrated for 105 days (September 1–December 15) for the southern polar region (>60°S dashes) and middle latitudes (35–55°S continuous line) in the period 1979–2006, showing major increase of O_3MD only over the polar region. In the lower panel, together with the O_3MD >60°S (solid line) is shown the temperature at 100 hPa (dashed line) which shows perfect correlation with ozone deficiencies over the same region and season (Balis *et al.*, 2009).

The situation revealed in the Antarctic at the Japanese Syowa station (69.0°S, 39.6°E) in their Dobson Umkehr measurements for derivation of the vertical ozone profile since 1977 is less optimistic in relation to ozone recovery (Miyagawa and Sasaki, 2009; Miyagawa *et al.*, 2009). In 2006 the ozone hole extended to the largest size ever observed and the lowest record of TOC since the beginning of observations in 1961 (114 DU on October 17) was observed at Syowa. Table 6.10 shows the linear trend for each layer.

Figure 6.21 shows time series of monthly mean deviation from averages over 1977–2000 and their 3-month mean (September to November); in these results natural variations such as solar activity, QBO, etc., had not been removed. Ozone

Table 6.10. Linear trend from 1987 to 2007 springtime periods in each Umkehr layer at the Japanese Antarctic research station at Syowa (Miyagawa *et al.*, 2009).

Layer	*Trend per decade* (%)	*Confidence*
1	4.5	3.5
2	−6.9	8.1
3	−21.7	14.9
4	−24.0	16.2
5	−8.9	8.1
6	−1.9	2.8
7	−5.7	2.9
8	−10.5	4.2
2 + 3	−14.3	11.5
8 + 9 + 10	−7.4	3.0

in the upper stratosphere shows a clear and steady decrease at a rate of 7.4%/decade (upper panel of Figure 6.21). It seems that the decrease starts around 1980. The sudden increase in 2002 is due to the major stratospheric sudden warming that suppressed ozone depletion in that year. By contrast, ozone in the lower stratosphere shows 2 to 3-year periodical variation, but also a clear decrease of 24.0%/decade from 1980 (lower panel of Figure 6.21). It will be noted that for this location, unlike the Northern Hemisphere locations mentioned above—Athens (38°N, 24°E), Belsk (52°N, 21°E), northern India (25°N–35°N)—there is no evidence of a reduction in ozone depletion starting around 1990 or so; these data, as indicated, only refer to a period during the existence of the ozone hole and are not based on annual averages.

Loyola *et al.* (2009b) using combined data from GOME, SCIAMACHY, and GOME-2 studied the size (area) of the Antarctic ozone hole in 2007 and compared it with the average size for the previous 12 years; they found no evidence of any significant recovery of the ozone layer (see Figure 6.22). MIPAS data have been analyzed by Papandrea *et al.* (2010) with the 2-D GMTR (Geo-fit Multi Target Retrieval) retrieval system in order to obtain fields of ozone and several molecular species related to ozone chemistry: HNO_3, N_2O, NO_2, N_2O_5, $ClONO_2$, COF_2, CFC-11, and CFC-12. The GMTR algorithm makes it possible to resolve the horizontal inhomogeneities of the atmosphere that are modeled using a 2-D discretization of the atmosphere. It is therefore especially suitable for analyzing

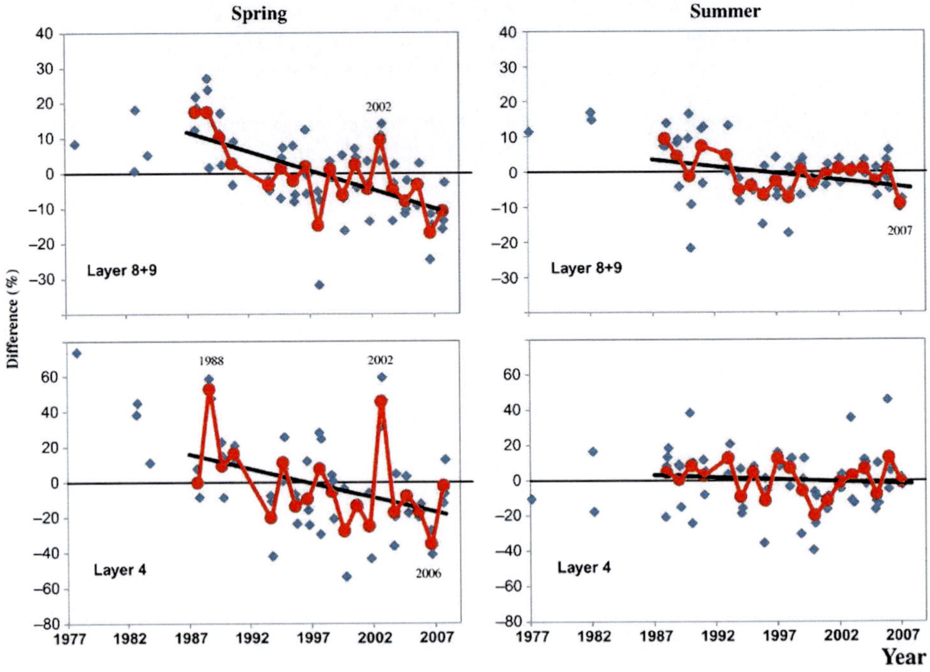

Figure 6.21. Long-term variation of ozone amount in layers 8 + 9 + 10 and 4 at Syowa. The gray line shows monthly deviations from the average over 1977–2000, and the red line shows 3-month (September to November) averages (Miyagawa and Sasaki, 2009).

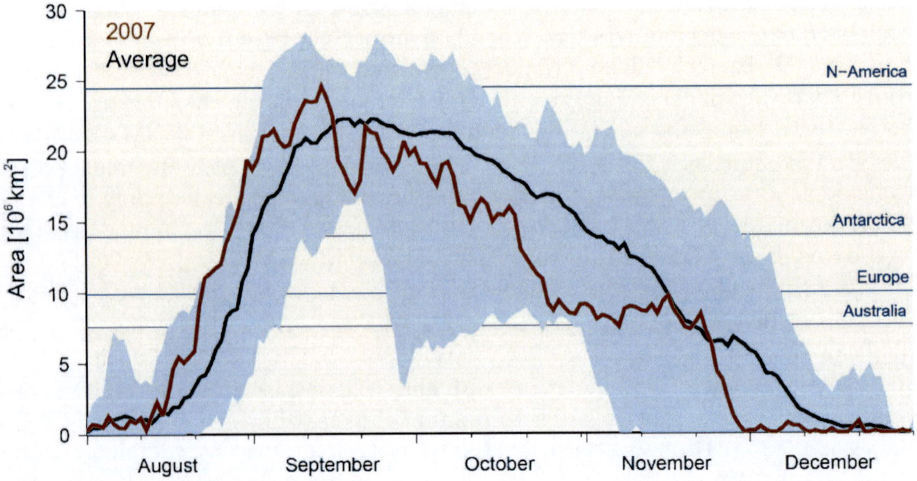

Figure 6.22. Size of the 2007 Antarctic ozone hole compared with the average size for the previous 12 years (Loyola *et al.*, 2009b).

portions of the atmosphere where strong gradients—such as those at the ozone hole—may be poorly reproduced by common horizontal homogeneous 1-D retrievals. The adopted strategy is well suited for refined analysis and correct monitoring of ozone recovery, as required by the Montreal Protocol and successive amendments.

Efstathiou (2005) studied characteristic features of the signature of quasi-stationary planetary waves in ozone over Antarctica during September of four successive years (2000–2003), using data from TOMS over 75.5°S. Planetary waves in total ozone seem to have been much stronger during 2002 than during the other years investigated. The latter agrees fairly well with the substantial reduction in the spatial extent of the polar vortex and the limited ozone hole size in 2002.

Evtushevsky *et al.* (2008a) studied the asymmetry of TOC and tropopause height during the Antarctic spring. Analysis of zonal anomalies in TOC and the thermal tropopause height shows that ozone depletion during the Antarctic spring has a strong influence on the height and sharpness of the tropopause over West Antarctica. The seasonal change of tropopause peak height occurs between 9 km (summer and autumn) and 13–14 km (winter and spring) over the Weddell Sea region, and in the range of 9–10 km at opposite longitudes in the Australian sector. Elevation of the tropopause is explained by the low-stratospheric temperature due to the seasonal temperature minimum in winter and ozone losses in spring. Accordingly, tropopause zonal asymmetry is determined by the strong east–west asymmetry of surface temperatures in winter and total ozone asymmetry in spring. Lowering of the elevated tropopause over the boundary of the Antarctic continent is explained by the influence of the cold continental troposphere both in winter and spring. Evtushevsky *et al.* (2008a) reported some details of the relative contribution of radiative (strong cooling of the lower stratosphere and continental troposphere) and dynamical (quasi-stationary planetary wave 1 in total ozone) processes to Antarctic tropopause formation. The sloping meridional profile of the tropopause and widening of the transition layer between the troposphere and stratosphere over West Antarctica suggest the possibility of enhanced air mixing not only vertically but also horizontally. This may contribute to the east–west asymmetry in Antarctic climate change.

Turning to the Northern Hemisphere, although there is nothing quite so dramatic as the ozone hole that occurs in the Antarctic spring there has nevertheless been a considerable loss of ozone in the spring in the Arctic. Average TOC departures from the pre-1976 values for February-March in the Arctic are shown in Figure 6.23 for the period 1990–2000.

Balis *et al.* (2009) calculated the zonal average O_3MD from the pre-1976 averages for January 1–April 15 (the first 105 days of the year) for three zones, 35–45°N, 45–60°N and >60°N (see Figure 6.24). The zone north of 60°N coincides with the Arctic stratospheric vortex within which conditions are favorable for ozone destruction if the temperature falls below about 78°C. Combining the three zones and considering the whole region north of 35°N one finds that the overall O_3MD for the first 105 days of each year increased from around 2,800 Mt in 1979–1984 to an average of around 7,400 Mt in the 1990s and exceeded 12,000 Mt in the springs of

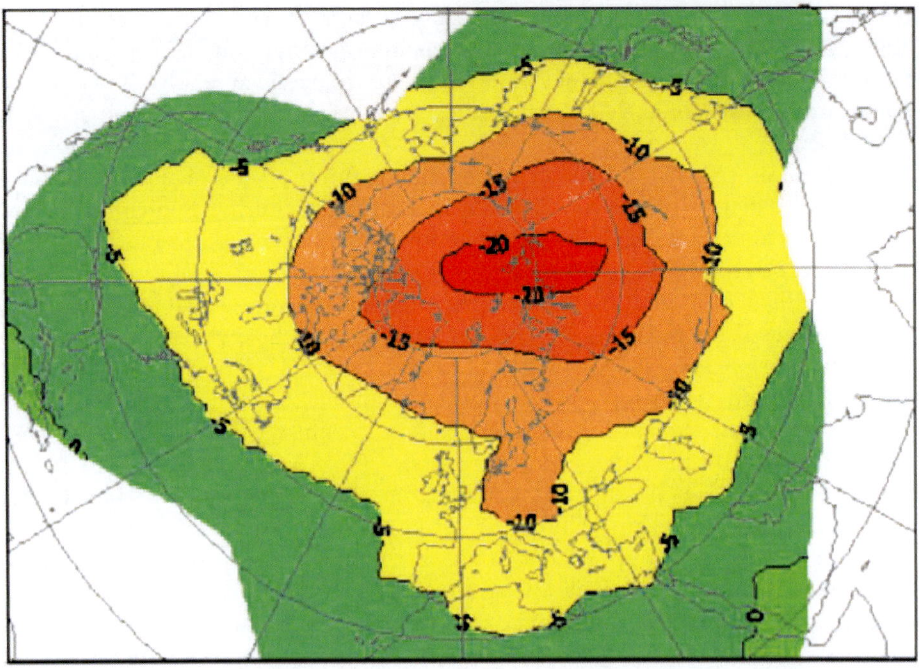

Figure 6.23. Average total ozone departures from pre-1976 values for the period 1990–2000, showing a large area coinciding with the winter–spring stratospheric vortex position, where ozone departures are between 10 and 20% even in 11-year averages (Balis *et al.*, 2009).

1993 and 1995. This value is comparable with the average O_3MD for the Antarctic springs of the 1990s. During the last seven years of this dataset the average $O_3MD > 35°N$ was around 7,200 Mt. The upper part of the mid-latitudes routinely contributes the largest contribution to the O_3MD north of 35°N; this is because of the longer time of exposure to sunlight during southward expansions of the Arctic vortex. Interannual variability of seasonally averaged temperatures in the lower stratosphere north of 60°N, an area where the polar vortex is usually situated, is the main parameter that drives the interannual variability of O_3MD over the polar regions of the Northern Hemisphere. This can be seen in Figure 6.25 where the O_3MD for the Arctic winter is plotted from 1978 to 2008 together with the O_3MD in the Antarctic winter which in turn, it will be recalled, is correlated with the temperature at 100 hPa (see the lower panel of Figure 6.20). This conforms to the earlier results obtained by McPeters and Labow (1996) for the Northern Hemisphere. The positive total ozone dependence in the difference between TOMS version 7 and Dobson/Brewer stations had been noted. This study covered the 30 ground-based instruments at the mid-latitudes (25°N–55°N). The TOMS–ground difference was considered as a function of TOMS total ozone. McPeters and Labow (1996) identified two TOC ranges with a distinct difference trend (their Figure 5): the difference increased by about 1% per 100 DU above a threshold of

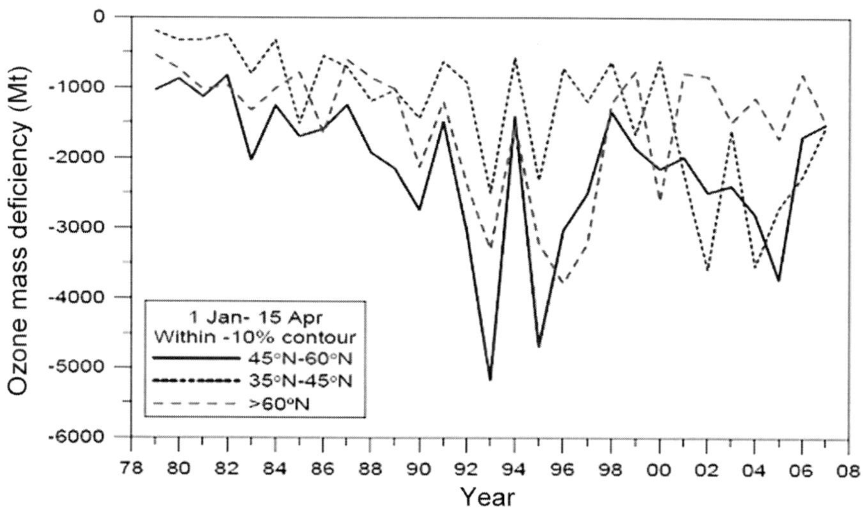

Figure 6.24. Integrated O_3MD for a period of 105 days within the −10% contour for latitudinal belts: 45–60°N (solid line); southern (lower) part of the mid-latitudes (35°–45°N) (dotted line); and >60°N (dashed line). It is seen that the contribution of the polar region to the O_3MD has been smaller during the last decade, when the contribution by the southern part of the northern mid-latitudes has rapidly increased (Balis *et al.*, 2009).

275 DU and 2.4% per 100 DU below that threshold. The authors assumed that the difference increase with total ozone was due to a combination of TOMS and Dobson errors.

Comparison of the O_3MD over the two hemispheres poleward from 35° reveals interesting differences. The values of the O_3MD integrated over the winter–spring seasons (January 1–April 15 for the Northern Hemisphere and September 1–December 15 for the Southern Hemisphere) from 1979–2007 are plotted in Figure 6.25. While values for the two hemispheres have been generally comparable, the magnitude of O_3MD in the Northern Hemisphere was less than that in the Southern Hemisphere, except in 1993 and 1995 when it was greater. The stratosphere of northern upper latitudes is warmer than in the Southern Hemisphere. The Arctic stratosphere is significantly chemically perturbed just as in the Antarctic; however, it warms up earlier in the spring by major meridional transport, so that by the time there is sunlight available to activate ozone losses, the PSCs have often already disappeared. That explains why the ozone loss in the Arctic has been less spectacular than in the Antarctic. The general conclusion from this figure is that there is, as yet, no evidence of a halt to the decline in the spring concentrations of ozone in the polar and mid-latitude regions.

The relationship between Arctic ozone loss and the volume of polar stratospheric clouds (PSCs) has been observed (Rex *et al.*, 2004, 2006). The surprising and to date unexplained qualities of this relationship are its compactness and its near linearity. The rate-determining step for activation in the early part of the winter

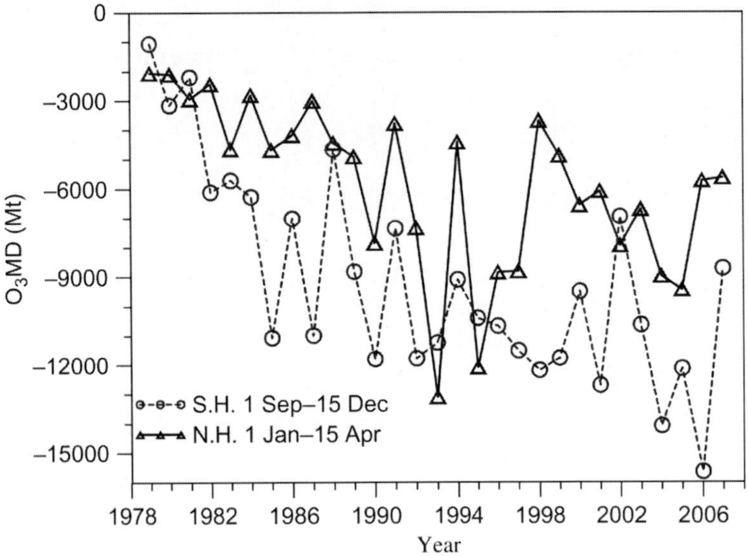

Figure 6.25. Total O_3MD (from average pre-1976 ozone levels) north of 35°N (January 1–April 15) and south of 35°S (September 1–December 15) (Balis *et al.*, 2009).

is the photolysis rate $J(HNO_3)$ as that limits the supply of $ClONO_2$ which reacts rapidly with HCl on PSCs. As a result the timescale for activation is weeks. The extent of activation thus depends on the continued presence of PSCs and, hence, on the time integral of the volume of PSCs. During the subsequent ozone loss period, ozone loss was found to depend on the competition between the photolysis of ClOOCl (leading to ozone loss) and the photolysis of HNO_3 (leading to deactivation). Both processes accelerate as the solar zenith angle decreases, and the integrated amount of ozone loss depends primarily on the extent of the initial chlorine activation—not on the speed of ozone loss.

One of the most striking features of Arctic ozone loss is its large interannual variability. For example, there were losses of $<10\%$ in 1998/1999 and $>65\%$ in 1999/2000 at around 18 km. Losses of 50% or more have been seen at similar altitudes in the Arctic in several winters since the early 1990s. Chemical losses in total ozone (i.e., the overall thickness of the ozone layer) in the Arctic vortex have varied between about 5 and 30% since the early 1990s. Year-to-year variability is apparently random and is driven by internal variability in the atmospheric system. This variability in ozone loss in the Arctic is to be contrasted with the Antarctic where nearly complete ozone loss has taken place in nearly all winters since the 1990s at altitudes between about 15 and 20 km.

Varotsos *et al.* (2012) by analyzing total ozone and temperature data over the Arctic region have found the following:

(1) Total ozone loss increased in the Arctic region in early 2011 and this was the largest loss of ozone above the Arctic ever observed.

(2) The Arctic TOC loss for December–April 2011 reached the levels of Antarctic TOC loss during 1989–1991.
(3) A highly extreme cold Arctic winter was observed during 2010/2011 which persisted longer than at any other time during the past 20 years.

The basic mechanism leading to rapid polar ozone loss is generally well understood (e.g., Newman *et al.*, 2007a, b). The air over the poles gradually cools at the start of the dark winter months and a westerly circumpolar flow of air (a polar vortex) is established. If the temperatures drop below a critical point, PSCs can form. HCl and $ClONO_2$ can react on the surface of PSCs, with these unreactive forms being converted to active forms (ClO, chlorine monoxide) which can rapidly destroy ozone. In the absence of further exposure to PSCs, the ClO formed continues to destroy ozone while it is gradually converted back to unreactive forms. Particularly large ozone losses can occur when airmasses become depleted in HNO_3 as PSCs fall to lower altitudes during prolonged cold periods, a process that delays the deactivation of active chlorine species.

Rex *et al.* (2004, 2006) reported a surprisingly compact and near linear relationship between column ozone loss and the volume of air where the temperature is low enough for PSC formation (V_{PSC}), when each quantity is integrated over each winter. The relationship can be used to infer the sensitivity of Arctic ozone loss to possible changes in stratospheric vortex temperatures and to test chemistry–transport and chemistry–climate models. An updated version of the relationship, using data for later years, was described by Harris *et al.* (2009) and the result is shown in Figure 6.26.

Sonbawne *et al.* (2009) made extensive observations of TOC over the Antarctic region at the Indian research station Maitri (70.76°S, 11.74°E) during two Antarctic summer periods of 2004–2005 and 2006–2007 using a ground-based hand-held sunphotometer/ozonometer (see Chapter 2) under clear-sky conditions to study short-term variations. Mean TOC during the 2004–2005 summer was 288.7 DU with a variability of 8% and the corresponding values during 2006–2007 summer were 280.4 DU and 11%. It was observed that both ozonometer and Dobson spectrophotometer measurements at this Antarctic station during January 2005 showed a short-term decrease in TOC of about 70 DU before regaining the original value. A simultaneous increase in precipitable water and decrease in surface temperature point to a possible interplay of chemistry and meteorological conditions. The occurrence of an intense solar proton event in January 2005 also suggests the possible role of precipitating charged particles in causing short-lived ozone decreases over this Antarctic region.

Kuttippurath *et al.* (2009) studied ozone depletion in the Arctic during the winter of 2007–2008 using the MIMOSA-CHIM model (Tripathi *et al.*, 2006, 2007). Kuttippurath *et al.* (2009) found that a sizable amount of ozone was depleted inside the vortex in the Arctic winter of 2007–2008 which appeared to be centered at 450–550 K. By considering modeled ozone loss at 475 K for the winters of 2004–2005, 2005–2006, 2006–2007, and 2007–2008, it became clear that interannual variations of ozone loss are very large in the Arctic and are well correlated with year-

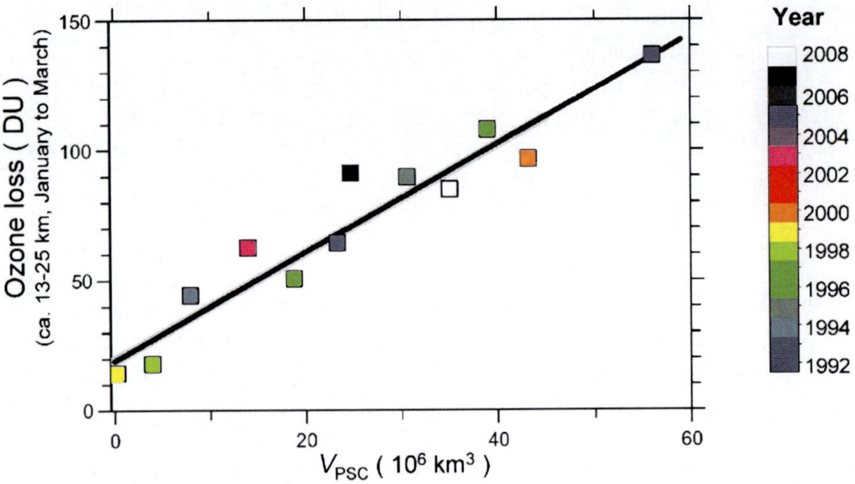

Figure 6.26. Integrated ozone loss as a function of V_{PSC} for 1992/1993 to 2007/2008. No values are shown for the winters 2000/2001, 2001/2002, 2003/2004, and 2005/2006 due to major warmings and/or lack of ozonesonde data. Ozone losses are derived from the Arctic ozonesonde network using the vortex average approach. V_{PSC} is derived from ECMWF analyses using the temperature of formation of nitric acid trihydrate (updated from Rex *et al.*, 2006) (Harris *et al.*, 2009).

to-year fluctuations in polar stratospheric temperatures. For instance, though the 2004–2005 winter was one of the coldest and witnessed significant amounts of ozone loss, the following winter (2005–2006) was one of the warmest and ozone loss was relatively small. Then the 2006–2007 and 2007–2008 Arctic winters were also characterized by very low temperatures inside the vortex and, hence, ozone loss was substantial. In addition, ozone loss in 2007–2008 was also the largest among the recent Arctic winters at 475 K. It has already been proved that there is a nearly linear relation between ozone loss and the volume of PSCs (Rex *et al.*, 2006; WMO, 2007), which are directly connected to polar temperatures as well. Therefore, observed and simulated ozone loss during recent winters are also in line with analyses done in the past (i.e., years with large and small ozone loss episodes). Specific change is hardly visible and large interannual variability masks any tangible trend in the data too. This is consistent with results obtained from other analyses (WMO, 2007) and is in accord with our present understanding of the trend of Arctic ozone loss (WMO, 2007).

Ground-based FTIR (Fourier Transform InfraRed) spectrometers are used to monitor several ozone-relevant species including chlorine compounds, as well as tracers such as HF or N_2O (see Section 3.3.3.3). One of these spectrometers has been operated continuously at the IRF (Institutet för Rymdfysik, the Swedish Institute of Space Physics) Kiruna (Sweden) since the winter of 1993–1994 (Blumenstock *et al.*, 2009). Atmospheric absorption spectra with high spectral resolution using the Sun as the source of radiation are recorded. Column amounts of several trace gases like O_3, N_2O, CH_4, HF, HCl, $ClONO_2$, ClO, NO, NO_2, and HNO_3 are derived. In addition, for strong absorbers (such as O_3) some profile information is retrieved, too. Long-term FTIR measurements in the polar vortex above Kiruna show significant ozone loss in the winters of 1995, 1996, 2000, 2003, 2005, and 2007. Fully consistent with that, chlorine activation and the removal of HNO_3 from the gas phase have also been observed in these winters. Comparison of observed ozone loss with data from the SAOZ (Système d'Analyse par Observation Zénithale) network shows very good agreement. Chlorine activation as well as ozone depletion correlates well with meteorological parameters such as stratospheric temperatures and volume of PSCs. In addition to studies using ozonesonde data, a correlation has been found between chlorine activation and the volume of PSCs (for details see Blumenstock *et al.*, 2009).

Studies of the area of the Antarctic ozone hole and of the O_3MD in the Arctic or Antarctic spring are based on the use of satellite data from TOMS and various other instruments which, in turn, are validated by point measurements with Dobson or other spectrophotometers. However, it is hard to know how and when the Antarctic ozone hole appears because it starts some time in July or in August when it is not possible to obtain TOC with the usual methods using sunlight (such as conventional Dobson spectrophotometers) or other ground-based spectrophotometers or satellite-flown instruments (such as TOMS, etc.). Consequently, the technique of making measurements of TOC during the Antarctic winter with a Dobson spectrophotometer at the Japanese Syowa Antarctic research station using moonlight instead of sunlight was developed (Chubachi and Kajiwara, 1988). It was found that ozone depletion started early in August at Syowa station between 1982 to 1994 (Chubachi, 1997). Further studies using Dobson spectrophotometers with moonlight as well as sunlight at three Antarctic stations and Macquarie Island—the South Pole (90°S), Arrival Heights (78°S, 167°E), Syowa Station (69°S, 40°E), and Macquarie Island (55°S, 159°E)—averaged for the years between 1993 and 2005 have been described by Chubachi (2009). The 10-day averages of the months are shown in Figure 6.27. This figure shows that the beginning of ozone depletion was observed at all these stations except Macquarie Island over the Antarctic in August. It also shows that the Antarctic ozone hole (defined as the area where TOC is below 220 DU; WMO, 2007) appeared in early September on average between 1993 and 2005. Macquarie Island was outside the ozone hole. For the three stations (South Pole, Arrival Heights, and Syowa), where the ozone hole appears, Chubachi (2009) also plotted the deviation from the polar night mean averaged over the periods 1980–1992 and 1993–2005 (see Figure 6.28). The magnitude of this deviation was greater in the later period indicating that the Montreal Protocol had, at that stage, not managed to achieve any reduction in ozone depletion in the ozone hole.

Kravchenko *et al.* (2009) made a detailed comparison between satellite-derived TOC values and results from ground-based measurements made at three Antarctic stations—Vernadsky (Ukraine; until February 6, 1996 this was the U.K.'s Faraday Station), Halley (U.K.), and Amundsen–Scott (South Pole, U.S.A.)—in Antarctica (Figure 6.29a) and Barrow (U.S.A.) in the Arctic (Figure 6.29b). These stations are equipped with Dobson spectrophotometers. The locations of these stations are

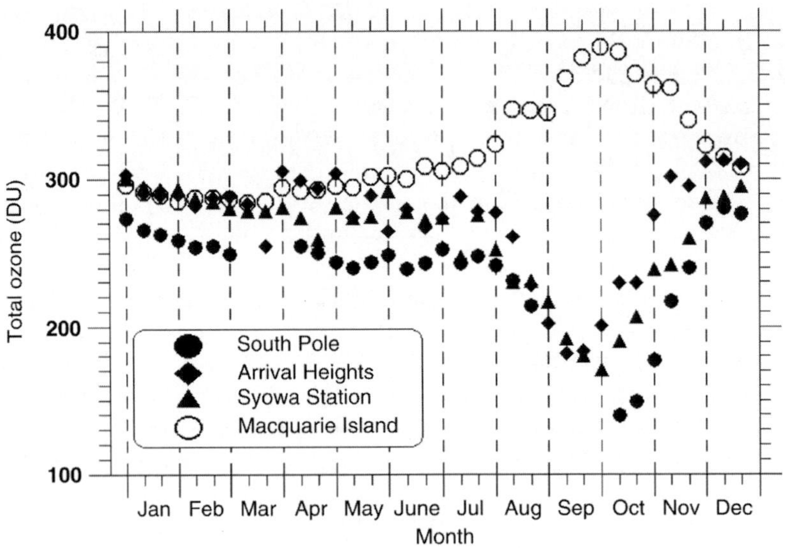

Figure 6.27. Seasonal changes of 10-day averaged TOC of each month at the South Pole, Arrival Heights, Syowa Station, and Macquarie Island averaged over the period between 1993 and 2005. The horizontal axis shows every 10-day period in each month,. The long tick marks on the horizontal axis and the dashed vertical lines indicate the first 10 days of each month. The next short tick marks indicate the middle 10 days and the third tick marks indicate the last 10 days of each month (Chubachi, 2009).

shown in Table 6.11. The reason for the inclusion of stations from both the Antarctic and the Arctic was to see if there is any difference between the two polar regions in terms of agreement between satellite and ground-based data. In the work of Kravchenko *et al.* (2009), EP-TOMS satellite data from 1996 to 2005 were used as the source of satellite data. Satellite daily TOC values were taken using version 8 of the algorithm introduced in 2004 and empirically corrected in 2007. The most persistent features of the relative EP-TOMS/Dobson difference are: (1) significant increase in dispersion during the period of the spring Antarctic ozone hole and (2) differing dependence on total ozone in trend tendency and significance for EP-TOMS and Dobson datasets. The results indicate the influence of specific conditions during the Antarctic ozone hole on the possible precision that could be achieved in assessments of Montreal Protocol effects in the ozone layer over this region.

It can be concluded that total ozone measurements in the polar regions, especially in Antarctica, remain influenced by total ozone dependence, which is probably most significant below 220 DU and above 220 DU for Dobson and EP-TOMS, respectively. These results give relative estimates; the individual contribution of TOMS and Dobson errors to total ozone dependence needs further investigation.

These conclusions are based on one decade of EP-TOMS measurements. However, Fioletov *et al.* (2008) demonstrated that systematic differences between the various satellite and ground-based instruments between 60°S and 60°N are

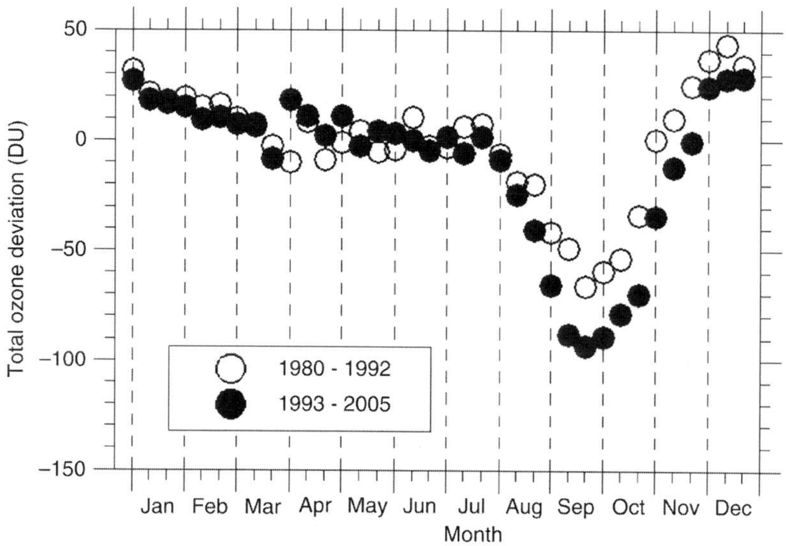

Figure 6.28. Seasonal change in the deviation of TOC from the polar night mean at the South Pole, Arrival Heights, and Syowa Station in the periods 1980–1992 (open circles) and 1993–2005 (filled circles). The horizontal axis shows every 10-day period in each month. The long tick marks on the horizontal axis and the dashed vertical lines indicate the first 10 days of each month. The next short tick marks indicate the middle 10 days and the third tick marks indicate the last 10 days of each month (Chubachi, 2009).

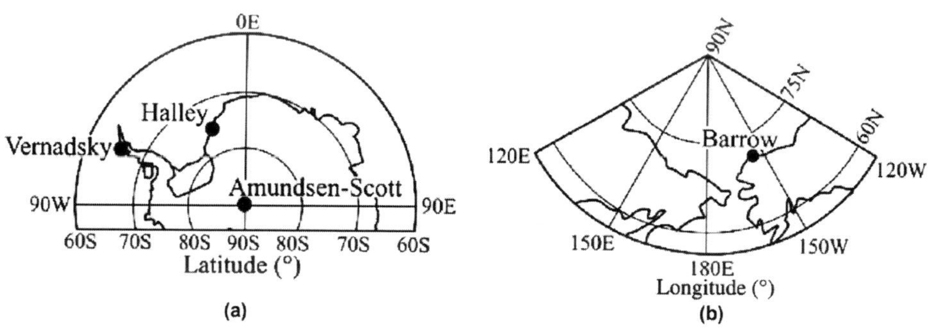

Figure 6.29. Geographical location of (a) Antarctic stations and (b) the Arctic station chosen for comparison of satellite EP-TOMS and ground-based measurements of total ozone content in 1996–2005 (Kravchenko *et al.*, 2009).

typically within ±3% (i.e., they are similar to differences at high latitudes). Another situation exists in the case of standard deviation; it ranges from 3.4% to 5.6% under conditions of normal TOC levels and increases to about 7% in the ozone hole compared to 2–3% at lower latitudes (Fioletov *et al.*, 2008). This increases uncertainty in the estimation of TOC change. Both dynamical influence and total ozone

Table 6.11. Geographical co-ordinates of the high-latitude Dobson stations considered by Kravchenko *et al.* (2009). The station numbers in column 2 correspond to the numbering given by the World Ozone and Ultraviolet Data Center (WOUDC; *www.woudc.org/data/Metadata/platform_e.html*).

Station	*Station number*	*Latitude*	*Longitude*
Vernadsky (Ukraine)	232	65°15′S	64°16′W
Halley (U.K.)	057	75°31′S	26°40′W
Amundsen–Scott (U.S.A.)	111	89°58′S	24°47′W
Barrow (U.S.A.)	199	71°19′N	156°36′W

dependence can contribute to the increase in standard deviation. In this relationship, satellite measurements exhibit much lower or statistically insignificant ΔTOC dependence on TOC and are more suitable for the assessment of Montreal Protocol effects on the ozone layer over Antarctica; what is more, satellite data offer higher spatial coverage relative to ground-based data in this region.

The relative percentage difference, ΔTOC, between empirically corrected EP-TOMS version 8 data and Dobson spectrophotometer observations from the four ground stations was analyzed both as a function of time (day of the year) and as a function of TOC itself. The relationship of ΔTOC to TOC is different depending on whether TOC_{TOMS} (the satellite values of TOC) or TOC_{Dobson} (the ground-based values of TOC) are used. Particular attention was given to the dependence of ΔTOC in the ranges of low (≤220 DU) and high (>220 DU) TOC ozone using the WMO threshold of 220 DU as defining the ozone hole (see Kravchenko *et al.*, 2009 for details).

In high-latitude regions, TOC measurements are complicated both from space and the ground because of the high solar zenith angle, snow ice, and cloud cover, as well as the unusually low TOC level inside the Antarctic ozone hole (Bernhard *et al.*, 2005; Bhartia and Wellemeyer, 2004; Kylling *et al.*, 2000; Labow *et al.*, 2004). Most of the Antarctic stations are located at the continental coast, where the ozone hole edge typically appears. Large dynamical distortions of the ozone hole edge result in frequent alternation of low (inside the hole) and high (outside the hole) TOC levels over the individual stations and can also lead to apparent changes during the observing day dependent on the solar position. The time of ozone records during the observational day may be different for different satellite and station instruments, and dynamical variations of the spatial distribution of TOC influence the agreement of data. Thus, it is to be expected that the magnitude of ΔTOC values would be larger under ozone hole conditions than at times when the ozone hole was not present or in the Arctic; this was, indeed, demonstrated in the work of Kravchenko *et al.* (2009), both from their studies of ΔTOC against day of the year (see Figure 2 of their paper) and against TOC (see Figure 3 and Tables 2 and 3 of their paper). Since spring ozone losses in Antarctica make a large contribution to global ozone changes, the assessment of data accuracy is especially important for the ozone hole.

The Polar Ozone and Aerosol Measurement (POAM III) satellite instrument provided 8 years (1998–2005) of Antarctic ozone profile measurements. These measurements showed that ozone in the vortex declined sharply in late winter/early spring in each of these years except 2002, which had an anomalous level of dynamical disturbance (Varotsos, 2002a, b), and 2004, which was also somewhat warmer than usual near 21 km in the vortex. To isolate chemical ozone loss, Hoppel *et al.* (2005) applied the Match technique (Rex *et al.*, 1998) to five years of data using the photochemical box model of Salawitch *et al.* (1993) and Canty *et al.* (2005) to calculate ozone photochemical loss. The model calculates ozone loss from chlorine and bromine reactions using JPL 2002/2006 kinetics (operated by the Jet Propulsion Laboratory). To simulate maximum expected loss, total reactive bromine (BrO_x) was set at 20 pptv, which equals the total inorganic bromine budget for the time period including shorter lived bromocarbons (Pfeilsticker *et al.*, 2000; Wamsley *et al.*, 1998; WMO, 2003). The abundance of reactive chlorine (ClO_x) was set to 3.7 ppbv. At a potential temperature of 469 K (~19 km altitude), 5-year average loss rates were found to increase slowly from ~2 ppbv/sunlit hour at the beginning of July to ~7 ppbv/sunlit hour at the beginning of September, and then decrease rapidly. Peak loss rates compare well with the maximum loss rates of ~6 ppbv/sunlit hour shown by Tripathi *et al.* (2007) for a Match analysis of Antarctic ozonesondes during 2003, and they are consistent with those observed at the South Pole. When Burkholder *et al.* (1990) ClOOCl cross-sections are used in the box model, the simulated loss rate increases by ~18%, which is still consistent with observations, when considering that ClO_x and. BrO_x may be overestimated in the model. When Huder and DeMore (1995) cross-sections are used, the September peak loss rate decreases to ~4.5 ppbv/sunlit hour, significantly less than the observed value of ~7 ppbv/sunlit hour. Finally, if Pope *et al.* (2007) cross-sections are used, the simulated loss rate is ~2.5 ppbv/sunlit hour, less than half the observed value. The Match results are, however, sensitive to the choice of meteorological analysis used for trajectory calculations. ECMWF trajectories yielded the smallest peak loss rates, which are expected to be the most accurate because of the higher spatiotemporal resolution compared to other meteorological data used in the study.

6.5.5 Antarctic ozone hole predictability; use of natural time series

Nowadays, it is widely recognized that the main problem in stratospheric ozone dynamics is ozone depletion, which has been most dramatically expressed by the formation of the spring TOC minimum in Antarctica, the so-called ozone hole, which we have already discussed (see Section 5.4). We have seen that most of the ozone loss in the ozone hole occurs in the lower stratosphere, and this is attributed to chemical ozone-destroying mechanisms. Considering this very important phenomenon, it is worth emphasizing the very complex and interactive nature of those processes that are responsible for stratospheric ozone formation and changes (Dameris, 2010a, b). Such changes depend on numerous interactions between heterogeneous chemical, photochemical, and transport processes, including close links between ozone and climate changes (Dameris *et al.*, 2007) as well as the impacts of such external factors as volcanic eruptions, extraterrestrial irradiance variations (solar activity, cosmic rays, etc.).

Undoubtedly, marked progress has been achieved in studying atmospheric ozone since the discovery of the Antarctic ozone hole. The interest of the scientific community in this issue is still strong due to the remarkable characteristic features observed during the last decade. For instance, the unprecedented event of major sudden stratospheric warming (SSW) (mainly caused by the propagation of very strong planetary waves) and the subsequent small-sized ozone hole that split into two holes in 2002 (see Section 6.5.1) attracted the attention of the ozone community (Allen *et al.*, 2003; Dowdy *et al.*, 2007; Glatthor *et al.*, 2004, 2005; Grytsai *et al.*, 2005; Lekien and Coulliette, 2007; Liu, Y. *et al.*, 2009; Manney *et al.*, 2005; Marchand *et al.*, 2005; Pazmino *et al.*, 2005; Siegmund *et al.*, 2005; von Savigny *et al.*, 2005; Wang *et al.*, 2005). In addition, a recent result shows that global total ozone variability obeys long-range power-law correlations with a (self-similarity) exponent α_{DFA} obtained from detrended fluctuation analysis (DFA) close to unity (Varotsos, 2005a, b).

Despite the afore-mentioned progress, a number of related problems remain unresolved so far, and there is a necessity for more comprehensive and integrative consideration of the complexity of all interactive processes including the chemistry and dynamics of the atmosphere. For instance, what caused the Southern Hemisphere to exhibit very strong planetary waves and, subsequently, an unprecedented major SSW in 2002 is not yet well understood. Another important question is to what extent are the dynamics of the Antarctic ozone hole governed by natural variability and how do they compare with the anthropogenic signature on this phenomenon. To reveal the hidden mechanisms behind these open questions we should probably work by considering Antarctic ozone as a complex non-linear system that may occasionally reach criticality. However, to address this question, detailed understanding of the evolution dynamics of ozone hole parameters over Antarctica is necessary.

Varotsos and Tzanis (2011) undertook the first study of dynamical evolution of the ozone hole area over Antarctica using a modern method of statistical physics: natural time analysis. This makes use of a new time domain termed "natural time", which was first introduced by Varotsos *et al.* (2002) in other geophysical applications. In particular, it has been found that novel dynamical features (e.g., non-linear) hidden behind time series in complex systems can emerge as a result of analyzing them in the new time domain of natural time, which conforms to the desire to reduce uncertainty and extract signal information as much as possible (Abe *et al.*, 2005). Analysis in natural time has the advantage that it enables study of the dynamical evolution of a complex system and identifies when the system enters a critical stage. Hence, natural time plays a key role in predicting impending major events in general. Relevant examples of data analysis in this new time domain have been presented in a large variety of fields including biology, Earth sciences, and physics. As a first example, we mention natural analysis of electrocardiograms (Varotsos, P.A. *et al.*,

2004, 2005a, 2007) which may herald a cardiac arrest. Second, natural time analysis can identify (Varotsos, P.A. *et al.*, 2003c, 2004, 2005a–c, 2006a–c) seismic electric signals (SESs) which are low-frequency electric signals that precede major earthquakes (Varotsos and Alexopoulos, 1977, 1984b, c; Varotsos and Lazaridou, 1991; Varotsos *et al.*, 1993). They are generated due to the migration of lattice defects (Varotsos, 1976; Varotsos and Alexopoulos, 1984d) when stress in the earthquake focal area reaches a critical value (Varotsos, 2007) before the rupture; thus, these signals are characterized by critical dynamics and, quite interestingly, also exhibit $\alpha_{\mathrm{DFA}} \approx 1$ (Varotsos, P.A. *et al.*, 2003a, b).

Among the physical quantities defined in the new time domain, the chosen variable for analysis of the time evolution of the ozone hole is the entropy S defined in natural time for the following reasons. Several time series most likely contain stochastic and deterministic components. Since the entropy is a concept equally applicable to deterministic as well as stochastic processes, its use is particularly appropriate for analysis of the temporal evolution of the ozone hole. Entropy, in general, has been used in time series analysis by other authors, but this approach differs essentially because the entropy that is employed is defined (Varotsos, P.A. *et al.*, 2004, 2011) in an entirely different time domain (i.e., the natural time domain).

Let us summarize the background of natural time χ. In a time series comprising N events, the natural time $\chi_k = k/N$ serves as an index (Varotsos *et al.*, 2002) for the occurrence of the k-th event. In the natural timescale these events are equally spaced in time. Let us study the evolution of the pair (χ_k, Q_k), where Q_k quantifies the magnitude of the k-th event. As an example, Figure 6.30 shows an artificial time series in conventional time and in natural time χ.

By defining

$$p_k = Q_k \Big/ \sum_{n=1}^{N} Q_n \tag{6.18}$$

entropy S in the natural time domain is defined (Varotsos, P.A. *et al.*, 2003a) as

$$\left.\begin{aligned} S &\equiv \sum_{k=1}^{N} p_k \chi_k \ln \chi_k - \left(\sum_{k=1}^{N} p_k \chi_k\right) \ln\left(\sum_{m=1}^{N} p_m \chi_m\right) \\ \text{or} \quad S &= \langle \chi \ln \chi \rangle - \langle \chi \rangle \ln \langle \chi \rangle, \end{aligned}\right\} \tag{6.19}$$

that is, the symbol $\langle\ \rangle$ stands for the average value of any quantity of natural time with respect to the distribution p_k.

It should be stressed that the entropy S occurring here is not simply a statistical entropy, but a dynamic one that takes into account the (temporal) order of the occurrence of the events. For critical dynamics S is smaller than the value S_u $(= \ln 2/2 - 1/4 \approx 0.0966)$ of a "uniform" distribution (as defined by Varotsos, P.A. *et al.*, 2002, 2003c, 2004, 2006a, b; e.g. when all p_k are equal or Q_k are independent and identically distributed random variables of finite variance). Thus, $S < S_u$. The same holds for the value of the entropy obtained when considering

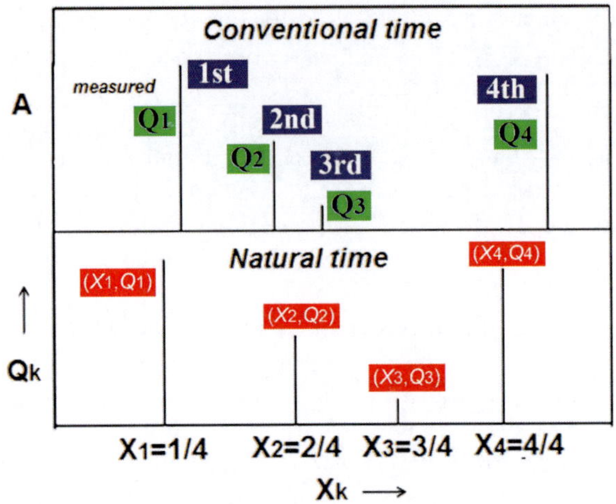

Figure 6.30. An artificial time series of events of a geophysical parameter in conventional time and in natural time χ (Varotsos and Tzanis, 2012).

time reversal $\hat{T}$ (the operator $\hat{T}$ is defined by $\hat{T}p_k = p_{N-k+1}$) which is labeled by S_-. Thus, for critical dynamics

$$S, S_- < S_u \tag{6.20}$$

It was found (Varotsos, P.A. *et al.*, 2005a) that, in general, S_- is different from S and, hence, S shows the breaking of time reversal symmetry, thus revealing the profound importance of considering the (true) time arrow in classifying the similar-looking signals of different dynamics. The difference $S - S_-$ will be labeled ΔS; this will also have a subscript (ΔS_i) meaning that the calculation is made (for each S and S_-) for a timescale i (= number of successive events).

Varotsos and Tzanis (2012) analyzed the time series A_k of the maximum daily ozone hole area (MDOHA) over Antarctica for each year during the period 1979–2009 using natural time analysis; A_k takes the role of Q_k in the above general analysis. The ozone hole area was determined from total ozone satellite measurements (TOMS and OMI) and data provided by NASA (*http://ozone watch.gsfc. nasa.gov/meteorology/annual_data.html*). It is defined to be that region of ozone values below 220 DU located south of 40°S. Linear interpolation has been used for 1995 for which there was no measurement. The procedure was as follows: a window of length i is sliding, each time by one year, through the whole time series. The entropies S and S_-, and their difference ΔS_i, are calculated each time. Thus, we form a new time series consisting of successive ΔS_i values.

The results for the entropy in natural time S_i for various window lengths (3 to 15 years)—sliding each time by one year through the whole time series for the period 1979–2009—are shown in Figure 6.31. It can be seen that for $i = 3$ to 7 years the value of S_i almost stabilizes during the last few years, but that for $i > 8$ years there is a gradual increase after around 1999. Note that all the S_i values are smaller than S_u,

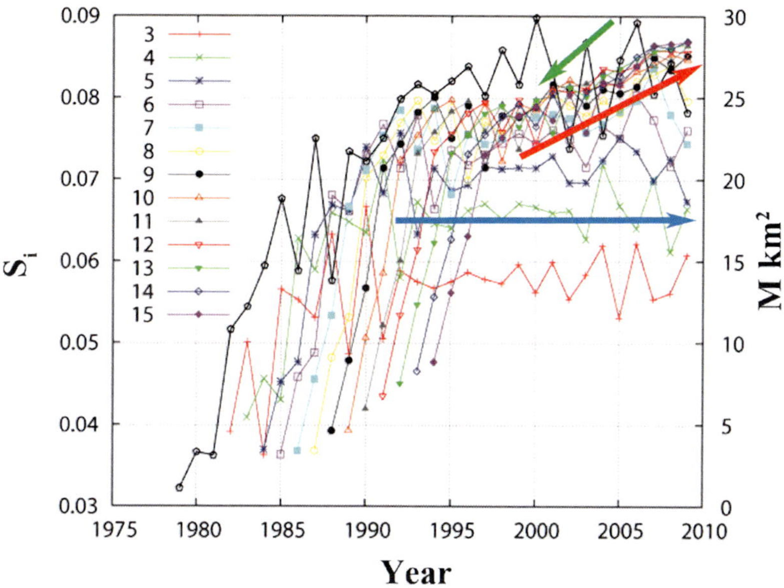

Figure 6.31. The entropy in natural time S_i for various window lengths $i = 3$ to 15 years (left scale) sliding each time by one year through the whole MDOHA time series (1979–2009). Data (right scale) are depicted by black pentagons (linear interpolation has been used for 1995 for which there was no measurement). For $3 \leq i \leq 7$ years, the S almost stabilizes during the last several years (see the blue arrow). For $i > 8$ years there is a gradual increase in S after around 1999 (see the red arrow). For the longer scales, around 2000 and 2001, $S \approx 0.080$, which is comparable to that deduced from a fractional Brownian motion with $\alpha_{\text{DFA}} \approx 1$ (see the green arrow). For critical dynamics $S < S_u$ $(= \ln 2/2 - 1/4 \approx 0.0966)$ of a uniform distribution and exhibits long-range correlations (Varotsos and Tzanis, 2012).

the value for a uniform distribution—a behavior which is consistent with the existence of long-range correlations (see Section 4.4). In addition, note that for longer timescales, the S_i values around 2000 and 2001 are close to or scatter around the value $S \approx 0.080$. This is comparable to the S value deduced from a fractional Brownian motion with α_{DFA} close to unity (see Figure 4 of Varotsos, P.A. *et al.*, 2006a, b).

The results for the entropy in natural time under time reversal $(S_-)_i$ for various window lengths (3 to 15 years) sliding each time by one year through the whole period (1979 to 2009) are presented in Figure 6.32. Here it can be seen that for all timescales $i = 3$ to 15 years the quantity $(S_-)_i$ almost stabilizes (except for a transient observed from 2002 to 2007) during the last few years. On the other hand, before 2002 the following precursory change is noticed. From 2000 to 2001, for all timescales the S_- values increase, except for $i = 13$, while from 2001 to 2002 all $(S_-)_i$ decrease, again except for $i = 13$; moreover, these $(S_-)_i$ values remain smaller than that (S_u) of a "uniform" distribution. Recall that, as mentioned above, both S and S_- values should be smaller than S_u for critical dynamics—see relation (6.20).

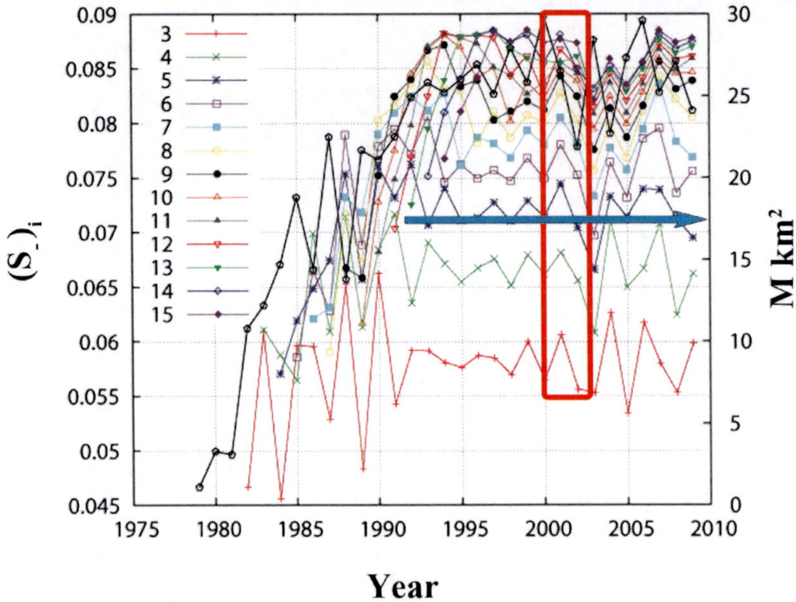

Figure 6.32. The entropy in natural time under time reversal $(S_-)_i$ for various window lengths $i = 3$ to 15 years (left scale) sliding each time by one year through the whole MDOHA time series from 1979 to 2009. Data (right scale) are depicted by the black pentagons (linear interpolation has been used for 1995 for which there was no measurement). For $3 \leq i \leq 15$ years, the $(S_-)_i$ almost stabilizes (except for a transient observed from 2002 to 2007) during the last several years (see the blue arrow). From 2000 to 2001, for all scales, the S_- values increase, except for $i = 13$, while from 2001 to 2002 all $(S_-)_i$ decrease, again except for $i = 13$ (see the brown rectangle). $(S_-)_i$ values remain smaller than that (S_u) of a uniform distribution. Recall that for critical dynamics, both S, S_- should be smaller than S_u (Varotsos and Tzanis, 2012).

The results for the entropy change $\Delta S_i = S_i - (S_-)_i$ under time reversal for various window lengths i (3 to 15 years) sliding each time by one year through the whole time series (1979–2009) are presented in Figure 6.33. Close inspection of this figure reveals that, before 2002, timescale invariance holds but only for short timescales. In particular, we observe that for timescales $i = 3$ to 7 years the values of ΔS_i almost coincide in 2000 and then decrease. On the other hand, for longer timescales $i = 10$ to 15, the values of ΔS_i clearly increase during the period 1990 to 2005.

In the complex system of the present case, the 11-year solar cycle—not surprisingly—seems to be of key importance, in the sense that both ranges of timescales (i.e., timescales shorter than 11 years as well as timescales comparable to or longer than 11 years) are appropriate for the detection of precursory changes. For example, for the former range it was found that in 2000 the ΔS_i values coincide for the timescales $i = 3$ to 7 years (see Figure 6.33).

As for the latter range, the results showed that for timescales exceeding 8 years the values of S_i exhibit a gradual increase after the year 1999. In other words, in

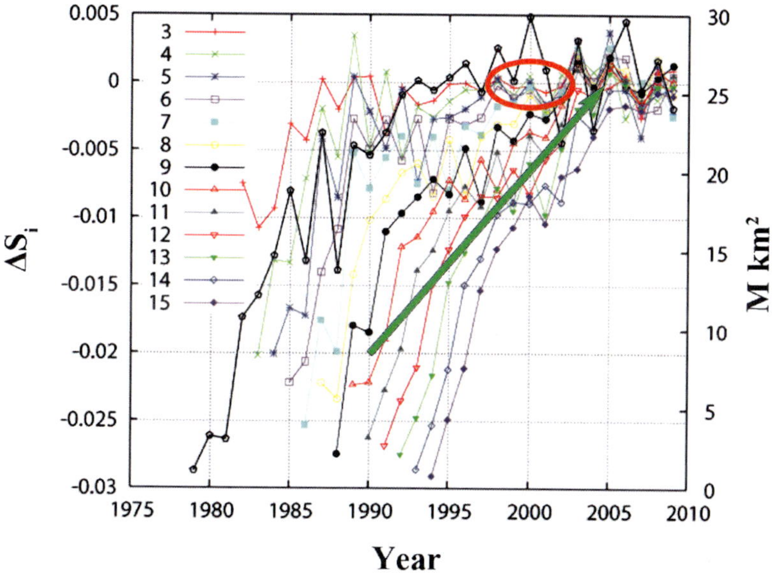

Figure 6.33. The entropy change in natural time under time reversal $\Delta S_i = S_i - (S_-)_i$ for various window lengths $i = 3$ to 15 years (left scale) sliding each time by one year through the whole time series of MDOHA from 1979 to 2009. Data (right scale) are depicted by black pentagons (linear interpolation has been used for 1995 for which there was no measurement). Before 2002, scale invariance holds but only for short scales. In particular, for scales $i = 3$ to 7 years the values of ΔS_i almost coincide at 2000 and then decrease (see the brown oval). For longer scales, ΔS_i clearly increases during 1990–2005 (see the green arrow) (Varotsos and Tzanis, 2012).

order to identify precursory effects, the investigation of the quantities S_i, $(S_-)_i$, and ΔS_i should be made at timescales surrounding (i.e., somewhat smaller or larger than) the 11-year period cycle which is of importance for the complex system under study. It should be emphasized that the signal for the year 2000 could be fairly associated with the phase transition in ozone hole size: in 2001 (largest ozone hole) and 2002 (smaller ozone hole).

Finally, Varotsos and Tzanis (2012) commented on the case of 1988, where the size of the Antarctic ozone hole was approximately equal to that observed in 2002 (before its split). In 1988, however, a quite strong but not major (without reversal of zonal mean zonal wind at 60°) SSW event occurred. That is why no break-up occurred in the diminutive Antarctic ozone hole in 1988. In addition, El Niño (5 to 7-year period) characteristic features in 1988 and 2002 were not similar (Varotsos, 2003a, b, 2004a, b). Precursory changes, if any, cannot be detected for this case due to the small number of observations of ozone hole size before 1988 (i.e., only 9 years of data are available).

This work can be summarized as follows. Three precursory changes have been identified:

(a) For timescales longer than 8 years, the entropy in natural time exhibits a gradual increase after around 1999.
(b) From 2000 to 2001, the entropy in natural time under time reversal shows an increase for all timescales (3 to 15 years) except for the timescale of almost 13 years.
(c) The values of entropy change in natural time almost coincide in 2000 for short timescales 3 to 7 years and then decrease.

The coexistence of these three facts is consistent with the approach of the system to a critical point (dynamic phase transition) almost two years before the ozone hole split over Antarctica.

Finally, it should be emphasized that natural time analysis does not constitute a specific theoretical model, but a general framework for analyzing experimental data.

6.6 LONG-TERM MONITORING OF THE OZONE LAYER

Although the Montreal Protocol now controls the production and emission of ozone-depleting substances, the timing of recovery of the ozone layer is unclear. Satellite sensors provide global measurements of ozone concentration which can be used to study the effects of implementation of the Montreal Protocol. However, a key issue in deriving long-term ozone trends from successive satellite instruments is intercomparability. Ground-based measurements offer continuous time series but only at a few locations. The combination of ground-based measurements with satellite data is therefore an effective means to evaluate satellite instrument intercomparability.

In order to assess the effectiveness of the measures taken in the framework of the Montreal Protocol, accurate and reliable measurements of atmospheric properties, such as ozone and temperature, are essential. In addition, co-ordinated *in situ* measurements provide and have provided a valuable contribution to our understandding of stratospheric processes. For instance, studies with ozonesondes have shown that ozone loss is not only related to enhanced chlorine levels, but also has a strong dependence on stratospheric temperature (Schulz *et al.*, 2000, 2001). Lidar measurements have, for example, been used to characterize polar stratospheric clouds (Massoli *et al.*, 2006) and to study geographic differences in observed stratospheric temperature and waves (Blum *et al.*, 2004). Nevertheless, ground-based and balloonborne measurements are only available at a few locations.

Satellite instruments are, in this respect, an excellent source of information as they can provide cost-effective and timely global coverage. However, one of the key issues when using different satellite instruments to derive long-term ozone trends is the degree to which the measurements of these instruments correlate properly. This is especially of concern when no simultaneous measurements are available due to coverage gaps in successive missions. A second major issue is the trade-off between geographic coverage (i.e., temporal and spatial resolutions) and the obtained accuracy for satellite retrievals (Meijer *et al.*, 2004).

The establishment of a long-term time series of ground-based measurements with known quality and comparison with satellite measurements is therefore a prerequisite to assess the quality of satellite data and to ensure good performance of individual sensors, as well as an appropriate combination of multisensor datasets. Ground-based and balloonborne measurements have already proved to be essential in the validation of various models (Reid *et al.*, 1998) and satellite retrievals (Cortesi *et al.*, 2007; Jiang *et al.*, 2007; Meijer *et al.*, 2003; Ridolfi *et al.*, 2007; Witte *et al.*, 2008).

The question of assessing the recovery of the Antarctic ozone hole was discussed by Newman *et al.* (2009). They identified two principal questions. The first was concerned with identifying the milestones of Antarctic ozone hole recovery as the levels of ODSs decline. The second was concerned with considering what might modify or change this recovery. Bodeker *et al.* (2007) defined three phases of ozone recovery:

(i) slowing of ozone decline, identified as the occurrence of a statistically significant reduction in the rate of decline in ozone because of changes (reductions) in EESC;
(ii) onset of ozone increases (turnaround), identified as the occurrence of statistically significant increases in ozone above previous minimum values due to declining EESC; and
(iii) full recovery of ozone from ODSs, identified as when ozone is no longer affected by ODSs.

In the absence of changes in the sensitivity of ozone to ODSs, this is likely to occur when EESC returns to pre-1980 levels.

There are many other factors apart from man-made ODSs affecting the ozone layer. There are, of course, natural fluctuations, the 11-year solar activity cycle, the equatorial Quasi-Biennial Oscillation of the lower-stratospheric zonal wind (QBO), and volcanic eruptions, all of which significantly affect the ozone layer. It is possible to predict the effects of the solar cycle and QBOs but volcanic eruptions cannot be predicted. Another important factor is climate change, due to increases in greenhouse gas concentrations, which influences stratospheric dynamics and chemistry, and therefore is expected to modify the speed of recovery of the ozone layer. Coupled 3-D models combine a general circulation model, which calculates winds and temperatures by numerically solving the equations of motion, with a representation of the photochemical processes that are important to the evolution of atmospheric constituents including stratospheric ozone. A better understanding of radiative, dynamical, and chemical processes and the interaction between climate change and atmospheric chemistry is highly necessary for reliable projections of future stratospheric ozone levels. Atmospheric models that consider the interaction of climate and atmospheric chemistry (i.e., so called chemistry–climate models, CCMs) have been developed recently to identify and quantify relevant processes affecting the ozone layer. Chemistry–climate models (CCMs) maintain consistency among the photochemical, radiative, and dynamical components of the model by transporting constituents using simulated meteorology and using the simulated fields for ozone and other gases to calculate radiative heating. Both 2-D models and CCMs contributed to the *Scientific Assessment of Ozone Depletion: 2006* (WMO, 2007).

Therefore, there are still open questions with regard to the beginning of ozone recovery and the timing of full recovery. Obviously, continuous measurements of TOC are needed to monitor the evolution of the stratospheric ozone layer, but extrapolation of these observations into the future needs to be undertaken on the basis of numerical models describing relevant atmospheric processes, and their mutual effects are needed as well. There are several components to the process of predicting the recovery of the ozone layer:

(a) The first stage involves projecting future concentrations of ODSs in the troposphere, based on available data on (mostly past) release of halocarbons and their various lifetimes in the troposphere. Surface observations have been combined with ODS emission estimates to develop scenarios of future ground levels of ODSs. For example, use of scenario Al (Daniel *et al.*, 2007) assumes adherence to the Montreal Protocol and projects a steady decline of most ODSs over the coming decades.
(b) The second stage involves projecting the migration of ODSs from the troposphere into the stratosphere and determining the resulting EESC.
(c) The third stage involves evaluating stratospheric ozone loss due to the calculated EESC. One needs to take care over the relation between EESC and ozone loss in the Antractic hole; this is because the ozone loss in the ozone hole tends to saturate (i.e., 100% loss—all the ozone between 15 km and 20 km has been destroyed in the ozone hole, see Section 6.5.1), resulting in no ozone change for either an increase or decrease of Antarctic EESC.

Any proposed model of this type needs to be evaluated and, if necessary, modified by using it with historical data to reproduce past and present TOC. Only models which have demonstrated their capability to reproduce recent developments adequately can be expected to provide reasonable estimates of future behavior.

A number of early studies predicted recovery of the Antarctic ozone hole in the middle of the 21st century (see Figure 1 of Newman *et al.*, 2009). Newman *et al.* (2006) projected the ozone hole area into the future by fitting the ozone hole area to Antarctic EESC and temperatures in the lower stratosphere, using values of the EESC from scenario A1. This projection indicated a leveling off in the period from the early 1990s to about 2015. After that the area was predicted to begin to decrease more clearly, returning to the 1980 situation around 2070. While the area of the ozone hole was used as a diagnostic by Newman *et al.* (2009), it is also possible to use ozone mass deficit O_3MD (Alvarez-Madrigal *et al.*, 2009), or ozone hole minimum (Polvani *et al.*, 2010) as a variable in regression. Hofmann *et al.* (1997) used an empirical model that was based on mid-latitude EESC and South Pole ozonesonde observations to predict recovery in 2050. Shindell *et al.* (1998a, b) used a GCM with parameterized chemistry to show ozone hole recovery in 2046. Newman and Rex (2007) used coupled chemistry–climate models (CCMs) to show recovery in approximately 2050 or somewhat later. Recent observational studies and improved model analysis suggest that ozone hole recovery will occur in approximately 2060–2075 (Austin and Wilson, 2006; Bodeker *et al.*, 2007; Eyring *et al.*, 2007; Waugh *et al.*, 2009; WMO, 2007).

There are a number of uncertainties associated with estimates of the recovery of the Antarctic ozone hole. These can be considered in three groups:

(1) future ODS levels;
(2) circulation changes; and
(3) climate change impact in the polar lower stratosphere.

But these three groups overlap to some extent. Future ODS levels can be affected by policy changes, such as the decision of the Meeting of Parties in the autumn of 2007 to accelerate the phase-out of HCFCs, although this decision was made as much because of the greenhouse gas effects of these compounds as because of their ozone-destroying properties. Reductions in these ODSs is estimated to bring forward recovery of the ozone hole by between about 4 years and 14 years (Newman *et al.*, 2009). As well as policy changes, the estimation of future ODS levels depends on estimates of future emissions and possible revisions of assumed values of the lifetimes of individual species (Montzka *et al.*, 2003b).

Brewer–Dobson (BD) circulation is a global-scale cell in the stratosphere in which air rises in the tropics and then moves polewards and downwards, mostly in the winter hemisphere. It was first proposed by Brewer (1949) and Dobson (1956) to explain water vapor and ozone measurements and is now generally accepted as the basic description of seasonal mean meridional mass transport in the stratosphere. It lifts air from the tropical tropopause into the middle to upper stratosphere, and then carries it southward in the upper stratosphere, where this air then descends over Antarctica over the course of the winter. An air parcel in this circulation takes an average of approximately 5.5 years to travel this distance, although the time taken by individual parcels varies between about 2 years to 8 years or more. Estimates of Antarctic EESC are based upon measurements of the mean age of air, estimates of the age spectrum width, and observed fractional release values in the lower stratosphere in the Northern Hemisphere. Newman *et al.* (2007a) explored uncertainties related to measurements and assumptions used to calculate Antarctic EESC and concluded that this led to uncertainty in the recovery to 1980 levels of ozone ranging between 2056 and of 2077. Another effect is that—as the concentration of greenhouse gases (GHGs) in the atmosphere increases—the mean age of air will decrease (i.e., Brewer–Dobson circulation will accelerate). Butchart *et al.* (2006) and Douglass *et al.* (2008) have shown that this faster circulation modifies the fractional release values, leading to a higher release for the same mean age. This modification of both the mean age and the release rate as a function of mean age adds considerable uncertainty to estimates of recovery of the Antarctic ozone hole.

6.6.1 Measurement of TOC and the OVP

To estimate the future evolution of the ozone layer confidently, the requirements of TOC measurement accuracy have increased. The TOMS retrieval algorithm and its data products have undergone three decades of progressive refinement. The most recent eighth version was introduced in 2004 (Bhartia and Wellemeyer, 2004; Labow *et al.*, 2004) and the latest empirical correction of the EP-TOMS dataset from 1996 to 2005 was made in 2007 (McPeters *et al.*, 2007).

In studying trends in TOC it is necessary to merge data from different satellite systems to form long-term datasets. Measurements from satellite instruments provide global coverage and are supplementary to selective ground-based observations (Loyola *et al.*, 2009a). To achieve this requires the combination and inter-calibration of data derived from different spaceborne instruments to produce a homogeneous long time series. Important factors such as spatial coverage, instrument drifts, record continuity, and long-term calibration stability need to be taken into consideration. This requires specific techniques and algorithms to produce consistent datasets that allow robust investigations including trend analysis. The harmonization of data from three European satellite-flown sensors—GOME (flown on ERS-2), SCIAMACHY (flown on ENVISAT), and GOME-2 (flown on MetOp-A)—using data from numerous ground-based observing instruments as well, to produce a continuous time series starting in June 1995, was discussed by Loyola *et al.* (2009b).

The Global Ozone Monitoring Experiment (GOME) on board the ESA satellite ERS-2 was launched in April 1995 (Burrows *et al.*, 1999), but global coverage was lost in June 2003. This data record was extended to May 2008 by measurements from the Scanning Imaging Absorption spectrometer for Atmospheric CHartographY (SCIAMACHY) on board the ESA satellite ENVISAT which was launched in February 2002 (Bovensmann *et al.*, 1999; Gottwald *et al.*, 2006), and GOME-2 on board EUMETSAT's MetOp-A satellite which was launched in October 2006 (Callies *et al.*, 2000). The data were merged based upon comparisons during the overlap period from July 2002 to May 2008 for GOME and SCIAMACHY, and from March 2007 to May 2008 for GOME and GOME-2, respectively. In both cases, GOME data are considered as the reference standard. The algorithm used for the adjustment of SCIAMACHY and GOME-2 is described in detail in Section 3 of Loyola *et al.* (2009a). The equator-crossing times in descending node for the satellites carrying these three instruments are at 10:30 AM local time for GOME, 10:00 AM local time for SCIAMACHY, and 09:30 AM local time for GOME-2. Thus, all three instruments provide atmospheric measurements within about one hour of each other. This relatively small time difference facilitates combination of retrieval results.

The GOME Data Processor (GDP) version 4.x, which is the operational algorithm for trace gas column retrievals from GOME, SCIAMACHY, and GOME-2, is an algorithm for the generation of TOC, NO_2, BrO, SO_2, H_2O, HCHO, and OClO. Operational GOME and GOME-2 TOC products (GDP 4.x) were used by Loyola *et al.* (2009a) and the SCIAMACHY TOC values that they used were retrieved using SDOAS, an adaptation of the algorithm GDOAS to the SCIAMACHY instrument

(Lerot *et al.*, 2009). The average agreement of GOME, SCIAMACHY, and GOME-2 total ozone products with ground-based and other satellite ozone column measurements is at the "percent level" (Balis *et al.*, 2007b, 2008). For the first time TOC values from three European satellite sensors—GOME (ERS-2), SCIAMACHY (ENVISAT), and GOME-2 (MetOp-A)—were combined and added up to a continuous time series starting in June 1995. However, this is not a static process. As data supplies from these instruments fail and as new TOC measuring satellite systems are launched this work will need to be updated to incorporate the new data into the long-term record, making sure that it is validated and properly harmonized with the earlier data if it is to be of use in studying long-term trends in TOC.

For the ground-based data incorporated with merged data from the three satellite-flown sensors by Loyola *et al.* (2009a), archived total ozone column measurements from the WMO Global Atmosphere Watch (GAW) network—routinely deposited at the World Ozone and Ultraviolet Radiation Data Center (WOUDC) in Toronto (Canada) (*http://www.woudc.org*)—were taken. The quality of the TOC values of each station and instrument that deposited data at WOUDC for any time period after 1995 was evaluated using a methodology described in great detail by Balis *et al.* (2007a). On that basis, 32 Brewer and 47 Dobson stations were considered suitable for comparison with GOME, SCIAMACHY, and GOME-2 total ozone column data. These 79 stations are identified in the Appendix of Loyola *et al.* (2009a). Monthly mean averages were then created using daily comparisons with two different sets (only coincident datasets were considered): the monthly mean and associated standard deviation of ground-based measurements and the equivalent ones for satellite measurements. Due to differences in the timespan and spatial resolution between the three satellite datasets considered, not all of these ground-based stations were utilized in comparisons of all three satellite instruments. In addition to the harmonized data from the three different satellite instruments, Loyola *et al.* (2009a) used modeled ozone data from a chemistry–climate model (CCM) in their studies of a long-term ozone dataset.

Kuttippurath *et al.* (2010b) used the passive tracer method to estimate ozone loss from ground-based measurements in the Antarctic. A sensitivity study shows that ozone depletion can be estimated within an accuracy of ~4%. The method was then applied to ground-based observations from Arrival Heights, Belgrano, Concordia, Dumont d'Urville, Faraday (now Vernadsky), Halley, Marambio, Neumeyer, Rothera, South Pole, Syowa, and Zhongshan for the diagnosis of ozone loss in the Antarctic. On average, the 10-day boxcar average of vortex mean ozone column loss deduced from ground-based stations was about $(55 \pm 5)\%$ for 2005–2009. Ozone loss computed from ground-based measurements is in very good agreement with those derived from satellite measurements (OMI and SCIAMACHY) and model simulations (REPROBUS and SLIMCAT), where differences are within ±3–5%. Historical ground-based total ozone observations in October showed that depletion started in the late 1970s, reached a maximum in the early 1990s, and stabilized afterwards due to saturation. There was no indication of ozone recovery at that stage. At southern mid-latitudes, a reduction of 20–50% was observed for a few days in October–November at the newly installed Rio Gallegos station (51.60°S, 69.32°W). Similar depletion of ozone was also observed episodically during vortex overpasses at Kerguelen (49.4°S, 70.3°E) in October–November and at Macquarie Island in July–August of recent winters. This illustrated the significance of measurements at the edges of Antarctica.

Extensive observations of TOC made over the Antarctic region at the Indian station Maitri (70.76°S, 11.74°E) during two summer periods of 2004–2005 and 2006–2007 using a ground-based ozonometer were examined to study short-term variations (Sonbawne *et al.*, 2009). Mean TOC during summer 2004–2005 was 288.7 DU with a variability of 8% and corresponding values during summer 2006–2007 were 280.4 DU and 11%. Both ozonometer and Dobson spectrophotometer measurements at this Antarctic station during January 2005 showed a short-term decrease in TOC by about 70 DU before regaining the original value. Simultaneous changes in precipitable water and surface temperature pointed to a possible interplay of chemistry and meteorological conditions. The occurrence of an intense solar proton event in January 2005 also suggested the possible role of precipitating charged particles in causing short-lived ozone decrease through production of OH radicals. Sunphotometer-derived TOC along with that from a Dobson spectrometer at a tropical urban Indian station, Pune (18°32′N, 73°51′E, 559 m amsl) for the period 1998–2003 have been investigated by Ernest Raj *et al.* (2009). On a monthly mean scale the agreement between the results from the ozonometer and Dobson was within ±3%. Ozone content at Pune during the winter months of January 1999 and 2002 were almost 6–7% lower than that observed during other years.

Schnadt Poberaj *et al.* (2009) described ozone measurements of the Global Atmospheric Sampling Program (GASP) performed from four commercial and one research aircraft in the late 1970s to compare them with the results of the MOZAIC project (see Section 1.7.4.3). The data were used to construct datasets of ozone concentrations in the upper troposphere/lower stratosphere (UT/LS) using the aircraft datasets (1975–1979 and 1994–2001) to study long-term changes between the 1970s and 1990s. They found that there were pronounced increases in ozone concentration in the upper troposphere over the Middle East and South Asia in the spring and summer seasons. Increases were also found over Japan, Europe, and the eastern parts of the United States depending on season. In the lower stratosphere the ozone concentration over northern mid-latitudes and high latitudes was found to be lower in the 1990s compared to the 1970s in all seasons of the year (as expected from various other data on stratospheric ozone depletion). Detailed comparisons with ozonesonde data were also made.

6.6.2 The use of models to predict ozone concentration

To be able to make predictions about the constitution of the atmosphere and, specifically, about the ozone layer, both atmospheric observations and atmospheric models are required. There are now multidecadal sets of atmospheric observations from balloon, aircraft, ground-based, and satelliteborne instruments. Analysis of such datasets leads to insights into atmospheric processes, and the insights lead to refinement of the conceptual basis of atmospheric models (Douglass, 2009). Any atmospheric model must be developed in such a way that it can be used to compute constituent distributions and evolution, thus making quantitative comparisons with observations possible.

The ozone concentration varies from day to day and from place to place and the main focus of predictive models concerned with climate modeling is on mean values. However, before we discuss this we should mention that there is some interest in making short-term (i.e., daily or monthly mean) predictions of ozone. For this purpose moving average, multiple linear regression, persistence, and neural network models are used (Aksoy *et al.*, 2009; Chattopadhyay, 2007; Chattopadhyay and Chattopadhyay-Bandyopadhyay, 2008; Chattopadhyay and Chattopadhyay, 2008; Unal *et al.*, 2011).

Turning to longer term modeling, models, in general, and atmospheric models, in particular, are made up of two parts. The first part is conceptual (as revealed through observations and theoretical development) and the second part is computational. The second part can be thought of as practical implementation of the conceptual model, and this is where there is a need for consideration of spatio-temporal resolution, the number of interacting species, the complexity of their photochemical interactions, and the practical limitations of computers.

The development of models from the early days of the 1970s until the present time has been reviewed by Douglass (2009). In the 1970s, following the original suggestion that man-made CFCs could lead to ozone depletion (Molina and Rowland, 1974) there were few observations, the conceptual model was not complete, and computational implementation of the model was primitive. At that time the models used were 1-D global average models, the one dimension being in the vertical direction, and the atmospheric photochemistry of ozone depletion was not well understood. One-dimensional models soon gave way to 2-D models; these were zonal models with the two dimensions allowing for altitude variations and latitudinal variations. These models were widely used in the 1980s and 1990s (sec WMO, 1995 and references therein). For example, Fleming *et al.* (2007) used a 2-D chemistry and transport model to study the effects of interannual dynamical variability on global total ozone for 1979–2004. Long-term meteorological data from the National Centers for Environmental Prediction/National Center for Atmospheric Research (NCEP/NCAR) reanalysis-2 project and the European Center for Medium Range Weather Forecasts (ECMWF) updated reanalysis (ERA-40) were used to construct yearly dynamical fields for use in the model. The simulations qualitatively resolved much of the seasonal and interannual variability observed in long-term global total ozone data, including fluctuations related to the Quasi-Biennial Oscillation (QBO). They performed a series of model experiments to examine the relative roles of interannual variability, changes in halogen and volcanic aerosol loading, and the 11-year solar cycle in controlling multidecadal ozone changes. Statistical regression was used to isolate these signals in observed and simulated time series. At northern mid-latitudes, simulated interannual dynamical variability acts to reinforce chemical ozone depletion caused by the enhanced aerosol loading following the eruption of

Mt. Pinatubo in 1991. However, at southern mid-latitudes, interannual variability masks the aerosol-induced chemical effect. The simulated solar cycle response to total ozone is generally consistent with observations and is primarily due to the direct photochemical effect. The halogen-induced total ozone trend for 1979–1996 derived from the model was in good agreement with that derived from observations in the tropics and the Northern Hemisphere. However, at southern mid-latitudes, the trend derived from the model was more sensitive to halogen loading than that derived from observations.

Two-dimensional models were successful in reproducing some aspects of observations (WMO, 1995 and references therein). However, the observations show that the assumption of zonal symmetry fails for some processes (e.g., those that require accurate representation of the Northern Hemisphere winter vortex) and the development of both conceptual and computational aspects of 3-D models continued.

In the 1990s, regression trend models for total column ozone included only QBO winds and ENSO ocean surface pressure as dynamical proxies. Trends derived from these regression models generally differ (are more negative) from 2-D chemical transport trends by about 2–5% decade^{-1} in the mid-latitudes during spring. After Ziemke *et al.* (1997) these regression models took into account additional dynamical proxies of total ozone in an effort to reduce errors in local ozone trends and reduce these model differences. These dynamical proxies included winds, relative vorticity, potential vorticity, temperatures, and geopotential heights. Inclusion of additional dynamical proxies improved statistics by reducing residuals and uncertainty regions in both zonal mean and zonally asymmetric trend models. Root-mean-square reductions, relative to a trend model with only QBO, solar, and trend terms, were as large as 50% in 14-year means in the Southern Hemisphere. For zonal mean or zonally asymmetric global trend models with one optional surrogate, a favorable choice was pre-filtered (at least deseasonalized and detrended) lower-stratospheric temperatures. Relative vorticity, potential vorticity, and geopotential height all exhibited similar relationships with total ozone, with the highest correlative behavior near 200 hPa (year-round at the mid-latitudes) and 10 hPa (high latitudes in winter–spring months). For models incorporating these latter proxies, combined 10 and 200 hPa (or similar) pressure levels were effective in reducing global residuals. ENSO, as a surrogate by itself or included with other dynamical proxies, had a comparatively small effect because of its episodic nature. Decadal variabilities in NCEP and microwave sounding unit channel 4 (MSU4) data as surrogates in trend models indicated maximal 1–3% decade^{-1} reductions everywhere in TOMS trends. Total ozone trends derived from the Goddard 2-D heterogeneous chemistry–transport model agreed favorably with trends in TOMS ozone, generally to within 2–3% decade^{-1} in both hemispheres.

Computational power has developed dramatically since the 1970s and this has made it possible for a computational model to represent many atmospheric processes directly. Douglass (2009) in an extensive review reflects that there are two ways that observations contribute to model development. First, observations lead directly to an improved conceptual model; as new processes are identified, the conceptual model is modified, and modifications are made to represent the new processes in the computational model. The second is that a discrepancy may be observed between observations and a simulation, thereby leading to the discovery of a new process that needs to be included in the model. As an example of the first, Douglass quotes the fact that computational models in 1985 could not account for the massive depletion of ozone in the Antarctic ozone hole in the Antarctic spring. This was because the mechanisms that lead to massive polar ozone destruction were unknown and, therefore, missing from the conceptual model and therefore from the computational model. Once the mechanisms responsible for the development of the ozone hole were understood, they could be incorporated into the conceptual model and, from that, into the computational model. Douglass also quotes a second example that concerns the stratospheric sulfate aerosol layer. Comparisons of observations of the concentrations of NO, NO_2, and HNO_3 in the lower stratosphere with model values revealed large discrepancies and suggested a missing component of the conceptual model. The role of the aerosol sulfate layer and the reaction on aerosol surfaces of water with N_2O_5, which is produced from NO_2 at night and photolyzed during the day, was recognized. This reaction leads to higher levels of HNO_3 and lower levels of NO and NO_2 and, in turn, to less importance of the nitrogen species in controlling ozone loss in the lower stratosphere (Fahey *et al.*, 1993). Once this was recognized as a necessary element in the conceptual model and implemented in the numerical model the simulated concentrations of nitrogen compounds compared favorably with observations (Kawa *et al.*, 1993). An example of the second way that observations contribute to model development (i.e., by finding a deficiency through comparisons of simulated fields with observations) is the observation that long-lived constituents are correlated with each other in the lower stratosphere.

CCMs are commonly used to simulate the past and future development of Earth's ozone layer. Fully coupled chemistry schemes calculate the chemical production and destruction of ozone interactively and ozone transport by simulated atmospheric flow. Due to the complexity of the processes acting on ozone it is not straightforward to disentangle the influence of individual processes on the temporal development of ozone concentrations. Methods that could quantify the influence of chemistry and transport on ozone concentration changes and could easily be implemented in CCMs and chemistry–transport models (CTMs) are necessary (see later in this section for a definition of CTM). Very recently Garny *et al.* (2011) demonstrated such a method in which ozone tendencies (i.e., the time rate of change of ozone) are partitioned into a contribution from ozone production and destruction (chemistry) and a contribution from transport of ozone (dynamics). In this scheme the influence of transport on ozone in a specific region is further divided into export of ozone out of that region and import of ozone from elsewhere into that region. For this purpose, Garny *et al.* (2011) used a diagnostic that disaggregates the ozone mixing ratio field into nine separate fields according to which of the nine predefined regions of the atmosphere the ozone originated. With this diagnostic the ozone mass fluxes between these regions are obtained. Furthermore, this method is used by Garny *et al.* (2011) to attribute long-term changes in ozone to chemistry and transport. They also quantify the relative change in ozone from one period to another that is due to changes in production or destruction rates, or due to changes in the import or export of ozone. As such, the diagnostics introduced by Garny *et al.* (2011) can be used to attribute changes in ozone on monthly, interannual, and long-term timescales to the responsible mechanisms.

Several CCMs have been developed and applied to investigate recent changes in stratospheric ozone concentrations (e.g., Eyring *et al.*, 2006) and to simulate future evolution of the ozone layer (e.g., Eyring *et al.*, 2007). The main features of 13 CCMs are summarized in Table 6.12 and their detailed descriptions are given in the literature cited in this table. These models were discussed by Eyring *et al.* (2006) to provide guidance for the interpretation of ozone predictions made by the same CCMs. The focus of the evaluation is on how well the fields and processes that are important for determining ozone distribution are represented in simulations of the recent past. The core period of the evaluation is from 1980 to 1999 but long-term trends are compared for an extended period (1960–2004). In general, the amplitude and phase of the annual cycle in total ozone was well simulated apart from the southern high latitudes. Most CCMs show reasonable agreement with observed total ozone trends and variability on a global scale, but a greater spread in the ozone trends in polar regions in spring, especially in the Arctic. It was argued that, despite the wide range of skills in representing different processes assessed, there is sufficient agreement between the majority of the CCMs and the observations that some confidence can be placed in their predictions.

Loyola *et al.* (2009a) concentrated on results from the CCM E39C-A (Stenke *et al.*, 2008), which is an upgraded version of the CCM E39C (Dameris *et al.*, 2005). Two model simulations were used with E39C-A, namely R1 and R2. The R1 simulation covers the period from 1960 to 2004 and the R2 model simulation covers the extended period from 1960 to 2050. Both model simulations included various natural and anthropogenic forcings like the 11-year solar cycle, the QBO, chemical and direct radiative effects of major volcanic eruptions, and the increase in greenhouse gas concentrations.

Another model, the three-dimensional Hamburg Model of the Neutral and Ionized Atmosphere (HAMMONIA), which treats atmospheric dynamics, radiation, and chemistry interactively from the Earth's surface up to the thermosphere (approximately 250 km) has been described by Schmidt *et al.* (2006). It is based on the ECHAM atmospheric general circulation model of the Max Planck Institute for Meteorology in Hamburg (Germany), which is extended to include important radiative and dynamical processes of the upper atmosphere and is coupled to a chemistry module containing 48 compounds. The model has been applied to study the effects of natural and anthropogenic climate forcing on the atmosphere, namely the 11-year solar cycle and a doubling of the present day concentration of carbon dioxide, respectively. The results include a temperature response to the solar cycle by 2 to 10 K in the mesopause region with the largest values occurring slightly above the summer mesopause. The concentration of ozone in the secondary maximum increases by up to 20% under solar maximum conditions. In the case of a doubling of CO_2 in the atmosphere the simulation indicates a cooling of the atmosphere everywhere above the tropopause but by the smallest values around the mesopause.

It was shown that the temperature response up to the mesopause is strongly influenced by changes in dynamics. During the Northern Hemisphere summer, dynamical processes alone would lead to an almost global warming of up to 3 K in the uppermost mesosphere.

Dikty *et al.* (2010a) investigated the latest version 1.07 of SABER (Sounding of the Atmosphere using Broadband Emission Radiometry) which measures tropical ozone at 1.27 μm and at 9.6 μm for the retrieval of temperature data with respect to daytime variations in the upper mesosphere. The processes involved are compared to daytime variations of the 3-D general circulation and chemistry model HAMMONIA. The results show good qualitative agreement for ozone. The amplitude of daytime variations is in both cases approximately 60% of the daytime mean. During equinox the daytime maximum ozone abundance for both observations and the model is higher than during solstice, especially above 0.01 hPa (approx. 80 km). The influence of tidal signatures either directly in ozone or indirectly via a temperature response above 0.01 hPa cannot be fully eliminated. Below 0.01 hPa (photo-)chemistry is the main driver for variations. We also use the HAMMONIA output of daytime variation patterns of several other different trace gas species (e.g., water vapor and atomic oxygen) to discuss the daytime pattern in ozone. In contrast to ozone, temperature data show little daytime variations between 65 and 90 km and their amplitudes are on the order of less than 1.5%. In addition, SABER and HAMMONIA temperatures show significant differences above 80 km.

The QBO is forced externally by linear relaxation ("nudging") of the simulated zonal winds in the equatorial stratosphere to a constructed QBO time series which follows observed equatorial zonal wind profiles (Giorgetta and Bengtsson, 1999). This assimilation is applied between 20°N and 20°S from 90 hPa up to the model top layer. The relaxation timescale is 7 days within the QBO core domain. Outside the core region the timescale depends on latitude and pressure (Giorgetta and Bengtsson, 1999). The influence of the 11-year solar cycle on photolysis is parameterized according to the intensity of 10.7 cm radiation from the sun (Lean *et al.* 1997, data available via *http://www.drao.nrc.ca/icarus/www/daily.html*). The impact of solar activity on short-wave radiative heating rates is considered on the basis of changes in the solar constant (Dameris *et al.*, 2005, Table 2).

Stolarski *et al.* (2006b) discussed the lessons learned about trends in stratospheric ozone from a 3-D chemical–transport model (CTM) taking into account the changing concentrations of CFCs, volcanic eruptions, the 11-year solar cycle, and the QBO. They used a CTM to simulate stratospheric composition for the past 30 years and the next 20 years using 50-year data of winds and temperatures from a general circulation model (GCM). The simulation included a representation of aerosol surface areas based on observations including volcanic perturbations from El Chichón in 1982 and Mount Pinatubo in 1991, and time-dependent mixing ratio boundary conditions for CFCs, halons, and other source gases such as N_2O and CH_4. A second CTM simulation was carried out for identical solar flux and boundary conditions but with constant background aerosol conditions. Statistical time series analysis was applied to both observed and simulated ozone concentrations to examine the capability of the analyses for the determination of trends in ozone

Table 6.12. Thirteen chemistry–climate models (CCMs) (Eyring *et al.*, 2006). Horizontal resolution is given in degrees latitude × longitude (grid point models). For spectral models the resolution is given as T30, T32, and T42 corresponding to the triangular truncation of the spectral domain with 30, 32, and 42 wave numbers, respectively. All the CCMs have a comprehensive range of chemical reactions. The time-varying concentrations of CO_2 as well as other GHGs are considered in the radiation schemes of all models.

Model	*Investigators*	*Underlying GCM*	*Domain/Resolution*	*Radiative feedback*
AMTRAC	J. Austin, R. J. Wilson	AM2 (Anderson *et al.*, 2004)	2° × 2.5°, 48 L, 0.0017 hPa	O_3, H_2O
CCSRNIES	H. Akiyoshi, T. Nagashima M. Yoshiki	CCSRNIES (Numaguti, 1993)	2.8° × 2.8° (T42), 34 L, 0.01 hPa	O_3, H_2O, CH_4, N_2O, CFCs
CMAM	J. McConnell, N. McFarlane, D. Plummer, J. Scinocca, T. Shepherd	CCCma AGCM3 (Scinocca and McFarlane, 2004)	3.75° × 3.75° (T32), 71 L, 0.0006 hPa	O_3, H_2O
E39C	M. Dameris, V. Eyring, V. Grewe, M. Ponater	ECHAM4 (Roeckner *et al.*, 1996)	3.75° × 3.75° (T30), 39 L, 10 hPa	O_3, CH_4, N_2O, H_2O, CFCs
GEOS CCM	A. Douglass, P. Newman, S. Pawson, R. Stolarski	GEOS-4 (Bloom *et al.*, 2005)	2° × 2.5°, 55 L, 0.01 hPa	O_3, H_2O, CH_4, N_2O, CFC-11, CFC-12
LMDZrepro	S. Bekki, F. Lott, F. Lefevre, M. Marchand	LMDz4 (Lott *et al.*, 2005)	2.5° × 3.75°, 50 L, 0.07 hPa	O_3, CH_4, N_2O, H_2O, CFC-11, CFC-12
MAECHAM4CHEM	C. Brühl, M. Giorgetta, E. Manzini, B. Steil	MAECHAM4 (Manzini *et al.*, 1997)	3.75° × 3.75° (T30), 39 L, 0.01 hPa	O_3, H_2O, CH_4
MRI	K. Shibata, M. Deushi	MRI/JMA98 (Shibata *et al.*, 1999)	2.8° × 2.8° (T42), 68 L, 0.01 hPa	O_3, CH_4, N_2O
SOCOL	E. Rozanov, M. Schraner	MAECHAM4 (Manzini *et al.*, 1997)	3.75° × 3.75° (T30), 39 L, 0.1 hPa	O_3, CH_4, N_2O, H_2O, CFCs

ULAQ	E. Manzini, G. Pitari	ULAQ-GCM (Pitari *et al.*, 1992)	$10° \times 22.5°$, 26 L, 0.04 hPa	O_3, H_2O, CH_4, N_2O, CFCs, HCFCs, aerosols
UMETRAC	N. Butchart, H. Struthers	UM (Pope *et al.*, 2000)	$2.5° \times 3.75°$, 64 L, 0.01 hPa	O_3, H_2O
UMSLIMCAT	M. Chipperfield, W. Tian	UM (Pope *et al.*, 2000)	$2.5° \times 3.75°$, 64 L, 0.01 hPa	O_3, CH_4, N_2O, H_2O
WACCM (v. 3)	B. Boville, R. Garcia, A. Gettelman, D. Kinnison, D. Marsh, F. Sassi	CAM (Collins *et al.*, 2004)	$4° \times 5°$, 66 L, 4.5×10^6 hPa	O_3, H_2O, N_2O, CH_4, CFC-11, CFC-12, NO, CO_2, O_2

due to CFCs and to separate these trends from the various natural effects in the atmosphere. The results demonstrated several difficulties associated with the interpretation of time series analyses of atmospheric ozone data. In particular, it showed that lengthening the dataset reduces the uncertainty in the derived trend due to interannual dynamic variability. It also showed that interannual variability can make it difficult to assess accurately the impact of a volcanic eruption on ozone concentrations. Such uncertainties make it difficult to obtain early proof of ozone recovery in response to decreasing chlorine.

For the past, Loyola *et al.* (2009a) used boundary conditions based on observational data. Detailed information about the experimental set-up is given by Stenke *et al.* (2008). For the future, some changes were made. QBO phases observed in the past were consistently continued, the solar activity signal (i.e., the intensity of 10.7 cm radiation) observed between 1977 and 2007 was continually repeated until 2050, and no allowance was made for major volcanic eruptions during this period. But the most important change was in relation to the sea surface temperatures (SSTs) that were used. In the R1 simulation the SSTs were prescribed as monthly means following the global sea ice and SST dataset HadISST1 derived from observations (Rayner *et al.*, 2003). To avoid discontinuity at the transition point from past to future, in R2 a continuous dataset derived from a coupled atmosphere–ocean model containing SST and sea ice data from 1960 to 2050 was used.

Fischer *et al.* (2008) described the use of the chemistry–climate model SOCOL (SOlar Climate Ozone Links) to study interannual to decadal variability of the stratosphere during the 20th century, including variations in ozone concentration. The forcings used included sea surface temperatures, sea ice, solar irradiance, stratospheric aerosols, QBO, changes in land properties, greenhouse gases, ozone-depleting substances, and emissions of carbon monoxides and nitrogen oxides. Transient simulations were in good agreement with observations, reconstructions, and reanalyses and allowed quantification of interannual to decadal variability during the 20th century (for details see Fischer *et al.*, 2008). The data were taken from the HadGEM1 model, provided by the U.K. Met Office's Hadley Center (Johns *et al.*, 2006). HadGEM1 used transient simulations with prescribed anthropogenic forcing as observed in the past and followed the SRES-A1B scenario in the future (for details on the simulations see Stott *et al.*, 2006).

It has been found that models underestimate O_3 abundance in the upper stratosphere and mesosphere (Baron *et al.*, 2009). This "ozone deficit" problem (e.g., Mlynczak *et al.*, 2000 and references therein) reveals that our knowledge of O_3 chemistry is still incomplete. One of the keys to solving this issue could lie in a better understanding of atmospheric cycles involving minor constituents that impact the O_3 balance (Baron *et al.*, 2009). Among these, the hydroperoxy radical (HO_2), a member of the odd hydrogen family or HO_x (HO_x = H, OH, HO_2), plays a major role in ozone loss processes above 40 km (Brasseur *et al.*, 1999; Jonsson, 2006). Currently, very few observations of HO_2 are available and they are primarily concerned with the stratosphere. The second Far-Infrared Spectrometer (FIRS-2), a balloonborne instrument, measured the HO_2 concentration from 30 to 50 km (Jucks *et al.*, 1998). The Microwave Limb Sounder (MLS) on board the Aura satellite has been performing HO_2 observations since September 2004 (Pickett *et al.*, 2008; Waters *et al.*, 2006), from which its abundance is retrieved in the altitude range 25–70 km (with version 2.2 of the Aura-MLS dataset). Comparisons between data from these two instruments show good agreement and also give results comparable to those from photochemical model simulations (Pickett *et al.*, 2008). The only mesospheric (50–80 km) observations available to date were obtained with a ground-based millimeter wavelength receiver, yielding HO_2 abundances smaller than model calculations by about 30% (Sandor and Clancy, 1998). The Sub-Millimeter Radiometer (SMR) aboard the Odin satellite has been providing data since October 2001 but, since HO_2 is not a primary target for the mission, the operational retrieval procedure is not well suited to retrieve its abundance and no retrieval was done prior to the work of Baron *et al.* (2009). It is worth noting that the second instrument of the Odin payload, the Optical Spectrograph and Infra-Red Imager System (OSIRIS), performs measurements from which atomic hydrogen and OH concentrations can be retrieved (Gattinger *et al.*, 2006a, b) so that the Odin satellite is thus potentially able to provide complete information on the H, HO, and HO_2 family.

A useful discussion of models is given by Douglass (2009). In an early attempt to produce a 3-D model suitable for ozone prediction Cunnold *et al.* (1975) used a limited set of photochemical reactions with a simplified general circulation model to produce a coupled 3-D model. The realism of the simulated ozone was demonstrated by comparing a global mean ozone profile with mid-latitude rocket measurements of ozone (Krueger and Minzer, 1973). The model profile compared reasonably well with the observations, but such simple comparisons fail to test many aspects of the simulation, and satellite data on ozone was not available at that time. The model included only hydrogen and nitrogen species, but did not include other processes now known to be important to stratospheric ozone loss.

Some datasets were important for comparison of results with 2-D models and for showing that 3-D models were required. The limb infrared monitor of the stratosphere (LIMS) instrument on Nimbus-7 provided profiles of temperature, ozone, water, nitric acid, and nitrogen dioxide for October 1978–May 1979 (Gille and Russell, 1984). The Total Ozone Mapping Spectrometer (TOMS) instrument on Nimbus-7 provided high horizontal resolution measurement of TOC.

One major problem in the development of 3-D general circulation models for atmospheric ozone is the treatment of photochemistry, including the need to represent temperature-dependent photochemical processes realistically. Rood *et al.* (1989) proposed that progress could be made using winds and temperatures from a data assimilation system. Temperatures from the assimilation system generally matched observed temperatures and horizontal winds were realistic. A model that solves constituent continuity equations using input winds for transport and input temperatures to calculate constituent production and loss is called a "chemistry–transport model" (CTM). The continuity equation is solved by process splitting (i.e., by solving the photochemical and transport contributions successively). The simulated ozone concentrations obtained by Rood *et al.* (1991) were compared with LIMS data but the agreement was only qualitative. Comparisons of data and simulated fields were not quantitative enough to be able to decide if differences were due to model photochemistry, input winds, or the numerical algorithms used for transport. The deficiencies of the model and subsequent work on model development are discussed by Douglass (2009). As computing facilities got better, various improved models were developed by different groups of workers.

The model deficiency was traced to the numerics of the transport scheme. The spectral transport scheme can generate negative mixing ratios and, although these are eliminated with a filling algorithm, the process affects different long-lived tracers differently, thus corrupting constituent correlations. Allen *et al.* (1991) describe implementation of an upstream-biased monotonic grid point transport scheme first described by van Leer (1974) in a CTM. In an upwind-biased scheme, advection takes place only in the direction of the wind and downwind boxes have no influence. As long as the Courant–Friedrichs–Levy (CFL) condition for the time step is satisfied, there are no undershoots, so the algorithm cannot produce negative mixing ratios. This contrasts with solution of the transport part of the continuity equation using centered differences, because such a scheme, like spectral transport, can produce negative mixing ratios and require filling near sharp gradients. Once the van Leer algorithm was implemented in the CTM, the calculated fields for long-lived tracers maintain constituent correlations. Note that this improvement to the numerical model did not necessitate a change to the conceptual model. A second simulation reported by Rood *et al.* (1991) was for winter 1989 and showed the first surprising result of this implementation. The CTM had the same conceptual model as for the 1979 simulation that was compared with LIMS. The meteorological assimilation system for 1989 was updated, but its quality was not radically different. The quality of the simulation was much improved, and the column ozone at high latitudes was shown to track day-to-day variability of TOMS observations (see Figure 11 in Rood *et al.*, 1991). The change to the numerical scheme that was implemented to maintain correlations between simulated long-lived species also led to a higher level of fidelity for comparisons of computed total ozone with observations.

The van Leer algorithm was attractive for use in global atmospheric models because it preserved sharp mixing ratio gradients, such as those found at the polar vortex boundary, but was monotonic (i.e., does not produce new maxima) and also maintained constituent correlations. However, it was necessary to take small time steps at high latitudes where the size of the grid boxes becomes small, particularly during periods of cross-polar flow. Lin and Rood (1996) developed a multi-dimensional upstream-biased flux form transport scheme that could accommodate arbitrarily long time steps: the flux form semi-Lagrangian (FFSL). This algorithm maintained the important features of the van Leer algorithm but allowed transported material to travel farther than a single grid box. The FFSL did not require a constraint on the time step, making the algorithm more practical for global calculations, but did have a constraint to ensure monotonic transport and conserve constituent correlations.

The remarkable result stemming from success with the new numerical transport algorithm for constituents was that this success inspired changes to the dynamical core of the GCM. The equations of motion that were solved in the general circulation model included material derivatives, and the approach to numerical solution of the equations of motion followed the approach used in the constituent transport algorithm. Douglass *et al.* (2003) reported on the features of tropical transport in a CTM using winds and temperatures from the Finite Volume General Circulation Model (FVGCM), a general circulation model developed at Goddard Space Flight Center, in collaboration with the National Center for Atmospheric Research. The CTM used the flux form semi-Lagrangian transport scheme (Lin and Rood 1996, 1997) to ensure accurate representation of transport by the resolved scale flow. Douglass *et al.* (2003) compared simulated fields of ozone, methane, and total reactive nitrogen with observations and found that transport was far more realistic in the CTM using meteorological fields from FVGCM than any simulation using assimilated fields.

Lamarque *et al.* (2010) presented a new dataset of gridded anthropogenic and biomass-burning emissions covering the historical period (1850–2000) in decadal increments at a horizontal resolution of 0.5° in latitude and longitude. The primary purpose of this inventory was to provide consistent gridded emissions of reactive gases and aerosols for use in chemistry model simulations needed by climate models for the Climate Model Intercomparison Program #5 (CMIP5) in support of the Intergovernmental Panel on Climate Change (IPCC) Fifth Assessment Report (AR5), which is due to be completed in 2013/2014. The best estimate for the 2000 inventory represents a combination of existing regional and global inventories to capture the best information available at this point; 40 regions and 12 sectors were used to combine the various sources. The historical reconstruction of each emitted compound, for each region and sector, was then forced to agree with the 2000 estimate, to ensure continuity between past and 2000 emissions. Simulations from two chemistry–climate models were used to test the ability of the emission dataset to capture long-term changes in atmospheric ozone, carbon monoxide, and aerosol distributions. The simulated long-term change in northern mid-latitude surface and mid-troposphere ozone was not quite as rapid as observed. However, stations outside this latitude band showed much better agreement.

Eleftheratos *et al.* (2011) performed a comparison between TOC values derived from solar backscatter ultraviolet (SBUV) satellite observations and those calculated from the chemical–transport model Oslo CTM2 for the period 2001–2007. Monthly mean total ozone amounts from improved model simulations were used to compute monthly, seasonal, and annual zonal means over 10° latitude zones, and compared with respective satellite retrievals over the Northern and Southern Hemispheres. The results showed that improved model simulations slightly underestimate total ozone over the Northern Hemisphere when compared with the satellites by 1.4% on average, and slightly overestimate total ozone over the southern extra-tropics, middle latitudes and high latitudes by 1.6% on average. The mean difference between model and satellite-derived total ozone columns from 75°S to 75°N was estimated to be about −0.3%. A linear regression analysis between model and satellite-derived total ozone data showed statistically significant correlations between the two datasets at all latitude zones (about +0.8 in the tropics and more than +0.9 over all other latitudes). The annual cycle of total ozone was shown to be well reproduced by the model at all latitudes.

Evaluation of atmospheric models is complex and we simply note a few key comparisons (Douglass, 2009).

There were NASA-sponsored model and measurement workshops held in 1992 (Prather and Remsberg, 1993) and 1997 (Park *et al.*, 1999). These workshops examined the many aspects of stratospheric models, identifying areas of success and failure for individual participating models and also finding some areas in which none of the model results matched the observations. Boering *et al.* (1996) used the known trend and stratospheric measurements of carbon dioxide to derive values for the mean age of stratospheric air (i.e., the mean time since the air had entered the stratosphere). Hall *et al.* (1999) used values of mean age derived from measurements to evaluate the models participating in the second workshop, showing that nearly all of the models produced "young" ages. The mean age is considered a key comparison for models, and values derived for observations are compared with values from 13 chemistry–climate models (Eyring *et al.*, 2006) and reported for models participating in the *Scientific Assessment of Ozone Depletion: 2006* (WMO, 2007). The simulated values for many models, including the NASA Goddard CCM (labeled GEOSCCM), were much more similar to observed values than for any simulated values included in the study reported by Hall *et al.* (1999). The tropical ascent rates were tested by examining the ascent of the "tape recorder" signal (i.e., the vertical propagation of seasonally varying anomalies in water vapor due to seasonal variation of the tropopause temperature; Mote *et al.*, 1996). Eyring *et al.* (2006) connect the mean age to the modeled ascent rate, and it is clear from Figure 8 in that paper that the ascent rate in the Goddard CCM closely matched the ascent rate derived from observations of stratospheric water vapor from the Halogen Occultation Experiment (HALOE) on the Upper Atmosphere Research Satellite (UARS).

There are differences among the predictions for ozone recovery among different models and it is a challenge to understand the causes of the variability and to determine which predictions are more reliable than the others. One key to the differences is found in simulated total inorganic chlorine. Since all the models use the same boundary conditions, the variability that is seen in simulated polar chlorine (Figures 6–8, WMO, 2007) results from differences in simulated transport and mixing. Douglass *et al.* (2008) compared an observed relationship between the mean age of air for a parcel and the amount of inorganic chlorine remaining in the parcel relative to its initial value with simulated relationships. The Goddard CCM-simulated relationship matched the observed relationship, suggesting that the simulated path through the height-dependent destruction rates for chlorofluorocarbons were realistic. Further investigation will determine if the differences among models and lack of fidelity to observations are due to issues with the conceptual model or due to issues with implementation in the computational model.

Global climate model (GCM) simulations for the IPCC Fourth Assessment Report differed in their treatment of ozone radiative forcing and, in particular, in future ozone radiative forcing (Miller *et al.*, 2006). For the 20th century, some simulations used the ozone dataset of Randel and Wu (1999) to constrain ozone radiative forcing, others used the dataset of Kiehl *et al.* (1999), while still others excluded ozone radiative forcing altogether. The Randel and Wu database is the output from a linear least squares regression model incorporating decadal trends, solar cycle terms, and QBO terms. Decadal trends are modeled using equivalent effective stratospheric chlorine (EESC; Daniel *et al.*, 1995). However, because only ozonesonde profiles from Syowa (69°S) are used to characterize the dependence of ozone on EESC poleward of 60°S, and because Syowa is close to the vortex edge and will occasionally sample extra-vortex air, ozone trends over Antarctica in this database are likely to be underestimated.

There was a similar range of treatments for 21st century stratospheric ozone changes; some models assumed that stratospheric ozone remained constant at present day values while others assumed a slow recovery to pre-industrial values. Still others excluded ozone radiative forcing completely. Those models that did not include future increases in ozone over southern high latitudes are unlikely to reliably simulate future changes in the southern annular mode (Perlwitz *et al.*, 2008) and hence changes in Antarctic surface climate (Thompson and Solomon, 2002).

Hassler *et al.* (2009) created a new global 3-D (latitude, pressure, time) zonal mean monthly mean ozone database, from 1979 to 2100. The database extends from 87.5°S to 87.5°N at a 5° latitude resolution and from 878 hPa to 0.0460 hPa in 70 pressure levels ~1 km apart. The full period is divided into two sub-periods: (1) observations from 1979 to 2006 obtained from several sources of high vertical resolution ozone profile measurements, and (2) extensions from 2007 to 2100 based on chemistry–climate model (CCM) projections. At each latitude and pressure level the CCM ozone time series are shifted and scaled to better match the measurements from 1997 to 2006 to ensure a smooth transition from observations to model output. This new database is expected to provide ozone boundary conditions suitable for long-term global climate model simulations.

Comparisons of simulated constituent fields with observations are clearly the key to improved modeling capability, for these comparisons lead to improvements in the conceptual model (i.e., the underlying set of physical and chemical processes), the computational model, and the numerical expression of the conceptual model that we must use for predictions. Discoveries such as the Antarctic ozone hole have led to dramatic changes in the conceptual model. The computational advances of the past two decades have made possible a much more physically realistic computational model. Detailed comparisons with observations placed physical constraints on the computational model. A good example of a physical constraint is the requirement that the numerical scheme maintain correlations between long-lived constituents. The change in numerical transport algorithms to meet this constraint inspired a change in the approach to solutions of the equations of motion in the general circulation model and, ultimately, to the development of an improved computational model.

A remaining challenge is for qualitative comparisons of simulated fields with observations to be replaced by quantitative comparisons that can be used to judge whether processes are adequately represented in various models and, eventually, to judge whether predictions from one model are more likely to be valid than those from another. Douglass *et al.* (1999) took a step towards this goal, judging the realism of various sets of meteorological fields in a CTM framework. Presently, the Chemistry–Climate Model Validation Activity (CCMVal), established by SPARC (Stratospheric Processes and their Role in Climate, a project of the World Climate Research Program) seeks to apply common process-oriented diagnostics to models participating in assessments. The program and its application are discussed by Eyring *et al.* (2005, 2006, 2007).

Instruments on NASA's Aura, the European Space Agency ENVISAT, and the Canadian Sci-Sat continue to produce high-quality global observations of ozone and other constituents. Global measurements of tropospheric constituents are informing the present generation of scientists. The conceptual model will be refined, perhaps transformed, as directed by comparisons of simulation with observation. The computational model will be improved as computational resources grow, and perhaps again be modified to accommodate physical constraints derived from observations.

Kopp *et al.* (2009) reported on the measurements of strato-mesospheric ozone concentrations that have been performed at the inner-tropical Mérida Atmospheric Research Station since May 2004. Due to the high altitude of this site more than 20 ozone profiles per day have been retrieved under favorable tropospheric conditions. The retrieved profiles have a measurement response larger than 70% in the vertical range 15–60 km, and vertical resolution is 5 km at best. According to Kopp *et al.* (2009). the major cause for stratospheric seasonal variations is seasonal change in Brewer–Dobson circulation which is reflected in variations of stratospheric mixing ratios and stratospheric column amounts.

In addition, the diurnal cycle of mesospheric ozone due to photochemical processes was also measured and compared to model calculations with a photochemical model by Kopp *et al.* (2009). The model was capable of reproducing the diurnal cycle of measurements of Novembers 2004–2007 very well, but the measurements also revealed interannual variations, especially in nighttime ozone concentrations, which were not reproduced by the model and whose cause is still unclear. Furthermore, in November 2006 an unusual decrease in mesospheric ozone was observed. Kopp *et al.* (2009) speculated that, since measurements were taken near the minimum of solar activity, variable solar irradiation was unlikely to cause the observed variations in mesospheric ozone.

While chemistry–climate models (CCMs) are able to reproduce many of the characteristics of global TOC and its long-term evolution, they fared less well in simulating the commonly used diagnostic of the area of the Antarctic ozone hole (i.e., the area within the 220 DU contour; Struthers *et al.*, 2009). Two possible reasons for this are: (1) the underlying global climate model (GCM) does not correctly simulate the size of the polar vortex, and (2) either or both the stratospheric chemistry scheme incorporated into the GCM and the model dynamics may result in systematic biases in calculated TOC fields such that the 220 DU contour is no longer appropriate for delineating the edge of the ozone hole. Both causes were examined by Struthers *et al.* in several CCMs with a view to developing better ozone hole area diagnostics. Analysis of the simulation of the polar vortex in the CCMs showed that the first of the two possible causes does play a role in some models. This in turn affects the ability of the models to simulate the large meridional gradients observed in total column ozone. The second of the two causes also strongly affects the ability of CCMs to track the observed size of the ozone hole. It was shown that applying a common algorithm to CCMs leads to better agreement observations.

6.6.3 Ozonesonde networks

Ozone soundings have been taken for 50 years or so and recently they have come to play a role in integrating model, aircraft, and ground-based measurements for better interpretation of atmospheric ozone losses (stratosphere) and pollution (troposphere).

Following the excitement surrounding the discovery of the Antarctic ozone hole, balloonborne ozone measurements from the South Pole station have been conducted each year since 1986 (see Hofmann, 2009 for a discussion). Antarctic balloon ozone soundings were also used to detect the effects of volcanic eruptions on stratospheric ozone. Solomon *et al.* (2005) used ozonesonde data from Syowa and the South Pole to show that following the eruption of Pinatubo (1992–1995) ozone depletion in the springtime lower stratosphere was enhanced compared to other times. Enhanced ozone reductions at 150 hPa, where it is generally not cold enough for extensive PSC formation, are related to sulfuric acid aerosol in the lower stratosphere (e.g., from the Mount Pinatubo eruption in 1991). The surfaces provided by the volcanic aerosol are adequate to enhance the heterogeneous chemical ozone loss process. Data from Syowa following the eruption of El Chichón in 1982 suggest that there may have been a small effect at the lower chlorine levels that would have existed, contributing to the non-linear onset of ozone depletion. There is no evidence for ozone depletion following the eruption of Agung in 1963 when chlorine levels were low.

A well-designed network of ozonesonde stations answers questions that cannot be solved by short campaigns or current satellite technology. Three strategic ozonesonde networks that were designed to answer specific questions about ozone chemistry and dynamics have been described by Thompson (2009)—one each for the high latitudes, the mid-latitudes, and the tropics. The first is Match, a project to track Arctic stratospheric ozone loss starting in the early 1990s (Harris *et al.*, 2002; Rex *et al.*, 1998; Streibel *et al.*, 2006); the second is the mid-latitude IONS (INTEX Ozone Nework Study, INTEX = Intercontinental Chemical Transport Experiment), which operated in 2004 and 2006 (Thompson *et al.*, 2007a); and the third is the tropical network, SHADOZ (Thompson *et al.*, 2003a).

6.6.3.1 Match

The Match network originated from the 1991–1993 European Arctic Stratospheric Ozone Experiment (EASOE) campaigns (Rex *et al.*, 1998; von der Gathen *et al.*, 1995) because there already was substantial coverage of mid and high-latitude sounding stations in the Northern Hemisphere. Several dozen of these stations agreed to launch ozonesondes on schedule over the period when ozone loss is greatest in the Arctic vortex, which is typically from December to mid-March.

The basic idea of Match is to determine stratospheric polar ozone losses by observing an airmass using ozonesondes at two stations over which the airmass passes; ozone loss can then be determined by the difference between the two ozonesonde profiles. The name "Match" comes from the attempt to match two ozonesoundings that correspond to the same airmass at different instants in its lifetime. The EASOE experiment included the launch of a large number (~1,400) of ozonesondes in the Arctic and mid-latitudes. This dataset was used afterwards to determine Arctic ozone losses for the first time with the Match method. Since the launches were not co-ordinated, matches were only obtained by chance. However, because of the large number of soundings made during EASOE enough matches were found for successful analysis, and in this way the Match method was established.

The method was then developed into an active method to eliminate chance and to be able to economize on the number of ozonesondes launched; it was made ready for the Second European Stratospheric and Mid-latitude Experiment (SESAME, 1998, 1999) in the winter of 1994–1995. The two main aims of the Match campaigns are (1) to measure chemical ozone loss in polar regions, and (2) to check our understanding of the underlying processes. The objective of Match—which is to probe (i.e., determine) the ozone content of large numbers of air parcels twice during their way through the atmosphere—is achieved by co-ordinating soundings in roughly the following way:

1. The trajectories (transport paths) of airmasses previously measured by ozonesondes are analyzed and future trajectories are forecast by meteorologists at the Free University (FU) in Berlin.
2. These trajectories are checked to see if such an airmass reaches the vicinity of one of the participating measuring sites (ozonesonde stations) within 10 days.
3. The staff at this ozonesonde station are informed and asked to launch an ozonesonde in order to examine this airmass for a second time.

Any decrease in ozone concentration within the time period of the two ozonesonde flights can then be attributed to chemical ozone depletion. Accumulated loss in the vertical (in K or altitude space) is computed. Column-integrated loss amounts within the vortex over the course of the Arctic winter are determined. The fractional loss is typically 10–15% of the total ozone column (the latter 350–475 DU). Factors affecting loss include temperatures, amount of polar stratospheric cloud formation, and duration of the Arctic vortex (Harris *et al.*, 2002).

Due to the large number of ozonesonde pairs, statistically significant ozone loss rates can be determined.

Since 1994–1995 the Alfred Wegener Institute for Polar and Marine Research (AWI) has co-ordinated the Match network campaign in most Arctic winters and in a single Antarctic winter in order to determine chemical ozone depletion in the stratosphere. Lists of Northern Hemispheric and Antarctic station names, corresponding staff members, and principal investigators are given on the AWI website. Recently, within the framework of the International Polar Year (IPY) both an Arctic and an Antarctic Match campaign took place. There was no Arctic Match campaign in the winter of 2008–2009 because of a major stratospheric warming event. This event prevented the conditioning of airmasses with chlorine radicals in February and March. A second Antarctic ozonesonde Match campaign took place between June 4 and October 31, 2007. About 300 ozonesondes were co-ordinated and launched at nine stations: Belgrano, Davis, Dome Concordia, Dumont d'Urville, Marambio, McMurdo, Neumeyer, South Pole, and Syowa.

The Match technique was used by Sasano and co-workers to determine ozone loss rates from data of the Improved Limb Atmospheric Spectrometer (ILAS) spaceborne sensor (Sasano *et al.*, 2000; Terao *et al.*, 2002).

The most important outcomes and implications of the Match method are:

- It has been demonstrated that Match is able to distinguish between dynamical and chemical variations (i.e., Match is able to determine chemical ozone losses). Vertical and time resolution are among the best reported.
- In the winter of 1995–1996 Match found a layer of air in which Antarctic-like conditions occurred (i.e., heavy denitrification prolonged ozone loss well into spring).
- Match found the effect of the chlorine nitrate trap (Rex *et al.*, 1997).
- Match data are very well suited to be compared with model outputs (in particular, with box model outputs). Two successful studies revealed that current photochemical models are not able to reproduce observed ozone loss rates quantitatively. It has also been shown that the use of tracer data from a 3-D model can improve the Match approach at higher altitudes.
- Match found a direct relationship between the minimum temperature that an air parcel encountered during a reasonable time period before the first sonde measurement and ozone loss. This allowed discovery of the threshold temperatures needed for ozone loss.

The Match campaigns were part of the Ozone Layer and UV Radiation in a Changing Climate (ORACLE-O3) project within the International Polar Year (IPY), which actually covered two years (from March 2007 to March 2009).

6.6.3.2 The mid-latitude summer ozone campaigns: IONS-04 and IONS-06

The IONS-04 and IONS-06 ozonesonde network campaigns (principal investigator: Anne M. Thompson) were associated with NASA's INTEX campaign. INTEX was an integrated atmospheric field experiment consisting of two phases. Phase A was conducted in the summer of 2004 over the central and eastern United States followed by Phase B in the spring of 2006 on the west coast of the United States and beyond (particularly, Mexico). The mission sought to answer questions about the transport and transformation of gases and aerosols on transcontinental/intercontinental scales and their impact on air quality and climate. The main constituents of interest were ozone and its precursors, aerosols and their precursors, and long-lived greenhouse gases. A particular focus of this study was to quantify and characterize the inflow and outflow of pollution over North America and its transformation during transport from and to distant continents. INTEX-A involved a NASA DC-8 aircraft with long-range and high-altitude capability and equipped with a comprehensive suite of *in situ* and remote-sensing instrumentation to provide a comprehensive suite of chemical, physical, and optical measurements involving gases and aerosols. This was complemented by several other aircraft and by several satellites making atmospheric chemistry measurements from space. Interaction with ongoing satellite measurement programs was an important goal of INTEX. The objectives of INTEX required the use of satellite data and were also aimed at helping to validate results from satellite-flown instruments. INTEX-A was operated in the summer of 2004 and INTEX-B in the spring of 2006.

The IONS program was similar in concept to the Match program, but whereas Match was concerned with monitoring ozone loss in the (northern) polar regions, IONS was concerned with tracking mid-latitude pollution plumes. Thus, IONS was concerned with mid-latitude tropospheric ozone, ozone precursors, and lightning strikes, whereas Match was concerned with stratospheric ozone losses. IONS-04 and IONS-06 ozone-sounding stations are identified in Table 1 of Thompson (2009) and soundings were made during the INTEX-A and INTEX-B campaigns. IONS-04 concentrated on pollution plumes originating in North America, following the trajectories of plumes towards Europe, and intercepting them over the Azores. IONS-06 involved studying the transport of pollution plumes from Mexico City northwards towards Texas. IONS data were used extensively for validation of data from satellites, with the ACE (Atmospheric Chemistry Experiment) comparing ozone retrievals to the 2004 campaign and the ozone sensors aboard Aura (launched in July 2004) to IONS-06 data. Aura sensors with enhanced tropospheric sensitivity include the Microwave Limb Sounder (MLS) (Jiang *et al.*, 2007; Schoeberl *et al.*, 2007) and the Tropospheric Emission Spectrometer (TES) (Osterman *et al.*, 2008). These sensors retrieve not just ozone but several other species, so that ozone and carbon monoxide correlations in pollution events can be followed simultaneously (Logan *et al.*, 2008; Thompson, 2009; Zhang *et al.*, 2006).

6.6.3.3 SHADOZ, a global tropical network

The SHADOZ (Southern Hemisphere ADditional OZonesondes) global tropical network (principal investigator: Anne M. Thompson) was conceived in the mid-1990s when the satellite community was making new tropospheric ozone products, and more advanced ozone sensors were being launched. Besides 1970 and 1980 instruments like Nimbus-7/TOMS (from 1978) and SAGE II (from 1984), ozone instruments on UARS and ESA platforms (e.g., GOME) began operations during this period. There was a critical shortage of validation data from the tropics and the Southern Hemisphere. A number of ozone-sounding stations were operating in the Southern Hemisphere tropics and subtropics, but with differing frequency and reporting procedures. Stations would stop and start to match campaign schedules and data distribution was uneven. With no long-standing project devoted to sondes and with no means of sharing data, few tropical stations could sustain ozone soundings. SHADOZ was designed to remedy this data deficiency by co-ordinating launches, by supplying additional sondes in some cases, and by providing a central archive location. Data are collected in a timely manner and made available to the SHADOZ and TOMS Science Teams, as well as to the scientific community as a whole. According to the SHADOZ website (*http://www.croc.gsfc.nasa.gov/shadoz/*) there are about a dozen active stations participating in SHADOZ.

SHADOZ data are quality-assured before distribution via a user-friendly, open access (no passwords) archive on the SHADOZ website (*http://www.croc.gsfc.nasa.gov/shadoz*; Thompson, 2009). The data are transmitted regularly to the WMO's World Ozone and Ultraviolet Data Centre (*woudc.org*) to further enhance distribution. An important factor in site selection for SHADOZ was having sufficient longitudinal coverage to span the standing wave 1 pattern in total ozone shown in Figure 3 of Kim *et al.* (1996). This phenomenon, seen by satellite (Fishman and Larsen, 1987) in the 1980s, refers to the 10–20 additional Dobson units of ozone column ($1\,DU = 2.69 \times 10^{16}\,cm^{-2}$) over the Atlantic and eastern Africa compared to the central Pacific where total ozone is a minimum. The question as to the stratospheric or tropospheric origin of the additional ozone was answered after several years of SHADOZ observations. Column-integrated stratospheric ozone was invariant longitudinally (Thompson *et al.*, 2003a, b), but an additional 15–20 DU appears, with magnitude depending on season, in the troposphere. Over the Atlantic the lower tropical tropopause layer (TTL) and high concentration of lower to mid-tropospheric ozone affected by biomass and biofuel burning, lightning, and possibly biogenic sources contribute to the thicker column.

The quality, spatial density, and frequency of SHADOZ sondes have led to many novel applications of the data. For example, Logan *et al.* (2003) used the early SHADOZ record to characterize ozone and temperature patterns produced by the QBO in the middle stratosphere. Stations closer to the equator (e.g., Nairobi and San Cristóbal) manifest stronger signals than those south of 5°S.

SHADOZ data are used to evaluate satellite data and assimilation models (e.g. Schoeberl *et al.*, 2007; Stajner *et al.*, 2008). Perhaps the most active research area to which SHADOZ data are being applied is to investigations of ozone and temperature in the TTL. The breadth of this activity has been aided by new satellite capabilities and by unresolved issues in the nature of troposphere-to-stratosphere transport of water vapor. Of particular interest is the role of convection in troposphere-to-stratosphere transport and the relative contributions of advection, radiative heating, and wave activity in determining TTL structure (Rex *et al.*, 2007).

Among the significant results of investigations based on SHADOZ observations are the following:

1. A small number of episodes in which very low O_3 layers are found in the upper troposphere (UT) (100–300 hPa) is observed at all SHADOZ stations except Nairobi (Kenya) and Irene (Argentina). When the frequency of the lowest ozone layers during the 2000–2004 period at Samoa, Ascension, and Natal is

compared to profiles at these sites in the 1986–1992 period, an increase in the occurrence of lowest ozone episodes is noted (Solomon *et al.* 2005). The shape of low-ozone layers follows convective outflow locations over regions of low surface ozone, raising a question about possible dynamical changes in the past 20 years.

2. Folkins and coworkers (Folkins *et al.*, 2000, 2002) were among the first to exploit ozone, temperature, and potential profiles from SHADOZ to demonstrate the importance of convective outflow of ozone in the TTL. Recently, Folkins *et al.* (2006) used a simple radiative detrainment model to connect CO and ozone from MLS with SHADOZ ozone and temperature, demonstrating the way in which mean upwelling and convection interact to establish the seasonal cycles of these parameters.
3. Randel *et al.* (2007) used SHADOZ and Aura/MLS data to delineate the seasonality of ozone and temperature in the TTL (16–19 km). Maximum and minimum ozone differ by almost a factor of 2. Ozone is nearly in phase with the annual temperature cycle and is associated with the mean upwelling rate of Brewer–Dobson circulation.
4. Takashima and Shiotani (2007) used 6 years of SHADOZ observations to characterize the seasonality of ozone and temperature in the tropical upper troposphere. Interactions are such that when the higher ozone longitude (corresponding to ozone maximum in wave 1) is at its greatest level during austral winter and spring, the temperature lapse rate is reduced; the opposite happens when ozone is at its seasonal minimum.
5. Wave influences in TTL ozone distribution were evaluated by Loucks (2007). The seasonal cycle of features involved in gravity or Kelvin wave forcing is strongest near the equator and more intense over the eastern Indian Ocean and Pacific than over the Atlantic or African SHADOZ sites.

6.6.4 Trends in TOC and tropopause properties

In Section 4.6 we considered quite extensively the relation between TOC and tropopause height, Z, with the decrease in TOC corresponding to an increase in Z varying from $-8.5\,\mathrm{DU\,km^{-1}}$ to $-11.2\,\mathrm{DU\,km^{-1}}$ in Greece and $-13\,\mathrm{DU\,km^{-1}}$ in Germany. We now move on to consider recent trends in these quantities and the correlation between them. Varotsos, C.A. *et al.* (2004) considered deseasonalized monthly means of TOC and Z over the period 1984–2002 with data from Athens. To subtract annual variation from both time series in order to get deseasonalized time series, smoothing of the 13-month running mean was applied to the raw data. Figure 6.34 shows the deseasonalized monthly means of TOC and Z, as well as their corresponding linear trends at Athens, during the period 1984–2002. As expected, high TOC is associated with a low tropopause and *vice versa*. The correlation coefficient between single monthly means of TOC and the tropopause was found to be -0.5, which is significant at the 99% confidence level. Linear regression analysis of the deseasonalized monthly mean TOC indicates a long-term decrease of (-7.54 ± 0.99) DU per decade (2σ).

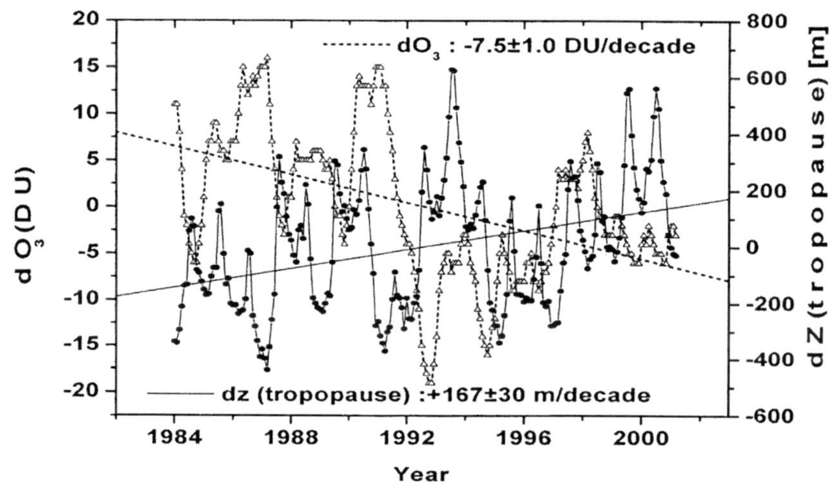

Figure 6.34. Time series of deseasonalized monthly means and the linear regression lines (2σ) of TOC (dO_3) and tropopause height (dZ) for 1984–2002 at Athens deduced from *in situ* observations (Varotsos, C.A. *et al.*, 2004).

Since TOC is highly variable due to various oscillations (like the QBO and NAO; Appenzeller *et al.*, 2000), it is worth mentioning that the application of more sophisticated trend analysis, which includes the effects of Pinatubo, solar cycle, QBO, etc. on TOC over Athens, gives approximately the same results for major long-term changes as linear regression analysis (Callis *et al.*, 1997; Varotsos, 2003a, b; Varotsos *et al.*, 2000). This was also the case for the Hohenpeissenberg time series (Steinbrecht *et al.*, 1998). There is no doubt, however, that the use of more sophisticated trend models should result in reduced trend estimate error.

Another interesting point in Figure 6.34 is the long-term increase—by about (167 ± 30) m (2σ) per decade—of Z at Athens for the period under consideration. However, if the $-9.7\,\mathrm{DU\,km^{-1}}$ correlation (linear fitting) between TOC and Z (considering both the summer and winter period) is representative on a long timescale, then the observed 167 m per decade increase in Z (shown in Figure 6.34) should be connected to a TOC decrease of 1.64 DU per decade. Thus, the observed increase in Z could explain about 22% of the observed decrease in TOC (−7.5 DU per decade) shown in Figure 6.34.

It is worth mentioning that 22% is more or less the same magnitude of mid-latitude ozone depletion as the 25% ozone decrease that relevant models have difficulty accounting for (Steinbrecht *et al.*, 1998). For example, 2-D model estimations by Jackman *et al.* (1996) give TOC trends of −3.5% per decade for northern mid-latitudes in spring, whereas observed trends are −5% per decade (WMO, 1995). Therefore, about 22% of the trend in TOC over Greece during 1984–2002 could be attributed to Z changes. It should be mentioned, however, that this part of the trend is only due to variations in the upper troposphere–lower stratosphere region and not attributable to all dynamical changes and

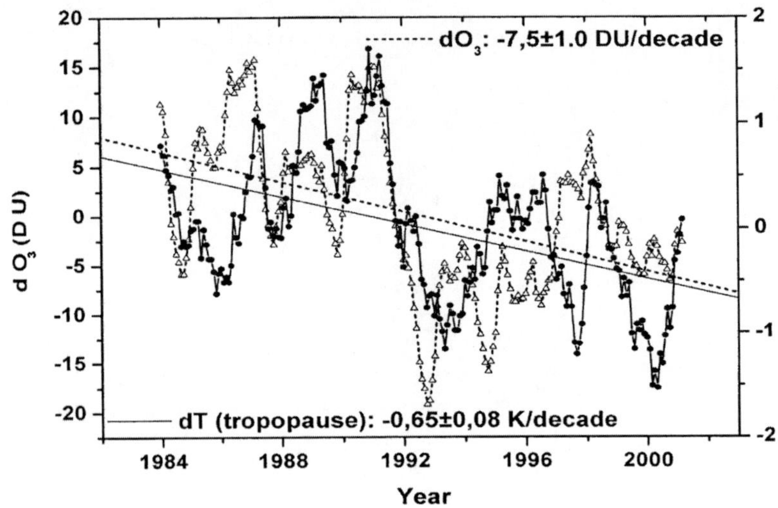

Figure 6.35. Time series of deseasonalized monthly means and the linear regression lines (2σ) of tropopause temperature ($\mathrm{d}T$) deduced from reanalysis data and TOC (dO_3) for 1984–2002 at Athens (Varotsos, C.A. *et al.*, 2004).

forcing on TOC. The above-mentioned results were also confirmed by observations obtained at Thessaloniki and Heraklion.

Because of the strong correlation between various meteorological indices, something akin to the latter result is obtained by using NAM-like or 330 K potential vorticity and meridional wind shear (from NCEP/NCAR reanalyses) dynamical proxies in statistical trend analyses of TOC during the winter–spring period (see also Steinbrecht *et al.*, 2001; WMO, 2003).

For comparison reasons Varotsos, C.A. *et al.* (2004) also studied the deseasonalized monthly means of TOC and tropopause temperature (calculated by using NCEP reanalysis data), as well as their corresponding linear trends, for the period 1984–2002, at Athens; the results are shown in Figure 6.35. It is clear that the trends shown in Figure 6.35 (deduced from reanalysis data) are in close agreement with those shown in Figure 6.34 (deduced from real observations).

Deseasonalized monthly means of 500 mb temperature and TOC over Athens during the period 1979–2002 show a correlation coefficient of -0.52 (significant at the 99% confidence level) and a regression coefficient of $(-2.7 \pm 0.4)\,\mathrm{DU\,K^{-1}}$ (2σ). In addition, linear regression gives an increase for the 500 mb temperature over Athens of $(+0.27 \pm 0.04)$ K per decade (2σ). This result is in qualitative agreement with the trend reported for tropospheric zonal mean temperatures at 40° to 50°N by Labitzke and van Loon (1995) (≈ 0.2 K per decade annual mean, ≈ 0.4 K per decade in winter). Combining the latter with the information obtained from Figure 6.30 supports the consideration that changing Z plays a crucial role in long-term TOC decrease. Increasing Z and decreasing tropopause temperature is qualitatively associated with the decreasing trend of ozone in the stratosphere. It should be noted that mid-tropospheric temperature is used as a proxy of temperature at the Earth's surface. Also, greenhouse gases warm the Earth's surface but cool the stratosphere radiatively and, therefore, affect ozone depletion. In addition, due to ozone depletion less heat would be absorbed in the stratosphere, whereas due to the increase in greenhouse gas content more heat would be trapped in the atmosphere and, thus, these effects may change the thermal structure of the tropopause region.

Furthermore, tropospheric warming pushes the tropopause up with the result that excess ozone in the stratosphere is destroyed photochemically. This goes to show that there is a close connection between increase in Z, tropospheric warming, and TOC depletion (Kondratyev, 1998b; Steinbrecht *et al.*, 1998; Varotsos, 2003b).

A similar analysis to that just described for Athens was carried out for Hohenpeissenberg (Germany) for 1967–2002 and for Belsk (Poland) for 1974–1996. The results showed that long-term increases in Z over Hohenpeissenberg and Belsk are of the same order of magnitude as that observed for Athens in Figure 6.29 (Varotsos, C.A. *et al.*, 2004). Analysis of radiosonde observations obtained for 1966–2000 for New Delhi (India, 28°N, 77°E) and for 1973–1998 at Thiruvananthapuram (India, 8°N, 76°E) showed that the long-term trend of Z is increasing in the range +0.6 to +1.0% per decade, while the long-term trend of tropopause temperature is decreasing in the range –0.5 to –0.9 K per decade.

These results should be combined with the recent findings of tropopause properties at high latitudes of the Northern Hemisphere. For instance, according to Highwood *et al.* (2000) the multidecadal radiosonde dataset (1965–1990) over the Arctic region shows that wintertime tropopause pressure has decreased by approximately 14 mb per decade, while that in other seasons has decreased by around 5 mb per decade. Tropopause temperature only shows a significant trend during winter, having decreased by 1.6 K per decade since 1965. Also, Zou *et al.* (2002) reported on ozone soundings taken on board the Chinese icebreaker *Xuelong* (at 75°N, 160°W), suggesting that the atmospheric ozone amount experienced high–low–high variation corresponding to low–high–low tropopause altitude.

The final question in connection with tropopause–TOC coupling and climate change is whether air circulation changes might be the result of GHG-induced Z variation and further climate change. On the basis of radiative balance, Santer *et al.* (2003) showed a mean mid-latitude tropopause pressure decrease of 3 hPa/decade resulting from changes in GHGs; this corresponds to a mid-latitude Z increase of about 120 m per decade, which is roughly consistent with observed changes discussed by Varotsos, C.A. *et al.* (2004). Therefore, if past dynamical changes actually represent a dynamical response to GHG-induced climate change, then these changes will increase in magnitude in the future (Kondratyev and Varotsos, 2000; Santer *et al.*, 2003). This would decrease future ozone levels and delay ozone recovery.

In this context, Gauss *et al.* (2003) analyzed the results of 11 chemistry–transport models and used them as input for radiative forcing calculations, mindful of the fact that radiative forcing due to changes in ozone is expected for the 21st century. They addressed future ozone recovery in the lower stratosphere and its impact on radiative forcing by applying two models that calculate both tropospheric and stratospheric changes. Their results suggested an increase in global mean tropospheric ozone between 11.4 and 20.5 DU for the 21st century, which corresponds to positive radiative forcing ranging from 0.40 to 0.78 $W m^{-2}$ on a global and annual average. They also found that the lower stratosphere contributes an additional 7.5–9.3 DU to the calculated increase in ozone column, augmenting radiative forcing by 0.15–0.17 $W m^{-2}$. As to the ability to scale radiative forcing to global mean ozone column change they suggested that—despite large variations between the 11 participating models—the calculated range for normalized radiative forcing is within 25%.

Let us break down the work described in this section into two parts: (a) the linear regression analyses of deseasonalized monthly mean TOC and Z at Athens for the 1984–2002 indicate that about 22% of the trend in TOC over Greece could be attributed to Z changes or, in other words, tropopause variations might be responsible for about 25% of the observed TOC trend over Greece. This part of the trend is only due to variations in the upper-troposphere/lower-stratosphere region and not attributable to all dynamical changes and forcing on TOC. (b) The observed changes in Z are consistent with recently derived modeled changes, which represent a dynamical response to GHG-induced climate change that would decrease future ozone levels and delay ozone recovery. Total ozone residual (TOR) data (what remains after removing natural periodicities) has been analyzed for the 1979 to 2004 period by Kulkarni *et al.* (2009). Tropospheric column ozone was observed to have a maximum during late summer of (48 ± 4) DU and a minimum during late winter of (30 ± 4) DU over the Indo-Ganges basin in India. The amplitude of the seasonal cycle was observed to be comparatively larger over the western part of the Indo-Ganges region $(\sim 51.0 \pm 2)$ DU than the central $(\sim 47.0 \pm 3)$ DU and eastern parts $(\sim 47.0 \pm 3)$ DU. Similarly, seasonal variation in tropospheric ozone was observed to be comparatively larger over the western part of the Indo-Ganges region (~22 DU) than the central (~18 DU) and eastern parts (~17 DU). The difference in tropospheric ozone amount over the eastern and western parts of this region also shows seasonal variation with a large difference (up to 4 DU) during the monsoon season. The monsoon system plays an important role in the seasonal variation of tropospheric ozone over different parts of Indo-Ganges region.

6.7 SCIENTIFIC ASSESSMENT OF OZONE DEPLETION 2010

Although the Montreal Protocol and its Amendments and Adjustments have successfully controlled the global production and consumption of ODSs over the last two decades, and the atmospheric abundances of nearly all major ODSs that were initially controlled are declining, nevertheless, ozone depletion will continue for many more decades because several key ODSs last a long time in the atmosphere after emissions end.

However, in contrast to the diminishing role of ODSs, changes in climate are expected to have an increasing influence on stratospheric ozone abundances in the coming decades. Moreover, ODSs and substitutes for ODSs are greenhouse gases;

we postpone consideration of these matters until Chapter 7 and in the present section consider only the question of ozone depletion.

Changes in the global atmospheric abundance of a substance are determined by the balance between its emissions to and removal from the atmosphere. Declines observed for ozone-depleting substances controlled under the Montreal Protocol are due to global emission reductions that have made emissions smaller than removals. As the majority of ODSs have been phased out, demand for hydrochlorofluorocarbon (HCFC) and hydrofluorocarbon (HFC) substitutes for the substances controlled under the Montreal Protocol has increased; these are also greenhouse gases. HCFCs deplete much less ozone per kilogram emitted than chlorofluorocarbons (CFCs), while HFCs are essentially non-ozone-depleting gases.

The 2010 Scientific Assessment (WMO, 2010a) reported on the tropospheric abundances and emissions of ODSs and their substitutes:

- "The amended and adjusted Montreal Protocol continues to be successful at reducing emissions (Figure 6.36) and thereby abundances of most controlled ozone-depleting substances in the lower atmosphere (troposphere), as well as abundances of total chlorine and total bromine from these ozone-depleting substances. By 2008, the total tropospheric abundance of chlorine from ODSs and methyl chloride had declined to 3.4 parts per billion (ppb) from its peak of 3.7 ppb. However, the rate of decline in total tropospheric chlorine by 2008 was only two-thirds as fast as was expected. This is because HCFC abundances increased more rapidly than expected, while CFCs decreased more slowly than expected. The discrepancy in CFC decreases is most likely because of emissions from 'banks' in existing applications such as refrigerators, air conditioners, and foams. The rapid HCFC increases are coincident with increased production in developing countries, particularly in East Asia. The rate of decline of total tropospheric bromine from controlled ODSs was close to that expected and was driven by changes in methyl bromide.
- Declines in CFCs made the largest contribution to the observed decrease in total tropospheric chlorine during the past few years and are expected to continue to do so through the rest of this century. Observations show that CFC-12 tropospheric abundances have decreased for the first time. The decline of methyl chloroform (CH_3CCl_3) abundances made a smaller contribution to the decrease in total chlorine than described in past Assessments, because this short-lived substance has already been largely removed from the atmosphere.
- Carbon tetrachloride (CCl_4) tropospheric abundances have declined less rapidly than expected. Emissions derived from data reported to the United Nations Environment Programme (UNEP) are highly variable and on average appear smaller than those inferred from observed abundance trends. Although the size of this discrepancy is sensitive to uncertainties in our knowledge of how long CCl_4 persists in the atmosphere (its 'lifetime'), the variability cannot be explained by lifetime uncertainties. Errors in reporting, errors in the analysis of reported data, and/or unknown sources are likely responsible for the year-to-year discrepancies.

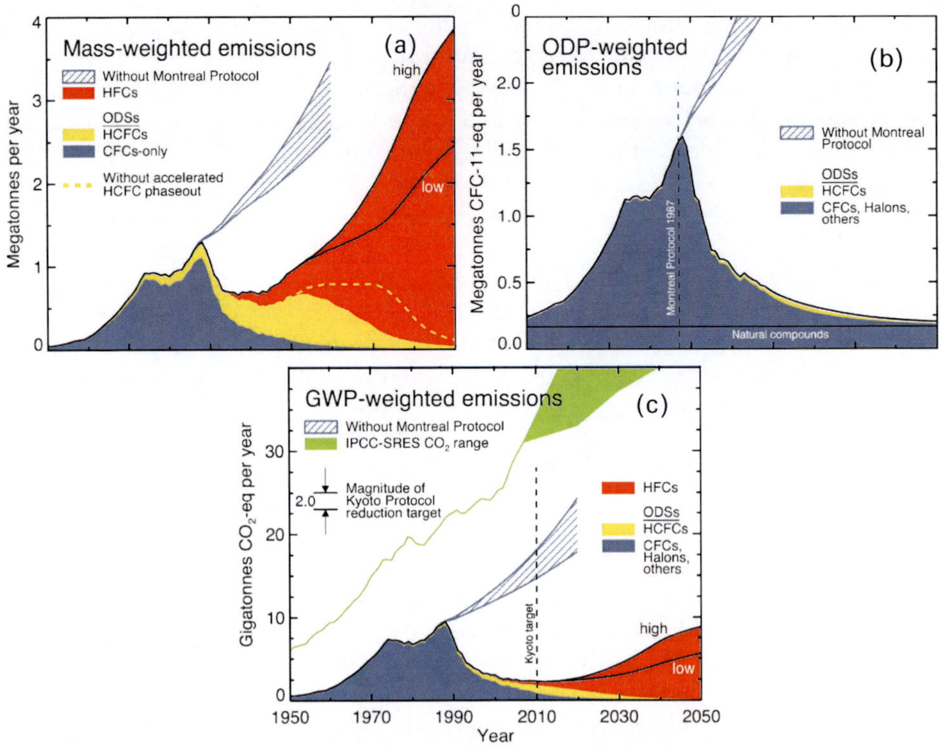

Figure 6.36. Emissions of ODSs and their substitutes. Global emissions of ODSs (CFCs, halons, HCFCs, and others) and their non-ozone-depleting substitutes (HFCs) from 1950 to 2050. Emissions are the total from developing and developed countries. The legends identify the specific groups of substances included in each panel. The high and low HFC labels identify the upper and lower limits, respectively, in global baseline scenarios. The blue hatched regions indicate the emissions that would have occurred in the absence of the Montreal Protocol with 2–3% annual production increases in all ODSs (WMO, 2010b). (a) Global mass-weighted emissions expressed as megatonnes per year. The yellow dashed line shows HCFC emissions calculated without the provisions of the 2007 accelerated HCFC phase-out under the Montreal Protocol. (b) Global ODP-weighted emissions expressed as megatonnes of CFC-11-equivalent per year. The emissions of individual gases are multiplied by their respective ODPs (CFC-11 = 1) to obtain aggregate equivalent CFC-11 emissions. The dashed line marks 1987, the year the Montreal Protocol was signed. (c) Global GWP-weighted emissions expressed as gigatonnes of CO_2-equivalent per year. The emissions of individual gases are multiplied by their respective GWPs (direct, 100-year time horizon; $CO_2 = 1$) to obtain aggregate equivalent CO_2 emissions. Shown for reference are emissions for the range of CO_2 scenarios from the IPCC's *Special Report on Emission Scenarios* (SRES). The CO_2 emissions for 1950–2007 are from global fossil fuel use and cement production. Beyond 2007, the shaded region for CO_2 reflects the maximum (A1B) and minimum (B2) SRES scenarios. The dashed line marks 2010, the middle year of the first commitment period of the Kyoto Protocol. Also shown is the magnitude of the reduction target of the first commitment period of the Kyoto Protocol, which is based on a 1990–2010 projection of global greenhouse gas emission increases and the reduction target for participating countries.

- Observations near the tropical tropopause suggest that several very short-lived industrial chlorinated chemicals, not presently controlled under the Montreal Protocol (e.g., methylene chloride, CH_2Cl_2; chloroform, $CHCl_3$; 1,2-dichloroethane, CH_2ClCH_2Cl; perchloroethylene, C_2Cl_4), reach the stratosphere. However, their contribution to stratospheric chlorine loading is not well quantified.
- Bromine from halons stopped increasing in the troposphere during 2005–2008. As expected, abundances of halon-1211 decreased for the first time during 2005–2008, while halon-1301 continued to increase but at a slower rate than in the previous Assessment.
- Tropospheric methyl bromide abundances continued to decline during 2005–2008, as expected due to reductions in industrial production, consumption, and emission. About half of the remaining methyl bromide consumption was for uses not controlled by the Montreal Protocol (quarantine and pre-shipment applications).
- Tropospheric abundances and emissions of some HCFCs are increasing faster now than four years ago. Abundances of HCFC-22, the most abundant HCFC, increased more than 50% faster in 2007–2008 than in 2003–2004, while HCFC-142b abundances increased about twice as fast as in 2003–2004. HCFC-141b abundances increased at a similar rate to that observed in 2003–2004. Total emissions of HCFCs are projected to begin to decline during the coming decade due to measures already agreed to under the Montreal Protocol.
- Tropospheric abundances and emissions of HFCs, used mainly as substitutes for CFCs and HCFCs, continue to increase. For example, abundances of HFC-134a, the most abundant HFC, have been increasing by about 10% per year in recent years. Abundances of other HFCs, including HFC-125, -143a, -32, and -152a, have also been increasing. Regional studies suggest significant HFC emissions from Europe, Asia, and North America."

The 2010 Scientific Assessment (WMO, 2010b) summarized the implications of total chlorine and bromine for ozone depletion:

- "Total chlorine has continued to decline from its 1990s peak values in both the troposphere and the stratosphere. Total tropospheric bromine is decreasing from its peak values, which occurred comparatively recently, while stratospheric bromine is no longer increasing.
- Relative declines in the sum of stratospheric chlorine and bromine from peak values are largest in midlatitudes and smallest in Antarctica (Figure 6.37). These declines are not as pronounced as observed in their tropospheric abundances. Differences between declines in the troposphere and different regions of the stratosphere are primarily associated with the time required for air to move from the troposphere to those regions. The relative declines are smallest in Antarctica primarily because the transport times to polar regions are the largest."

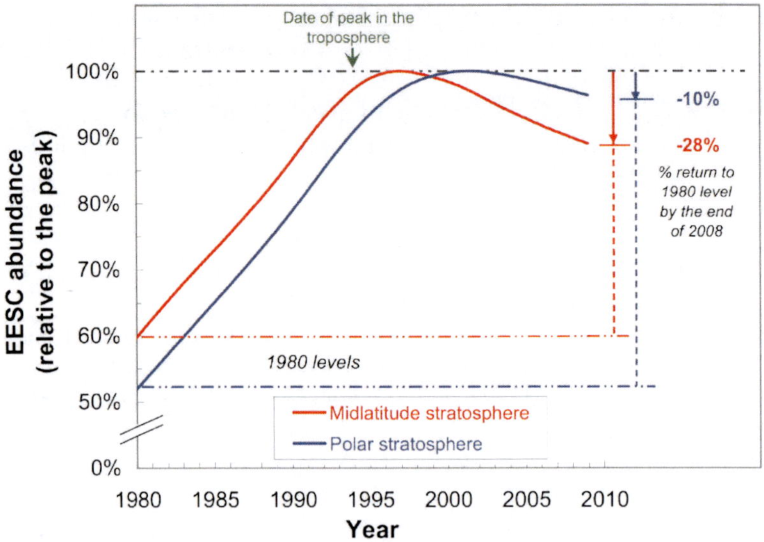

Figure 6.37. Stratospheric EESC derived for mid-latitude and polar stratospheric regions relative to peak abundances, plotted as a function of time. Peak abundances are ~1,950 ppt for the mid-latitude stratosphere and ~4,200 ppt for the polar stratosphere. Percentages shown to the right indicate the observed change in EESC by the end of 2008 relative to the change needed for EESC to return to its 1980 abundance. A significant portion of the 1980 EESC level is from natural emissions (WMO, 2010b).

The Montreal Protocol and its Amendments and Adjustments have made large contributions toward reducing global greenhouse gas emissions (Figure 6.36). In 2010, the decrease of annual ODS emissions under the Montreal Protocol is estimated to be about 10 Gt of avoided CO_2-equivalent emissions per year, which is about five times larger than the annual emissions reduction target for the first commitment period (2008–2012) of the Kyoto Protocol.

The sum of HFCs currently used as ODS replacements contributes about 0.4 Gt of CO_2-equivalent per year to total global CO_2-equivalent emissions, while the HCFCs contribute about 0.7 Gt. CO_2-equivalent emissions of HFCs are increasing by about 8% per year and this rate is expected to continue to grow, while the contribution from HCFCs is expected to start decreasing in the next decade.

Emissions of HFC-23, a byproduct of HCFC-22 production, contributed about 0.2 Gt of CO_2-equivalent per year in 2006–2008. HFC-23 is a particularly potent greenhouse gas with a lifetime of about 220 years. Its emissions have increased in the past decade despite global emission reduction measures, including those covered by the Kyoto Protocol's Clean Development Mechanism projects.

Total chlorine has continued to decline from its 1990s' peak values in both the troposphere and the stratosphere. Total tropospheric bromine is decreasing from its peak values, which occurred comparatively recently, while stratospheric bromine is no longer increasing.

Relative declines in the sum of stratospheric chlorine and bromine from peak values are largest in the mid-latitudes and smallest in Antarctica (Figure 6.37). These declines are not as pronounced as observed in their tropospheric abundances. Differences between declines in the troposphere and different regions of the stratosphere are primarily associated with the time required for air to move from the troposphere to those regions. Relative declines are smallest in Antarctica primarily because the transport times to polar regions are the largest.

7

Atmospheric ozone and climate

7.1 INTRODUCTION

Numerous interactive processes are responsible for climate formation and change (Brasseur, 1997; Houghton *et al.*, 1996; Kondratyev, 1989, 1998b; Zillman, 1997). It is therefore difficult to single out the impact of an individual cause of climate change. Nevertheless, many attempts have been undertaken to assess climatic the impacts of various processes and phenomena, including processes in the stratosphere with special emphasis on the role of stratospheric ozone variability (Kondratyev, 1998b, 1999b). In this context, climate change affects the variability of the ozone layer through changes in atmospheric circulation, chemical composition, and temperature. However, changes to the ozone layer affect climate through radiative processes (e.g., changes in temperature gradients) which consequently modify atmospheric dynamics. Therefore, there is a strong link between climate change and ozone layer variability. The problem becomes more complex when assuming that the processes involved in this coupling often exhibit non-linear behavior (Fleming *et al.*, 2009; WMO, 2007).

Long ago a hypothesis was suggested concerning the important role of the stratosphere in climate change (stratospheric climate regulation). Although it was soon shown that such a hypothesis should be considered as a gross exaggeration, there are no doubts that the relation between the stratosphere and climate deserves serious attention, especially in the context of anthropogenically induced ozone depletion (Bojkov, 1997; Hansen *et al.*, 1997; Kondratyev, 1981, 1985, 1998b; *Ozone Depletion* ..., 1989; Schmidt *et al.* 2010; SPARC, 1998; Stolarski *et al.* 2010; Uchino, 1998; Varotsos, 2004a; Wang and Isaksen, 1995; WMO, 2007). This is why a WCRP core project (COPES) "Stratospheric Processes and Their Role in Climate" (SPARC) was approved and accomplished (Chanin, 1993, 1998; Chanin and Geller, 1998; Geller, 1998; Shine *et al.*, 1995; SPARC, 1997). The following key scientific issues were identified (Phillips, 1997):

- The influence of the stratosphere on climate
- The physics and chemistry associated with stratospheric ozone decrease
- Stratospheric variability and monitoring
- UV irradiation changes.

Later on (SPARC, 1998), the principal SPARC research topics were classified into the three following categories: "Stratospheric Indicators of Climate Change", "Stratospheric Processes and Their Relation to Climate", and "Modeling Stratospheric Effects on Climate".

As far as stratospheric processes and their relation to climate is concerned the following important topics were emphasized:

- Gravity wave processes and parameterization of the effects of unresolved internal gravity waves in global numerical atmospheric models
- Stratosphere–troposphere exchange
- Chemistry–climate of the upper troposphere–lower stratosphere
- Changes of ozone and their influence on climate
- The Quasi-Biennial Oscillation and its possible role in coupling the stratosphere and the troposphere.

To study upper-tropospheric and lower-stratospheric ozone, aerosols, and climate the following priorities should be taken into account:

- Observations of stratospheric aerosols and polar stratospheric clouds (PSCs)
- Heterogeneous chemistry of the upper troposphere
- Gas phase processes
- Chemical–radiative coupling in the upper troposphere and lower stratosphere.

According to the SPARC Implementation Plan, 2009, SPARC contributions to implementation of the WCRP (COPES) Strategic Framework are briefly as follows (*http://www.atmosp.physics.utoronto.ca/SPARC/index.html*):

1. Major challenges
   1.1 Present and near future overarching themes
      (a) Detection, attribution and prediction of stratospheric changes
      (b) Stratospheric chemistry and climate
      (c) Stratosphere–troposphere coupling
   1.2 SPARC overarching activities that should be maintained beyond 2013
      (a) Chemistry–Climate Model Validation Activity (CCMVal)
      (b) Assessments of key uncertainties in measurements
      (c) Linking various scientific communities
2. Detailed Science Objectives
   2.1 Detailed Objectives to 2013
      (a) Chemistry and Climate
      (b) Detection, attribution and prediction of stratospheric changes
      (c) Stratosphere–troposphere dynamical coupling
   2.2 Key Science Questions Beyond 2013 as far as Currently Conceivable

3. Detailed Contributions to Unifying COPES Themes
   (a) Modeling
   (b) Observations (including some cross-cutting with modeling; i.e., compared with simultaneous model runs)
   (c) Climate System Science (including some cross-cutting with modeling and observations)
4. Contributions to Implementing COPES Cross-cutting Initiatives (including overlapping with COPES themes)
   (a) Atmospheric Chemistry and Climate (AC&C)
   (b) Anthropogenic Climate Change
   (c) Seasonal Prediction
   (d) Decadal Prediction
   (e) International Polar Year
5. SPARC Legacy and Outlook beyond 2013.

Nowadays, it has become clear that understanding basic climate features is impossible without consideration of the complexity of all interactive processes including the chemistry and dynamics of the atmosphere (Wang *et al.*, 1996), especially in connection with studying atmospheric greenhouse effect changes (Wuebbles, 1996). The spatiotemporal variability of optically active minor gas and aerosol components in the atmosphere controls radiative flux divergence (including the greenhouse effect) while, on the other hand, climate dynamics is a principal factor of minor component changes (the necessity to take into account biospheric processes makes the situation prohibitively complex).

An important step forward has been made in this context by Wang, C. *et al.* (1998) who studied climate–chemistry interactions and their uncertainty using a coupled global-scale modeling system consisting of a global climate model and an atmospheric chemistry model. This system was designed to predict, as functions of time, latitude, and altitude, the zonally averaged concentrations of the major chemically and radiatively important trace species in the atmosphere. The biogeochemical models and climate sub-models in this system are fully interactive. The nine-level 2-D statistical–dynamical climate sub-model has a horizontal resolution of 7.826°. The biogeochemical sub-model has 25 chemical species including CO_2, CH_4, N_2O, O_3, CO, H_2O, NO_x, HO_x, SO_2, sulfate aerosol, and chlorofluorocarbons. The model does not include non-methane hydrocarbon chemistry, which leads to underestimation of ozone production, especially in polluted areas. The continuity equations for $CFCl_3$, N_2O, O_3, CO, NO, CO_2, NO_2, N_2O_5, HNO_3, CH_4, CH_2O, SO_2, and H_2SO_4 include convergence due to transport (advection, eddy diffusion, and convection), whereas the remaining very reactive atoms, free radicals, or molecules are assumed to be unaffected by transport (photochemical steady states) due to their very short lifetimes.

There are 41 gas phase and 12 heterogeneous reactions in the chemistry model. The gas phase chemistry has three major parts: tropospheric O_3–HO_x–NO_x–CO–CH_4 reactions, tropospheric SO_2–sulfate reactions, and stratospheric chlofluorocarbon and N_2O removal reactions (Table 7.1). Dissolution of soluble

Table 7.1. Gaseous phase chemical reactions included in the model (Wang, C. *et al.*, 1998).

Number	*Reaction*
R1	$O_3 + h\nu \rightarrow O(^1D) + O_2$
R2	$O(^1D) + H_2O \rightarrow 2OH$
R3	$O(^1D) + N_2 \rightarrow O + N_2$
R4	$O(^1D) + O_2 \rightarrow O + O_2$
R5	$CO + OH \rightarrow H + CO_2$
R6	$H + O_2 + M \rightarrow HO_2 + M$
R7	$HO_2 + NO \rightarrow OH + NO_2$
R8	$NO_2 + h\nu \rightarrow NO + O$
R9	$O + O_2 + M \rightarrow O_3 + M$
R10	$HO_2 + O_3 \rightarrow OH + 2O_2$
R11	$OH + O_3 \rightarrow HO_2 + O_2$
R12	$NO + O_3 \rightarrow NO_2 + O_2$
R13	$NO_2 + OH + M \rightarrow HNO_3 + M$
R14	$NO_2 + O_3 \rightarrow NO_3 + O_2$
R15	$NO_3 + NO_2 + M \rightarrow N_2O_5 + O_2$
R16	$HO_2 + HO_2 \rightarrow H_2O_2 + O_2$
R17	$H_2O_2 + h\nu \rightarrow 2OH$
R18	$H_2O_2 + OH \rightarrow HO_2 + H_2O$
R19	$HO + HO_2 \rightarrow H_2O + O_2$
R20	$HO + HO \rightarrow H_2O + O$
R21	$HO + HO + M \rightarrow H_2O_2 + M$
R22	$CH_4 + OH \rightarrow CH_3 + H_2O$
R23	$CH_3 + O_2 + M \rightarrow CH_3O_2 + M$
R24	$CH_3O_3 + NO \rightarrow CH_3O + NO_2$
R25	$CH_3O + O_2 \rightarrow CH_2O + HO_2$
R26	$CH_3O_2 + HO_2 \rightarrow CH_3O_2H + O_2$

Number	*Reaction*
R27	$CH_3O_2H + h\nu \rightarrow CH_3O + OH$
R28	$CH_3O_2H + OH \rightarrow CH_3O_2 + H_2O$
R29	$CH_2O + h\nu \rightarrow CHO + H$
R30	$CH_2O + OH \rightarrow CHO + H_2O$
R31	$CHO + O_2 \rightarrow CO + HO_2$
R32	$SO_2 + OH + M \rightarrow HOSO_2 + M$
R33	$HOSO_2 + O_2 \rightarrow HO_2 + SO_3$
R34	$SO_3 + H_2O \rightarrow H_2SO_4$
R35	$CFCl_3 + O(^1D) \rightarrow$ products
R36	$CFCl_3 + h\nu \rightarrow$ products
R37	$CF_2Cl_2 + O(^1D) \rightarrow$ products
R38	$CF_2Cl_2 + h\nu \rightarrow$ products
R39	$N_2O + h\nu \rightarrow N_2 + O(^1D)$
R40	$N_2O + O(^1D) \rightarrow 2NO$
R41	$N_2O + O(^1D) \rightarrow N_2 + O_2$

gases in precipitating water is included in the wet deposition term in the model (Table 7.2). Emissions of CO_2, CH_4, N_2O, $CFCl_3$, CF_2Cl_2, CO, NO, and SO_2 are taken into account as functions of latitude. With the exception of two chlorofluorocarbons (CFCs), all emissions are separated into natural and anthropogenic emissions.

Predicted concentrations of chemical species in the chemistry sub-model are used interactively to calculate radiative forcing (RF) in the climate sub-model. The results obtained indicate that model predictions of the surface trends of several key species are close to observations over the past 10–20 years. The model provides reasonable simulations of seasonal variability and good agreement with observations in high latitudes. Predicted vertical distributions of climate-relevant species as well as seasonal variations are also in good agreement with observations. Wang, W.C. *et al.* (1998) emphasized that the differences involve not only higher concentrations of the major long-lived trace gases such as CO_2 (745 ppm), N_2O (410 ppb), and CH_4

Table 7.2. Aqueous phase chemical reactions included in the model (Wang, C. *et al.*, 1998).

Number	*Reaction*
R42	$H_2SO_4(g) \leftrightarrow H_2SO_3(aq)$
R43	$H_2SO_4(aq) \leftrightarrow HSO_4^- + H^+$
R44	$HNO_3(g) \leftrightarrow HNO_3(aq)$
R45	$HNO_3(aq) \leftrightarrow NO_3^- + H^+$
R46	$CH_2O(g) \leftrightarrow CH_2O(aq)$
R47	$SO_2(g) \leftrightarrow SO_2(aq)$
R48	$SO_2(aq) + H_2O \leftrightarrow HSO_3 + H^+$
R49	$HSO_3^- \leftrightarrow SO_2^- + OH^-$
R50	$H_2O_2(g) \leftrightarrow H_2O_2(aq)$
R51	$HO(g) \leftrightarrow HO(aq)$
R52	$HO_2(g) \leftrightarrow HO_2(aq)$
R53	$HO_2(aq) \leftrightarrow H^+ + O_2^-$

(4.4 ppm) but also about 20% lower concentrations of the major tropospheric oxidizer (OH free radical), and almost double the current concentrations of the short-lived air pollutants: CO and NO_x. A very important result is quantification of the two-way feedback between climate dynamics and atmospheric chemistry. Wang, W.C. *et al.* (1998) underlined, however, that there are still many uncertainties in the model and therefore further studies are necessary.

Ozone in the stratosphere and in the troposphere affects climate in various ways (Lübken *et al.*, 2010). In the stratosphere, ozone is a strong absorber of ultraviolet solar radiation and of upwelling thermal emission while, in the troposphere, ozone (although it absorbs ultraviolet radiation) plays a role as a greenhouse gas (Wang and Isaksen, 1995). During recent years substantial progress has been achieved in making assessments of the contributions made by the dynamics of the ozone content in the stratosphere and troposphere to climate change (Mahlman *et al.*, 1994; Shine, 1986; van Dorland and Fortuin, 1994; Varotsos, 2004b).

Brasseur *et al.* (2006) presented an assessment of how expected climate changes (2100 versus 2000) should affect the chemical composition of the troposphere by using a global chemistry–transport model of the atmosphere (the Model for Ozone and Related Tracers, version 2: MOZART-2) driven by prescribed surface emissions and by meteorological fields provided by the ECHAM5/Max Planck Institute Ocean Model (MPI-OM-1) coupled atmosphere–ocean model. One of the principal results obtained by Brasseur *et al.* (2006) was that ozone changes resulting from climate change alone are negative in a large fraction of the troposphere because of enhanced photochemical destruction by water vapor. In addition, Brasseur *et al.* (2006) suggested that predictions depend strongly on future trends in atmospheric methane levels, which are not well established. They claimed that changes in the emissions of NO_x by bacteria in soils and of non-methane hydrocarbons by vegetation associated with climate change could have a significant impact on future ozone levels. The effects of ozone depletion on global biogeochemical cycles, *via* increased UV-B radiation at the Earth's surface, constitute another important issue. Zepp *et al.* (2003) reported on the role of UV-B in global climate change, based on the detailed interactions between ozone depletion and climate change, which are central to the prediction and evaluation of future Earth environmental conditions. They pointed out that elevated UV-B has significant effects on the terrestrial biosphere with important implications for the cycling of carbon, nitrogen, and other elements. For example, increased UV has been shown to induce carbon monoxide production from dead plant matter in terrestrial ecosystems, nitrogen oxide production from Arctic and Antarctic snowpacks, and halogenated substances from several terrestrial ecosystems. In particular, Zepp *et al.* (2003) suggested that the fact that UV effects on the decomposition of dead leaf material confirm that these effects are complex. Decomposition can be retarded, accelerated, or remain unchanged. Zepp *et al.* (2003) emphasized that it has been difficult to relate the effects of UV on decomposition rates to leaf litter chemistry, as this is very variable. However, new evidence shows that UV has effects on some fungi, bacterial communities, and soil fauna that could play roles in decomposition and nutrient cycling. They focused on an important new result that not only is nitrogen cycling in soils significantly perturbed by increased UV-B, but that these effects persist for over a decade. As nitrogen cycling is temperature dependent, this finding clearly links the impacts of ozone depletion to the ability of plants to use nitrogen in a warming global environment. There are many other potential interactions between UV and climate change impacts on terrestrial biogeochemical cycles that remain to be quantified.

Zepp *et al.* (2003) also discussed the fact that UV-B strongly influences aquatic carbon, nitrogen, sulfur, and metal cycling that affects a wide range of life processes. In this context, UV-B accelerates the decomposition of colored dissolved organic matter (CDOM) entering the sea *via* terrestrial run-off, thus having important effects on oceanic carbon cycle dynamics. Since UV-B influences the distribution of CDOM, there is an impact of UV-B on estimates of oceanic productivity based on remote sensing of ocean color. Thus, oceanic productivity estimates based on remote sensing require estimates of CDOM distributions. In this regard UV-B transforms dissolved organic matter into dissolved inorganic carbon and nitrogen, including carbon dioxide and ammonium, and into organic substances that are either more or less readily available to micro-organisms. The extent of these transformations is correlated with the loss of UV absorbance by organic matter. Changes in aquatic primary productivity and decomposition due to climate-related changes in circulation and nutrient supply, which occur concurrently with increased UV-B exposure, have synergistic influences on the penetration of light into aquatic ecosystems. New research has confirmed that UV affects the biological availability of iron, copper, and other trace metals in aquatic environments, thus potentially affecting the growth of phytoplankton and other micro-organisms that are involved in carbon and nitrogen cycling. Finally, Zepp *et al.* (2003) pointed out that there are several instances where UV-B modifies the air–sea exchange of trace gases which in turn alters atmospheric chemistry, including the carbon cycle.

Calculations have shown (WMO, 1995, 1999) that the total ozone content decrease observed in the 1980s and 1990s ought to have determined the decrease of RF for the surface–atmosphere system in middle and high latitudes. Being globally averaged over 10 years, such a decrease turns out to be similar in magnitude, but opposite in sign, to the RF enhancement due to CFC concentration growth during the same time period. Since regional inhomogeneities of the RF distribution are substantial in this case (especially in high latitudes), specification of these estimates based on the use of global 3-D models is very urgent (Raes *et al.*, 2010).

Assessment of the possible effect of anthropogenic sulfate aerosol in the Northern Hemisphere (in the Southern Hemisphere the effect is negligibly small) under clear-sky conditions has led to a cooling estimate of about $1\,\mathrm{W\,m^{-2}}$ (on average, for the Northern Hemisphere), which should be compared to a greenhouse warming of about $2.5\,\mathrm{W\,m^{-2}}$ for the last century. No doubt, estimates of the climate-forcing contribution of aerosol are very preliminary and require further verification, the more so because the more important role of aerosol, as a factor affecting the radiative properties of clouds, may be substantial (Kondratyev, 1999a, b; Perlwitz and Miller, 2010).

7.2 RADIATIVE-FORCING CALCULATIONS

Most of the climatic assessments of ozone impacts have been concentrated on radiative forcing (RF) calculations (e.g., Hansen *et al.*, 1997; Kondratyev, 1998b; Portmann *et al.*, 1997; Solomon and Daniel, 1996; SPARC, 1998; Stevenson *et al.*, 1998; Zwick, 1998).

New assessments of the RF caused by decreasing ozone content in the lower stratosphere at middle and high latitudes suggested the the possibility of considerable negative RF (cooling) for the surface–troposphere system, contrary to positive greenhouse RF due to the increasing amount of CFCs. However, initial calculations of anthropogenic climate changes did not take into account recent growth in ozone content in the troposphere, which should cause greenhouse warming.

7.2.1 Estimates of changes in RF from pre-industrial times to the present

According to Shindell *et al.* (2003a, b) comparison with pre-industrial (PI) simulations gives a global annual average RF due to tropospheric ozone increases of $0.30\,\mathrm{W\,m^{-2}}$ with standard assumptions for PI emissions. Locally, forcing reaches more than 0.8 W m^{-2} in parts of the northern subtropics during spring and summer, and is more than 0.6 W m^{-2} through nearly all the northern subtropics and mid-latitudes during summer. An alternative PI simulation with soil NO_x emissions reduced by two-thirds and emissions of isoprene, paraffins, and alkenes from vegetation increased by 50% gives a forcing of 0.33 W m^{-2}. Given the large uncertainties in PI ozone amounts, the true value may lie well outside this range. Of course, regional variations in forcing due to inhomogeneous gases such as tropospheric ozone are much larger than that for long-lived GHGs, so that this comparison only tells part of the story. The results suggest that alternative scenarios for slowing global warming via controls on tropospheric ozone and methane (Hansen *et al.*, 2000) may be quite useful (though the relative importance of CO_2 is likely to increase in the 21st century).

Simulations of PI to present day (PD) changes performed by Shindell *et al.* (2006) show tropospheric ozone burden increases of 11% while the stratospheric burden decreases by 18%. The resulting tropopause RF values are 0.06 W m^{-2} from stratospheric ozone and 0.40 W m^{-2} from tropospheric ozone. Global mean mass-weighted OH decreases by 16% from the PI to the PD. The stratosphere–troposphere exchange (STE) of ozone also decreased substantially during this time, by 14%. Comparison of the PD with a simulation using 1979 pre-ozone hole conditions for the stratosphere shows a much larger downward flux of ozone into the troposphere in 1979, resulting in a substantially greater tropospheric ozone burden than that seen in the PD run. This implies that reduced STE due to Antarctic ozone depletion may have offset as much as 2/3 of the tropospheric ozone burden increase from PI to PD. However, the model overestimates the downward flux of ozone at high southern latitudes, so this estimate is likely an upper limit. According to Shindell *et al.* (2006) projected future ozone recovery leads to an enhanced influx of ozone from the stratosphere, again with potentially large impacts on tropospheric ozone burden. As the influx is confined to middle to high latitudes, however, the impact on RF is not large, nor do the STE changes dramatically affect surface pollution. These will continue to be governed primarily by emissions, which dominate over the effects of climate change on tropospheric pollution at low levels globally and are important at all levels in the tropics.

The calculated pre-industrial global fields of monthly mean ozone concentration and TOC for June 1850 are presented in Figure 7.1–7.3 (Berntsen *et al.*, 1997). Under background conditions two maxima in surface ozone concentrations of about 15 ppb are located on the west coasts of continents in the tropics due to *in situ* production over continents and easterly flow in the trade winds. The third maximum over northern India and the Himalayas was caused by a combination of downward mixing of stratospheric air (probably due to the influence of the mountain range on the dynamic process in the GCM) and *in situ* production. In the upper troposphere, ozone concentrations are generally low (5–30 ppb) in the tropics (in particular, over the Pacific). Although net chemical ozone production is negative at higher latitudes, ozone concentrations increase towards the poles in both hemispheres because of stratosphere–troposphere exchange (Pan *et al.*, 2010).

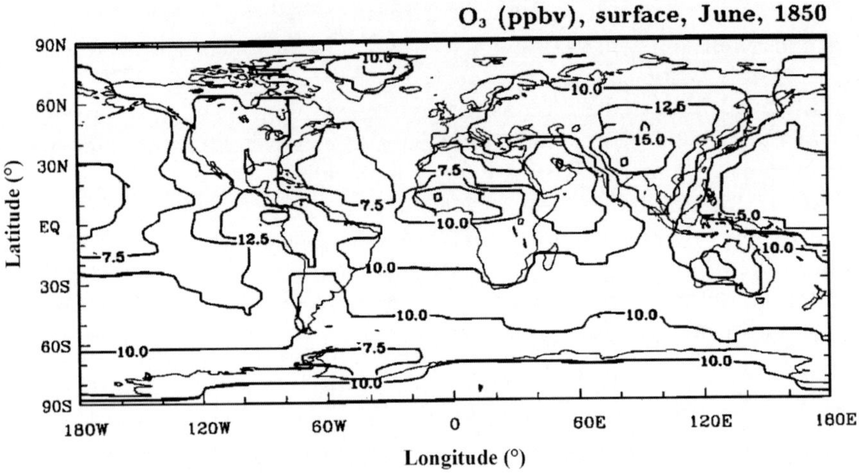

Figure 7.1. Calculated monthly concentration of ozone in the surface layer (i.e., surface to ≈ 500 m) for June 1850. Isolines are at 5, 7.5, 10, 12.5, 15, and 20 ppbv (Berntsen *et al.*, 1997).

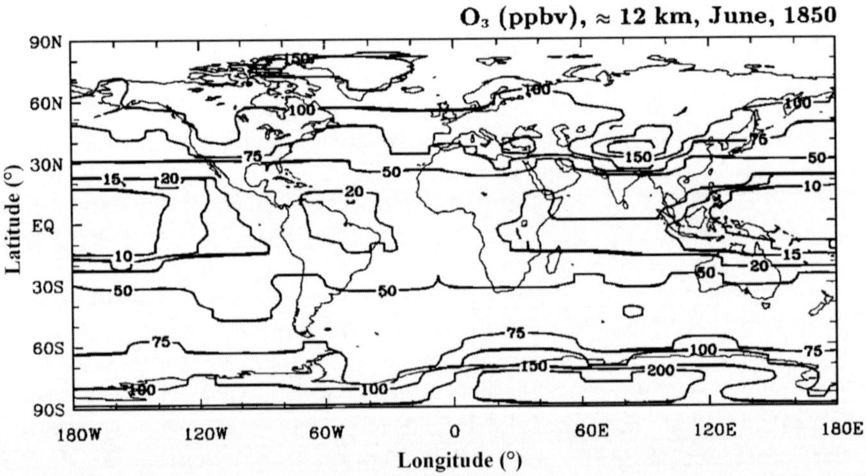

Figure 7.2. Calculated monthly concentration of ozone in the upper troposphere ($L = 7$, $\approx$10.3–13.7 km) for June 1850. Isolines are at 10, 15, 20, 50, 75, 100, 150, and 200 ppbv (Berntsen *et al.*, 1997).

The increase in CH_4, CO, and NO_x concentrations since pre-industrial times led to increased ozone production in the CTM. Table 7.3 illustrates the ozone budget in 1850 and 1990. The year 1850 is very close to the beginning of the industrial era. Calculations indicate a positive net chemical ozone production in the atmospheric boundary layer (ABL) and in the upper troposphere (UT), but a net chemical loss in the middle troposphere (MT, 1.4–7.4 km) because of the low NO_x concentration in

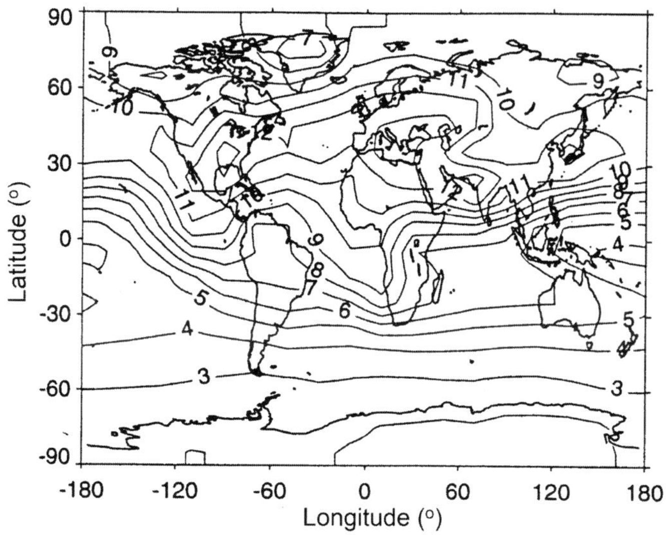

Figure 7.3. Calculated change in annual mean tropospheric ozone in Dobson units between 1850 and 1990 (Berntsen *et al.*, 1997).

this altitude region. Berntsen *et al.* (1997) underlined, however, the preliminary nature of these assessments, since a number of processes have not been taken into account (including heterogeneous processes on cirrus cloud ice particles and carbonaceous aerosols). Nevertheless, a comparison with ozonesonde data indicates, in general, satisfactory results. The TOC changes considered between 1850 and 1990 show a clear increase due to ABL ozone over industrialized regions in the Northern Hemisphere.

Radiative transfer calculations were made by Berntsen *et al.* (1997), taking into account the effects of clouds and thermal adjustment in the stratosphere. Calculations indicate that global and annual-averaged RF at the tropopause are equal to $0.28\,W\,m^{-2}$ (Reading model) and $0.31\,W\,m^{-2}$ (OsloRad model) which is in the lower part of the IPCC (Houghton *et al.*, 1996) estimated range and coincides with the assessment made by van Dorland *et al.* (1997). The latter conclusion is also true of comparison with the negative RF by sulfate aerosol (which is similar in magnitude to ozone RF, but displaced southward in source regions at northern mid-latitudes). Tropospheric ozone increase resulted in cooling the lower stratosphere by up to 0.9 K, with possibly half of this cooling occurring in the past two to three decades.

Within ACCENT, a European Network of Excellence, ozone changes between 1850 and 2000 were assessed for the troposphere and the lower stratosphere (up to 30 km) by a variety of seven chemistry–climate models and three chemistry–transport models. Gauss *et al.* (2006) reported on ACCENT stating that changes in atmospheric ozone have occurred since the pre-industrial era as a result of increasing anthropogenic emissions. In particular, modeled ozone changes were

Table 7.3. Net chemical production of ozone (PO_3) in the planetary boundary layer (PBL) (Tg/6 months) and the upper troposphere (UT) and net upward flux of ozone from the PBL (Berntsen *et al.*, 1997).

Latitude	*Net PO_3 in PBL*[a]		*Net PO_3 in UT*[b]		*Net upward flux from PBL*[c]	
	Summer	*Winter*	*Summer*	*Winter*	*Summer*	*Winter*
			1850 Case			
90°–32°N	49	9.9	4.9	−0.17	−56	−62
32°–0°N	45	32	82	53	−8.7	−42
0°–32°S	41	44	65	48	−27	−60
32°–90°S	0.12	1.9	0.74	0.74	−30	−44
			1990 Case			
90°–32°N	176	25	83	21	−95	−114
32°–0°N	149	154	189	152	31	−26
0°–32°S	75	106	126	110	−32	−75
32°–90°S	6.9	5.5	7.2	3.8	−37	−55

Northern Hemisphere summer and Southern Hemisphere winter are defined as April–September, and Northern Hemisphere winter and Southern Hemisphere summer are defined as October–March.
[a] Model layers 1 and 2; <1.5 km altitude.
[b] Model layers 6 and above; >7 km altitude.
[c] Through top of layer 2.

taken as input for detailed calculations of RF. When consideration is restricted to changes in chemistry (constant climate) the modeled global mean tropospheric ozone column increase since pre-industrial times ranges from 7.9 DU to 13.8 DU among the 10 participating models, while stratospheric column reduction lies between 14.1 DU and 47.9 DU in models considering stratospheric chemistry. According to their findings, the resulting RF is strongly dependent on the location and altitude of the modeled ozone change and varies between 0.26 W m^{-2} and 0.53 W m^{-2} due to ozone change in the troposphere and −0.25 W m^{-2} and +0.12 W m^{-2} due to stratospheric ozone change. In addition, they found that changes in ozone and other greenhouse gases since pre-industrial times have altered climate. Six out of the ten participating models performed an additional calculation that took both chemical and climate change into account. The isolated effect of climate change results in enhancement of the tropospheric ozone column in all models, ranging from 1% to 37%, while stratospheric reduction becomes slightly less severe in most models. In the three chemistry–climate models with detailed tropospheric and stratospheric chemistry the inclusion of climate change increases the resulting RF due to tropospheric ozone change by up to 0.08 $W m^{-2}$, while RF due to stratospheric ozone change is reduced by up to 0.14 $W m^{-2}$. Considering tropospheric and stratospheric change combined, the total ozone column change is negative while the resulting net RF is positive.

7.2.2 Detailed studies of changes in RF in recent decades

According to Hansen *et al.* (1997) there has been net negative RF from all ozone changes, since 1979, which is larger in magnitude than changes in all the non-CO_2 long-lived gases over the same period. This means that future ozone layer recovery in the coming decades will be accompanied by enhancement of GHG warming by greenhouse gases (GHGs). For the time being, it is assumed that the observed cooling of the stratosphere is mostly a result of ozone depletion (SPARC, 1998).

The observed downward trend of global ozone since 1979 began to level off in the mid-to-late 1990s in response to stratospheric chlorine amounts regularizing and at some latitudes has shown signs of recovery. As the ozone layer "recovers", it is important to understand how temperatures in the lower stratospheric (LS) will respond, as these temperatures are related to RF from the troposphere and global warming at the surface. S. Zhou and C. Long, in an unpublished paper presented at the 2008 Quadrennial Ozone Symposium in Tromsø, reported on the relationship between lower-stratospheric temperatures and total ozone variations based on nearly 30 years of NCEP Reanalysis-2 and SBUV(/2) Version 8 ozone data (1979–2007). From this study it was found that the global cooling trend in the LS has been halted since the beginning of the 21st century. But, while extra-tropical temperature trends have leveled off, the tropics continue to show a cooling trend. These different latitudinal characteristics of LS temperature were diagnosed in terms of various factors including dynamics, ozone, and other greenhouse gases.

The impact of increasing GHGs on the recovery of stratospheric ozone was examined by Waugh *et al.* (2009) using simulations of the Goddard Earth Observing System Chemistry–Climate Model. In this model, GHG-induced climate change has a large impact on ozone evolution and when O_3 recovery milestones are reached. The two distinct milestones of "O_3 returning to historical values" and "O_3 being no longer significantly influenced by ozone depleting substances (ODSs)" can be reached at very different dates, and which occurs first varies between regions. GHG-induced cooling in the upper stratosphere causes O_3 to increase and O_3 returns to 1980 or 1960 values several decades before O_3 is no longer significantly influenced by ODSs. In contrast, transport changes in the tropical and southern mid-latitude lower stratosphere cause O_3 to decrease. Here O_3 never returns to 1980 values, even when anthropogenic ODSs have been removed from the atmosphere. O_3 returning to 1960 (or 1980) values should not necessarily be interpreted as O_3 recovery from the effects of ODSs.

The recovery of stratospheric ozone as ozone-depleting substances decline through the 21st century can be expected to be affected by climate change. T. G. Shepherd in an unpublished paper presented at the 2008 Quadrennial Ozone

Symposium in Tromsø reported on three simulations with the Canadian Middle Atmoshere Model (CMAM) from 1950 to 2100 to investigate the possibility of a circulation change (i.e., change in the Brewer–Dobson circulation) which would affect the ozone distribution as well as tropopause height. The finding was that the recovery of global ozone essentially follows chlorine loading, suggesting that there is no "super-recovery" in global ozone. However, strengthened Brewer–Dobson circulation (seen also in a decrease in the age of air) most notably in the Northern Hemisphere was noted. This leads to a persistent decrease in tropical ozone of about 2%, and to a dynamical "super-recovery" of about 4% in northern mid-latitudes and about 8% in the Arctic. The changes occur principally in the lowermost stratosphere and appear to be associated with changes in subtropical wave drag in the lower stratosphere. Tropical changes were consistent with recent observations of ozone and temperature decreases just above the tropopause.

When considering the effect of the observed ozone changes on radiative forcing (RF) and climate, it is of course necessary to take into account both the decrease of the ozone content in the stratosphere and its increase in the troposphere. Moreover, specific features of the vertical ozone profile should be taken into account. In this connection, Wang *et al.* (1993) analyzed the results of observations of the vertical ozone profile at seven stations at middle and high latitudes (from Tateno, 36°N, to Resolute, 75°N) to study ozone variations that have taken place during the last few decades and to assess climate changes in the context of increased concentrations of other GHGs: CO_2, CH_4, CFC-11, CFC-12, and N_2O.

The results obtained by Wang *et al.* (1993) show that observed changes in the vertical ozone profile (especially in the lower stratosphere and upper troposphere) can markedly affect the climate (see also Hassler *et al.*, 2009). While the causes of the ozone decrease in the lower stratosphere are known and connected with increasing concentrations of CFCs, there are still many uncertainties with regard to the increasing trend of tropospheric ozone at Northern Hemisphere middle and high latitudes (between 2 and 2.5% per year in the spring and autumn troposphere), although, it is probably determined by the effect of such ozone precursors as CO, NO_x, and hydrocarbons (much effort is needed to solve this problem, especially bearing in mind the importance of forecasts of tropospheric ozone trends).

Table 7.4 shows the results of calculations of RF due to short-wave (SW) and long-wave (LW) radiation in the surface–atmosphere system, determined by changes in the ozone content over Hohenpeissenberg (47°48′N, 11°0′E) and Payern (46.82°N, 6.94°E) for the period 1971–1980 to 1981–1990. Calculations were made with the use of two techniques: at a fixed air temperature (FT) and by means of prescribed fixed dynamic (FD) heating (which takes account of the temperature profile).

Comparison of the data for the troposphere and stratosphere makes it possible to analyze the effect of the vertical ozone profile on the formation of RF. As can be seen, the increase in ozone content in the troposphere determines positive RF for both SW and LW radiation in July, with the troposphere contributing most to the formation of total RF.

Table 7.4. Radiative forcing (W m^{-2}) due to changes in ozone content for the period 1972–1990 (Kondratyev and Varotsos, 1995b).

		Stratosphere		*Troposphere*		*Total*	
Month	*Heating*	*SW*	*LW*	*SW*	*LW*	*SW*	*LW*
			Hohenpeissenberg				
January	FT	0.207	−0.019	0.062	0.160	0.270	0.140
	FD	0.207	−0.103	0.062	0.074	0.270	0.057
July	FT	0.084	−0.004	0.098	0.392	0.182	0.388
	FD	0.084	−0.026	0.098	0.369	0.182	0.366
			Payern				
January	FT	0.141	−0.027	0.056	0.186	0.198	0.158
	FD	0.141	−0.193	0.056	0.018	0.198	−0.008
July	FT	0.346	−0.045	0.027	0.07	0.373	0.024
	FD	0.346	−0.306	0.027	−0.194	0.373	−0.237

FT stands for fixed air temperature, FD stands for fixed dynamic heating.

Table 7.5 contains estimates characterizing the relative contributions of ozone and other GHGs to RF. The data in this table, again, demonstrate the marked contribution of tropospheric ozone dynamics to the formation of RF.

The strong sensitivity of ozone-induced RF to lower-stratospheric temperature points to the need for studies of climatic consequences of ozone content variations based on the use of atmospheric general circulation models that adequately account for the radiation–dynamics interaction.

Schwartzkopf and Ramaswamy (1993) pointed out that observations from satelliteborne and ground-based instruments indicate that significant changes in atmospheric ozone concentrations took place during the 1980s. Decreases in total ozone in middle and high latitudes were measured by the TOMS instrument, while Stratospheric Aerosol and Gas Experiment (SAGE) data indicate ozone depletion in the lower stratosphere at all latitudes, including the tropics. On the other hand, observations at several locations in the middle latitudes of the Northern Hemisphere have shown significant decadal increases in tropospheric ozone. Earlier calculations led to the conclusion that the decadal changes in lower-stratospheric ozone produce substantial negative RF of the surface–troposphere system at middle and high latitudes. The magnitude of ozone forcing is sensitive to both total ozone loss and the altitude profile of ozone depletion (see also Sections 4.6.2 and 5.7).

Observations from satellites, ozonesondes, and total column ozone measurements have provided strong evidence that downward trends in stratospheric ozone

Table 7.5. The relative roles of contributions of ozone and other greenhouse gases to the formation of RF ($W\,m^{-2}$) (Kondratyev and Varotsos, 1995b).

	Changes in ozone content				*Combined effect of ozone and other greenhouse gases*	
	Total		*Troposphere*			
Season	*FT*	*FD*	*FT*	*FD*	*FT*	*FD*
			Hohenpeissenberg			
January	0.410	0.327	0.222	0.136	0.794	0.710
July	0.570	0.548	0.490	0.467	1.050	1.027
			Payern			
January	0.356	0.190	0.242	0.074	0.797	0.630
July	0.397	0.136	0.097	−0.167	0.879	0.617
			Resolute			
Winter	−0.038	−0.085	0.012	−0.036	0.230	0.183
Summer	0.194	−0.009	0.052	−0.153	0.620	0.415
			Goose Bay			
Winter	0.025	−0.066	0.013	0.012	0.379	0.288
Summer	0.043	−0.218	−0.153	−0.415	0.497	0.235
			Churchill			
Winter	0.013	−0.086	−0.022	−0.122	0.209	0.110
Summer	−0.097	−0.425	−0.507	−0.839	0.155	−0.174
			Edmonton			
Winter	0.080	−0.069	−0.008	−0.158	0.275	0.127
Summer	0.110	−0.002	0.078	−0.031	0.361	0.252
			Tateno			
Winter	0.239	0.096	0.084	−0.060	0.491	0.348
Summer	0.304	0.287	0.295	0.279	0.585	0.569

have decreased and, in some cases, have flattened out, starting towards the end of the 1990s. At mid-latitudes in the upper stratosphere and in the 18–25 km altitude region the decreases have been shown to be consistent with reductions in equivalent effective stratospheric chlorine (EESC) levels. From the tropopause to 18 km altitude ozone has increased slightly and numerical models have approximately simulated ozone change by changes in atmospheric transport rates. Moreover, ozone is no longer decreasing in the Antarctic springtime ozone hole region. It has been suggested that this is because of ozone loss saturation effects (i.e., virtually all the ozone over the altitude range of approximately 14–21 km in spring has been destroyed—see Figure 5.4). However, a recent study has shown that—after the effects of loss saturation and year-to-year dynamically induced changes in the vortex have been accounted for—the behavior of the underlying ozone is approximately consistent with changes in EESC.

Schwartzkopf and Ramaswamy (1993) undertook new estimations of the sensitivity of RF to ozone changes with the use of SAGE I and II data on ozone vertical profiles up to 17 km (below this level the presence of clouds complicates remote sounding) as well as data on the increase of tropospheric ozone.

The following parameters were fixed: vertical temperature and humidity profiles, cloudiness, surface temperature, and albedo. Three scenarios for annual-averaged profiles within the 17–50 km layer for decadal ozone losses in tropical (4.5°N) and middle (40.5°N) latitudes (similar in each latitude belt) were chosen:

i. No ozone depletion below 17 km (S1);
ii. A linear decrease with altitude of the percentage of ozone depletion between 17 km and the tropopause, where the depletion becomes zero (S2);
iii. The depletion percentage between the tropopause and 17 km is equal to the SAGE depletion percentage at 17 km (S3).

An additional standard profile (S4) was also used which employs percentage ozone depletion adjusted so that TOC is equal to that obtained using profile S1. Tables 7.6 and 7.7 illustrate some of the results obtained by Schwartzkopf and Ramaswamy (1993). These results show that the amount of stratospheric ozone RF during the 1980s depends principally on two factors: total ozone content depletion and the altitude profile of ozone change. In the tropics, where stratospheric ozone change profiles are available almost to the troposphere, the principal question is the quantity of lower-stratospheric ozone loss. An important conclusion is that the

Table 7.6. Decadal stratospheric column ozone change in DU, stratospheric ozone forcing in $W m^{-2}$, and stratospheric RF gradient (SRFG) in $W m^{-2}$ $(DU)^{-1}$ for standard tropical (January, 4.5°N) profiles. The SRFG derived here is valid over a stratospheric column ozone decrease of 0–10 DU (Schwartzkopf and Ramaswamy, 1993).

Profile	*S1*	*S2*	*S3*	*S4*
Column change	−7.34	−8.07	−8.79	−7.34
Forcing	−0.052	−0.071	−0.090	−0.052
SRFG	0.0071	0.0087	0.0102	0.0071

Table 7.7. Same as Table 7.6, except for the mid-latitudes (January, 40.5°N). The SRFG derived here is valid over stratospheric total ozone content decrease of 0–30 DU (Schwartzkopf and Ramaswamy, 1993).

Profile	*S1*	*S2*	*S3*	*S4*
Column change	−10.38	−15.55	−20.72	−10.38
Forcing	−0.035	−0.084	−0.134	−0.078
SRFG	0.0034	0.0054	0.0065	0.0075

magnitude of decadal Northern Hemisphere stratospheric ozone RF would increase significantly if ozone depletion occurred in the tropics, as implied by SAGE observations (see also Hegglin and Shepherd, 2009).

In mid-latitudes, stratospheric ozone forcing depends strongly on the altitude profile of decadal ozone change in the mid-latitude lower stratosphere, especially in the layers nearest the tropopause (accurate measurements of ozone change in the vicinity of the tropopause are necessary to arrive at a precise estimate of stratospheric ozone RF). If the observed increase of tropospheric ozone in various regions of the Northern Hemisphere are hemispherically representative then one can conclude that substantial positive tropospheric ozone RF occurred during the 1980s, of magnitude possibly comparable to that produced by decadal stratospheric ozone decreases. Although the results considered require further confirmation (uncertainty in both the magnitude and the sign of decadal total ozone forcing is rather high), it may be concluded that changes of total ozone content and vertical ozone profiles (both in the stratosphere and troposphere) belong to significant climate-forming processes. This is why various radiative transfer calculations have been made during recent years to assess ozone-induced RF variability (see, e.g., Berntsen *et al.*, 1997; Deser and Phillips, 2009; Kondratyev, 1998b; Lelieveld and van Dorland, 1995; Roelofs *et al.*, 1997; van Dorland *et al.*, 1997; Wang and Isaksen, 1995; Wang and Sze, 1980, Wang *et al.*, 1993, 1998), which have been based on various assumptions.

An important aspect of the climatic impact of ozone is its interactive nature with regard to the contribution of other greenhouse gases (Kondratyev, 1998b). To analyze the stratospheric contribution to such interactivity Schneider *et al.* (1993) assessed the response of stratospheric ozone to increased levels of CO, N_2O, and CFCs (a similar analysis is of course needed for tropospheric ozone). The computations show that CO_2 doubling leads to a 3.1% increase in global ozone content, which results in an associated increase in solar heating, reducing the maximum temperature drop (due to CO_2 doubling) by about 3°C. The CFC increase from continuous emissions at the 1985 rate causes a 4.5% loss of ozone. For the combined perturbation (due to CO_2 doubling and CFC increase) a net loss of total ozone of 1.3% has been calculated. All ozone perturbations calculated by Schneider *et al.* (1993) exhibit an asymmetry between Northern and Southern Hemispheres, with

Table 7.8. Atmospheric composition changes. In the CH_4 and N_2O columns are reported the factors by which the reference mixing ratios have been multiplied. In Cases B and C, the feedback (FB) referred to is temperature feedback introduced by ozone (Nicoli and Visconti, 1982).

Version	CO_2 (ppm)	Cl_y (ppb)	CH_4	N_2O
A	320	2.3	1	1
B	320	10 (FB)	1	1
C	320	10 (no FB)	1	1
D	640	2.3	1	1
E	640	10	1	1
F	320	2.3	2	1
G	320	2.3	1	2
H	640	10	1	2
I	640	10	2	2

relative ozone losses being larger in the high latitudes of the Southern Hemisphere because of the larger eddy mixing and larger horizontal advection than is the case in the northern hemisphere.

It became abundantly clear at an early stage of studying the ozone impact on climate that climate changes, GHGs, and ozone are highly coupled. This was demonstrated by Nicoli and Visconti (1982). who applied a 1-D radiative–convective–photochemical model of the atmosphere (0–70 km) that took account of 60 chemical and 25 photochemical reactions to simulate the impact of N_2O, CH_4, CO_2, and Cl_y emissions to the atmosphere on TOC and surface air temperature (SAT). Calculations were made for a cloudless atmosphere and solar zenith angle of 60°. Table 7.8 characterizes various versions of prescribed trace gas dynamics where "1" or "2" stand for nominal or doubled concentration values: FB and NFB (with and without taking account of feedback, respectively). Table 7.9 illustrates the results of numerical modeling: SAT change (K) and TOC changes (%).

CO_2 doubling leads to tropospheric warming and stratospheric cooling of the Cl_y mixing ratio by 10 ppb results in TOC depletion equal to 9% (NFB) or 10.9% (FB). Nitrous oxide doubling leads to ozone depletion and weak warming, whereas the result of CH_4 doubling leads to TOC increase and climate warming.

Methane emissions weaken the Cl_y impact on ozone due to Cl_y-induced HCl formation. Enhancement of the greenhouse effect due to combined N_2O and CH_4 emissions lead to surface air temperature increase of about 2.3 K (instead of 1.67 K in the case of CO_2 doubling). When the simple additive effect of the four components

Table 7.9. SAT and TOC changes (Nicoli and Visconti, 1982).

Version	*SAT changes* (K)	*Total ozone changes* (%)
B		−9.3
C	No FB	−10.9
D	1.88	2.2
E	1.67	−9
F	0.27	1.7
G	0.35	−2.7
H	2.1	−9.7
I	2.3	−4.7

is considered TOC depletion is 11.9%, while interactive consideration of their impacts results in only 4.7% ozone depletion (Table 7.9).

Increasing atmospheric CH_4 concentration results in not only greenhouse effect enhancement, but also tends to increase two radiatively–chemically important gases: tropospheric O_3 through enhanced tropospheric chemical activities and stratospheric H_2O due to stratospheric CH_4 oxidation. These increases lead to RF increase and thus enhance the CH_4 future warming effect.

To examine the indirect effects of the increase of atmospheric CH_4 induced by climate–chemistry interaction Wang, W.C. *et al.* (1998) conducted model sensitivity experiments. A concrete aim of modeling has been to analyze the climate implications of changes in tropospheric O_3 and CH_4, when increases in moisture and temperature caused by increasing GHGs (CO_2, N_2O, CH_4, and CFCs) are considered.

A principal conclusion made by Wang, W.C. *et al.* (1998) is that tropospheric O_3 can substantially increase in response to CH_4 increase, which is, however, in sharp contrast to O_3 decrease, when increases in moisture and temperature are included. In addition, CH_4 increase can be substantially reduced when enhanced OH associated with global warming is taken into account. Wang, W.C., *et al.* (1998) predicted chemical composition changes (shown in Table 7.10) to calculate relevant RF changes (Table 7.11). The values of tropospheric O_3 and stratospheric H_2O in the last two columns of Table 7.10 are annual and global mean changes; Cases I and II correspond to situations without and with global warming, respectively.

As can be seen from Table 7.11, calculated annual and global mean total RF without global warming (Case I) is 3.92 $W\,m^{-2}$ with CO_2 increase contributing 2.59 $W\,m^{-2}$ while CH_4 accounts for 0.65 $W\,m^{-2}$. The contributions of increases of tropospheric O_3 and stratospheric H_2O associated with CH_4 increase are 0.19 and 0.03 $W\,m^{-2}$, respectively. Numerical modeling has shown that the effect of global

Table 7.10. Minor gas concentrations used in RF calculations (Wang, W.C. *et al.*, 1998).

Year	*CO_2* (ppmv)	*N_2O* (ppmv)	*$CFCl_3$* (ppbv)	*CF_2Cl_2* (ppbv)	*Cases*	*CH_4* (ppmv)	*Troposphere mean O_3* (%)	*Stratosphere H_2O* (ppmv)
1990	354	0.31	0.28	0.48		1.72	—	—
2050	539	0.37	0.55	1.07	I	3.24	9.6	0–3
					II	2.69	−1.4	0–2

Table 7.11. Global and annual RF ($W\,m^{-2}$) (Wang, W.C. *et al.*, 1998).

Case	*CO_2*	*N_2O*	*$CFCl_3$, CF_2Cl_2*	*CH_4*	*Troposphere O_3*	*Stratosphere H_2O*	*Total*
I	2.59	0.19	0.28	0.6	0.19	0.030	3.92
II	2.59	0.19	0.28	0.44	−0.04	0.019	3.47

warming (Case II) is to decrease zonal mean forcing in all latitudes with the maximum value of about $0.5\,W\,m^{-2}$ at low latitudes. Annual and global mean total RF (Table 7.11) decreases to $3.47\,W\,m^{-2}$ (an 11.5% reduction) with CH_4, tropospheric O_3, and stratospheric H_2O contributing 0.44, −0.04, and $0.02\,W\,m^{-2}$, respectively. There are much larger latitudinal effects, however. For example, RF due to O_3 changes in the low latitudes decreases from a positive $0.2\,W\,m^{-2}$ to a negative $0.1\,W\,m^{-2}$.

Wang, W.C. *et al.* (1998) emphasized that the results considered indicate the importance of including climatic changes in tropospheric chemistry in studying CH_4–O_3 coupling. Undoubtedly, further studies with 3-D models are necessary.

Another numerical 2-D zonally averaged interactive dynamical radiative–photochemical model of the atmosphere including aerosol physics was used by Dyominov and Zadorozhny (2008) to investigate mechanisms of the impact of anthropogenic discharges of greenhouse gases on expected recovery of the Earth's ozone layer. A forecast of changes to global ozone abundance in the Earth's atmosphere caused by continuing anthropogenic discharges in the atmosphere of greenhouse gases (CO_2, CH_4, and N_2O) and ozone-depleting gases was made for the period from 1980 to 2050. Model calculations showed that expected cooling of the stratosphere caused by increases in greenhouse gases, most importantly CO_2, essentially influences the ozone layer through the temperature dependence of gas phase reaction rates and through enhancement of polar ozone depletion via increased PSC formation. The influence of greenhouse gases on the ozone layer through the temperature dependence of gas phase reaction rates was shown to be the main, extremely important mechanism that determines the influence of greenhouse gases on the speed of recovery of the global ozone abundance in the Earth's atmosphere. The impact of greenhouse gases on global ozone through the modification of polar stratospheric clouds makes an insignificant contribution to the dynamics of recovery of global ozone. This mechanism begins to be more effective in comparison with the first one in about 2020 and only at northern polar latitudes and only for a limited period from the end of February to the beginning of the April.

Rosenfield and Schoeberl (1986) made calculations of radiative flux divergence (RFD) in the stratosphere and RFD-induced atmospheric circulation using observed temperature and ozone fields in the Southern Hemisphere for September–October periods of 1979, 1980, and 1984. The principal purpose of numerical modeling was to analyze limitations imposed by the stratospheric radiative regime at the end of winter/beginning of spring on the formation of atmospheric circulation due to RFD sensitivity to temperature and ozone variations. Their results indicate that: (1) during September–October there was an enhancement of total radiative warming of the lower stratosphere near the South Pole ($RFD > 0$); (2) an interannual depletion trend of total radiative cooling in the lower stratosphere at mid and subpolar latitudes took place due to a decrease in mean zonal temperature; (3) temperature changes because of observed ozone variations were highly improbable; (4) RFD-induced atmospheric circulation in the lower stratosphere in spring reveals a trend toward upward motion enhancement near the pole in 1979–1980 but a weakening of downward motions in mid and subpolar latitudes in 1979–1984 (this indicates annual enhancement of the upward branch of the circulation cell near the pole and the downward branch in middle latitudes in September and October (this conclusion agrees with observed TOC changes); (5) assessments of mass circulation show that for there to be upward motions near the pole there has to be an additional source of stratospheric warming that leads to a temperature increase of about 0.5 K/day (this source should have been more powerful in 1984 than in 1980).

Forster and Shine (1997) made new calculations of RF with improved radiative transfer schemes and temperature trends from stratospheric ozone changes (Nimbus-7 SBUV and SAGE I, II data). The results are shown in Table 7.12. These results clearly indicate the importance of ozone-induced RF.

Van Dorland *et al.* (1997) carried out numerical modeling of the global distribution of RF due to tropospheric ozone and sulfate aerosols. The 3-D concentration field for both components and their precursors were calculated on a $10^\circ \times 10^\circ$ horizontal grid (with 10 isobaric layers in the vertical) with the use of a chemistry–transport model of the troposphere (MOGUNTIA) for the pre-industrial case (1850), their recent contemporary atmosphere (1990), and for the future (2050).

There was significant growth in the concentration of tropospheric ozone (TrOZ) since pre-industrial times (e.g., Cartalis and Varotsos, 1994). According to assessments by van Dorland *et al.* (1997), increases up to 19 DU are found in July in Europe. As far as the seasonal cycle of zonal mean ozone is concerned the highest change (of about 14 DU) was found at northern mid-latitudes (40°–60°N) in summer with the principal contribution from Europe and North America. A second zonal mean maximum (of about 8 DU) occurs in September and October, mainly due to

Table 7.12. Adjusted annual global average net RF ($W m^{-2}$) from various ozone changes and greenhouse gas changes over different time periods (Forster and Shine, 1997).

Ozone data	*1979–1991*	*1979–1996*	*1964–1996*
SBUV	-0.13 ± 0.02	-0.22 ± 0.03	-0.26 ± 0.05
SAGE A (no depletion below 15.5 km)	-0.08 ± 0.02	-0.14 ± 0.03	-0.16 ± 0.05
SAGE B (constant percent depletion between 15.5 km and the tropopause)	-0.11 ± 0.02	-0.19 ± 0.03	-0.22 ± 0.05
SAGE B normalized to SBUV total ozone	-0.14 ± 0.02	-0.23 ± 0.03	-0.28 ± 0.05
Dobson	-0.11 ± 0.03	-0.18 ± 0.04	-0.21 ± 0.06
Well-mixed greenhouse gases	$+0.45 \pm 0.05$	$+0.70 \pm 0.05$	$+1.7 \pm 0.10$

biomass burning in the tropics of the Southern Hemisphere (Africa and South America). More detailed information was discussed by Levy *et al.* (1997).

The results of RF calculations made by van Dorland *et al.* (1997) indicate (the extended Morcrette radiative transfer model was used) that global annual average RF over the industrial period (1985–1990) is equal to (0.38 ± 0.15) and $-0.36\,W\,m^{-2}$ for tropospheric ozone (a TrOZ increase of about 8 DU) and sulfate aerosol changes, respectively (i.e., an approximate balance between these two RF components takes place). Van Dorland *et al.* (1997) emphasized, however, that both constituents have characteristic seasonal cycles, and their RF dependences on several radiative parameters are quite different, being characterized by different space and time variability. Local RF values achieve $1.0\,W\,m^{-2}$ over northern Africa and Saudi Arabia. The pattern of sulfate RF shows a higher variability compared to ozone.

It is quite natural that ozone-induced RF changes are at maximum downwind of industrialized regions and biomass-burning areas, which results in some asymmetry between the two hemispheres. Peak values of about $1\,W\,m^{-2}$ were found over north Africa and Saudi Arabia. Sulfate aerosol increases are to a larger degree confined to industrial areas in the Northern Hemisphere and to biomass-burning regions in Africa and South America than is the case of TrOZ, due to the shorter lifetime in the atmosphere. This leads to large variability in space and time with peak values up to $-5.5\,W\,m^{-2}$ over Europe and the United States in July, and $-4\,W\,m^{-2}$ over South East Asia in January.

A significant difference between RF due to ozone and sulfate changes exists because ozone is optically active in both the long-wave and short-wave spectral regions (on average, long-wave RF is about 80% of the total), whereas sulfate aerosols mainly scatter short-wave radiation. Therefore, ozone RF is a maximum over areas with high surface temperatures and its changes are strongest in the vicinity of the tropopause (in particular, in summer) and over regions with large convective activity.

In general, clouds shield albedo and in-cloud absorption effects on trace gas retrievals. For total ozone column retrieval the most important cloud effect is the albedo effect; because of the high albedo of clouds, the total ozone slant column is larger than that without clouds. To correct for these effects cloud fraction and cloud pressure are the two most important parameters. Therefore, Wang *et al.* (2008) developed the cloud algorithm FRESCO in which cloud is assumed to be a Lambertian reflector with an albedo of 0.8. The FRESCO effective cloud fraction and cloud pressure have been used in cloud corrections for ozone and NO_2 retrievals from GOME, SCIAMACHY, and GOME-2 measurements by many users.

Van Dorland *et al.* (1997) pointed out that clouds mask the greenhouse effect of tropospheric ozone, but enhance short-wave forcing. The net effect is a decrease of about 13% for the global and annual mean. Since normalized short-wave RF is linearly dependent on the product of surface albedo and $\mu_0 = \sec\theta$ ($\theta =$ solar zenith angle), the ratio of long-wave to short-wave RF is smallest over cold bright surfaces and in the case of large insolation (like Antarctica in January). The dependence of short-wave RF due to sulfate aerosol on surface albedo and angle of incidence is quite different from that of ozone.

Van Dorland *et al.* (1997) developed analytical fits for both ozone and sulfate RF which take into account the most important relevant radiative parameters, which are surface temperature for long-wave RF, surface albedo and solar zenith angle for short-wave RF, and cloudiness for RF in both long-wave and short-wave spectral regions.

In the case of ozone RF, normalized short-wave forcing can be written as:

$$\frac{\Delta F_{\text{sw}}}{\Delta O_3} = S_0 f_d \mu_0 \left(a + \frac{b}{\mu_0} \right) \left(R_s + \frac{c}{\mu_0} \right) \tag{7.1}$$

where normalized forcing is defined as the forcing per DU of total ozone; S_0 is the solar constant; f_d is the day fraction; $\mu_0 = \cos\theta$; and R_s is surface albedo. The best fit was obtained for $a = 7.3 \times 10^{-5}\ \text{DU}^{-1}$, $b = 0$, and $c = 0.075$. The constant $b = 0$ value means that tropospheric RF due to ozone changes is predominantly determined by diffuse radiation (c/μ_0) and not by direct radiation.

The trend for normalized mean RF values in the Northern Hemisphere over the period 1850–1990 was highest in July (about 0.53 W m^{-2} per 10 DU) and minimal in January (0.37 W m^{-2} per 10 DU). Annual variation is caused mainly by the seasonal cycle of surface temperature.

Table 7.13 (van Dorland *et al.*, 1997) illustrates the trends for ozone (long wave and short wave) and for sulfate aerosols (short wave) over the period 1850–1990 and consequent total RF (clear sky and cloudy sky) for January and July at three locations and for the Northern Hemisphere. The locations selected are in Europe (55°N, 15°E), in the United States (35°N, 27.5°E), and in South East Asia (25°N, 115°E), where RF is considerable and reaches a maximum in some cases. These data clearly demonstrate the impacts of various factors on RF, including cloud contribution.

Calculations of the current combined zonal mean RF due to well-mixed greenhouse gases, tropospheric ozone, and sulfate made by van Dorland *et al.*

Table 7.13. January and July clear-sky and average cloudiness sensitivities of ozone and sulfate aerosol and consequent RF for the industrial period 1850–1990 (van Dorland *et al.*, 1997).

	Month	*Location*			*Northern Hemisphere*		
		Europe	*U.S.A.*	*SE Asia*	*Mean*	*Ocean*	*Land*
Cloud cover (%)	January	62	49	59	43	45	40
	July	42	37	53	45	50	41
			Ozone				
$\tilde{a}_{LW,clr}$ ($W\,m^{-2}$/10 DU)	January	0.27	0.27	0.34	0.39	0.46	0.37
	July	0.42	0.45	0.51	0.51	0.54	0.48
$\tilde{a}_{SW,clr}$ ($W\,m^{-2}$/10 DU)	January	0.03	0.06	0.06	0.05	0.04	0.05
	July	0.09	0.10	0.08	0.09	0.07	0.10
$\tilde{a}_{TOT,clr}$ ($W\,m^{-2}$/10 DU)	January	0.30	0.33	0.40	0.43	0.51	0.37
	July	0.51	0.55	0.59	0.59	0.61	0.58
$\tilde{a}_{TOT,cld}$ ($W\,m^{-2}$/10 DU)	January	0.27	0.30	0.33	0.37	0.42	0.33
	July	0.45	0.46	0.47	0.53	0.53	0.52
ΔO_3 (DU)	January	10.9	11.7	12.4	10.1	9.5	10.7
	July	19.2	18.3	11.8	11.6	9.7	13.5
			Sulfate				
$\tilde{a}_{SW,clr}$ ($W\,m^{-2}$/0.01 OD)	January	−0.20	−0.24	−0.26	−0.22	−0.25	−0.20
	July	−0.38	−0.24	−0.21	−0.23	−0.27	−0.23
$\tilde{a}_{SW,cld}$ ($W\,m^{-2}$/0.01 OD)	January	−0.10	−0.12	−0.15	−0.16	−0.17	−0.15
	July	−0.32	−0.22	−0.17	−0.19	−0.18	−0.19
ΔSO_4 (0.01 OD)	January	10.8	9.7	27.4	4.2	2.3	6.1
	July	17.6	25.7	9.2	3.0	0.8	5.0
			Ozone + Sulfate				
ΔF_{clr} ($W\ m^{-2}$)	January	−1.82	−1.91	−6.64	−0.45	−0.07	−0.82
	July	−5.60	−5.15	−1.17	0.03	0.39	−0.31
ΔF_{cld} ($W\ m^{-2}$)	January	−0.77	−0.78	−3.74	−0.26	0.02	−0.54
	July	−4.59	−4.61	−0.96	0.07	0.38	−0.22

Here $\tilde{a}_{LW,clr}$, $\tilde{a}_{LW,cld}$, $\tilde{a}_{TOT,clr}$, are long-wave, short-wave, and total clear-sky sensitivities, respectively. $\tilde{a}_{TOT,cld}$ is total sensitivity under average cloudiness conditions (for ozone per 10 DU and for sulfate per 0.01 optical depth). Also shown are calculated O_3 and SO_4 changes and total RF (ΔF_{clr} is clear-sky forcing, and ΔF_{cld} is forcing for average cloudiness). OD denotes optical depth.

(1997) indicate that RF values are positive with the lowest values less than 1 $W m^{-2}$ in March and April northward of about 50°N. Maximum RF values (of about 2.5 $W m^{-2}$) are located in the northern subtropics (mainly in summer). There are, however, various spots where there are negative RF values. Such a spot is South East Asia, which in January is characterized by total RF up to $-2 W m^{-2}$. Four spots of negative RF exist in July over Western Europe, the United States, South East Asia, and Central Africa, being surrounded by large areas of relatively strong positive RF.

In accordance with the MOGUNTIA chemistry–transport model the maximum TrOZ changes by 2050 occur around 25°N, showing a southward shift compared to the 1850–1990 changes. Ozone increases in the southern United States/Mexico, India/Pakistan, and China regions up to 20, 14, and 10 DU, respectively, were obtained (of course, these results depend on the choice of ozone-producing and depleting gas dynamics).

Assessment of future RF dynamics indicates that by 2050 there will be mean annual global RF of 0.28 $W m^{-2}$ (with RF = 0.34 $W m^{-2}$ in the Northern Hemisphere)—an ozone increase for the next half century equal to about 70% of the change since pre-industrial times. Because of enhanced industrial activities in central America and in southern Asia the regions of maximum ozone-induced change will shift southward (with expected peak values of 0.9 $W m^{-2}$) as compared to the 1850–1990 period.

Maximum ozone RF is expected to occur in August, instead of July, when conditions are optimal for the photochemical processes of tropospheric ozone formation and for the long-wave forcing component. In contrast with the reduced O_3 trend over the period 1990–2050, compared to the period 1850–1990, sulfate changes are expected to increase in the future, resulting in global average RF = 0.59 $W m^{-2}$ (India and South East Asia being the principal contributors).

Van Dorland *et al.* (1997) pointed out that, compared to current RF due to well-mixed greenhouse gases, the tropospheric ozone contribution to the greenhouse effect is about 15%. In 2050 the contribution of ozone to total positive RF since pre-industrial times is expected to be somewhat smaller, but still significant. Increases in the sulfate burden act to (over) compensate positive RF on a regional scale.

Similar results of RF calculations were obtained by Haywood *et al.* (1998). According to their results the global annual mean adjusted RF, including clouds, ranges from +0.25 to +0.32 $W m^{-2}$ with about 80% of the contribution from the long-wave component (see also Velders *et al.* 2007).

Numerical modeling of the global 3-D distribution of tropospheric ozone changes and respective RF was considered by Berntsen *et al.* (1997). The ozone concentration field was calculated with the use of a chemical tracer model (CTM) developed at the University of Oslo, whereas RF caused by the increase of ozone was computed by two independent radiative transfer models: the University of Reading model (Reading) and the University of Oslo/Norwegian Institute for Air Research model (OsloRad).

The principal purpose was to study the impact of increased emissions of NO_x, CO, and CH_4, on the RF of climate, by introducing the ozone perturbation calculated by the CTM in two different radiative transfer models. The CTM was run with emissions according to 1850 and 1990 conditions, respectively. The CTM has a resolution of 8° latitude by 10° longitude and nine vertical levels below 10 hPa. A comprehensive photochemical scheme with about 50 chemical components and 110 gas phase reactions was incorporated to simulate the photochemistry of the troposphere. In the stratosphere, concentrations of ozone, NO_x, and HNO_3 are fixed according to the observed distribution of these species. The model uses pre-calculated meteorological input data with 8-hour resolution from a 1-year climate simulation with the NASA GIS general circulation model, which allows synoptic-scale variability to be resolved.

Ozone production in the troposphere is caused by the oxidation of NO to NO_2 by HO_2 or RO_2 radicals and the subsequent photolysis of NO_2 through the reactions:

$$NO + HO_2/RO_2 \rightarrow NO_2 + OH/RO \tag{7.2}$$

$$NO_2 + h\nu \rightarrow NO + O \tag{7.3}$$

$$O + O_2 + M \rightarrow O_3 + M \tag{7.4}$$

Since ozone production in the free troposphere is generally limited by the abundance of NO_x, Berntsen *et al.* (1997) emphasized the necessity to prescribe realistic emission inventories for NO_x and CO (Table 7.14).

Numerical modeling by Berntsen *et al.* (1997) indicates significant increases in upper-tropospheric ozone concentrations at northern mid-latitudes (15–45 ppbv, depending on season) at about 10 km altitude (see also Table 7.3). Berntsen *et al.* (1997) pointed out that tropospheric ozone increase is caused mainly by enhanced *in situ* production due to transport of ozone precursors from the atmospheric boundary layer (ABL), with a similar contribution from the increased transport of ozone produced in the ABL. Because of enhanced global mean concentrations of HO_2 by 40% and an OH concentration increase by 6%, the lifetime of ozone in the troposphere decreased by about 35%. The calculated increase of surface ozone concentration in Europe is found to be in good agreement with observations.

Figures 7.4 and 7.5 characterize the global distribution of net RF and its components according to the Reading and OsloRad model results. An important conclusion from these data is that regional and seasonal, as well as hemispheric, differences in the RF due to changes in tropospheric ozone are higher than corresponding differences in the RF due to changes in the concentration of well-mixed greenhouse gases (see, e.g., Myhre and Stordal, 1997; Myhre *et al.*, 2008). This indicates differences between the climatic impacts of ozone and well-mixed greenhouse gases (Wang *et al.*, 1995). Berntsen *et al.* (1997) emphasised in this context that detailed experiments with a general circulation model are required to establish the extent of the differences.

Besides RF spatial variability, it is very important to take into account its temporal changes, including seasonal variation (Bernsten *et al.*, 1997). Another important question concerns the impact of tropospheric ozone increase on

Table 7.14. Emission rates applied in the model study of Berntsen *et al.* (1997).

Species	*1850*	*1990*
NO_x ($Tg\,N\,yr^{-1}$)		
Surface	11.6	37.8
Lightning	5.2	5.2
Aircraft	0.0	0.6
CO ($Tg\,yr^{-1}$)		
Total	290	1,400
Anthropogenic	0	810
Biomass burning	100	400
Natural sources[a]	190	190
Isoprene ($Tg\,yr^{-1}$)	267	205
Other NMHCs[b] ($Tg\,yr^{-1}$)	175.7 (64.8[c])	175.7

[a] Natural sources include emissions from vegetation, soils, wildfires, and ocean.
[b] Other NMCHs includes ethane, butane, hexane, ethene, propene, and *m*-xylene.
[c] Sensitivity experiment with no anthropogenic NMCH emissions.

stratospheric temperature (Dall'Amico *et al.*, 2010; Forster and Shine, 1997; Ramaswamy *et al.*, 1996).

From the discussion in this section it is clear that both tropospheric and stratospheric chemistry need to be taken into account, because tropospheric ozone increase since pre-industrial times has moderated lower-stratospheric ozone depletion in low latitudes, while stratospheric ozone depletion during the last three decades has influenced tropospheric ozone increase in high latitudes. In this context it is important to investigate further the transport of ozone through the tropopause. In the long run it should become possible to consider RF from combined tropospheric and stratospheric changes. Clearly, further model studies will be needed to investigate the contribution from climate change to changes in ozone, both in the troposphere and in the stratosphere, with focus on changes in temperatures, water vapor concentrations, and stratospheric circulation. Regarding RF calculations of this type it will be important in the future to distinguish clearly between RF due to ozone changes resulting from changing emissions and from changing climate, which under certain circumstances should be considered as a climate feedback rather than a contribution to RF.

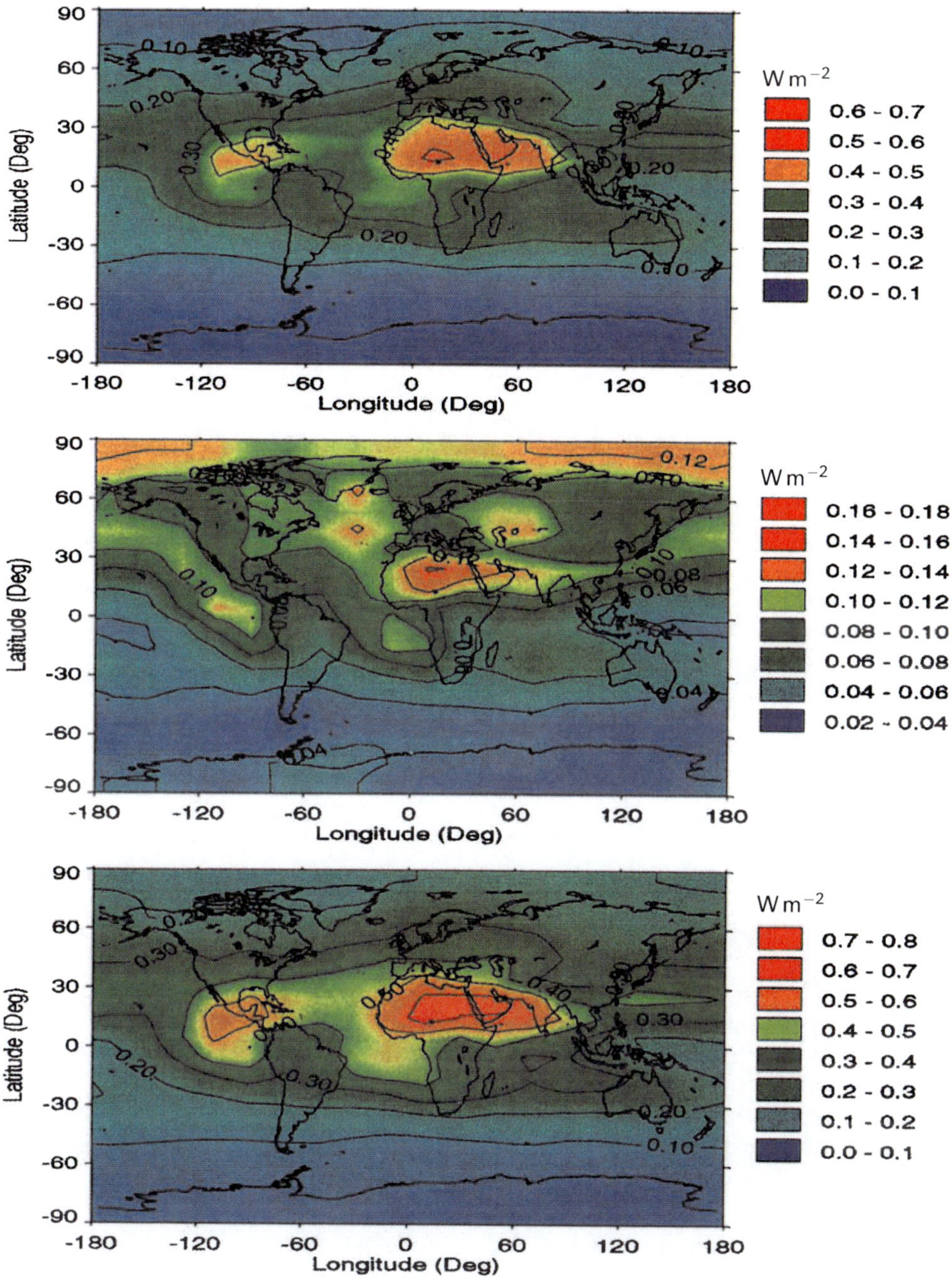

Figure 7.4. RF (calculated using the Reading model) due to the change in tropospheric ozone since pre-industrial times according to the chemistry–transport model (CTM). Annual mean cloudy-sky adjusted forcing is shown: (top) thermal infrared, (middle) short wave, (bottom) net (Berntsen *et al.*, 1997).

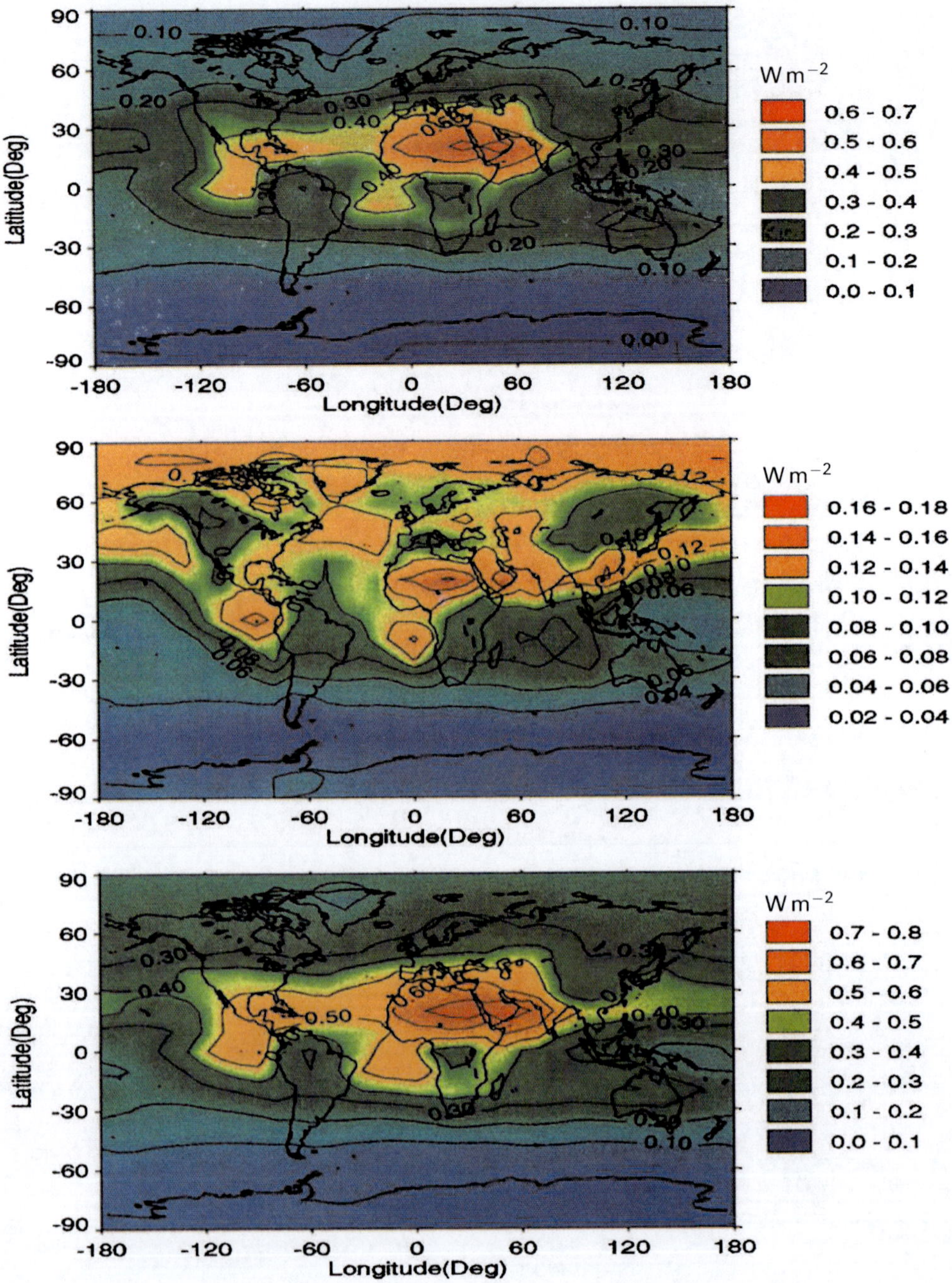

Figure 7.5. As Figure 7.4, but for the OsloRad model (Berntsen *et al.*, 1997).

It is crucial to stress the point that quantifying the combined effects of ozone depletion and changes in other greenhouse gases and radiatively and chemically active atmospheric constituents is of great importance to human welfare. Achieving this goal requires evaluating the impact of stratospheric ozone depletion and its recovery on the tropospheric climate as well as elucidating the effects of climate change on the evolution of ozone itself. This, in turn, requires the understanding and quantification of the long-term sensitivity of the climate system to significant perturbations in the radiation budget of the atmosphere associated with human activities. Detecting and quantifying such effects also requires a quantitative understanding of the role of natural events, such as volcanoes and solar variability, on the composition and evolution of the atmosphere and, ultimately, of the effects of such events on the climate change signal throughout the active atmosphere and at the surface. With the advent of climate modeling as a key tool for studying and predicting the evolution of the climate system, the linked concepts of RF and climate sensitivity have come into wide use as a means to understand and quantify key aspects of modeling results. These concepts are reviewed by McFarlane (2008) and their relevance to understanding and quantifying the radiative impact of ozone and ozone depleting substances on the RF of the climate system is discussed.

7.2.3 Contribution of the transport sector

The contribution to RF due to increases in tropospheric ozone arising from the transport sector has been studied by Balkanski *et al.* (2010), Dalsøren *et al.* (2010), Fuglestvedt *et al.* (2008), and Myhre *et al.* (2011) using several different models. Ozone changes from three transport sectors have been simulated for the period from pre-industrial times to the present. RF has been calculated with one set of long-wave and short-wave radiative transfer schemes. For the year 2000, the simulated RF was found to be dominated by the road (31 $mW\,m^{-2}$) rather than maritime shipping (24 $mW\,m^{-2}$) and aircraft (17 $mW\,m^{-2}$) sectors in four of the models. Four of the six models have quite similar latitudinal variation in RF for the three transport sectors, but two differ considerably from the rest. Total global and annual mean RF from the transport sector is slightly below 0.1 $W\,m^{-2}$. Variation between the six models was largest for RF due to ozone from aircraft. By contrast, Jonson *et al.* (2009), using results from five global atmospheric chemistry models, calculated RF for 2000 due to changes in O_3 and CH_4 (and CH_4-induced stratospheric water vapor) as a result of emissions of short-lived gases—oxides of nitrogen (NO_x), carbon monoxide, and non-methane hydrocarbons—from three transport sectors (road, maritime shipping, and aircraft). In addition, Jonson *et al.* (2009) quantified the uncertainties involved, using results from these models plus other published data. They showed that RF due to short-term O_3 changes is positive and highest for road transport. The net RF of O_3 and CH_4 combined (i.e., including the impact of CH_4 on ozone and stratospheric water vapor) was found, using all five models, to be positive for the road ransport ($+16 \pm 13\,mW\,m^{-2}$ with one standard deviation) and air ($+6 \pm 5\,mW\,m^{-2}$) traffic sectors but negative for shipping ($-18 \pm 0\,mW\,m^{-2}$).

Within the EU project QUANTIFY (Quantifying the Climate Impact of Global and European Transport Systems) an ensemble of chemistry–transport models and chemistry–climate models was applied to calculate and compare the impacts of emissions from road traffic, aircraft, and shipping on atmospheric composition. Within the framework of the QUANTIFY project the maximum impact of emissions from the transport sector on the ozone column calculated by the models was found to be an increase of about 4 DU. This impact was greatest over the subtropical North Atlantic. The contribution to zonal mean ozone perturbation can be as high as 15% (5.5 ppbv). In the upper troposphere the impact amounts up to 8 ppbv. Ships contribute more than 80% to maximum ozone perturbation by the transport sector in the boundary layer during the Northern Hemisphere summer. The reason is that the relatively pristine marine boundary layer over the Atlantic is quite sensitive to perturbations. In the zonal mean these changes in the boundary layer can exceed 10%. The efficiency of NO_x emission to produce ozone is largest for aircraft emissions owing to the long lifetime of NO_x reservoir species in the tropopause region. Road emissions have a strong impact on ozone in the continental boundary layer and are more efficiently transported upward into the free troposphere than ship emissions, especially in summer when convection over the land is more intense and extends higher than over the oceans. In the upper troposphere in summer the direct impact of upward-transported road emissions is comparable to that from (local) aircraft emissions. The oxidation capacity in the boundary layer is most strongly affected by ship emissions, which have a significant influence on the lifetime of many trace gases including methane. Ship emissions perturb the zonal mean OH concentration by 10–16%. The associated reduction of the lifetime of methane is estimated to be of the order of 3.5–4%. For road traffic emissions this is about 1.5% and for aircraft emissions about 1%. The contributions of emissions from the transport sector to acidification and eutrophication were greatest near to the west coast of Europe and the east coasts of the U.S. and Asia where overall emissions are largest. Model results have also been evaluated against available observations (e.g., from instrumented aircraft campaigns and ozone soundings).

The growing effect of emissions, including significant emissions of ozone precursors, from the transport sector on climate was considered by Berntsen and Fuglestvedt (2008). The impact of all emissions from different transport modes on ozone, methane lifetime, and the long-term "primary mode" ozone effect was calculated with the Oslo CTM2 model (Berntsen and Fuglestvedt, 2008). Due to very different mixtures of precursors (NO_x, CO, and volatile organic compounds (VOCs)) and differences in chemical background conditions where the emission takes place, it was found that the impact is significantly different between transport modes. RF from the road transport, shipping, aviation, and rail subsectors was comprehensively analyzed using both past and forward-looking perspectives. It was also found that, since pre-industrial times, transport has contributed 31% of total man-made O_3 forcing, respectively. A forward-looking perspective showed that the current emissions from transport are likely to be responsible for 16% of integrated net forcing over the next 100 years from all current man-made emissions. The dominating contributor to positive forcing (warming) is CO_2, followed by tropospheric O_3.

Road transport is the largest contributor to warming. The transport sector also exerts cooling through reduced methane lifetime and atmospheric aerosol effects. Shipping causes net cooling, except on future timescales of several centuries.

7.3 OZONE-INDUCED CLIMATIC IMPACTS

Ozone–climate interaction studies are still at an early stage of development. Only a limited number of attempts have been made to examine the sensitivity of climate dynamics to ozone changes (Gillet *et al.*, 2009).

Interest in the impact of ozone on climate was greatly stimulated by the discovery of the springtime ozone hole over the Antarctic continent in the mid-1980s (see also Sections 5.4 and 6.5). As mentioned above, the initial results of numerical modeling show that TOC depletion leads to climate cooling, while TrOZ increase results in climate warming. Bintanja *et al.* (1997) undertook a new investigation of the climate response due to changes in radiative fluxes caused by prescribed changes in stratospheric and tropospheric ozone. They used for this purpose a simplified climate model which combined a zonal energy balance atmosphere model and an advection–diffusion zonal ocean model. Radiative fluxes were calculated at 33 vertical levels. Such a coupled model makes it possible to simulate latitudinal and seasonal variations in zonal mean surface air temperature and average lower (12–22 km) and higher (22–100 km) stratospheric temperatures. Two types of numerical experiments were performed: (1) experiments in which the model is perturbed by (uniform) ozone changes after which the model is allowed to reach a (quasi) steady state; and (2) transient climate experiments in which the model is forced by temporally varying ozone concentrations (in the form of observed trends). These experiments include altitude, latitude, and seasonally varying ozone changes.

Bintanja *et al.* (1997) demonstrated the model's capability to simulate present day climate successfully, which is an important prerequisite of perturbed climate simulations. Intercomparison between calculated and observed annual mean SAT (surface air temperature) values as well as annual SAT variations indicates that, generally, the difference between model results and observations is small. Simulated temperatures over Antarctica are a few degrees too low, while the annual SAT range is somewhat underestimated in Northern Hemisphere mid-latitudes. There is also agreement in the main essential features of mean meridional energy transport in the atmosphere and ocean. Seasonal and latitudinal distributions of snow cover, surface and planetary albedo, surface radiation fluxes, turbulent fluxes, and ocean temperatures were found to be simulated with reasonable accuracy.

Calculations of quasi-equilibrium climate sensitivity due to uniform changes in ozone were made with the use of three versions of the surface energy balance model (Bintanja *et al.*, 1997): (1) ALL, in which all feedback mechanisms are operative; (2) AT, in which the albedo–temperature (α-T) feedback is switched off by keeping the amount of snow and sea ice and the snow and ice albedo at their present day values; and (3) NTF (no tropospheric feedback), in which the troposphere, Earth

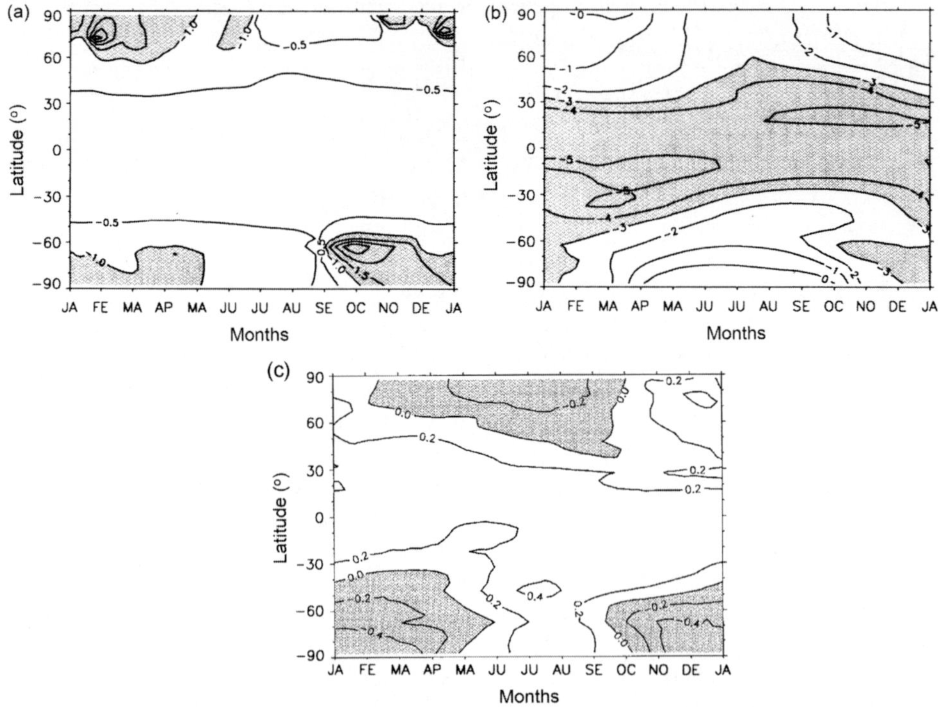

Figure 7.6. Meridional and seasonal distribution of the change in (a) surface air temperature, (b) STL temperature, and (c) STH temperature resulting from STL ozone reduction of 50% (Bintanja *et al.*, 1997).

surface, and ocean are kept fixed at their present day state—only the stratosphere is allowed to change. RF at the tropopause in the NTF version is equivalent to the concept of adjusted forcing (AF), which is RF at the tropopause with an adjustment for stratospheric temperature included. In addition to adjusted forcing, Bintanja *et al.* (1997) also considered initial forcing, which is the initial change in net radiation at the tropopause without temperature adjustments.

Figure 7.6 depicts the latitudinal and seasonal distribution of differences in SAT, mean lower stratosphere (STL) (i.e., the 198.6–41.4 hPa layer), and mean higher stratosphere temperature between the perturbed state and the present day state in the event of a 50% reduction in STL ozone concentration. Figure 7.7 demonstrates differences in some key radiation quantities. As emphasized by Bintanja *et al.* (1997), the main result of STL ozone reduction is the decrease in short-wave absorption in the STL, in particular in the polar summer (as shown earlier by Ramanathan and Dickinson, 1979 and Ramanathan *et al.*, 1976, 1985). This is due to high insolation combined with high surface reflectivity in polar areas, and to the fact that absolute ozone changes in polar regions (~66 DU) are larger than those in the tropics. Negative tropopause RF (Figure 7.7c) results in global decrease in SAT (Figure

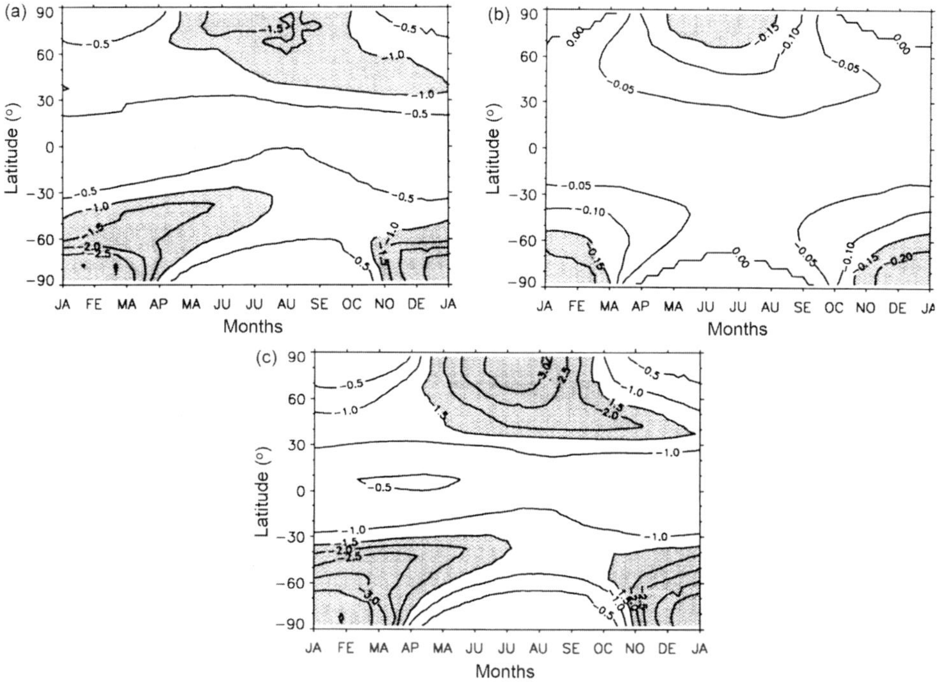

Figure 7.7. (a) Meridional and seasonal distribution of the adjusted change in net radiation ($W m^{-2}$) at the tropopause, resulting from STL ozone reduction of 50%; (b) initial change in short-wave radiative divergence in the STL ($K day^{-1}$); and (c) adjusted change in long-wave radiation ($W m^{-2}$) at the tropopause (Bintanja *et al.*, 1997).

7.6a) with strong poleward enhancement in winter due to the α-T feedback. Global average cooling is about 3.5°C in the lower stratosphere with maximum values in the tropics and equals 0.46°C at the surface with maximum cooling in the polar winter (due to the α-T feedback).

Bintanja *et al.* (1997) emphasized that the albedo–temperature feedback is consistently stronger in the case of tropospheric and lower-stratospheric ozone perturbations than in the case of, for instance, CO_2 perturbations. This can be attributed mainly to differences in the meridional gradient in tropopause RF.

Similar (50%) ozone reduction in the higher stratosphere (STH) (i.e., the 41.4–0.003 hPa layer) results in a much smaller effect on tropopause RF because of smaller long-wave adjusted forcing and SAT decrease is consequently much smaller.

Since tropospheric ozone increase enhances the atmospheric greenhouse effect this leads to increased SAT (in particular in the polar regions) and to lower stratospheric temperatures. Figure 7.8 illustrates the meridional distribution of the mean annual SAT change for various scenarios, whereas Figure 7.9 shows the meridional distribution of the change in mean annual short-wave, long-wave, and net radiation of the tropopause.

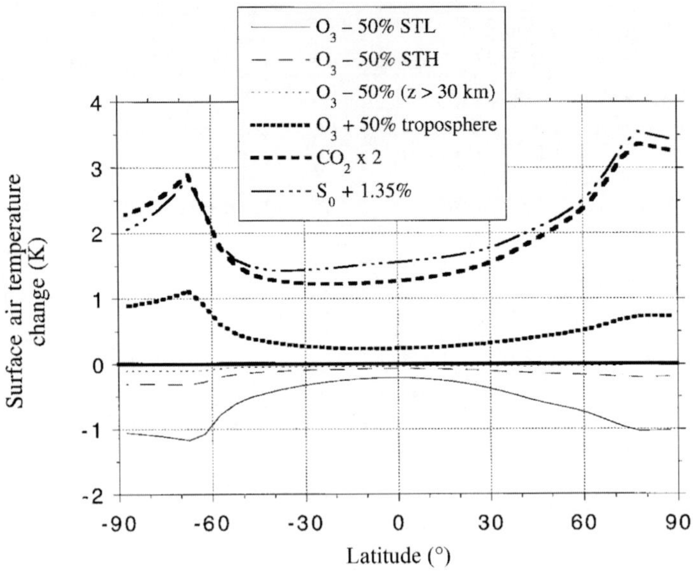

Figure 7.8. Meridional distribution of the change in annual mean surface air temperature for the various scenarios (Bintanja *et al.*, 1997).

Data on the meridional and seasonal distribution of the change in STL temperature ($K\,yr^{-1}$) for the TOC(z) scenario and TOT(φ, z, t) scenario (φ is latitude, z is height, and t is time) are presented in Figure 7.10.

The assessments obtained by Bintanja *et al.* (1997) for the transient response of the climate system to a number of more or less realistic ozone perturbation scenarios (based on recent ozone trend observations) indicated that the combined result of TOC decrease and TrOZ increase is a small global average cooling from -0.001 to $-0.003\,K\,yr^{-1}$, which compensates for about 10% of greenhouse warming due to greenhouse gases. Relative changes in ozone induced-temperature variations are strongest in the case of tropospheric ozone (0.011–$0.015\,K\,DU^{-1}$) with the principal contribution from the upper mean global troposphere and the lowest from the lower stratosphere (0.008–$0.010\,K\,DU^{-1}$) and the higher stratosphere ($0.001\,K\,DU^{-1}$). Combined mean global sensitivity (taking account of both stratospheric cooling and tropospheric warming) is only about 0.002–$0.003\,K\,DU^{-1}$.

In view of the simplified climate model used, Bintanja *et al.* (1997) emphasized that their model can provide qualitative insights into the fundamental processes that determine the sensitivity of climate for ozone changes (see also Pawsson *et al.*, 2008). In this context, Rind *et al.* (2008) used the new Goddard Institute for Space Studies Global Climate Middle Atmosphere Model 3 to investigate various aspects of solar cycle influence on the troposphere/stratosphere system. Three different configurations of sea surface temperatures were used to help determine whether the tropospheric response is due to forcing from above (UV variations impacting the stratosphere) or below (total solar irradiance changes acting through the surface

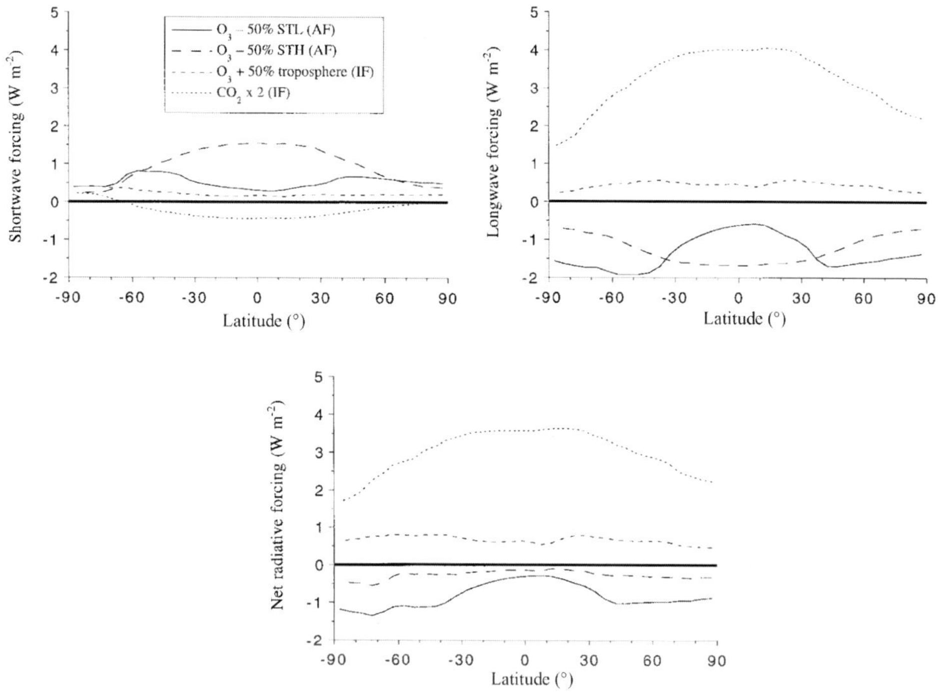

Figure 7.9. Meridional distribution of the change in annual mean short-wave (top left), long-wave (top right), and net (bottom) radiation at the tropopause for the various scenarios. Tropopause forcing includes stratospheric temperature adjustment (adjusted forcing) for ozone changes in the stratosphere, whereas initial tropopause forcing is shown for tropospheric ozone changes and for CO_2 doubling (Bintanja *et al.*, 1997).

temperature field). The results showed that the stratospheric response is highly repeatable and significant. The predominant tropospheric response consists of warming in the troposphere, with precipitation decreases south of the equator and in the Northern Hemisphere subtropics and mid-latitudes, with increases north of the equator especially over southern Asia.

Roelofs *et al.* (1997) discussed results from the tropospheric chemistry–climate European Center Hamburg Model by comparing two simulations that consider a pre-industrial and a contemporary emission scenario. Due to these results, photochemical O_3 production from anthropogenically emitted precursors contributes about 30% to the present day tropospheric O_3 content, which is roughly equal to natural photochemical production. Transports of stratospheric ozone into the troposphere contribute about 40%. Table 7.15 shows more detailed information on tropospheric source and sink terms for ozone and the tropospheric burdens of O_3 from stratospheric origin (O_3s), O_3 from photochemical formation in the troposphere (O_3t), and total O_3 as well as dry deposition (Tg O_3 per year) and net transport.

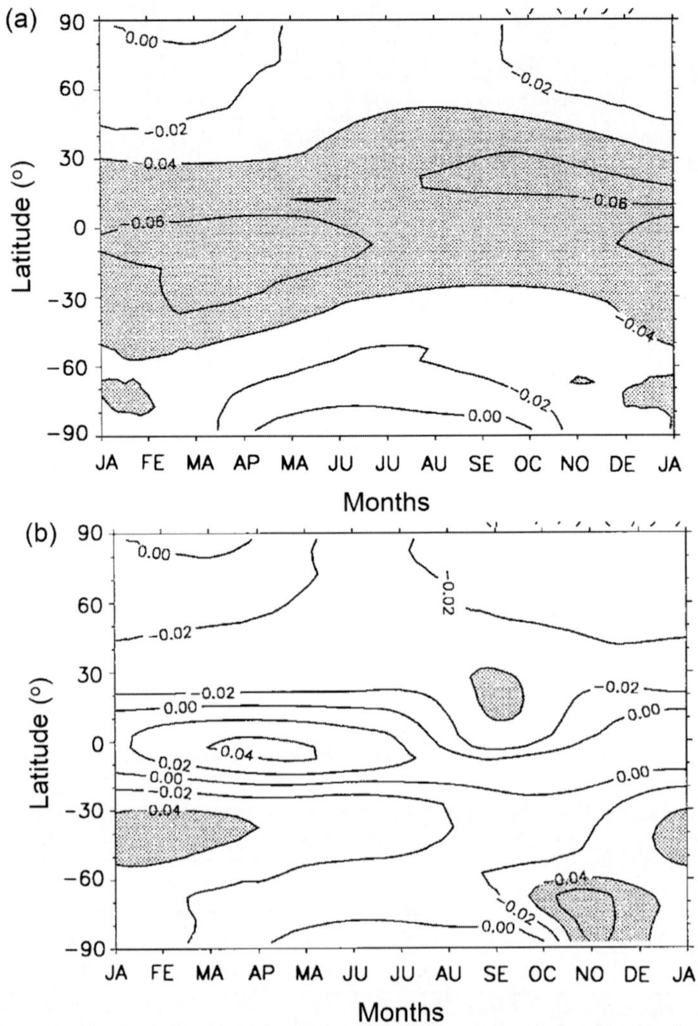

Figure 7.10. Meridional and seasonal distribution of the change in STL temperature (K yr^{-1}) for (a) the TOT(z) scenario and (b) the TOT(φ, z, t) scenario (bottom) (Bintanja *et al.*, 1997).

Photochemical ozone production predominantly takes place through the reactions $NO + HO_2$ and $NO + CH_3O_2$, whereas the principal destruction mechanisms are the reactions $O_3 + OH$, $O_3 + HO_2$ and $O(^1D) + H_2O$. Additional reactions through which relatively small amounts of O_3 are produced/destroyed are $HNO_3 + OH$, N_2O_5 decomposition and NO_3 photolysis (see also Sections 4.5, 6.5.4). The increase of ozone precursors (NO_x, CH_4, and CO) results in photochemical production increase from 1,732 to 3,415 Tg O_3 yr^{-1}, but the increase in photochemical destruction is somewhat smaller. Therefore, at the present time the

Table 7.15. Simulated global budget of tropospheric ozone ($Tg\,yr^{-1}$) (Roelofs *et al.*, 1997).

Components	*Present day*	*Pre-industrial*	*Anthropogenic*
Net STE O_3	459	622	−163
Chemical production	3,415	1,732	1,683
Chemical destruction	−3,340	−2,055	−1,285
Net chemical production/destruction	75	−323	398
Dry deposition	−534	−296	−238
O_3s	110	108	2
O_3t	161	82	79
Total ozone	271	190	81

model calculates net production of O_3 in the troposphere to dominate, while in the pre-industrial case net destruction of ozone prevails. Net O_3 transport into the troposphere ($459\,Tg\,O_3\,yr^{-1}$) is the result of downward transport of O_3s ($930\,Tg\,O_3\,yr^{-1}$) and upward transport of O_3t ($471\,Tg\,O_3\,yr^{-1}$) (Roelofs and Lelieveld, 1997). In the pre-industrial simulation the downward transport of O_3s was $889\,Tg\,O_3\,yr^{-1}$ and the upward transport of O_3t was $267\,Tg\,O_3\,yr^{-1}$.

Calculations of total RF based on the results mentioned above led Roelofs *et al.* (1997) to global and annual average RF by tropospheric ozone changes of $0.42\,W\,m^{-2}$ ($0.51\,W\,m^{-2}$ in the Northern Hemisphere and $0.33\,W\,m^{-2}$ in the Southern Hemisphere). The respective long-wave and short-wave RF components are 0.35 and 0.16 (NH), 0.24 and 0.09 (SH), 0.29 and $0.13\,W\,m^{-2}$ (global). Zonally averaged values indicate the presence of an RF maximum near 25°N, especially in the Northern Hemisphere summer which coincides with peak TOC change. The geographical distributions of total RF at the tropopause due to tropospheric O_3 increases during summer; and winter seasons indicate the existence of maxima over northern Africa and the Middle East.

Fuglestvedt *et al.* (1998) pointed out that through production of tropospheric O_3, NO_x emissions lead to positive RF and warming. But, by affecting the concentration of OH radicals, NO_x also reduces the levels of CH_4, thereby giving negative forcing and cooling. An important fact in this context is that the lifetime of NO_x varies from hours to days, giving large spatial variations in the levels of NO_x. Fuglestvedt *et al.* (1998) used a 3-D global chemical tracer model and a radiation transfer model to study the role of NO_x emissions for the RF of climate. For this purpose geographical regions were selected so as to represent different chemical and physical conditions; the chemical and radiative effects of reducing NO_x emissions by 20% in each region were studied. The results indicate large geographical differences in the effects of NO_x on O_3 as well as variations in the annual profile of changes due

Table 7.16. Changes in NO_x emissions due to a 20% reduction in the emissions in each region and the calculated changes in lifetime ($\Delta\tau_0$) and steady-state concentration of methane. These changes are also normalised to the magnitude of change in NO_x emissions (Fuglestvedt *et al.*, 1998).

Region	E_{NO_x} ($Tg\,N\,yr^{-1}$)	ΔE_{NO_x} (Tg N yr-1)	$\Delta\tau_0$ (years)	$\Delta\tau_0/\Delta E_{NO_x}$	ΔCH_4 (ppbv)	$\Delta CH_4/\Delta E_{NO_x}$
1. Australia	0.84	−0.169	0.023	−0.136	7.27	−43.1
2. S-Europe	1.97	−0.393	0.012	−0.031	3.85	−9.8
3. M-Europe	2.22	−0.444	0.01	−0.023	3.22	−7.3
4. Scandinavia	0.21	−0.043	0.00106	−0.025	0.342	−8.0
5. Southeast Asia	1.18	−0.236	0.035	−0.148	11.5	−48.6
6. U.S.A.	7.53	−1.506	0.048	−0.032	15.7	−10.4
All regions	13.96	−2.79	0.136	−0.049	43.8	−15.7

to non-linearities in O_3 chemistry and differences in convective activity. The effect of NO_x emissions on methane depends on localization of emissions.

Calculated ozone and methane RFs are of similar magnitude but of opposite sign. The methane effect takes place on a global scale with a decay of approximately a decade, while the ozone effect is regional in character and lasts for a few weeks. Table 7.16 illustrates changes in NO_x emissions due to a 20% reduction in the emission in each region, whereas Table 7.17 contains values of total RF.

Sexton *et al.* (1996) conducted numerical climate modeling using the Hadley Centre atmospheric GCM (HADAM2a) for the period from late 1948 to the end of 1994 with the aim of analyzing the impacts of different specified atmospheric forcing distributions. Two ensemble calculations were made for varying ozone, one of which ($SUL + O_3$) took into account increasing sulfate aerosols as well as a linearly decreasing trend in ozone concentrations in the lower stratosphere from 1975 based on estimated and observed TOC decreases. Ozone loss was assumed to occur within about 12 km on a fixed tropopause, with twice the rate of loss in the lower 7 km than the upper part. Another ensemble ($SST + O_3$) was only forced with decreasing stratospheric ozone since 1975. During this period, observations showed a decrease of mean annual temperature at the 50 hPa level of about 1.3°C, with three short periods of large warming due to volcanic eruptions. The observed cooling is very well simulated only when CO_2 increases and stratospheric ozone reductions are taken into account.

Similar results were obtained by Sexton *et al.* (1996) in the case of modeling the zonal mean temperature distribution between 1961–1975 and 1987–1994. The observed changes are dominated by stratospheric and uppermost tropospheric cooling as well as warming lower in the troposphere. The cooling effects of the

Table 7.17. Total RF ($10^{-2}\,W\,m^{-2}$) due to ozone changes for different regions (Fuglestvedt *et al.*, 1998).

		NH	*SH*	*Global*
South East Asia	January	−0.439	−0.360	−0.400
	July	−0.432	−0.764	−0.598
	Annual	−0.522	−0.600	−0.561
U.S.A.	January	−0.324	−0.020	−0.171
	July	−1.963	−0.037	−1.000
	Annual	−1.021	−0.021	−0.521
Scandinavia	January	−0.001	0.000	−0.000
	July	−0.057	0.000	−0.028
	Annual	−0.025	0.000	−0.012
Combined	January	−0.911	−1.100	−1.001
	July	−3.513	−1.172	−2.342
	Annual	−2.162	−1.161	−1.662

stratospheric ozone hole in the Antarctic are very evident between 300 hPa and 100 hPa. As far as the global land air surface temperature is concerned, the observed warming trend of 0.5°C per decade is underestimated in all simulations. The numerical modelling considered clearly shows that inclusion of stratospheric ozone depletion is an important anthropogenic forcing when considering changes in the vertical temperature structure of the troposphere and stratosphere, and there may be an effect at the surface as well. Sexton *et al.* (1996) emphasised, however, that there is considerable uncertainty in the magnitude and distribution of the forcing due to stratospheric ozone, so there is an urgent need for the ozone dataset to be improved.

Austin and Butchart (1997) discussed results from the first long integration of the UKMO (U.K. Meteorological Office) 49-level Unified Model with fully interactive stratospheric chemistry. Photochemistry is active in the 400–1 hPa layer and outside this layer constituents are transported passively. (The evolution of 32 species is considered using 15 advected tracers).The photochemistry scheme includes all the processes known to affect stratospheric constituents, except for reactions on sulfate aerosol. The results of a 19-month integration, starting with observations on January 15, 1992, indicate that the simulated chemical behavior of the atmosphere was reasonably consistent with observations, particularly in the Southern Hemisphere. However, model TOC was significantly underpredicted in the Northern Hemisphere, especially during the spring. The principal cause of the discrepancy was probably due to poor meridional transport.

In the past few decades, tropospheric westerly winds in the Southern Hemisphere have been observed to accelerate on the poleward side of the surface wind maximum. This has been attributed to the combined anthropogenic effects of increasing greenhouse gases and decreasing stratospheric ozone and is predicted to continue by the Intergovernmental Panel on Climate Change/Fourth Assessment Report (IPCC/AR4) models. Son *et al.* (2008) examined the predictions of chemistry–climate model validation (CCMVal) models, focusing in particular on the fact that, unlike the AR4 models, the CCMVal models have a fully interactive stratospheric chemistry. Son *et al.* (2008) stated that, owing to the expected disappearance of the ozone hole in the first half of the 21st century, the CCMVal models predict that tropospheric westerlies in Southern Hemisphere summer will decelerate, on the poleward side, in contrast with the prediction of most IPCC/AR4 models.

Rind *et al.* (1997) analyzed the influence of ozone changes on global-warming simulation (in the case of CO_2 doubling) using the GISS Global Climate/Middle Atmosphere Model (GCMAM). When ozone is allowed to respond photochemically, stratospheric cooling is reduced by 20%, with little effect on the troposphere. Shindell *et al.* (1997b) discussed application of the GCMAM to investigate the response of ozone to a doubled CO_2 climate, and the resulting changes on atmospheric radiation and dynamics (interactive ozone was considered in the 200–0.1 hPa layer). The results show that the ozone response is opposite to temperature changes, so that cooling in the upper stratosphere results in increased ozone abundances, while lower-stratospheric warming reduces ozone in that region. The increased overhead column also reduces the amount of UV reaching the lower stratosphere. The net result is ozone decrease of up to 7% in the lower stratosphere. Shindell *et al.* (1997b) emphasized the large differences between their results and those of other models in extratropical latitudes (probably due to stronger tropospheric warming in the GCMAM). When parameterized heterogeneous chemistry on PSCs was included, it was found that the Antarctic ozone hole becomes significantly larger and of longer duration in the doubled CO_2 atmosphere. An ozone hole of variable strength, roughly one quarter to two thirds that of the current Antarctic hole, forms in the Arctic as a result of overall cooling in the polar lower stratosphere combined with the reduction of sudden warmings in the doubled CO_2 atmosphere.

Perlwitz *et al.* (1997) pointed out that there was an increase in the number of winter months with a strong circumpolar vortex (CPV) in the stratosphere compared with the number of months with mean CPVs. To examine the potential causes of this trend, numerical climate modeling was undertaken with the use of the general circulation model ECHAM 3/LSG/T21 developed at DKRZ (Hamburg). The principal conclusion is that the GHG scenario (i.e., enhancement of the greenhouse effect) is in accordance with the observed winter mean trend of recent decades both in stratospheric and tropospheric circulation. At the same time this mode was also a naturally occurring feature in the unforced control run. However, these trends in the control run lasted at most for 40 years in one direction. After this time the trend reversed, which had not happened in the case of the GHG experiment (such a persistent trend never occurred in the 700-year control run). Thus, the increased greenhouse effect enhances and stabilizes the natural baroclinic mode towards a stronger polar night vortex.

Perlwitz *et al.* (1997) emphasized that the TOC decrease since 1979 is another potential candidate for stratospheric changes. The results of calculations indicate, however, that the observed changes in stratospheric circulation in winter are not consistent with forcing by reduced ozone. In spring, GHG forcing and ozone act in the same direction: both keep the CPV persistent. The results suggest that ozone anomalies do not actually change the circulation except in spring. As far as winter circulation is concerned, there are circulation changes that determine the ozone distribution, but not *vice versa*.

Krivolutsky (1997) suggested, on the basis of 1-D photochemical modeling, that a possible cause of the Antarctic ozone hole might be a decrease in air density and, therefore, the formation of an oxygen hole leading to transformed photochemistry. Bazilevskaya *et al.* (1997) established an empirical interrelationship between the long-term trend of ozone depletion and solar and galactic cosmic rays. Linear regression analysis indicates the influence of solar proton events on TOC of the order of several DU during years in which there are strong solar proton events (SPEs) and periodic forcing by galactic cosmic rays (GCRs) with the amplitude of several DU.

Christiansen *et al.* (1997) applied the Arpège climate model (a 3-D spectral model, T42, extending from the surface up to about 80 km; Déqué *et al.*, 1994) in perpetual January mode to assess the sensitivity of atmospheric general circulation to large perturbations in stratospheric ozone concentration. Three numerical experiments were carried out: uniform reduction of the mixing ratio by 50% and 70% and an ozone hole-type reduction with removal of ozone from 500 hPa and upward until TOC reduces to 50% of the unperturbed value. Values of sea ice cover extent, sea surface temperatures, surface albedo, and solar radiation insolation were prescribed. The model was integrated for 300 days and averages were taken over the last 100 days (enough to reach a new statistical equilibrium). Since relaxation to a fixed temperature profile was used as the upper boundary condition, the full dynamical model response could only be evaluated below 1 hPa ($\sim$50 km) level. In accordance with Fels *et al.* (1980), to break down the response of the model into its radiative, thermal, and dynamical parts both radiative drive and temperature change were considered under the fixed dynamical heating (FDH) approximation as well as the full dynamics. In all three experiments both the FDH and the full dynamical feedback version were compared with the control run. An important feature of the model is that it includes a prognostic equation for the ozone mixing ratio, thus ensuring that the sources and sinks of ozone will be in balance with the dynamics in the control run.

Christiansen *et al.* (1997) pointed out that in all experiments the main effect of the dynamics was a weakening of diabatic meridional circulation and an associated latitudinal smoothing of the temperature response. The weakening of the diabatic meridional circulation is in agreement with a reduction in wave forcing associated with an extended tongue of low refractive index close to the tropopause. The FDH approximation, in general, represents the pattern of the temperature response well, but the strength of the response is altered when dynamical effects are allowed. In the lower stratosphere this adjustment is 25–50% of the fixed dynamical heating response, but at the 10 hPa level it reaches 100%. In both uniform reduction experiments the FDH temperature response is negative everywhere except in the polar night region, where a small heating is observed. In the lower stratosphere the cooling is strongest in the tropical region at 20 hPa. Here the cooling is −12 K and −25 K in the 50% and 75% reduction, respectively. In the upper stratosphere and in the mesosphere the response is large with values up to −35 K and −75 K, respectively. Responses in the ozone hole-type experiment differ in several aspects from the uniform reduction experiment (Christiansen *et al.*, 1997).

Mean zonal wind is very stable to uniform reductions, but the strength of the winter stratospheric jet increases drastically in the Antarctic ozone hole-type experiment. In the control experiment, variability in the winter hemispheres is oscillatory in nature, and there is no sign in the 300-day experiment that the oscillation is merely a transient phenomenon, but a firm conclusion about the persistence and periodicity of the oscillation requires a simulation of longer duration. Since the timescale of the oscillation is close to one year, phase locking with the annual cycle is often employed.

The results obtained by Christiansen *et al.* (1997) do not confirm the hypothesis suggested by Kiehl and Boville (1988) that the climate response to ozone depletion is non-linear over the range 0–75% in the sense that only ozone reduction above a certain threshold value would lead to significant climate response. For instance, the dynamical response in the 75% reduction experiment is almost identical to the dynamical response in the 50% reduction experiment. If the threshold exists then it must be higher than 75%.

Christiansen *et al.* (1997) emphasized that they have not considered long-term interactions in the climate system (such as the atmosphere–ocean interaction). This is why the troposphere, as a whole, is relatively insensitive to imposed changes in ozone concentrations. The surface–troposphere temperature response is generally of the order of 1–2 K or smaller, whereas in the stratosphere similar changes are of the order of 5–10 K or more.

Sinha and Toumi (1997) made some assessments which indicate that a surface temperature–lightning–O_3 feedback, currently absent in general circulation models, could significantly affect anthropogenic climate change. They pointed out that lightning is a major source of NO_x in the troposphere and, thus, of tropospheric ozone. On the other hand, Toumi *et al.* (1996) suggested that tropospheric ozone, especially in the tropics, may increase substantially as a result of more frequent lightning flashes under warmer climate conditions, thus acting as a positive radiation feedback on climate change. Sinha and Tuomi (1997) undertook a study to quantify the extent of this forcing and the associated positive climate feedback.

In the context mentioned, Sinha and Toumi (1997) carried out numerical modeling that incorporated tropospheric ozone–surface temperature parameterization of the last glacial maximum and of a climate in which greenhouse gas concentrations had doubled. A 1-D radiative–convective model was applied, in which CO_2, CH_4, tropospheric O_3, and—in the case of paleoclimate simulation—surface albedo were varied both independently and in combination.

Table 7.18 shows the results of the model runs driven by changes in climate variables since the last glacial maximum. The initial state to which perturbations were applied was that of the present day climate. The parameter $\varepsilon = \Delta T_s/\Delta F$, where ΔT_s is surface temperature change, ΔF is the applied RF ($W\,m^{-2}$) at the tropopause

Table 7.18. Results from perturbing the climate according to paleoclimate data (Sinha and Toumi, 1997).

Case	*Variable(s) changed*	*Change in variable*	ΔT_s (K)	E_i (K/(W m^{-2}))
	Single-parameter perturbations			
1	CO_2	355 to 175 ppmv	−2.22	0.796
2	CH_4	1,720 to 300 ppbv	−0.66	0.572
3	R_s	0.139(0.331[a]) to 0.208(0.358[a])	−5.25	0.534
4	O_3 (no LFB[b])	−49%	−0.31	0.554
5	O_3 (with LFB)	−72%[c]	−0.48	0.559
	Effect of ozone on paleoclimate simulations			
6	CO_2, CH_4, R_s	As Cases 1, 2, and 3	−8.13	0.579
7	CO_2, CH_4, R_s, O_3 (no LFB)	As Cases 1, 2, 3, and 4	−8.46	0.602
8	CO_2, CH_4, R_s, O_3 (with LFB)	As Cases 1, 2, 3, and 5	−8.69	0.618

[a] Planetary albedo.
[b] LFB denotes lightning–ozone feedback.
[c] Taken from the equilibrium solution of Case 8.

(which is called in accordance with IPCC terminology "instantaneous RF"), and ε is the instantaneous climate sensitivity parameter. The upper part of Table 7.18 characterizes the relative importance of individual parameters (Cases 1 to 5). As can be seen, the largest surface cooling is due to surface albedo perturbations, while the highest T_s sensitivity to a given climate forcing (ε value) takes place in the case of CO_2. The results for the full simulations (Cases 6 to 8) show that removing industrial tropospheric O_3 (−50%) increases the temperature drop by 0.33 K (compare Cases 6 and 7), whereas the decrease in O_3 due to LFB (a further −22%) adds an additional cooling of 0.23 K (Case 7 versus Case 8). Thus, according to Sinha and Toumi (1997) simulation of ozone feedback has an effect that is roughly two thirds of that due to industrialization alone. Overall, the effect of changes in tropospheric ozone from both sources increases the temperature drop by around 4–7% (i.e., plays only a minor role in determining ΔT_s between the last glacial maximum and the contemporary climate). The dominant contributions belong to surface albedo and CO_2 changes.

Table 7.19 illustrates the situation in the case of global warming due to GHG increases. Being much smaller than the warming due to CO_2 doubling (Case 9, $\Delta T_s = 2.20$ K), $\Delta T_s = 0.23$ in Case 11 of tropospheric O_3 change still illustrates the significance of taking LFB into account. This in contrast to the ice age

Table 7.19. Results from perturbing the climate according to projected greenhouse gas emission data (Sinha and Toumi, 1997).

Case	*Variable(s) changed*	*Change in variable*[a]	ΔT_s (K)	ε (K/(W m^{-2}))
9	CO_2	355 to 175 ppmv	2.20	0.804
10	CH_4	1,720 to 300 ppbv	0.40	0.556
11	O_3	+34%[b]	0.23	0.869
12	CO_2, CH_4	As Cases 9 and 10	2.71	0.785
13	CO_2, CH_4, O_3	As Cases 9, 10, and 11	2.95	0.779

[a] In all cases water vapor feedback is included.
[b] Taken from the equilibrium results of Case 13.

simulation, indicating that tropospheric O_3 can significantly affect the magnitude of any warming episodes. The lightning-ozone feedback augments the surface temperature change by around 10%, when CO_2 and CH_4 concentrations are doubled.

Sinha and Toumi (1997) emphasized a number of uncertainties in the simulation results. In particular, the lack of explicit dynamics in the model together with the restriction to a single dimension excluded treatment of some potentially non-linear processes. There are also some other sources of significant uncertainties.

Saunois *et al.* (2008) reported on the response of tropospheric ozone to the dynamics of the West African monsoon, as well as surface emissions and NO_x production by lightning (LNO_x). For this purpose they used an idealized vertical–meridional zonally symmetrical version of the Meso-NH model, which is a non-hydrostatic mesoscale model intended to be applicable to all scales ranging from large (synoptic) scales to small (large eddy) scales and can be coupled with an atmospheric chemistry model (Fillipi *et al.*, 2011). An O_3–NO_x–VOC chemical scheme was added to the dynamical model, including surface emissions and a parameterization of LNO_x production. The model shows that ozone precursors emitted at the surface are uplifted by deep convection and then advected in the upper branches of Hadley cells on both sides of the Inter Tropical Convergence Zone (ITCZ). The NO_x produced by lightning promotes chemical ozone production in the middle and upper troposphere as a result of the oxidation of CO and VOCs. Analysis of the convective and chemical tendencies shows that the ozone minimum at the ITCZ is induced by venting of ozone-poor airmasses into the upper troposphere. The 2-D model suffers from limitations due to the absence of exchange with higher latitudes and ventilation in the zonal direction. Despite these restrictions, sensitivity simulations show that the LNO_x source and biogenic VOCs are necessary to create the meridional gradient of ozone observed by the Measurements of OZone and water vapor by in-service AIrbus airCraft (MOZAIC—Section 1.7.4.3) aircraft in the southern Hadley cell. The LNO_x source is also required to maintain the meridional ozone gradient up to 24°N in the northern Hadley cell. The modeled meridional gradient of O_3 in the upper troposphere ranges from 0.22 to 0.52 ppbv/deg without the LNO_x source and from 0.60 to 1.08 ppbv/deg with the LNO_x source in the southern and the northern cells, respectively.

A number of new results in the field of ozone–climate simulations were discussed during the First SPARC General Assembly (SPARC, 1997) and in the SPARC Implementation Plan, 2009 (*http://www.atmosp.physics.utoronto.ca/SPARC/index.html*).

Rind *et al.* (SPARC, 1997) undertook a new effort to simulate the doubled CO_2 climate with the use of the GISS (Goddard Space Flight Center) GCMAM in a series of experiments, in which sea surface temperatures were allowed to adjust to radiation influences. The integration for 58 years indicates global warming by 5.1°C at the surface, while the stratosphere cools by up to 10°C. When ozone is allowed to respond photochemically, stratospheric cooling is reduced by 20%, with little effect on the troposphere. The effect of planetary wave energy increases in the stratosphere, producing dynamical warming at high latitudes, weakens in the lower stratosphere, when O_3 chemistry on PSC impacts are included.

Discussing the specific consequences of including ozone chemistry in a coupled chemistry–climate model, Shindell *et al.* (1997b) analyzed the response of ozone to a doubled CO_2 climate. The results indicate that the ozone response is opposite to temperature changes, so that cooling in the upper stratosphere results in increased ozone abundances, while lower-stratospheric warming reduces ozone in that region. In contrast to the other models, a net TOC decrease (0–2%) at the mid-latitudes was obtained due to large tropospheric warming. Taking account of the impact of parameterized heterogeneous chemistry on PSCs results in simulation of a significantly larger Antarctic ozone hole of longer duration in the doubled CO_2 atmosphere. In addition, an ozone hole of variable strength, roughly one quarter to two thirds that of the current Antarctic hole, forms in the Arctic in four out of five years as a result of overall cooling in the polar stratosphere combined with fewer sudden warmings in the doubled CO_2 atmosphere.

Graf *et al.* (1995) came to the conclusion that observed winter trends in atmospheric circulation since 1957 are characterized in the stratosphere by intensification of the polar night vortex and in the troposphere by intensification of westerlies over the North Atlantic. Moreover, not only has the mean winter strength of the stratospheric polar vortex increased, but there was a tremendous increase in the number of winter months with a strong polar vortex in the stratosphere. The ratio of the number of winter months experiencing a strong versus weak polar night vortex increased from 1.1 (1957/1958–1971/1972) to 6.8 (1981/1982–1995/1996). Perlwitz *et al.* (1997) explored two possible explanations of this phenomenon: (1) the effect of increasing greenhouse gas concentrations; (2) the effect of decreasing stratospheric ozone concentration. Since observational time series are too short for this kind of analysis, simulation modeling was undertaken with the use of the coupled climate model ECHAM3/LSG/T21 developed at the Climate Center of the University of Hamburg. The principal results obtained by Perlwitz *et al.* (1997) indicate that:

- the observed changes in stratospheric circulation in winter are consistent with greenhouse gas forcing. In winter they are not consistent with forcing by reduced ozone. However, the ozone distribution depends on changes in stratospheric circulation.
- In spring, greenhouse gas forcing and ozone act in the same direction. Both keep the polar night vortex persistent.

Thus, the results suggest that ozone anomalies do not actively change the circulation except in spring.

Baumgaertner *et al.* (2010) investigated the effects of strengthened stratospheric/mesospheric residual circulation on the transport of NO produced by energetic particle precipitation (EPP). During periods of high geomagnetic activity, EPP is responsible for wintertime ozone loss in the polar middle atmosphere between 1 and 6 hPa. However, as climate change is expected to increase the strength of Brewer–Dobson circulation including extratropical downwelling, enhancements of EPP NO_x concentrations are expected to be transported to lower altitudes in extratropical regions, becoming more significant in the ozone budget. They also considered changes in mesospheric residual circulation. They used simulations with the chemistry–climate model system EMAC to compare present day effects of EPP NO_x with expected effects in a climate change scenario for 2100. In years of strong geomagnetic activity, similar to that observed in 2003, additional polar ozone loss of up to 0.4 pmol/mol at 5 hPa is found in the Southern Hemisphere. However, this would be approximately compensated by ozone enhancement originating from stronger poleward transport of ozone from lower latitudes caused by strengthened Brewer–Dobson circulation, as well as by slower photochemical ozone loss reactions in a stratosphere cooled by increased greenhouse gas concentrations. In the Northern Hemisphere the EPP NO_x effect appears to lose importance due to the different nature of climate change-induced circulation changes.

A new important step in the assessment of our current understanding of ozone climate impact was made by the *Scientific Assessment of Ozone Depletion: 2006* (WMO, 2007) and earlier in the Fifth Session of the Scientific Steering Group on Stratospheric Processes and Their Role in Climate, which took place at Port Jefferson, N.Y., November 17–20, 1997 (Chanin, 1998). The principal purpose of the session was discussion of the results and perspectives of making quantitative estimates of the impact of stratospheric processes on climate; in particular, the impacts of optically active components of the stratosphere such as ozone and aerosols, which are present in both the upper troposphere and lower stratosphere. A number of scientific problems are of interest in this context, especially stratosphere–troposphere exchange, the influence of ozone and aerosols on climate change, and many others. Despite the great significance of the upper troposphere and lower stratosphere for climate formation, relevant observations are not sufficient as yet and numerical modeling requires further development.

The dynamics of the upper troposphere and the chemistry of the lower stratosphere are first-priority problems. One of the concrete tasks of observation data analysis is to study the dependence of global-warming (GW) characteristics entering the stratosphere on the intensity, temporal change, and small-scale structure of convection in the troposphere as well as to verify satellite information on convection induced by gravity waves. Special attention should be paid to studying the interaction between the microphysics of cirrus clouds and chemical, radiation, and dynamic processes.

In this regard, Yoshiki and Sato (2000) examined GW in the lower polar stratosphere by analyzing operational radiosonde observations for 10 years. Both the potential energy and kinetic energy of gravity waves were found to vary annually and reach maxima in winter in the Arctic and in spring in the Antarctic. Sato and Yoshiki (2008) analyzed intensive radiosonde observations at Syowa Station (69.0°S, 39.6°E) to examine GW characteristics in the Antarctic lower stratosphere. Clear signals of GW whose phases propagate upward, suggesting downward energy propagation, were detected in June and October when the polar night jet (PNJ) was present. On the other hand, downward phase propagation (i.e., upward energy propagation) components were dominant in all months. It was also found that there is a spectral peak around the inertial frequency in a wide range of vertical wavenumbers in December when the background wind is weak, whereas large spectral densities are distributed over lower frequency regions in June and October. It was also found that upward and downward-propagating wave packets observed simultaneously in the same height regions have similar horizontal wavelengths and phase velocities. This fact suggested that these GWs are generated from the same source with a similar mechanism. When the wave packets were observed, both the local Rossby number and the residual in the non-linear balance equation—estimated using NCEP–NCAR reanalysis data—are large around the PNJ situated slightly south of Syowa Station. Therefore, it is likely that observed inertia–gravity waves are generated by spontaneous adjustment around the geostrophically unbalanced PNJ and propagate toward Syowa Station.

From the viewpoint of stratosphere troposphere exchange (STE) the following priorities were defined: STE quantitative characteristics and their annual variations for various minor gas components (MGCs) including ozone, water vapor, and others; the mesoscale structure of MGC spatial distributions; processes in the tropical tropopause layer and water vapor; the causes of model failures in the UT/LS layer. To solve these and other problems it is particularly necessary to obtain more complete and reliable data for: tropopause height; MGC concentration in the UT/LS layer; wind field in the tropics (in view of its significance for MGC transport); concentrations of water vapor and its condensates within the tropopause layer. Intercomparison between observation data and numerical modeling results is especially important.

There is still not enough information on the lower stratosphere to explain the destruction of ozone due to Cl and Br compounds. Further analysis is necessary of the impact of volcanic eruptions (such as El Chichón and Pinatubo) with regard to stratospheric aerosol influence on temperature and ozone concentration in the lower stratosphere. The impact of subsonic and supersonic aviation on the upper troposphere/lower stratosphere (including climate change) still remains unclear (Borrmann *et al.*, 2010; Köhler *et al.*, 2008).

Various observations have shown stabilization or a weak increase in the stratospheric ozone layer since the late 1990s. Recent coupled chemistry–climate model simulations have predicted that the stratospheric ozone layer will likely return to pre-1980 levels in the middle of the 21st century, as a result of the decline of ODSs under the 1987 Montreal Protocol. Since the ozone layer is an important component in determining the stratospheric and tropospheric–surface energy balance, the recovery of the ozone layer may have a significant impact on the tropospheric–surface climate. Hu *et al.* (2011), using multimodel ensemble results from both the IPCC Fourth Assessment Report (IPCC/AR4) models and coupled chemistry–climate models, showed that the troposphere is warmed more than would be expected without ozone recovery, suggesting an enhancement of tropospheric warming due to ozone recovery. It was predicted that enhanced tropospheric warming would be most significant in the upper troposphere, with a global mean warming magnitude of 0.41 K for 2001–2050. They also predicted that relatively large enhanced warming would occur in the extratropics and polar regions in summer and autumn in both hemispheres while enhanced warming would be stronger in the Northern Hemisphere than in the Southern Hemisphere. Enhanced warming was also predicted at the surface, with the strongest enhancement predicted to be located in the Arctic in boreal winter. Global annual mean enhancement of surface warming of about 0.16 K for 2001–2050 was also predicted.

The upper troposphere/lower stratosphere is that part of the atmosphere which is characterized by low temperatures and high ozone concentration. The processes that are responsible for ozone changes here are far from being clear (especially with regard to the relationship between *in situ* formation and stratosphere–troposphere exchange). Special attention should be paid to specific features of processes in the upper troposphere and lower stratosphere (rapid mixing in the upper troposphere, but slow mixing in the lower stratosphere; different NO_x concentrations in these layers; etc.). An important aspect of the problems under discussion is the influence of the Quasi-Biennial Oscillation (QBO) in the stratosphere on the troposphere and its role in STE (Hsu and Prather, 2009). The trends of stratospheric temperatures and vertical ozone concentration profiles are especially interesting as stratospheric indicators of climate change. In this context, the impact of ozone depletion on the cooling trend in the lower stratosphere in middle and high latitudes deserves serious attention.

A rather controversial situation exists in the upper stratosphere where not only the amplitude but even the sign of temperature variations changes depending on the latitude. A number of problems are relevant to water vapor dynamics in the lower stratosphere including: further accumulation of observation data on water vapor climatology in the lower stratosphere; studies of those processes that are responsible for water vapor changes; filtering out long-term trends of water vapor content; development of models that are able to adequately simulate water vapor and impacts on water vapor changes (Johny *et al.*, 2009; Noel *et al.*, 2010).

The intercomparison of 13 simulation models in the GRIPS Project indicates that: (1) all models realistically simulate the latitude and altitude structure of the mean zonal state of the atmosphere, although certain biases take place with regard to temperature calculations, location and intensity of basic jetstreams, etc., the causes of which have not been understood as yet; (2) there are sometimes significant disagreements between calculated and observed planetary waves in the troposphere and stratosphere of the Northern Hemisphere; (3) the quality of simulations is fairly low when there are low-frequency oscillations in the tropical stratosphere (this is especially true for QBO simulations).

A number of field programs are being directed toward solving these and other problems. A substantial contribution is being made by satellite remote-sensing data from UARS/HALOE, SAGE II, SAGE III, among others. The *International Workshop on Stratospheric Change and Its Role in Climate*, which was held in Tsukuba (Japan), October 20–22, 1997, was devoted to discussing SPARC problems in the context of the ATMOS-C1 satellite observation program (Uchino, 1998). During the first stage of this program (1995–1997) the following aims were reached: (1) monitoring stratospheric variability; (2) investigating chemical processes in the stratosphere; (3) observing and numerical modeling of stratospheric changes and their impact on tropospheric climate.

The principal aims of the ATMOS-C1 mission were remote sensing of the optically active minor components of the stratosphere as well as monitoring stratospheric ozone and climate parameters. A specific task is the study of stratospheric temperature decrease due to ozone depletion at various levels (adequate information on vertical ozone concentration profile changes is very important in this respect, as are the dynamics of stratospheric warming). Priorities include: interaction between gravity waves and convection; climatic consequences of water vapor changes in the stratosphere; interrelationships between the frequency of hurricanes in the Atlantic and El Niño events; consideration of cloud and ozone contributions in predictions of anthropogenic climate change; polar ozone depletion and climate change; interactions between the stratosphere and troposphere, between processes in the high and mid-latitude stratosphere; small-scale processes and ozone change near the UT/LS boundary.

An important step in studying ozone as a climate gas was marked by the papers presented at the *European Geophysical Society Assembly* in Vienna (*Annales Geophysicae*, 1998). For instance, Christiansen presented a series of general circulation model experiments with prescribed ozone perturbations in different vertical intervals. The results were discussed by Christiansen *et al.* (1997) who emphasized that climate is particularly sensitive to ozone changes occurring in the vicinity of the tropopause. The large value of climate sensitivity and the corresponding large positive feedback found for perturbations in the upper stratosphere should be contrasted with the relatively weak feedbacks found both for ozone perturbations in the lower stratosphere, in the troposphere, and in a doubled CO_2 experiment (*Annales Geophysicae*, 1998).

Fortuin *et al.* (*Annales Geohysicae*, 1998) demonstrated the importance of using adequate observational ozone data to simulate the climatic influence of ozone changes. It was found that the systematically lower ozone values of the new climatology around the tropopause leads to lower temperatures there, accompanied by increased convection and high cloudiness, especially in the tropics. The influence of the ozone hole over Antarctica—in contrast to the ozone spring maximum of the previous climatology—primarily involves cooling and enhancement of the polar vortex not only in October but also in November when the ozone hole has almost recovered.

Ponater *et al.* (1999) pointed out that aircraft-induced NO_x emissions lead to significant modifications of atmospheric ozone concentrations and the resulting climate impact is also important, even for present day subsonic air traffic (a comprehensive 3-D model of the atmosphere–ocean system was applied for these assessments). However, it was emphasized that the relation between RF and climate change (in terms of surface temperature change) is not as straightforward as usually found in greenhouse gas experiments. Even in the case of present day aircraft emission levels there is considerable non-linear dependence on the global mean temperature response from RF. RF seems to be a bad predictor of future aircraft scenarios.

The ozone impact on climate connected with the consequences of subsonic and supersonic aircraft flights was discussed by Brunner *et al.* (2003). Fortuin *et al.* (1995) assessed RF in northern mid-latitudes due to aircraft emissions up to 1990. The results indicate local RF at the tropopause which should be negative in summer (-0.5 to $0.0\,W\,m^{-2}$) and either negative or positive in winter (-0.3 to $0.2\,W\,m^{-2}$). Moreover, the indirect effect of contrails has to be added, which for the North Atlantic Flight Corridor covers the range from -0.2 to $0.3\,W\,m^{-2}$ in summer and 0.0 to $0.3\,W\,m^{-2}$ in winter. Fortuin *et al.* (1995) emphasized that major positive contributions come from contrails, stratospheric water vapor in winter, and ozone in summer. The direct effect of NO_2 is negligible and the contribution of CO_2 is relatively small. Sausen *et al.* (1997) pointed out that NO_x emissions from conventional subsonic aircraft significantly enhance (up to a factor of 2) the atmospheric NO_x concentration at levels near the tropopause of the northern extratropics. A number of model calculations indicate that enhanced atmospheric NO_x concentrations result in an increase in ozone concentration, in particular at altitudes between 400 and 50 hPa (i.e., in the vicinity of the tropopause) where RF induced by ozone increment reaches a maximum; an important aspect of the problem is the necessity to consider factors influencing ozone chemistry in aircraft plumes (Moulik and Milford, 1999).

Sausen *et al.* (1997) conducted several sensitivity experiments with the general circulation model ECHAM 4 to calculate the impact of RF changes on monthly averaged mean zonal air temperature. ECHAM 4 is a global spectral model at T21 horizontal resolution. Diabatic processes are calculated on a Gaussian grid of approximately $5.6^\circ \times 5.6^\circ$. The model has 19 vertical levels from the surface to the 10 hPa level. Land surface temperature and snow cover are calculated interactively, whereas sea surface temperature is prescribed according to a climatological annual cycle.

A set of five model simulations of 30 model years each in the annual cycle mode were performed: (1) control calculations (for climatological ozone distributions) and four versions of ozone perturbations (ΔO_3); (2) $\Delta O_3'$ perturbations due to the MOGUNTIA model; (3) $5 \times \Delta O_3'$; (4) $\Delta O_3''$ due to the CTMK model; (5) $5 \times \Delta O_3''$. Results indicate that in July (when both ozone increase and solar radiation in the Northern Hemisphere are largest) the temperature response due to NO_x emissions reaches its maximum. Although, at first glance, the signal looks rather chaotic (with warming and cooling extremes of similar magnitude), statistical tests (Student *t*-test) have revealed certain changes at the 95% significance level. Consistent features were discovered, such as: significant warming in the Northern Hemisphere troposphere, warming in the upper tropical troposphere, and cooling of the northern extratropical stratosphere. Sausen *et al.* (1997) emphasized, however, that—since atmospheric temperature changes are usually spatially coherent and the area with significant changes is small—the total response might be little more than noise.

Meijide *et al.* (2010) applied the fingerprint technique to separate signal from noise and demonstrate that ozone changes due to aircraft emissions could lead to a statistically significant change in meridional temperature profile. The magnitude of the signal appears to depend non-linearly on the magnitude of ozone increase. Zonal mean temperature changes are in the range of ± 0.2 K, which is about 5–10% of the response of the same model's simulation for doubling CO_2 in the upper troposphere. However, the signal due to ozone changes is less coherent. Sausen *et al.* (1997) pointed out some limitations of the results: (1) prescribing SST can lead to the temperature signal being dampened; (2) using the MOGUNTIA and CTMK models can lead to the temperature signal being enhanced.

Wyser and Ström (1998) emphasized that aircraft exhaust may reduce the crystal size in natural cirrus clouds and assessed the change in cloud RF as a result of such a size reduction by assuming a constant ice water content. Applying a 1-D model to compute the radiative transfer for an atmospheric column showed that negative short-wave cloud forcing is enhanced with smaller crystals as they mainly increase the reflectivity of clouds. The change in long-wave cloud RF is always positive although its magnitude depends strongly on the ice water path. Net RF depends on mean crystal size, surface albedo, and ice water content. It appears that there is a range of diameters between 15 and 25 μm where the response to reduction in crystal size is fairly insensitive. Below this range the change is negative, while above this range the change is positive. In regions where there is dense air traffic, the magnitude of the change in cloud forcing could be on the order of $0.3\,W\,m^{-2}$ under the assumption of a 20% decrease in mean crystal size from about 30 μm to 20 μm. Wyser and Ström (1998) emphasized that, although aircraft exhaust has the potential to affect climate, the results should be taken with caution as they are based on parameterized optical properties for cirrus clouds.

The substantial impact of solar activity on the ozone layer means that it is possible that climate change is due to solar activity-induced ozone variations (Kondratyev, 1998b). Studies into this possibility were carried out by Haigh (1994, 1996, 1998). Haigh (1998) applied the 3-D spectral (T42) UGCM model with 19 vertical levels (up to 10 hPa) that was developed within the U.K. Universities "Global Atmospheric Modelling Programme (UGAMP)" to simulate the impact of solar variability on climate. Two modeling versions were considered: (1) control run I represented minimum activity conditions during the solar 11-year cycle; (2) runs IIa and IIc represented solar maximum. In both solar maximum runs, solar irradiance is increased by 0.08%. The difference between the two solar maximum runs lies in the specification of the increase (from run I) of stratospheric ozone amounts.

The results obtained confirm the earlier conclusion of Haigh (1996) that changes in stratospheric ozone may provide a mechanism whereby small changes in solar irradiance can cause a measurable impact on climate. It has been proposed (Haigh, 1996) that changes in the thermal structure of the lower stratosphere could cause a strengthening of low-latitude upper-tropospheric easterly winds. New calculations (Haigh, 1998) indicate that this easterly acceleration can come about as a result of the tropopause lowering and deceleration of poleward meridional winds in the upper portions of Hadley cells. A model experiment—in which a very different change in ozone structure was assumed—produced results that are consistent with this hypothesis.

Varotsos (1989) studied the association between the solar cycle and TOC, suggesting that global TOC does not show any evident connection during the period 1958–1984 with the 10.7 cm solar flux (F10.7). However, when the data are separated according to the east or west phase of the Quasi-Biennial Oscillation (QBO) in the equatorial stratosphere, the following connection is found: when the QBO is in its west phase, global total ozone is positively correlated with the solar cycle; the opposite holds for the east phase of the QBO. Labitzke and van Loon (1997b) studied the annual mean and interannual variability of TOC (TOMS data for 1979–1993) to assess spatial correlations of ozone with the 11-year sunspot cycle. They found that correlations reach their highest and most statistically significant values between 5° and 30° latitude in either hemisphere, but especially in the summer hemisphere. The equatorial regions where ozone is produced show comparatively small statistically insignificant correlations; and the subpolar regions where ozone is most plentiful show little or no correlation with the solar cycle.

Note that stratospheric processes are only poorly represented in IPCC/AR4. For example, climatological or simple specifications of time-varying ozone concentrations are imposed and the QBO is absent. Dall'Amico *et al.* (2010) investigated the impact of an improved stratospheric representation using two sets of perturbed simulations with the Hadley Centre coupled ocean–atmosphere model HadGEM1 with natural and anthropogenic forcings for the 1979–2003 period. In the first set of simulations, the usual zonal mean ozone climatology with superimposed trends was replaced with a time series of observed zonal mean ozone distributions that included interannual variability associated with the solar cycle, the QBO, and volcanic eruptions. In addition to this, the second set of perturbed simulations included a scheme in which the stratospheric zonal wind in the tropics is relaxed to appropriate zonal mean values obtained from European Center for Medium Range Weather Forecasts (ECMWF) updated reanalysis (ERA-40), thus forcing a QBO. The contribution of the solar cycle signal in the ozone field to this improved representation of stepwise cooling due to volcanic eruptions is discussed. The improved ozone field in addition

Table 7.20. Climate impacts of 20% reduction in NO_x (Fuglestvedt *et al.*, 1998).

Region	$\Delta(CH_4)$ (ppb)	$\Delta F(CH_4)$ (10^{-2} W m^{-2})	$\Delta F(CH_4)/\Delta E_{NO_x}$ (10^{-2} W m^{-2}/Tg N yr^{-1})	$\Delta F(O_3)/\Delta E_{NO_x}$ (10^{-2} W m^{-2}/(Tg N yr^{-1}))
Australia	7.27	0.317	−1.88	—
S. Europe	3.85	0.168	−0.43	—
E. Europe	3.22	0.141	−0.32	—
Scandinavia	0.34	0.015	−0.35	0.29
SE Asia	11.5	0.501	−2.12	2.37
U.S.A.	15.7	0.684	−0.45	0.35

to the QBO resulted in improved simulation of observed trends, both globally and at tropical latitudes.

An important contribution to studying the climatic implications of ozone and ozone-related processes was marked by the papers presented at the *International Ozone Symposium* in 1996 (Bojkov and Visconti, 1998). As a continuation of earlier investigations into tropospheric ozone as a climate gas, Wang and Isaksen (1995) and Fuglestvedt *et al.* (1998) assessed the impacts of reduced NO_x emissions on RF through changes in tropospheric O_3 and CH_4. For this purpose they selected six geographical regions representing different climatic and physical conditions and prescribed a 20% reduction of NO_x emissions for these regions. Table 7.20 illustrates some of the results. In the table $\Delta(CH_4)$ and $\Delta F(CH_4)$ are changes (due to NO_x reduction) of CH_4 concentration and radiance forcing, respectively. A sensitivity factor is defined as $\alpha = \dfrac{\Delta\bar{O}_3}{\Delta E_{NO_x}}$ where $\Delta\bar{O}_3$ (ppbv yr^{-1}) is global mean ocean ozone change and ΔE_{NO_x} (Tg N yr^{-1}) is the change in NO_x emissions. Fuglestvedt *et al.* (1998) emphasized that changes in NO_x emissions will not only affect O_3 but also OH and, thereby, also the level of CH_4. The following sequence of events will take place:

$$\Delta E_{NO_x} \rightarrow \Delta[OH]_0 \rightarrow \Delta[CH_4]_0 \rightarrow \Delta[OH] \rightarrow \Delta[CH_4] \rightarrow \Delta[OH] \rightarrow \text{etc.}$$

The results obtained by Fuglestvedt *et al.* (1998) indicate large variations in the sensitivity of chemical responses and RFs due to changes in O_3 and CH_4 between the regions. An important conclusion is that O_3 and CH_4 forcings are of opposite sign and generally of similar magnitude. Although the calculated O_3 forcing due to a 20% reduction in NO_x emissions is noticeable on a regional basis, in terms of global annual averages the effects are negligible. The general conclusion is that emissions of NO_x have complex and potentially important impacts on climate through atmosphere–chemistry interactions.

Table 7.21. Long-wave, short-wave, and net RF due to stratospheric ozone; global, Northern Hemispheric, and Southern Hemispheric averages, for the four seasons (Dec., Jan., Feb.—DJF, etc.) and the yearly average (YA) (Myhre *et al.*, 1998).

Forcing	*DJF*	*MAM*	*JJA*	*SON*	*YA*
			Global average		
Long wave	−0.148	−0.177	−0.173	−0.167	−0.167
Short wave	0.188	0.217	0.235	0.212	0.213
Net	0.039	0.040	0.061	0.044	0.046
			Northern Hemisphere average		
Long wave	−0.127	−0.252	−0.246	−0.151	−0.194
Short wave	0.135	0.272	0.329	0.192	0.232
Net	0.008	0.020	0.082	0.041	0.038
			Southern Hemisphere average		
Long wave	−0.170	−0.103	−0.100	−0.184	−0.139
Short wave	0.240	0.163	0.141	0.232	0.194
Net	0.071	0.060	0.040	0.048	0.055

Myhre *et al.* (1998) calculated RF and stratospheric temperature change due to changes in stratospheric ozone during the period from 1960 to 2004 using a 2-D stratospheric chemistry–transport model and taking into account both the short-wave and long-wave components of RF. Table 7.21 shows some of the results. More detailed results revealed (in accordance with earlier results) that RF at the tropopause is very sensitive to the altitude of stratospheric ozone changes.

While ozone reductions in the upper stratosphere give positive RF (especially at low latitudes), ozone reductions in the lower stratosphere give negative RF. As is typical for atmospheric processes in general, the final results are small differences of much larger values (Table 7.21). Net RF is positive for all seasons and for the yearly average. RF in 1996 with 1969 as a reference year is calculated to be $0.05\,W\,m^{-2}$. Previous RF calculations based on observed changes in ozone from the SAGE satellites found a negative forcing of $-0.1\,W\,m^{-2}$ (SAGE data probably overestimate ozone reduction, which is much smaller according to ozonesonde data). Myhre *et al.* (1998) calculated that, on the global scale, heating of the troposphere–surface system due to ozone reductions in the upper and middle stratosphere nearly balances the cooling due to reductions in lower-stratospheric ozone. This is why RF due to stratospheric ozone is probably very small on the global scale.

Wauben *et al.* (1998) investigated the sensitivity of RF (net downward radiation at the tropopause level) due to changes in the global distribution of ozone with reference to ozone changes from aviation. The global distribution of RF was computed for a horizontal grid (10° longitude × about 8° latitude) for January and July. The results indicate that RF is generally most sensitive to ozone changes just below the tropopause (increasing from the surface up to the tropopause and rapidly decreasing in the stratosphere), with the highest RF sensitivity in the tropics and strong seasonal dependence in the polar regions. Sensitivity is characterized not only by large latitudinal variations, but also longitudinal ones. The assessment of ozone and relevant RF changes due to aircraft NO_x emissions show that RF attains maximum values of 0.13 $W m^{-2}$ in the Mediterranean in July, whereas maximum values over the northern subtropical region of the Atlantic (0.07 $W m^{-2}$) take place in January. These maxima are located a bit more equatorward of the regions of maximum ozone perturbation.

Wauben *et al.* (1998) emphasized that aircraft emissions do not only affect atmospheric composition by changing the ozone concentration, but other components are changed as well (see also Section 5.2). The impacts of aerosols and contrails on RF are especially interesting in this context. There are serious uncertainties, however, about the accuracy of assessments of aircraft-induced RF changes.

Using TOMS total ozone data from Nimbus-7 and Meteor-3 satellites for the period from 1979 to 1994, Vargin (1998) investigated the non-zonal global structure of TOC (especially for the Southern Hemisphere) and calculated the non-zonal atmospheric heating rates due to ozone absorption of solar UV radiation in the Hartley (200–300 nm), Huggins (300–350 nm), and Chappuis (450–750 nm) bands (similar calculations were made earlier by Bojkov, 1976). The principal conclusion made by Vargin (1998) was that large-scale ozone disturbances were regularly observed in September during the whole period considered within the 55°S–70°N latitude belt. The maximum amplitudes of the first TOC harmonics in the Southern Hemisphere reached about 100 DU (September 1988). The strong QBO contribution to long-term TOC variations at high latitudes (55°S–70°S) was observed. The non-zonal part of ozone-induced solar radiative heating in September 1988 was about 2 K per day in the 40–45 km layer and 0.2 K per day at the 20 km altitude.

Christiansen and Guldberg (1998) conducted a general circulation model study of the dynamical and radiative effects of ozone perturbations, including uniform 50% and 75% ozone reductions, and a hole-type reduction concentrated in the lower part of the stratosphere and the upper part of the troposphere (see also Christiansen *et al.*, 1997). An earlier study by Ramaswamy *et al.* (1996) revealed possible cooling of the lower stratosphere due to dynamical feedback from latitudes where no ozone perturbations are imposed. This strengthened the belief that ozone depletion may be responsible for most of the observed cooling at all latitudes in the lower stratosphere. Christiansen and Guldberg (1998) emphasised in this context that changes in the dynamical structure of the stratosphere may alter the stratospheric temperature response and, thereby, RF on the troposphere. Hence, the importance of studying the role of dynamical feedback in RF on the troposphere–surface system.

To study the contribution to feedback made by various aspects of ozone impacts a comparison was made between pure radiative effects (no thermal or dynamical adjustment), the effect of thermal adjustment alone (the fixed dynamical heating or FDH approximation), and the effect of the full dynamical system (Arpège model). The principal conclusions made by Christiansen and Guldberg are that: (1) the major effect of dynamical adjustment is a weakening of meridional circulation and a smoothing of the temperature response in the meridional direction (these results question the validity of the FDH approach); (2) dynamical temperature adjustment in the lower stratosphere plays an important role in the latitudinal distribution of RF; (3) in agreement with previous studies thermal adjustment in the lower stratosphere changes the sign of the forcing.

For example, models consistently project a decrease in tropical lower-stratospheric ozone associated with increased tropical upwelling. Such a decrease in lower stratospheric tropical ozone has in fact been observed (Randel and Wu, 2007), but it was attributed to climate change, not to CFCs, and so is not expected to reverse in the future. Using the CCMVal-1 model simulation archive, Son *et al.* (2008) showed that recovery of the Antarctic ozone hole should lead to a reversal of the observed Southern Annular Mode (SAM) trend over the next half century. Such a reversal is not predicted by IPCC/AR4 models and even those with imposed ozone recovery do not predict as large a change in the SAM trend as was found in the high top CCMs (Figure 7.11). This demonstrates the importance of a fully coupled representation of ozone and climate in a stratosphere-resolving model.

In the early days of climate models used to estimate rises in temperature due to increases in carbon dioxide concentration little attention was paid to ozone. However, stratospheric ozone recovery in the Southern Hemisphere is expected to drive pronounced trends in atmospheric temperature and circulation from the stratosphere to the troposphere in the 21st century; therefore, ozone changes need to be accounted for in future climate simulations (Karpechkol *et al.*, 2010). Many climate models do not include interactive ozone chemistry and rely on prescribed ozone fields, which may be obtained from coupled chemistry–climate model (CCM) simulations. However, CCMs vary widely in their predictions of ozone evolution, complicating the selection of ozone boundary conditions for future climate simulations. In order to assess which models might be expected to better simulate future ozone evolution, and thus provide more realistic ozone boundary conditions, Karpechkol *et al.* (2010) assessed the ability of 12 CCMs to simulate observed ozone climatology and trends and ranked these models according to their errors averaged across the individual diagnostics chosen. According to their analysis no one model performs better than the others in all the diagnostics; however, combining errors in individual diagnostics into one metric of model performance allowed them to rank the models objectively. The multimodel average showed better overall agreement with observations than any individual model. Based on this analysis it was concluded that the multimodel average ozone projection presents the best estimate of future ozone evolution and they recommended it for use as a boundary condition in future climate simulations.

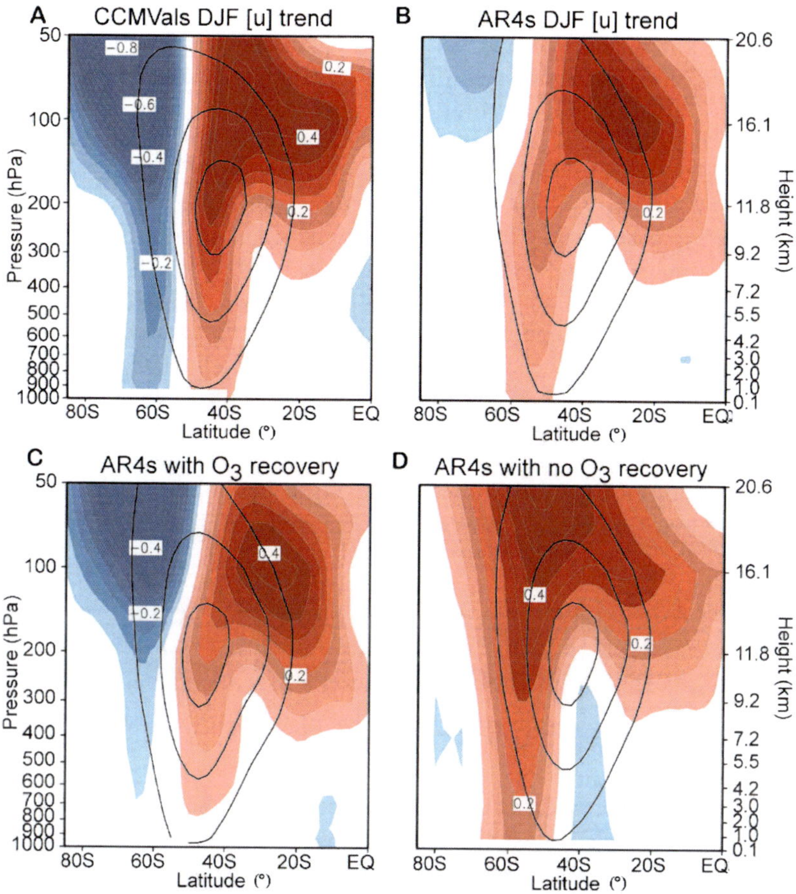

Figure 7.11. Trends in December to February (DJF) zonal mean zonal wind. Multimodel mean trends between 2001 and 2050 are shown for CCMVal models (a), AR4 models (b), AR4 models with prescribed ozone recovery (c), and AR4 models with no ozone recovery (d). Shading and contour intervals are $0.05\,ms^{-1}$ $decade^{-1}$. Deceleration and acceleration are indicated in blue and red, respectively, and trends weaker than $0.05\,ms^{-1}$ $decade^{-1}$ are omitted. Superimposed black solid lines are DJF zonal mean zonal wind averaged from 2001 to 2010, with a contour interval of $10\,ms^{-1}$, starting at $10\,ms^{-1}$. EQ, equator (Son *et al.*, 2008).

7.3.1 The health impacts of changes in ozone concentration

An important issue attracting a great deal of interest in the international community concerns the health impacts from climate change-induced changes in ozone variability.

In this context, Hogrefe *et al.* (2004) studied the health impacts from climate change-induced changes in ozone levels in 85 U.S. cities. The results obtained indicate that the climate change scenario would produce higher ambient ozone levels, with an average increase of 2.8 ppb in daily average ozone (range −0.1 to

6.4 ppb). Daily 1-hour and 8-hour maxima increased for all 85 cities, with an average increase of 4.6 and 4.2 ppb, respectively. Elevated ozone concentrations from global warming in the 2050s are estimated to produce a 0.25% increase (95% confidence interval 0.14, 0.36%) in daily mortality, averaged across the 85 cities. Similar results were recently reported by Selin *et al.* (2009) by using the MIT Emissions Prediction and Policy Analysis (EPPA) model. In this study they compared the economic effects of air pollution for the present day and for 2050 specified by different chemistry–climate models, including the NCAR Community Climate Model and the GEOS-Chem global tropospheric chemistry model, in order to assess a range of possible outcomes. Among their findings is that climate change and local air pollution are closely linked. Climate change can influence the magnitude and severity of local air pollution and changes in the amount of air pollutants can have implications for climate change. For example, increases in temperature and stagnation due to climate warming may increase the severity and frequency of ozone and aerosol pollution episodes. Caps on cooling aerosols, such as sulfate, and ozone precursors, such as NO_x and VOCs, imposed to control local air pollution can influence the global climate and can have ancillary climate benefits.

7.4 CONCLUSIONS ON TROPO-STRATOSPHERIC VARIABILITY

The presence of ozone in the atmosphere results in a number of problems which should be considered in the context of global environmental change. Of course, the most important problem is conservation of the ozone layer in the stratosphere for the protection of life on the Earth against harmful ultraviolet solar radiation. The cessation of CFC emissions into the atmosphere (in accordance with the Montreal Protocol) is a prerequisite to solving this problem. A number of important unsolved scientific questions remain unanswered, which indicates the necessity for further study of those photochemical processes in the stratosphere that are responsible for ozone layer dynamics. Studies of the impacts of solar activity on the ozone layer, especially in connection with the potential role of stratospheric processes in solar climatic interrelations, play a special role in this context. Heterogeneous chemical reactions on the surfaces of aerosol particles of various (natural and anthropogenic) origins and the impact of stratospheric aviation constitute further important aspects of stratospheric ozone studies.

The impact of ozone on human health and on ecosystems as well as in connection with global climate changes (tropospheric ozone as a greenhouse gas) are further examples of the significance of tropospheric ozone studies.

The fact that all these problems are interrelated (Bojkov and Visconti, 1998; Kondratyev, 1989, 1998a; Schneider *et al.*, 1993; SPARC, 1998) demonstrates the necessity to apply a systematic approach to solving the relevant problems. Since ozone problems are located at the "boundary" between the World Climate Research Program (WCRP) and the International Geosphere–Biosphere Program (IGBP) it would make sense to work out a special interdisciplinary ozone program embracing the specific ozone problems that need to be solved by developing a better global ozone observing system and making further progress in numerical simulation modeling.

The *Scientific Assessment of Ozone Depletion: 2006* (WMO, 1999) comprehensively analyzes present day results and future perspectives. Therefore, we shall limit ourselves to just a few brief comments relevant to some unsolved problems.

7.4.1 Stratospheric ozone dynamics and its determining factors

This complicated problem is connected with the necessity of explaining the numerous dynamic and chemical processes that determine changes in total ozone (TOC) under conditions of anthropogenic forcings due to emissions to the atmosphere of ozone-destroying substances (ODSs) such as CFCs. To do this requires, apart from chemical processes (principally, heterogeneous chemical reactions on atmospheric aerosol particles), the stratosphere–troposphere interaction and air exchange between high–middle and middle–tropical latitude belts to be taken into account.

Atmospheric chemical constituents such as ozone play important roles in determining the radiative balance of the atmosphere. Although numerical models such as chemistry–transport models (CTMs) are powerful tools for analyzing the chemical and transport processes of atmospheric chemical constituents, current models typically do not fully succeed in simulating the distribution of chemical constituents. The performance of the chemical data assimilation system was assessed by assuming that CTMs provide a perfect representation of chemical constituents. The optimized assimilation system reproduced the global distribution of chemical constituents well. Using a developed chemistry–climate coupling data assimilation system, the impact of assimilating chemical species on the performance of meteorological reanalysis was also demonstrated. By assimilating ozone into CTMs and considering the radiative interaction between CTMs and GCMs, analysis errors in both chemical and meteorological fields were greatly reduced.

As far as observational data are concerned, there is a need to: (1) monitor ozone at high spatiotemporal resolutions and, most importantly, ODSs (ClO and NO_x), aerosol, water vapor, temperature, and other meteorological parameters; (2) obtain adequate information to meet the needs of numerical modeling of stratospheric variability. The global dynamics of the ozone layer should be addressed by looking at the concentrations of CFCs and their substitutes (HCFCs), possible intensification of stratospheric aviation, and probable changes in temperature and circulation in the stratosphere (stratospheric climate) under the influence of increasing GHG concentrations (the interactivity of climatic dynamics and ozone layer dynamics should of course be borne in mind).

Solution of these and other problems should be supported by the EOS CHEM program, which will provide global databases on the concentrations of O_3, ClO, NO_3, $ClONO_2$, HCl, HNO_3, N_2O_5, CFCs, H_2O, CH_4, as well as on aerosol and temperature at a vertical resolution of 2–3 km (MLS) and 1 km (HIRDLS). Stratosphere–troposphere exchange will be studied by using the combined data of aircraft measurements at a vertical resolution of ~100 m along horizontal routes from 100 km to 500 km and of satellite remote sounding.

During the EOS CHEM program (Schoeberl *et al.*, 2006), two additional developments appeared: (1) observations of the UV radiation flux and substantiation of requirements for these observations; (2) importance of including measurements of concentrations of OH or other HO_x radicals in the program by modifying the MLS instruments in some way. The hydroxyl problem is urgent because of its important role in atmospheric chemistry for several reasons: (a) OH controls the conversion of CH_4 to H_2O; (b) reactions in the presence of HO_x are most important as mechanisms for ozone loss in the upper and lower stratosphere; (c) reactions in the presence of OH determine the rate of oxidation of gaseous sulfur compounds (SO_2 and OCS) to give sulfate aerosol; (d) OH competes with heterogeneous chemical reactions to control the transformation of radicals and "reserve" components in the NO_y and Cl_x systems of families. So, for example, the reaction $OH + NO_2 \rightarrow HNO_3$ competes both with the inverse reaction ($OH + HNO_3 \rightarrow H_2O + NO_3$) with reactions

$$ClO + NO_2 \rightarrow ClONO_2 \quad \text{and} \quad OH + HCl \rightarrow Cl + H_2O \tag{7.5}$$

Since the content of OH in models of chemical processes is expressed, as a rule, through concentrations of other gaseous components, measurement data are needed that characterize the dependence of OH on the concentration of such components under a wide range of real conditions.

Though the ozone residing in the middle and upper stratosphere contributes relatively little to TOC and, therefore, weakly affects the near-surface UV radiation flux, it is very interesting as one of the factors responsible for the formation of the temperature field and atmospheric circulation in the stratosphere, which may affect the structure and composition of the lower stratosphere. In this connection, the following two unsolved problems are most urgent: (1) the differences that remain between the data of observations and the results of numerical modeling for the zone in which photochemical processes dominate at altitudes above 40 km; (2) poorly studied heterogeneous chemical processes in the lower stratosphere (23–35 km).

7.4.2 Tropospheric processes

The combined use of direct measurement and remote sounding data in association with numerical modeling will open up prospects for solving most urgent problems. Here are some of the most important unsolved problems.

Discovering the concentrations of most substantial GHGs (H_2O, CO_2, CH_4, N_2O, HCFCs, O_3) and their determining factors. While long-lived gaseous components such as CO_2, CH_4, N_2O, and HCFCs are rather uniformly mixed in the atmosphere, representative information can be obtained from direct surface observational data, but in the case of water vapor and tropospheric ozone (characterized by strong spatiotemporal variabilities) remote sounding data are important.

Discovering the concentrations of oxidants in the troposphere (particularly, ozone and hydroxyl). Oxidizers such as ozone, hydroxyl, and others (particularly, H_2O_2) form in the troposphere as a result of a complicated set of photochemical reactions in the presence of nitrogen oxides, hydrocarbons, and water vapor, the relative roles of which remain poorly studied, though the primary importance of heterogeneous chemistry of the atmosphere. Not only are observation data on numerous trace gases and aerosols needed but also problem-oriented field experiments that particularly take into account studies of the gas exchange between the atmosphere and the biosphere, the atmosphere and the ocean. The correlation approaches of direct measurements of trace gas fluxes in the atmosphere can play a special role in this context. Satellite remote sounding should become an important source of information on the global fields of precursors (and indicators) of oxidants in the troposphere—such as NO_x CO, hydrocarbons, water vapor—which are short-lived and strongly variable.

A substantial source of ozone and NO_x for the troposphere is by transport from the stratosphere, but quantitative data on transport from one to the other are disputable. Since the gas exchange between the stratosphere and the troposphere takes place apparently via mesoscale processes, direct aircraft measurements at a high spatial resolution are very important. Satellite RS data will supplement them with respect to spatial representativity and accounting for the role of large-scale processes.

As far as NO_2 is concerned, Gruzdev (2009) analyzed diurnal and annual variations in column NO_2, column NO_2 decrease related to the Pinatubo eruption, column NO_2 change during the 11-year cycle of solar activity (SA), and linear trends in column NO_2. This was done by using the data of ground-based spectrometric measurements of stratospheric column NO_2 at stations of the Network for the Detection of Atmospheric Composition Change (NDACC). It was found that annual linear trend estimates of column NO_2 are predominantly positive in the middle and low latitudes of the Southern Hemisphere, whereas they are predominantly negative in the middle and low latitudes of the Northern Hemisphere. The maximum values of trends were about 10% per decade by module; seasonal trend estimates can differ from annual ones.

7.5 NEW CLIMATE RESEARCH ASPECTS DEDUCED FROM GLOBAL OZONE DYNAMICS RESEARCH AND REMOTE SENSING

7.5.1 Climate modeling and atmospheric ozone

As we have seen in various sections of this book, remote sensing is playing an increasingly important role in ozone and climate research and, as the archives of remote-sensing observations extend over longer periods of time, we are obtaining increasing evidence of trends in the climate system. Moreover, a growing array of more accurate sensors and measurements is providing more and more information about the Earth's atmospheric constituents (e.g., ozone) and providing additional data for input to climate models. A great deal of effort in climate studies goes into modeling the concentrations of atmospheric constituents in an effort to predict future climatic conditions. These models start by choosing a set of various atmospheric field variables—temperature, pressure, velocity, concentrations of ozone, water vapor, carbon dioxide and other greenhouse gases, aerosols, etc.—the values of which are specified at a set of points on a regular grid. Then a set of equations (algebraic, differential, and integral equations) representing the physics of the relations among these field variables is established. A set of starting values of these field variables and for various different assumptions about the evolution of forcing conditions is taken as a given, and then numerical solutions of these equations are found for some subsequent time. The forcing conditions assumed are usually attenuation of the ozone layer and increases in carbon dioxide and in the equivalent carbon dioxide contribution representing all the (non-H_2O) greenhouse gases in the atmosphere. These models, which have evolved from weather forecast models, involve massive computing facilities and there are a number of laboratories around the world running such models. The IPCC makes extensive use of the outputs from such models in producing its reports, which are intended to guide policy makers at the national and international level, although there are people who are critical of the IPCC and its heavy reliance on modeling projections.

Fluctuations in the atmosphere limit the validity of long-term weather forecasts to 1 or 2 weeks. Since climate models have been developed from weather forecast models and are based on the same physical mechanisms and equations that are used in weather forecast models, one might wonder whether there is any hope of being able to make meaningful climate forecast predictions over the scale of a century or two if one has no hope of being able to make reliable weather forecasts for more than a few weeks ahead. However, we recall that in looking at climate we are looking at an average of the weather over a reasonably long time. In looking at climate change we are looking at long-term changes in this average. Viewed over a century or two we see that climatic parameters (temperature, rainfall, etc.) are basically stable and vary only slowly. It is the nature of these slow long-term variations that are of concern to us in climate studies after local short-term fluctuations have been smoothed out. It is the stability of long-term components that makes climate prediction possible. In a climate model we consider the effect of various forcing conditions on a given mean state. Despite randomness or chaos at the small-scale level (see Section 7.5.5) we can nevertheless make useful and meaningful calculations at the macroscopic level based on the transition from one mean (or quasi-equilibrium) state to another. Thus, it is common to assume in climate prediction that the climate system and its components (such as the ozone layer) are in equilibrium with its forcing (e.g., CFC abundance); that means, as long as its forcing is constant and slowly varying components alter only slightly in the timescale considered, the mean state of the climate system will be stable and that, if there is a change in the forcing, the mean state will change until it is again in balance with the forcing. We shall consider this assumption below (see Section 7.5.5). The forcing that is commonly considered is that due to the presence of increasing quantities of greenhouse gases in the atmosphere.

The basic method to make such scenarios feasible involves the use of numerical climate models that simulate interactive processes in the "atmosphere–ocean–land surface–cryosphere–biosphere" climatic system. Because there are many such models, the serious difficulty arises as to which is the best model to choose and so climatologists attempt to draw the most reliable conclusions from the outputs of all available models. This difficulty becomes much more serious when taking into account the fact that big uncertainties in fundamental climate-forming factors (e.g., ozone depletion, melting of the ice sheets and the permafrost, carbon cycle feedbacks, the role of clouds, biogeochemical cycles, etc.) still remain (Cracknell and Varotsos, 2011).

As just mentioned, the climate system consists of subsystems that are interacting continuously with each other, not only naturally but also through human intervention. Thus, in order to reduce the level of existing uncertainties, modeling of the interaction between nature and society is urgently required with long-term, non-linear changes in the climate system taken into account. In addition, various advances in different branches of modern physics, which are now occasionally used in climate research, would systematically need to be employed to explore currently unknown features involved in the non-linear dynamics of climate (e.g., in the ozone layer). Some of these advances are briefly discussed in the following subsections (Cracknell and Varotsos, 2011).

When discussing current methodological challenges and the most recent advances to better evaluate the prospects of, and to foster, climate and ozone research, the following key scientific priorities should be taken into account:

- The many atmospheric chemical processes whose rates vary non-linearly with atmospheric parameters such as concentration, temperature, and pressure.
- The characteristic features of spatiotemporal fluctuations of various atmospheric variables such as aerosols, greenhouse gases, and oxides of nitrogen which are closely related to the greenhouse effect and ozone dynamics.
- Low-frequency variability and long-term memory in a number of climatic variables.
- Plausible causes of the occurrence of long-range correlations associated with the long-term memory of climatic variables.

7.5.2 Role of phase transitions in climate system dynamics

The climate system is a dynamic system that is often characterized by phase transitions that—according to Landau's theory (or similarly to Ehrenfest classes)—can be continuous (i.e., second-order) transitions or first-order transitions (Cracknell, 1971; Cracknell and Varotsos, 2011; Landau and Lifshitz, 1958).

In a first-order transition the first derivatives of the appropriate thermodynamic potential have discontinuities at the transition, so that such a transition is accompanied by latent heat and a change in volume. For a continuous transition the first derivatives remain continuous, so that there is no latent heat or change in volume, and only higher order derivatives, such as compressibility or specific heat, are discontinuous.

Many important phase transitions fall in the category of first-order phase transitions, including the solid/liquid and liquid/gas transitions, which therefore involve latent heat and, consequently, are of great importance in climate change and ozone science, since heat balance plays an important role in the climate system and its components. Despite the apparent simplicity of the idea of latent heat, there are some aspects of its involvement in the climate that are not necessarily fully understood or properly taken into account. We mention one curious effect: hot water can in fact freeze faster than initially cold water under a wide range of experimental conditions, which is nowadays known as Mpemba's effect (Jeng, 2005). Although this phenomenon (crucial for ozone depletion) had been known for centuries and was described by Aristotle (see Bacon, 1620; Webster, 1923; see also Adam and Tannery, 1902; Descartes, 1637; Kuhn, 1970), it was not introduced to the modern scientific community until 1969, when a Tanzanian secondary school student named Erasto Mpemba rediscovered it. However, it is still not known exactly why this happens. Further experiments into the causes of the Mpemba effect will enable us to address several hydrospheric and atmospheric open problems. For instance, the way thin laminar ice clouds (frequently observed in the tropical tropopause layer) form and their ability to dehydrate ascending air are of crucial importance to the water vapor budget of the stratosphere (Immler *et al.*, 2008). Also, the rate of ice nucleation in clouds is not easily determined and large discrepancies exist between model predictions and the actual ice crystal concentration measured in clouds (Chen *et al.*, 2008). It is also very important for the formation and evolution of PSCs (see Section 5.7).

7.5.3 Nambu dynamics and ozone–climate modeling

One interesting possible development in climate and ozone modeling, which is being pursued principally at the Freie Universität in Berlin, involves the use of Nambu dynamics (Névir and Sommer, 2009a, b). For almost four decades it has been known that—while Hamiltonian mechanics is based on the flows generated by a smooth Hamiltonian (i.e., a Hamiltonian that has derivatives of all orders)—Nambu dynamics involves multiple Hamiltonians. Nambu dynamics stems from a paper by Nambu (1973) on the generalization of Hamiltonian dynamics for ordinary differential equations obeying Liouville's theorem of non-divergent motion in phase space. Twenty years later this concept was picked up in ideal fluid mechanics by Névir and Blender (1993) to establish a Nambu field approach for partial differential equations. Here, the non-divergent vorticity equation is written using the conservation laws of kinetic energy and enstrophy. Enstrophy is a quantity directly related to the kinetic energy in the flow model that corresponds to dissipation effects in the fluid and can be described as the integral of the square of the vorticity.

In climate and ozone sciences there is a constant search to try to construct better models for global forecasting and simulations. To achieve this, the primary demand is to start from the most general equations (a non-hydrostatic compressible system characterized by a turbulent multicomponent nature and consisting of dry air, water vapor, liquid water, and an ice phase often with embedded particles) and then to determine a model set of equations with net conservation of mass, total energy, and enstrophy. To reach this target a suitable approach is to apply a Nambu formulation of a model equation set, starting from the approach of Névir (1998) which is not limited to an ideal theory (Gassmann and Herzog, 2008; Névir and Sommer, 2009a, b).

In other words, Nambu field theory is generalized to establish a unified energy vorticity theory of ideal fluid mechanics. Using this approach, the degeneracy of the corresponding non-canonical Poisson bracket—a characteristic property of Hamiltonian fluid mechanics—can be replaced by a non-degenerate bracket. The description of a system by Nambu mechanics is substantially more elegant and intuitive when compared with conventional Hamiltonian dynamics. It should be emphasized that Nambu dynamics has proved to be an appropriate tool to develop a quasi-geostrophic baroclinic model in potential quasi-geostrophic vorticity co-ordinates, by employing potential enstrophy and total energy as integrals of motion (Salmon, 2007). In barotropic fluids, helicity is the quantity that is considered to be a conserved quantity. Recently, Bihlo (2008) incorporated dissipative terms (see the next subsection) into the theory and showed that Rayleigh–Bénard convection (the convection pattern occurring in the first 2 km of the Earth's atmosphere) can be represented as a Nambu-metriplectic problem.

7.5.4 Dissipation-induced instabilities in ozone and climate fields

If H stands for mechanical energy in the global climate system and Q stands for all other non-mechanical contributions to the energy (chemical, thermal, etc.), then total energy $U = H + Q$ is conserved (if the system is isolated). All instabilities can be divided into two classes: (i) those that are accounted for by the Hamiltonian description ($\mathrm{d}H = 0$), and (ii) those that are due to dissipation ($\mathrm{d}H < 0$). Thus, instabilities are intrinsically either Hamiltonian or dissipation induced (Cracknell and Varotsos, 2011).

However, recalling the Thomson–Tait–Chetayev theorem—which states that if a system with unstable potential energy is stabilized by gyroscopic forces then this stability is lost after the addition of an arbitrarily small dissipation (Krechetnikov and Marsden, 2007)—we conclude that there is interplay between stability and dissipation in the global climate system. As a simple example, the double spherical (conical) pendulum has two relative equilibria: the straight stretched-out branch and the cowboy branch (the classical two-body Kepler problem; Goldstein, 1956) (see Figure 7.12). The former is stable, while the latter has an indefinite second variation and, thus, is susceptible to dissipation-induced instabilities. Nowadays, the separation of positional forces (which are proportional to displacements) into dissipative and non-dissipative depends on whether or not the global climate system is considered closed or open.

According to the physical definition of dissipative forces—which are proportional to velocities (i.e., gyroscopic forces)—a set of non-conservative forces acting on a mechanical system with a relative equilibrium is called dissipative if—under the action of these forces and in the absence of forces that work against these non-conservative forces—the total mechanical energy of the whole physical system decreases in order to maintain relative equilibrium. The dissipative forces between a given pair of particles are like those produced by a dashpot. This means that the dissipative force on a given particle due to another particle is proportional to the relative approaching velocity. Note that this type of force is completely different

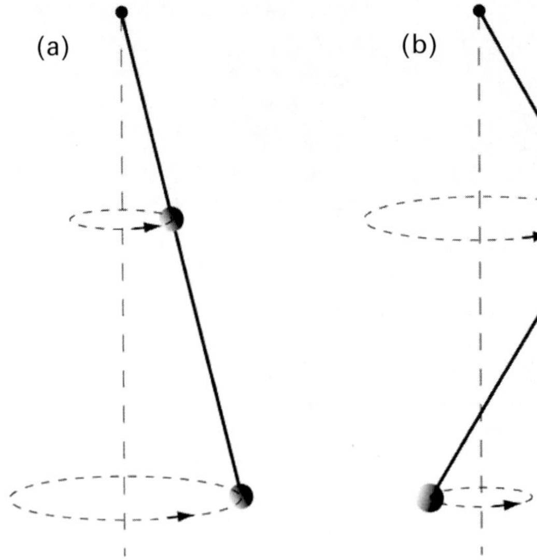

Figure 7.12. (a) Stretched-out and (b) cowboy solutions of the double spherical (conical) pendulum problem (reprinted with permission from Krechetnikov and Marsden, 2007, © 2007 American Physical Society).

from the viscous frictional force applied to each particle (which depends on its absolute velocity). This last type of dissipative force does not conserve momentum.

This allows us to define the generalized notion of dissipation-induced instability: a conservative system with a spectrally stable relative equilibrium is said to suffer from dissipation-induced instability if the introduction of dissipative forces destabilizes this equilibrium in the Lyapunov sense (Pesin, 1977) (e.g., baroclinic instability).

Baroclinic instability (Krechetnikov and Marsden, 2007) is a large-scale instability of westerly winds in the mid-latitudes, when the basic equilibrium state is subject to vertical shear. The basic state is maintained against dissipative effects by an external source of energy. The basic state results from a temperature gradient between the subtropical and polar regions, which causes a pressure gradient overhead. The latter is balanced by the Coriolis force to form the geostrophic flow known as Westerlies. The instability of these baroclinic zonal currents is known to occur as a result of the release of available potential energy from sloping density surfaces.

It is worth mentioning that the lack of theoretical understanding of dissipation-induced instabilities led to the flight failure of the first American satellite—*Explorer I*—launched on January 31, 1958 shortly after the Russian *Sputniks* (launched on October 4 and November 3, 1957). Specifically, *Explorer I* was long and narrow like a pencil, spinning around its long axis (axis of minimum moment of inertia) at 750 revolutions per minute, and not around the axis of maximum moment of inertia (despite the advice of Stanford University astronomer Ronald Bracewell). Thus, an instability, caused by its antennae flexing, dissipated a small amount of rotational energy, with the result that once *Explorer I* had made just one Earth orbit, it flipped over and from then on it windmilled. In contrast, *Sputnik I* (a 55 cm diameter sphere that weighed 83 kg and had four antennae) was spinning around its maximum moment of inertia (Cracknell and Varotsos, 2007b; Krechetnikov and Marsden, 2007).

7.5.5 Deterministic, chaotic, or stochastic ozone–climate time series?

As a result of the discussion in Section 7.5.1 the question might be raised as to whether climate models are, or should be, deterministic, chaotic, or stochastic. The difference between these three situations is very simply (Cracknell and Varotsos, 2011):

- In *deterministic* mechanisms one cause corresponds to one effect.
- In *stochastic* processes one cause leads to more than one effect and probabilities are needed to predict one particular effect.
- In *chaotic* systems the ability to predict a certain plausible result is extremely low, so low that the probability of any particular result is close to zero (i.e., there is no prediction because the sensitivity of the system is very high). In other words, chaos is when the present determines the future, but the approximate present does not define the approximate future.

Models generate time series according to pre-specified rules while real data always possess a stochastic component due to omnipresent dynamical noise. In this context, Wold (1938) proved that any (stationary) time series can be decomposed into two different parts. The first (deterministic) part can be exactly described by a linear combination of its own past; the second part is a moving average component of a finite order. Hence, it may seem superfluous to ask whether a time series generated by natural processes is deterministic, chaotic, or stochastic.

As mentioned above (Section 7.5.1) the chaotic character of atmospheric–oceanic–lithospheric–cryospheric–biospheric dynamics limits the reliability of climate projections. Although being of quite different physical origin, time series arising from chaotic systems share with those generated by stochastic processes several properties that make them almost indistinguishable: (1) a wide-band power spectrum, (2) a delta-like autocorrelation function, (3) irregular behavior of measured signals, etc. (Rosso *et al.*, 2007). Thus, reliable prediction of global climate change or of one of its components (e.g., the ozonosphere) is impossible without consideration of the complexity of all interactive processes (Varotsos, 2002a; Varotsos *et al.*, 2000).

Nowadays, the most satisfactory analysis of the climate system is that in which its underlying mechanisms can be modeled from first principles. For example, the approach of general circulation models (GCMs) allows a deeper understanding of causes and effects, but the conclusions drawn from their simulations will be revised as these models are continuously improved; this constitutes a weakness of the approach by the IPCC (as pointed out particularly strongly in chapter 2 of Lovelock,

2009). An alternative modeling path is given by data-driven (statistical) techniques. According to this philosophy, the evolution of the climate system is studied by recording time series, which are sets of observations collected at regular intervals of time. Withn this framework Verdes (2007), using two independent driving force reconstruction techniques, showed that the combined effect of greenhouse gases and aerosol emissions has been the main external driver of global climate during the past few decades.

7.6 WMO/UNEP SCIENTIFIC ASSESSMENT 2010

The WMO/UNEP's *Scientific Assessment of Ozone Depletion: 2010* (WMO, 2010a) considered developments since the previous assessment (WMO, 2007). We quote from the Executive Summary of the 2010 Assessment (WMO, 2010b). Regarding the Antarctic it observed that "The Antarctic ozone hole is the clearest manifestation of the effect of ODSs on the ozone layer. The depletion far exceeds natural variability and has occurred without exception since 1980. The ozone hole also provides the most visible example of how ozone depletion affects surface climate.

- Springtime Antarctic total column ozone losses (the ozone hole), first recognizable around 1980, continue to occur every year. Although the ozone losses exhibit year-to-year variations that are primarily driven by year-to-year changes in meteorology, October mean column ozone within the vortex has been about 40% below 1980 values for the past fifteen years. The average erythemal ("sunburning") UV measured at the South Pole between 1991 and 2006 was 55–85% larger than the estimated values for the years 1963–1980.
- Doubts raised since the previous Assessment regarding our understanding of the cause of the Antarctic ozone hole have been dispelled. New laboratory measurements on the key chemistry involved in polar ozone depletion have reaffirmed that past changes in ODSs are indeed the cause of the ozone hole. This is also supported by quantification of the chemicals responsible for the ozone hole via field observations.
- There is increased evidence that the Antarctic ozone hole has affected the surface climate in the Southern Hemisphere. Climate models demonstrate that the ozone hole is the dominant driver of the observed changes in surface winds over the Southern Hemisphere mid and high latitudes during austral summer. These changes have contributed to the observed warming over the Antarctic Peninsula and cooling over the high plateau. The changes in the winds have also been linked to regional changes in precipitation, increases in sea ice around Antarctica, warming of the Southern Ocean, and a local decrease in the ocean sink of CO_2.
- The trends in the summertime winds in the Southern Hemisphere are not expected to persist over the next few decades. This is because of the expected offsetting influences on the surface winds of increasing greenhouse gases and the recovering ozone hole.

- Observed Antarctic springtime column ozone does not yet show a statistically significant increasing trend. Year-to-year variability, due to meteorology, is much larger than the expected response to the small ODS decreases in the Antarctic vortex to date. This is consistent with simulations using chemistry–climate models (CCMs).
- The evolution of Antarctic springtime column ozone over the rest of the century is expected to be dominated by the decline in ODS abundance. CCM simulations show that greenhouse gas changes have had, and will continue to have, a small impact on the ozone hole compared to the effects of the ODS changes. There are some indications that small episodic Antarctic ozone holes may occur even through the end of the century. In spring and early summer, Antarctica will continue to experience excess surface UV."

Turning to the Arctic and to the general global situation, the 2010 Assessment concluded that "As a result of the controls introduced by the Montreal Protocol and its Amendments and Adjustments, it is expected that the decline in ODSs will lead to an increase in stratospheric ozone abundances. However, it will be challenging to attribute ozone increases to the decreases in ODSs during the next few years because of natural variability, observational uncertainty, and confounding factors, such as changes in stratospheric temperature or water vapor. A feature of this Assessment is the coordinated use by the community of chemistry–climate models (CCMs) with integrations covering the period from 1960–2100, which has allowed more detailed study of the long-term changes in the stratosphere and of the relative contributions of ODSs and greenhouse gases (GHGs).

- Average total ozone values in 2006–2009 remain at the same level as the previous Assessment, at roughly 3.5% and 2.5% below the 1964–1980 averages respectively for 90°S–90°N and 60°S–60°N. Midlatitude (35°–60°) annual mean total column ozone amounts in the Southern Hemisphere [Northern Hemisphere] over the period 2006–2009 have remained at the same level as observed during 1996–2005, at ~6% [~3.5%] below the 1964–1980 average.
- The ozone loss in Arctic winter and spring between 2007 and 2010 has been variable, but has remained in a range comparable to the values prevailing since the early 1990s. Substantial chemical loss continues to occur during cold Arctic winters.
- Robust linkages between Arctic stratospheric ozone depletion and tropospheric and surface climate trends have not been established, as expected from the smaller ozone depletion compared with the Antarctic.
- Chemistry–climate models reproduce both the latitudinal and vertical structure of the observed ozone trends in both northern and southern midlatitudes during the past periods of increase of the ODSs, confirming our basic understanding of ozone change. Simulations agree with observations that the last decade has shown flattening of the time series of global total ozone.
- Analyses based on surface and satellite measurements show that erythemal UV irradiance over midlatitudes has increased since the late 1970s. This is in qualitative agreement with the observed decrease in column ozone, although

other factors (mainly clouds and aerosols) have influenced long-term changes in erythemal irradiance. Clear-sky UV observations from unpolluted sites in mid-latitudes show that since the late 1990s, UV irradiance levels have been approximately constant, consistent with ozone column observations over this period.

- New analyses of both satellite and radiosonde data give increased confidence in changes in stratospheric temperatures between 1980 and 2009. The global-mean lower stratosphere cooled by 1–2 K and the upper stratosphere cooled by 4–6 K between 1980 and 1995. There have been no significant long-term trends in global-mean lower stratospheric temperatures since about 1995. The global-mean lower-stratospheric cooling did not occur linearly but was manifested as downward steps in temperature in the early 1980s and the early 1990s. The cooling of the lower stratosphere includes the tropics and is not limited to extra-tropical regions as previously thought.
- The evolution of lower stratospheric temperature is influenced by a combination of natural and human factors that has varied over time. Ozone decreases dominate the lower stratospheric cooling since 1980. Major volcanic eruptions and solar activity have clear shorter-term effects. Models that consider all of these factors are able to reproduce this temperature time history.
- Changes in stratospheric ozone, water vapor, and aerosols all radiatively affect surface temperature. The radiative forcing of climate in 2008 due to stratospheric ozone depletion ($-0.05 \pm 0.1\,\mathrm{W\,m^{-2}}$) is much smaller than the positive radiative forcing due to the CFCs and HCFCs largely responsible for that depletion (about $+0.3\,\mathrm{W\,m^{-2}}$). For context, the current forcing by CO_2 is approximately $+1.7\,\mathrm{W\,m^{-2}}$. Radiative calculations and climate modeling studies suggest that the radiative effects of variability in stratospheric water vapor ($\pm \sim 0.1\,\mathrm{W\,m^{-2}}$ per decade) can contribute to decadal variability in globally averaged surface temperature. Climate models and observations show that major volcanic eruptions (e.g., Mt. Pinatubo in 1991, roughly $-3\,\mathrm{W\,m^{-2}}$) can cool the surface for several years.
- The global middle and upper stratosphere are expected to cool in the coming century, mainly due to CO_2 increases. Stratospheric ozone recovery will slightly offset the cooling. HFCs could warm the tropical lower stratosphere and tropopause region by about 0.3°C if stratospheric abundances reach the 1 ppb level.
- Emerging evidence from model simulations suggests that increasing greenhouse gases lead to an acceleration of the stratospheric circulation usually referred to as the Brewer–Dobson circulation. Such an acceleration could have important consequences, particularly decreases in column ozone in the tropics and increases in column ozone elsewhere. However, responsible mechanisms remain unclear and observational evidence for the circulation increase is lacking.
- Global ozone is projected to increase approximately in line with the ODS decline, and the increase is accelerated by cooling of the upper stratosphere. Global ozone is not very sensitive to circulation changes, so high confidence can be placed in this projection.

- The evolution of ozone in the Arctic is projected to be more sensitive to climate change than in the Antarctic. The projected strengthening of the stratospheric Brewer–Dobson circulation is expected to significantly increase lower stratospheric ozone in the Arctic, augmenting the GHG-induced ozone increase from upper stratospheric cooling and hastening the return to 1980 levels.
- GHG-induced temperature and circulation changes are projected to hasten the return of midlatitude total column ozone to 1980 levels by several decades, rising well above 1980 levels by the end of the century. The effect is most pronounced in northern midlatitudes, where it would result in clear-sky surface UV radiation levels well below 1980 values by the second half of the century. In southern midlatitudes, the effect of circulation changes is projected to be weaker and ozone is also influenced by depletion in the Antarctic, where the return to 1980 levels occurs much later."

References and bibliography

Abbas, M.M., Kunde, V.G., Mumma, M.J., Kostink, T., Buhl, D., and Frerking, M.A. (1979). Stratospheric sounding by infrared heterodyne spectroscopy, *Journal of Geophysical Research*, **84**, 2681–2690.

Abe, S., Sarlis, N.V., Skordas, E.S., Tanaka, H.K., and Varotsos, P.A. (2005). Origin of the usefulness of the natural-time representation of complex time series, *Physical Review Letters*, **94**, 170601.

Acarreta, J.R., de Haan, J.F., and Stammes, P. (2004). Cloud pressure retrieval using the O_2–O_2 absorption band at 477 nm, *Journal of Geophysical Research*, **109**, doi: 10.1029/2003JD003915.

Adam, C.E. and Tannery, P. (1902). *Oeuvres de Descartes*, Vol. VI, Leopold (erf), Paris, p. 238 (trans. F.C. Frank).

Aikin, A.C. (1989). Mesospheric ozone behaviour determined from satellite data, in: R.D. Bojkov and P. Fabian (Eds.), *Proceedings of the Quadrennial Ozone Symposium 1988, Gottingen, Germany*, Deepak Publications, Hampton, VA., pp. 388–390.

Aksoy, B., Incecik, S., Topcu, S., Demirhan Bari, D., Kahya, C., Acar, Y., Ozunlu, M., and Ekici, M. (2009). Total ozone over Ankara and its forecasting using regression models, *International Journal of Remote Sensing*, **30**, 4387–4400.

Alexandris, D., Varotsos, C., Kondratyev, K.Ya., and Chronopoulos, G. (1999). On the altitude dependence of solar effective UV, *Physics and Chemistry of the Earth* C, **24**, 515–517.

Allen, D.J., Douglass, A.R., Rood, R.B., and Guthrie, P.D. (1991). Application of a monotonic upstream-biased transport scheme to three-dimensional constituent transport calculations, *Monthly Weather Review*, **119**, 2456–2464.

Allen, D.R., Bevilacqua, R.M., Nedoluha, G.E., Randall, C.E., and Manney, G.L. (2003). Unusual stratospheric transport and mixing during the 2002 Antarctic winter, *Geophysical Research Letters*, **30**, 1599–1602.

Allen, G., Vaughan, G., Bower, K.N., Williams, P.I., Crosier, J., Flynn, M., Connolly, P., Hamilton, J.F., Lee, J.D., Saxton, J.E. *et al.* (2008). Aerosol and trace-gas measurements in the Darwin area during the wet season, *Journal of Geophysical Research*, **113**, doi: 10.1029/2007JD008706.

Alvarez-Madrigal, M., Pérez-Peraza, J., and Velasco, V.M. (2009). The influence of cosmic rays on the size of the Antarctic Ozone Hole, *Current Development in Theory and Applications of Wavelets*, **3**, 233–248.

Ambrosio, M., Antolini, R., Auriemma, G., Baker, R., Baldini, A., Barbarino, G.C., Barish, B.C., Battistoni, G., Bellotti, R., Bemporad, C. *et al.* (1997). Seasonal variations in the underground muon intensity as seen by MACRO, *Astroparticle Physics*, **7**, 109–124.

Ancellet, G., Papayannis, A., Pelon, J., and Megie, G. (1988). Tropospheric ozone lidar measurements, in: I.S. Isakson (Ed.), *NATO Workshop on Tropospheric Ozone 1987, Lillehammer, Norway*, D. Reidel, Dordrecht, The Netherlands.

Andersen, S.B., Weatherhead, E.C., Stevermer, A., Austin, J., Brühl, C., Fleming, E.L., de Grandpré, J., Grewe, V., Isaksen, I., Pitari, G. *et al.* (2006). Comparison of recent modeled and observed trends in total column ozone, *Journal of Geophysical Research*, **111**, D02303, doi: 10.1029/2005JD006091.

Anderson, J.L., Balaji, V., Broccoli, A.J., Cooke, W.F., Delworth, T.L., Dixon, K.W., Donner, L.J., Dunne, K.A., Freidenreich, S.M., Garner, S.T. *et al.* (2004). The new GFDL global atmosphere and land model AM2/LM2: Evaluation with prescribed SST simulations. *Journal of Climate*, **17**, 4641–4673.

Anderson, J.G., Margitan, J.J., and Stedman, D.H. (1977). Atomic chlorine and the chlorine monoxide radical in the stratosphere: Three in situ observations, *Science*, **198**, 501–503.

Anderson, J.G., Grassl, H.J., Shetter, R.E., and Margitan, J.J. (1980). Stratospheric free chlorine measured by balloon-borne in situ resonance fluorescence, *Journal of Geophysical Research*, **85**, 2869–2887.

Anderson, J., Russell, J.M., Solomon, S., and Deaver, L.E. (2000). Halogen Occultation Experiment confirmation of stratospheric chlorine decreases in accordance with the Montreal Protocol, *Journal of Geophysical Research*, **105**, 4483–4490.

Andronache, C. and Chameides, W. L. (1998). Interactions between sulphur and soot emissions from aircraft and their role in contrail formation, 2: Development. *Journal of Geophysical Research*, **103**, 10787–10802.

Angell, J.K. (1980). Temperature and ozone variations in the stratosphere, *Pure and Applied Geophysics*, **118**, 378–386.

Angell, J.K. (1986). The close relation between Antarctic total-ozone depletion and cooling of the Antarctic low stratosphere. *Geophysical Research Letters*, **13**, 1240–1243.

Annales Geophysicae, 1998, Part III: Space and Planetary Science (Supplement III to Volume 16). European Geophysical Society, Vienna, pp. C941–C952.

Antón, M., Loyola, D., López, M., Vilaplana, J.M., Bañón, M., Zimmer, W., and Serrano, A. (2009). Comparison of GOME-2/MetOp total ozone data with Brewer spectroradiometer data over the Iberian Peninsula, *Annales Geophysicae*, **27**, 1377–1386.

Antón, M., Vilaplana, J.M., Kroon, M., Serrano, A., Parias, M., Cancillo, M.L., and de la Morena, B.A. (2010). The empirically corrected EP-TOMS total ozone data against Brewer measurements at El Arenosillo (Southwestern Spain), *IEEE Transactions on Geoscience and Remote Sensing*, **48**, 3039–3045.

Appenzeller, C., Weiss, A.K., and Staehelin, J. (2000). North Atlantic Oscillation modulates total ozone winter trends. *Geophysical Research Letters*, **27**, 1131–1134.

Ariya, P.A., Jobson, B.T., Sander, R., Niki, H., Harris, G.W., Hopper, J.F., and Anlauf, K.G. (1998). Measurements of C_2–C_7 hydrocarbons during the Polar Sunrise Experiment 1994: Further evidence for halogen chemistry in the troposphere. *Journal of Geophysical Research*, **103**, 13169–13180.

Arnone, R.A., Ladner, Sh., La Violette, P.E., Brock, J.C., and Rochford, P.A. (1998). Seasonal and interannual variability of surface photosynthetically available radiation in the Arabian Sea, *Journal of Geophysical Research*, **103**, 7735–7748.

Asimakopoulos, D., Deligiorgi, D., Drakopoulos, C., Helmis, C., Kokkori, K., Lalas, D., Sikiotis, D., and Varotsos, C. (1992). An experimental-study of nighttime air-pollutant transport over complex terrain in Athens, *Atmospheric Environment*, **26**, 59–71.

Atrasheuski, Y., Krasouski, A., Liudchik, A., and Turishev, L. (2006). Development of ozone monitoring instruments in Belarus, *TECO-2006: WMO Technical Conference on Meteorological and Environmental Instruments and Methods of Observation, IOM, '94, Geneva, Switzerland, December 4–6, 2006* (*http://www.wmo.int/pages/prog/ www/IMOP/ publications/IOM-94-TECO$_2$006/P2%2801%29_Krasouski_ Belarus.pdf*).

Ausloos, M. and Ivanova, K. (2001). Power-law correlations in the southern-oscillation-index fluctuations characterizing El Niño, *Physical Review E*, **63**, 047201, Part 2.

Austin, J., and Butchart, N. (1997). Simulations of stratospheric constituents in a climate model with interactive ozone, *Proceedings of the SPARC General Assembly, December 2–6, 1996, Melbourne, Australia, Vols. I, II, WCRP-99 (WMO/TD-No. 814)*, Geneva, Switzerland, pp. 449–452.

Austin, J. and Butchart, N. (1998). Simulations of stratospheric constituents in a climate model with interactive ozone, *Proceedings of the First SPARC General Assembly. Vol. 2, WCRP-99 (WMO/TD-N814)*, pp. 449–452.

Austin, J. and Wilson R.J. (2006). Ensemble simulations of the decline and recovery of stratospheric ozone. *Journal of Geophysical Research*, **111**, D16314, doi: 10.1029/2005JD006907.

Bacmeister, J.T., Kuell, V., Offenmann, D., Riese, M., and Elkins, J.W. (1999). Intercomparison of satellite and aircraft observations of ozone, CFC-11 and NO_y using trajectory mapping. *Journal of Geophysical Research*, **104**, 16375–16390.

Bacon, F. (1620 [1869 Edition]). *Novum Organum, Vol. VIII: The Works of Francis Bacon* (J. Spedding, R.L. Ellis, and D.D. Heath, Eds.), Houghton, Mifflin & Co., Boston, 235 pp.

Baker, M.B. (1997). Cloud microphysics and climate. *Science*, **276**, 1072–1078.

Balatsko, L.M., Dziomin, V.S., Krasouski, A.N., Liudchik, A.M., and Pakatashkin, V.I. (2008). Ozone optical track analyzer TrIO-1 and its metrological certification features, *Measurement Techniques*, **51**, 1139–1142.

Balatsko, L.M., Dziomin, V.S., Krasouski, A.N., Liudchik, A.M., and Pakatashkin, V.I. (2010a). Theory of multiwave differential optical absorption spectroscopy surface ozone measurement with open path meters, *International Journal of Remote Sensing*, **31**, 523–529.

Balatsko, L., Dziomin, V., Krasouski, A., Liudchik, A., and Pakatashkin , V. (2010b). Multiwave differential optical absorption spectroscopy surface ozone measurement with open path meters, in: Milind Kr. Sharma (Ed.), *Advances in Measurement Systems*, InTech, Rijeka, Croatia, Chapter 22, pp. 537–560.

Balis, D., Kroon, M., Koukouli, M.E., Brinksma, E.J., Labow, G., Veefkind, J.P., and McPeters, R.D. (2007a). Validation of Ozone Monitoring Instrument total ozone column measurements using Brewer and Dobson spectrophotometer ground-based observations. *Journal of Geophysical Research*, **112**, D24546, doi: 10.1029/2007JD008796.

Balis, D., Lambert, J.C., van Roozendael, M., Spurr, R., Loyola, D., Livschitz, Y., Valks, P., Amiridis, V., Gerard, P., Granville, J. *et al.* (2007b). Ten years of GOME/ERS2 total ozone data.The new GOME data processor (DP) version 4, 2: Ground-based validation and comparisons with TOMS V7/V8. *Journal of Geophysical Research*, **112**, D07307, doi: 10.1029/2005JD006376.

Balis, D., Koukouli, M., Loyola, D., Valks, P., and Hao, N. (2008). *Validation of GOME-2 Total Ozone Products (OTO/O3, NTO/O3) Processed with GDP4.2* (Technical Report SAF/O3M/AUTH/GOME-2VAL/RP/01, Revision 1B; *http://www.wdc.dlr.de/sensors/gome2/SAF-O3M-AUTH-GOME-2VAL-RP-01.pdf*).

Balis, D, Bojkov, R., Tourpali, K., and Zerefos, C. (2009). Characteristics of the ozone decline over both hemispheres, *International Journal of Remote Sensing*, **30**, 3887–3895.

Balkanski, Y., Myhre, G., Gauss, M., Rädel, G., Highwood, E., and Shine, K.P. (2010). Direct radiative effect of aerosols emitted by transport: From road, shipping, and aviation. *Atmospheric Chemistry and Physics*, **10**, 4477–4489.

Barlett, L.M. and Vaughan, G. (1998). Ozone and NO_2 measurements by ground-based spectrometer at Aberystwyth since 1991, in: R.D. Bojkov and G. Visconti (Eds.), *Proceedings of the XVIII Quadrennial Ozone Symposium, L'Aquila, Italy, September 12–21, 1996*, Consorzio Parco Scientifico e Tecnologico d'Abruzzo, L'Aquila, Italy, Vol. 1, 9–12.

Barnes, R.A. and Simeth, P.G. (1986). Design of a rocket-borne radiometer for stratospheric ozone measurements. *Review of Scientific Instruments*, **57**, 544–551.

Baron, P., Dupuy, E., Urban, J., Murtagh, D.P., Eriksson, P., and Kasai, Y. (2009). HO_2 measurements in the stratosphere and the mesosphere from the sub-millimeter limb sounder Odin/SMR, *International Journal of Remote Sensing*, **30**, 4195–4208.

Barras, E., Ruffieux, D., and Hocke, K. (2009). Stratospheric ozone profiles over Switzerland measured by SOMORA, ozonesonde and MLS/AURA satellite. *International Journal of Remote Sensing*, **30**, 4033–4041.

Barratt-Brown, E.P. (1991). Building a monitoring and compliance regime under the Montreal Protocol, *Yale Journal of International Law*, **16**, 519–565.

Barrett, P.H., Bollinger, L.M., Cocconi, G., Eisenberg, Y., and Greisen, K. (1952). Interpretation of cosmic-ray measurements far underground. *Reviews of Modern Physics*, **24**, 133–178.

Barrie, L.A. and Platt, U. (1997). Arctic tropospheric chemistry: Overview to Tellus special issue. *Tellus*, **49B**, 450–454.

Barrie, L. A., Platt, U., and Shepson, P. (1998). Surface ozone depletion at the polar sunrise fueled by sea-salt halogens. *IGACtivities Newsletter*, **14**, 4–7.

Basher, R.E. (1982a). *Review of the Dobson Spectrophotometer and Its Accuracy* (WMO Global Ozone Research and Monitoring Project, Report No. 13). WMO, Geneva, Switzerland, 90 pp. (*http://www.esrl.noaa.gov/gmd/ozwv/dobson/papers/report13/4th.html*).

Basher, R.E. (1982b). Units for column amounts of ozone and other atmospheric gases. *Quarterly Journal of the Royal Meteorological Society*, **108**, 460–462.

Basher, R.E. (1995). *Survey of WMO-Sponsored Dobson Spectrophotometer Intercomparisons* (WMO Global Ozone and Monitoring Project Report No. 19). WMO, Geneva, Switzerland.

Bass, A.M. and Paur, R.J. (1985). The ultraviolet cross section of ozone, 1: The measurements, in: C.S. Zerefos and A. Ghazi (Eds.), *Atmospheric Ozone: Proceedings of the Quadrennial Ozone Symposium, Halkidiki, Greece*, D. Reidel, Hingham, MA, pp. 606–610.

Batchelor, R.L., Strong, K., Lindenmaier, R.L., Mittermeier, R., Fast, H., Drummond, J.R., and Fogal, P.F. (2009). A new Bruker IFS 125HR FTIR spectrometer for the Polar Environment Atmospheric Research Laboratory at Eureka, Canada: Measurements and comparison with the existing Bomem DA8 spectrometer. *Journal of Atmospheric and Oceanic Technology*, **26**, 1328–1340.

Batchelor, R.L., Kolonjari, F., Lindenmaier, R., Mittermeier, R.L., Daffer, W., Fast, H., Manney, G., Strong, K., and Walker, K.A. (2010). Four Fourier transform spectrometers and the Arctic polar vortex: Instrument intercomparison and ACE-FTS validation at Eureka during the IPY springs of 2007 and 2008, *Atmospheric Measurement Techniques*, **3**, 51–66.

Bates, D.R. and Nicolet, M. (1950). The photochemistry of atmospheric water vapour, *Journal of Geophysical Research*, **55**, 301–327.

Bates, T.S., Huebert, B.S., Gras, J.L., Griffith, F.R., and Durkee, P.A. (1998). International Global Atmospheric Chemistry (IGAC) project's First Aerosol Characterization Experiment (ACE-I) overview. *Journal of Geophysical Research*, **103**, 16297–16318.

Baumert, K.A., Herzog, T., and Pershing, J. (2005). *Navigating the Numbers: Greenhouse Gas Data and International Climate Policy*, World Resources Institute, Washington, 132 pp.

Baumgaertner, A. J. G., Jöckell, P., Dameris, M., and Crutzen, P.J. (2010). Will climate change increase ozone depletion from low-energy-electron precipitation? *Atmospheric Chemistry and Physics*, **10**, 9647–9656.

Baumgardner, D., Miake-Lye, R.C, Anderson, M.R., and Brown, R.C. (1998). An evaluation of the temperature, water vapor, and vertical velocity structure of aircraft contrails. *Journal of Geophysical Research*, **103**, 8727–8736.

Becker, G., Muller, R., McKenna, D.S., Rex, M., and Carslaw, K.S. (1998). Ozone loss rates in the Arctic stratosphere in the winter 1991/1992: Model calculations compared with Match results. *Geophysical Research Letters*, **25**, 4325–4328.

Bazilevskaya, G., Krivolutsky, A., Ivanova, I., Nazarova, M., Perejaslova, N., and Petrenko, I. (1997). Cosmic rays and ozone variability, *Proceedings of the SPARC General Assembly, December 2–6, 1996, Melbourne, Australia* (Vols. I and II, WMO/TD-No. 814), WMO, Geneva, Switzerland, pp. 469–472.

Beer, R. (2006). TES on the Aura mission: Scientific objectives, measurements and analysis overview, *IEEE Transactions on Geoscience and Remote Sensing*, **44**, 1102–1105.

Beer, R., Glavich, T.A., and Rider, D.M. (2001). Tropospheric emission spectrometer for the Earth Observing System's Aura satellite, *Applied Optics*, **40**, 2356–2367.

Bègue, N., Bencherif, H., Sivakumar, V., Kirgis, G., Mze, N., and Leclair de Bellevue, J. (2010). Temperature variability and trends in the UT-LS over a subtropical site: Réunion (20.8°S, 55.5°E), *Atmospheric Chemistry and Physics*, **10**, 8563–8574.

Begum, D.A. (1989). Studies of the total ozone field around the subtropical jet stream. PhD thesis, University of Wales.

Begum, D.A. (1993). Climatological study of the total ozone field around the subtropical jet stream. *International Journal of Climatology*, **13**, 915–921.

Bekoryukov, V.I. (1991). Spatial and seasonal features of climate distribution of total ozone content and partial pressure of ozone. *Russian Meteorology and Hydrology*, **11**, 47–54.

Benedick, R.E. (1991). *Ozone Diplomacy: New Directions in Safeguarding the Planet*. Harvard University Press, Cambridge, MA, 300 pp.

Benedick, R.E. (2007). *Avoiding Gridlock on Climate Change*, National Academy of Sciences: Issues in Science and Technology, pp. 37–40.

Benedick, R.E. (2009). Science inspiring diplomacy: The improbable Montreal Protocol, in: C. Zerefos, G. Contopoulos, and G. Skalkeas (Eds.), *Proceedings of the Symposium for the 20th Anniversary of the Montreal Protocol*, Springer-Verkag, Berlin, pp. 13–19.

Bernath, P.F., McElroy, C.T., Abrams, M.C., Boone, C.D., Butler, M., Camy-Peyret, C., Carleer, M., Clerbaux, C., Coheur, P.-F., Colin, R. *et al.* (2005). Atmospheric Chemistry Experiment (ACE): Mission overview, *Geophysical Research Letters*, **32**, L15S01, doi: 10.1029/2005GL022386.

Bernhard, G., Evans, R.D., Labow, G.J., and Oltmans, S.J. (2005). Bias in Dobson total ozone measurements at high latitudes due to approximations in calculations of ozone absorption coefficients and air mass. *Journal of Geophysical Research*, **110**, D10305, doi: 10.1029/2004JD005559.

Berntsen, T. and Fuglestvedt, J. (2008). Global temperature responses to current emissions from the transport sectors, *Proceedings of the National Academy of Sciences of the U.S.A.*, **105**, 19154–19159.

Berntsen, T. and Isaksen, I.S.A. (1997). A global three-dimensional chemical transport model for troposphere, 1: Model description and CO and ozone results. *Journal of Geophysical Research*, **102**, 21239–21280.

Berntsen, T., Isaksen, I.S.A., Myhre, G., Fuglestvedt, J.S., Stordal, F., Larsen, D.A., Freckleton, R.S., and Shine, K.P. (1997). Effects of anthropogenic emissions on tropospheric ozone and its radiative forcing. *Journal of Geophysical Research*, **102**, 28101–28126.

Bertaux, J.L., Megie, G., Widemann, T., Chassefiere, E., Pellinen, R., Kyrölä, E., Korpela, S., and Simon, P. (1991). Monitoring of ozone trend by stellar occultations: The GOMOS instrument, *Advances in Space Research*, **11**, 237–242.

Bertaux, J.L., Kyrölä, E., and Wehr, T. (2000a). Stellar occultation technique for atmospheric ozone monitoring: GOMOS on Envisat, *Earth Observation Quarterly*, **67**, 17–20.

Bertaux, J.L., Quemerais, E., Lallement, R., Lamassoure, E., Schmidt, W., and Kyrölä, E. (2000b). Monitoring solar activity on the far side of the Sun from sky reflected Lyman alpha radiation. *Geophysical Research Letters*, **27**, 1331–1334.

Bertaux, J.L., Hauchecorne, A., Dalaudier, F., Cot, C., Kyrölä, E., Fussen, D., Tamminen, J., Leppelmeier, G.W., Sofieva, V., Hassinen *et al.* (2004). First results on GOMOS/Envisat, *Advances in Space Research*, **33**, 1029–1035.

Bertaux, J.L., Kyrölä, E., Fussen, D., Hauchecorne, A., Dalaudier, F., Sofieva, V., Tamminen, J., Vanhellemont, F., Fanton d'Andon, O., Barrot, G. *et al.* (2010). Global ozone monitoring by occultation of stars: An overview of GOMOS measurements on ENVISAT, *Atmospheric Chemistry and Physics*, **10**, 12091–12148.

Bevilacqua, R.M., Hoppell, K., Hornstein, J., Lucke, R., Shettle, E.P., Ainsworth, T.L., Debrestian, D., Fromm, M., Lumpe, J., Krigman, S. *et al.* (1995). First results from POAM II: The dissipation of the 1993 Antarctic ozone hole. *Geophysical Research Letters*, **22**, 909–912.

Bhartia, P.K. (2009). Role of satellite measurements in the discovery of the stratospheric ozone depletion. Statement from the Executive Secretary for the Vienna Convention and the Montreal Protocol. Twenty years of ozone decline, in: C. Zerefos, G. Contopoulos, and G. Skalkeas (Eds.), *Proceedings of the Symposium for the 20th Anniversary of the Montreal Protocol*, Springer-Verlag, Berlin, pp. 5–8.

Bhartia, P.K. and McPeters, R.D. (2011). The SBUV Ozone Record: The good, the bad, and the ugly, *SPARC/IO$_3$C/WMO-IGACO-O3/UV Workshop on Past Changes in the Vertical Distribution of Ozone, January 25–27*, WMO, Geneva, Switzerland.

Bhartia, P.K. and Wellemeyer, C. (2002). OMI Ozone Product, in: P.K. Bhartia (Ed.), *TOMS-V8 Total O Algorithm in OMI Algorithm Theoretical Basis Document, II*, Editor , NASA GSFC, Greenbelt, MD, pp. 15–31.

Bhartia, P.K. and Wellemeyer, C.W. (2004). *TOMS-V8 Algorithm Theoretical Basis Document* (*http://toms.gsfc.nasa.gov/version8/v8toms_atbd.pdf* [last accessed June 2, 2009]).

Bhartia, P.K., Klenk, K.F., Wong, C.K., Gordon, D., and Fleig, A.J. (1984). Intercomparison of Nimbus-7 SBUV/TOMS total ozone data sets with Dobson and M83 results. *Journal of Geophysical Research*, **89**, 5239-5247.

Bhartia, P.K., Heath, D.F., and Fleig, A.F. (1985). Observation of anomalously small ozone densities in south polar stratosphere during October 1983 and 1984, *Symposium on Dynamics and Remote Sensing of the Middle Atmosphere, 5th Scientific Assembly, August 1985*, International Association of Geomagnetism and Aeronomy, Prague, Czechoslovakia.

Bhartia, P.K. and Wellemeyer, C. (2002). TOMS-V8 Total O_3 Algorithm, in: *OMI ATBD, II, OMI Ozone Products* (ATBD-OMI-02), NASA Goddard Space Flight Center, Greenbelt, MD, pp. 15–31.

Bhartia, P.K., Wellemeyer, C.G., Taylor, S.L., Nath, N., and Gopolan, A. (2004). Solar Backscatter Ultraviolet (SBUV) Version 8, in: C.S. Zerefos (Ed.), *Ozone, Vol. I, Proceedings of the XX Quadrennial Ozone Symposium, June 1–8, Kos, Greece*, International Ozone Commission, Athens, Greece, pp. 295–296.

Bhatt. P.P., Remsberg, E.E., Gordley, L.L., McInerney, J.M., Brackett, V.G., and Russell, J.M. III (1999). An evaluation of the quality of Halogen Occultation Experiment ozone profiles in the lower stratosphere. *Journal of Geophysical Research*, **104**, 9261–9276.

Bianchini, G., Cortesi, U., Palchetti, L., and Pascale, E. (2004). SAFIREA (Spectroscopy of the Atmosphere by Far-Infrared Emission—Airborne): Optimized instrument configuration and new assessment of improved performance, *Applied Optics*, **43**, 2962–2977.

Bianchini, G., Boscaleri, A., Carli, B., Mencaraglia, F., Palchetti, L., and Pascale, E. (2006). Infrared Balloon Experiment: Improved instrumental configuration and assessment of instrument performance, *Applied Optics*, **45**, 1041–1051.

Bigelow, D.S., Slusser, J.R., Beaubien, A.E., and Gibson, J.H. (1998). The USDA ultraviolet radiation monitoring program. *Bulletin of the American Meteorological Society*, **79**, 601–615.

Biggs, R.H. and Joyner, M.E.B. (1994). *Stratospheric Ozone Depletion/UV-B Radiation in the Biosphere* (NATO ASI Series, Global Environmental Change Vol. 18), Springer-Verlag, Berlin, 358 pp.

Bihlo, A. (2008). Rayleigh–Bènard convection as a Nambumetriplectic problem. *Journal of Physics*, **41A**, 292001, doi: 10.1088/1751–8113/41/29/292001.

Bintanja, R., Fortuin, J.P., and Kelder, H. (1997). Simulation of the meridionally and seasonally varying climate response caused by changes in ozone concentrations. *Journal of Climate*, **10**, 1288–1311.

Bisling, P.G.F., Glauer, J., Köhler, S., Weitkamp, C., Knaps, A., Möllmann-Coers, M., Geiss, H., Voltz-Thomas, A., and Kley, D. (1998). Dispersion, stratification and transport of tropospheric ozone observed by lidar and tracer experiments, in: R.D. Bojkov and G. Visconti (Eds.), *Stratospheric Ozone: Proceedings of the XVIII Quadrennial Ozone Symposium, L'Aquila, Italy, September 12–21, 1996*, Consorzio Parco Scientifico e Tecnologico d'Abruzzo, L'Aquila, Italy, Vol. 2, pp. 319–322.

Bloom, S., da Silva, A., Dee, D., Bosilovich, M., Chern, J.D., Pawson, S., Schubert, S., Sienkiewicz, M., Stajner, I., Tan, W.W. *et al.* (2005). Documentation and validation of the Goddard Earth Observing System (GEOS) data assimilation system—Version 4, in: *Global Modeling Data Assimilation 104606* (Tech. Rep. Ser. 26), NASA GSFC, Greenbelt, MD.

Blum, U., Fricke, K.H., Baumgarten, G., and Schöch, A. (2004). Simultaneous lidar observations of temperatures and waves in the polar middle atmosphere on both sides of the Scandinavian mountains: A case study on 19/20 January 2003, *Atmospheric Chemistry and Physics Discussions*, **4**, 969–989.

Blumenstock, T., Mikuteit, S., Hase, H., Boyd, I., Calisesi, Y., De-Clercq, C., Lambert, J.-C., Koopman, R., McDermid, S., Oltmans, S. *et al.* (2004). Comparison of MIPAS O_3 profiles with ground-based measurements, *Proceedings Second Atmospheric Chemistry Validation of ENVISAT Workshop (ACVE-2)* (ESA SP-562), ESA European Space Research Institute (ESRIN), Frascati, Italy, pp. 157–163.

Blumenstock, T., Hase, F., Kramer, I., Mikuteit, S., Fischer, H., Goutail, F., and Raffalski, U. (2009). Winter to winter variability of chlorine activation and ozone loss as observed by ground-based FTIR measurements at Kiruna since winter 1993/94, *International Journal of Remote Sensing*, **30**, 4055–4064.

Bodeker, G.E., Waugh, D.W., Akiyoshi, H., Braesicke, P., Eyring, V., Fahey, D.W., Manzini, E., Newchurch, M.J., Portmann, R.W., Robock *et al.* (2007). The ozone layer in the 21st Century, *Scientific Assessment of Ozone Depletion: 2006* (Global Ozone Research and Monitoring Project Report No. 50), WMO, Geneva, Switzerland, 572 pp.

Boering, K.A., Wofsy, S.C., Daube, B.C., Schneider, H.R., Loewenstein, M., and Podolske, J.R. (1996). Stratospheric mean ages and transport rates from observations of carbon dioxide and nitrous oxide. *Science*, **274**, 1340–1343.

Bojkov R.D. (1968). The ozone variations during the solar eclipse of 20 May 1966. *Tellus*, **20**, 417–421.

Bojkov, R.D. (1969). Differences in Dobson spectrometer and filter ozonometer measurements on total ozone. *Journal of Applied Meteorology*, **8**, 362–368.

Bojkov, R.D. (1976). Global net heating rates due to absorption by ozone. *Proceedings IAMAP/WMO Symposium on Atmospheric Ozone, Dresden*, WMO, Geneva, Switzerland, Vol. 2, pp. 375–381.

Bojkov, R.D. (1986a) Surface ozone during the second half of nineteenth century, *Journal of Climate and Applied Meteorology*, **25**, 343–352.

Bojkov, R.D. (1986b). The 1979–1985 ozone decline in the Antarctic as reflected in ground based observations. *Geophysical Research Letters*, **13**, 1236–1239.

Bojkov, R.D. (1988). Ozone changes at the surface and in the free troposphere, in: I.S.A. Isaksen (Ed.), *Tropospheric Ozone*, D. Reidel, Amsterdam. pp. 83–96.

Bojkov, R.D. (1997). Global ozone change and possible climate implications, in: C.S. Zerefos and A.F. Bais (Eds.), *Solar Ultraviolet Radiation* (NATO ASI Series 1: Global Environmental Change Vol. 52), Springer-Verlag, Berlin, pp. 303–324.

Bojkov, R.D. (1998). Ozone variations in the Northern Polar region, *Meteorology and Atmospheric Physics*, **38**, 117–130.

Bojkov, R. D. (1999). International assessment of ozone depletion: 1998. *WMO Bulletin*, **48**(1), 35–44.

Bojkov, R.D. and Balis, D.S. (2009). The history of total ozone measurements: The early search for signs of a trend and an update, in: C. Zerefos, G. Contopoulos, and G. Skalkeas (Eds.), *Proceedings of the Symposium for the 20th Anniversary of the Montreal Protocol*, Springer-Verlag, Berlin, pp. 73–110.

Bojkov, R.D. and Fioletov, V.E. (1995). Estimating the global ozone characteristics during the last 30 years, *Journal of Geophysical Research*, **100**, 16537–16551.

Bojkov, R.D. and Fioletov, V.E. (1996). Total ozone variations in the tropical belt: An application for quality of ground based measurements, *Meteorology and Atmospheric Physics*, **58**, 223–240.

Bojkov, R.D. and Fioletov, V.E. (1998). Latitudinal and regional total ozone from quality controlled ground-based data, in: R.D. Bojkov and G. Visconti (Eds.), *Stratospheric Ozone: Proceedings of the XVIII Quadrennial Ozone Symposium, L'Aquila, Italy,*

September 12–21, 1996, Consorzio Parco Scientifico e Tecnologico d'Abruzzo, L'Aquila, Italy, Vol. 2, pp. 13–16.

Bojkov, R.D. and Reinsel, G.C. (1985). Trends in tropospheric ozone concentrations, in: C.S. Zerefos and A. Ghazi (Eds.), *Atmospheric Ozone: Proceedings of the Quadrennial Ozone Symposium, Halkidiki, Greece, September 3–7, 1984*, D. Reidel, Dordrecht, The Netherlands, pp. 775–781.

Bojkov, R.D. and Visconti, G. (1998). *Atmospheric Ozone: Proceedings XVIII Quadrennial Ozone Symposium, L'Aquila, Italy, September 12–21, 1996*, Consorzio Parco Scientifico e Tecnologico d'Abruzzo, L'Aquila, Italy (two volumes).

Bojkov, R.D., Mateer, C.L., and Hansson, A.L. (1988). Comparison of ground-based and total ozone mapping spectrometer measurements used in assessing the performance of the global ozone observing system. *Journal of Geophysical Research*, **93**, 9525–9533.

Bojkov, R., Bishop, L., Hill, W.J., Reinsel, G.C., and Tiao, G.C. (1990). A statistical trend analysis of revised dobson total ozone data over the northern-hemisphere, *Journal of Geophysical Research*, **95**, 9785–9807.

Bojkov, R.D., Komhyr, W.D., Lapworth, A., and Vanicek, K. (1993a). *Handbook for Dobson Ozone Data Re-evaluation* (WMO Global Ozone Research and Monitoring Project, Report No. 29, WMO/TDK-No. 597), WMO, Geneva, Switzerland.

Bojkov, R.D., Zerefos, C.S., Balis, D.S., Ziomas, I.C., and Bais, A.F. (1993b). Record low total ozone during northern winter of 1992 and 1993. *Geophysical Research Letters*, **20**, 1351–1354.

Bojkov, R.D., Fioletov, V.E., and Shalamyansky, A.M. (1994). Total ozone changes over Eurasia since 1973 based on reevaluation filter ozonometer data. *Journal of Geophysical Research*, **99**, 22985–22999.

Bojkov, R.D., Bishop, L., and Fioletov, V.E. (1995). Total ozone trends from quality controlled ground-based data (1964–1994), *Journal of Geophysical Research*, **100**, 25867–25876.

Bojkov, R.D., Balis, D.S., and Zerefos, C.S. (1998). Characteristics of the ozone decline in the Northern polar and middle latitudes during the winter–spring, *Meteorology and Atmospheric Physics*, **69**, 119–135.

Boone, C.D., Nassar, R., Walker, K.A., Rochon, Y., McLeod, S.D., Rinsland, C.P., and Bernath, P.F. (2005). Retrievals for the atmospheric chemistry experiment Fourier-transform spectrometer, *Applied Optics*, **44**, 7218–7231.

Borchi, F. and Pommereau, J.-P. (2007). Evaluation of ozonesondes, HALOE, SAGE II and III, ODIN-OSIRIS and SMR, and ENVISAT-GOMOS, -SCIAMACHY and -MIPAS ozone profiles in the tropics from SAOZ long duration balloon measurements in 2003 and 2004, *Atmospheric Chemistry and Physics*, **7**, 2671–2690.

Borrell, P., Varotsos, C.A., and Kondratyev, K.Ya. (1998). Key problems of chemistry and photochemistry of the troposphere: To the completion of the first phase of EUROTRAC Project and future perspectives, *Ecological Chemistry*, **7**, 1–12.

Borrmann, S., Kunkel, D., Weigel, R., Minikin, A., Deshler, T., Wilson, J.C., Curtius, J., Volk, C.M., Homan, C.D., Ulanovsky, A. *et al.* (2010). Aerosols in the tropical and subtropical UT/LS: In-situ measurements of submicron particle abundance and volatility, *Atmospheric Chemistry and Physics*, **10**, 5573–5592.

Bortoli, D., Silva, A.M., Costa, M.J., Domingues, A.F., and Giovanelli, G. (2009a). Measurements of stratospheric ozone and nitrogen dioxide at Evora, Portugal, *International Journal of Remote Sensing*, **30**, 4209–4226.

Bortoli, D., Silva, A.M., Costa, M.J., Domingues, A.F., and Giovanelli, G. (2009b). Monitoring of atmospheric ozone and nitrogen dioxide over the south of Portugal by ground-based and satellite observations. *Optics Express*, **17**, 12944–12959.

Bose, S., Ghosh, A.B., Lal, M., Maitza, A., and Sharma, R.C. (1998). Ground based observations of ozone and nitrogen dioxide during the solar eclipse 1995, in: R.D. Bojkov and G. Visconti (Eds.), *Proceedings of the XVIII Quadrennial Ozone Symposium, L'Aquila, Italy, September 12–21, 1996*, Consorzio Parco Scientifico e Tecnologico d'Abruzzo, L'Aquila, Italy, Vol. 2, pp. 891–894.

Bovensmann, H., Burrows, J.P., Buchwitz, M., Frerick, J., Noel, S., Rozanov, V.V., Chance, K.V. and Goede, A.P.H. (1999). SCIAMACHY: Mission objectives and measurement modes, *Journal of the Atmospheric Sciences*, **56**, 127–150.

Bowman, K. P. (1989). Global patterns of the quasi-biennial oscillation in total ozone. *Journal of the Atmospheric Sciences*, **46**, 3328–3343.

Boyd, I.S., Bodeker, G.E., Connor, B.J., Swart, D.P.J., and Brinksma, E.J. (1998). An assessment of ECC ozonesondes operated using 1% and 0.5% KI cathode solutions at Lauder, New Zealand, *Geophysical Research Letters*, **5**, 2409–2412.

Bramstedt, K., Gleason, J., Loyola, D., Thomas, W., Bracher, A., Weber, M., and Burrows, J.P. (2003). Comparison of total ozone from the satellite instruments GOME and TOMS with measurements from the Dobson network 1996–2000. *Atmospheric Chemistry and Physics*, **3**, 1409–1419.

Brandt, J., Christensen, J., Frohn, L.M., and Berkowicz, R. (2003). Air pollution forecasting from regional to urban street scale: Implementation and validation for two cities in Denmark. *Physics and Chemistry of the Earth*, **28**, 335–344.

Brasseur, G. (Ed.) (1997). *The Stratosphere and Its Role in the Climate System*. Springer-Verlag, Berlin, 368 pp.

Brasseur, G.P., Müller, J.-F., and Granier, C. (1996). Atmospheric impact of NO_x emissions by subsonic aircraft: A three-dimensional model study, *Journal of Geophysical Research*, **101**, 1423–1428.

Brasseur, G., Amanatidis, G.T., and Angeletti, G. (Eds.) (1998a). European scientific assessment of the atmospheric effects of aircraft emissions, *Atmospheric Environment*, **32**, 2327–2422.

Brasseur, G.P, Cox, R.A., Hauglestaine, D., Isaksen, I., Lelieveld, J., Lister, D.H., Sausen, R., Schumann, U., Wahner, A., and Wiesen, P. (1998b). European satellite assessment of the atmospheric effects of aircraft emissions. *Atmospheric Environment*, **32**, 2329–2418.

Brasseur, G.P., Hauglustaine, D.A., Walters, S., Rasch, P.J., Müller, J.-F., Granier, C., and Tie, X.X. (1998c). MOZART: A global chemical transport model for ozone and related chemical tracers, Part I: Model description. *Journal of Geophysical Research*, **103**, 28265–28289.

Brasseur, G.P., Orlando, J.J., and Tyndall, G.S. (1999). *Atmospheric Chemistry and Global Change, 2nd Edition*, Oxford University Press, Oxford, U.K.

Brasseur, G.P., Schultz, M., Granier, C., Saunois, M., Diehl, T., Botzet, M., Roeckner, E., and Walters, S. (2006). Impact of climate change on the future chemical composition of the global troposphere. *Journal of Climate*, **19**, 3932–3951.

Brauers, T., Dom, H.-P., Koch, H., Kraus, A.B., and Plass-Duelmer, C. (1998). Meteorological aspects, ozone and solar radiation measurements during POPCORN 1994. *Journal of Atmospheric Chemistry*, **31**, 33–52.

Brewer, A.W. (1949). Evidence for a world circulation provided by measurements of helium and water vapour distribution in the stratosphere. *Quarterly Journal of the Royal Meteorological Society*, **75**, 351–363.

Brewer, A.W. (1973). A replacement for the Dobson spectrophotometer? *Pure and Applied Geophysics*, **106**, 919–927.

Brewer, A.W. and Milford, J.R. (1960). The Oxford–Kew ozonesonde. *Proceedings of the Royal Society of London, A*, **256**, 470–495.

Bronnimann, S. (2005). Interactive comment on "Detection and measurement of total ozone from stellar spectra, Paper 2: Historic data from 1935–1942" by R. E. M. Griffin, *Atmospheric Chemistry and Physics Discussions*, **5**, S4045–S4048.

Bronnimann, S., Vogler, C., Staehelin, J., Stolarski, R., and Hansen, G. (2007). Total ozone observations during the past 80 years, *Advances in Global Change Research*, **33**, 129–140, doi: 10.1007/978-1-4020-6766-2_8.

Bruhl, C., Drayson, R.S., Russell J.M. III, Crutzen, P.J., McInerney, J.M., Purcell, P.N., Claude, H., Gernandt, H., McGee, T.J., McDermid, I.S. *et al.* (1996). Halogen Occultation Experiment ozone channel validation, *Journal of Geophysical Research*, **101**, 10217–10240.

Brune, W.I, Toohey, D.W., Anderson, J.G., Starr, W.L, Vedder, J.F., and Danielsen, E.F. (1988). In situ northern mid-latitude observations of ClO, O_3, and BrO in the wintertime lower stratosphere, *Science*, **242**, 558–562.

Brunner, D., Staehelin, J., Maeder, J.A., Wohltmann, I., and Bodeker, G.E. (2006). Variability and trends in total and vertically resolved stratospheric ozone based on the CATO ozone data set, *Atmospheric Chemistry and Physics*, **6**, 4985–5008.

Brunner, D., Staehelin, J., Rogers, H.L., Köhler, M.O., Pyle, J.A., Hauglustaine, D., Jourdain, L., Berntsen, T.K., Gauss, M., Isaksen, I.S.A. *et al.* (2003). An evaluation of the performance of chemistry transport models by comparison with research aircraft observations, Part 1: Concepts and overall model performance. *Atmospheric Chemistry and Physics*, **3**, 1609–1631.

Bühler, S., Eyring, V., Wehr, T., and Kunzi, K. (1996). Limits imposed on millimeter and submillimeter wave limb sounders by continuum emissions, in: R.D. Bojkov and G. Visconti (Eds.), *Proceedings of the XVIII Quadrennial Ozone Symposium, L'Aquila, Italy, September 12–21, 1996*, Consorzio Parco Scientifico e Tecnologico d'Abruzzo, L'Aquila, Italy, Vol. 2, pp. 883–886.

Burkholder, J.B., Orlando, J.J., and Howard, C.J. (1990). Ultraviolet-absorption cross sections of Cl_2O_2 between 210 and 410 nm, *Journal of Physical Chemistry*, **94**, 687–695.

Burnett, C.R. and Burnett, E.B. (1985). Atmospheric hydroxyl response to the partial solar eclipse of May 1984, *Geophysical Research Letters*, **12**, 263–266.

Burrows, J.P., Weber, M., Buchwitz, M., Rozanov, V.V., Ladstädter-Weissenmayer, A., Richter, A., de Beek, R., Hoogen, R., Bramstedt, K., Eichmann, K.U. *et al.* (1999). The Global Ozone Monitoring Experiment (GOME): Mission concept and first scientific results, *Journal of the Atmospheric Sciences*, **56**, 151–175.

Butchart, N., Scaife, A.A., Bourqui, M., de Grandpré, J., Hare, S.H.E., Kettleborough, J., Langematz, U., Manzini, E., Sassi, F., Shibata *et al.* (2006). Simulations of anthropogenic change in the strength of the Brewer–Dobson circulation, *Climate Dynamics*, **27**, 727–741.

Calisesi, Y., Soebijanta, V.T., and van Oss, R. (2005). Regridding of remote soundings: Formulation and application to ozone profile comparison, *Journal of Geophysical Research*, **110**, D23306, doi: 10.1029.2005.JD006122.

Callies, J., Corpaccioli, E., Eisinger, M., Hahne, A., and Lefebvre, A. (2000). GOME-2: Metop's second generation sensor for operational ozone monitoring, *ESA Bulletin*, **102**, 28–36.

Callis, L. and Natarajan, M. (1986). The Antarctic ozone minimum: Relationship to odd nitrogen, odd chlorine, the final warming, and the 11-year solar cycle, *Journal of Geophysical Research*, **91**, 10771–10796.

Callis, L.B., Natarajan, M., Lambeth, J.D., and Boughner, R.E. (1997). On the origin of midlatitude ozone changes: Data analysis and simulations for 1979–1993, *Journal of Geophysical Research*, **102**, 1215–1228.

Cammas, J.P., Jakoby-Koaly, S., Suhre, K., Rosset, R., and Marenco, A. (1998). Atlantic subtropical potential vorticity barrier as seen by Measurements of Ozone by Airbus In-Service Aircraft (MOZAIC) flights, *Journal of Geophysical Research*, **103**, 25681–25693.

Camp, C.D., Roulston, M.S., and Yung, Y.L. (2003). Temporal and spatial patterns of the interannual variability of total ozone in the tropics. *Journal of Geophysical Research*, **108**, doi: 10.1029/2001JD001504.

Campos, T.L., Weinheimer, A.S., Zheng, J., Montzka, O.D., Walega, J.G., Grahek, F.E., Yay, S.A., Collins, J.E. Jr., Wade, L.O., Sachse, G.W. *et al.* (1998). Measurements of NO and NO_x emission indices during SUCCESS, *Geophysical Research Letters*, **25**, 1713–1716.

Canty, T., Rivière, E.D., Salawitch, R.J., Berthet, G., Renard, J.B., Pfeilsticker, K., Dorf, M., Butz, A., Bosch, H., Stimpfle, R.M. *et al.* (2005). Nighttime OClO in the winter Arctic vortex, *Journal of Geophysical Research*, **110**, D01301.

Canziani P.O., Compagnucci, R.H., Bischoff, S.A., and Legnani, W.E. (2002). A study of impacts of tropospheric synoptic processes on the genesis and evolution of extreme total ozone anomalies over southern South America, *Journal of Geophysical Research*, **107**, 4741, doi: 10.1029/2001JD000965.

Carlotti, M., Paruolo, G., Alboni, F., Carli, B., Mencaraglia, F., and Ridolfi, M. (1995). *Determination of Atmospheric Pressure and Temperature Distributions from MIPAS Limb Emission Line Spectra* (ESA Contract Report No. 142956), European Space Agency, Noordwijk, The Netherlands.

Cartalis, C. and Varotsos, C. (1994). Surface ozone in Athens, Greece, at the beginning and at the end of the twentieth century, *Atmospheric Environment*, **28**, 3–8.

Chan, L.Y., Liu, H.Y., Lam, K.S., and Wang, T. (1998). Observations of total column ozone and vertical ozone distribution in subtropical Hong Kong, in: R.D. Bojkov and G. Visconti (Eds.), *Proceedings of the XVIII Quadrennial Ozone Symposium, L'Aquila, Italy, September 12–21, 1966*, Consorzio Parco Scientifico e Tecnologico d'Abruzzo, L'Aquila, Italy, Vol. 1, pp. 107–110.

Chandra, S. and Stolarski, R.S. (1991). Recent trends in stratospheric total ozone: Implications of dynamical and El Chichon perturbations, *Geophysical Research Letters*, **18**, 2277–2280.

Chandra, S. and Varotsos, C.A. (1995). Recent trends of the total column ozone: Implications for the Mediterranean region, *International Journal of Remote Sensing*, **16**, 1765–1769.

Chandra, S., Varotsos, C., and Flynn, L. E. (1996). The mid-latitude total ozone trends in the Northern Hemisphere, *Geophysical Research Letters*, **23**, 555–558.

Chandra, S., Ziemke, J.R., Bhartia, P.K., and Martin, R.V. (2002). Tropical tropospheric ozone: Implications for dynamics and biomass burning, *Journal of Geophysical Research*, **107**, 4188–4198.

Chandra, S., Ziemke, J.R., and Martin, R.V. (2003). Tropospheric ozone at tropical and middle latitudes derived from TOMS/MLS residual: Comparison with a global model, *Journal of Geophysical Research*, **108**, 4291, doi: 10.1029/2002JD002912.

Chandra, S., Ziemke, J.R., Duncan, B.N., Diehl, T.L., Livesey, N.J., and Froidevaux, L. (2009). Effect of the 2006 El Niño on tropospheric ozone and carbon monoxide:

Implications for dynamics and biomass burning, *Atmospheric Chemistry and Physics*, **9**, 4239–4249.

Chanin, M.-L. (Ed.) (1993). *The Role of the Stratosphere in Global Change* (NATO ASU Series), Springer-Verlag, Heidelberg, 425 pp.

Chanin, M.-L. (1998). Fifth Session of the Scientific Steering Group on Stratospheric Processes and Their Role in Climate, Port Jefferson, NY, November 17–20, 1997, *SPARC Newsletter*, **10**, 1–5.

Chanin, M.-L. and Geller, M. (1998). Report on the WCRP IJSC Meeting, *SPARC Newsletter*, **11**, 1–2.

Chapman, S. (1930). On ozone and atomic oxygen in the upper atmosphere, *Philosophical Magazine*, **10**, 369–383.

Chapman, S. (1932). A theory of upper-atmosphere ozone, *Quarterly Journal of the Royal Meteorological Society*, **58**, 11–13.

Charlton, A.J. and Polvani, L.M. (2007). A new look at stratospheric sudden warming, Part I: Climatology and modelling benchmarks, *Journal of Climate*, **20**, 449–469.

Charyulu, D.V., Sivakumar, V., Bencherif, H., Kirgis, G., Hauchecorne, A., Keckhut, P., and Narayana Rao, D. (2007). 20-year LiDAR observations of stratospheric sudden warming over a mid-latitude site, Observatoire de Haute Provence (OHP; 44°N, 6°E): Case study and statistical characteristics, *Atmospheric Chemistry and Physics Discussions*, **7**, 15739–15779.

Chattopadhyay, G. and Chattopadhyay, S. (2010). Univariate approach to the monthly total ozone time series over Kolkata, India: Autoregressive integrated moving average (ARIMA) and autoregressive neural network (AR-NN) models, *International Journal of Remote Sensing*, **31**, 575–583.

Chattopadhyay, S. (2007). Prediction of mean monthly total ozone time series: Application of radial basis function network, *International Journal of Remote Sensing*, **28**, 4037–4046.

Chattopadhyay, S. and Bandyopadhyay, B. (2007). Artificial neural network with backpropagation learning to predict mean monthly total ozone in Arosa, Switzerland, *International Journal of Remote Sensing*, **28**, 4471–4482.

Chattopadhyay, S. and Chattopadhyay-Bandyopadhyay, G. (2008). Forecasting daily total ozone concentrationa comparison between neurocomputing and statistical approaches, *International Journal of Remote Sensing*, **29**, 1903–1916.

Chattopadhyay, S. and Chattopadhyay, G. (2008). A factor analysis and neural network-based validation of the Varotsos–Cracknell theory on the 11-year solar cycle, *International Journal of Remote Sensing*, **29**, 2775–2786.

Chen, Z., Ivanov, P.C., Hu, K., and Stanley, H.E. (2002). Effect of nonstationarities on detrended fluctuation analysis. *Physical Review E*, **65**, 041107.

Chen, Z., Hu, K., Carpena, P., Bernaola-Galvan, P., Stanley, H.E., and Ivanov, P.C. (2005). Effect of nonlinear filters on detrended fluctuation analysis. *Physical Review E*, **71**, 011104.

Chen, Z.M., Wang, H.L., Zhu, L.H., Wang, C.X., Jie, C.Y., and Hua, W. (2003). Aqueous-phase ozonolysis of methacrolein and methyl vinyl ketone: A potentially important source of atmospheric aqueous oxidants, *Atmospheric Chemistry and Physics*, **8**, 2255–2265.

Chernikov, A.A., Borisov, Y.A., and Zvyagintsev, A.M. (2000). Tendencies in the ozone layer measurements by satellite instrument TOMS and ground-based network, *Earth Research from Space*, **6**, 23–32.

Chimonas, G. and Hines, C.O. (1970). Atmospheric gravity waves induced by a solar eclipse, *Journal of Geophysical Research*, **75**, 875–878.

Chimonas, G. and Hines, C.O. (1971). Atmospheric gravity waves induced by a solar eclipse, 2, *Journal of Geophysical Research*, **76**, 7003–7005.

Chiou, E.W., Chu, W.P., and Thomason, L.W. (1998a). Optimization of SAGE III spectral channels for water vapor retrieval. *Ninth Conference on Satellite Meteorology and Oceanography, May 25–29, 1998, Paris, France*, pp. 579–582.

Chiou, E.W., Chu, W.P., and Thomason, L.W. (1998b). Simulation studies for the retrieval of water vapor profiles from the SAGE III experiment. *Tenth Symposium on Meteorological Observations and Instrumentation, January 11–16, 1998, Phoenix, AZ.*

Christiansen, B. and Guldberg, A. (1998). Dynamical and radiative effects of idealized ozone perturbations, in: R.D. Bojkov and G. Visconti (Eds.), *Proceedings of the XVIII Quadrennial Ozone Symposium, L'Aquila, Italy, September 12–21, 1996*, Consorzio Parco Scientifico e Tecnologico d'Abruzzo, L'Aquila, Italy, Vol. 2, pp. 805–808.

Christiansen, B., Guldberg, A., Hansen, A.W., and Riishøjgaard, L.P. (1997). On the response of a three-dimensional general circulation model to imposed changes in the ozone distribution. *Journal of Geophysical Research*, **102**, 13051–13077.

Christodoulakis, J., Tzanis, C., and Varotsos, C. (2008). Technical report: Standardization of the Athens Dobson spectrophotometer versus Reference Dobson spectrophotometer 064, *International Journal of Remote Sensing*, **29**, 1917–1920.

Chu, W.P., McCormick, M.P., Lenoble, J., Brogniez, C., and Pruvost, P. (1989). SAGE II inversion algorithm, *Journal of Geophysical Research*, **94**, 8339–8352.

Chubachi, S. (1984). Preliminary result of ozone observations at Syowa Station from February, 1982 to January, 1983. *Memoirs of National Institute of Polar Research Japan*, Special Issue, **34**, 13–20.

Chubachi, S. (1985). A special ozone observation at Syowa Station, Antarctica, from February 1982 to January 1983, in: C.S. Zerefos and A. Ghazi (Eds.), *Proceedings of the Quadrennial Ozone Symposium held in Halkidiki, Greece, September 3–7, 1984*, D. Reidel, Dordrecht, The Netherlands, pp. 285–289.

Chubachi, S. (1986). On the cooling of the stratospheric temperature at Syowa Antarctica, *Geophysical Research Letters*, **13**, 1221–1223.

Chubachi, S. (1997) Annual variation of total ozone at Syowa Station, Antarctica, *Journal of Geophysical Research*, **102**, 1349–1354.

Chubachi, S. (2009). Seasonal start of the Antarctic ozone hole derived by the observations with Dobson spectrophotometers, *International Journal of Remote Sensing*, **30**, 3907–3916.

Chubachi, S. and Kajiwara, R. (1988). Total ozone by lunar Dobson observation at Syowa, Antarctica, *Geophysical Research Letters*, **15**, 905–906.

Cicerone, R.J., Stolarski, R.S., and Walters, S. (1974). Stratospheric ozone destruction by man-made chlorofluoromethanes, *Science*, **185**, 1165–1167.

Cionni, I., Eyring, V., Lamarque, J.F., Randel, W.J., Stevenson, D.S., Wu, F., Bodecker, G.E., Shepherd, T.G., Shindell, D.T., and Waugh, D.W. (2011). Ozone database in support of CMIP5 simulations: Results and corresponding radiative forcing, *Atmospheric Chemistry and Physics Discussions*, **11**, 10875–10933.

Clarisse, L., Coheur, P.F., Prata, A.J., Hurtmans, D., Razavi, A., Phulpin, T., Hadji-Lazaro, J., and Clerbaux, C. (2008). Tracking and quantifying volcanic SO_2 with IASI: The September 2007 eruption at Jebel at Tair, *Atmospheric Chemistry and Physics*, **8**, 7723–7734.

Claude, H., Kohler, U., and Steinbrecht, W. (1998). New trend analyses of the homogenized ozone records at Hohenpeissenberg, in: R.D. Bojkov and G. Visconti (Eds.), *Proceedings of the XVIII Quadrennial Ozone Symposium, L'Aquila, Italy, September 12–21, 1996*, Consorzio Parco Scientifico e Tecnologico d'Abruzzo, L'Aquila, Italy, Vol. 1, 21–24.

Clerbaux, C., Boynard, A., Clarisse, L., George, M., Hadji-Lazaro, J., Herbin, H., Hurtmans, D., Pommier, M., Razavi, A., Turquety, S. *et al.* (2010). Monitoring of atmospheric composition using the thermal infrared IASI/MetOp sounder, *Atmospheric Chemistry and Physics*, **9**, 6041–6054.

Collette, C. and Ausloos M. (2004). Scaling analysis and evolution equation of the North Atlantic oscillation index fluctuations, *International Journal of Modern Physics, C*, **15**, 1353–1366.

Collins, W.D., Rasch, P.J., Boville, B.A., Hack, J.J., McCaa, J.R., Williamson, D.L., Kiehl, J.T., Briegleb, B., Bitz, C., Lin, S.-J. *et al.* (2004). *Description of the NCAR Community Atmosphere Model (CAM3)* (Technical Note NCAR-TN-464 + STR), National Center for Atmospheric research, Boulder, CO, 226 pp.

Collins, W.J., Derwent, R.G., Garnier, B., Johnson, C.E., and Sanderson, M.G. (2003). Effect of strastosphere–troposphere exchange on the future tropospheric ozone trend. *Journal of Geophysical Research*, **108**, 8528, doi: 10.1029/2002JD002617.

Connor, B.J., Parrish, A., Tsou, J.-J., and McCormick, M.P. (1995). Error analysis for the ground-based microwave ozone measurements during STOIC, *Journal of Geophysical Research*, **100**, 9283–9291.

Considine, D.B., Stolarski, R.S., Hollandsworth, S.M., Jackman, C.H., Fleming, E.L. (1999). A Monte Carlo uncertainty analysis of ozone trend predictions in a two-dimensional model. *Journal of Geophysical Research*, **104**, 1749–1766.

Cooper, M. (2004). Validation of SABER temperature measurements using ground-based instruments. *International Geoscience and Remote Sensing Symposium (IGARSS '04), September 20–24, 2004*, Vol. 6, pp. 4099–4101.

Corlett, G.K. and Monks, P.S. (2001). A comparison of total column ozone values derived from the Global Ozone Monitoring Experiment (GOME), the TIROS Operatioonal Vertical Sounder (TOVS), and the Total Ozone Mapping spectrometer (TOMS), *Journal of the Atmospheric Sciences*, **58**, 1103–1116.

Cortesi, U., Blom, C.E., Camy-Peyret, C., Chance, K., Davies, J., Goutail, F., Kuttippurath, J., McElroy, C.T., Mencaraglia, F., Oelhaf, H. *et al.* (2004). MIPAS ozone validation by stratospheric balloon and aircraft measurements, *Proceedings of the Second Atmospheric Chemistry Validation of ENVISAT Workshop (ACVE-2), May 3–7, 2004* (ESA SP-562), ESA European Space Research Institute (ESRIN), Frascati, Italy.

Cortesi, U., Bianchini, G., Carli, B., Palchetti, L., Trambusti, M., and Ade, P.A.R. (2005). SAFIRE-A (spectroscopy of the atmosphere by using far-infrared emission-airborne): Assessment of measurement capabilities and future developments, *Advances in Space Research*, **36**, 888–893.

Cortesi, U., Lambert, J.C., De Clerco, C., Bianchini, G., Blumenstock, T., Bracher, A., Castelli, E., Catoire, V., Chance, K.V., De Maziere, M. *et al.* (2007). Geophysical validation of MIPAS-ENVISAT operational ozone data, *Atmospheric Chemistry and Physics*, **7**, 4807–4867.

Cox, R.A. (1999). Ozone and peroxy radical budgets in the marine boundary layer: Modeling the effect of NO_x, *Journal of Geophysical Research*, **104**, 8047–8056.

Cracknell, A.P. (1971). The application of Landau's theory of continuous phase transitions to magnetic phase transitions. *Journal of Physics C: Solid State Physics*, **4**, 2488–2500.

Cracknell, A.P. and Varotsos, C.A. (1994). Ozone depletion over Scotland as derived from Nimbus-7 TOMS measurements, *International Journal of Remote Sensing*, **15**, 2659–2668.

Cracknell, A.P. and Varotsos, C.A. (1995). The present status of the total ozone depletion over Greece and Scotland: A comparison between Mediterranean and more northerly latitudes, *International Journal of Remote Sensing*, **16**, 1751–1763.

Cracknell, A.P. and Varotsos, C.A. (2007a). Editorial and cover: Fifty years after the first artificial satellite: From Sputnik 1 to ENVISAT, *International Journal of Remote Sensing*, **28**, 2071–2072.

Cracknell, A.P. and Varotsos, C.A. (2007b). The IPCC Fourth Assessment Report and the fiftieth anniversary of Sputnik, *Environmental Science and Pollution Research*, **14**, 384–387.

Cracknell, A.P. and Varotsos, C.A. (2009). The contribution of remote sensing to the implementation of the Montreal Protocol and the monitoring of its success, *International Journal of Remote Sensing*, **30**, 3853–3873.

Cracknell, A.P. and Varotsos, C.A. (2011). New aspects of global climate-dynamics research and remote sensing, *International Journal of Remote Sensing*, **32**, 579–600.

Cristofanelli, P., Bonasoni, P., Bonafè, U., Calzolari, F., Duchi, R., Marinoni, A., Roccato, F., Vuillermoz, E., and Sprenger, M. (2009). Influence of lower stratosphere/upper troposphere (LS/UT) transport events on surface ozone at the Everest-Pyramid GAW Station (Nepal): First year of analysis, *International Journal of Remote Sensing*, **30**, 4083–4097.

Cronn, D.R., Bamesberger,W.L., Menzia, F.A., Waylett, S.F., and Waylett, A.S. (1986). Atmospheric trace gas trends at Palmer Station, Antarctica: 1982–1985, *Geophysical Research Letters*, **13**, 1272–1275.

Crutzen, P.J. (1969). Determination of parameters appearing in the oxygen–hydrogen atmosphere, *Annales de Géophysique*, **25**, 275–279.

Crutzen, P.J. (1970). The influence of nitrogen oxides on the atmospheric ozone content, *Quarterly Journal of the Royal Meteorological Society*, **96**, 320–325.

Crutzen, P.J. (1971). Ozone production rates in an oxygen, hydrogen, nitrogen oxide atmosphere, *Journal of Geophysical Research*, **76**, 7311–7327.

Crutzen, P.J. (1974). Estimates of possible future ozone reductions from continued use of fluoro–chloro–methanes CF_2Cl_2, $CFCl_3$, *Geophysical Research Letters*, **1**, 205–208.

Crutzen, P.J. (1998). The Bulletin interview, *WMO Bulletin*, **47**(2), 111–123.

Crutzen, P.J. and Arnold, F. (1986). Nitric acid cloud formation in the cold Antarctic stratosphere: A major cause for the springtime "ozone hole", *Nature*, **324**, 651–655.

Crutzen, P.J., Grooß, J.-U., Bruhl, C., Muller, R., and Russell J.M. III (1995). A reevaluation of the ozone budget with HALOE UARS data: No evidence for the ozone deficit, *Science*, **268**, 705–708.

Crutzen, P.J., Lawrence, M.G., and Poschl, U. (1999). On the background photochemistry of tropospheric ozone, *Tellus A*, **51**, 123–146.

Cunnold, D.M. and Wang, H.J. (1998). The effects of temperature uncertainties on the interpretation of ozone trends in the upper stratosphere, in: R.D. Bojkov, and G. Visconti (Eds.), *Proceedings of the XVIII Quadrennial Ozone Symposium, L'Aquila, Italy, September 12–21, 1996*, Consorzio Parco Scientifico e Tecnologico d'Abruzzo, L'Aquila, Italy, Vol. 1, pp. 111–114.

Cunnold, D., Alyea, F., Phillips, N., and Prinn, R. (1975). A three-dimensional dynamical–chemical model of atmospheric ozone, *Journal of the Atmospheric Sciences*, **32**, 170–194.

Cunnold, D.M., Chu, W.P., Barnes, R.A., McCormick, M.P., and Veiga, R.E. (1989). Validation of SAGE II ozone measurements, *Journal of Geophysical Research*, **94**, 8447–8460.

Daehler, M., Bevilacqua, R.M., Kriebel-Hamilton, D.L., Goldizen, D., Olivero, J.J., Hartmann, G.K., Degenhardt, W., Richards, M.L., Connor, B.J., and Aellig, C.P.

(1998). Validation of Millimeter-wave Atmospheric Sounder (MAS) ozone measurements, *Journal of Geophysical Research*, **103**, 19615–19639.

Dahlback, A., Hansen, G., Svenoe, T., Opedal, L., and Toennessen, F. (1998). Monitoring of total ozone by means of multi-channel UV filter instruments, in: R.D. Bojkov and G. Visconti (Eds.), *Proceedings of the XVIII Quadrennial Ozone Symposium, L'Aquila, Italy, September 12–21, 1996*, Consorzio Parco Scientifico e Tecnologico d'Abruzzo, L'Aquila, Italy, Vol. 2, pp. 887–890.

Dall'Amico, M., Gray, L.J., Rosenlof, K.H., Scaife, A.A., Shine, K.P., and Stott, P.A. (2010). Stratospheric temperature trends: impact of ozone variability and the QBO, *Climate Dynamics*, **34**, 381–398.

Dalsøren, S.B., Magnus, S.E., Myhre, G., Endresen, O., Isaksen, I.S.A., and Fuglestvedt, J.S. (2010). Impacts of the large increase in international ship traffic 2000–2007 on tropospheric ozone and methane, *Environmental Science and Technology*, **44**, 2482–2489.

Dameris, M. (2010a). Climate change and atmospheric chemistry: How will the stratospheric ozone layer develop? *Angewandte Chemie*, International Edition, **49**, 8092–8102.

Dameris, M. (2010b). Depletion of the ozone layer in the 21st century, *Angewandte Chemie*, International Edition, **49**, 489–491.

Dameris, M., Grewe, V., Khler, I., Sausen, R., Bruhl, C., Grooß, J.-U., and Steil, B. (1998). Impact of aircraft NO_x-emissions on tropospheric and stratospheric ozone. Part II: 3-D model results. *Atmospheric Environment*, **32**, 3185–3200.

Dameris, M., Grewe, V., Ponater, M., Deckert, R., Eyring, V., Mager, F., Matthes, S., Schnadt, C., Stenke, A., Steil, B. *et al.* (2005). Long-term changes and variability in a transient simulation with a chemistry–climate model employing realistic forcing, *Atmospheric Chemistry and Physics*, **5**, 2121–2145.

Dameris, M., Matthes, S., Deckert, R., Grewe, V., and Ponater, M. (2006). Solar cycle effect delays onset of ozone recovery, *Geophysical Research Letters*, **33**, L03806-1–L03806-4, doi: 10.1029/2005GL024741.

Dameris, M., Peter, T., Schmidt, U., and Zellner, R. (2007). The ozone hole and its causes, *Chemie in Unserer Zeit*, **41**, 152–168.

Daniel, J.S. and Solomon, S. (1998). On the climate forcing of carbon monoxide, *Journal of Geophysical Research*, 103, 13245–13260.

Daniel, J.S., Solomon, S., and Albritton, D.L. (1995). On the evaluation of halocarbon radiative forcing and global warming potentials, *Journal of Geophysical Research*, **100**, 1271–1285.

Daniel, J.S., Solomon, S., Portmann, R.W., and Garcia, R.R. (1999). Stratospheric ozone destruction: The importance of bromine relative to chlorine, *Journal of Geophysical Research*, **104**, 23871–23880.

Daniel, J.S., Velders, G.J.M., Douglass, A.R., Forster, P.M.D., Hauglustaine, D.A., Isaksen, I.S.A., Kuijpers, L.J.M., McCulloch, A., Wallington, T.J., Ashford, P. *et al.* (2007). Halocarbon scenarios, ozone depletion potentials, and global warming potentials. Chapter 8 in *Scientific Assessment of Ozone Depletion: 2006* (Global Ozone Research and Monitoring Project Report No. 50), WMO, Geneva, Switzerland, 572 pp.

Danilin, M.Y., Ko, M.K.W., Froidevaux, L., Santee, M.L., Lyjak, L.V., Bevilacqua, R.M., Zawodny, J.M., Sasano, Y., Irie, H., Kondo, Y. *et al.* (2002). Trajectory hunting as an effective technique to validate multiplatform measurements: Analysis of the MLS, HALOE, SAGEII, ILAS, and POAM-II data in October–November 1996, *Journal of Geophysical Research*, **107**, 4420, doi: 10.1029/2001JD002012.

Davies, W.E., Vaughan, G., and O'Connor, F.M. (1998). Observation of the near-zero ozone concentration in the upper troposphere in midlatitudes, *Geophysical Research Letters*, **25**, 1173–1176.

De Backer, H. (2009). Time series of daily erythemal UV doses at Uccle Belgium, *International Journal of Remote Sensing*, **30**, 4145–4151.

De Backer, H., DeMuer, D., Schoubs, E., and Allaart, M. (Eds.) (1998). A new pump correction profile for Brewer Mast ozonesondes, in: R.D. Bojkov and G. Visconti (Eds.), *Proceedings of the XVIII Quadrennial Ozone Symposium, L'Aquila, Italy, September 12–21, 1996*, Consorzio Parco Scientifico e Tecnologico d'Abruzzo, L'Aquila, Italy, Vol. 2, pp. 891–894.

Debrestian, D., Lumpe, J., Bevilacqua, R.M., Shettle, E.P., Hornstein, J.S., and Olivero, J.J. (1997). POAM II observations of polar mesospheric clouds in the Southern Hemisphere, *Advances in Space Research*, **19**, 587–590.

De Clercq, C. and Lambert, J.-C. (2006). A forward model of limb infrared emission spectra in a two-dimensional atmosphere, in: *Proceedings of Atmospheric Science Conference, Frascati, Italy, May 8–12, 2006* (ESA SP-628), ESA, Noordwijk, The Netherlands.

Degórska, M. and Rajewska-Wiech, B. (1989). The effect of stray light on total ozone measurements at Belsk, Poland, in: R.D. Bojkov and G. Visconti (Eds.), *Proceedings of the Quadrennial Ozone Symposium, 1988, Göttingen, Germany*, Deepak Publications, Hampton, VA, pp. 759–761.

de Laat, A.T.J., Zachariasse, M., Roelofs, G.J., van Velthoven, P., Dickerson, R.R., Rhoads, K.P., Oltmans, S.J., and Lelieveld, J. (1999). Tropospheric O_3 distribution over the Indian Ocean during spring 1995 evaluated with a chemistry–climate model. *Journal of Geophysical Research*, **104**, 13881–13895.

DeLuisi, J.J., Longenecker, D.U., Mateer, C.L., and Wuebbles, D.J. (1989). An analysis of Northern mid-latitude Umkehr measurements corrected for stratospheric aerosol for 1979–1986. *Journal of Geophysical Research*, **94**, 9837–9846.

DeLuisi, J., Petropavloskikh, I., Mateer, C., and Thomason, L. (1998). An evaluation of the uncertainties in estimated stratospheric aerosol errors to retrieved Umkehr ozone profiles, in: R.D. Bojkov and G. Visconti (Eds.), *Atmospheric Ozone: Proceedings of the XVIII Quadrennial Ozone Symposium, L'Aquila, Italy, September 12–21, 1996*, Consorzio Parco Scientifico e Tecnologico d'Abruzzo, L'Aquila, Italy, Vol. 1, 259–262.

De Mazière, M., Van Roozendael, M., Hermans, C., Simon, P.C., Demoulin, P., Roland, G., and Zander, R. (1998). Quantitative evaluation of the post-Mount Pinatubo NO_2 reduction and recovery, based on 10 years of Fourier transform infrared and UV–visible spectroscopic measurements at Jungfraujoch, *Journal of Geophysical Research*, **103**, 10849–10858.

Déqué, M., Dreveton, C., Braun, A., and Cariolle, D. (1994). The ARPEGE/IFS atmosphere model: A contribution to the French community climate modelling, *Climate Dynamics*, **10**, 249–266.

Derwent, R.G., Jenkin, M.E., Saunders, S.M., and Pilling, M.S. (1998). Photochemical ozone creation potential for organic compounds in north-west Europe calculated with master chemical mechanism, *Atmospheric Environment*, **32**, 2429–2441.

Descartes, R. (1637). *Discours de la méthode: Les météores*, Ian Marie, Leyden, 31 pp.

Deser, C. and Phillips, A.S. (2009). Atmospheric circulation trends, 1950–2000: The relative roles of sea surface temperature forcing and direct atmospheric radiative forcing, *Journal of Climate*, **22**, 396–413.

Dethof, A. and Helm, E.V. (2004). Ozone assimilation in the ERA-40 reanalysis project, *Quarterly Journal of the Royal Meteorological Society*, **130**, 2851–2872.

De Toma, G., London, J., Chen, L., Rottman, G.J., Woods, T.N., and Froidevaux, L. (1998). Solar cycle changes in ultraviolet irradiance and corresponding changes in mesospheric (45 to 65 km) ozone and temperature, in: R.D. Bojkov and G. Visconti (Eds.), *Atmospheric Ozone: Proceedings of the XVIII Quadrennial Ozone Symposium, L'Aquila, Italy, September 12–21, 1996*, Consorzio Parco Scientifico e Tecnologico d'Abruzzo, L'Aquila, Italy, Vol. 1, pp. 259–262.

de Zafra, R.L., Jaramillo, M., Parrish, A., Solomon, P., Connor, B., and Barnett, J. (1987). High concentrations of chlorine monoxide at low altitudes in the Antarctic spring stratosphere: Diurnal variation, *Nature*, **328**, 408–411.

Dhomse, S., Weber, M., Wohltmann, I., Rex, M., and Burrows, J.P. (2006). On the possible causes of recent increases in northern hemisphere total ozone from a statistical analysis of satellite data from 1979–2003, *Atmospheric Chemistry and Physics*, **6**, 1165–1180.

Dials, M., Gille, J.C., Barnett, J.J., and Whitney, J.G. (1998). Description of the High Resolution Dynamics Limb Sounder (HIRDLS) instrument, *Proceedings SPIE*, **3437**, 84–91.

Dikty, S., Schmidt, H., Weber, M., von Savigny, C., and Mlynczak, M.G. (2010a). Daytime ozone and temperature variations in the mesosphere: A comparison between SABER observations and HAMMONIA model, *Atmospheric Chemistry and Physics*, **10**, 8331–8339.

Dikty, S., Weber, M., von Savigny, C., Sonkaew, T., Rozanov, A., and. Burrows, J.P. (2010b). Modulations of the 27 day solar rotation signal in stratospheric ozone from Scanning Imaging Absorption Spectrometer for Atmospheric Cartography (SCIAMACHY) (2003–2008), *Journal of Geophysical Research*, **115**, D00I15, doi: 10.1029/2009JD012379.

Dingmin, L., Shine, K.P., and Gray, L.S. (1995). The role of ozone-induced diabatic heating anomalies in the quasi-biennial oscillation, *Quarterly Journal of the. Royal Meteorological Society*, **121**, Part B, N524, 937–944.

Dobber, M., Stammes, P., Levelt, P., Kleipool, Q., and Voors, R. (2006). In-flight calibration of GOME-2 level-1 data using the Ozone Monitoring Instrument. *Proceedings of the First EPS/MetOp RAO Workshop, ESA-ESRIN, Frascati, Italy, May 15–17, 2006* (ESA SP-618), ESA European Space Research Institute (ESRIN), Frascati, Italy

Dobson, G.M.B. (1931). A photoelectric spectrophotometer for measuring the amount of atmospheric ozone. *Proceedings of the Royal Society, London*, **43**, 324–339.

Dobson, G.M.B. (1956). Origin and distribution of the polyatomic molecules in the atmosphere, *Proceedings of the Royal Society, London*, **236A**, 187–193.

Dobson, G.M.B. (1957a). *Observer's Handbook for the Ozone Spectrophotometer* (Annals of the International Geophysical Year, V, Part I), Pergamon Press, Oxford, U.K., pp. 46–89.

Dobson, G.M.B. (1957b). *Adjustment and Calibration of the Ozone Spectrophotometer* (Annals of the International Geophysical Year, V, Part I), Pergamon Press, Oxford, U.K., pp. 90–113.

Dobson, G.M.B. (1963a). Note on the measurement of ozone in the atmosphere, *Quarterly Journal of the Royal Meteorological Society*, **89**, 46–89.

Dobson, G.M.B. (1963b). *Exploring the Atmosphere*, Oxford University Press, Oxford, 200 pp.

Dobson, G.M.B. (1968). Forty years' research on atmospheric ozone at Oxford: A history, *Applied Optics*, **7**, 387–405.

Dobson, G.M.B. (1973). The laminated structure of the ozone in the atmosphere. *Quarterly Journal of the Royal Meteorological Society*, **99**, 599–607.

Dobson, G.M.B. and Harrison, D.N. (1926). Observations of the amount of ozone in the Earth's atmosphere and its relation to other geophysical conditions. *Proceedings of the Royal Society*, **110**, 660–693.

Dobson, G.M.B. and Normand, C.W.B. (1957a). The Dobson spectrophotometer, *Annals of the International Geophysical Year*, **16**, 46–191.

Dobson, G.M.B. and Normand, C.W.B. (1957b). Observer's handbook for the ozone spectrophotometer, *Annals of the International Geophysical Year*, **5**, 46–89.

Dobson, G.M.B. and Normand, C.W.B. (1962). Determination of the constants etc. used in the calculation of the amount of ozone from spectrophotometer measurements and of the accuracy of the results, *Annals of the International Geophysical Year*, **16**, 161–191.

Domine, F. and Xueref, I. (2001). Evaluation of depth profiling using laser resonant desorption as a method to measure diffusion coefficients in ice. *Analytical Chemistry*, **73**, 4348–4353.

Donahue, N.M., Kroll, J H., Anderson, J G., and Demerjian, K.L. (1998). Direct observation of OH production from the ozonolysis of olefins, *Geophysical Research Letters*, **25**, 59–62.

Doran, J.C., Barnard, J.C., Arnott, W.P., Cary, R., Coulter, R., Fast, J.D., Kassianov, E.I., Kleinrnan, L., Laulainen, N.S., Martin, T. *et al.* (2007). The T1–T2 study: Evolution of aerosol properties downwind of Mexico City, *Atmospheric Chemistry and Physics*, **7**, 1585–1598.

Dorokhov, V.M., Khattatov, V.U., Klimova, T.N., Pankratov, A.V., and Torgovichev, V.A. (1989). An automated spectral complex for measurements of atmospheric ozone and UV radiation, in: R.D. Bojkov and P. Fabian (Eds.), *Proceedings of the Quadrennial Ozone Symposium, 1988, Gottingen, Germany*, Deepak Publications, Hampton, VA, pp. 766–771.

Douglass, A.R. (2009). Global observations: The key to model development and improved assessments, in: C. Zerefos, G. Contopoulos, and G. Skalkeas (Eds.), *Proceedings of the Symposium for the 20th Anniversary of the Montreal Protocol*, Springer-Verlag, Berlin, pp. 251–259.

Douglass, A.R. and Stanford, J. (1982). Model of the Antarctic sink for stratospheric water vapor, *Journal of Geophysical Research*, **87**, 5001–5009.

Douglass, A.R., Weaver, C.J., Rood, R., and Coy, L. (1996). A three dimensional simulation of the ozone annual cycle using winds from a data assimilation system, *Journal of Geophysical Research*, **101**, 1463–1474.

Douglass, A.R., Prather, M.J., Hall, T.M., Strahan, S.E., Rasch, P.J., Sparling, L.C., Cox, L., and Rodriguez, J.M. (1999). Choosing meteorological input for the Global Modeling Initiative assessment of high-speed aircraft, *Journal of Geophysical Research*, **104**, 27545–27564.

Douglass, A.R., Schoeberl, M.R., and Rood, R.B. (2003). Evaluation of transport in the lower tropical stratosphere in a global chemistry and transport model, *Journal of Geophysical Research*, **108**, 4259, doi: 10.1029/2002JD002696.

Douglass, A.R., Stolarski, R.S., Schoeberl, M.R., Jackman, C.H., Gupta, M.L., Newman, P.A., Nielsen, J.E., and Fleming, E.L. (2008). Relationship of loss, mean age of air and the distribution of CFCs to stratospheric circulation and implications for atmospheric lifetimes, *Journal of Geophysical Research*, **113**, D14309, doi: 10.1029/2007JD009575.

Dowdy, A.J., Vincent, R.A., Tsutsumi, M., Igarashi, K., Murayama, Y., Singer, W., Murphy, D.J., and Riggin, D.M. (2007). Polar mesosphere and lower thermosphere dynamics: Response to sudden stratospheric warmings, *Journal of Geophysical Research*, **112**, D17105, doi: 10.1029/2006JD008127.

Dragani, R., Redaelli, G., Mariotti, A., Rudakov, V.V., MacKenzie, A.R., Stefanutti, L., and Visconti, G. (2002). High resolution stratospheric tracer fields reconstructed with Lagrangian techniques: A comparative analysis of predictive skill, *Journal of the Atmospheric Sciences*, **59**, 1943–1958.

Dubey, M.K., Smith, G.P., Hartley, W.S., Kinnson, D.E., and Cornell, P.S. (1997). Rate parameter uncertainty effects in assessing stratospheric ozone depletion by supersonic aviation, *Geophysical Research Letters*, **24**, 2737–2740.

Dubrovský, M. and Kalvová, J. (1998). The daily total ozone: The mean annual cycle and correlation with meteorological conditions, in: R.D. Bojkov and G. Visconti (Eds.), *Atmospheric Ozone: Proceedings of the XVIII Quadrennial Ozone Symposium, L'Aquila, Italy, September 12–21, 1996*, Consorzio Parco Scientifico e Tecnologico d'Abruzzo, L'Aquila, Italy, Vol. 1, pp. 33–36.

Dudhia, A., Jay, V.L., and Rodgers, C.D. (2002). Microwindow selection for high-spectral-resolution sounders, *Applied Optics*, **41**, 3665–3673.

Dunlea, E.J., Herndon, S.C., Nelson, D.D., Volkamer, R.M., Lamb, B.K., Allwine, E.J., Grutter, M., Ramos Villegas, C.R., Marquez, C., Blanco, S. *et al.* (2006). Technical note: Evaluation of standard ultraviolet absorption ozone monitors in a polluted urban environment, *Atmospheric Chemistry and Physics*, **6**, 3163–3180.

Dupuy, E., Walker, K.A., Kar, J., Boone, C.D., McElroy, C.T., Bernath, P.F., Drummond, J.R., Skelton, R., McLeod, S.D., Hughes, R.C. *et al.* (2009). Validation of ozone measurements from the Atmospheric Chemistry Experiment (ACE), *Atmospheric Chemistry and Physics*, **9**, 287–343.

Dütsch, H.U. (1970). Atmospheric ozone: A short review, *Journal of Geophysical Research*, **75**, 1707–1712.

Dütsch, H.U. and Staehelin, J. (1989). Discussion of the 60 year total ozone record at Arosa based on measurements of the vertical-distribution and a meteorological parameter, *Planetary and Space Science*, **37**, 1587–1599.

Dyominov, I.G. and Zadorozhny, A.M. (2008). Greenhouse gases and future long-term changes in the stratospheric temperature and the ozone layer, *International Journal of Remote Sensing*, **29**, 2749–2774.

Ebel, A. (2001). Evaluation and reliability of meso-scale air pollution simulations, *Lecture Notes in Computer Science*, **2179**, 255–263.

Ebel, A., Hass, H., Jakobs, H.J., Laube, M., Memmesheimer, M., Oberreuter, A., Geiss, H., and Kuo, Y.-H. (1991). Simulation of ozone intrusion caused by a tropopause fold and cut-off low, *Atmospheric Environment*, **25A**, 2131–2144.

Ebel, A., Friedrich, R., and Rodhe, H. (1997). *Tropospheric Modeling and Emission Estimation*, Springer-Verlag, Berlin, Vol. 24, 450 pp.

EC (2003). *Ozone–Climate Interactions* (EC Air Pollution Report 81, EUR 20623), Office for Official Publications of the EC, Luxembourg, Belgium, 143 pp.

Efstathiou, M.N. (2005). The planetary waves in total ozone over Antarctica, *International Journal of Remote Sensing*, **26**, 3389–3393.

Efstathiou, M.N. and Varotsos, C.A. (2010). On the altitude dependence of the temperature scaling behaviour at the global troposphere, *International Journal of Remote Sensing*, **31**, 343–349.

Efstathiou, M.N., Varotsos, C., and Kondratyev, K.Ya. (1998). An estimation of the surface solar ultraviolet irradiance during an extreme total ozone minimum, *Meteorology and Atmospheric Physics*, **68**, 171–176.

Efstathiou, M.N., Varotsos, C.A., Singh, R.P., Cracknell, A.P., and Tzanis, C. (2003). On the longitude dependence of total ozone trends over middle-latitudes, *International Journal of Remote Sensing*, **24**, 1361–1367.

Efstathiou, M.N., Tzanis, C., and Varotsos, C.A. (2009). Long-term memory dynamics of total ozone content, *International Journal of Remote Sensing*, **30**, 3897–3905.

Eisele, H. and Trickl, T. (1998). Lidar sounding of tropospheric ozone at Garmisch Partenkirchen., in: R.D. Bojkov and G. Visconti (Eds.), *Atmospheric Ozone: Proceedings of the XVIII Quadrennial Ozone Symposium, L'Aquila, Italy, September 12–21, 1996*, Consorzio Parco Scientifico e Tecnologico d'Abruzzo, L'Aquila, Italy, Vol. 1, pp. 351–354.

Elansky, N.F., Glushchenco, Y.U.V., Gruzdev, A.N., and Elokhov, A.S. (1983). Measurements of the ozone content in the atmosphere from an aircraft during the solar eclipse on July 31, 198?, *Izvestia Atmospheric and Oceanic Physics*, **19**, 153–156.

Elansky, N.F., Postylyakov, O.V., Mitin, I.V., Bais, A.F., and Zerefos, C.S. (1998). The extended Umkehr method for retrieving ozone profiles from Brewer observations, in: R.D. Bojkov and G. Visconti (Eds.), *Proceedings of the XVIII Quadrennial Ozone Symposium, L'Aquila, Italy, September 12–21, 1996, September 12–21, 1996*, Consorzio Parco Scientifico e Tecnologico d'Abruzzo, L'Aquila, Italy, Vol. 1, pp. 119–122.

Eleftheratos, K., Zerefos, C.S., Gerasopoulos, E., Isaksen, I.S.A., Rognerud, B., Dalsøren, S., and Varotsos, C. (2011). A note on the comparison between total ozone from Oslo CTM2 and SBUV satellite data, *International Journal of Remote Sensing*, **32**, 2535–2545.

Elokhov, A.S. and Gruzdev, A.N. (1991). Total ozone and nitrogen dioxide measurements at Molodezhnaya and Mizny Antarctic stations in spring 1987–autumn 1988, *Optics of the Atmosphere and Ocean*, **4**, 1006–1009.

Elokhov, A.S. and Gruzdev, A.N. (1992). Total ozone and total NO_2 latitudinal distribution derived from measurements in the Atlantic Ocean in May 1988, in: R.D. Hudson (Ed.), *Proceedings of the Quadrennial Ozone Symposium, Charlottesville, VA* (NASA Conference Publication), NASA, Washington, D.C., pp. 695–698.

Engelen, R.J. and Stephens, G.L. (1997). Infrared radiative transfer in the 9.6-μm band: Application to TIROS operational vertical sounder ozone retrieval, *Journal of Geophysical Research*, **102**, 6925–6939.

Ellsaesser, H.W. (Ed.) (1992). *Global 2000 Revisited: Mankind's Impact on Spaceship Earth*, Paragon House, New York, 436 pp.

Ellsaesser, H.W. (Ed.) (1994). The unheard arguments: A rational view of stratospheric ozone, *21st Century Science and Technology*, **7**(23), 37–45.

Ernest Raj, P., Sonbawne, S.M., Dani, K.K., Saha, S.K., Pandithurai, G., and Devara, P.C.S. (2009). Changes observed in sun photometer derived total column ozone and possible implications on surface reaching UV radiation over a tropical Indian station, *International Journal of Remote Sensing*, **30**, 4153–4165.

Errera, Q. and Fonteyn, D. (2001). Four-dimensional variational chemical assimilation of CRISTA stratospheric measurements, *Journal of Geophysical Research*, **106**, 12253–12265.

Eskes, H. (2004). Stratospheric ozone: Satellite observations, data assimilation and forecasts, in: *Proceedings of the Seminar on Recent Developments in Data Assimilation for Atmosphere and Ocean, September 8–12, 2003*, European Centre for Medium-Range Weather Forecasts, Reading, U.K., pp. 341–360.

ESA (2003). *Proceedings ENVISAT Validation Workshop, December 9–13, 2002, ESRIN, Frascati, Italy* (CD-ROM, SP-531), ESA-ESTEC, Noordwijk, The Netherlands.

ESA (2004). *Proceedings Second Atmospheric Chemistry Validation of ENVISAT Workshop (ACVE-2)* (ESA SP-562), ESA European Space Research Institute (ESRIN), Frascati, Italy.

ESA (2006) *Proceedings Third Atmospheric Chemistry Validation of ENVISAT Workshop (ACVE-3), ESRIN, Frascati, Italy, December 4–7, 2006* (ESA SP-642), ESA European Space Research Institute (ESRIN), Frascati, Italy.

ESA (2008). *PREMIER: Candidate Earth Explorer Core Missions* (Reports for Assessment, ESA SP-1313(5)), Mission Science Division, ESA-ESTEC, Noordwijk, The Netherlands, 121 pp.

Evans, R.D., Komhyr, W.D., and Grass, R.D. (1994). An automated optical wedge calibrator for the Dobson Ozone Spectrophotometer, in: A.R.D. Hudson (Ed.), *Proceedings of the Quadrennial Ozone Symposium, Charlottesville, VA, June 4–13, 1992* (NASA Conference Publication), NASA GSFC, Greenbelt, MD, pp. 749–753.

Evans, R.G., McConville, R.G, Oltmans, S., Petropavlovskikh, I., and Quincy, D. (2009). Measurement of internal stray light within Dobson ozone spectrophotometers, *International Journal of Remote Sensing*, **30**, 4247–4258.

Evans, W.F.J., Asbridge, I.A., Kerr, J.B., Mateer, C.L., and Olafson, R.A. (1980). The effects of SO_2 on Dobson and Brewer total ozone measurements, *Proceedings of the Quadrennial International Ozone Symposium, Boulder, CO, August 4–9, 1980, Volume 1*, International Association of Meteorology and Atmospheric Physics, Boulder, CO, pp 48–56.

Evtushevsky, O.M., Grytsai, A.V., Klekociuk, A.R., and Milinevsky, G.P. (2008a). Total ozone and tropopause zonal asymmetry during the Antarctic spring, *Journal of Geophysical Research*, **113**, D00B06, doi: 10.1029/2008JD009881.

Evtushevsky, O., Milinevsky, G., Grytsai, A., Kravchenko, V., Grytsai, Z. and Leonov, M. (2008b). Comparison of ground-based Dobson and satellite EP-TOMS total ozone measurements over Vernadsky station, Antarctica, 1996–2005, *International Journal of Remote Sensing*, **29**, 2675–2683.

Eyring, V., Harris, N.R.P., Rex, M., Shepherd, T.G., Fahey, D.W., Amanatidis, G.T., Austin, J., Chipperfield, M.P., Dameris, M., Forster, P.M. *et al.* (2005). A strategy for process-oriented validation of coupled chemistry–climate models, *Bulletin of the American Meteorological Society*, **86**, 1117–1133.

Eyring, V., Butchart, N., Waugh, D.W., Akiyoshi, H., Austin, J., Bekki, S., Bodeker, G., Boville, B., Brühl, C., Chipperfield, M. *et al.* (2006). Assessment of temperature, trace species, and ozone in chemistry-climate model simulations of the recent past. *Journal of Geophysical Research*, **111**, D22308, doi: 10.1029/2006JD007327.

Eyring, V., Waugh, D.W., Bodeker, G.E., Cordero, E., Akiyoshi, H., Austin, J., Beagley, S.R., Boville, B., Braesicke, P., Brühl, C. *et al.* (2007). Multi-model projections of stratospheric ozone in the 21st century, *Journal of Geophysical Research*, **112**, D16303, doi: 10.1029/2006JD008332.

Fabry, C. and Buisson, H. (1913). L'Absorption de l'ultraviolet par l'ozone et la limite du spectre solaire, *Journal de Physique (Paris)*, **III**, 196–206.

Fabry, C. and Buisson, H. (1921). Etude de l'extremité ultraviolette du spectre solaire, *Journal de Physique (Paris)*, **VI**, 197–226.

Fadnavis, S. and Beig, G. (2010). Solar cycle variability in middle atmospheric ozone over tropics, *International Journal of Remote Sensing*, **31**, 565–573.

Fahey, D.W. and Hegglin, M.I. (2011). Twenty questions and answers about the ozone layer: 2010 update, *Scientific Assessment of Ozone Depletion: 2010*, World Meteorological Organization, Geneva, Switzerland, 72 pp.

Fahey, D.W., Kawa, S.R., Woodbridge, E.L., Tin, P., Wilson, J.C., Jonsson, H.H., Dye, J.E., Baumgardner, D., Borrmann, S., Toohey, D.W. *et al.* (1993). In situ measurement constraining the role of sulfate aerosols in midlatitude ozone depletion, *Nature*, **363**, 509–514.

Farman, J.C. (1989). Measurements of total ozone using Dobson spectrometers: Some comments on their history, *Planetary and Space Science*, **37**, 1601–1604.

Farman, J.C., Gardiner, G., and Shanklin, J.D. (1985). Large losses of total ozone in Antarctica reveal seasonal ClO_x/NO_x interaction, *Nature*, **315**, 207–210.

Feder, T. (2008). Passenger jets collect data for research on climate change and pollution, *Physics Today*, **61**(7), 18–19.

Feister, U. (1994). Comparison between Brewer spectrometer, M124 filter ozonometer and Dobson spectrophotometer, in: A.R.D. Hudson (Ed.), *Proceedings of the Quadrennial Ozone Symposium, Charlottesville, VA, June 4–13, 1992* (NASA Conference Publication 3266), NASA, Washington, D.C., pp. 770–773.

Feister, U. and Grewe, R. (1995). Higher UV radiation inferred from low ozone levels at northern mid-latitudes in 1992 and 1993, *Global and Planetary Change*, **11**, 25–34.

Feister, U., Grasnick, K.H., and Jakobi, G. (1989). Surface ozone and solar radiation, in: R.D. Bojkov and P. Fabian (Eds.), *Proceedings of the Quadrennial Ozone Symposium, 1988, Göttingen, Germany*, Deepak Publications, Hampton, VA, pp. 441– 446.

Felker, S.R., Moody, J.L., Wimmers, A.J., Osterman, G., and Bowman, K. (2011). A multisensor upper tropospheric ozone product (MUTOP) based on TES ozone and GOES water vapor: Derivation, *Atmospheric Chemistry and Physics*, **11**, 6515–6527.

Felker, S.R., Moody, J. L., Wimmers, A. J., Osterman, G., and Bowman, K. (2011). A multisensor upper troposphere ozone product (MUTOP) based on TES ozone and GOES water vapor: Derivation, *Atmospheric Chemistry and Phyics*, **11**, 6515–6517.

Fels, S.B., Mahlman, J.D., Schwarzkopf, M.D., and Sinclair, R.W. (1980). Stratospheric sensitivity to perturbations in ozone and carbon dioxide radiative and dynamical response, *Journal of the Atmospheric Sciences*, **37**, 2265–2297.

Fenn, M.A., Browell, E.V., Butler, C.F., Grant, W.B., Kooi, S.A., Clayton, M.B., Gregory, G.L., Newell, R.E., Zhu, Y., Dibb, J.E. *et al.* (1999). Ozone and aerosol distributions and air mass characteristics over the South Pacific during the burning season, *Journal of Geophysical Research*, **104**, 16197–16212.

Feretis, E., Theodorakopoulos, P., Varotsos, C., Efstathiou, M., Tzanis, C., Xirou, T., Alexandrou, N., and Aggelou, M. (2002). On the plausible association between environmental conditions and human eye damage, *Environmental Science and Pollution Research*, **9**, 163–165.

Ferm, M., De Santis, F., and Varotsos, C. (2005). Nitric acid measurements in connection with corrosion studies, *Atmospheric Environment*, **39**, 6664–6672.

Ferm, M., Watt, J., O'Hanlon, S., De Santis, F., and Varotsos, C. (2006). Deposition measurement of particulate matter in connection with corrosion studies, *Analytical and Bioanalytical Chemistry*, **384**, 1320–1330.

Fernandez, W., Castro, V., and Hidalgo, H. (1993). Air temperature and wind changes in Costa Rica during the total solar eclipse of July 11, 1991, *Earth, Moon and Planets*, **63**, 133–147.

Fierli, F., Hauchecorne, A., and Nedeljkovic, D. (1998). Relationship between PCS events and temperature measured by ALOMAR RMR lidar at Andoya (69°N, 16°E), in: R.D. Bojkov and G. Visconti (Eds.), *Atmospheric Ozone: Proceedings of the XVIII Quadrennial Ozone Symposium, L'Aquila, Italy, September 12–21, 1996*, Consorzio Parco Scientifico e Tecnologico d'Abruzzo, L'Aquila, Italy, Vol. 1, pp. 503–506.

Fillipi, J-B., Bosseur, F., Pialat, X., Santoni, P-A., Strada, S., and Man, C. (2011). Simulation of coupled fire/atmosphere interaction with the MesoNH-ForeFree models, *Journal of Combustion*, **2011**, 1–13.

Fioletov, V.E. and Shepherd, T.G. (2005). Summertime total ozone variations over middle and polar latitudes, *Geophysical Research Letters*, **32**, L04807, doi: 10.1029/2004GL022080.

Fioletov, V.E., Bojkov, R.D., Bhartia, P.K., McPeters, R., Miller, A.J., and Planet, W. (1998). On total ozone trends derived from solar backscatter ultraviolet (SBUV) and SBUV/2 measurements, in: R.D. Bojkov and G. Visconti (Eds.), *Atmospheric Ozone: Proceedings of the XVIII Quadrennial Ozone Symposium, L'Aquila, Italy, September 12–21, 1996*, Consorzio Parco Scientifico e Tecnologico d'Abruzzo, L'Aquila, Italy, Vol. 1, pp. 37–40.

Fioletov, V.E., Kerr, J.B., Wardle, D.I., Davies, J., Hare, E.W., McElroy, C.T., and Tarasick, D.W. (1997). Long-term ozone decline over the Canadian Arctic to early 1997 from ground-based and balloon observations, *Geophysical Research Letters*, **24**, 2705–2708.

Fioletov, V.E., Griffeen, E., Kerr, J.B., Wardle, D.I., and Uchino, O. (1998). Influence of volcanic sulphur dioxide on spectral irradiance as measured by Brewer spectrophotometers, *Geophysical Research Letters*, **25**, 1655–1658.

Fioletov, V., Kerr, J., Hare, E., Labow, G., and McPeters, R. (1999). An assessment of the world ground-based total ozone network performance from the comparison with satellite data, *Journal of Geophysical Research*, **104**, 1737–1747.

Fioletov, V.E., Bodeker, G.E., Miller, A.J., McPeters, R.D., and Stolarski, R. (2002). Global and zonal total ozone variations estimated from ground-based and satellite measurements: 1964–2000, *Journal of Geophysical Research*, **107**, 4647, doi: 10.1029/2001JD001350.

Fioletov, V.E., Tarasick, D.W., and Petropavlovskikh, I. (2006). Estimating ozone variability and instrument uncertainties from SBUV(/2), ozonesonde, Umkehr, and SAGE II measurements: Short-term variations, *Journal of Geophysical Research*, **111**, D02305, doi: 10.1029/2005JD006340.

Fioletov, V.E., Labow, G., Evans. R., Hare, E.W., Köhler, U., McElroy, C.T., Miyagawa, K., Redondas, A., Savastiouk, V., Shalamyansky, A.M. *et al.* (2008). Performance of the ground-based total ozone network assessed using satellite data, *Journal of Geophysical Research*, **113**, D14313, doi: 10.1029/2008JD009809.

Fiore, A.M., Jacob, D.S., Logan, J.A., and Yin, J.H. (1998). Long-term trends in ground level ozone over the contiguous United States, 1980–1995, *Journal of Geophysical Research*, **103**, 1471–1480.

Fischer, H. and Oelhaf, H. (1996). Remote sensing of vertical profiles of atmospheric trace constituents with MIPAS limb-emission spectrometers, *Applied Optics*, **35**, 2787–2796.

Fischer, H., Nikitas, C., Parchatka, K., Zenker, T., Harris, G.W., Matuska, P., Schmitt, R., Mihelcic, D., Muesgen, D., Paetz, H.-W. *et al.* (1998). Trace gas measurements during the Oxidizing Capacity of the Tropospheric Atmosphere campaign 1993 at Izaña, *Journal of Geophysical Research*, **103**, 13505–13518.

Fischer, H., Blom, C.E., Oelhaf, H., Carli, B., Carlotti, M., Delbouille, L., Ehhalt, D., Flaud, J.-M., Isaksen, I., Lopez-Puertas, M. *et al.* (2000). ENVISAT-MIPAS, the Michelson Interferometer for Passive Atmospheric Sounding: An instrument for atmospheric chemistry and climate research, in: C. Readings and R.A. Harris (Eds.), *ESA SP-1229*, European Space Agency, Noordwijk, The Netherlands

Fischer, H., Birk, M. , Blom, C.E., Carli, B., Carlotti, M., von Clarmann, T., Delbouille, L., Dudhia, A., Ehhalt, D., Endemann, M. *et al.* (2007). MIPAS: An instrument for atmospheric and climate research, *Atmospheric Chemistry and Physics Discussions*, **7**, 8795–8893.

Fischer, A.M., Schraner, M., Rozanov, E., Kenzelmann, P., Schnadt Poberaj, C., Brunner, D., Lustenberger, A., Luo, B.P., Bodeker, G.E., Egorova, T. *et al.* (2008). Interannual-to-decadal variability of the stratosphere during the 20th century: Ensemble simulations with a chemistry–climate model, *Atmospheric Chemistry and Physics*, **8**, 7755–7777.

Fish, D.J., Jones, R.L., Freshwater, R.A., Roscol, H.K., and Oldham, D.J. (1994). Measurements of stratospheric composition using a Star Pointing Spectrometer, *Proceedings of the Quadrennial Ozone Symposium, Charlottesville, VA* (NASA Conference Publication 3266), NASA, Washington, D.C., pp. 671–674.

Fisher, D.A., Hales, C.H., Filkin, D.L., Ko, M.K.W., Dak Sze, N., Connell, P.S., Wuebbles, D.J., Isaksen, I.S.A., and Stordal, F. (1990a). Model calculations of the relative effects of CFCs and their replacements on stratospheric ozone, *Nature*, **344**, 508–512.

Fisher, D.A., Hales, C.H., Wang, W-C., Ko, M.K.W., and Dak Sze, N. (1990b). Model calculations of the relative effects of CFCs and their replacements on global warming, *Nature*, **344**, 513–516.

Fisher, D.A., Ko, M., Wuebbles, D., and Isaksen, I. (1990c). Evaluating ozone depletion potentials, *Nature*, **348**, 203–204.

Fisher, H., Nikitas, C., Parchatka, K., Zenker, T., Harris, G.W., Matuska, P., Schmitt, R., Mihelcic, D., Muesgen, D., Paetz, H.-W. *et al.* (1998). Trace gas measurements during the Oxidizing Capacity of the Tropospheric Atmosphere campaign 1993 at Izaña, *Journal of Geophysical Research*, **103**, 13505–13518.

Fishman, J. and Larsen, J.C. (1987). Distribution of total ozone and stratospheric ozone in the tropics: Implications for the distribution of tropospheric ozone, *Journal of Geophysical Research*, **92**, 6627–6634.

Fix, A., Ehret, G., Flentje, H., Poberaj, G., Gottwald, M., Finkenzeller, H., Bremer, H., Bruns, M., Burrows, J.P., Kleinbohl, A. *et al.* (2005). SCIAMACHY validation by aircraft remote measurements: Design, execution, and first results of the SCIA-VALUE mission, *Atmospheric Chemistry and Physics*, **5**, 1273–1289.

Fleig, A.J., Klenk, K.F., Bhartia, P.K., and Gordon, D. (1982). *User's Guide for the Total Ozone Mapping Spectrometer (TOMS) Instrument First Year Ozone-T Dataset* (NASA RP-1096), NASA, Washington, D.C., 50 pp.

Fleming, E.L., Jackman, C.H., Weisenstein, D.K., and Ko, K.K.W. (2007). The impact of interannual variability on multidecadal total ozone simulations, *Journal of Geophysical Research*, **112**, 10310–10330.

Fleming, S.W., Hudson, P., and Quilty, E.J. (2009). Interpreting nonstationary environmental cycles as amplitude-modulated (AM) signals, *Canadian Journal of Civil Engineering*, **36**, 720–731.

Flentje, H., Claude, H., Elste, T., Gilge, S., Köhler, U., Plass-Dülmer, C., Steinbrecht, W., Thomas, W., Werner, A., and Fricke, W. (2010). The Eyjafjallajökull eruption in April 2010: Detection of volcanic plume using in-situ measurements, ozone sondes and lidar-ceilometer profiles, *Atmospheric Chemistry and Physics*, **10**, 10085–10092.

Flury, T., Hocke, K., Kaempfer, N., and Haefele, A. (2008). Sudden stratospheric warming and its effect on ozone above Bern, *ISSI Workshop, October 22, 2008*, Institute of Applied Physics, University of Bern, Switzerland.

Flury, T., Hocke, K., Haefele, A., Kaempfer, N., and Lehmann, R. (2009). Ozone depletion, water vapor increase, and PSC generation at midlatitudes by the 2008 major stratospheric warming, *Journal of Geophysical Research*, **114**, doi: 10.1029/2009JD011940.

Flynn, L.E., Labow, G.J., Beach, R.A., Rawlins, M.A., and Flitner, D.E. (1996). Estimation of ozone with total ozone portable spectroradiometer instruments, I: Theoretical model and error analysis, *Applied Optics*, **35**, 6076–6083.

Flynn, L.E., McNamara, D., Beck, C.T., Petropavlovskikh, I., Beach, E., Pachepsky, Y., Li, Y.P., Deland, M., Huang, L.-K., Long, C.S. *et al.* (2009). Measurements and products from the Solar Backscatter Ultraviolet (SBUV/2) and Ozone Mapping and Profiler Suite (OMPS) instruments, *International Journal of Remote Sensing*, **30**, 4259–4272.

Fogal, P.F., McElroy, C.T., Goldman, A., and Murray, D.G. (1989). Tunable diode laser heterodyne spectrophotometry of ozone, in: R.D. Bojkov and P. Fabian (Eds.), *Proceedings of the Quadrennial Ozone Symposium, 1988, Göttingen, Germany*, Deepak Publications, Hampton, VA, pp. 732–735.

Folkins, I., Oltmans, S.J., and Thompson, A.M. (2000). Tropical convective outflow and near-surface equivalent potential temperatures, *Geophysical Research Letters*, **27**, 2549–2552.

Folkins, I., Braun, C., Thompson, A.M., and Witte, J. (2002). Tropical ozone as in indicator of deep convective outflow, *Journal of Geophysical Research*, **107**, D13, doi: 10.1029/2001JD001178.

Folkins, I., Bernath, P., Boone, C., Walker, K., Thompson, A.M., and Witte, J.C. (2006). The seasonal cycles of O_3, CO and convective outflow at the tropical tropopause, *Geophysical Research Letters*, **33**, L16802, doi: 10.1029/2006GL026602.

Fonteyn, D., Bonjean, S., Chabrillat, S., Daerden, F., and Errera, Q. (2003). 4D-VAR chemical data assimilation of ENVISAT chemical products (BASCOE): Validation support issues, *Proceedings of the ENVISAT Validation Workshop* (ESA SP-531), European Space Agency, Noordwijk, The Netherlands.

Forster, P.M. de F. and Shine, K.P. (1997). Radiative forcing and temperature trends from stratospheric ozone changes, *Journal of Geophysical Research*, **102**, 10841–10855.

Forster, P.M. de F. and Tourpali, K. (2001). Effect of tropopause height changes on the calculation of ozone trends and their radiative forcing, *Journal of Geophysical Research*, **106**, 12241–12251.

Fortuin, J.P.F. and Kelder, H.M. (1998). An ozone climatology based on ozonesonde and satellite measurements, *Journal of Geophysical Research*, **103**, 31709–31734.

Fortuin, J.P. F., Van Dorland, R., Wauben, W.M.F., and Kelder, H. (1995). Greenhouse effects of aircraft emissions as calculated by a radiative transfer model, *Annales Geophysicae*, **13**, 413–418.

Founda, D., Melas, D., Lykoudis, S., Lisaridis, I., Gerasopoulos, E., Kouvarakis, G., Petrakis, M., and Zerefos, C. (2007). The effect of the total solar eclipse of 29 March 2006 on meteorological variables in Greece, *Atmospheric Chemistry and Physics*, **7**, 5543–5553.

Friedl-Vallon, F., Maucher, G., Seefeldner, M., Trieschmann, O., Kleinert, A., Lengel, A., Keim, C., Oelhaf, H., and Fischer, H. (2004). Design and characterisation of the balloon-borne Michelson Interferometer for Passive Atmospheric Sounding (MIPAS-B2), *Applied Optics*, **43**, 3335–3355.

Frisk, U., Hagstroem, M., Ala-Laurinaho, J., Andersson, S., Berges, J.-C., Chabaud, J.-P., Dahlgren, M., Emrich, A., Florén, H.-G., Florin, G. *et al.* (2003). The ODIN satellite, I: Radiometric design and test, *Astronomy and Astrophysics*, **402**, L27–L34, doi: 10.1051/0004-6361:20030335.

Froidevaux, L., Waters, J.W., Read, W.G., Elson, L.S., Flower, D.A., and Jarnot, R.F. (1994). Global ozone observations from the UARS MLS: An overview of zonal-mean, *Journal of the Atmospheric Sciences*, **51**, 2846–2866.

Froidevaux, L., Read, W.G., Lungu, T.A., Cofield, R.E., Fishbein, E.F., Flower, D.A., Jarnot, R.F., Ridenoure, B.P., Shippony, Z., Waters, J.W. *et al.* (1996). Validation of UARS Microwave Limb Sounder ozone measurements, *Journal of Geophysical Research*, **101**, 10017–10060.

Froidevaux, L., Livesey, N.J., Read, W.G., Salawitch, R.J., Waters, J.W., Drouin, B.J., MacKenzie, I.A., Pumphrey, H.C., Bernath, P., Boone, C. *et al.* (2006). Temporal decrease in upper atmospheric chlorine, *Geophysical Research Letters*, **33**, L23812, doi: 10.1029/2006GL027600.

Froidevaux, L., Jiang, Y.B., Lambert, A., Livesey, N.J., Read, W.G., Waters, J.W., Browell, E.V., Hair, J.W., Avery, M.A., McGee, T.J. *et al.* (2008). Validation of Aura Microwave Limb Sounder stratospheric ozone measurements, *Journal of Geophysical Research*, **113**, D15S20, doi: 10.1029/2007JD008771.

Fromm, M.D., Lumpe, J.D., Bevilacqua, R.M., Shettle, E.P., Hornestein, J., Massie, S.T., and Fricke, K.H. (1997). Observations of Antarctic polar stratospheric clouds by POAM II: 1994–1996, *Journal of Geophysical Research*, **102**, 23659–23672.

Fuglestvedt, J.S., Berntsen, T.K., Isaksen, I.S.A., Mao, H., Liang, X.-Z., and Wang, W.-C. (1998). Impacts of reduced NO_x emissions on radiative forcing through changes in tropospheric O_3 and CH_4, in: R.D. Bojkov and G. Visconti (Eds.), *Atmospheric Ozone*, Consorzio Parco Scientifico e Tecnologico d'Abruzzo, L'Aquila, Italy, Vol. 2, pp. 809–812.

Fuglestvedt, J.S., Berntsen, T.K., Isaksen, I.S.A., Mao, H., Liang, X.-Z. and Wang, W.-C. (1999). Climatic forcing of nitrogen oxides through changes in tropospheric ozone and methane: Global 3D model studies, *Atmospheric Environment*, **33**, 961–977.

Fuglestvedt, J.S., Berntsen, T., Myhre, G., Rypdal, K., and Skeie, R.B. (2008). Climate forcing from the transport sectors, *Proceedings of the National Academy of Sciences*, **105**, 454–458.

Fujimoto, T., Uchino, O., and Nagai, T. (1998). Four years intercomparison of MRI ozone lidar, ozonesonde and SAGE II data, in: R.D. Bojkov and G. Visconti (Eds.), *Atmospheric Ozone: Proceedings of the XVIII Quadrennial Ozone Symposium, L'Aquila, Italy, September 12–21, 1996*, Consorzio Parco Scientifico e Tecnologico d'Abruzzo, L'Aquila, Italy, Vol. 1, pp. 127–130.

Fujiwara, M., Kita, K., Ogawa, T., Komala, N., Saraspriya, S., Suripto, A., and Sano, T. (1998). Total ozone enhancement in September and October 1994 in Indonesia, in: R.D. Bojkov and G. Visconti (Eds.), *Proceedings of the XVIII Quadrennial Ozone Symposium, L'Aquila, Italy, September 12–21, 1996*, Consorzio Parco Scientifico e Tecnologico d'Abruzzo, L'Aquila, Italy, Vol. 1, pp. 363–366.

Furrer, R., Dohler, W., Kirsch, H.J., Plessing, P., and Gorsdorf, U. (1992). Evidence for vertical ozone redistribution since 1967, *Journal of Atmospheric and Terrestrial Physics*, **54**, 1423–1445.

Fusco, A.C. and Logan, J.A. (2003). Analysis of 1970–1995 trends in tropospheric ozone at Northern Hemisphere midlatitudes with the GEOS-CHEM model, *Journal of Geophysical Research*, **108**, ACH4-1–ACH4-25.

Fussen, D. (1998). A critical analysis of the Stratospheric Aerosol and Gas Experiment II spectral inversion algorithm, *Journal of Geophysical Research*, **103**, 8455–8464.

Fussen, D., Vanhellemont, F., Dodion, J., Bingen, C., Walker, K.A., Boone, C.D., McLeod, S.D., and Bernath, P.F. (2005). Initial intercomparison of ozone and nitrogen dioxide number density profiles retrieved by the ACE-FTS and GOMOS occultation experiments, *Geophysical Research Letters*, **32**, L16S02, doi: 10.1029/2005GL022468.

Galmarini, S., Steyn, D.G., and Ainslie, B. (2004). The scaling law relating world point-precipitation records to duration, *International Journal of Climatology*, **24**, 533–546.

Ganguly, N.D. and Iyer, K.N. (2009). Monitoring the success of implementation of Montreal Protocol in India with satellite measurements, *International Journal of Remote Sensing*, **30**, 3927–3941.

Gardner, M.W. and Dorling, S.R. (1998). Artificial neural networks (the multilayer perceptron): A review of applications in the atmospheric sciences, *Atmospheric Environment*, **32**, 2627–2636.

Garny, H., Grewe, V., Dameris, M., Bodeker, G.E., and Stenke, A. (2011). Attribution of ozone changes to dynamical and chemical processes in CCMs and CTMs, *Geoscientific Model Development Discussion*, **4**, 1–43.

Gassmann, A. and Herzog, H-J. (2008). Towards a consistent numerical compressible non-hydrostatic model using generalized Hamiltonian tools, *Quarterly Journal of the Royal Meteorological Society*, **134**, 1597–1613.

Gattinger, R.L., Boone, C.D., Walker, K.A., Degenstein, D.A., Bernath, P.F., and Llewellyn, E.J. (2006a). Comparison of Odin-OSIRIS OH $A^2\Sigma^+$–$X^2\Pi$ 0–0 mesospheric observations and ACE-FTS water vapor observations, *Geophysical Research Letters*, **33**, L15808, doi: 10.1029/2006GL026,425.

Gattinger, R.L., Degenstein, D.A., and Llewellyn, E.J. (2006b). Optical Spectrograph and Infra-Red Imaging System (OSIRIS) observations of mesospheric OH $A^2\Sigma^+$–$X^2\Pi$ 0–0 and 1–1 band resonance emissions, *Journal of Geophysical Research*, **111**, D13303, doi: 10.1029/2005JD006369.

Gauss, M., Myhre, G., Pitari, G., Prather, M.J., Isaksen, I.S.A., Berntsen, T.K., Brasseur, G.P., Dentener, F.J., Derwent, R.G., Hauglustaine, D.A. *et al.* (2003). Radiative forcing in the 21st century due to ozone changes in the troposphere and the lower stratosphere, *Journal of Geophysical Research*, **108**, 4292.

Gauss, M., Isaksen, I.S.A., Lee, P.S., and Sovde, O.A. (2006). Impact of aircraft NO_x emissions on the atmosphere: Tradeoffs to reduce the impact, *Atmospheric Chemistry and Physics*, **6**, 1529–1548,

Geer, A.J., Lahoz, W.A., Bekki, S., Bormann, N., Errera, Q., Eskes, H.J., Fonteyn, D., Jackson, D.R., Juckes, M.N., Massart, S. *et al.* (2006). The ASSET intercomparison of ozone analyses: Method and first results, *Atmospheric Chemistry and Physics*, **6**, 5445–5474.

Geller, M.A. (1998). Presentation of SPARC at the WCRP Conference: SPARC future prospects, *SPARC Newsletter*, **10**, 6–10.

Georgiadis, T., Orciari, R., Bonasoni, P., Fortezza, F., Alberti, L., Leoncini, G., Venieri, L., Gnami, V., Montanari, T., and Rambelli, E. (1998). Airborne measurements for the study of vertical layering of photochemical ozone over a coastal area, in: R.D. Bojkov and G. Visconti (Eds.), *Atomospheric Ozone: Proceedings of the XVIII Quadrennial Ozone Symposium, L'Aquila, Italy, September 12–21, 1996*, Consorzio Parco Scientifico e Tecnologico d'Abruzzo, L'Aquila, Italy, Vol. 1, pp. 367–370.

Geraci, M.J. and Luers, J.K. (1978). An evaluation of electrochemical concentration cell sonde measurements of atmospheric ozone, *Journal of Applied Meteorology*, **17**, 1521–1527.

Gernandt, H., Goersdorf, U., Claude, H., and Varotsos, C.A. (1995). Possible impact of polar stratospheric processes on mid-latitude vertical ozone distributions, *International Journal of Remote Sensing*, **16**, 1839–1850.

Ghude, S.D., Kulkarni, P.S., Beig, G., Jain, S.L., and Arya, B.C. (2009). Global distribution of tropospheric ozone and its precursors: A view from the space, *International Journal of Remote Sensing*, **30**, 485–495.

Gille, J.C. and Russell, J.M. (1984). The limb infrared monitor of the stratosphere: Experiment description, performance and results, *Journal of Geophysical Research*, **89**, 5125–5140.

Gille, J.C., Barnett, J.J., Arter, P., Barker, M., Bernath, P.F., Boone, C.D., Cavanaugh, C., Chow, J., Coffey, M., Craft, J. *et al.* (2008). High resolution dynamics limb sounder: experiment overview, recovery, and validation of initial temperature data, *Journal of Geophysical Research*, **113**, doi: 10.1029/2007JD008824.

Gillet, N.P., Scinocca, J.F., Plummer, D.A., and Reader, M.C. (2009). Sensitivity of climate to dynamically-consistent zonal asymmetries in ozone, *Geophysical Research Letters*, **36**, L10809, doi: 10.1029/2009GL037246.

Giorgetta, M.A. and Bengtsson, L. (1999). Potential role of the quasi-biennial oscillation in the stratosphere–troposphere exchange as found in water vapor in general circulation model experiments, *Journal of Geophysical Research*, **104**, 6003–6019.

Glaccum, W., Lucke, R., Bevilacqua, R.M., Shettle, E.P., Hornstein, J.S., Chen, D.T., Lumpe, J.D., Krigman, S.S., Debrestian, D.J., Fromm, M.D. *et al.* (1996). The Polar Ozone and Aerosol Measurement (POAM II) instrument, *Journal of Geophysical Research*, **101**, 14479–14487.

Glatthor, N., von Clarmann, T., Fischer, H., Grabowski, U., Hopfner, M., Kellmann, S., Kiefer, M., Linden, A., Milz, M., Steck *et al.* (2004). Spaceborne ClO observations by the Michelson Interferometer for Passive Atmospheric Sounding (MIPAS) before and during the Antarctic major warming in September/October 2002, *Journal of Geophysical Research*, **109**, D11307.

Glatthor, N., von Clarmann, T., Fischer, H., Funke, B., Grabowski, U., Hopfner, M., Kellmann, S., Kiefer, M., Linden, A., Milz, M. *et al.* (2005). Mixing processes during the Antarctic vortex split in September–October 2002 as inferred from source gas and ozone distributions from ENVISAT-MIPAS, *Journal of the Atmospheric Sciences*, **62**, 787–800.

Glatthor, N., von Clarmann, T., Fischer, H., Funke, B., Gil-Lopez, S., Grabowski, U., Hopfner, M., Kellmann, M.S., Linden, A., Lopez-Puertas, M. *et al.* (2006). Retrieval of stratospheric ozone profiles from MIPAS/ENVISAT limb emission spectra: A sensitivity study, *Atmospheric Chemistry and Physics*, **6**, 2767–2781.

Godin, S., Megie, G., and Pelon, J. (1989). Systematic lidar measurements of the stratospheric ozone vertical distribution, *Geophysical Research Letters*, **16**, 547–550.

Godin, S., Carswell, A.I., Donovan, D.P., Claude, H., Steinbrecht, W., McDermid, I.S., McGee, T.J., Gross, M.R., Nakane, H., Swart, D.P.J. *et al.* (1999). Ozone differential absorption lidar algorithm intercomparison, *Applied Optics*, **38**, 6225–6236.

Godin-Beekmann, S. (2009). International multi-instruments ground-based networks: Recent developments within the Network for the Detection of Atmospheric Composition Changes, in: C. Zerefos, G. Contopoulos, and G. Skalkeas (Eds.), *Twenty Years of Ozone Decline: Proceedings of the Symposium for the 20th Anniversary of the Montreal Protocol*, Springer-Verlag, Berlin, pp. 135–156.

Godzik, B. (1998). Ground level ozone concentration in the Krakow region, southern Poland, *Environmental Pollution*, **98**, 273–280.

Gogosheva, T.Z., Petkov, B., and Kristev, D. (2002). Measurement of the UV radiation and total ozone during solar eclipse on 11 August 1999, *Geomagnetism and Aeronomy*, **42**, 274–278.

Goldstein, H. (1956). *Classical Mechanics*, Addison-Wesley, Reading, MA.

Gonzalez, M. (2009). Statement from the Executive Secretary for the Vienna Convention and the Montreal Protocol, in: C. Zerefos, G. Contopoulos, and G. Skalkeas (Eds.), *Twenty Years of Ozone Decline: Proceedings of the Symposium for the 20th Anniversary of the Montreal Protocol*, Springer-Verlag, Berlin, pp. 5–8.

Goto, K., Hondoh, T., and Higashi, A. (1986). Determination of diffusion-coefficients of self-interstitials in ice with a new method of observing climb of dislocations by X-ray topography, *Japanese Journal of Applied Physics*, **25**, 351–357.

Gottwald, M., Bovensmann, H., Lichtenberg, G., Noël, S., Von Bargen, A., Slijkhuis, S., Piters, A., Hoogeveen, R., Von Savigny, C., Buchwitz, M. *et al.* (2006). *SCIAMACHY: Monitoring the Changing Earth's Atmosphere*, DLR, Institut für Methodik der Fernerkundung (IMF), Germany, 168 pp.

Goutail, F., Lefèvre, F., Kuttippurath, J., Pazmiño, A., Pommereau, J.-P., Chipperfield, M., Feng, W., Van Roozendael, M., Eriksen, P., Stebel, K. *et al.* (2010). Total ozone loss during the 2009/2010 Arctic winter and comparison to previous years, *Geophysical Research Abstracts*, **12**, EGU2010-3725-2.

Graf, H.-F., Perlwitz, J., Kirchner, I., and Schult, I. (1995). On the interrelationship between recent climate trends, ozone changes and increased greenhouse gas forcing, in: W.-C. Wang and I.S.A. Isaksen (Eds.), *Global Environmental Change* (NATO ASI Series, Subseries I), NATO, Vol. 32, pp. 163–179.

Grant, W.B. (Ed.) (1989). *Ozone Measuring Instruments for the Stratosphere, Vol. 1: Collected Works in Optics*, Optical Society of America, Washington, D.C.

Grant, W.B., Pierce, R.B., Oltmans, S.J., and Browell, E.V. (1998). Seasonal evolution of total and gravity wave induced laminae in ozonesonde data in the tropics and subtropics, *Geophysical Research Letters*, **25**, 1863–1866.

Gratien, A., Picquet-Varrault, B., Orphal, J., Doussin, J.-F., and Flaud, J.-M. (2010). New laboratory intercomparison of the ozone absorption coefficients in the mid-infrared (10 μm) and ultraviolet (300–350 nm) spectral regions, *Journal of Physical Chemistry A*, **114**, 10045–10048.

Grewe, V. (2007). Impact of climate variability on tropospheric ozone, *Science of the Total Environment*, **374**, 167–181.

Grewe, V., Dameris, M., Hein, R., Kohler, I., and Sausen, R. (1999). Impact of future subsonic aircraft NO_x emissions on the atmospheric composition, *Geophysical Research Letters*, **26**, 47–50.

Grewe, V., Shindell, D.T., and Eyring, V. (2004). The impact of horizontal transport on the chemical composition in the tropopause region: Lightning NO_x and streamers, *Advances in Space Research*, **33**, 1058–1061.

Griffin, R.E.M. (2006). Detection and measurement of total ozone from stellar spectra, 2: Historic data from 1935–1942, *Atmospheric Chemistry and Physics*, **6**, 2231–2240.

Griggs, M. (1964). Design of the bubbler ozone detector, *Journal of Aircraft*, **1**, 306–308.

Grund, C.J., Banta, R.M., George, J., Howell, J.N., Post, M.J., and Richter, R.A. (1998). High-resolution Doppler lidar for boundary layer and cloud research, *Journal of Atmospheric and Oceanic Technology*, **18**, 376–393.

Gruzdev, A.N. (2009). Latitudinal structure of variations and trends in stratospheric NO_2, *International Journal of Remote Sensing*, **30**, 4227–4246.

Grytsai, A., Grytsai, Z., Evtushevsky, A., and Milinevsky, G. (2005). Interannual variability of planetary waves in the ozone layer at 65 degrees S, *International Journal of Remote Sensing*, **26**, 3377–3387.

Grytsai, Z.I., Evtushevsky, A.M., Leonov, N.A., and Milinevsky, G.P. (2000). Comparison of ground-based and TOMS-EP total ozone data for Antarctica and northern midlatitude stations (1996–1999), *Physics and Chemistry of the Earth (B)*, **25**, 459–461.

Guillas, S., Stein, M.L., Wuebbles, D.J. and Xia, J. (2004). Using chemistry–transport modeling in statistical analysis of stratospheric ozone trends from observations, *Journal of Geophysical Research*, **109**, D22303, doi: 10.1029/2004JD005049, 12 pp.

Guirlet, M., Kibane-Dawe, I., and Harris, N. (1998). Third European Stratospheric Experiment on Ozone (THESEO), *SPARC Newsletter*, **10**, 24.

Guld, T., Piesch, C., Blom, C.E., Fisher, H., Fergg, F., and Wildgruber, G. (1994). The airborne MIPAS infrared emission experiment, *Proceedings of the First International Airborne Remote Sensing Conference and Exhibition*, Environmental Research Institute of Michigan, Ann Arbor, MI, Vol. II, pp. 301–311.

Gushchin, G.P. (1963). Universal ozonometer, *Proceedings of the Main Geophysical Observatory of Leningrad*, **141**, 83–98 [in Russian].

Gushchin, G.P., Sokolenko, S.A., Dudko, B.G., and Lagutina, V.V. (1985a). Ozonometer M-124, *Proceedings of the Main Geophysical Observatory of Leningrad*, **499**, 60–67 [in Russian].

Gushchin, G.P., Sokolenko, S.A., and Kovalyov, V.A. (1985b). Total-ozone measuring instruments used at the USSR station network, in: C.S. Zerefos and A. Ghazi (Eds.), *Atmospheric Ozone*, Reidel, Dordrecht, The Netherlands, pp. 543–546.

Hadji-Lazaro, J., Clerbaux, C., and Megie, G. (1998). Simultaneous measurements of the concentrations of ozone, carbon monoxide and methane in the tropopshere at the global scale from a satellite, in: N. Chaumerliac, G. Megie, N. Papineau, and G. Angeletti (Eds.), *Proceedings of the Advanced Study Course on Tropospheric Chemistry and Space Observation* (EU Report No. 65).

Haigh, J.D. (1994). The role of stratospheric ozone in modulating the solar radiative forcing on climate, *Nature*, **370**, 544–546.

Haigh, J.D. (1996). The impact of solar variability on climate, *Science*, **272**, 981–984.

Haigh, J.D. (1998). Stratospheric ozone and the impact of solar variability on climate, *Proceedings of the First SPARC General Assembly* (WCRP-99, WMO/TD-N814), WMO, Geneva, Switzerland, Vol. 2, pp. 12–15.

Hales, J. (1996). Scientific background for AMS Policy Statement on Atmospheric Ozone, *Bulletin of the American Meteorological Society*, **77**, 1245–1253.

Hall, T.M., Waugh, D.W., Boering, K.A., and Plumb, R.A. (1999). Evaluation of transport instratospheric models, *Journal of Geophysical Research*, **104**, 18815–18839.

Hampson, J. (1964). *Photochemical Behavior of the Ozone Layer* (Technical Note 1627/64), CARDE, Valcartier, Quebec.

Hansen, G. and Chipperfield, M.P. (1999). Ozone depletion at the edge of the Arctic polar vortex 1996/1997, *Journal of Geophysical Research*, **104**, 1846–1897.

Hansen, J., Sato, M., and Ruedy, R. (1997). Radiative forcing and climate response, *Journal of Geophysical Research*, **102**, 6831–6864.

Hansen, J., Sato, M., Ruedy, R., Lacis, A., and Oinas, V. (2000). Global warming in the twenty-first century: An alternative scenario, *Proceedings of the National Academy of Sciences*, **97**, 9875–9880.

Hanson, D.R. and Mauersberger, K. (1988). The vapor pressures of HNO_3-H_2O solutions, *Journal of Physical Chemistry*, **92**, 6167–6170.

Hare, E. and Fioletov, V. (1998). An examination of the total ozone data in the World Ozone and Ultraviolet Radiation Data Center, in: R.D. Bojkov and G. Visconti (Eds.), *Proceedings of the XVIII Quadrennial Ozone Symposium, L'Aquila, Italy, September 12–21, 1996*, Consorzio Parco Scientifico e Tecnologico d'Abruzzo, L'Aquila, Italy, Vol. 1, pp. 45–48.

Harris, N. and Hudson, R. (1998). *SPARC/IOC/GAW Assessment of Trends in the Vertical Distribution of Ozone* (SPARC Report No. 1) (*http://www.atmosp.physics.utoronto.ca/SPARC/SPARCReport1/index.html*).

Harris, N.R.P. and Rowland, F.S. (1986). Trends in total ozone at Arosa, *EoS, Transactions of the American Geophysical Union*, **67**, 875.

Harris, N.R.P., Kibane-Dawe, I., and Amanatidis, G.T. (Eds.) (1998). Air Pollution Research Report 66, *Polar Stratospheric Ozone 1997: Proceedings of the Fourth European Symposium, September 22–26, 1997, Schliersee, Bavaria, Germany* (EUR 18032), Office for Official Publications of the EC, Brussels, Belgium, 772 pp.

Harris, N.R.P., Rex, M., Goutail, F., Knudsen, B.M., Manney, G.L., Müller, R., and von der Gathen, P. (2002). Comparison of empirically derived ozone losses in the Arctic vortex, *Journal of Geophysical Research*, **107**, doi: 10.1029/2001JD000482.

Harris, N.R.P., Kyrö, E., Staehelin, J., Brunner, D., Andersen, S.-B., Godin-Beekmann, S., Dhomse, S., Hadjinicolaou, P., Hansen, G., Isaksen, I. *et al.* (2008). Ozone trends at northern mid- and high latitudes: A European perspective, *Annales Geophysicae*, **26**, 1207–1220.

Harris, N.R.P., Lehmann, R., Rex, M., and von der Gathen, P. (2009). Understanding the relation between Arctic ozone loss and the volume of polar stratospheric clouds, *International Journal of Remote Sensing*, **30**, 4065–4070.

Harvey, V.L., Randall, C.E., Manney, G.L., and Singleton, C.S. (2008). Low-ozone pockets observed by EOS-MLS, *Journal of Geophysical Research*, **113**, D17112, doi: 10.1029/2007JD009181.

Hase, F. (2000). *Inversion von Spurengasprofilen aus hochaufgelösten bodengebundenen FTIR-Messungen in Absorption* (Wissenschaftliche Berichte, FZK Report No. 6512), Forschungszentrum Karlsruhe, Germany.

Hase, F., Hannigan, J.W., Coffey, M.T., Goldman, A., Höpfner, M., Jones, N.B., Rinsland, C.P., and Wood, S.W. (2004). Intercomparison of retrieval codes used for the analysis of high-resolution, ground-based FTIR measurements, *Journal of Quantitative Spectroscopy and Radiative Transfer*, **87**, 25–52.

Hassinen, S., Tamminen, J., Tanskanen, A., Leppelmeier, G., Mälkki, A., Koskela, T., Karhu, J.M., Lakkala, K., Veefkind, P., Krotkov, N. *et al.* (2008). Description and validation of the OMI very fast delivery products, *Journal of Geophysical Research*, **113**, D16S35, doi: 10.1029/2007JD008784.

Hassler, B., Bodeker, G.E., Cionni, I., and Dameris, M. (2009). A vertically resolved, monthly mean, ozone database from 1979 to 2100 for constraining global climate model simulations, *International Journal of Remote Sensing*, **30**, 4009–4018.

Haywood, J.M., Schwarzkopf, M.D., and Ramaswamy, V. (1998). Estimates of radiative forcing due to modeled increases in tropospheric ozone, *Journal of Geophysical Research*, **103**, 16999–17007.

Heath, D.F. (1988). Non-seasonal changes in total column ozone from satellite observations, 1970–86, *Nature*, **332**, 219–227.

Heath, D.F., Wei, Z., Fowler, W.K., and Nelson, V.W. (1993). Comparison of spectral radiance calibrations of SBUV-2 satellite ozone monitoring instruments using integrating sphere and flat-plate diffuser technique, *Metrologia*, **30**, 259–264.

Hedin, A. (1991). Extension of the MSIS thermosphere model into the middle and lower atmosphere, *Journal of Geophysical Research*, **96**, 1159–1172.

Heese, B., Barthel, K., and Hov, O. (1992). A comparison of total ozone data from satellite and ground-based observations at northern latitudes, *Journal of Geophysical Research*, **97**, 3825–3830.

Hegglin, M.I. and Shepherd, T.G. (2009). Large climate-induced changes in ultraviolet index and stratosphere-to-troposphere ozone flux, *Nature Geoscience*, **2**, 687–691.

Hegglin, M.I., Boone, C.D., Manney, G.L., Shepherd, T.G., Walker, K.A., Bernath, P.F., Daffer, W.H., Hoor, P., and Schiller, C. (2008). Validation of ACE-FTS satellite data in the upper troposphere/lower stratosphere (UTLS) using non-coincident measurements, *Atmospheric Chemistry and Physics*, **8**, 1483–1499.

Hegglin, M.I., Gettelman, A., Hoor, P., Krichevsky, R., Manney, G.L., Pan, L.L., Son, S.-W., Stiller, G., Tilmes, S., Walker, K.A. *et al.* (2010). Multimodel assessment of the upper troposphere and lower stratosphere: Extratropics, *Journal of Geophysical Research*, **115**, D00M09, doi: 10.1029/2010JD013884.

Hein, R., Dameris, M., Schnadt, C., Land, C., Grewe, V., Köhler, I., Ponater, M., Sausen, R., Steil, B., Landgraf, J., and Brühl, C. (2001). Results of an interactively coupled atmospheric chemistry–general circulation model: Comparison with observations, *Annales Geophysicae*, **19**, 435–457.

Helas, G., Andreae, M.O., Fontan, J., Cros, B., and Delmas, R. (1989). Ozone measurements in equatorial Africa during DECAFE 88, in: R.D. Bojkov and P. Fabian (Eds.), *Proceedings of the Quadrennial Ozone Symposium, 1988, Göttingen, Germany*, Deepak Publications, Hampton, VA, pp. 430–432.

Henriksen, K. and Stannes, K. (1989). Solar irradiance and ozone column density in polar regions derived with an UV sensitive double monochromator, in: R.D. Bojkov and P. Fabian (Eds.), *Proceedings of the Quadrennial Ozone Symposium, 1988, Göttingen, Germany*, Deepak Publications, Hampton, VA, pp. 407–410.

Hering, W.S. (1964). *Ozone Observations over North America*, Vol. 1, Air Force Cambridge Research Laboratories, Bedford, MA.

Hering, W.S. and Borden, T.R. (1964). *Ozone Observations over North America*, Vol. 2, Air Force Cambridge Research Laboratories, Bedford, MA.

Herman, J.R., Bhartia, P.K., Krueger, A.J., McPeters, R.D., Wellemeyer, C.G., Seftor, C.J., Barry, G.J., Schlesinger, M., Torres, O., Labow, G. *et al.* (1996). *Meteor-3 Total Ozone Mapping Spectrometer (TOMS) Data Products User's Guide* (NASA RP-1393), NASA, Washington, D.C.

Herman, J.R., Bhartia, P.K., Torres, O., Hsu, C., Seftor, C., and Celarier, E. (1997). Nimbus TOMS/absorbing aerosols, *Journal of Geophysical Research*, **102**, 16911–16922.

Herman, J.R., Krotkov, N., Celarier, E., Larko, D., and Labow, G. (1999). Distribution of UV radiation at the Earth's surface from TOMS-measured UV-backscattered radiances, *Journal of Geophysical Research*, **104**, 12059–12076.

Herman, J.R., Larko, D., Celarier, E., and Ziemke, J. (2001). Changes in global UV reflectivity from clouds and aerosols, *Journal of Geophysical Research*, **106**, 5353–5368.

Hervig, M. and McHugh, M. (1999). Cirrus detection using HALOE measurements, *Geophysical Research Letters*, **26**, 719–722.

Highwood, E.J., Shine, K.P., Hurley, M.D., and Wallington, T.J. (1999). Estimation of direct radiative forcing due to non-methane hydrocarbons, *Atmospheric Environment*, **33**, 759–767.

Highwood, E.J., Hoskins, B.J., and Berrisford, P. (2000). Properties of the Arctic tropopause, *Quarterly Journal of the Royal Meteorological Society*, **126**, 1515–1532.

Hilsenrath, E., Cebula, R.P., Hynes, S.J., and Caffrey, R.T. (1991). Implications of Space Shuttle flight on the calibration of instruments observing atmospheric ozone and the solar irradiance, *Metrologia*, **28**, 301–308.

Hilsenrath, E., Bhartia, P.K., Cebula, R.P., and Wellemeyer, C.G. (1997). Calibration and intercalibration of backscatter ultraviolet (BUV) satellite ozone data, *Advances in Space Research*, **19**, 1345–1353.

Hoegger, B., Levrat, G., Staehelin, J., Schill, H., and Ribordy, P. (1992). Recent developments of the Light Climatic Observatory ozone measuring station of the Swiss Meteorological Institute (LKO) at Arosa, *Journal of Atmospheric and Terrestrial Physics*, **54**, 497–505.

Hofmann, D. (2009). International balloon measurements for ozone research, in: C. Zerefos, G. Contopoulos, and G. Skalkeas (Eds.), *Twenty Years of Ozone Decline: Proceedings of the Symposium for the 20th anniversary of the Montreal Protocol*, Springer, Berlin, pp. 157–172.

Hofmann, D.J., Rosen, J.M., Harder, J.A., and Rolf, S.R. (1986). Ozone and aerosol measurements in the springtime Antarctic stratosphere in 1985, *Geophysical Research Letters*, **13**, 1253–1255.

Hofmann, D.J., Oltmans, S.J., Harris, J.M., Johnson, B.J., and Lathrop, J.A. (1997). Ten years of ozonesonde measurements at the south pole: Implications for recovery of springtime Antarctic ozone, *Journal of Geophysical Research*, **102**, 8931–8943.

Hofmann, D., Johnson, B.J., and Oltmans, S.J. (2009). Twenty-two years of ozonesonde measurements at the South Pole, *International Journal of Remote Sensing*, **30**, 3995–4008.

Hoffmann, P., Singer, W., Keuer, D., Hocking, W.K., Kunze, M.K., and Murayama, Y. (2007). Latitudinal and longitudinal variability of mesospheric winds and temperature during stratospheric warming events, *Journal of Atmospheric and Solar Terrestrial Physics*, **69**, 2355–2366.

Hogrefe, C., Lynn, B., Civerolo, K., Ku, J.-Y., Rosenthal, J., Rosenzweig, C., Goldberg, R., Gaffin, S., Knowlton, K., and Kinney, P.L. (2004). Simulating changes in regional air pollution over the eastern United States due to changes in global and regional climate and emissions, *Journal of Geophysical Research*, **109**, D22301, doi: 10.1029/2004JD004690.

Hoinka, K.P. (1998). Statistics of the global tropopause pressure, *Monthly Weather Review*, **126**, 3303–3325.

Hoinka, K.P. (1999). Temperature, humidity, and wind at the global tropopause, *Monthly Weather Review*, **127**, 2248–2265.

Hoinka, K.P., Claude, H., and Köhler, U. (1996). On the correlation between tropopause pressure and ozone above Central Europe, *Geophysical Research Letters*, **23**, 1753–1756.

Hollandsworth, S.M., McPeters, R., Flynn, L., Planet, W., Miller, A.J., and Chandra, S. (1995). Ozone trends deduced from combined Nimbus-7 SBUV and NOAA-11 SBUV/2 data, *Geophysical Research Letters*, **22**, 905–908.

Holm, E.V., Untch, A., Simmons, A, Saunders, R., Bouttier, F., and Andersson, E. (1999). *The SODA EU Project 1996–1999* (final report from ECMWF, Abstract), Office for Official Publications of the EC, Luxembourg, Belgium, 12 pp.

Holton, J.R., Haynes, P.H., McIntyre, M.E., Douglass, A.R., Rood, R.B., and Pfister, L. (1995). Stratosphere–troposphere exchange, *Reviews of Geophysics*, **33**, 403–439.

Holzworth, G.C. (1972). *Mixing Heights, Wind Speeds and Potential for Urban Air Pollution throughout the Contiguous United States* (PB-207-102), Office of Air Programs, Research Triangle Park, NC.

Hondoh, T., Goto, A., Hoshi, R., Ono, T., Anzai, H., Kawase, R., Pinienta, P., and Mae, S. (1989). Utilization of high speed X-ray topography for determining diffusion coefficients of point defects in nearly perfect crystals, *Review of Scientific Instruments*, **60**, 2494–2497.

Hood, L.L. and Zhou, S. (1998). Stratospheric effects of 27-day solar ultraviolet variations: An analysis of UARS MLS ozone and temperature data, *Journal of Geophysical Research*, **103**, 3629–3637.

Hood, L.L., McPeters, R.D., McCormack, J.P., Flynn, L.E., Hollandsworth, S.M., and Gleason, J.F. (1993). Altitude dependence of stratospheric ozone trends based on Nimbus 7 SBUV data, *Geophysical Research Letters*, **20**, 2667–2670.

Hoor, P., Borken-Kleefeld, J., Caro, D., Dessens, O., Endresen, O., Gauss, M., Grewe, V., Hauglustaine, D., Isaksen, I.S.A., Jöckel, P. *et al.* (2009). The impact of traffic emissions on atmospheric ozone and OH: Results from QUANTIFY, *Atmospheric Chemistry and Physics*, **9**, 3113–3136.

Hopfner, M., Stiller, G.P., Kuntz, M., von Clarmann, T., Echle, G., Funke, B., Glatthor, N., Hase, F., Kemnitzer, H., and Zorn, S. (1998). The Karlsruhe optimized and precise radiative transfer algorithm, Part II: Interface to retrieval applications, *SPIE Proceedings*, **3501**, 186–195.

Hopfner, M., Blom, C.E., Echle, G., Glatthor, N., Hase, F., and Stiller, G. (2001). Retrieval simulations for MIPAS-STR measurements, in: W.L. Smith (Ed.), *IRS 2000: Current Problems in Atmospheric Radiation; Proceedngs of the International Radiation Symposium, St. Petersburg, Russia, July 24–29, 2000*, Deepak Publications, Hampton, VA, S.1121– 1124.

Hopfner, M., Oelhaf, H., Wetzel, G., Friedl-Vallon, F., Kleinert, A., Lengel, A., Maucher, G., Nordmeyer, H., Glatthor, N., Stiller, G. *et al.* (2002). Evidence of scattering of tropospheric radiation by PSCs in mid-IR limb emission spectra: MIPAS-B observations and KOPRA simulations, *Geophysical Research Letters*, **29**, 1278, doi: 10.1029/2001GL014443.

Hoppel, K., Bevilacqua, R., Canty, T., Salawitch, R., and Santee, M. (2005). A measurement/model comparison of ozone photochemical loss in the Antarctic ozone hole using Polar Ozone and Aerosol Measurement observations and the Match technique, *Journal of Geophysical Research*, **110**, D19304.

Hoskins, B.J. (2003). Climate change at cruising altitude? *Science*, **301**, 469–470.

Houghton, J.T., Meiro Filho, L.G., Callander, B.A., Harris, N., Kattenberg, A., and Maskell, K. (Eds.) (1996) *Climate Change 1995: The Science of Climate Change*, Cambridge University Press, Cambridge, U.K., 584 pp.

Høv, Ø., Kley, D., Volz-Thomas, A., Beck, J., Grennfelt, P., and Penkett, S.A. (1997). An overview of tropospheric ozone research, in: Ø. Høv (Ed.), *Tropospheric Ozone Research*, Springer-Verlag, Berlin, pp. 1–34.

Hsu, J. and Prather, M.J. (2009). Stratospheric variability and tropospheric ozone, *Journal of Geophysical Research*, **114**, D06102, doi: 10.1029/2008JD010942.

Huber, M., Blumthaler, M., and Ambach, W. (1998). Ground based total atmospheric ozone determination with an absolutely calibrated UV spectrometer, in: R.D. Bojkov and G. Visconti (Eds.), *Atmospheric Ozone: Proceedings of the XVIII Quadrennial Ozone Symposium, L'Aquila, Italy, September 12–21, 1996*, Consorzio Parco Scientifico e Tecnologico d'Abruzzo, L'Aquila, Italy, Vol. 2, pp. 911–914.

Huder, K.J. and DeMore, W.B. (1995). Absorption cross sections of the ClO dimer, *Journal of Physical Chemistry*, **99**, 3905–3908.

Hudson, R.D., Andrade, M.F., Follette, M.B., and Frolov, A.D. (2003). The total ozone fields separated into the meteorological regimes, Part 1: Defining the regimes, *Journal of the Atmospheric Sciences*, **60**, 1669–1677.

Hudson, R.D., Andrade, M.F., Follette, M.B., and Frolov, A.D. (2006). The total ozone field separated into meteorological regimes, Part II: Northern Hemisphere mid-latitude total ozone trends, *Atmospheric Chemistry and Physics*, **6**, 5183–5191.

Hu, K., Ivanov, P.C., Chen, Z., Carpena, P., and Stanley, H.E. (2001). Effect of trends on detrended fluctuation analysis, *Physical Review E*, **64**, 011114, Part 1.

Hu, Y., Xia, Y., and Fu, Q. (2011). Tropospheric temperature response to stratospheric ozone recovery in the 21st century, *Atmospheric Chemistry and Physics*, **11**, 7687–7699.

Huthwelker, T., Ammann, M., and Peter, T. (2006). The uptake of acidic gases on ice, *Chemical Reviews*, **106**, 1375-1444.

Ialongo, I., Casale, G.R., and Siani, A.M. (2008). Comparison of total ozone and erythemal UV data from OMI with ground-based measurements at Rome station, *Atmospheric Chemistry and Physics*, **8**, 3283–3289.

Immler, F., Kruger, K., Fujiwara, M., Verver, G., Rex, M., and Schrems, O. (2008). Correlation between equatorial Kelvin waves and the occurrence of extremely thin ice clouds at the tropical tropopause. *Atmospheric Chemistry and Physics*, **8**, 4019–4026.

ICG (2006). *http://www.fz-juelich.de/icg/icg-2/josie/ozone_sondes/kc79/*, Institut für Chemie und Dynamik der Geosphäre, Berlin.

Ionov D.V., Timofeyev, Yu.M., and Shalamyansky, A.M. (2002). Comparisons of the satellite (GOME, TOMS) and the ground-based total ozone measurements, *Earth Research from Space*, **3**, 10–19 [in Russian].

Ionov, D.V., Goutail, F., Pazmiño, A., and Pommereau, J.-P. (2008). Validation of satellite total ozone and NO_2 data with ground-based SAOZ network, *Geophysical Research Abstracts*, **10**, EGU2008-A-01547.

Isaksen, I.S.A. (1988). *Tropospheric Ozone: Regional and Global Scale Interactions*, Reidel, Dordrecht, The Netherlands, 425 pp.

IPCC (2005). *Safeguarding the Ozone Layer and the Global Climate System: Issues Related to Hydrofluorocarbons and Perfluorocarbons* (edited by the Intergovernmental Panel on Climate Change), Cambridge University Press, Cambridge, U.K.

Jackman, C.H., Fleming, E.L., Chandra, S., Considine, D.B., and Rosenfield, J.E. (1996). Past, present and future modeled ozone trends with comparison to observed trends, *Journal of Geophysical Research*, **101**, 753–767.

Jacob, D.J., Leovy, C., Schoeberl, M., McNeal, R.J., and Gleason, J. (1995). Atmospheric trace gas measurements for the year 2000 and beyond, *The Earth Observer*, **7**, 31–38.

Jacobi, C., Kurschner, D., Muller, H.G., Pancheva, D., Mitchell, N.J., and Naujokat, B. (2003). Response of the mesopause region dynamics to the February 2001 stratospheric warming, *Journal of Atmospheric and Solar-Terrestrial Physics*, **65**, 843–855.

Jacovides, C.P., Varotsos, C., Kaltsounides, N.A., Petrakis, M., and Lalas, D.P. (1994). Atmospheric turbidity parameters in the highly polluted site of Athens basin, *Renewable Energy*, **4**, 465–470.

Jaeglé, L., Jacob, D.J., Brune, W.H., Tan, D., Faloona, I.C, Weinheimer, A.S., Ridley, B.A, Campos, T.L., and Sachse, G.W. (1998a). Sources and chemistry of NO_x and production of ozone in the upper troposphere over the United States, *Geophysical Research Letters*, **25**, 1705–1708.

Jaeglé, L., Jacob, D.J., Brune, W.H., Tan, D., Faloona, I.C., Weinheimer, A.S., Ridley, B.A., Campos, T.L., and Sachse, G.W. (1998b). Sources of HO_x and production of ozone in the upper troposphere over the United States, *Geophysical Research Letters*, **25**, 1709–1712.

James, P.M. (1998). An interhemispheric comparison of ozone minihole climatologies, *Geophysical Research Letters*, **25**, 301–304.

James, P.M., Peters, D., and Greisiger, K.M. (1997). A study of ozone mini-hole formation using a tracer advection model driven by barotropic dynamics. *Meteorology and Atmospheric Physics*, **64**, 107–121.

Jaross, G., Krueger, A.J., Cebula, R.P., Seftor, C., Hartman, U., Haring, R., and Burchfield, D. (1995). Calibration and postlaunch performance of the Meteor-3/ TOMS instrument, *Journal of Geophysical Research*, **100**, 2985–2995.

Jeng, M. (2005). Hot water can freeze faster than cold?!? arXiv:physics/0512262v1 [physics.histph], 29 December 2005.

Jeuken, A.B.M., Eskes, H.J., van Velthoven, P.F.J., Kelder, H.M., and Hólm, E.V. (1999). Assimilation of total ozone satellite measurements in a three-dimensional tracer transport model. *Journal of Geophysical Research*, **104**, 5551–5563.

Jiang, Y., Yung, Y.L., and Zurek, R.W. (1996). Decadal evolution of the Antarctic ozone hole. *Journal of Geophysical Research*, **101**, 8985–8999.

Jiang, Y.B., Froidevaux, L., Lambert, A., Livesey, N.J., Read, W.G., Waters, J.W., Bojkov, B., Leblanc, T., McDermid, I.S., Godin-Beekmann, S. *et al.* (2007). Validation of Aura microwave limb sounder ozone by ozonesonde and lidar measurements, *Journal of Geophysical Research*, **112**, D24S34, doi: 10.1029/2007JD008776.

Jing, P., Cunnold, D.M., Wang, H.J., and Yang, E.S. (2004). Isentropic croos-tropopause ozone transport in the Northern Hemisphere. *Journal of the Atmospheric Sciences*, **61**, 1068–1078.

Jobson, B.T., McKeen, S.A., Parrish, D.D., Fehsenfeld, F.C, Blake, D.R., Goldstein, A.H., Schauffler, S.M., and Elkins, J.W. (1999). Trace gas mixing ratio variability versus lifetime in the troposphere and stratosphere: Observations. *Journal of Geophysical Research*, **104**, 16091–16115.

Johns, T.C., Durman, C.F., Banks, H.T., Roberts, M.J., McLaren, A.J., Ridley, J.K., Senior, C.A., Williams, K.D., Jones, A., Rickard, G.J. *et al.* (2006). The new Hadley Centre Climate Model (HadGEM1): Evaluation of coupled simulations, *Journal of Climate*, **19**, 1327–1353.

Johnson, D.G., Jucks, K.W., Traub, W.A., and Chance, K.V. (1995). The Smithsonian stratospheric far-infrared spectrometer and data reduction system, *Journal of Geophysical Research*, **100**, 3091–3106.

Johnston, H.S. (1971). Reduction of stratospheric ozone by nitrogen oxide catalysts from supersonic transport exhaust, *Science*, **173**, 517–522.

Johny, C.J., Sarkar, S.K., and Punyaseshudu, D. (2009). Spatial and temporal variation of water vapour in upper troposphere and lower stratosphere over Indian region, *Current Science*, **97**, 1735–1741.

Joiner, J., Lee, H.-T, Strow, L.L., Bhartia, P.K., Hannon, S., Miller, A.S., and Rokke, L. (1998). Radiative transfer in the 9.6 μm HIRS ozone channel using collocated SBUV-determined ozone abundances, *Journal of Geophysical Research*, **103**, 19213–19230.

Jones, P.D., Osborn, T.J., Briffa, K.R., Folland, C.K., Horton, E.B., Alexander, L.V., Parker, D., and Rayner, N.A. (2001). Adjusting for sampling density in grid box land and ocean surface temperature time series, *Journal of Geophysical Research*, **106**, 3371–3380.

Jonson, J.E., Tarrasón, L., Klein, H., Vestreng, V., Cofala, J., and Whall, C. (2009). Effects of ship emissions on European ground-level ozone in 2020, *International Journal of Remote Sensing*, **30**, 4099–4110.

Jonsson, A. (2006). Modelling the middle atmosphere and its sensitivity to climate change. PhD thesis, Stockholm University, Department of Meteorology, Stockholm, Sweden.

Jost, H., Peter, R., and Kampher, N. (1998). Comparison of ground-based millimetre wave ozone profiles with Dobson Umkehr, in: R.D. Bojkov and G. Visconti (Eds.), *Atmospheric Ozone: Proceedings of the XVIII Quadrennial Ozone Symposium, L'Aquila, Italy, September 12–21, 1996*, Consorzio Parco Scientifico e Tecnologico d'Abruzzo, L'Aquila, Italy, Vol. 1, pp. 131–134.

Jucks, K.W., Johnson, D.G., Chance, K.V., Traub, W.A., Margitan, J.J., Osterman, G.B., Salawitch, R.J., and Sasano, Y. (1998). Observations of OH, HO_2, H_2O, and O_3 in the upper stratosphere: Implications for HO_x photochemistry. *Geophysical Research Letters*, **25**, 3935–3938.

Kahya, C., Demirhan, D., Topcu, S., and Incecik, S. (2005). An examination of the laminae characteristics in ozone profiles in Eastern and South Eastern Europe, *International Journal of Remote Sensing*, **26**, 3455–3466.

Kalabokas, P.D. and Repapis, C.C. (2004). A climatological study of rural surface ozone in central Greece, *Atmospheric Chemistry and Physics*, **4**, 1139–1147.

Kane, R.P., Sahai, Y., and Casiccia, C. (1998). Latitude dependence of the quasi-biennial oscillation and quasi-triennial oscillation characteristics of the total ozone measured by TOMS, *Journal of Geophysical Research*, **103**, 8477–8490.

Kantelhardt, J.W., Zschiegner, S.A., Koscielny-Bunde, E., Havlin, S., Bunde, A., and Stanley, H.E. (2002). Multifractal detrended fluctuation analysis of nonstationary time series, *Physica A*, **316**, 87–114.

Kapitza, A.P. (1996). Controversies in the theory of ozone hole formation, *Optics of the Atmosphere and Ocean*, **9**, 1164–1167 [in Russian].

Karol, I.L., Ozolin, Y.E., and Rozanov, E.V. (1992). Effect of space rocket launches on ozone, *Annales Geophysicae*, **10**, 810–814.

Karol, I.L., Shalamyansky, A.M., Solomatnikova, A.A., and Titova, E.A. (2009). Winter ozone transport variations and the Montreal Protocol impact as revealed by the total ozone ground-based measurements over the Russian territory in 1973–2005, in: C. Zerefos, G. Contopoulos, and G. Skalkeas (Eds.), *Proceedings of the Symposium for the 20th anniversary of the Montreal Protocol*, Springer-Verlag, Berlin, pp. 349–356.

Karpechkol, A.Yu., Gillett, N.P., Hassler, B., Rosenlof, K.H., and Rozanov, E. (2010). Quantitative assessment of Southern Hemisphere ozone in chemistry–climate model simulations, *Atmospheric Chemistry and Physics*, **10**, 1385–1400.

Katsambas, A., Varotsos, C.A., Veziryianni, G., and Antoniou, C. (1997). Surface solar ultraviolet radiation: A theoretical approach of the SUVR reaching the ground in Athens, Greece, *Environmental Science and Pollution Research*, **4**, 69–73.

Kawa, S.R., Fahey, D.W., Wilson, J.C., Schoeberl, M.R., Douglass, A.R., Stolarski, R.S., Woodbridge, E.L., Jonsson, H., Lait, L.R., Newman, P.A. *et al.* (1993). Interpretation of NO_x/NO_y observations from AASE-II using a model of chemistry along trajectories, *Geophysical Research Letters*, **20**, 2507–2510.

Kawabata, K., Ogawa, H., Yonekura, Y., Suzuki, H., Suzuki, M., Iwasaka, Y., Shibata, T., and Sakai, T. (1997). Ground-based radiometry of stratomesospheric ozone employing a superconductive receiver, *Journal of Geophysical Research*, **102**, 1371–1377.

Kaye, J. (1997). Summary of space-based observations of atmospheric chemistry, *SPARC Newsletter*, **8**, 21—27.

Kaye, J. (1998). Summary of Total Ozone Mapping Spectrometer (TOMS) science team meeting, *The Earth Observer*, **10**(3), 15–22.

Kaye, J.A. and Miller, T.L. (1996). The ATLAS series of shuttle missions, *Geophysical Research Letters*, **23**, 2285–2288.

Kaye, J.A. and Readings, M. (1998). Report of the Global Observing Systems Space Panel, *Third Session, Paris, France, May 27–30, 1997* (WMO/TD 845), WMO, Geneva, Switzerland.

Kaye, J.A. and Rind, D. (1998). Summary of chemistry/climate modeling meeting, *The Earth Observer*, **10**(3), 27–36.

Kazantzidis, A., Bais, A.F., Zempila, M.M., Meleti, C., Eleftheratos, K., and Zerefos, C.S. (2009). Evaluation of ozone column measurements over Greece with NILUUV multichannel radiometers, *International Journal of Remote Sensing*, **30**, 4273–4281.

Keckhut, P., McDermid, S., Swart, D., McGee, T., Godin-Beekmann, S., Adriani, A., Barnes, J., Baray, J.-L., Bencherif, H., Claude, H. *et al.* (2004). Review of ozone and temperature lidar validations performed within the framework of the Network for the Detection of Stratospheric Change, *Journal of Environmental Monitoring*, **6**, 721–733.

Keim, C., Blom, C.E., von der Gathen, P., Gulde, T., Hopfner, M., Liu, G.Y., Oulanovski, A., Piesch, C., Ravegnani, F., Sartorius, C. *et al.* (2004). Validation of MIPAS-ENVISAT by correlative measurements of MIPAS-STR, *Proceedings of ACVE-2 Meeting, May 3–7, 2004, Frascati, Italy* (ESA SP-562), ESA, Noordwijk, The Netherlands.

Kerr, J.B., Asbridge, I.A., and Evans, W.F.J. (1988). Intercomparison of total ozone measured by the Brewer and Dobson spectrophotometers at Toronto, *Journal of Geophysical Research*, **93**, 11129–11140.

Kerr, J.B., McElroy, C.T., and Wardle, D.W. (1998). The Brewer Instrument Calibration Centre 1984–1996, in: R.D. Bojkov and G. Visconti (Eds.), *Atmospheric Ozone: Proceedings of the XVIII Quadrennial Ozone Symposium, L'Aquila, Italy, September 12–21, 1996*, Consorzio Parco Scientifico e Tecnologico d'Abruzzo, L'Aquila, Italy, Vol. 2, pp. 915–918.

Kerridge, B.J., Goutail, F., Bazureau, A., Wang, D.-Y., Bracher, A., Weber, M., Bramstedt, H., Siddans, R., Latter, B.G., Reburn, W.J. *et al.* (2004). MIPAS ozone validation by satellite intercomparison, *Proceedings Second Atmospheric Chemistry Validation of ENVISAT Workshop (ACVE-2), ESRIN, Frascati, Italy* (ESA SP-562), ESA European Space Research Institute (ESRIN), Frascati, Italy.

Kiehl, J.T. and Boville, B.A. (1988). The radiative–dynamical response of a stratospheric–tropospheric general circulation model to changes in ozone, *Journal of the Atmospheric Sciences*, **45**, 1798–1817.

Kiehl, J.T., Schneider, T.L., Portmann, R.W., and Solomon, S. (1999). Climate forcing due to tropospheric and stratospheric ozone, *Journal of Geophysical Research*, **104**, 31239–31254.

Kiesewetter, G., Sinnhuber, B.-M., Vountas, M., Weber, M., and Burrows, J.P. (2010). A long-term stratospheric ozone data set from assimilation of satellite observations: High-latitude ozone anomalies, *Journal of Geophysical Reseach*, **115**, D10307, doi: 10.1029/2009JD013362.

Kikuchi, K., Nishibori, T., Ochiai, S., Ozeki, H., Irimajiri, Y., Kasai, Y., Koike, M., Manabe, T., Mizukoshi, K., Murayama, Y. *et al.* (2010). Overview and early results of the Superconducting Submillimeter-Wave Limb-Emission Sounder (SMILES), *Journal of Geophysical Research*, **115**, D23306, doi: 10.1029/2010JD014379.

Kilbane-Dawe, I., Harris, N.R.P., Pyle, J.A., Rex, M., Lee, A.M. and Chipperfield M.P. (2001). A comparison of Match ozonesonde-derived and 3D model ozone loss rates in the Arctic polar vortex during the winters of 1994/95 and 1995/96, *Journal of Atmospheric Chemistry*, **39**, 123–138.

Kim, J.H. and Newchurch, M.J. (1998). Biomass-burning influence on tropospheric ozone over New Guinea and South America, *Journal of Geophysical Research*, **103**, 1455–1462.

Kim, J.H., Hudson, R.D., and Thompson, A.M. (1996). A new method of deriving time-averaged tropospheric column ozone over the tropics using TOMS radiances: Intercomparison and analysis using TRACE A data, *Journal of Geophysical Research*, **101**, 24317–24330.

Kirk-Davidoff, D.B., Anderson, J.G., Hintsa, E.J., and Keith, D.W. (1999). The effect of climate change on ozone depletion through stratospheric H_2O, *Nature*, **402**, 399–401.

Kiss, P., Müller, R., and Janosi, I.M. (2007). Long-range correlations of extrapolar total ozone are determined by the global atmospheric circulation, *Nonlinear Processes in Geophysics*, **14**, 435–442.

Kleinert, A., Aubertin, G., Perron, G., Birk, M., Wagner, G., Hase, F., Nett, H., and Poulin, R. (2007). MIPAS Level 1B algorithms overview: Operational processing and characterisation, *Atmospheric Chemistry and Physics*, **7**, 1395–1406.

Kleinman, L.I., Daum, P.H., Lee, J.H., Lee, Y-N., Weinstein-Lloyd, J., and Springston, S.R. (1998). Photochemistry of O_3 and related components over southern Nova Scotia, *Journal of Geophysical Research*, **103**, 13519–13529.

Kleipool, Q.L., de Jonge, A.R.W., Whyborn, N.D., van den Broek, M., Schrijver, H., Goede, A.P.H., Urban, J., Wohlgemuth, J., Eyring, V., Kullmann, H. *et al.* (1998). Chlorine measured by ASUR over the Arctic during the SESAME campaigns, in: R.D. Bojkov and G. Visconti (Eds.), *Proceedings of the XVIII Quadrennial Ozone Symposium, L'Aquila, Italy, September 12–21, 1996*, Consorzio Parco Scientifico e Tecnologico d'Abruzzo, L'Aquila, Italy, Vol. 2, pp. 535–538.

Klenk, K.F., Bhartia, P.K., Fleig, A.J., Kaveeshwar, V.G., McPeters, R.D., and Smith P.M. (1982). Total ozone determination from the backscattered ultraviolet (BUV) experiment, *Journal of Applied Meteorology*, **21**, 1672–1684.

Knudsen, B.M., Larsen, N., Mikkelson, I.S., Morcrette, J.-J., Braathen, G.O., Kyr, E., Fast, H., Gernandt, H., Kanzawa, H., Nakane, H. *et al.* (1998). Ozone depletion in and below the Arctic vortex for 1997, *Geophysical Research Letters*, **25**, 627–630.

Ko, M.K.W., Sze, N.D., Scott, C., Rodríguez, J.R., Weisenstein, D.K., and Sander, S.P. (1998). Ozone depletion potential of CH_3Br, *Journal of Geophysical Research*, **103**, 28187–28195.

Kobayashi, J. and Toyama, Y. (1966). On various methods of measuring the vertical distribution of atmospheric ozone (III): Carbon iodine type chemical ozone sonde, *Papers in Meteorology and Geophysics*, **17**, 113–126.

Koehler, J. and Hajost, S.A. (1990). The Montreal Protocol: A dynamic agreement for protecting the ozone layer, *Ambio*, **19**, 82–86.

Koelemeijer, R.B.A., Stammes, P., Hovenier, J.W., and de Haan, J.F.A. (2001). Fast method for retrieval of cloud parameters using oxygen A band measurements from the Global Ozone Monitoring Experiment, *Journal of Geophysical Research*, **106**, 3475–3490.

Köhler, U. (1999). A comparison of the new filter ozonometer MICROTOPS II with Dobson and Brewer spectrometers at Hohenpeissenberg, *Geophysical Research Letters*, **26**, 1385–1388.

Köhler, U. and Claude, H. (1998). Homogenized ozone records at Hohenpeissenberg, in: R. D. Bojkov and G. Visconti (Eds.), *Proceedings of the XVIII Quadrennial Ozone Symposium, L'Aquila, Italy, September 12–21, 1996*, Consorzio Parco Scientifico e Tecnologico d'Abruzzo, L'Aquila, Italy, Vol. 1, 57–60.

Köhler, M.O., Rädel, G., Dessens, O., Shine, K.P., Rogers, H.L., Wild, O., and Pyle, J.A. (2008). Impact of perturbations to nitrogen oxide emissions from global aviation, *Journal of Geophysical Research*, **113**, D11305, doi: 10.1029/2007JD009140.

Kolev, N., Tatarov, B., Grigorieva, V., Donev, E., Simeonov, P., Umlensky, V., Kaprielov, B., and Kolev, I. (2005). Aerosol lidar and in situ ozone observations in PBL over Bulgaria during solar eclipse on 11 August 1999, *International Journal of Remote Sensing*, **26**, 3567–3584.

Komhyr, W.D. (1969). Electrochemical concentration cells for gas analysis, *Annales de Geophysique*, **25**, 203–210.

Komhyr, W.D. (1980). *Operations Handbook: Ozone Observations with a Dobson Spectrophotometer* (WMO Global Ozone Research and Monitoring Project, Report No. 6), WMO, Geneva, Switzerland, 125 pp.

Komhyr, W.D. (1987). Ozonemessungen bis 40 km Hohe mit ECC-Ozonsonden, *Promet Meteorologische Fortbildung*, **17**, 29.

Komhyr, W.D. and Evans, R.D. (1980). Dobson spectrophotometer total ozone measurement errors caused by interfering absorbing species such as SO_2, NO_2 and photochemically produced O_3 in polluted air, *Geophysical Research Letters*, **7**, 157–160.

Komhyr, W.D. and Harris, T.B. (1971). *Development of the ECC Ozonesonde* (NOAA Technical Report ERL 200-APCL 18), U.S. Department of Communications, NOAA Environmental Research Laboratory, Boulder, CO.

Komhyr, W.D., Gammon, R.H., Harris, T.B., Waterman, L.S., Conway, T.J., Taylor, W.R. and Thoning, K.W. (1985). Global atmospheric CO distribution and variations from 1968–1982 NOAA/GMCC CO flask sample data, *Journal of Geophysical Research*, **90**, 5567–5596.

Komhyr, W.D., Grass, R.D., and Leonard, R.K. (1989a). Dobson Spectrophotometer 83: A standard for total ozone measurements, 1962–1987, *Journal of Geophysical Research*, **94**, 9847–9861.

Komhyr, W.D., Oltmans, S.J., Franchois, P.R., Evans, W.F.J., and Matthews, W.A. (1989b). The latitudinal distribution of ozone to 35 km altitude from ECC ozonesonde observations, 1985–1987, in: R.D. Bojkov and P. Fabian (Eds.), *Proceedings of the Quadrennial Ozone Symposium 1988, Göttingen, Germany*, Deepak Publications, Hampton, VA., pp. 147–150.

Komhyr, W.D., Grass, R.D., Evans, R.D., Leonard, R.K., Quincy, D.M., Hofmann, D.J., and Koenig, G.L. (1994). Unprecedented 1993 ozone increase over the United States from Dobson spectrophotometer observations, *Geophysical Research Letters*, **21**, 201–204.

Komhyr, W.D., Barnes, R.A., Brothers, G.B., Lathrop, J.A., and Opperman, D.P. (1995). Electrochemical concentration cell ozonesonde performance evaluation during STOIC 1989, *Journal of Geophysical Research*, **100**, 9231–9244.

Kondratyev, K.Ya. (Ed.) (1972). *Explorations of the Environment from Manned Spacecraft*, Gidrometeoizdat, Leningrad, 297 pp. [in Russian].

Kondratyev, K.Ya. (1981). *Stratosphere and Climate*, VINITI, Moscow, 223 pp. [in Russian].

Kondratyev, K.Ya. (1985). Key problems of environmental-studies: International Geosphere–Biosphere Program, *Zeitschrift fur Meteorologie*, **35**, 309–313.

Kondratyev, K.Ya. (1989). *Ozone Dynamics: Progress in Science and Technology, Geomagnetism and Upper Atmosphere*, VINITI, Moscow, Vol. 11, p. 212 [in Russian].

Kondratyev, K.Ya. (1997). A new stage of middle atmosphere satellite remote sensing, *Studying the Earth from Space*, **2**, 127–144 [in Russian].

Kondratyev, K.Ya. (1998a). Aerosol and climate: Some results and perspectives of remote sensing. 1. Complex nature of climate change and variability of aerosol properties, *Ecol. Chemistry*, **7**(2), 73–85 [in Russian].

Kondratyev, K.Ya. (1998b). *Multidimensional Global Change*, Wiley/Praxis, Chichester, U.K., 761 pp.

Kondratyev, K.Ya. (1999a). *The Climatic Effect of Aerosols and Clouds*, Springer/Praxis, Heidelberg, Germany/Chichester, U.K., 272 pp.

Kondratyev, K.Ya. (1999b). *Ecodynamics and Geopolicy: Global Problems*, Russian Academy of Sciences, St. Petersburg, 1036 pp. [in Russian].

Kondratyev, K.Ya. and Galindo, I. (1997). *Volcanic Activity and Climate*, A. Deepak, Hampton, VA, 382 pp.

Kondratyev, K.Ya. and Varotsos, C.A. (1995a). Volcanic eruptions and global ozone dynamics, *International Journal of Remote Sensing*, **16**, 1887–1895.

Kondratyev, K.Ya. and Varotsos, C.A. (1995b). Atmospheric ozone variability in the context of global change, *International Journal of Remote Sensing*, **16**, 1851–1881.

Kondratyev, K.Ya. and Varotsos, C.A. (1995c). Atmospheric greenhouse effect in the context of global climate change, *Nuovo Cimento della Società Italiana di Fisica C-Geophysics and Space Physics*, **18**, 123–151.

Kondratyev, K.Ya. and Varotsos, C.A. (1996a). Global total ozone dynamics: Impact on surface solar ultraviolet radiation variability and ecosystems. I. Global ozone dynamics and environmental safety, *Environmental Science and Pollution Research*, **3**, 153–157.

Kondratyev, K.Ya. and Varotsos, C.A. (1996b). Global total ozone dynamics: Impact on surface solar ultraviolet radiation variability and ecosystems, *Environmental Science and Pollution Research*, **3**, 205–209.

Kondratyev, K.Ya. and Varotsos, C.A. (1997). A review of greenhouse effect and ozone dynamics in Greece, in: C. Varotsos (Ed.), *Atmospheric Ozone Dynamics: Observations in the Mediterranean Region* (NATO ASI Series, Series I, Vol. 53). Springer-Verlag, Berlin, pp. 175–228.

Kondratyev, K.Ya. and Varotsos, C.A. (2000). *Atmospheric Ozone Variability: Implications for Climate Change, Human Health and Ecosystems*, Springer/Praxis, Heidelberg, Germany/Chichester, U.K., 617 pp.

Kondratyev, K.Ya. and Varotsos, C.A. (2001a). Global tropospheric ozone dynamics, Part I: Tropospheric ozone precursors, *Environmental Science and Pollution Research*, **8**, 57–62.

Kondratyev, K.Ya. and Varotsos, C.A. (2001b). Global tropospheric ozone dynamics, Part II: Numerical modelling of tropospheric ozone variability, *Environmental Science and Pollution Research*, **8**, 113–119.

Kondratyev, K.Ya. and Varotsos, C.A. (2002). Remote sensing and global tropospheric ozone observed dynamics, *International Journal of Remote Sensing*, **23**, 159–178.

Kondratyev, K.Ya, Varotsos, C.A., and Cracknell, A.P. (1994). Total ozone amount trend at St. Petersburg as deduced from Nimbus-7 TOMS observations, *International Journal of Remote Sensing*, **15**, 2669–2677.

Kondratyev, K.Ya., Pokrovsky, O.M., and Varotsos, C.A. (1995). Atmospheric ozone trends and other factors of surface ultraviolet radiation variability, *Environmental Conservation*, **22**, 259–261.

Kondratyev, K.Ya., Buznikov, A.A., and Pokrovsky, O.M. (1996). *Global Change and Remote Sensing*, Wiley/Praxis, Chichester, U.K., 396 pp.

Kondratyev, K.Ya., Nakajima, T., Sumi, A., and Tanaka, T. (1998). Priorities of global change and the development of remote sensing in Japan, *International Journal of Remote Sensing*, **19**, 1259–1282.

Kondratyev, K.Ya., Grigoriev, A.A., and Varotsos, C.A. (2002). *Environmental Disasters*, Springer/Praxis, Heidelberg, Germany/Chichester, U.K., 484 pp.

Kopp, G., Belova, A., Diez y Riega, V.E., Groß, J., Hochschild, G., Hoffmann, P., Murtagh, D., Raffalski, U., and Urban, J. (2007). Intercomparison of Odin–SMR ozone profiles with ground-based millimetre-wave observations in the Arctic, the mid-latitudes, and the tropics, *Canadian Journal of Physics*, **85**, 1097–1110.

Kopp, G., Calderón, S.M., Gross, J., Hochschild, G., Hoffmann, P., Notholt, J. and Sinnhuber, M. (2009). Inner-tropical ozone measurements at the Mérida Atmospheric Research Station (MARS) using ground-based microwave radiometry, *International Journal of Remote Sensing*, **30**, 4019–4032.

Koscielny-Bunde, E., Bunde, A., Havlin, S., Roman, H.E., Goldreich, Y., and Schellnhuber, H.J. (1998). Indication of a universal persistence law governing atmospheric variability, *Physical Review Letters*, **81**, 729–732.

Kosmidis, E., Zerefos, C.S., Balis, D., and Bais, A.F. (1998). Testing of the new Brewer Umkehr algorithm, in: R.D. Bojkov and G. Visconti (Eds.), *Proceedings of the XVIII*

Quadrennial Ozone Symposium, L'Aquila, Italy, September 12–21, 1996, Consorzio Parco Scientifico e Tecnologico d'Abruzzo, L'Aquila, Italy, Vol. 1, pp. 139–142.

Kostopoulos, D., Varotsos, P.A., and Mourikis, S. (1975). Conductivity of crystalline NaI, *Canadian Journal of Physics*, **53**, 1318–1320.

Koutoulaki, K., Rodgers, C.D., and Taylor, F.W. (1998). Non-LTE effects in the ISAMS ozone data, in: R.D. Bojkov and G. Visconti (Eds.), *Proceedings of the XVIII Quadrennial Ozone Symposium, L'Aquila, Italy, September 12–21, 1996*, Consorzio Parco Scientifico e Tecnologico d'Abruzzo, L'Aquila, Italy, Vol. 2, pp. 919–922.

Kramer, H.J. (2002). *Observation of the Earth and Its Environment: Survey of Missions and Sensors*, Springer-Verlag, 1510 pp.

Krasnopoler, A. and George, S.M. (1998). Infrared resonant desorption of H_2O from ice multilayers, *Journal of Physical Chemistry B*, **102**, 788–794.

Kramarova, N.A. and Kuznetsov, G.I. (2009). An investigation of long-term variability of ozone 3-dimensional fields in the tropics connected with the 11-year solar cycle and peculiarities of the general atmospheric circulation, *International Journal of Remote Sensing*, **30**, 6483–6496.

Krasouski, A.N. and Liudchik, A.M. (2004). Further development of the extraterrestrial constant calibration for days when ozone changes, in: C.S. Zerefos (Ed.), *Proceedings of the XX Quadrennial Ozone Symposium, June 1–8, 2004, Kos, Greece*, International Ozone Commission, Athens, Greece, pp. 569–570.

Krasouski, A.N., Liudchik, A.M., and Lukyanava, N.F. (2010). Ozone anomaly statistics over Europe over a period of instrumental observation, *International Journal of Remote Sensing*, **31**, 513–521.

Kravchenko, V., Evtushevsky, A., Grytsai, A., Milinevsky, G., and Shanklin, J. (2009). Total ozone dependence of the difference between the empirically corrected EP-TOMS and high-latitude station datasets, *International Journal of Remote Sensing*, **30**, 4283–4294.

Krechetnikov, R. and Marsden, J.E. (2007). Dissipation-induced instabilities in finite dimensions, *Reviews of Modern Physics*, **79**, 519–553.

Krivolutsky, A.A. (1997). Possible nature of spring ozone decreation and its sensitivity to long-term variability, *Proceedings of the First SPARC General Assembly* (WCRP-99, WMO/TD-N814), WMO, Geneva, Switzerland, Vol. 2, pp. 465-468.

Križan, P. and Laštovička, J. (2005). Trends in positive and negative ozone laminae in the Northern Hemisphere, *Journal of Geophysical Research*, **110**, D10107, doi: 10.1029/2004JD005477.

Križan, P. and Laštovička, J. (2006). Ozone laminae: Comparison of the Southern and Northern Hemisphere, and tentative explanation of trends, *Journal of Atmospheric and Solar-Terrestrial Physics*, **68**, 1962–1972.

Krotkov, N.A., Bhartia, P.K., Herman, J.R., Fioletov, V., and Kerr, J. (1998). Satellite estimation of spectral surface UV irradiance in the presence of tropospheric aerosols, 1: Cloud-free case, *Journal of Geophysical Research*, **103**, 8779–8793.

Krueger, A.J. (1986). Rocket techniques for measurements of ozone and related neutral minor constituents, in: R. A. Goldberg (Ed.), *Handbook for MAP*, International Council of Scientific Unions, Urbana, IL, Vol. 19, pp. 29–55.

Krueger, A.J. and Minzer, R. (1973). *A Proposed Mid-latitude Ozone Model for the U.S. Standard Atmosphere* (NASA X-651-73-72), NASA GSFC, Greenbelt, MD.

Krueger, B.C., Neuber, R., El Naggar, S., and Walther, H. (1989). Measurements of ozone profiles by a lidar method during the Arctic winter and spring 1988, in: R.D. Bojkov and P. Fabian (Eds.), *Proceedings of the Quadrennial Ozone Symposium, 1988, Göttingen, Germany*, Deepak Publications, Hampton, VA, pp. 160–163.

Krzyścin, J.W. (2002). Long-term changes in ozone mini-hole event frequency over the Northern Hemisphere derived from ground-based measurements, *Internatiional Journal of Climatology*, **22**, 1425–1439.

Krzyścin, J.W. (2006). Change in ozone depletion rates beginning in the mid 1990s: Trend analyses of the TOMS/ SBUV merged total ozone data, 1978–2003, *Annales Geophysicae*, **24**, 493–502.

Krzyścin, J.W. and Rajewska-Więch, B. (2009). Trends in the ozone vertical distribution from the Umkehr observations at Belsk, 1963–2007, *International Journal of Remote Sensing*, **30**, 3917–3926.

Krzyścin, J.W., Degorska, M. and Rajewska-Więch, B. (1998). Seasonal acceleration of the rate of total ozone decreases over central Europe: impact of tropopause height changes, *Journal of Atmospheric and Solar-Terrestrial Physics*, **60**, 1755–1762.

Krzyścin, J.W, Degorska, M., and Rajewska-Więch, B. (2001). Impact of interannual meteorological variability on total ozone in northern middle latitudes: A statistical approach, *Journal of Geophysical Research*, **106**, 17953–17960.

Krzyścin, J.W., Jarosławski, J. and Rajewska-Więch, B. (2005). Beginning of the ozone recovery over Europe? Analysis of the total ozone data from the ground-based observations, 1964–2004, *Annales Geophysicae*, **23**, 1–11.

Krzyścin, J.W., Jarosławski, J., Rajewska-Więch, B., Sobolewski, P.S., and Froidevaux, L. (2008). Comparison of ozone profiles by the Umkehr measurements taken at Belsk (52°N, 21°E) with the Aura Microwave Limb Sounder overpasses: 2004-2006, *Journal of Geophysical Research*, **113**, D15S09, doi: 10.1029/2007JD008586.

Kuhn, T. (1962). Structure of Scientific Revolutions (First Edition). University of Chicago Press, Chicago, IL, 168 pp.

Kuhn, T.S. (1970). *The Structure of Scientific Revolutions* (Second Edition), University of Chicago Press, Chicago, IL.

Kulinich, B.V., Evtushevsky, A.M., Leonov, N.A., and Milinevsky, G.P. (2005). Seasonal change of difference between the ground-based and EP-TOMS satellite ozone measurements, *Ukrainian Antarctic Journal*, **3**, 5–9.

Kulkarni, P.S., Jain, S.L., Ghude, S.D., Arya, B.C., Dubey, P.K., and Shahnawaz (2009). On some aspects of tropospheric ozone variability over the Indo-Gangetic (IG) basin, India, *International Journal of Remote Sensing*, **30**, 4111–4122.

Kulmälä, M. and Wagner, P.E. (Eds.) (1996). *Nucleation and Atmospheric Aerosols*, Elsevier Science, Oxford, U.K., 967 pp.

Kunzi, K.F. and Rubin, G. (1988). The use of microwave sensors in a global ozone monitoring network, *International Radiation Symposium, August 18–24, 1988, Lille, France.*

Kurosu, T., Rozanov, V.V., and Burrows, J.P. (1997). Parameterization schemes for terrestrial water clouds in the radiative transfer model GOMETRAN, *Journal of Geophysical Research*, **102**, 21809–21823.

Kurylo, M.J. (2009). The role of airborne science in the study of polar ozone, in: C. Zerefos, G. Contopoulos, and G. Skalkeas (Eds.), *Twenty Years of Ozone Decline: Proceedings of the Symposium for the 20th Anniversary of the Montreal Protocol*, Springer-Verlag, Berlin, pp. 173–182.

Kurylo, M.J. and Zander, R.J. (2001). The NDSC: Its status after ten years of operation, *Proceedings of the Quadrennial Ozone Symposium 2000, Hokkaido University, Sapporo, Japan*, National Space Development Agency of Japan, Tokyo (NASDA), pp. 167–168.

Kuttippurath, J., Bremer, H., Burrows, J., Kleinboehl, A., Kuellmann, H., Kuenzi, K., Notholt, J., Sinnhuber, M., von Savigny, C., Lautie, N. *et al.* (2007). Intercomparison of ozone profile measurements from ASUR, SCIAMACHY, MIPAS, OSIRIS, and SMR, *Journal of Geophysical Research*, **112**, D09311, doi: 10.1029/2006JD007830.

Kuttippurath, J., Godin-Beekmann, S., Lefèvre, F., and Pazmiño, A. (2009). Ozone depletion in the Arctic winter 2007/08, *International Journal of Remote Sensing*, **30**, 4071–4082.

Kuttippurath, J., Godin-Beekmann, S., Lefèvre, F., and Goutail, F. (2010a). Spatial, temporal, and vertical variability of polar stratospheric ozone loss in the Arctic winters 2004/2005–2009/2010, *Atmospheric Chemistry and Physics*, **10**, 9915–9930.

Kuttippurath, J., Goutail, F., Pommereau, J.-P., Lefèvre, F., Roscoe, H.K., Pazmiño, A., Feng, W., Chipperfield, M.P., and Godin-Beekmann, S. (2010b). Estimation of Antarctic ozone loss from ground-based total column measurements, *Atmospheric Chemistry and Physics*, **10**, 6569–6581.

Kylling, A, Dahlback, A., and Mayer, B. (2000). The effect of clouds and surface albedo on UV irradiances at a high latitude site, *Geophysical Research Letters*, **27**, 1411–1414.

Kyrölä, E., Tamminen, J., Leppelmeier, G.W., Sofieva, V., Hassinen, S., Bertaux, J.L., Hauchecorne, A., Dalaudier, F., Cot, C., Korablev, O. *et al.* (2004). GOMOS on Envisat: An overview, *Advances in Space Research*, **33**, 1020–1028.

Kyrölä, E., Tamminen, J., Leppelmeier, G.W., Sofieva, V., Hassinen, S., Seppälä, A., Verronen, P.T., Bertaux, J.L., Hauchecorne, A., Dalaudier, F. *et al.* (2006). Nighttime ozone profiles in the stratosphere and mesosphere by the Global Ozone Monitoring by Occultation of Stars on Envisat, *Journal of Geophysical Research*, **111**, D24306, doi: 10.1029/2006JD007193.

Kyrölä, E., Tamminen, J., Sofieva, V., Bertaux, J.L., Hauchecorne, A., Dalaudier, F., Fussen, D., Vanhellemont, F., Fanton d'Andon, O., Barrot, G. *et al.* (2010a). GOMOS O_3, NO_z, and NO_3 observations in 2002–2008, *Atmospheric Chemistry and Physics*, **10**, 7723-7738.

Kyrölä, E., Tamminen, J., Sofieva, V., Bertaux, J.L., Hauchecorne, A., Dalaudier, F., Fussen, D., Vanhellemont, F., Fanton d'Andon, O., Barrot, G. *et al.* (2010b). Retrieval of atmospheric parameters from GOMOS data, *Atmospheric Chemistry and Physics Discussions*, **10**, 10145–10217.

Labitzke, K. and van Loon, H. (1995). A note on the distribution of trends below 10 hPa: The extratropical northern hemisphere, *Journal of the Meteorological Society of Japan*, **73**, 883–889.

Labitzke, K. and van Loon, H. (1997a). The signal of the 11-year sunspot cycle in the upper troposphere–lower stratosphere, *Space Science Reviews*, **80**, 393–410.

Labitzke, K. and van Loon, H. (1997b). Total ozone and the 11-yr sunspot cycle, *Journal of Atmospheric and Solar-Terrestrial Physics*, **59**, 9–19.

Labow, G.J., Flynn, L.E., Rawlins, M.A., Beach, R.A., and, Simmons, C.A. (1996). Estimation of ozone with total ozone portable spectroradiometer instruments, II: Practical and comparisons, *Applied Optics*, **35**, 6084–6089.

Labow, G.J., McPeters, R.D., and Bhartia, P.K. (2004). A comparison of TOMS, SBUV and SBUV/2 Version 8 total column ozone data with data from ground stations, In: C.S. Zerefos (Ed.), *Proceedings of the XX Quadrennial Ozone Symposium, June 1–8, 2004, Kos, Greece*, International Ozone Commission, Athens, Greece, pp. 123–124.

Lacis, K., Wuebbles, D.J., and Logan, J.A. (1990). Radiative forcing of climate by changes in the vertical distribution of ozone, *Journal of Geophysical Research*, **95**, 9971–9981.

Lamago, D., Dameris, M., Schnadt, C., Eyring, V., and Brühl, C. (2003). Impact of large solar zenith angles on lower stratospheric dynamical and chemical processes in a coupled chemistry–climate model, *Atmospheric Chemistry and Physics*, **3**, 1981–1990.

Lamarque, J.-F., Granier, C., Bond, T., Eyring, V., Heil, A., Kainuma, M., Lee, D., Liousse, C., Mieville, A., Riahi, K. *et al.* (2009). Gridded emissions in support of IPCC AR5, *IGAC Newsletter*, **41**, 12–18.

Lamarque, J.-F., Bond, T.C., Eyring, V., Granier, C., Heil, A., Klimont, Z., Lee, D., Liousse, C., Mieville, A., Owen, B. *et al.* (2010). Historical (1850–2000) gridded anthropogenic and biomass burning emissions of reactive gases and aerosols: Methodology and application, *Atmospheric Chemistry and Physics*, **10**, 7017–7039.

Lambert, J.C., van Roozendael, M., Mazière, M.D., Simon, P.C., Pommereau, J.P., Goutail, F., Sarkissian, A., and Gleason, J.F. (1999). Investigation of pole-to-pole performances of spaceborne atmospheric chemistry sensors with the NDSC, *Journal of the Atmospheric Sciences*, **56**, 176–193.

Lambert, J.C., van Roozendael, M., Simon, P.C., Pommereau, J.P., Goutail, F., Gleason, J.F., Andersen, S.B., Arlander, D.W., Bui Van, N.A., Claude, H. *et al.* (2000). Combined characterization of GOME and TOMS total ozone measurements from space using ground-based observations from the NDSC, *Advances in Space Research*, **26**, 1931– 1940.

Land, C., Feichter, J., and Sausen, R. (2002). Impact of vertical resolution on the transport of passive tracers in the ECHAM4 model, *Tellus (B)*, **54**, 344–360.

Landau, L.D. and Lifshitz, E.M. (1958). *Statistical Physics*, Pergamon, London.

Landgraf, J. and Crutzen, P.S. (1998). An efficient method for online calculations of photolysis and heating rates, *Journal of the Atmospheric Sciences*, **55**, 863–878.

Laštovička, J. (1998). Second IAGA/IAMAS Workshop: Solar activity forcing of the middle atmosphere, *SPARC Newsletter*, **10**, 26–27.

Lat, S., Subbaraya, B.H., Acharya, Y.B., Ramani, S.V., Borchers, R., and Fabian, P. (1989). Vertical distribution of ozone over Hyderabad using a balloon-borne suntracking photometer, in: R. D. Bojkov and P. Fabian (Eds.), *Proceedings of the Quadrennial Ozone Symposium, 1988, Göttingen, Germany*, Deepak Publications, Hampton, VA, pp. 136–139.

Lawrence, M.G., Crutzen, P.J., Rasch, P.J., Eaton, B.E., and Mahowald, N.M. (1999) A model for studies of tropospheric photochemistry: Description, global distributions, and evaluation, *Journal of Geophysical Research*, **104**, 26245–26277.

Lean, J., Rottmann, G., Kyle, H., Woods, T., Hickey, J., and Puga, L. (1997). Detection and parameterisation of variations in solar mid- and near-ultraviolet radiation (22–400 nm), *Journal of Geophysical Research*, **102**, 29939–29956.

Lee, C., Richter, A., Weber, M., and Burrows, J.P. (2008). SO_2 retrieval from SCIAMACHY using the Weighting Function DOAS (WFDOAS) technique: Comparison with standard DOAS retrieval, *Atmospheric Chemistry and Physics Discussions*, **8**, 10817–10839.

Lee, S.H., Akimoto, H., Nakane, H., Kurnosenko, S., and Kinjo, Y. (1998). Lower tropospheric ozone trend observed in 1989–1997 at Okinawa, Japan, *Geophysical Research Letters*, **25**, 1637–1640.

Lekien, F. and Coulliette, C. (2007). Chaotic stirring in quasi-turbulent flows, *Philosophical Transactions of the Royal Society A*, **365**, 3061–3084.

Lelieveld, J. and van Dorland, R. (1995). Ozone chemistry changes in the troposphere and consequent radiative forcing of climate, in: W.-C. Wang and I.S.A. Isaksen (Eds.), *Atmospheric Ozone as a Climate Gas* (NATO ASI Ser. I), Springer-Verlag, New York, Vol. 32, pp. 227–258.

Lerot, C., van Roozendael, M., van Geffen, J., van Gent, J., Fayt, C., Spurr, R., Lichtenberg, G., and von Bargen, A. (2009). Six years of total ozone column measurements from SCIAMACHY nadir observations, *Atmospheric Measurement Techniques*, **2**, 87–98.

Lerot, C., van Roozendael, M., Lambert, J.-C., Granville, J., van Gent, J., Loyola, D., and Spurr, R. (2010). The GODFIT algorithm: A direct fitting approach to improve the accuracy of total ozone measurements from GOME, *International Journal of Remote Sensing*, **31**, 543–550.

Levelt, P.F., Allaart, M.A.F., and Kelder, H.M. (1996). On the assimilation of total-ozone satellite data, *Annales Geophysicae*, **14**, 1111–1118.

Levelt, P.F., Hilsenrath, E., Leppelmeier, G.W., van den Oord, G.H.J., Bhartia, P.K., Tamminen, J., de Haan, J.F., and Veefkind, J.P. (2006). Science objectives of the Ozone Monitoring Instrument, *IEEE Transactions on Geoscience and Remote Sensing*, **44**, 1199–1208.

Levy, H. II, Kasibhatla, P.S., Moxim, W.J., Klonecki, A.A., Hirsch, A.I., Oltmans, S.J., and Chameides, W.L. (1997). The global impact of human activity on tropospheric ozone, *Geophysical Research Letters*, **24**, 791–794.

Li, J., Schmidt, C.C. Nelson, J.P., Schmit, T.J., and Menzel, W.P. (2001). Estimation of total atmospheric ozone from GOES sounder radiances with high temporal resolution, *Journal of Atmospheric and Oceanic Technology*, **18**, 157–168.

Li, J.L., Cunnold, D.M., Wang, H.J., Yang, E.S., and Newchurch, M.J. (2002). A discussion of upper stratospheric ozone asymmetries and SAGE trends, *Journal of Geophysical Research*, **107**, 4705, doi: 10.1029/2001JD001398.

Liang, L., Horowitz, L.W., Jacob, D.S., Wang, Y., Fiore A.M., Logan, J.A., Gardner, G.M., and Mungerd, M. (1998). Seasonal budgets of reactive nitrogen species and ozone over the United States, and export fluxes to the global atmosphere, *Journal of Geophysical Research*, **103**, 13435–13450.

Lin, S.J. and Rood, R.B. (1996). Multidimensional flux-form semi-Lagrangian transport schemes, *Monthly Weather Review*, **124**, 2046–2070.

Lin, S.J. and Rood, R.B. (1997). A finite-volume integration scheme for computing pressure-gradient forces in general vertical coordinates, *Quarterly Journal of the Royal Meteorological Society*, **123**, 1749–1762.

Lin, X., Trainer, M., and Hsie, E.-Y. (1998). A modeling study of tropospheric species during the North Atlantic Regional Experiment (NARE), *Journal of Geophysical Research*, **103**, 13593–13613.

Liu, C.X., Liu, Y., Cai, Z.N., Gao, S.T., Lu, D.R., and Kyrölä, E. (2009). A Madden–Julian Oscillation-triggered record ozone minimum over the Tibetan Plateau in December 2003 and its association with stratospheric "low-ozone pockets", *Geophysical Research Letters*, **36**, L15830.

Liu, X., Chance, K., Sioris, C.E., Spurr, R.J.D., Kurosu, T.P., Martin, R.V., and Newchurch, M.J. (2005). Ozone profile and tropospheric ozone retrievals from the Global Ozone Monitoring Experiment: Algorithm description and validation, *Journal of Geophysical Research*, **110**, D20307, doi: 10.1029/2005JD006240.

Liu, X., Bhartia, P.K., Chance, K., Spurr, R.J.D., and Kurosu, T.P. (2010a). Ozone profile retrievals from the Ozone Monitoring Instrument, *Atmospheric Chemistry and Physics*, **10**, 2521–2537.

Liu, X., Bhartia, P.K., Chance, K., Froidevaux, L., Spurr, R.J.D., and Kurosu, T.P. (2010b). Validation of Ozone Monitoring Instrument (OMI) ozone profiles and stratospheric ozone columns with Microwave Limb Sounder (MLS) measurements, *Atmospheric Chemistry and Physics*, **10**, 2539–2549.

Liu, Y., Liu, C.X., Wang, H.P., Tie, X.X., Gaos, S.T., Kinnison, D., and Brasseur, G. (2009). Atmospheric tracers during the 2003–2004 stratospheric warming event and impact of ozone intrusions in the troposphere, *Atmospheric Chemistry and Physics*, **9**, 2157–2170.

Liudchik, A.M. (2009). Variational method for determining the extraterrestrial constant for a solar ultraviolet spectrometer ozonometer, *Journal of Applied Spectroscopy*, **76**, 273–279.

Liudchik, A. and Krasouski, A. (2002). Extending Langley calibration to days when ozone changes, *Stratospheric Ozone 2002: Proceedings of the Sixth European Symposium, September 2–6, 2002, Goteborg, Sweden*, pp. 293-296.

Liudchik, A.M. and Krasouski, A.N. (2010). A new calculation technique of the extraterrestrial parameter for column ozone observations made by spectrophotometers, *International Journal of Remote Sensing*, **31**, 323–328.

Livesey, N.J., Filipiak, M.J., Froidevaux, L., Read, W.G., Lambert, A., Santee, M.L., Jiang, J.H., Pumphrey, H.C., Waters, J.W., Cofield, R.E. *et al.* (2008). Validation of Aura Microwave Limb Sounder O_3 and CO observations in the upper troposphere and lower stratosphere, *Journal of Geophysical Research*, **113**, D15S02, doi: 10.1029/2007JD008805.

Livingston, F.E. and George, S.M. (1999). Effect of HNO_3 and HCl on HDO diffusion on crystalline D_2O ice multilayers, *Journal of Physical Chemistry B*, **103**, 4366–4376.

Livingston, F.E. and George, S.M. (2001). Diffusion kinetics of HCl hydrates in ice measured using infrared laser resonant desorption depth-profiling, *Journal of Physical Chemistry A*, **105**, 5155–5164.

Livingston, F.E., Whipple, G.C., and George, S.M. (1997). Diffusion of HDO into single-crystal (H_2O)-O-16 ice multilayers: Comparison with (H_2O)-O-18, *Journal of Physical Chemistry B*, **101**, 6127–6131.

Livingston, F.E., Whipple, G.C., and George, S.M. (1998). Surface and bulk diffusion of HDO on ultrathin single-crystal ice multilayers on Ru(001), *Journal of Chemical Physics*, **108**, 2197–2207.

Livingston, F.E., Smith, J.A., and George, S.M. (2000). Depth-profiling and diffusion measurements in ice films using infrared laser resonant desorption, *Analytical Chemistry*, **72**, 5590–5599.

Livingston *et al.* (2001). Personal communication.

Livingston, F.E., Smith, J.A., and George, S.M. (2002). General trends for bulk diffusion in ice and surface diffusion on ice, *Journal of Physical Chemistry A*, **106**, 6309–6318.

Logan, J.A. (1999a). An analysis of ozonesonde data for the troposphere: Recommendations for testing three-dimensional models and development of a gridded climatology for tropospheric ozone, *Journal of Geophysical Research*, **104**, 16115–16149.

Logan, J.A. (1999b). An analysis of ozonesonde data for the lower stratosphere: Recommendations for testing models, *Journal of Geophysical Research*, **104**, 16151–16170.

Logan, J.A., Jones, D.B.A., Megretskaia, I.A., Oltmans, S.J., Johnson, B.J., Vömel, H., Randel, W.J., Kimani, W., and Schmidlin, F.J. (2003). Quasibiennial oscillation in tropical ozone as revealed by ozonesonde and satellite data, *Journal of Geophysical Research*, **108**, 4244–4266.

Logan, J.A., Megretskaia, L., Nassar, R., Murray, L.T., Zhang, L., Bowman, K.W., Worden, H.M., and Luo, M. (2008). Effects of the 2006 El Niño on tropospheric composition as revealed by data from the Tropospheric Emission Spectrometer (TES), *Geophysical Research Letters*, **35**, L03816, doi: 10.1029/2007GL031698.

London, J. and Liu, S.C. (1992). Long-term tropospheric and lower stratospheric ozone variations from ozonesonde observations, *Journal of Atmospheric and Terrestrial Physics*, **54**, 599–625.

Lott, F., Fairhead, L., Hourdin, F., and Levan, P. (2005). The stratospheric version of LMDz: Dynamical climatologies, Arctic oscillation, and impact on the surface climate, *Climate Dynamics*, **25**, doi: 10.1007/s00382-005-064.

Loucks, A.L. (2007). Evaluation of dynamical sources of ozone laminae in the tropical troposphere and tropical tropopause layer, MS thesis, Penn State University.

Loughman, R.P., Flittner, D.E., Herman, B.M., Bhartia, P.K., Hilsenrath, E., and McPeters, R.D. (2005). Description and sensitivity analysis of a limb scattering ozone retrieval algorithm, *Journal of Geophysical Research*, **110**, doi: 10.1029/2004JD005429.

Lovelock, J.E. (1970). Air pollution and climate change, *Atmospheric Environment*, **5**, 403–411.

Lovelock, J.E. (1971). Atmospheric fluorine compounds as indicators of air movements, *Nature*, **230**, 379.

Lovelock, J.E. (1972). Atmospheric turbidity and CCl_3F concentrations in rural Southern England and Southern Ireland, *Atmospheric Environment*, **6**, 917–925.

Lovelock, J.E. (2000). *Homage to Gaia: The Life of an Independent Scientist*, Oxford University Press, Oxford, U.K.

Lovelock, J.E. (2009). *The Vanishing Face of Gaia: A Final Warning*, Allen Lane, London.

Lovelock, J.E., Maggs, R.J., and Wade, R.J. (1973). Halogenated hydrocarbons in and over the Atlantic, *Nature*, **241**, 194–196.

Loyola, D.G., Coldewey-Egbers, R.M., Dameris, M., Garny, H., Stenke, A., Van Roozendael, M., Lerot, C., Balis, D., and Koukouli, M. (2009a). Global long-term monitoring of the ozone layer: A prerequisite for predictions, *International Journal of Remote Sensing*, **30**, 4295–4318.

Loyola, D., Erbertseder, T., Balis, D., Lambert, J-C., Spurr, R., Van Roozendael, M., Valks, P., Zimmer, W., Meyer-Arnek, J., and Lerot, C. (2009b). Operational monitoring of the Antarctic Ozone Hole: Transition from GOME and SCIAMACHY to GOME-2, in: C. Zerefos, G. Contopouolos, and G. Stalkeas (Eds.), *Twenty Years of Ozone Decline: Proceedings of the Symposium for the 20th Anniversary of the Montreal Protocol*, Springer-Verlag, Berlin, pp. 213–236.

Low, P.S. (1998). Tropospheric ozone: The missing GHG in the UNFCCC equation, *Linkages Journal*, **3**, 32–34.

Lu, H., McCartney, S.A., and Sadtchenko, V. (2007). Fast thermal desorption spectroscopy study of H/D isotopic exchange reaction in polycrystalline ice near its melting point, *Journal of Chemical Physics*, **127**, 184701, doi: 10.1063/1.2786101.

Lu, H.P., McCartney, S.A., and Sadtchenko, V. (2009). H/D exchange kinetics in pure and HCl doped polycrystalline ice at temperatures near its melting point: Structure, chemical transport, and phase transitions at grain boundaries, *Journal of Chemical Physics*, **130**, 054501, doi: 10.1063/1.3039077.

Lubin, D. and Frederick, J.E. (1990). Column ozone measurements from Palmer Station, Antarctica: Variations during the austral springs of 1988 and 1989, *Journal of Geophysical Research*, **95**, 13883–13889.

Lubin, D., Jensen, E.H., and Gies, H.P. (1998). Global surface ultraviolet radiation climatology from TOMS and ERBE data, *Journal of Geophysical Research*, **103**, 26061–26091.

Lübken, F.-J., Austin, J., Langematz, U., and Oberheide, J. (2010). Introduction to special section on climate and weather of the Sun–Earth system, *Journal of Geophysical Research*, **115**, D00119, doi: 10.1029/2009JD013784.

Lucke, R.L., Korwan, D., Bevilacqua, R.M., Hornstein, J.S., Shettle, E.P., Chen, D.T., Daehler, M., Lumpe, J.D., Fromm, M.D., Debrestian, D. *et al.* (1999). The Polar

Ozone and Aerosol Measurement (POAM III) instrument and early validation results, *Journal of Geophysical Research*, **104**, 18785–18799.

Lumpe, J.D., Bevilacqua, R.M., Hoppel, K.W., and Randall, C.E. (2002). POAM III retrieval algorithm and error analysis retrieval, *Journal of Geophysical Research*, **107**, 4575–4607, doi: 10.1029/2002JD002137.

MacKenzie, I.A., Harwood, R.S., Stott, P.A., and Watson, G.C. (1999). Radiative–dynamic effects of the Antarctic ozone hole and chemical feedback, *Quarterly Journal of the Royal Meteorological Society*, **125**, Part B, 2171–2203.

Mäder, J., Koch, G., and Stähelin, J. (2002). Modeling of ozone measurements monthly means or daily means? *European Geophysical Society XXVII General Assembly, Nice, France, April 21–26, 2002.*

Madronich, S. (1998). WMO Ad Hoc Scientific Steering Committee on UV monitoring: activities and scientific issues, *SPARC Newsletter*, **10**, 19–26.

Maduro, R.A. and Schauerhammer, R. (1992). *The Holes in the Ozone Scare: The Scientific Evidence that the Sky Isn't Falling*, 21st Century Science Associates, Washington, D,C.

Mailhot, J., Strapp, J.W., MacPherson, J.I., Benoit, R., Blair, S., Donaldson, N.R., Froude, F., Benjamin, M., Zawadzki, I., and Rogers, R.R. (1998). The Montreal-96 Experiment on Regional Mixing and Ozone (MERMOZ): An overview and some preliminary results, *Bulletin of the American Meteorological Society*, **79**, 439–442.

Maillard-Barras, E., Ruffieux, D., and Hocke, K. (2009). Stratospheric ozone profiles over Switzerland measured by SOMORA, ozonesonde and MLS/AURA satellite, *International Journal of Remote Sensing*, **30**, 4033–4041.

Mahlman, J.D., Pinto, J.P., and Umscheid, L.J. (1994). Transport, radiative, and dynamical effects of the Antarctic ozone hole: A GFDL «SKYHI» model experiment, *Journal of the Atmospheric Sciences*, **51**, 489–508.

Manney, G.L., Froidevaux, L., Waters, J.W., Zurek, R.W., Gille, J.C., Kumer, J.B., Mergenthaler, J.L., Roche, A.E., Oneill, A., and Swinbank, R. (1995). Formation of low-ozone pockets in the middle stratospheric anticyclone during winter, *Journal of Geophysical Research*, **100**, 13939–13950.

Manney, G.L., Froidevaux, L., Santee, M.L., Zurek, R.W., and Waters, J.W. (1997). MLS observations of Arctic ozone loss in 1996–97, *Geophysical Research Letters*, **24**, 2697–2700.

Manney, G.L., Sabutis, J.L., Allen, D.R., Lahoz, W.A., Scaife, A.A., Randall, C.E., Pawson, S., Naujokat, B., and Swinbank, R. (2005). Simulations of dynamics and transport during the September 2002 Antarctic major warming, *Journal of the Atmospheric Sciences*, **62**, 690–707.

Mantis, H.T., Zerefos, C.S., Bais, A., Ziomas, I., and Kelessis, A. (1986). The Northern Hemisphere ozone minimum in 1982–1983, *Archives for Meteorology, Geophysics and Bioclimatology*, Series B, **36**, 135–145.

Manzini, E., McFarlane, N.A., and McLandress, C. (1997). Impact of the Doppler spread parameterization on the simulation of the middle atmosphere circulation using the MA/ECHAM4 general circulation model, *Journal of Geophysical Research*, **102**, 25751–25762.

Marchand, M., Bekki, S., Pazmiño, A., Lefèvre, F., Godin-Beekmann, S., and Hauchecorne, A. (2005). Model simulations of the impact of the 2002 Antarctic ozone hole on the midlatitudes, *Journal of the Atmospheric Sciences*, **62**, 871–884.

Marchand, P., Riou, S., and Ayotte, P. (2006). Diffusion kinetics for methanol in polycrystalline ice, *Journal of Physical Chemistry*, **110**, 11654–11664.

Martens, P. (1998). *Health and Climate Change: Modelling the Impacts of Global Warming and Ozone Depletion*, Earthscan, London, 224 pp.

Massoli, P., Maturilli, M., and Neuher, R. (2006). Climatology of Arctic polar stratospheric clouds as measured by lidar in Ny-Ålesund, Spitsbergen (79°N, 12°E), *Journal of Geophysical Research*, **111**, D09206, doi: 10.1029/2005JD005840.

Mast, G.M. and Saunders, H.E. (1962). Research and development of the instrumentation of ozone sensing, *Instrument Society of America Transactions*, **1**, 325–328.

Mateer, C.L. (1971). Estimation of total ozone from satellite measurements of backscattered ultraviolet earth radiance, *Journal of the Atmospheric Sciences*, **28**, 1307–1311.

Mateer, C.L. and DeLuisi, J.J. (1985). A comparison of ozone profiles derived from standard Umkehr and short Umkehr measurements from 15 stations, in: C. S. Zerefos and A. Ghazi (Eds.), *Atmospheric Ozone: Proceedings of the Quadrennial Ozone Symposium, Halkidiki, Greece*, D. Reidel, Dordrecht, The Netherlands, pp. 290–294.

Mateer, C.L. and DeLuisi, J.J. (1992). A new Umkehr inversion algorithm, *Journal of Atmospheric and Terrestrial Physics*, **54**, 537–556.

Mateer, C.L. and Dütsch, H.U. (1964). *Uniform Evaluation of Umkehr Observations from the World Ozone Network, Part I: Proposed Standard Evaluation Technique*, National Center for Atmospheric Research, Boulder, CO, 105 pp.

Mauersberger, K., Anderson, S., Murphy, D., and Morton, J. (1985). Precise ozone measurements using a mass spectrometer beam system, in: C.S. Zerefos and A. Ghazi (Eds.), *Atmospheric Ozone: Proceedings of the Quadrennial Ozone Symposium, Halkidiki, Greece*, D. Reidel, Hingham, MA, pp. 493–498.

Mauldin L.E. III, Zaun, N.H., McCormick, M.P., Guy, J.H., and Vaughn, W.R. (1985). Stratospheric Aerosol and Gas Experiment II instrument: A functional description, *Optical Engineering*, **24**, 307–312.

Mauldin, L.E. III, McCormick, M.P. Jr., McMaster, L.R., and Vaughan, W.R. (1986). The stratospheric aerosol and gas experiment II (SAGE II) design and in-orbit performance, in: *Instrumentation for Optical Remote Sensing from Space: Proceedings of the Meeting, Cannes, France, November 27–29, 1985* (A87-1964707-19), Society of Photo-Optical Instrumentation Engineers, Bellingham, WA, pp. 104–111.

Mauzerall, D.L., Logan, J.A., Jacob, D.S., Anderson, B.E., Blake, D.R., Bradshaw, J.D., Heikes, B., Sachse, G.W., Singh, H., and Talbot, B. (1998). Photochemistry in biomass burning plumes and implications for tropospheric ozone over the tropical Southern Atlantic, *Journal of Geophysical Research*, **103**, 8401–8424.

Mayer, M. and Putz, E. (1998). Stratospheric trace gas measurements by means of DOAS at an Alpine monitoring station in Austria, in: R.D. Bojkov and G. Visconti (Eds.), *Atmospheric Ozone: Proceedings of the XVIII Quadrennial Ozone Symposium, L'Aquila, Italy, September 12–21, 1996*, Consorzio Parco Scientifico e Tecnologico d'Abruzzo, L'Aquila, Italy, Vol. 1, 65–68.

McCormack, J.P., Miller, A.J., Nagatani, R., and Fortuin, J.P.F. (1998). Interannual variability in the spectral distribution of extratropical total ozone, *Geophysical Research Letters*, **25**, 2153–2156.

McCormick, M.P. (1987). SAGE II: An overview, *Advances in Space Research*, **7**, 219–226.

McCormick, M.P. (1998). Summary of the SAGE III (Stratospheric Aerosol and Gas Experiment III) science team meeting, *The Earth Observer*, **10**(3), 7–14.

McCormick, M.P., Steele, H.M., Hamill, P., Chu, W.P., and Swissler, T.J. (1982). Polar stratospheric cloud sightings by SAM 11, *Journal of the Atmospheric Sciences*, **39**, 1387–1397.

McCormick, M.P., Veiga, R.E., and Chu, W.P. (1992). Stratospheric ozone profile and total ozone trends derived from SAGE I and SAGE II data, *Geophysical Research Letters*, **19**, 269–272.

McDermid, I.S., Godin, S., and Lindquist, L.O. (1990). Ground based lidar DIAL system for long term measurements of stratospheric ozone, *Applied Optics*, **29**, 3603–3612.

McDermid, I.S., Bergwerff, J.B., Bodeker, G., Boyd, I.S., Brinksma, E.J., Connor, B.J., Farmer, R., Gross, M.R., Kimvilakani, P., Matthews, W.A *et al.* (1998). OPAL: Network for the detection of stratospheric change Ozone Profiler Assessment at Lauder, New Zealand. II. Intercomparison of revised results, *Journal of Geophysical Research*, **103**, 28693–28699.

McElroy, C.T., Mateer, C.L., Kerr, J.B., and Wardle, D.I. (1989). Umkehr observations made with the Brewer ozone spectrophotometer, in: R.D. Bojkov and P. Fabian (Eds.), *Proceedings of the Quadrennial Ozone Symposium, 1988, Göttingen, Germany*, Deepak Publications, Hampton, VA, pp. 441–446.

McElroy, C.T., Kerr, J.B., Wardle, D.I., McArthur, L.J.B., Shah, G.M., Garneau, M., McClean, S.G., Thirsk, R., Davies, J.A., Evans, W.F.J. *et al.* (1991). SPEAM-I (SunPhotometer Earth Atmosphere Measurement) observations of high-altitude ozone from STS41-G, *Canadian Journal of Physics*, **69**, 1123–1127.

McElroy, C.T., Hare, E.W., and Kerr, J.B. (1994). Ozone trends estimated from Umkehr measurements made at Edmonton, Alberta, Canada, in: R. D. Hudson (Ed.), *Ozone in the Troposphere and Stratosphere: Proceedings of the Quadrennial Ozone Symposium, Charlottesville, VA* (NASA CP-3266), NASA, Washington, D.C., pp. 199–202.

McElroy, C.T., Hahn, J.F., and Hare, E. (1998). Determining changes in ozone at high-altitude from Brewer Umkehr observations made at Canadian stations, in: R.D. Bojkov and G. Visconti (Eds.), *Proceedings of the XVIII Quadrennial Ozone Symposium, L'Aquila, Italy, September 12–21, 1996*, Consorzio Parco Scientifico e Tecnologico d'Abruzzo, L'Aquila, Italy, Vol. 1, pp. 147–150.

McElroy, C.T., Nowlan, R.C., Drummond, R.J., Bernath, F.P., Barton, V.D., Dufour, G.D., Midwinter, C., Hall, B.R., Ogyu, A., Ullberg, A. *et al.* (2007). The ACE-MAESTRO instrument on SCISAT: Description, performance, and preliminary results, *Applied Optics*, **46**, 4341–4356.

McElroy, M.B., Salawitch, R.J., Wofsky, S.C., and Logan, J.A. (1986). Reductions of Antarctic ozone due to synergistic interactions of chlorine and bromine, *Nature*, **321**, 759–762.

McFarlane, N.A. (2008). Connections between stratospheric ozone and climate: Radiative forcing, climate variability and change, *Atmosphere-Ocean*, **46**, 139–158.

McHugh, M., Magill, B., Walker, K.A., Boone, C.D., Bernath, P.F., and Russell J.M. III (2005). Comparison of atmospheric retrievals from ACE and HALOE, *Geophysical Research Letters*, **32**, L15S10, doi: 10.1029/2005GL022403.

McInturff, R. (1978). *Stratospheric Warmings: Synoptic, Dynamic and General-circulation Aspects* (NASA RP-1017), NASA, National Meteorological Center, Washington, D.C., 50 pp.

McKenzie, R., Connor, B., and Bodeker, G. (1999). Increased summertime UV radiation in New Zealand in response to ozone loss, *Science*, **285**, 1709–1711.

McKenzie, R.L. and Johnston, P.V. (1982). Seasonal variations in stratospheric NO_2 at 45°S, *Geophysical Research Letters*, **9**, 1255–1258.

McLinden, C.A., Tegtmeier, S., and Fioletov, V. (2009). Technical note: A SAGE-corrected SBUV zonal-mean ozone data set, *Atmospheric Chemistry and Physics*, **9**, 7963–7972.

McNamara, F.D., Beck, C.T., Petropavlovskikh, I., Beach, E., Pachepsky, Y., Li, Y.P., Deland, M., Huang, L-K., Long, C.S., Tiruchirapalli, R. *et al.* (2009). Measurements and products from the Solar Backscatter Ultraviolet (SBUV/2) and the Ozone Mapping and Profiler Suite (OMPS) instruments, *International Journal of Remote Sensing*, **30**, 4259–4272.

McPeters, R.D. and Labow, G.J. (1996) An assessment of the accuracy of 14.5 years of Nimbus 7 TOMS version 7 ozone data by comparison with the Dobson network, *Geophysical Research Letters*, **23**, 3695–3698.

McPeters, R.D., Miles, T., Flynn, L.E., Wellemeyer, C.G., and Zawodny, J. (1994). A comparison of SBUV and SAGE II ozone profiles: Implications for ozone trend, *Journal of Geophysical Research*, **99** 20513–20524.

McPeters, R.D., Hollandsworth, S.M., Flynn, L.E., and Herman, J.R. (1996). Long term trends derived from the 16-year combined Nimbus 7/Meteor 3 TOMS version 7 record, *Geophysical Research Letters*, **23**, 3699–3702.

McPeters, R.D., Bhartia, P.K., Krueger, A.J., Herman, J.R., Wellemeyer, C.G., Seftor, C.J., Jaross, G., Torres, O., Moy, L., Labow, G. *et al.* (1998). *Earth Probe Total Ozone Mapping Spectrometer (TOMS) Data Products User's Guide* (NASA TP-1998-206895), NASA GSFC, Greenbelt, MD, 70 pp.

McPeters, R., Taylor, S., Jaross, G., Haffner, D., Labow, G., and Kowalewski, M. (2007). *Empirically Corrected TOMS Earth Probe Dataset* (*http://toms.gsfc.nasa.gov/news/Corrected_EP _TOMS_README.pdf* [last accessed June 2, 2009]).

Mees, J., Crewell, S., Nett, H., de Lange, G., van de Stadt, H., Kuipers, J.J., and Panhuyzen, R.A. (1995). ASUR: An airborne SIS-receiver for atmospheric measurements at 625 to 720 GHz, *IEEE Transactions on Microwave Theory and Techniques*, **43**, 2543–2548.

Megie, G., Allain, J.Y., Chanin, M.L., and Blamont, J.E. (1977). Vertical profile of stratospheric ozone by lidar sounding from the ground, *Nature*, **270**, 329–331.

Megie, G., Chanin, M.L., Ehhalt, D., Fraser, P., Frederic, J.F., Gille, J.C., McCormick, M.P., and Schoeberl, M. (1998). Global trends, *Scientific Assessment of Stratospheric Ozone Depletion: 1998* (WMO Global Ozone Research and Monitoring Project, Report No. 20), WMO, Geneva, Switzerland.

Meier, A., Blumenstock, T., and Nakajima, H. (1998). Regular isotopic observation of stratospheric ozone and its implications for the ozone formation theory, *Proceedings of the Fourth European Symposium on Stratospheric Ozone* (Air Pollution Research Report 66), Office for Official Publications of the EC, Brussels, Belgium, pp. 216–219.

Meijer, Y.J., van der A, R.J., Van Oss, R.F., Swart, D.P.J., Kelder, H.M., and Johnston, P.V. (2003). Global ozone monitoring experiment ozone profile characterization using interpretation tools and lidar measurements for intercomparison, *Journal of Geophysical Research*, **108**, doi: 10.1029/2003JD003498.

Meijer, Y.J., Swart, D.P.J., Allaart, M., Andersen, S.B., Bodeker, G., Boyd, I., Braathen, G., Calisesi, Y., Claude, H., Dorokhov, V. *et al.* (2004). Pole-to-pole validation of Envisat GOMOS ozone profiles using data from ground-based and balloon sonde measurements, *Journal of Geophysical Research*, **109**, D23305, doi: 10.1029/2004JD004834.

Meijide, A., Cardenas, L.M., Bol, R., Bergstermann, A., Goulding, K., Well, R., Vallejo, A., and Scholefield, D. (2010). Dual isotope and isotopomer measurements for the understanding of N_2O production and consumption during denitrification in an arable soil, *European Journal of Soil Science*, **61**, 364–374.

Meleti, C., Bais, A.F., Kazadzis, S., Kouremeti, N., Garane, K., and Zerefos, C. (2009). Factors affecting solar UV irradiance measured since 1990 at Thessaloniki, Greece, *International Journal of Remote Sensing*, **30**, 4167–4179.

Mercer, J.L., Kröger, C., Nardi, B., Johnson, B.J., Chipperfield, M.P., Wood, S.W., Nichol, S.E., Santee, M.L., and Deshler, T. (2007). Comparison of measured and modeled ozone above McMurdo Station, Antarctica, 1989–2003, during austral winter/spring, *Journal of Geophysical Research*, **112**, D19307, 12 pp., doi: 10.1029/2006JD007982.

Michaels, P.J., Singer S.F., Knapperberger, O.P.E., Kerr, J.B., and McElroy, C.T. (1994). Analysing ultraviolet-B radiation: Is there a trend? *Science*, **264**, 1341–1343.

Miller, A.J., Tiao, G.C., Reinsel, G.C., Wuebbles, D., Bishop, L., Kerr, J., Nagatani, R.M., DeLuisi, J.J., and Mateer, C.L. (1995). Comparisons of observed ozone trends in the stratosphere through examination of Umkehr and balloon ozonesonde data, *Journal of Geophysical Research*, **100**, 11209–11217.

Miller, A.J., Nagatani, R.M., Flynn, L.E., Kondragunta, S., Beach, E., Stolarski, R., McPeters, R.D., Bhartia, P.K., DeLand, M.T., Jackman, C.H. *et al.* (2002). A cohesive total ozone data set from the SBUV(/2) satellite system, *Journal of Geophysical Research*, **107**, doi: 10.1029/2001JD000853.

Miller, A.J., Cai, A., Tiao, G., Wuebbles, D.J., Flynn, L.E., Yang, S.-K., Weatherhead, E.C., Fioletov, V., Petropavlovskikh, I., Meng, X-L. *et al.* (2006). Examination of ozonesonde data for trends and trend changes incorporating solar and Arctic oscillation signals, *Journal of Geophysical Research*, **111**, D13305, doi: 10.1029/2005JD006684.

Miller, R.L., Schmidt, G.A., and Shindell, D.T. (2006). Forced annular variations in the 20th century Intergovernmental Panel on Climate Change Fourth Assessment Report models, *Journal of Geophysical Research*, **111**, D18101, doi: 10.1029/2005JD006323.

Miller, W.C., Hare, G., Strain, D.C., George, K.P., Stickney, M.E., and Beckman, A.O. (1949). A new spectrophotometer employing a glass Fery prism, *Journal of the Optical Society of America*, **39**, 377–386.

Mims, F.M. and Mims, E.R. (1993). Fluctuations in column ozone during the total solar eclipse of July 11, 1991, *Geophysical Research Letters*, **20**, 367–370.

Miyagawa, K. and Sasaki, T. (2009). Vertical ozone profile by Umkehr measurements at Syowa Station, *International Journal of Remote Sensing*, **30**, 4043–4053.

Miyagawa, K., Sasaki, T., Nakane, H., Petropavlovskikh, I., and Evans, R.D. (2009). Reevaluation of long-term Umkehr data and ozone profiles at Japanese stations, *Journal of Geophysical Research*, **114**, D07108.

Mlch, P. and Halenka, T. (1998). On the intercomparison of the laminar structures in ozone profiles over Europe and Canada, in: R.D. Bojkov and G. Visconti (Eds.), *Proceedings of the XVIII Quadrennial Ozone Symposium, L'Aquila, Italy, September 12–21, 1996*, Consorzio Parco Scientifico e Tecnologico d'Abruzzo, L'Aquila, Italy, Vol. 1, pp. 151–154.

Mlynczak, M.G. (1997). Energetics of the mesosphere and lower thermosphere and the Saber Experiment, *Advances in Space Research*, **20**, 1177–1183.

Mlynczak, M.G., Garcia, R.R., Roble, R.G., and Hagan, M. (2000). Solar energy deposition rates in the mesosphere derived from airglow measurements: Implications for the ozone model deficit problem, *Journal of Geophysical Research*, **105**, 17525–17538.

Molemaker, M.J. and de Arellano, J.V.-G. (1998). Control of chemical reactions by convective turbulence in the boundary layer, *Journal of the Atmospheric Sciences*, **55**, 568–579.

Molina, L.T. and Molina, M.J. (1987). Production of the Cl_2O_2 from the self-reaction of the ClO radical, *Journal of Physical Chemistry*, **91**, 433–436.

Molina, M. and Rowland, F.S. (1974). Stratospheric sink for chlorofluoromethanes: Chlorine atom-catalysed destruction of ozone, *Nature*, **249**, 810–812.

Molina, L.T., Molina, M.J., Stachnick, R.A., and Tom, R.D. (1985). An upper limit to the rate of the HCl + $ClONO_2$ reaction, *Journal of Physical Chemistry*, **89**, 3779–3781.

Molina, M., Zaelke, D., Sarma, K.M., Andersen, S.O., Ramanathan, V., and Kaniaru, D. (2009). Reducing abrupt climate change risk using the Montreal Protocol and other regulatory actions to complement cuts in CO_2 emissions, *Proceedings of the National Academy of Sciences of the U.S.A.*, **106**, 20616–20621.

Monod, A. and Carlier, P. (1999). Impact of clouds on the tropospheric ozone budget: Direct effect of multiphase photochemistry of soluble organic compounds, *Atmospheric Environment*, **33**, 4431–4446.

Montzka, S.A., Butler, J.H., Myers, R.C., Thompson, T.M., Swanson, T.H., Clarke, A.D., Lock, L.T., and Elkins, J.W. (1996). Decline in the tropospheric abundance of halogen from halocarbons: Implications for stratospheric ozone depletion, *Science*, **272**, 1318–1322.

Montzka, S.A., Butler, J.H., Hall, B.D., Mondeel, D.J., and Elkins, J.W. (2003a). A decline in tropospheric organic bromine, *Geophysical Research Letters*, **30**, 1826, doi: 10.1029/2003GL017745.

Montzka, S.A., Fraser, P., Butler, J.H., Connell, P.S., Cunnold, D.M., Daniel, J.S., Derwent, R.G., Lal, S., McCulloch, A., Oram, D.E. *et al.* (2003b). Controlled substances and other source gases, *Scientific Assessment of Ozone Depletion: 2002* (WMO Global Ozone Research and Monitoring Project Report No. 47), WMO, Geneva, Switzerland, 87 pp.

Moody, J.L., Munger, J.W., Goldstein, A.H., Jacob, D.S., and Wofsy, S.C. (1998). Harvard forest regional-scale air mass composition by Patterns in Atmospheric Transport History (PATH), *Journal of Geophysical Research*, **103**, 13181–13194.

Moreau, G., Robert, C., and Rigaud, P. (1989). A new multipass ozonometer, in: R.D. Bojkov and P. Fabian (Eds.), *Proceedings of the Quadrennial Ozone Symposium, 1988, Göttingen, Germany*, Deepak Publications, Hampton, VA, pp. 755–758.

Moreau, G., Robert, C., Catoire, V., Chartier, M., Camy-Peyret, C., Huret, N., Pirre, M., Pomathiod, L., and Chalumeau, G. (2005). SPIRALE: A multispecies in situ balloon-borne instrument with six tunable diode laser spectrometers, *Applied Optics*, **44**, 5972– 5989.

Morris, G.A., Schoeberl, M.R., Sparling, L.C., Newman, P.A., Lait, L.R., Elson, L., Waters, J., Suttie, R.A., Roche, A., Kumer, J. *et al.* (1995). Trajectory mapping and applications to data from the Upper Atmosphere Research Satellite, *Journal of Geophysical Research*, **100**, 16491–16505.

Morris, G.A., Kawa, S.R., Douglas, A.R., Schoeberg, M.R., Froidevaux, L., and Waters, J. (1998). Low-ozone pockets explained, *Journal of Geophysical Research*, **103**, 3595–3610.

Morris, G.A., Gleason, J.F., Russel, J.M. III, Schoeberl, M.R., and McCormick, M.P. (2002). A comparison of HALOE v19 with SAGE II v6.00 ozone observations using trajectory mapping, *Journal of Geophysical Research*, **107**, 4177–4186.

Morys, M., Mims, F.M. III, and Anderson, S.E. (1996). Design, calibration and performance of MICROTOPS II hand-held ozonometer, *12th International Symposium on Photobiology*, Solar Light Co., Vienna.

Mote, P.W., Rosenlof, K.H., McIntyre, M.E., Carr, E.S., Gille, J.C., Holton, J.R., Pumphrey, H.C., Russell, J.M. III, and Waters, J.W. (1996). An atmospheric tape recorder: The imprint of tropical tropopause temperatures on stratospheric water vapor, *Journal of Geophysical Research*, **101**, 3989–4006.

Moulik, M.D. and Milford, J.B. (1999). Factors influencing ozone chemistry in subsonic aircraft plumes, *Atmospheric Environment*, **33**, 869–880.

Mount, G.H., Sanders, R.W., Schmeltekopf, A.L., and Solomon, S. (1987). Visible spectroscopy at McMurdo Station, Antarctica, 1: Overview and daily variations of NO_2 and O_3, austral spring, *Journal of Geophysical Research*, **92**, 8320–8328.

Munro, R., Siddans, R., Reburn, W.J., and Kerridge, B.J. (1998). Direct measurements of tropospheric ozone from space, *Nature*, **392**, 168–171.

Murata, I., Sato, K., Okano, S., and Tomikawa, Y. (2009). Measurements of stratospheric ozone with a balloon-borne optical ozone sensor, *International Journal of Remote Sensing*, **30**, 3961–3966.

Murtagh, D., Frisk, U., Merino, F., Ridal, M., Jonsson, A., Stegman, J., Eriksson, G.W.P., Jimenez, C., Megie, G., de la Noee, J. *et al.* (2002). An overview of the Odin atmospheric mission, *Canadian Journal of Physics*, **80**, 309–319.

Myhre, G. and Stordal, F. (1997). Role of spatial and temporal variations in the computation of radiative forcing and GWP, *Journal of Geophysical Research*, **102**, 11181–11200.

Myhre, G., Stordal, F., Rognerud, B., and Isaksen, I.S.A. (1998). Radiative forcing due to stratospheric ozone, in: R.D. Bojkov and G. Visconti (Eds.), *Atmospheric Ozone: Proceedings of the XVIII Quadrennial Ozone Symposium, L'Aquila, Italy, September 12–21, 1996*, Consorzio Parco Scientifico e Tecnologico d'Abruzzo, L'Aquila, Italy, Vol. 2, 813–816.

Myhre, G, Shine, K.P., Rädel, G., Gauss, M., Isaksen, I.S.A., Tang, Q., Prather, M.J., Williams, J.E., van Velthoven, P., Dessens, O. *et al.* (2011). Radiative forcing due to changes in ozone and methane caused by the transport sector, *Atmospheric Environment*, **45**, 387–394.

Mze, N., Hauchecorne, A., Bencherif, H., Dalaudier, F., and Bertaux, J.-L. (2010). Climatology and comparison of ozone from ENVISAT/GOMOS and SHADOZ/balloon-sonde observations in the southern tropics, *Atmospheric Chemistry and Physics*, **10**, 8025–8035.

Nair, H., Allen, M., Froidevaux, L., and Zurek, R.W. (1998). Localized rapid ozone loss in the northern winter stratosphere: An analysis of UARS observations, *Journal of Geophysical Research*, **103**, 1555–1571.

Nair, P.J., Godin-Beekmann, S., Pazmiño, A., Hauchecorne, A., Ancellet, G., Petropavlovskikh, I., Flynn, L.E., and Froidevaux, L. (2010). Coherence of long-term stratospheric ozone vertical distribution time series used for the study of ozone recovery at a northern mid-latitude station, *Atmospheric Chemistry and Physics Discussions*, **10**, 28519–28564.

Nair, P.J., Godin-Beekmann, S., Pazmiño, A., Hauchecorne, A., Ancellet, G., Petropavlovskikh, I., Flynn, L.E., and Froidevaux, L. (2011). Coherence of long-term stratospheric ozone vertical distribution time series used for the study of ozone recovery at a northern mid-latitude station, *Atmospheric Chemistry and Physics*, **11**, 4957–4975.

Nakane, H., Sasano, Y., Hayashida-Amano, S., Sugimoto, N., Matsui, I., and Minato, A. (1992a). Comparison of ozone profiles obtained with NIES DIAL and SAGE II measurements. *Journal of Meteorological Society of Japan*, **71**, 153–159.

Nakane, H., Hayashida, S., Sasano, Y., Sugimoto, N., Matsui, I., and Minato, A. (1992b). Vertical profiles of temperature and ozone observed during DYANA campaign with NIES ozone lidar system at Tsukuba, *Journal of Geomagnetism and Geoelectricity*, **44**, 1071–1083.

Nambu, Y. (1973). Generalized Hamiltonian dynamics, *Physical Review*, **D7**, 2405–2412.

NASA/NOAA (2002). Ozone Hole Press Release 02-185, September 30, 2002 (*http://toms.gsfc.nasa.gov/*).

Nash, E.R., Newman, P.A., Rosenfield, J.E., and Schoeberl, M.R. (1996). An objective determination of the polar vortex using Ertel's potential vorticity, *Journal of Geophysical Research*, **101**, 9471–9478.

Nehme, R. (2006). Diploma-thesis University of Duisburg-Essen; P. Behr, R. Nehme, R. Zellner, in preparation.

Neuendorffer, A.C. (1996). Ozone monitoring with TIROS-N operational vertical sounders, *Journal of Geophysical Research*, **101**, 18807–18828.

Névir, P. (1998). *Die Nambu-Felddarstellung der Hydro-Thermodynamik und ihre Bedeutung für die dynamische Meteorologie* (Habilitatiosschrift FB Geowissenschaften), Freie Universität, Berlin.

Névir, P. and Blender, R. (1993). A Nambu representation of incompressible hydrodynamics using helicity and enstrophy, *Journal of Physics*, **26A**, L1189–L1193.

Névir, P. and Sommer, M. (2009a). A conservative scheme for the shallow-water system on a staggered geodesic grid based on a Nambu representation, *Quarterly Journal of the Royal Meteorological Society*, **135**, 485–494.

Névir, P. and Sommer, M. (2009b). Energy–vorticity theory of ideal fluid mechanics, *Journal of the Atmospheric Sciences*, **66**, 2073–2084.

Newchurch, M.J., Yang, E.S., Cunnold, D.M., Reinsel, G.C., Zawodny, J.M., and Russell, J.M. III (2003). Evidence for slowdown in stratospheric ozone loss: First stage of ozone recovery. *Journal of Geophysical Research*, **108**, 4507, doi: 10.1029/2003JD003471.

Newell, R.E., Browell, E.V., Davis, D.D., and Liu, Sh.E. (1997). Western Pacific tropospheric ozone and potential vorticity: Implications for Asian pollution, *Geophysical Research Letters*, **24**, 2733–2736.

Newman, P.A. and Nash, E.P. (2005). The unusual Southern Hemisphere stratosphere winter of 2002, *Journal of the Atmospheric Sciences*, **62**, 614–628.

Newman, P.A. and Rex, M. (2007). *Scientific Assessment of Ozone Depletion: 2006* (Global Ozone Research and Monitoring Project, Report No. 50), World Meteorological Organization, Geneva, Switzerland.

Newman, P.A. and Schoeberl, M.R. (1986). October Antarctic temperature and total ozone trends form 1979–1985, *Geophysical Research Letters*, **13**, 1206–1209.

Newman, P.A., Rex, M., Austin, J., Braathen, G.O., Canziani, P.O., Carslaw, K.S., Forster, P.M. de F., Godin-Beekmann, S., Knudsen, B.M., Kreher, K. *et al.* (2002). Polar stratospheric ozone: Past and future, *Scientific Assessment of Ozone Depletion: 2002* (WMO Global Ozone Research and Monitoring Project Report No. 47), WMO, Geneva, Switzerland, 104 pp.

Newman, P.A., Kawa, S.R., and Nash, E.R. (2004). On the size of the Antarctic ozone hole, *Geophysical Research Letters*, **31**, L21104, doi: 10.1029/2004GL020596.

Newman, P.A., Nash, E.R., Kawa, S.R, Montzka, S.A., and Schauffler, S.M. (2006). When will the Antarctic ozone hole recover? *Geophysical Research Letters*, **33**, L12814, doi: 10.1029/2005GL025232.

Newman, P.A., Daniel, J.S., Waugh, D.W., and Nash, E.R. (2007a). A new formulation of equivalent effective stratospheric chlorine (EESC), *Atmospheric Chemistry and Physics*, **7**, 4537–4552.

Newman, P.A., Rex, M., Canziani, P.O., Carslaw, K.S., Drdla, K., Godin-Beekmann, S., Golden, D.M., Jackman, C.H., Kreher, K., Langematz, U. *et al.* (2007b). Polar ozone: Past and present, *Scientific Assessment of Ozone Depletion: 2006* (WMO Global Ozone Research and Monitoring Project Report No. 50), WMO, Geneva, Switzerland, 572 pp.

Newman, P.A., Oman, L.D., Douglass, A.R., Fleming, E.L., Frith, S.M., Hurwitz, M.M., Kawa, S.R., Jackman, C.H., Krotkov, N.A., Nash, E.R. *et al.* (2008). What would have happened to the ozone layer if chlorofluorocarbons (CFCs) had not been regulated? *Atmospheric Chemistry and Physics Discussions*, **8**, 20565–20606.

Newman, P.A., Oman, L.D., Douglass, A.R., Fleming, E.L., Frith, S.M., Hurwitz, M.M., Kawa, S.R., Jackman, C.H., Krotkov, N.A., Nash, E.R. *et al.* (2009). What would have happened to the ozone layer if chlorofluorocarbons (CFCs) had not been regulated? *Atmospheric Chemistry and Physics*, **9**, 2113–2128.

Nicoli, M.P. and Visconti, G. (1982). Impact of coupled perturbations of atmospheric trace gases on world's climate and ozone, *Pure and Applied Geophysics*, **120**, 626–641.

NOAA (2001). *NOAA Operational Sounding Products from Advanced-TOVS Polar Orbiting Environmental Satellites* (Technical Report NESDIS 102), Washington, D.C., 63 pp.

NOAA (2009). *KLM User's Guide: Satellite and Data Description of NOAA's Polar-orbiting Satellites from NOAA-15 and Later* (February 2009 Revision), National Oceanic and Atmospheric Administration, Silver Springs, MD.

Noel, S., Bramstedt, K., Rozanov, A., Bovensmann, H., and Burrows, J.P. (2010). Water vapour profiles from SCIAMACHY solar occultation measurements derived with an onion peeling approach, *Atmospheric Measurement Techniques*, **3**, 523–535.

Noxon, J.F., Whipple, E.C. Jr., and Hyde, R.S. (1979). Stratospheric NO_2, 1: Observation method and behavior at mid-latitude, *Journal of Geophysical Research*, **84**, 5047–5065.

Numaguti, A. (1993). Dynamics and energy balance of the Hadley circulation and the tropical precipitation zones: Significance of the distribution of evaporation, *Journal of the Atmospheric Sciences*, **50**, 1874–1887.

Oelhaf, H., Friedl-Vallon, F., Kleinert, A., Lengel, A., Maucher, G., Nordmeyer, H., Wetzel, G., Zhang, G., and Fischer, H. (2003). ENVISAT validation with MIPAS-B, in: *Proceedings of ENVISAT Validation Workshop, December 9–13, 2002, ESRIN, Frascati, Italy* (CD-ROM, vol. SP-531), ESA Publications Division, ESTEC, Noordwijk, The Netherlands.

Ogawa, T. and Komala, N. (1989). Diurnal and seasonal variations of the tropospheric ozone in tropical Asia, in: R.D. Bojkov and P. Fabian (Eds.), *Proceedings of the Quadrennial Ozone Symposium, 1988, Göttingen, Germany*, Deepak Publications, Hampton, VA, pp. 437–440.

Ogawa, T., Koike, M., and Suzuki, K. (1989). Observation of vertical ozone profiles with the EXOS-C backscattered UV spectrometer, in: R.D. Bojkov and P. Fabian (Eds.), *Proceedings of the Quadrennial Ozone Symposium, 1988, Göttingen, Germany*, Deepak Publications, Hampton, VA, pp. 214–217.

Okano, S., Okabayashi, M., and Gernandt, H. (1996). Observations of ozone profiles in the upper stratosphere using a UV sensor on board a light-weight high-altitude balloon, *Memoirs of National Institute of Polar Research*, Special Issue, **51**, 225–231.

Olafson, R.A. and Asbridge, I.A. (1981). Stray light in the Dobson spectrophotometer and its effect on ozone measurements, in J. London (Ed.), *Proceedings of the Quadrennial International Ozone Symposium*, International Association of Meteorology and Atmospheric Physics (IAMAP), Boulder, CO, pp. 46–47.

Olsen, M.A., Douglass, A.R., and Schoeberl, M.R. (2003). A comparison of Northern and Southern Hemisphere cross-tropopause ozone flux, *Geophysical Research Letters*, **30**, 1412, doi: 10.1029/2002GL016538.

Olsen, M.A., Douglass, A.R., Gille, J.C., Nardi, B., Yudin, V.A., Kinnison, D.E., and Khosravi, R. (2008). HIRDLS observations and simulation of a lower stratospheric intrusion of tropical air to high latitudes, *Geophysical Research Letters*, **35**, L21813, doi: 10.1029/2008GL035514.

Oltmans, S.J., Lebone, A.S., Scheel, H.E., Harris, J.M., Levy, H. II, Galbally, I.E., Brunke, E.-G., Meyer, C.P., Lathorp, J.A., Johnson, B.S. *et al.* (1998). Trends of ozone in the troposphere, *Geophysical Research Letters*, **25**, 143–146.

OMI (2009). *Ozone Monitoring Instrument (OMI) Data User's Guide*, OMI Team (*http://disc.sci.gsfc.nasa.gov/Aura/additional/documentation/README.OMI_DUG.pdf*).

Orphal, J. (2003). A critical review of the absorption cross-sections of O_3 and NO_2 in the ultraviolet and visible, *Journal of Photochemistry and Photobiology A: Chemistry*, **157**, 185–209.

Orsolini, Y.J., Cariolle, D., and Deque, M. (1995). Filamentation and layering of an idealised tracer by observed winds in the lower stratosphere, *Geophysical Research Letters*, **22**, 839–842.

Osprey, S., Barnett, J., Smith, J., Adamson, P., Andreopoulos, C., Arms, K.E., Armstrong, R., Auty, D.J., Ayres, D.S., Baller, B. *et al.* (2009). Sudden stratospheric warmings seen in MINOS deep underground muon data, *Geophysical Research Letters*, **36**, L05809, doi: 10.1029/2008GL036359.

Osterman, G.B. (1999). The budget and partitioning of reactive nitrogen species in the Arctic stratosphere, *Geophysical Research Letters*, **26**, 1157–1160.

Osterman, G.B., Kulawik, S.S., Worden, H.M., Richards, A.D., Fisher, B.M., Eldering, A., Shephard, M.W., Froidevaux, L., Labow, G., Luo, M. *et al.* (2008). Validation of Tropospheric Emission Spectrometer (TES) measurements of the total, stratospheric and tropospheric column abundance of ozone, *Journal of Geophysical Research*, **113**, DI5S16, doi: 10.1029/2007JD008801.

Ozone Depletion, Greenhouse Gases, and Climate Change (1989). National Academy Press, Washington, D.C., 136 pp.

Palchetti L., Bianchini, G., Pellegrini, M., Esposito, F., Restieri, R., and Pavese, G. (2004). Radiometric performances of the Fourier transform spectrometer for the radiation explorer in the far infrared (REFIR) space mission, in: R. Meynart, S.P. Neeck, and H. Shimoda (Eds.), *Sensor, Systems, and Next Generation Satellites, VIII: Proceedings of the Society of Photo-Optical Instrumentation Engineers*, **5570**.

Pan, L., Solomon, S., Randel, W., Lamarque, J.F., Hess, P., Gille, J., Chiou, E.W., and McCormick, M.P. (1997). Hemispheric asymmetries and seasonal variations of the lowermost stratospheric water vapor and ozone derived from SAGE II data, *Journal of Geophysical Research*, **102**, 28177-28184.

Pan, L.L., Bowman K.P., Atlas, E.L., Wofsy, S.C., Zhang, F.Q., Bresch, J.F., Ridley, B.A., Pittman, J.V., Homeyer, C.R., Romashkin, P. *et al.* (2010). The Stratosphere–Troposphere Analyses of Regional Transport 2008 Experiment, *Bulletin of the American Meteorological Society*, **91**, 327–342.

Papandrea, E., Arnone, E., Brizzi, G., Carlotti, M., Castelli, E., Dinelli, B.M., and Ridolfi, M. (2010). Two-dimensional tomographic retrieval of MIPAS/ENVISAT measurements of ozone and related species, *International Journal of Remote Sensing*, **31**, 477–483.

Park, C.B., Nakane, H., Sugimoto, N., Matsui, I., Sasano, Y., Fujinuma, Y., Ikeuchi, I., Kurokawa, J.-I., and Furuhashi, N. (2006). Algorithm improvement and validation of National Institute for Environmental Studies ozone differential absorption lidar at the Tsukuba Network for Detection of Stratospheric Change complementary station, *Applied Optics*, **45**, 3561–3576.

Park, J.H., Ko, M.K.W., Jackman, C.H., Plumb, R.A., Kaye, J.A., and Sage, K.H. (Eds.) (1999). *Models and Measurements Intercomparison II* (NASA/TM-199-209554), NASA, Washington, D.C.

Parnes, S. (1992). Israeli students begin work on Techsat 1 satellite, *Space News*, **34**.

Parrish, A., Connor, B.J., and Tsou, J. J. (1998). Ground-based microwave measurements of ozone profiles following the Pinatubo eruption, in: R.D. Bojkov and G. Visconti (Eds.), *Atmospheric Ozone: Proceedings of the XVIII Quadrennial Ozone Symposium, L'Aquila, Italy, September 12–21, 1996*, Consorzio Parco Scientifico e Tecnologico d'Abruzzo, L'Aquila, Italy, Vol. 1, pp. 155–158.

Parrish, D.D., Trainer, M., Holloway, J.S., Yee, J.E., Warshawsky, M.S., Fehsenfeld, F.C., Forbes, G.L., and Moody, J.L. (1998). Relationships between ozone and carbon monoxide at surface sites in the North Atlantic region, *Journal of Geophysical Research*, **103**, 1357–1376.

Patil, S.D. and Revadekar, J.V. (2008). Extremes in total ozone content over the northern India, *International Journal of Remote Sensing*, **30**, 2389–2397.

Paul, J., Fortuin, J.P.F., and Kelder, H. (1998). An ozone climatology based on ozonesonde and satellite measurements, *Journal of Geophysical Research*, **103**, 31709–31734.

Pawson, S., Stolarski, R.S., Douglass, A.R., Newman, P.A., Nielsen, J.E., Frith, S.M., and Gupta, M.L. (2008). Goddard Earth Observing System chemistry–climate model simulations of stratospheric ozone-temperature coupling between 1950 and 2005, *Journal of Geophysical Research*, **113**, D12103.

Pazmino, A.F., Godin-Beckmann, S., Ginzburg, M., Bekki, S., Hauchecorne, A., Piacentini, R.D., and Quel, E.J. (2005). Impact of Antarctic polar vortex occurrences on total ozone and UVB radiation at southern Argentinean and Antarctic stations during 1997–2003 period, *Journal of Geophysical Research*, **110**, D03103.

Pelon, J. and Megie, G. (1982). Ozone monitoring in the troposphere and lower stratosphere: Evaluation and operation of a ground-based lidar station, *Journal of Geophysical Research*, **87**, 4947–4955.

Pelon, J., Godin, S., and Megie, G. (1986). Upper stratospheric (30–50 km) lidar observations of the ozone vertical distribution, *Journal of Geophysical Research*, **91**, 8667–8671.

Peng, C.K., Buldyrev, S.V., Havlin, S., Simons, M., Stanley, H.E., and Goldberger, A.L. (1994). Mosaic organization of DNA nucleotides, *Physical Review E*, **49**, 1685–1689.

Peng, C.K., Havlin, S., Stanley, H.E., and Goldberger, A.L. (1995). Quantification of scaling exponents and crossover phenomena in nonstationary heartbeat time series, *Chaos*, **5**, 82–87.

Penner, J.E., Lister, D.H., Griggs, D.J., Dokken, D.J., and McFarland, M. (Eds.) (1999). *Aviation and the Global Atmosphere* (a special report of IPCC WG I and III), Cambridge University Press, Cambridge, U.K., 373 pp.

Perez, P. and Trier, A. (2001). Prediction of NO and NO_2 concentrations near a street with heavy traffic in Santiago, Chile, *Atmospheric Environment*, **35**, 1783–1789.

Perliski, L. and London, J. (1989). Satellite observed long-term averaged seasonal and spatial ozone variations in the stratosphere, *Planetary and Space Science*, **37**, 1509–1525.

Perliski, L.M. and London, J. (2002). Satellite observed long-term averaged seasonal and spatial ozone variations in the stratosphere, *Planetary and Space Science*, **37**, 1509–1525.

Perlwitz, J.P. and Miller, R.L. (2010). Cloud cover increase with increasing aerosol absorptivity: A counterexample to the conventional semi-direct aerosol effect, *Journal of Geophysical Research*, **115**, 1–23.

Perlwitz, J., Graf, H.-F., and Kirchner, I. (1997). Increasing greenhouse effect and ozone trends, *Proceedings of the First SPARC General Assembly* (WCRP-99, WMO/TD-N814), WMO, Geneva, Switzerland, Vol. 2, pp. 461–464.

Perlwitz, J., Pawson, S., Fogt, R.L., Nielsen, J.E., and Neff, W.D. (2008). Impact of stratospheric ozone hole recovery on Antarctic climate, *Geophysical Research Letters*, **35**, L08714, doi: 10.1029/2008GL033317.

Perov, S.P. and Tishin, S.V. (1989). Diurnal variations of ozone in the mesosphere, in: R.D. Bojkov and P. Fabian (Eds.), *Proceedings of the Quadrennial Ozone Symposium, 1988, Göttingen, Germany*, Deepak Publications, Hampton, VA, pp. 399–402.

Pesin, Y.B. (1977). Characteristic Lyapunov exponents and smooth ergodic theory, *Russian Mathematical Surveys*, **32**, 55–114.

Petelina, S.V., Llewellyn, E.J., Walker, K.A., Degenstein, D.A., Boone, C.D., Bernath, P.F., Haley, C.S., von Savigny, C., Lloyd, N.D., and Gattinger, R.L. (2005). Validation of ACE-FTS stratospheric ozone profiles against Odin/OSIRIS measurements, *Geophysical Research Letters*, **32**, L15S06, doi: 10.1029/2005GL022377.

Peter, R., Caliseri, Y., and Kampfer, N. (1998). Variability of middle atmospheric ozone abundances derived from continuous ground-based millimeter wave measurements, in: R.D. Bojkov and G. Visconti (Eds.), *Proceedings of the XVIII Quadrennial Ozone Symposium, L'Aquila, Italy, September 12–21, 1996*, Consorzio Parco Scientifico e Tecnologico d'Abruzzo, L'Aquila, Italy, Vol. 2, pp. 559–562.

Petrenko, V.F. and Whitworth, R.W. (1999). *Physics of Ice*, Oxford University Press, New York.

Petropavlovskikh, I., Ahn, C., Bhartia, P.K., and Flynn, L.E. (2005a). Comparison and covalidation of ozone anomalies and variability observed in SBUV(/2) and Umkehr northern midlatitude ozone profile estimates, *Geophysical Research Letters*, **32**, L06805.

Petropavloskikh, I., Bhartia, P.K., and Deluisi, J.J. (2005b), New Umkehr ozone profile retrieval algorithm optimized for climatological studies, *Geophysical Research Letters*, **32**, L16808.

Petropavlovskikh, I. , Evans, R., McConville, G., Miyagawa, K., and Oltmans, S. (2009). Effect of the out-of-band stray light on the retrieval of the Umkehr Dobson ozone profiles, *International Journal of Remote Sensing*, **30**, 6461–6482.

Petropavlovskikh, I., McConville, G., Evans, R., Oltmans, S., Quincy, D., Lantz, K., and Disterhoft, P. (2010). Sensitivity of Dobson and Brewer Umkehr ozone profile retrievals to the choice of the ozone cross section, *March 23–25, 2010 Second WMO O_3 X-Section Meeting* (*http://igaco-o3.fmi.fi/ACSO/presentations_2010/ground-based/OTM_2010_Petropavlovskikh.pdf*).

Pfeilsticker, K., Sturges, W., Bösch, H., Camy-Peyret, C., Chipperfield, M., Engel, A., Fitzenberger, R., Müller, M., Payan, S., and Sinnhuber, B.-M. (2000). Lower stratospheric organic and inorganic bromine budget for the Arctic winter 1998/99, *Geophysical Research Letters*, **27**, 3305–3308.

Phillips, C. (Ed.) (1997). *Stratospheric Processes and Their Role in Climate* (a WCRP project), SPARC Office, Verrières-le-Buisson, France, 8 pp.

Pickett, H.M., Drouin, B.J., Canty, T., Salawitch, R.J., Fuller, R.A., Perun, V.S., Livesey, N.J., Waters, J.W., Stachnik, R.A., Sander, S.P. *et al.* (2008), Validation of Aura Microwave Limb Sounder OH and HO_2 measurements, *Journal of Geophysical Research*, **113**, D16S30, doi: 10.1029/2007JD008775.

Pierce, R.B., Fairlie, T.D., Remsberg, E.E., Russel, J.M. III, and Grose, W.L. (1997). HALOE observations of the Arctic vortex during the 1997 spring: Horizontal structure in the lower stratosphere, *Geophysical Research Letters*, **24**, 2705–2708.

Piesch, C., Gulde, T., Sartorius, C., Friedl-Vallon, F., Seefeldner, M.,Wölfel, M., Blom, C.E., and Fischer, H. (1996). Design of a MIPAS instrument for high-altitude aircraft, *Proceedings of the Second International Airborne Remote Sensing Conference and Exhibition*, Environmental Research Institute of Michigan, Ann Arbor, MI, Vol. II, pp. 199–208.

Pirre, M., Pisso, I., Marécal, V., Catoire, V., Mébarki, Y., and Robert, C. (2008). Intrusion of recent air in midlatitude stratosphere revealed by in situ tracer measurements and trajectory calculations, *Journal of Geophysical Research*, **113**, D11302, doi: 10.1029/2007JD009188.

Pitari, G., Palermi, S., Visconti, G., and Prinn, R.G. (1992). Ozone response to a CO_2 doubling: Results from a stratospheric circulation model with heterogeneous chemistry, *Journal of Geophysical Research*, **97**, 5953–5962.

Pitts, M.C., Thomason, L.W., Zawodny, J.M., Wenny, B.N., Livingston, J.M., Russell, P.B., Yee, J.-H., Swartz, W.H., and Shetter, R.E. (2005). Ozone observations by the Gas and Aerosol Measurement Sensor during SOLVE II, *Atmospheric Chemistry and Physics Discussions*, **5**, 9953–9992.

Platt, U. and Stutz, J. (2008). *Differential Optical Absorption Spectroscopy: Principles and Applications*, Springer-Verlag, Heidelberg, 597 pp.

Polvani, M.L., Waugh, W.D., Correa, J.P.G., and Seok-Woo Son (2010). Stratospheric ozone depletion: The main driver of 20th century atmospheric circulation changes in the Southern Hemisphere, *Journal of Climate*, **24**, 795–812, doi: 10.1175/2010JCLI3772.1.

Pommereau, J.P. (2009). The European Arctic Ozone Campaigns, in: C. Zerefos, G. Contopoulos, and G. Skalkeas (Eds.), *Proceedings of the Symposium for the 20th Anniversary of the Montreal Protocol*, Springer-Verlag, Berlin, pp. 201–212.

Pommereau, J.P. and Goutail, F. (1988a). O_3 and NO_2 ground based measurements by visible spectrometry during Arctic winter and spring 1988, *Geophysical Research Letters*, **15**, 891–895.

Pommereau, J.P. and Goutail, F. (1988b). Stratospheric O_3 and NO_2 observations at the Southern Polar Circle in summer and fall 1988, *Geophysical Research Letters*, **15**, 895–897.

Pommereau, J.P. and Piquard, J. (1994a). Observations of the vertical distribution of stratospheric OClO, *Geophysical Research Letters*, **21**, 1231–1234.

Pommereau, J.P. and Piquard, J. (1994b). Ozone and nitrogen dioxide vertical distributions by UV–visible solar occultation from balloons, *Geophysical Research Letters*, **21**, 1227–1230.

Pommereau, J.P. and Schmidt, U. (1991). CHEOPS III: An ozone research programme in the Arctic winter 1989/90, *Geophysical Research Letters*, **18**, 759–762.

Ponater, M., Sausen, R., Feneberg, B., and Roeckner, E. (1999). Climate effect of ozone changes caused by present and future air traffic, *Climate Dynamics*, **15**, 631–642.

Pope, F.D., Hansen, J.C., Bayes, K.D., Friedl, R.R., and Sander, S.P. (2007). Ultraviolet absorption spectrum of chlorine peroxide, ClOOCl, *Journal of Physical Chemistry*, **111**, 4322–4332.

Pope, V., Gallani, M., Rowntree, P., and Stratton, R. (2000). The impact of new physical parametrizations in the Hadley Centre climate model HadCM3, *Climate Dynamics*, **16**, 123–146.

Portmann, R.W., Solomon, S., Fishman, J., Olson, J.R., Kiehl, J.T., and Briegleb, B. (1997). Radiative forcing of the Earth's climate system due to tropical tropospheric ozone production, *Journal of Geophysical Research*, **102**, 9409–9417.

Pougatchev, N.S., Connor, B.J., and Rinsland, C.P. (1995). Infrared measurements of the ozone vertical distribution above Kitt Peak, *Journal of Geophysical Research*, **100**, 16689–16698.

Pougatchev, N.S., Connor, B.J., Jones, N.B., and Rinsland, C.P. (1998). Ground-based infrared measurements of the ozone vertical distribution in the troposphere and lower stratosphere, in: R.D. Bojkov and G. Visconti (Eds.), *Proceedings of the XVIII Quadrennial Ozone Symposium*, Vol. 1, pp. 167–170.

Prather, M.J., Midgley, P., Rowland, F.S., and Stolarski, R. (1996). The ozone layer: The road not taken, *Nature*, **381**, 551–554.

Prather, M.J. (1998). Time scales in atmospheric chemistry: Coupled perturbations to N_2O, NO_y and O_3, *Science*, **279**, 1339–1342.

Prather, M.J. (2002). Lifetimes of atmospheric species: Integrating environmental impacts, *Geophysical Research Letters*, **29**, 2063, doi: 10.1029/2002GL016299.

Prather, M.J. and Remsberg, E.E. (Eds.) (1993). *The Atmospheric Effects of Stratospheric Aircraft: Report of the 1992 Models and Measurements Workshop* (NASA RP-1292). NASA, Washington, D.C., Vols. 1–3, 764 pp.

Proffitt, M.H. and McLaughlin, R.J. (1983). Fast response dual-beam UV-absorption ozone photometer suitable for use on stratospheric balloons, *Review of Scientific Instruments*, **54**, 1719–1728.

Pszenny, A. and Brasseur G. (1997). Tropospheric ozone: An emphasis on IGAC research, *Global Change Newsletter*, **30**, 2–10.

Pundt, I. and Pommereau, J.-P. (1998). Investigation of stratospheric bromine and iodine oxides using the SAOZ balloon sonde, in: R.D. Bojkov and G. Visconti (Eds.), *Proceedings of the XVIII Quadrennial Ozone Symposium, L'Aquila, Italy, September 12–21, 1996*, Consorzio Parco Scientifico e Tecnologico d'Abruzzo, L'Aquila, Italy, Vol. 2, pp. 575–578.

Pyle, J.A., Shepherd, T., Bodeker, G.E., Canziani, P., Dameris, M., Forster, P.M., Gruzdev, A., Muller, R., Muthama, N., Pitari, G. *et al.* (2005). *Ozone and Climate: A Review of Interconnections* (special report on safeguarding the ozone layer and global climate system, IPCC/TEAP), Cambridge University Press, Cambridge, U.K., pp. 83-132.

Pyle, J.A., Harris, N.R.P., Farman, J.C., Arnold, F., Braathen, G., Cox Faucon, R.A., Jones, R.L., Megie, G., O'Neill, A., Platt, U. *et al.* (1994). An overview of the EASOE campaign, *Geophysical Research Letters*, **21**, 1191–1194.

Raes, F., Liao, H., Chen, W.T., and Seinfeld, J.H. (2010). Atmospheric chemistry–climate feedbacks, *Journal of Geophysical Research*, **115**, D12121, doi: 10.1029/2009JD013300.

Rafanelli, C., De Simone, S., Damiani, A., Lund Myhre, C., Edvardsen, K., Svenoe, T., and Benedetti, E. (2009). Stratospheric ozone during the arctic winter: Brewer measurements in Ny-Ålesund, *International Journal of Remote Sensing*, **30**, 4319–4330.

Ramanathan, V. and Dickinson, R.E. (1979). The role of stratospheric ozone in the zonal and seasonal radiative energy balance of the earth–troposphere system, *Journal of the Atmospheric Sciences*, **36**, 1084–1104.

Ramanathan, V., Callis, L.B., and Boughner, R.E. (1976). Sensitivity of surface temperature and atmospheric temperature to perturbations in the stratospheric concentration of ozone and nitrogen dioxide, *Journal of the Atmospheric Sciences*, **33**, 1092–1112.

Ramanathan, V., Cicerone, R.J., Singh, H.B., and Kiehl, J.T. (1985). Trace gas trends and their potential role in climate change, *Journal of Geophysical Research*, **90**, 5547–5566.

Ramaswamy, V. and Bower, M.M. (1994). Effects of changes in radiatively active species upon the lower stratospheric temperatures, *Journal of Geophysical Research*, **99**, 18909–18921.

Ramaswamy, V., Schwarzkopf, M.D., and Randel, W.J. (1996). Fingerprints of ozone depletion in the spatial and temporal pattern of recent lower-stratospheric cooling, *Nature*, **382**, 616–618.

Randall, C.E., Rusch, D.W., Bevilacqua, R.M., Lumpe, J., Ainsworth, T.L., Debrestian, D., Fromm, M., Krigman, S.S., Hornstein, J.S., Shettle, E.P. *et al.* (1995). Preliminary results from POAM-II: Stratospheric ozone at high northern latitudes, *Geophysical Research Letters*, **22**, 2733–2736.

Randall, C.E., Rusch, D.W., Bevilacqua, R.M., Hoppel, K.W., Lumpe, J.D., Shettle, E., Thompson, E., Deaver, L., Zawodny, J., Kyro, E. *et al.* (2003). Validation of POAM III ozone: Comparisons with ozonesonde and satellite data, *Journal of Geophysical Research*, **108**, 6–17.

Randel, W.J. and Cobb, J.B. (1994). Coherent variations of monthly mean total ozone at low stratospheric temperature, *Journal of Geophysical Research*, **99**, 5433–5447.

Randel, W.J. and Wu. F. (1996). Isolation of the ozone QBO in SAGE II data by singular value decomposition, *Journal of the Atmospheric Sciences*, **53**, 2546–2559.

Randel, W.J. and Wu, F. (1999). A stratospheric ozone trends data set for global modeling studies, *Geophysical Research Letters*, **26**, 3089–3092.

Randel, W.J. and Wu, F. (2007). A stratospheric ozone profile data set for 1979–2005: Variability, trends, and comparisons with column ozone data, *Journal of Geophysical Research*, **112**, D06313.

Randel, W.J., Park, M., Wu, F., and Livesey, N. (2007). A large annual cycle in ozone above the tropical tropopause linked to the Brewer–Dobson circulation, *Journal of the Atmospheric Sciences*, **64**, 4479–4488.

Randel, W.J., Shine, K.P., Austin, J., Barnett, J., Claud, C., Gillett, N.P., Keckhut, P., Langematz, U., Lin, R., Long, C. *et al.* (2009). An update of observed stratospheric temperature trends, *Journal of Geophysical Research*, **114**, D02107, doi: 10.1029/2008JD01042.

Rao, T.N., Kirkwood, S., Arvelius, J., von der Gathen, P., and Kivi, R. (2003). Climatology of UTLS ozone and the ratio of ozone and potential vorticity over northern Europe, *Journal of Geophysical Research*, **108**, 4703, doi: 10.1029/2003JD003860.

Raspollini, P., Belotti, C., Burgess, A., Carli, B., Carlotti, M., Ceccherini, S., Dinelli, B.M., Dudhia, A., Flaud, J.-M., Funke, B. *et al.* (2006). MIPAS level 2 operational analysis, *Atmospheric Chemistry and Physics*, **6**, 5605–5630.

Rathman, W., Monks, P.S., Llewellyn Jones, D., and Burrows, J.P. (1997). A preliminary comparison between TOVS and GOME level 2 ozone data, *Geophysical Research Letters*, **24**, 2191–2194.

Ravishankara, A.R., Turnipseed, A.A., Jensen, N.R., Barone, S., Mills, M., Howard, C.J., and Solomon, S. (1994). Do hydrofluorocarbons destroy stratospheric ozone? *Science*, **263**, 71–75.

Ravishankara, A.R., Hancock, G., Kawasaki, M., and Matsumi, Y. (1998). Photochemistry of ozone: Surprises and recent lessons, *Science*, **280**, 60–61.

Rayner, N.A., Parker, D.E., Horton, E.B., Folland, C.K., Alexander, L.V., Rowell, D.P., Kent, E.C., and Kaplan, A. (2003). Global analyses of sea surface temperature, sea ice, and night marine air temperature since the late nineteenth century, *Journal of Geophysical Research*, **108**, doi: 10.1029/2002JD002670.

Rayner, N.A., Brohan, P., Parker, D.E., Folland, C.K., Kennedy, J., Vanicek, M., Ansell, T., and Tett, S.F.B. (2006). Improved analyses of changes and uncertainties in sea-surface temperature measured in situ since the mid-nineteenth century: HadSST2 data set, *Journal of Climate*, **19**, 446–469.

Reburn, W.J., Siddans, R., Munro, R., and Kerridge, B.J. (1998). Upper troposphere/lower stratosphere constituent retrieval simulations for AMAS, in: R.D. Bojkov and G. Visconti (Eds.), *Proceedings of the XVIII Quadrennial Ozone Symposium, L'Aquila, Italy, September 12–21, 1996*, Consorzio Parco Scientifico e Tecnologico d'Abruzzo, L'Aquila, Italy, Vol. 2, pp. 583–586.

Redaelli, G. (1997). Lagrangian techniques for the analysis of stratospheric measurements, PhD thesis, University of L'Aquila, Italy.

Redaelli, G., Cortesi, U., Bianchini, G., Castelli, E., Dinelli, B.M., Grassi, B., Mencaraglia, F., Taddei, A., and Visconti, G. (2006). Multitechnique comparison of MIPAS O_3 measurements with correlative data obtained by FT-FIR measurements during the ENVISAT stratospheric aircraft and balloon campaign, *Proceedings of the Third Atmospheric*

Chemistry Validation of ENVISAT Workshop (ACVE-3), ESA-ESRIN, Frascati, Italy, December 4–7, 2006 (ESA SP-642), ESA European Space Research Institute (ESRIN), Frascati, Italy.

Regener, V.H. (1960). On a sensitive method for the recording of atmosphere ozone, *Journal of Geophysical Research*, **65**, 3975–3977.

Regener, V.H. (1964). Measurement of atmospheric ozone with the chemiluminescent method, *Journal of Geophysical Research*, **69**, 3795–3800.

Reichler, T., Dameris, M., and Sausen, R. (2003). Determining the tropopause height from gridded data, *Geophysical Research Letters*, **30**, 2042–2046.

Reid, S.J. and Vaughan, G. (1991). Lamination in ozone profiles in the lower stratosphere, *Quarterly Journal of the Royal Meteorological Society*, **117**, 825–844.

Reid, S.J., Vaughan, G., and Kyro, E. (1993). Occurrence of ozone laminae near the boundary of the stratospheric polar vortex, *Journal of Geophysical Research*, **98**, 8883–8890.

Reid, S.J., Vaughan, G., Mitchell, N.J., Prichard, I.T., Smit, H.J., Jorgensen, T.S., Varotsos, C., and De Backer, H. (1994). Distribution of the ozone laminae during EASOE and the possible influence of inertia–gravity waves, *Geophysical Research Letters*, **21**, 1479–1482.

Reid, S.J., Rex, M., von der Gathen, P., Floisand, I., Stordial, F., Carver, G.D., Beck, A., Reimer, E., Kruger-Carstensen, R., De Haan, L.L. *et al.* (1998). A study of ozone laminae using diabatic trajectories, contour advection and photochemical trajectory model simulations, *Journal of Atmospheric Chemistry*, **30**, 187–207.

Reinsel, G.C., Tiao, G.C., Deluisi, J.J., Basu, S., and Carriere, K. (1989). Trend analysis of aerosol-corrected UMKEHR ozone profile data through 1987, *Journal of Geophysical Research*, **94**, 16373–16386.

Reinsel, G.C., Tam, W.-K., and Ying, L.H. (1994a). Comparison of trend analyses for Umkehr data using new and previous inversion algorithms, *Geophysical Research Letters*, **21**, 1007–1010.

Reinsel, G.C., Tiao, G.C., Wuebbles, D.J., Kerr, J.B., Miller, A.J., Nagatani, R.M., Bishop, L., and Ying, L.H. (1994b). Seasonal trend analysis of published ground-based and TOMS total ozone data through 1991, *Journal of Geophysical Research*, **99**, 5449–5464.

Reinsel, G.C., Weatherhead, E.C., Tiao, G.C., Miller, A.J., Nagatani, R.M., Wuebbles, D.J., and Flynn, L.E. (2002). On detection of turnaround and recovery in trend for ozone, *Journal of Geophysical Research*, **107**, doi: 10.1029/2001JD000500.

Reinsel, G.C., Miller, A.J., Weatherhead, E.C., Flynn, L.E., Nagatani, R.M., Tiao, G.C., and Wuebbles, D.J. (2005). Trend analysis of total ozone data for turnaround and dynamical contributions, *Journal of Geophysical Research*, **110**, D16306, doi: 10.1029/2004JD004662.

Reithmeier, C. and Sausen, R. (2002). ATTILA: Atmospheric Tracer Transport in a Lagrangian Model, *Tellus B*, **54**, 278–299.

Rempel, A.W. and Wettlaufer, J.S. (2003). Isotopic diffusion in polycrystalline ice, *Journal of Glaciology*, **49**, 397-406.

Retalis A., Cartalis, C., Tombrou, M., and Varotsos, C. (1997). Atmospheric soundings in support of the definition of the tropopause region in the south-eastern Mediterranean region, in: C. Varotsos (Ed.), *Atmospheric Ozone Dynamics Observations in the Mediterranean Region* (NATO ASI Series, Vol. 1), Springer-Verlag, Berlin, pp. 281–284.

Revadekar, J.V. and Patil, S.D. (2009). Recent extremes in total ozone content over the northern parts of India in view of Montreal Protocol, *International Journal of Remote Sensing*, **30**, 3967–3974.

Rex, M., Harris, N.R.P., von der Gathen, P., Lehman, R., Braathen, G.O., Reimer, E., Beck, A., Chipperfield, M.P., Alfier, R., Allaart, M. *et al.* (1997). Prolonged stratospheric ozone loss in the 1995–96 Arctic winter, *Nature*, **389**, 835–838.

Rex, M., von der Gathen, P., Harris, N.R.P., Lucic, D., Knudsen, B.M., Braathen, G.O., Reid, S.J., De Backer, H., Claude, H., Fabian, R. *et al.* (1998). In situ measurements of stratospheric ozone depletion rates in the Arctic winter 1991/1992: A Lagrangian approach, *Journal of Geophysical Research*, **103**, 5843–5853.

Rex, M., Salawitch, R.J., Santee, M.L.,Waters, J.W., Hoppel, K., and Bevilacqua, R. (2003). On the unexplained stratospheric ozone losses during cold Arctic Januaries, *Geophysical Research Letters*, **30**, doi: 10.1029/2002GL016008.

Rex, M., Salawitch, R.J., von der Gathen, P., Harris, N.R.P., Chipperfield, M.P., and Naujokat, B. (2004). Arctic ozone loss and climate change, *Geophysical Research Letters*, **31**, LO4116, doi: 10.1029/2003GL018844.

Rex, M., Salawitch, R.J., Deckelmann, H., von der Gathen, P., Harris, N.R.P., Chipperfield, M.P., Naujokat, B., Reimer, E., Allaart, M., Andersen, S.B. *et al.* (2006). Arctic winter 2005: Implications for stratospheric ozone loss and climate change, *Geophysical Research Letters*, **33**, L23808, doi: 10.1029/2006GL026731.

Rex, M., Lehmann, R., Wohlmann, I., Kremser, S., Krueger, K., Dameris, M., and Langematz, U. (2007). Ascent rates, dehydration and vertical diffusion in the tropical tropopause region and lower stratosphere, in: *Reunion Island Symposium* (unpublished proceedings).

Ridley, B.A., Walega, S.G., Lamarque, J.-F., Grahek, F.E., Trainer, M., Hubler, G., Lin, X., and Fehsenfeld, F.C. (1998). Measurements of reactive nitrogen and ozone to 5-km altitude in June 1990 over the southeastern United States, *Journal of Geophysical Research*, **103**, 8365–8388.

Ridolfi, M., Blum, U., Carli, B., Catoire, V., Ceccherini, S., De Clercq, C., Fricke, K.H., Iarlori, M., Kerridge, B., Keckhut, P. *et al.* (2007). Geophysical validation of temperature retrieved by the ESA processor from MIPAS/ENVISAT atmospheric limb-emission measurements, *Atmospheric Chemistry and Physics*, **7**, 4459–4487.

Rind, D. and Lonergan, D. (1995). Modeled impacts of stratospheric ozone and water vapor perturbations with implications for high-speed civil transport aircraft, *Journal of Geophysical Research*, **100**, 7381–7396.

Rind, D., Shindell, D., Lonergan, D., and Balachandran, N.K. (1997). Climate change and the middle atmosphere: The doubled CO_2 climate revised, *Proceedings of the First SPARC General Assembly* (WCRP-99, WMO/TD-N814),WMO Geneva, Switzerland, Vol. 2, pp. 453-456.

Rind, D., Lean, J., Lerner, J., Lonergan, P., and Leboissitier, A. (2008). Exploring the stratospheric/tropospheric response to solar forcing, *Journal of Geophysical Research*, **113**, D24103, doi: 10.1029/2008JD010114.

Rinsland, C.P., Connor, B.J., Jones, N.B., Boyd, I., Matthews, W.A., Goldman, A., Murcray, F.J., Murcray, D.G., David, S.J., and Pougatchev, N.S. (1996). Comparison of infrared and Dobson total ozone columns measured from Lauder, New Zealand, *Geophysical Research Letters*, **23**, 1025–1028.

Rinsland, C.P., Jones, N.B., Connor, B.J., Logan, J.A., Pougatchev, N.S., Goldman, A., Murcray, F.J., Stephen, T.M., Pine, A.S., Zander, R. *et al.* (1998). Northern and southern hemisphere ground-based infrared measurements of tropospheric carbon monoxide and ethane, *Journal of Geophysical Research*, **103**, 28197–28217.

Rinsland, C.P., Salawitch, R.J., Gunson, M.R., Solomon, S., Zander, R., Mahieu, E., Goldman, A., Newchurch, M.J., Irion, F.W., and Chang, A.Y. (1999). Polar stratospheric descent of NO_y and CO and Arctic denitrification during winter 1992–1993, *Journal of Geophysical Research*, **104**, 1847–1861.

Roan, S.L. (1989). *Ozone Crisis: The 15-year Evolution of a Sudden Global Emergency*, John Wiley & Sons, New York, 270 pp.

Rodgers, C.D. (1976). Retrieval of atmospheric-temperature and composition from remote measurements of thermal-radiation, *Reviews of Geophysics*, **14**, 609–624.

Rodgers, C.D. (1990). Characterization and error analysis of profiles retrieved from remote sounding measurements, *Journal of Geophysical Research*, **95**, 5587–5595.

Rodgers, C.D. (2000). *Inverse Methods for Atmospheric Sounding: Theory and Practice*, World Scientific, Singapore, 240 pp.

Rodgers, C.D. and Connor, B.J. (2003). Intercomparison of remote sounding instruments, *Journal of Geophysical Research*, **108**, 4116–4130.

Roeckner, E., Arpe, K., Bengtsson, L., Christoph, M., Claussen, M., Dümenil, L., Esch, M., Giorgetta, M., Schlese, U., and Schulzweida, U. (1996). *The Atmospheric General Circulation Model ECHAM-4: Model Description and Simulation of Present-day Climate* (Report No. 218), Max-Planck-Institut für Meteorologie, Hamburg, Germany.

Roelofs, G.J. and Lelieveld, J. (1997). Model study of the influence of cross-tropopause O_3 transports on tropospheric O_3 levels, *Tellus B*, **49**, 38–55.

Roelofs, G.-J., Lelieveld, J., and van Dorland, R. (1997). A three-dimensional chemistry/general circulation model simulation of anthropogenically derived ozone in the troposphere and its radiative climate forcing, *Journal of Geophysical Research*, **102**, 23389–23401.

Roelofs, G.J., Kentarchos, A.S., Trickl, T., Stohl, A., Collins, W.J., Crowther, R.A, Hauglustaine, D., Klonecki, A., Law, K.S., Lawrence, M.G., von Kuhlmann, R., and van Weele, M. (2003). Intercomparison of tropospheric ozone models: Ozone transport in a complex tropopause folding event, *Journal of Geophysical Research*, **108**, 8529, doi: 10.1029/2003JD003462.

Rood, R.B., Allen, D.J., Baker, W.E., Lamich, D.J., and Kaye, J.A. (1989). The use of assimilated stratospheric data in constituent transport calculations, *Journal of the Atmospheric Sciences*, **46**, 687–701.

Rood, R.B., Douglass, A.R., Kaye, J.A., Geller, M.A. Chi, Y., Allen, D.J., Larson, E.M., Nash, E.R., and Nielsen, J.E. (1991). Three-dimensional simulations of wintertime ozone variability in the lower stratosphere, *Journal of Geophysical Research*, **96**, 5055–5071.

Roots, O., Roose, A., Eerme, K., and Teinemaa, E. (2009). Developing long-term monitoring of ozone in Estonia: The mandate of the Montreal Protocol, *International Journal of Remote Sensing*, **30**, 4181–4194.

Rosenfield, J.E., Frith, S.M., and Stolarski, R.S. (2005). Version 8 SBUV ozone profile trends compared with trends from a zonally averaged chemical model, *Journal of Geophysical Research*, **110**, D12302, doi: 10.1029/2004JD005466.

Rosenfield, J.R. and Schoeberl, M.R. (1986). Computation of stratospheric heating rates and the diabatic circulation for the Antarctic spring, *Geophysical Research Letters*, **13**, 1339–1342.

Rosso, O.A., Larrondo, H.A., Martin, M.T., Plastino, A., and Fuentes, M.A. (2007). Distinguishing noise from chaos, *Physical Review Letters*, **99**, 154102-1–154102-4.

Rothman, L.S., Jacquemart, D., Barbe, A., Chris Bennerc, D., Birk, M., Brown, L.R., Carleer, M.R., Chackerian C. Jr., Chance, K., Coudert, L.H. *et al.* (2005). The HITRAN 2004 molecular spectroscopic database, *Journal of Quantitative Spectroscopy and Radiative Transfer*, **96**, 139–204.

Rowland, F.S. (2006). Stratospheric ozone depletion, *Philosophical Transactions of the Royal Society, B: Biological Sciences*, **361**, 769–790.

Rowland, F.S. (2009). Stratospheric ozone depletion, in: C. Zerefos, G. Costopoulos, and G. Skalkeas (Eds.), *Twenty Years of Ozone Decline: Proceedings of the Symposium for the 20th Anniversary of the Montreal Protocol*, Springer-Verlag, Berlin, pp. 23–66.

Rowland, F.S. and Molina, M.J. (1974). Stratospheric ozone destruction catalyzed by chlorine atoms from photolysis of chlorofluoromethanes, *EoS, Transactions of the American Geophysical Union*, **55**, 1153.

Rowland, F.S., Sato, S., Khwaja, H., and Elliot, S.M. (1986). The hydrolysis of chlorine nitrate and its possible atmospheric significance, *Journal of Physical Chemistry*, **90**, 1985–1988.

Rowland, F.S, Harris N.R.P, Bojkov R.D., and Bloomfield, P. (1989). Statistical error analyses of ozone trends: Winter depletion in the northern hemisphere, in: R.D. Bojkov and P. Fabian (Eds.), *Ozone in the Atmosphere: Proceedings of the Quadrennial Ozone Symposium, 1988, Göttingen, Germany*, Deepak Publications, Hampton, VA, pp. 71–75.

Rozanov, V., Diebel, D., Spurr, R.J.D., and Burrows, J.P. (1997). GOMETRAN: A radiative transfer model for the satellite project GOME—the plane-parallel version, *Journal of Geophysical Research*, **102**, 16683–16695.

Rozanov, V., Kurosu, T., and Burrows, J.P. (1998). Retrieval of atmospheric constituents in the UV–visible: A new quasi-analytical approach for the calculation of weighting functions, *Journal of Quantitative Spectroscopy and Radiative Transfer*, **60**, 277–299.

Rozanov, V., Buchwitz, M., Eichmann K.U., de Beek, R., and Burrows, J.P. (2002). SCIATRAN—a new radiative transfer model for geophysical applications in the 240–2400 nm spectral region: The pseudo-spherical version, *Advances in Space Research*, **29**, 1831–1835.

Rubin, G. and Kunzi, K.F. (1989). Ground-based microwave sensor to retrieve stratospheric ozone profiles, in: R.D. Bojkov and P. Fabian (Eds.), *Proceedings of the Quadrennial Ozone Symposium, 1988, Göttingen, Germany*, Deepak Publications, Hampton, VA, pp. 751–754.

Russell, J.M. III, Gordley, L.L., Park, J.H., Drayson, S.R., Hesketh, W.D., Cicerone, R.J., Tuck, A.F., Frederick, J.E., Harries, J.E., and Crutzen, P.J. (1993). The Halogen Occultation Experiment, *Journal of Geophysical Research*, **98**, 10777–10798.

Sahoo, A., Sarkar, S., Singh, R.P., Kafatos, M., and Summers, M.E. (2005). Declining trend of total ozone column over the northern parts of India, *International Journal of Remote Sensing*, **26**, 3433–3440.

Salawitch, R.J. (1998). Ozone depletion: A greenhouse warming connection, *Nature*, **392**, 551–552.

Salawitch, R.J., Wofsky, S.C., Gottlieb, E.W., Lait, L.R., Newman, P.A., Schoeberl, M.R., Loewenstein, M., Podolske, J.R., Strahan, S.E., Proffitt, M.H. *et al.* (1993). Chemical loss of ozone in the Arctic polar vortex in the winter of 1991–1992, *Science*, **261**, 1146–1149.

Salmon, R. (2007). A general method for conserving energy and potential enstrophy in shallow water models, *Journal of the Atmospheric Sciences*, **16**, 515–531.

Sander, R. (1999). Modeling atmospheric chemistry: Interactions between gas-phase species and liquid cloud/aerosol particles, *Surveys in Geophysics*, **20**, 1–31.

Sandor, B.J. and Clancy, R.T. (1998). Mesospheric HO_x chemistry from diurnal microwave observations of HO_2, O_3, and H_2O, *Journal of Geophysical Research*, **103**, 13337–13351.

Santee, M.L., Manney, G.L., Livesey, N.J., Froidevaux, L., MacKenzie, I.A., Pumphrey, H.C., Read, W.G., Schwartz, M.J., Waters, J.W., and Harwood, R.S. (2005). Polar processing and development of the 2004 Antarctic ozone hole: First results from MLS on Aura, *Geophysical Research Letters*, **32**, L12817, doi: 10.1029/2005GL022582.

Santer, B.D., Wehner, M.F., Wigley, T.M.L., Sausen, R., Meehl, G.A., Taylor, K.E., Ammann, C., Arblaster, J., Washington, W.M., Boyle, J.S., and Brüggemann, W. (2003). Contributions of anthropogenic and natural forcing to recent tropopause height changes, *Science*, **301**, 479–483.

Sarkissian, A., Vaughan, G., Roscoe, H.K., Bartlett, L.M., O'Connor, F.M., Drew, D.G., Hughes, P.A., and Moore, D.M. (1997). Accuracy of measurements of total ozone by a SAOZ ground-based zenith sky visible spectrometer, *Journal of Geophysical Research*, **102**, 1379–1390.

Sarma, K.M. (1998). Protection of the ozone layer: A success story of UNEP, *Linkages Journal*, **3**(3), 6–10.

Sarma, K.M. and Taddonio, K.N. (2009). The role of financial assistance by the Multilateral Fund in Technology Change to Protect the Ozone Layer, in: C. Zerefos, G. Contopoulos, and G. Skalkeas (Eds.), *Twenty Years of Ozone Decline: Proceedings of the Symposium for the 20th Anniversary of the Montreal Protocol*, pp. 441–458.

Sasano, Y., Bodeker, G., and Kreher, K. (1998). Preliminary ILAS observations of O_3, HNO_3 and N_2O in the Arctic stratosphere during the winter of 1996/97, *SPARC Newsletter*, **10**, 18–19.

Sasano, Y., Terao, Y., Tanaka, H.L., Yasunari, T., Kanzawa, H, Nakajima, H., Yokota, T., Nakane, H., Hayashida, S., and Saitoh, N. (2000). ILAS observations of chemical ozone loss in the Arctic vortex during early spring 1997, *Geophysical Research Letters*, **27**, 213–216.

Sathishkumar, S., Sridharan, S., and Jacobi, C. (2009). Dynamical response of low-latitude middle atmosphere to major sudden stratospheric warming events, *Journal of Atmospheric and Solar-Terrestrial Physics*, **71**, 857–865.

Sato, K, and Yoshiki, M. (2008). Gravity wave generation around the polar vortex in the stratosphere revealed by 3-hourly radiosonde observations at Syowa Station, *Journal of the Atmospheric Sciences*, **65**, 3719–3735.

Saunois, M., Mari, C., Thouret, V., Cammas, J.P., Peyrille, P., Lafore, J.P., Sauvage, B., Volz-Thomas, A., Nedelec, P., and Pinty, J.P. (2008). An idealized two-dimensional approach to study the impact of the West African monsoon on the meridional gradient of tropospheric ozone, *Journal of Geophysical Research*, **113**, D07306.

Sausen, R. and Santer, B.D. (2003). Use of changes in tropopause height to detect human influences on climate, *Meteorologische Zeitschrift*, **12**, 131–136.

Sausen, R., Feneberg, B., and Ponater, M. (1997). Climatic impact of aircraft-induced ozone changes, *Geophysical Research Letters*, **24**(10), 1203–1206.

Schauffler, S.M., Atlas, E.L., Donnelly, S.G., Andrews, A., Montzka, S.A., Elkins, J.W., Hurst, D.F., Romashkin, P.A., Dutton, C.S., and Stroud, V. (2003). Chlorine budget and partitioning during the Stratospheric Aerosol and Gas Experiment (SAGE) III Ozone Loss and Validation Experiment (SOLVE), *Journal of Geophysical Research*, **108**, 4173, doi: 10.1029/2001JD002040.

Schmidt, G.A., Ruedy, R., Miller, R.L., and Lacis, A.A. (2010). The attribution of the present-day total greenhouse effect, *Journal of Geophysical Research*, **115**, D20106, doi: 10.1029/2010JD014287.

Schmidt, H., Brasseur, G.P., Charron, M., Manzini, E., Giorgetta, M.A., Diehl, T., Fomichev, V.I., Kinnison, D., Marsh, D., and Walters, S. (2006). The HAMMONIA chemistry–climate model: Sensitivity of the mesopause region to the 11-year solar cycle and CO_2 doubling, *Journal of Climate*, **19**, 3903–3931.

Schnadt, C., Dameris, M., Ponater, M., Hein, R., Grewe, V., and Steil, B. (2002). Interaction of atmospheric chemistry and climate and its impact on stratospheric ozone, *Climate Dynamics*, **17**, 501–517.

Schnadt, C., Poberaj, C., Staehelin, J., Brunner, D., Thouret, V., DeBacker, H., and Stübi, R. (2009). Long-term changes in UT/LS ozone between the late 1970s and the 1990s deduced from the GASP and MOZAIC aircraft programs and from ozonesondes, *Atmospheric Chemistry and Physics Discussions*, **9**, 2435–2499.

Schneider, H.R., Ko, M.R.W., Shia, R.-L., and Sze, N.-D. (1993). A two-dimensional model with coupled dynamics, radiative transfer, and photochemistry, 2: Assessment of the response of stratospheric ozone to increased levels of CO_2, N_2O, CH_4, and CFC, *Geophysical Research Letters*, **98**, 20444–20449.

Schoeberl, M.R. and Hartmann, D.L. (1991). The dynamics of the polar vortex, *Science*, **251**, 46–52.

Schoeberl, M.R. and Krueger, A. (1983). Medium scale disturbances in total ozone during southern hemisphere summer, *Bulletin of the American Meteorological Society*, **64**, 1358–1362.

Schoeberl, M.R. and Rodriguez, J.M. (2009). The rise and fall of dynamical theories of the ozone hole, in: C. Zerefos, G. Costopoulos, and G. Skalkeas (Eds.), *Twenty Years of Ozone Decline: Proceedings of the Symposium for the 20th Anniversary of the Montreal Protocol*, Springer-Verlag, Berlin, pp. 263–272.

Schoeberl, M.R., Krueger, A.J., and Newman, P.A. (1986). The morphology of Antarctic total ozone as seen by TOMS, *Geophysical Research Letters*, **13**, 1217–1220.

Schoeberl, M., Douglass, A.R., Hilsenrath, E., Bhartia, P.K., Beer, R., Waters, J.W., Gunson, M.R., Froidevaux, L., Gille, G.C., Barnett, J.J. *et al.* (2006). Overview of the EOS Aura Mission, *IEEE Transactions on Geoscience and Remote Sensing*, **44**, 1066–1072.

Schoeberl, M.R., Ziemke, J.R., Bojkov, R., Livesey, N., Duncan, B., Strahan, S., Froidevaux, L., Kulawik, S., Bhartia, P.K., Chandra, S. *et al.* (2007). A trajectory based estimate of the tropospheric column ozone column using the residual method, *Journal of Geophysical Research*, **112**, D24S49, doi: 10.1029/ 2007JD008873.

Schotland, R.M. (1974). Errors in the lidar measurements of atmospheric gases by differential absorption, *Journal of Applied Meteorology*, **13**, 71–77.

Schubert, S.D. and Munteanu, M.-J. (1998). An analysis of tropopause pressure and total ozone correlations, *Monthly Weather Review*, **116**, 569–582.

Schultz, M.G., Jacob, D.J., Wang, Y., Logan, J.A., Atlas, E.L., Blake, D.R., Blake, N.J., Bradshaw, J.D., Browell, E.V., Fenn, M.A. *et al.* (1999). On the origin of tropospheric ozone and NO_x over the tropical South Pacific, *Journal of Geophysical Research*, **104**, 5829–5843.

Schulz, A., Rex, M., Steger, J., Harris, N.R.P., Braathen, G.O., Reimer, E., Alfier, R., Beck, A., Alpers, M., Cisneros, J. *et al.* (2000). Match observations in the Arctic winter 1996/97: High stratospheric ozone loss rates correlate with low temperatures deep inside the polar vortex, *Geophysical Research Letters*, **27**, 205–208.

Schulz, A., Rex, M., Harris, N.R.P., Braathen, G.O., Reimer, E., Alfier, R., Kilbanedawe, I., Eckermann, S., Allaart, M., Alpers, M. *et al.* (2001). Arctic ozone loss in threshold conditions: Match observations in 1997/1998 and 1998/1999, *Journal of Geophysical Research*, **106**, 7495–7503.

Schulz, A., Rex, M., Steger, J., Harris, N.R.P., Braathen, G.O., Reimer, E., Alfier, R., Beck, A., Alpers, M., Cisneros, J. *et al.* (2010). Match observations in the Arctic winter 1996/97: High stratospheric ozone, *Atmospheric Chemistry and Physics*, **10**, 3099–3105.

Schwartzkopf, M.D. and Ramaswamy, V. (1993). Radiative forcing due to ozone in the 1980s: Dependence on altitude of ozone change, *Geophysical Research Letters*, **20**, 205–208.

Scinocca, J.F. and McFarlane, N.A. (2004). The variability of modeled tropical precipitation, *Journal of the Atmospheric Sciences*, **61**, 1993–2015.

Seftor, C.J., Hsu, N.C., Herman, J.R., Bhartia, P.K., Torres, O., Rose, W.I., Schneider, D.J., and Krotkov, N. (1997). Detection of volcanic ash clouds from Nimbus-7/TOMS reflectivity data, *Journal of Geophysical Research*, **102**, 16749–16759.

Seidel, D.J. and Randel, W.J. (2006). Variability and trends in the global tropopause estimated from radiosonde data, *Journal of Geophysical Research*, **111**, D21101, doi: 10.1029/2006JD007363.

Seinfeld, J.H. and Pandis, S.N. (1998). *Atmospheric Chemistry and Physics*, John Wiley & Sons, New York, 1326 pp.

Sekiguchi, Y. (1986). Antarctic ozone change correlated to the stratospheric temperature field, *Geophysical Research Letters*, **13**, 11202–11205.

Sekiyama, T.T., Shibata, K., Deushi, M., Kodera, K., and Lean, J.L. (2006). Stratospheric ozone variation induced by the 11-year solar cycle: Recent 22-year simulation using 3-D chemical transport model with reanalysis data, *Geophysical Research Letters*, **33**, L17812, doi: 10.1029/2006GL026711.

Selin, N.E., Wu, S., Nam, K.M., Reilly, J.M., Paltsev, S., Prinn, R.G., and Webster, M.D. (2009). Global health and economic impacts of future ozone pollution, *Environmental Research Letters*, **4**, 1–9.

Sembhi, H., Remedios, J.J., and Raspollini, P. (2006). Validation of cloud-free tropical UTLS MIPAS ozone and water vapour, *Proceedings of ASC 2006, ESA-ESRIN, Frascati, Italy, May, 8–12, 2006*, ESA European Space Research Institute (ESRIN), Frascati, Italy.

SESAME (1998). Special issue, *Journal of Atmospheric Chemistry*, **30**.

SESAME (1999). Special issue, *Journal of Atmospheric Chemistry*, **32**.

Sexton, D.M.H., Karoly, D.J., Folland, C.K, Rowell, D.P., and Parker, D.E. (1996). GCM simulation of the climate response to observed ozone decreases, *Proceedings of the First SPARC General Assembly* (WCRP-99, WMO/TD-N814), WMO, Geneva, Switzerland, Vol. 2, pp. 177–180.

Shalamyansky, A.M. (1993). Ozonometric network of CIS, *Meteorology and Hydrology*, **9**, 100–104 [in Russian].

Shephard, M.W., Worden, H.M., Cady-Pereira, K.E., Lampel, M., Luo, M.-Z., Bowman, K.W., Sarkissian, E., Beer, R., Rider, D.M., Tobin, D.C. *et al.* (2008). Tropospheric Emission Spectrometer nadir spectral radiance comparisons, *Journal of Geophysical Research*, **113**, D15S05, doi: 10.1029/2007JD008856.

Shi, G.-Y., Wang, W.-C., and Ko, M.R.W. (1986). Radiative heating due to stratospheric aerosols over Antarctica, *Geophysical Research Letters*, **13**, 1335–1338.

Shibata, K., Yoshimura, H., Ohizumi, M., Hosaka, M., and Sugi, M. (1999). A simulation of troposphere, stratosphere and mesosphere with an MRI/JMA98 GCM, *Papers in Meteorology and Geophysics*, **50**, 15–53.

Shindell, D.T., Wong, S., and Rind, D. (1997a). Interannual variability of the Antarctic ozone hole in a GCM, Part 1: The influence of tropospheric wave variability, *Journal of the Atmospheric Sciences*, **54**, 2308–2319.

Shindell, D.T., Rind, D., and Lonergan, P. (1997b). The doubled CO_2 climate including ozone chemistry, *Proceedings of the First SPARC General Assembly* (WCRP-99, WMO/TD-N814), WMO, Geneva, Switzerland, Vol. 2, pp. 457–460.

Shindell, D.T., Rind, D., and Lonergan, P. (1998a). Climate change and the middle atmosphere, Part IV: Ozone response to doubled CO_2, *Journal of Climate*, **11**, 895–918.

Shindell, D.T., Rind, D., and Lonergan, P. (1998b). Increased polar stratospheric ozone losses and delayed eventual recovery owing to increasing greenhouse-gas concentrations, *Nature*, **392**, 589–592.

Shindell, D.T., Rind, D., Balachandran, N., Lean, J., and Lonergan, P. (1999). Solar cycle variability, ozone, and climate, *Science*, **284**(5412), 305–308.

Shindell, D.T., Faluvegi, G., and Bell, N. (2003a). Preindustrial-to-present-day radiative forcing by tropospheric ozone from improved simulations with the GISS chemistry–climate GCM, *Atmospheric Chemistry and Physics*, **3**, 1675–1702.

Shindell, D.T., Schmidt, G.A., Miller, R.L., and Mann, M.E. (2003b). Volcanic and solar forcing of climate change during the preindustrial era, *Journal of Climate*, **16**, 4094–4107.

Shindell, D.T., Faluvegi, G., Unger, N., Aguilar, E., Koch, D.M., Bauer, S.E., and Miller, R.L. (2006). Simulations of preindustrial, present-day, and 2100 conditions in the NASA GISS composition and climate model G-PUCCINI, *Atmospheric Chemistry and Physics Discussions*, **6**, 4795–4878.

Shine, K.P. (1986). On the modeled thermal response of the Antarctic stratosphere to a depletion of ozone, *Geophysical Research Letters*, **13**, 1331–1334.

Shine, K.P., Briegleb, A.S., Grossman, A.S., Hauglestaine, D., Mao, H., Ramaswamy, V., Schwartzkopf, M.D., van Dorland, R., and Wang, W.-C. (1995). Radiative forcing due to changes in ozone: A comparison of different codes, in: W.-C. Wang and I.S.A. Isaksen (Eds.), *Atmospheric Ozone as a Climate Gas* (NATO ASI Ser. I), Springer-Verlag, New York, Vol. 32, pp. 373–396.

Shine, K.P., Freckleton, R.S., and Forster, P.M. de F. (1998). Comment on "Climate forcing by stratospheric ozone depletion calculated from observed temperature trends" by Zhong *et al.*: Authors' reply, *Geophysical Research Letters*, **25**, 663–665.

Shiotani, M. (1992). Annual, Quasi-Biennial, and El Niño–Southern Oscillation (ENSO) time-scale variations in equatorial total ozone, *Journal of Geophysical Research*, **97**, 7625–7633.

Shobha, K., Flynn, L.E., Shettle, E.P., and Beach, E. (2004). Analysis and validation of Version 8 SBUV/2 total and profile ozone data, *Proceedings of the XX Quadrennial Ozone Symposium, June 1–8, 2004, Kos, Greece,* International Ozone Commission, Athens, Greece, pp. 565–566.

Siddans, R., Munro, R., Reburn, W.J., Kerridge, B.J., Ellison, B., Moyna, B., Urban, J., Küllman, H., and Kunzi, K. (1998). Stratospheric constituent retrievals from airborne 500 GHz measurements in SESAME Phase III, in: R.D. Bojkov and G. Visconti (Eds.), *Atmospheric Ozone: Proceedings of the XVIII Quadrennial Ozone Symposium, L'Aquila, Italy, September 12–21, 1996*, Consorzio Parco Scientifico e Tecnologico d'Abruzzo, L'Aquila, Italy, Vol. 2, pp. 599–602.

Siegenthaler, A., Peter, R., and Kampfer, N. (1998). Ground-based millimeter wave observations of stratospheric chlorine monoxide and ozone above the Jungfraujoch in the winter 1996, in: R.D. Bojkov and G. Visconti (Eds.), *Atmospheric Ozone: Proceedings of the XVIII Quadrennial Ozone Symposium, L'Aquila, Italy, September 12–21, 1996*, Consorzio Parco Scientifico e Tecnologico d'Abruzzo, L'Aquila, Italy, Vol. 2, pp. 603–606.

Siegmund, P., Eskes, H., and van Velthoven, P. (2005). Antarctic ozone transport and depletion in austral spring 2002, *Journal of the Atmospheric Sciences*, **62**, 838–847.

Singh, R.P., Sarkar, S., and Singh, A. (2002). Effect of El Niño on inter-annual variability of ozone during the period 1978–2000 over the Indian subcontinent and China, *International Journal of Remote Sensing*, **23**, 2449–2456.

Sinha, A. and Toumi, R. (1997). Tropospheric ozone, lightning, and climate change, *Journal of Geophysical Research*, **102**, 10667–10671.

Smit, H.G.J. and Strater, W. (2004). *Jülich Ozone Sonde Intercomparison Experiment 2000 (JOSIE-2000)* (TD No. 1225), World Meteorological Organization Global Atmospheric Watch (WMO/GAW), Geneva, Switzerland.

Sofieva, V.F., Kyrölä, E., Verronen, P.T., Seppälä, A., Tamminen, J., Marsh, D.R., Smith, A.K., Bertaux, J.-L., Hauchecorne, A., Dalaudier, F. *et al.* (2009). Spatio-temporal observations of the tertiary ozone maximum, *Atmospheric Chemistry and Physics*, **9**, 4439–4445.

Solomon, S. (1988). The mystery of the Antarctic Ozone Hole, *Reviews of Geophysics*, **26**, 131–148.

Solomon, S. (1999). Stratospheric ozone depletion: A review of concepts and history, *Reviews of Geophysics*, **37**, doi: 10.102911999RG900008.

Solomon, S. and Albritton, D. L. (1992). Time-dependent ozone depletion potentials for short- and long-term forecasts, *Nature*, **357**, 33–37.

Solomon, S. and Daniel, J.S. (1996). Impact of the Montreal Protocol and its amendments on the rate of change of global radiative forcing, *Climate Change*, **32**, 7–17.

Solomon, S., Garcia, R.R., Rowland, F.S., and Wuebbles, D.J. (1986). On the depletion of Antarctic ozone, *Nature*, **321**, 755–758.

Solomon, S., Mount, G.H., Sanders, R.W., and Schmeltekopf, A.L. (1987). Visible spectroscopy at McMurdo Station, Antarctica, 2: Observations of OClO, *Journal of Geophysical Research*, **92**, 8329–8338.

Solomon, S., Mills, M., Heidt, L.E., Pollock, W.H., and Tuck, A.F. (1992). On the evaluation of ozone depletion potentials, *Journal of Geophysical Research*, **97**, 825–842.

Solomon, S., Portmann, R.W., Sasaki, T., Hofmann, D.J., and Thompson, D.W.J. (2005). Four decades of osonesonde measurements over Antarctica, *Journal of Geophysical Research*, **110**, D21311, doi: 10.1029/2005JD005917.

Solomonov, S.V., Kropotkina, E.P., and Rozanov, S. (2003). Study of vertical ozone distributions in the stratosphere and mesosphere at the millimeter wavelengths, *Radiophysics and Quantum Electronics*, **46**, 8–9.

Son, S.W., Polvani, L.M., Waugh, D.W., Akiyoshi, H., Garcia, R., Kinnison, D., Pawson, S., E. Rozanov, E., Shepherd, T.G., and Shibata, K. (2008). The impact of stratospheric ozone recovery on the Southern Hemisphere Westerly Jet, *Science*, **320**, 1486–1489.

Sonbawne, S.M., Ernest Raj, P., Devara, P.C.S., and Dani, K.K. (2009). Variability in sun photometer derived summer-time total column ozone over the Indian station "Maitri" in Antarctic region, *International Journal of Remote Sensing*, **30**, 4331–4341.

Sonnemann, G.R., Grygalashvyly, M., and Berger, U.A. (2006). A case study of stratospheric warming event in January 2001 on the basis of a new 3D model LIMA of dynamics and chemistry of the middle atmosphere, *Journal of Atmospheric Solar and Terrestrial Physics*, **68**, 2012–2025.

Sonnemann, G.R. Hartogh, P., Jarchow, C., Grygalashvyly, M., and Berger, U. (2007). On the winter anomaly of the night-to-day ratio of ozone in the middle to upper mesosphere in middle to high latitudes, *Advances in Space Research*, **40**, 846–854.

Sørensen, J.H. and Nielsen, N.W. (2001). Intrusion of stratospheric ozone to the free troposphere through tropopause folds: A case study, *Physics and Chemistry of the Earth, Part B: Hydrology, Oceans and Atmosphere*, **26**, 801–806.

Spang, R., Remedios, J.J., and Barkley, M.P. (2004). Colour indices for the detection and differentiation of cloud types in infra-red limb emission spectra, *Advances in Space Research*, **33**, 1041–1047.

Spänkuch, D. and Schulz, E. (1995). Diagnosing and forecasting total column ozone by statistical relations, *Journal of Geophysical Research*, **100**, 18873–18885.

Spänkuch, D., Dhler, W., Güldner, J., and Schulz, E. (1998). Estimation of the amount of tropospheric ozone in a cloudy sky by ground-based Fourier-transform infrared emission spectroscopy, *Applied. Optics*, **37**, 3133–3142.

SPARC (1997). *Stratospheric Processes and Their Role in Climate: Proceedings of the SPARC General Assembly, December 2–6, 1996, Melbourne, Australia* (WCRP-99, WMO/TD-No. 814), WMO, Geneva, Switzerland, Vols. I, II, 672 pp.

SPARC (1998). *Stratospheric Processes and Their Role in Climate: Implementation Plan* (WCRP-1059, WMO/TD-No. 914), WMO, Geneva, Switzerland, 131 pp.

SPARC (2001). Report on the SPARC/IOC Workshop on Understanding Ozone Trends by N.R.P. Harris and R.D. Hudson, *SPARC Newsletter*, **17**, 10–13.

Spearman, C. (1904). The proof and measurement of association between two things, *American Journal of Psychology*, **15**, 72–101.

Speuser, W., Sahand, S., and Schurath, U. (1989). A novel fast response chemiluminescence sonde for routine soundings of stratospheric ozone up to 1.5 mb, in: R.D. Bojkov and P. Fabian (Eds.), *Proceedings of the Quadrennial Ozone Symposium, 1988, Göttingen, Germany*, Deepak Publications, Hampton, VA, pp. 747–750.

Spurr, R., Van Roozendael, M., Fayt, C., and Lambert, J.-C. (2004). The GODFIT (GOME Direct FITting) algorithm: A new approach for total column retrieval, *Proceedings of the ENVISAT ERS Symposium, September 6–10, 2004, Salzburg, Austria* (*http://earth.esa.int/workshops/salzburg04*).

Spurr, R., Loyola, D., Thomas, W., Balzer, W., Mikusch, E., Aberle, B., Slijkhuis, S., Ruppert, T., Van Roozendael, M., Lambert, J.-C. *et al.* (2005). GOME level 1-to-2 data processor version 3.0: A major upgrade of the GOME/ERS-2 total ozone retrieval algorithm, *Applied Optics*, **44**, 7196–7209.

Spurr, R., Van Roozendael, M., Loyola, D., Lerot, C., van Geffen, J., van Gent, J., Fayt, C., Lambert, J.-C., and Tepper, R. (2007). *GOME Direct Fitting Phase-B (GODFIT-2)* (ATBD ESA ITT 18817/04/I-OL.ATBD-02), European Space Agency, Noordwijk, The Netherlands.

Srivastava, G.P., Pakkir, M.P.M., and Balwalli, R.R. (1982). Ozone concentration measurements near the ground at Raichur during the solar eclipse of 1980, *Proceedings of Indian Natural Sciences Academy*, **A48**, 138–142.

Staehelin, J., Kegel, R., and Harris, N.R.P. (1998a). Trend analysis of the homegenized total ozone series at Arosa (Switzerland), 1926–1996, *Journal of Geophysical Research*, **103**, 8385–8400.

Staehelin, J., Renaud, A., Bader, J., McPeters, R., Viatte, P., Hoegger, B., Bugnion, V., Giroud, M., and Schill, H. (1998b). Total ozone series at Arosa (Switzerland): Homogenization and data comparison, *Journal of Geophysical Research*, **103**, 5827–5841.

Staehelin, J., Harris, N.R.P., Appenzeller, C., and Eberhard, J. (2001). Ozone trends: A review, *Reviews of Geophysics*, **39**, 231–290.

Staehelin, J., Kerr, J., Evans, R., and Vanicek, K. (2003). *Comparison of Total Ozone Measurements of Dobson and Brewer Spectrophotometers and Recommended Transfer Functions* (Technical Report WMO TD N. 1147, No. 149), WMO, Geneva, Switzerland.

Staehelin, J., Vogler, C., and Brönnimann, S. (2009). The long history of ozone measurements: Climatological information derived from long ozone records, in: C. Zerefos, G. Contopoulos, and G. Skalkeas (Eds.), *Twenty Years of Ozone Decline: Proceedings of the Symposium for the 20th anniversary of the Montreal Protocol*, Springer-Verlag, Berlin, pp. 119–131.

Stajner, I., Wargan, K., Pawson, S., Hayashi, H., Chang, L.P., Hudman, R.C., Froidevaux, L., Livesey, N., Levelt, P.F., Thompson, A.M. *et al.* (2008). Assimilated ozone from EOS-

Aura: Evaluation of the tropopause region and tropospheric columns, *Journal of Geophysical Research*, **113**, D16S32, doi: 10.1029/2007JD008863.

Stanley, E. (1999). Scaling, universality, and renormalization: Three pillars of modern critical phenomena, *Reviews of Modern Physics*, **71**, 358–366.

STAR (2001). *Scientific and Technical Aerospace Reports, Vol. 39*, Scientific and Technical Information Program Office, NASA Langley Research Center, Cleveland, OH.

Steffanutti, L., Sokolov, L., McKenzie, A.R., Peter, Th., Khattatov, V.H., and Balestri, S. (1999). The MSS *Geophysica* as a platform for the Airborne Polar Experiment, *Journal of Atmospheric and Oceanic Technology*, **16**, 1303–1312.

Steffanutti, L., Mackenzie, A.R., Santacesaria, V., Adriani, A., Balestri, S., Borrmann, S., Khattatov, V., Mazzinghi, P., Mitev, V., Rudakov, V. *et al.* (2004). The APE-THESEO tropical campaign: An overview, *Journal of Atmospheric Chernistry*, **48**, 1–33.

Steil, B., Dameris, M., Brühl, C., Crutzen, P.J., Grewe, V., Ponater, M., and Sausen, R. (1998). Development of a chemistry module for GCMs: First results of a multiannual integration, *Annales Geophysicae*, **16**, 205–228.

Steinbrecht, W., Claude, H., Köhler, U., and Hoinka, K.P. (1998). Correlations between tropopause height and total ozone: Implications for long-term changes, *Journal of Geophysical Research*, **103**, 19183–19192.

Steinbrecht, W., Claude, H., Köhler, U., and Winkler, P. (2001). Interannual changes of total ozone and northern hemisphere circulation patterns, *Geophysical Research Letters*, **28**, 1191–1194.

Steinbrecht, W., Claude, H., Schönenborn, F., McDermid, I.S., Leblanc, T., Godin, S., Song, T., Swart, D.P.J., Meijer, Y.J., Bodeker, G.E. *et al.* (2006a). Long-term evolution of upper stratospheric ozone at selected stations of the Network for the Detection of Stratospheric Change (NDSC), *Journal of Geophysical Research*, **111**, D10308, doi: 10.1029/2005JD006454.

Steinbrecht, W., Hassler, B., Brühl, C., Dameris, M., Giorgetta, M.A., Grewe, V., Manzini, E., Matthes, S., Schnadt, C., Steil, B. *et al.* (2006b). Interannual variation patterns of total ozone and lower stratospheric temperature in observations and model simulations, *Atmospheric Chemistry and Physics*, **6**, 349–374.

Steinbrecht, W., Claude, H., Schönenborn, F., Mcdermid, I.S., Leblanc, T., Godin, S., Beekmann, S., Keckhut, P., Hauchecorne, A., Van Gijsel, J.A.E *et al.* (2009). Ozone and temperature trends in the upper stratosphere at five stations of the Network for the Detection of Atmospheric Composition Change, *International Journal of Remote Sensing*, **30**, 3875–3886.

Stenchikov, G.L., Kirchner, I., Roback, A., Graf, H.-F., Antuña, J.C., Grainger, R.G., Lambert, A., and Thomason, L. (1998). Radiative forcing from the 1991 Mount Pinatubo volcanic eruption, *Journal of Geophysical Research*, **103**, 13837–13857.

Stenke, A., Dameris, M., Grewe, V., and Garny, H. (2008). Implications of Lagrangian transport for coupled chemistry–climate simulations, *Atmospheric Chemistry and Physics*, **9**, 5489–5504.

Stevenson, D.C., Johnson, C.E., Collins, W.J., Derwent, R.G., Shine, K.P., and Edwards, J.M. (1998). Evolution of tropospheric ozone radiative forcing, *Geophysical Research Letters*, **25**, 3819–3828.

Stohl, A., Cooper, O.R., and James, P. (2004). A cautionary note on the use of meteorological analysis fields for quantifying atmospheric mixing, *Journal of the Atmospheric Sciences*, **61**, 1446–1453.

Stolarski, R.S. and Cicerone, R.J. (1974). Stratospheric chlorine: A possible sink for ozone, *Canadian Journal of Chemistry*, **52**, 1610–1615.

Stolarski, R.S. and Schoeberl, M.R. (1986). Further interpretation of satellite measurement of Antarctic total ozone, *Geophysical Research Letters*, **13**, 1210–1212.

Stolarski, R.S., Krueger, A.J., Schoeberl., M.R., McPeters, R.D., Newman, P.A., and Alpert, J.C. (1986). Nimbus-7 satellite measurements of the springtime Antarctic ozone decrease, *Nature*, **322**, 808–811.

Stolarski, R.S., Bloomfield, P., McPeters, R.D., and Herman, J.R. (1991). Total ozone trends deduced from NIMBUS 7 TOMS data, *Geophysical Research Letters*, **18**, 1015–1018.

Stolarski, R., Bojkov, R., Bishop, L., Zerefos, C., Staehelin, J., and Zawodny, J. (1992). Measured trends in stratospheric ozone, *Science*, **256**, 342–349.

Stolarski, R.S., Douglass, A.R., Gupta, M., Newman, P.A., Pawson, P., Schoeberl, M.R., and Nielsen, J.E. (2006a). An ozone increase in the Antarctic summer stratosphere: A dynamical response to the ozone hole, *Geophysical Research Letters*, **33**, doi: 10.1029/2006GL026820.

Stolarski, R.S., Douglass, A.R., Steenrod, S., and Pawson, S. (2006b). Trends in stratospheric ozone: Lessons learned from a 3D chemical transport model, *Journal of the Atmospheric Sciences*, **63**, 1028–1041.

Stolarski, R.S., Douglass, A.R., Newman, P.A., Pawson, S., and Schoeberl, M.R. (2010). Relative contribution of greenhouse gases and ozone-depleting substances to temperature trends in the stratosphere: A chemistry–climate model study, *Journal of Climate*, **23**, 28–42.

Stott, P., Jones, G., Lowe, J., Thorne, P., Durman, C., Johns, T., and Thelen, J. (2006). Transient climate simulations with the HadGEM1 climate model: Causes of past warming and future climate change, *Journal of Climate*, **19**, 2763–2782.

Strand, O.N. and Westwater, E.R. (1968). Statistical estimation of the numerical solution of a Fredholm integral equation of the first kind, *Journal of the ACM*, **15**, 100–114.

Streibel, M., Rex, M., von der Gathen, P., Lehmann, R., Harris, N.R.P., Braathen, G.O., Reimer, E., Deckelmann, H., Chipperfield, M., Millard, G. *et al.* (2006). Chemical ozone loss in the Arctic winter 2002/2003 determined with Match, *Atmospheric Chemistry and Physics*, **6**, 2783–2792.

Struthers, H., Bodeker, G.E., Austin, J., Bekki, S., Cionni, I., Dameris, M., Giorgetta, M.A., Grewe, V., Lefèvre, F., Lott, F. *et al.* (2009). The simulation of the Antarctic ozone hole by chemistry–climate models, *Atmospheric Chemistry and Physics*, **9**, 6363–6376.

Stübi, R., Bugnion, V., Giroud, M., Jeannet, P., Viatte, P., Hoegger, B., and Staehelin, J. (1998). Long term ozone balloon sounding series at Payerne: Homogenization method and problems, in: R.D. Bojkov and G. Visconti (Eds.), *Proceedings of the XVIII Quadrennial Ozone Symposium, L'Aquila, Italy, September 12–21, 1996,* Consorzio Parco Scientifico e Tecnologico d'Abruzzo, L'Aquila, Italy, Vol. 1, pp. 179–182.

Stübi, R., Levrat, G., Hoegger, B., Viatte, P., Staehelin, J., and Schmidlin, F.J. (2008). In-flight comparison of Brewer–Mast and electrochemical concentration cell ozonesondes, *Journal of Geophysical Research*, **113**, D13302, doi: 10.1029/2007JD009091.

Stuhler, H. and Frohn, A. (1998). The production of NO by hypersonic flights, *Atmospheric Environment*, **32**, 3153–3155.

Sussmann, R. and Schaefer, K. (1997). Infrared spectroscopy of tropospheric trace gases: Combined analysis of horizontal and vertical column abundances, *Applied Optics*, **36**, 735–741.

Sutton, R. (1994). Lagrangian flow in the middle atmosphere, *Quarterly Journal of the Royal Meteorological Society*, **120**, 1299–1321.

Svendby, T.M. and Dahlback, A. (2002). Twenty years of revised Dobson total ozone measurements in Oslo, Norway, *Journal of Geophysical Research*, **107**, 4369–4387.

Syrovatkina, O.A., Karol, I.L., Shalamyanskii, A.M., and Klyagina, L.P. (2008). Interannual variations in total ozone fields at high latitudes of the Northern Hemisphere, November to March 1998–2005, *Russian Meteorology and Hydrology*, **33**(2), 91–97, doi: 10.3103/S1068373908020040.

Syvorotkin, V.L. (1998). Earth's outgassing destroys the ozonosphere, *Earth and Universe*, **1**, 21-27 [in Russian].

Tabazadeh, A. and Turco, R.P. (1993). A model for heterogeneous chemical processes on the surfaces of ice and nitric-acid trihydrate particles, *Journal of Geophysical Research*, **98**, 12727–12740.

Taddei, A., Redaelli, G., Grassi, B., and Visconti, G. (2006). Selfconsistency analysis of MIPAS data using the Trajectory Hunting Technique (THT), *Proceedings of the Atmospheric Science Conference, May 8–12, 2006, ESA-ESRIN, Frascati, Italy*, ESA European Space Research Institute (ESRIN), Frascati, Italy.

Takashima, H. and Shiotani, M. (2007). Ozone _variation in the tropical tropopause layer as seen from ozonesonde data, *Journal of Geophysical Research*, **112**, D11123, doi: 10.102912006JD008322.

Takigawa, M., Takahashi, M., and Akigoshi, H. (1999). Simulation of ozone and other chemical species using a Center for Climate System Research/National Institute for Environmental Studies atmospheric GCM with coupled stratospheric chemistry, *Journal of Geophysical Research*, **104**, 14003–14018.

Talkner, P. and Weber, R.O. (2000). Power spectrum and detrended fluctuation analysis: Application to daily temperatures, *Physical Review E*, **62**, 150–160.

Tamminen, J., Kyrölä, E., Sofieva, V.F., Laine, M., Bertaux, J.L., Hauchecorne, A., Dalaudier, F., Fussen, D., Vanhellemont, F., Fanton-d'Andon, O. *et al.* (2010). GOMOS data characterization and error estimation, *Atmospheric Chemistry and Physics Discussions*, **10**, 6755–6796.

Tatarov, B., Nakane, H., Park, Ch.B, Sugimoto, N., and Matsui, I. (2009). Lidar observation of long-term trends and variations of stratospheric ozone and temperature over Tsukuba, Japan, *International Journal of Remote Sensing*, **30**, 3951–3960.

Taupin, F.G., Bessafi, M., Baldy, S., and Bremaud, P.J. (1999). Tropospheric ozone above the southwestern Indian Ocean is strongly linked to dynamical conditions prevailing in the tropics, *Journal of Geophysical Research*, **104**, 8057–8066.

Taylor, F.W. (2011). Review article: Forty years of satellite meteorology at Oxford, *Quarterly Journal of the Royal Meteorological Society*, **137**, 277–285.

Tegtmeier, S., Fioletov, V.E., and Shepherd, T.G. (2010). A global picture of the seasonal persistence of stratospheric ozone anomalies, *Journal of Geophysical Research*, **115**, D18119, doi: 10.1029/2009JD013011.

Terao, Y., and Logan, J.A. (2007). Consistency of time series and trends of stratospheric ozone as seen by ozonesonde, SAGE II, HALOE, and SBUV(/2), *Journal of Geophysical Research*, **112**, D06310, doi: 10.1029/2006JD007667.

Terao, Y., Sasano, Y., Nakajima, H., Tanaka, H.L., and Yasunari, T. (2002). Stratospheric ozone loss in the 1996/1997 Arctic winter: Evaluation based on multiple trajectory analysis for double-sounded air parcels by ILAS, *Journal of Geophysical Research*, **107**, 8210, doi: 10.1029/2001JD000615.

Thomas, R.W.L. and Holland, A.C. (1977). Ozone estimates derived from Dobson direct sun measurements: Effect of atmospheric temperature variations and scattering, *Applied Optics*, **16**, 613–618.

Thompson, A.M. (2009). An overview of strategic ozone sounding networks: Insights into ozone budgets, UT/LS processes and tropical climate signatures, in: C. Zerefos,

G. Contopoulos, and G.. Skalkeas (Eds.), *Twenty Years of Ozone Decline: Proceedings of the Symposium for the 20th Anniversary of the Montreal Protocol*, Springer-Verlag, Berlin, pp. 237–249.

Thompson, A.M., Witte, J.C., McPeters, R.D., Oltmans, S.J., Schmidlin, F.J., Logan, J.A., Fujiwara, M., Kirchhoff, V.W.J.H., Posny, F., Coetzee, G.J.R. *et al.* (2003a). Southern Hemisphere Additional Ozonesondes (SHADOZ) 1998–2000 tropical ozone climatology, 1: Comparison with Total Ozone Mapping Spectrometer (TOMS) and ground-based measurements, *Journal of Geophysical Research*, **108**(2), 8238, doi: 10.1029/2001JD000967.

Thompson, A.M., Witte, J.C., Oltmans, S.J., Schmidlin, F.J., Logan, J.A., Fujiwara, M., Kirchhoff, V.W.J.H., Posny, F., Coetzee, G.J.R., Hoegger, B. *et al.* (2003b). Southern Hemisphere Additional Ozonesondes (SHADOZ) 1998–2000 tropical ozone climatology, 2: Tropospheric variability and the zonal wave-one, *Journal of Geophysical Research*, **108**(2), 8241, doi: 10.1029/2002JD002241.

Thompson, A.M., Stone, J.B., Witte, J.C., Miller, S.K., Pierce, R.B., Chatfield, R.B., Oltmans, S.J., Cooper, O.R., Loucks, A.L., Taubman, B.F. *et al.* (2007a). Intercontinental transport experiment ozonesonde network study (IONS, 2004), 1: Summertime upper tropospheric/lower stratosphere ozone over northeastern North America, *Journal of Geophysical Research*, **112**, DI2S12, doi: 10.1029/2006JD007441.

Thompson, A.M., Stone, J.B., Witte, J.C., Miller, S.K., Oltmans S.J., Kucsera, T.L., Ross, K.L., Pickering, K.E., Merrill, J.T., Forbes, G. *et al.* (2007b). Intercontinental transport experiment ozonesonde network study (IONS, 2004), 2: Tropospheric ozone budgets and variability over northeastern North America, *Journal of Geophysical Research*, **112**, D12S13, doi: 10.1029/2006JD007670.

Thompson, A.M., Witte, J.C., Smit, H.G.J., Oltmans, S.J., Johnson, B.J., Kirchhoff, V.W.J.H., and Schmidlin, F.J. (2007c). Southern Hemisphere Additional Ozonesondes (SHADOZ) 1998–2004 tropical ozone climatology, 3: Instrumentation, station variability, evaluation with simulated flight profiles, *Journal of Geophysical Research*, **112**, D03304, doi: 10.1029/2005JD007042.

Thompson, A.M., Yorks, J.E., Miller, S.K., Witte, J.C., Dougherty, K.M., Morris, G.A., Baumgardner, D., Ladino, L., and Rappenglück B. (2008). Tropospheric ozone sources and wave activity over Mexico City and Houston during MILAGRO/Intercontinental Transport Experiment Ozone Network Study, 2006 (IONS-06), *Atmospheric Chemistry and Physics*, **8**, 5113–5125.

Thompson, A.M., Oltmans, S.J., Tarasick, D.W., von der Gathen, P., Smit, H.G.J., and Witte, J.C. (2011). Strategic ozone sounding networks: Review of design and accomplishments, *Atmospheric Environment*, **45**, 2145–2163.

Thompson, D.W.J. and Solomon, S. (2002). Interpretation of recent southern hemisphere climate change, *Science*, **296**, 895–899.

Thouret, V., Marenco, A., Nedelec, P., and Grouhel, C. (1998a). Ozone climatologies at 9–12 km altitude as seen by the MOZAIC airborne program between September 1994 and August 1996, *Journal of Geophysical Research*, **103**, 25653–25679.

Thouret, V., Marenco, A., Logan, J.A., Nedelec, P., and Grouhel, C. (1998b). Comparisons of ozone measurements from the MOZAIC airborne program and the ozone sounding network at eight locations, *Journal of Geophysical Research*, **103**, 25965–25720.

Thuburn J. and Craigh, G.C. (1997). GCM tests of theories for the height of the tropopause, *Journal of the Atmospheric Sciences*, **54**, 869–882.

Tiao, G.C., Reinsel, G.C., Pedrick, J.H., Allenby, G.M., Mateer, C.L., Miller, A.J., and Deluisi, J.J. (1986). A statistical trend analysis of ozonesonde data, *Journal of Geophysical Research*, **91**, 3121–3136.

Timmreck, C, Graf, H.F., and Kirchner, I. (1999). A one and half year interactive MA/ ECHAM4 simulation of Mount Pinatubo aerosol, *Journal of Geophysical Research*, **104**, 9337–9359.

Titova, E.A. and Karol, I.L. (2010). Analysis of the influence that climatic variability has on the formation of the total-ozone-content field at extratropical latitudes of the Northern Hemisphere, *Izvestiya Atmospheric and Oceanic Physics*, **46**, 635–642, doi: 10.1134/ S0001433810050087.

Toon, O.B., Hamill, P., Turco, R.P., and Pinto, J. (1986). Condensation of HNO_3 and HCl in the winter polar stratosphere, *Geophysical Research Letters*, **13**, 1284–1287.

Torres, O. and Bhartia, P.K. (1999). Impact of aerosol absorption on total ozone retrieval from satellite measurements of backscattered ultraviolet radiation, *Journal of Geophysical Research*, **104**, 21569–21578.

Torres, O., Herman, J.R., Bhartia, P.K., and Ahmad, Z. (1995). Properties of Mount Pinatubo aerosols as derived from Nimbus-7 total ozone mapping spectrometer measurements, *Journal of Geophysical Research*, **100**, 14043–14055.

Toumi, R., Haigh, J.D. and Law, K.S. (1996). A tropospheric ozone–lightning climate feedback, *Geophysical Research Letters*, **23**, 1037–1040.

Tourpali, K., Tie, X.X., Zerefos, C.S., and Brasseur, G. (1997). Decadal evolution of total ozone decline: Observations and model results, *Journal of Geophysical Research*, **102**, 23955–23962.

Tripathi, O.P., Godin-Beekmann, S., Lefèvre, F., Marchand, M., Pazmiño, A., Hauchecorne, A., Goutail, F., Schlager, H., Volk, C.M., Johnson, B. *et al.* (2006). High resolution simulation of recent Arctic and Antarctic stratospheric chemical ozone loss compared to observations, *Journal of Atmospheric Chemistry*, **55**, 205–226.

Tripathi, O.P., Godin-Beekmann, S., Lefèvre, F., Pazmiño, A., Hauchecorne, A., Chipperfield, M., Feng, W., Millard, G., Rex, M., Streibel, M. *et al.* (2007). Comparison of polar ozone loss rates simulated by one-dimensional and three-dimensional models with Match observations in recent Antarctic and Arctic winters, *Journal of Geophysical Research*, **112**, D12307, doi: 10.1029/2006JD008370.

Tsitomeneas, S., Brekoulakis, A., and Carabatos-Nedelec, C. (2003). Detection of environmental parameters of the atmosphere using laser radiation and telecommunication techniques, *International Society for Optical Engineering (SPIE), Proceedings*, **5131**, 235–238.

Tsou, J.J., Connor, B.J., Parrish, A., McDermid, I.S., and Chu, W.P. (1995). Ground-based microwave monitoring of middle atmosphere ozone: Comparison to lidar and Stratospheric and Gas Experiment II satellite observations, *Journal of Geophysical Research*, **100**, 3005–3016.

Tsou, J.J., Connor, B.J., Parrish, A, Pierce, R.B., Boyd, I.S., Bodeker, G.E., Chu, W.P., Russell, J.M., Swart, D.P.J., and McGee, T.J. (2000). NDSC millimeter wave ozone observations at Lauder, New Zealand, 1992–1998: Improved methodology, validation, and variation study, *Journal of Geophysical Research*, **105**, 24263– 24281.

Tuck, A., Watson, R., and Toon, B. (Eds.) (1989). The Airborne Antarctic Ozone Experiment (AAOE), *Journal of Geophysical Research* (special issue), **94**, 11179–11737, 16434–16858.

Tung, K.K. (1986). On the relationship between the thermal structure of the stratosphere and the seasonal distribution of ozone, *Geophysical Research Letters*, **13**, 1308–1311.

Tung, K.K., Ko, M.K.W., Rodriguez, J.M., and Sze, N.D. (1986). Are Antarctic ozone variations a manifestation of dynamics or chemistry? *Nature*, **333**, 811–814.

Tzanis C. (2005). Ground-based observations of ozone at Athens, Greece during the solar eclipse of 1999, *International Journal of Remote Sensing*, **26**, 3585–3596.

Tzanis, C. (2009a). On the relationship between total ozone and temperature in the troposphere and the lower stratosphere, *International Journal of Remote Sensing*, **30**, 6075–6084.

Tzanis, C. (2009b). Total ozone observations at Athens, Greece by satellite-borne and ground-based instrumentation, *International Journal of Remote Sensing*, **30**, 6023–6033.

Tzanis, C. and Varotsos, C.A. (2008). Tropospheric aerosol forcing of climate: A case study for the greater area of Greece, *International Journal of Remote Sensing*, **29**, 2507–2517.

Tzanis, C., Varotsos, C., and Viras, L. (2008). Impacts of the solar eclipse of 29 March 2006 on the surface ozone concentration, the solar ultraviolet radiation and the meteorological parameters at Athens, Greece, *Atmospheric Chemistry and Physics*, **8**, 425–430.

Tzanis, C., Christodoulakis, J., Efstathiou, M., and Varotsos, C. (2009a). Comparison of the Athens Dobson spectrophotometer with World Standard Instruments, *International Journal of Remote Sensing*, **30**, 3943–3950.

Tzanis, C., Varotsos, C.A., Asimakopoulos, D.N., and Cracknell, A.P. (2009b). Surface ultraviolet radiation and ozone content as indicators of environment quality, *International Journal of Remote Sensing*, **30**, 4123–4143.

Tzortziou, M.A., Krotkov, N.A., Cede, A., Herman, J.R., and Vasilkov, A. (2008). A new technique for retrieval of tropospheric and stratospheric ozone profiles using sky radiance measurements at multiple view angles: Application to a Brewer spectrometer, *Journal of Geophysical Research*, **113**, D06304, doi: 10.1029/2007JD009093.

Uchino, O. (1998). Tsukuba International Workshop on stratospheric change and its role in climate and on the ATMOS-C1 satellite mission, Tsukuba, Japan, October 20–22, 1997, *SPARC Newsletter*, **10**, 28–32.

Uchino, O., Maeda, M., Yamamura, H., and Hirono, M. (1983). Observation of stratospheric vertical ozone distribution by a XeCl lidar, *Journal of Geophysical Research*, **88**, 5273–5280, doi: 10.1029/JC088iC09p05273.

Unal, Y.S., Incecik, S., Topcu, S., and Oztopal, A. (2010). Daily peak ozone forecast in Istanbul, *International Journal of Remote Sensing*, **31**, 551–561.

UNEP (1977). *Report of the UNEP Meeting of Experts Designated by Governments, Intergovernmental and Nongovernmental Organizations on the Ozone Layer* (UNEP/WG/7/25/Rev1, annex, March 8), United Nations Environment Program (UNEP), Zurich, Switzerland.

UNEP (1992). *Urban Air Pollution in Megacities in the World Health Organization, United Nations Environmental Programme*, Blackwell, Oxford, U.K.

UNEP (1994). *Scientific Assessment of Ozone Depletion* (WMO Global Ozone Research and Monitoring Project Report No. 37), WMO, Geneva, Switzerland.

UNEP (2005). *Production and Consumption of Ozone Depleting Substances under the Montreal Protocol, 1986–2004*, Ozone Secretariat, UNEP, Nairobi, 79 pp.

Urban, J., Eyring, V., Kullmann, H., Kunzi, K., Mellmann, G., Wohlgemuth, J., Goede, A., de Jonge, A., Kleipool, Q., Mees, J. *et al.* (1998). Observations of stratospheric HlO, HCl, O_3, N_2O, and HO_2 at high latitudes during the winters of 1995 and 1996 with the Airborne-Submillimeter-SISRadiometer, in: R.D. Bojkov and G. Visconti (Eds.), *Atmospheric Ozone: Proceedings of the XVIII Quadrennial Ozone Symposium, L'Aquila, Italy, September 12–21, 1996*, Consorzio Parco Scientifico e Tecnologico d'Abruzzo, L'Aquila, Italy, Vol. 2, pp. 615–618.

Urban, J., Lautie, N., Le Flochemoen, E., Jimenez, C., Eriksson, P., de la Noee, J., Dupuy, E., Ekstrom, M., El Amraoui, L., Frisk, U. *et al.* (2005). Odin/SMR limb observations of stratospheric trace gases: Level 2 processing of ClO, N_2O, HNO_3, and O_3, *Journal of Geophysical Research*, **110**, D14307, doi: 10.1029/2004JD005741.

van Dorland, R. and Fortuin, J.P.F. (1994). Simulation of the observed stratospheric temperature trends 1967–1987 over Antarctica due to ozone hole deepening, in: J. Van Ham and Co-editors, *Non-CO_2 Greenhouse Gases*, Kluwer Academic, Dordrecht, The Netherlands, pp. 237–245.

van Dorland, R., Dentener, F.J., and Lelieveld, J. (1997). Radiative forcing due to tropospheric ozone and sulfate aerosols, *Journal of Geophysical Research*, **102**, 28079–28100.

Van Gijsel, J.A.E., Swart, D.P.J., Baray, J-L., Claude, H., Fehr, T., Von Dergathe, T.P., Godin-Beekmann, S., Hansen, G.H., Leblanc, T., Mcdermid, I.S. *et al.* (2009). Global validation of ENVISAT ozone profiles using lidar measurements, *International Journal of Remote Sensing*, **30**, 3987–3994.

Vanicek, K. (1998). Differences between Dobson and Brewer observations of total ozone at Hradec, Kralove, in: R.D. Bojkov and G. Visconti (Eds.), *Atmospheric Ozone: Proceedings of the XVIII Quadrennial Ozone Symposium, L'Aquila, Italy, September 12–21, 1996*, Consorzio Parco Scientifico e Tecnologico d'Abruzzo, L'Aquila, Italy, Vol. 1, pp. 81–84.

van Leer, B. (1974). Towards the ultimate conservative difference scheme, II: Monotonicity and conservation combined in a second-order scheme, *Journal of Computational Physics*, **14**, 361–370.

Van Roozendael, M., Lambert, J.C., Simon, P.C., Hansen, G., DeMuer, D., Schoubs, E., Koopman, R., Vanderwoerd, H., Barbe, A., Claude, H. *et al.* (1998a). Ground-based validation of GOME total ozone measurements by means of Dobson, Brewer and GUV instruments, in: R.D. Bojkov and G. Visconti (Eds.), *Proceedings of the XVIII Quadrennial Ozone Symposium, L'Aquila, Italy, September 12–21, 1996*, Consorzio Parco Scientifico e Tecnologico d'Abruzzo, L'Aquila, Italy, Vol. 2, pp. 665–668.

Van Roozendael, M., Peeters, P., Roscoe, H.K., Backer, H.D., Jones, A., Vaughan, G., Goutail, F., Pommereau, J.P., Kyrö, E., Wahlstrøm, C. *et al.* (1998b). Validation of ground-based UV–visible measurements of total ozone by comparison with Dobson and Brewer spectrophotometers, *Journal of Atmospheric Chemistry*, **29**, 55–83.

Van Roozendael, M., Loyola, D., Spurr, R., Balis, D., Lambert, J.-C., Livschitz, Y., Valks, P., Ruppert, T., Kenter, P., Fayt, C. *et al.* (2006). Ten years of GOME/ERS-2 total ozone data—The new GOME data processor (GDP) version 4, 1: Algorithm description, *Journal of Geophysical Research*, **111**, D14311, doi: 10.1029/2005JD006375.

Van Roozendael, M., Bernath, P., and Murtagh, D. (Eds.) (2009). GOMOS (Global Ozone Monitoring by Occultation of Stars): Data processing, geophysical validation and scientific results, *Atmospheric Chemistry and Physics*, Special Issue.

Vargin, P.N. (1998). Climatological structure of nonzonal heating rates in the stratosphere caused by large-scale inhomogeneity of total ozone, in: R.D. Bojkov and G. Visconti (Eds.), *Atmospheric Ozone: Proceedings XVIII Quadrennial Ozone Symposium, L'Aquila, Italy, September 12–21, 1996*, Consorzio Parco Scientifico e Tecnologico d'Abruzzo, L'Aquila, Italy, Vol. 2, pp. 817–820.

Varotsos, C.A. (1989). Comment on connection between the 11-year solar cycle, the Q.B.O. and the total ozone, *Journal of Atmospheric and Terrestrial Physics*, **51**, 367–370.

Varotsos, C.A. (1994). Solar ultraviolet radiation and total ozone, as derived from satellite and ground-based instrumentation, *Geophysical Research Letters*, **21**, 1787–1790.

Varotsos, C.A. (1995a). Observational evidence for chemical ozone depletion over the Arctic in winter 1991–92, *Nature*, **375**, 131–134.

Varotsos, C.A. (1995b). On the correction of the total ozone content over Athens, Greece as deduced from satellite-observations, *International Journal of Remote Sensing*, **16**, 1771–1776.

Varotsos, C.A. (Ed.) (1997). *Atmospheric Ozone Dynamics: Observations in the Mediterranean Region* (NATO ASI Series, Global Environmental Change, Vol. 53), Springer-Verlag, Berlin, 336 pp.

Varotsos, C.A. (2002a). The southern hemisphere ozone hole split in 2002, *Environmental Science and Pollution Research*, **9**, 375–376.

Varotsos, C.A. (2002b). A note on "what caused Antarctica to exhibit quite strong planetary waves in 2002", *COSPAR Symposium on Natural and Manmade Hazards Using Space Technology, Houston, TX, October 9-19*.

Varotsos, C.A. (2003a). What is the lesson from the unprecedented event over Antarctica in 2002? *Environmental Science and Pollution Research*, **10**, 80–81.

Varotsos, C.A. (2003b). Why did a "no-ozone-hole" episode occur in Antarctica, *EoS, Transactions of the American Geophysical Union*, **84**, 183.

Varotsos, C.A. (2004a). Atmospheric pollution and remote sensing: Implications for the southern hemisphere ozone hole split in 2002 and the northern mid-latitude ozone trend, *Advances in Space Research*, **33**, 249–253.

Varotsos, C.A. (2004b). The extraordinary events of the major, sudden stratospheric warming, the diminutive Antarctic ozone hole and its split in 2002, *Environmental Science and Pollution Research*, **11**, 405–411.

Varotsos, C.A. (2005a). Airborne measurements of aerosol, ozone, and solar ultraviolet irradiance in the troposphere, *Journal of Geophysical Research*, **110**, doi: 1029/2004JD005397.

Varotsos, C.A. (2005b). Power-law correlations in column ozone over Antarctica, *International Journal of Remote Sensing*, **26**, 3333–3342.

Varotsos, C.A. (2005c). Modern computational techniques for environmental data: Application to the global ozone laye, *Lecture Notes in Computer Science*, **3516**, 504–510.

Varotsos, C.A. and Cracknell, A.P. (1993). Ozone depletion over Greece as deduced from Nimbus-7 TOMS measurements, *International Journal of Remote Sensing*, **14**, 2053–2059.

Varotsos, C.A. and Cracknell, A.P. (1994). Three years of total ozone measurements over Athens obtained using the remote sensing technique of a Dobson spectrophotometer, *International Journal of Remote Sensing*, **15**, 1519–1524.

Varotsos, C.A. and Cracknell, A.P. (2004). New features observed in the 11-year solar cycle, *International Journal of Remote Sensing*, **25**, 2141–2157.

Varotsos, C.A. and Cracknell, A.P. (2007). Technical note: Validation of ENVISAT (SCIAMACHY) versus Dobson and TOMS atmospheric ozone measurements in Athens, Greece—Input for the upcoming IPY campaign, *International Journal of Remote Sensing*, **28**, 2073–2075.

Varotsos, C.A. and Kirk-Davidoff, D. (2006). Long-memory processes in ozone and temperature variations at the region 60°S–60°N, *Atmospheric Chemistry and Physics*, **6**, 4093–4100.

Varotsos, C.A. and Kondratyev, K.Ya. (1995). Ozone dynamics over Greece as derived from satellite and in situ measurements, *International Journal of Remote Sensing*, **16**, 1777–1798.

Varotsos, C.A. and Kondratyev, K.Ya. (1998a). Total ozone dynamics in mid-latitudes of the Northern Hemisphere, *Doklady Akademii Nauka*, **359**(6), 105–107 [in Russian].

Varotsos, C.A. and Kondratyev, K.Ya. (1998b). On potential interconnection between vertical ozone concentration profile in mid-latitudes and processes in the polar stratosphere, *Doklady Akademii Nauka*, **360**(1), 11–14 [in Russian].

Varotsos, C.A. and Tzanis, C. (2012). A new tool for the study of the ozone hole dynamics over Antarctica, *Atmospheric Environment*, **47**, 428–434.

Varotsos, C.A. and Zellner, R. (2010). A new modeling tool for the diffusion of gases in ice or amorphous binary mixture in the polar stratosphere and the upper troposphere, *Atmospheric Chemistry and Physics*, **10**, 3099–3105.

Varotsos, C.A., Helmis, C., and Cartalis, C. (1992). Annual and semiannual waves in ozone as derived from SBUV vertical global ozone profiles, *Geophysical Research Letters*, **19**, 926–928.

Varotsos, C., Kalabokas, P., and Chronopoulos, G. (1994a). Association of the laminated vertical ozone structure with the lower-stratospheric circulation, *Journal of Applied Meteorology*, **33**, 473–476.

Varotsos, C.A., Kalabokas, P., and Chronopoulos, G. (1994b). On the laminated structure of ozone in the subtropical atmosphere, in: A.R.D. Hudson (Ed.), *Proceedings of the Quadrennial Ozone Symposium, Charlottesville, VA, June 4–13, 1992* (NASA CP-3266). NASA, Washington, D.C.

Varotsos C., Kalabokas, P., and Chronopoulos, G. (1994c). Stratosphere troposphere ozone exchange at Athens, Greece, *Toxicological and Environmental Chemistry*, **44**, 211–216.

Varotsos, C.A., Chronopoulos, G.J., Katsikis, S., and Sakellariou, N.K. (1995a). Further evidence of the role of air-pollution on solar ultraviolet-radiation reaching the ground, *International Journal of Remote Sensing*, **16**, 1883–1886.

Varotsos, C.A., Cracknell, A.P., Sakellariou, N., Katsikis, S., Chronopoulos, G., and Kassomenos P. (1995b). On the SO_2, NO_2 interferences in total ozone measurements made with the Dobson spectrophotometer No. 118 in Athens, *International Journal of Remote Sensing*, **16**, 1805–1814.

Varotsos, C.A., Kondratyev, K.Ya., and Katsikis, S. (1995c). On the relationship between total ozone and solar ultraviolet radiation at St Petersburg, Russia, *Geophysical Research Letters*, **22**, 3481–3484.

Varotsos, C.A., Gkhosh, S., Katsikis, S., Kondratyev, K.Ya., and Cracknell, A.P. (1997a). Comparison of the results of observations of the total ozone content using data from satellite and ground observations (for the example of Athens), *Earth Observation and Remote Sensing*, **14**, 531–538.

Varotsos, C.A., Kondratyev, K.Ya., and Chronopoulos, G. (1997b). On the total ozone depletion in middle latitudes of the northern hemisphere due to satellite observations over Athens, Greece, *Doklady Russian Academy of Science*, **355**(1), 104–105 [in Russian].

Varotsos, C.A., Katsikis, S., and Cracknell, A.P. (1998). On the influence of stray light on the total ozone measurements made with Dobson spectrophotometer No. 118 in Athens, Greece, *International Journal of Remote Sensing*, **19**, 3307–3315.

Varotsos, C.A., Alexandris, D., and Chronopoulos, G. (1999). On the role of the lower-stratospheric circulation to the vertical ozone structure, *Physics and Chemistry of the Earth*, **24**, 481-485.

Varotsos, C.A., Kondratyev, K.Ya., and Cracknell, A.P. (2000). New evidence for ozone depletion over Athens, Greece, *International Journal of Remote Sensing*, **21**, 2951–2955.

Varotsos, C.A., Alexandris, D., Chronopoulos, G., and Tzanis, C. (2001a). Aircraft observations of the solar ultraviolet irradiance throughout the troposphere, *Journal of Geophysical Research*, **106**, 14843–14854.

Varotsos, C.A., Kondratyev, K.Ya., and Efstathiou, M. (2001b). On the seasonal variation of the surface ozone in Athens, Greece, *Atmospheric Environment*, **35**, 315–320.

Varotsos, C.A., Efstathiou, M.N., and Kondratyev, K.Ya. (2003). Long-term variation in surface ozone and its precursors in Athens, Greece: A forecasting tool, *Environmental Science and Pollution Research*, **10**, 19–23.

Varotsos, C.A., Cartalis, C., Vlamakis, A., Tzanis, C., and Keramitsoglou, I. (2004). The long-term coupling between column ozone and tropopause properties, *Journal of Climate*, **17**, 3843–3854.

Varotsos, C.A., Ondov, J., and Efstathioiu, M. (2005). Scaling properties of air pollution in Athens, Greece and Baltimore, Maryland, *Atmospheric Environment*, **39**, 4041–4047.

Varotsos, C.A., Ondov, J.M., Cracknell, A.P., Efstathiou, M.N., and Assimakopoulos, M.-N. (2006). Long-range persistence in global Aerosol Index dynamics, *International Journal of Remote Sensing*, **27**, 3593–3603.

Varotsos, C.A., Assimakopoulos, M.N., and Efstathiou, M. (2007). Technical note: Long-term memory effect in the atmospheric CO_2 concentration at Mauna Loa, *Atmospheric Chemistry and Physics*, **7**, 629–634.

Varotsos, C.A., Cracknell, A.P., Tzanis, C., Tsitomeneas, S., and Viras, L. (2008a). Association of the vertical ozone structure with the lower-stratospheric circulation, *International Journal of Remote Sensing*, **29**, 2685–2695.

Varotsos, C.A., Milinevsky, G., Grytsai, A., Efstathiou, M., and Tzanis, C. (2008b). Scaling effect in planetary waves over Antarctica, *International Journal of Remote Sensing*, **29**, 2697–2704.

Varotsos, C.A., Efstathiou, M., and Tzanis, C. (2009). Scaling behaviour of the global tropopause, *Atmospheric Chemistry and Physics*, **9**, 677–683.

Varotsos, C.A., Cracknell, A.P., and Tzanis, C. (2010). Major atmospheric events monitored by deep underground muon data, *Remote Sensing Letters*, **1**, 169–178.

Varotsos, C.A., Cracknell, A.P., and Tzanis, C. (2012). The exceptional ozone depletion over the Arctic in January–March 2011, *Remote Sensing Letters*, **3**, 343–352.

Varotsos, P. (1976). Comments on the formation entropy of a Frenkel defect in BaF_2 and CaF_2, *Physical Review B*, **15**, 938.

Varotsos, P. (1978). Estimate of pressure dependence of dielectric constant in alkali halides, *Physica Status Solidi B*, **90**, 339–343.

Varotsos, P. (1980). On the temperature variation of the bulk modulus of mixed alkali halides, *Physica Status Solidi B*, **99**, K93–K96.

Varotsos, P. (1981). Determination of the composition of the maximum conductivity or diffusivity in mixed alkali halides, *Journal of the Physics and Chemistry of Solids*, **42**, 405–407.

Varotsos, P. (2007). Comparison of models that interconnect point defect parameters in solids with bulk properties, *Journal of Applied Physics*, **101**, 123503, doi: 10.1063/1.2745359.

Varotsos, P. and Alexopoulos, K. (1977). Estimation of migration enthalpy and entropy for cation vacancy motion in alkali halides with NaCl type structure, *Physical Review B*, **15**, 2348–2351.

Varotsos, P.A. and Alexopoulos, K. (1978). Curvature in conductivity plots of silver-halides as a consequence of anharmonicity, *Journal of the Physics and Chemistry of Solids*, **39**, 759–761.

Varotsos, P. and Alexopoulos, K. (1979). On the possibility of the enthalpy of a Shottky defect decreasing with increasing temperature, *Journal of Phys. C: Solid State*, **12**, 761–764.

Varotsos, P. and Alexopoulos, K. (1980). Migration entropy for the bound fluorine motion in alkaline-earth fluorides, *Journal of the Physics and Chemistry of Solids*, **41**, 443–446.

Varotsos, P. and Alexopoulos, K. (1981). Migration parameters for the bound fluorine motion in alkaline-earth fluorides II, *Journal of the Physics and Chemistry of Solids*, **42**, 409–410.

Varotsos, P. and Alexopoulos, K. (1984a). Connection between the formation volume and formation Gibbs energy in noble gas solids, *Physical Review B*, **30**, 7305–7306.

Varotsos, P. and Alexopoulos, K. (1984b). Physical properties of the variations of the electric field of the earth preceding earthquakes, I, *Tectonophysics*, **110**, 73–98.

Varotsos, P. and Alexopoulos, K. (1984c). Physical-properties of the variations of the electric-field of the earth preceding earthquakes, II: Determination of epicenter and magnitude, *Tectonophysics*, **110**, 99–125.

Varotsos, P. and Alexopoulos, K. (1984d). Connection between the formation volume and formation Gibbs energy in noble-gas solids, *Physical Review B*, **30**, 7305–7306.

Varotsos, P. and Alexopoulos, K. (1986). Thermodynamics of point defects and their relation with the bulk properties, in: S. Amelinckx, R. Gevers, and J. Nihoul (Eds.), *Theory of Metal Oxidation*, North Holland, Amsterdam, The Netherlands, 474 pp.

Varotsos, P. and Lazaridou, M. (1991). Latest aspects of earthquake prediction in Greece based on seismic electric signals, *Tectonophysics*, **188**, 321–347.

Varotsos, P., Alexopoulos, K., and Nomicos, K. (1982). Comments on the pressure variation of the Gibbs energy for bound and unbound defects, *Physica Status Solidi B*, **111**, 581–590.

Varotsos, P., Alexopoulos, K., Varotsos, C., and Lazaridou, M. (1985). Interconnection of point defect parameters in BaF_2, *Physica Status Solidi A*, **88**, 137–140.

Varotsos, P., Varotsos, C.A., Hatjicontis, V., and Lazaridou M. (1986). On a plausible explanation of the connection of point-defect parameters with the melting-point, *Journal of the Physics and Chemistry of Solids*, **47**, 79–82.

Varotsos, P., Alexopoulos, K., and Lazaridou, M. (1993). Latest aspects of earthquake prediction in Greece based on seismic electric signals, II, *Tectonophysics*, **224**, 1–37.

Varotsos, P., Eftaxias, K., Lazaridou, M., Antonopoulos, G., Makris, J., and Poliyiannakis, J. (1996). Summary of the five principles suggested by Varotsos *et al.* [1996] and the additional questions raised in this debate, *Geophysical Research Letters*, **23**, 1449–1452.

Varotsos, P., Sarlis, N.V., and Lazaridou, M. (1999). Interconnection of defect parameters and stress-induced electric signals in ionic crystals, *Physical Review B*, **59**, 24–27.

Varotsos, P.A., Sarlis, N.V., and Skordas, E.S. (2002). Long-range correlations in the electric signals that precede rupture, *Physical Review E*, **66**, 011902, doi: 10.1103/PhysRevE.66.011902.

Varotsos, P.A., Sarlis, N.V., and Skordas, E.S. (2003a). Long-range correlations in the electric signals that precede rupture: Further investigations, *Physical Review E*, **67**, 21109–21121.

Varotsos, P.A., Sarlis, N.V., and Skordas, E.S. (2003b). Attempt to distinguish electric signals of a dichotomous nature, *Physical Review E*, **68**, 031106, doi: 10.1103/PhysRevE.68.031106.

Varotsos, P.A., Sarlis, N.V., and Skordas, E.S. (2003c). Electric fields that "arrive" before the time-derivative of the magnetic field prior to major earthquakes, *Physical Review Letters*, **91**, 148501.1–148501.4.

Varotsos, P.A., Sarlis, N.V., Skordas, E.S., and Lazaridou, M.S. (2004). Entropy in the natural time-domain, *Physical Review E*, **70**, 011106.

Varotsos, P.A., Sarlis, N.V., Skordas, E.S., and Lazaridou, M.S. (2005a). Natural entropy fluctuations discriminate similar-looking electric signals emitted from systems of different dynamics, *Physical Review E*, **71**, 011110, doi: 10.1103/PhysRevE.71.011110.

Varotsos, P.A., Sarlis, N.V., Tanaka, H.K., and Skordas, E.S. (2005b). Some properties of the entropy in the natural time, *Physical Review E*, **71**, 032102, doi: 10.1103/PhysRevE.71.032102.

Varotsos, P.A., Sarlis, N.V., Tanaka, H.K., and Skordas, E.S. (2005c). Similarity of fluctuations in correlated systems: The case of seismicity, *Physical Review E*, **72**, 041103.

Varotsos, P.A., Sarlis, N.V., Skordas, E.S., Tanaka, H.K., and Lazaridou, M.S. (2006a). Entropy of seismic electric signals: Analysis in the natural time under time reversal, *Physical Review E*, **73**, 031114, doi: 10.1103/PhysRevE.73.031114.

Varotsos, P.A., Sarlis, N.V., Skordas, E.S., Tanaka, H.K., and Lazaridou, M.S. (2006b). Attempt to distinguish long range temporal correlations from the statistics of the increments by natural time analysis, *Physical Review E*, **74**, 021123, doi: 10.1103/PhysRevE.74.021123.

Varotsos, P.A., Sarlis, N.V., Skordas, E.S., and Lazaridou, M.S. (2007). Identifying sudden cardiac death risk and specifying its occurrence time by analyzing electrocardiograms in natural time, *Appied Physics Letters*, **91**, 064106, doi: 10.1063/1.2768928.

Varotsos, P.A., Sarlis, N.V., and Skordas, E.S. (2011). *Natural Time Analysis: The New View of Time*, Springer/Praxis, Heidelberg, Germany/Chichester, U.K., 476 pp., ISBN 97836421 64484.

Vay, S.A., Anderson, B.E., Sache, G.W., Collins, J.E. Jr., Podolske, J.R., Twohy, C.H., Gandrud, B., Chan, K.R., Baughcum, S.L. and Wallio, H.A. (1998). DC-8-based observations of aircraft CO, CH_4, N_2O, and H_2O(g) emission indices during SUCCESS, *Geophysical Research Letters*, **25**, 1717–1720.

Veefkind, J.P., de Haan, J.F., Brinksma, E.J., Kroon, M., and Levelt, P.F. (2006). Total ozone from the ozone monitoring instrument (OMI) using the DOAS technique, *IEEE Transactions on Geoscience and Remote Sensing*, **44**, 1239–1244.

Velders, G.J.M., Andersen, S.O., Daniel, J.S., Fahey, D.W., and McFarland, M. (2007). The importance of the Montreal Protocol in protecting climate, *Proceedings of the National Academy of Sciences of the U.S.A.*, **104**, 4814–4819.

Verdes, P.F. (2007). Global warming is driven by anthropogenic emissions: A time series analysis approach, *Physical Review Letters*, **99**, 048501-1–048501-4.

Viatte, C., Schneider, M., Redondas, A., Hase, F., Eremenko, M., Chelin, P., Flaud, J.M., Blumenstock, T., and Orphal, J. (2010). Comparison of ground-based FTIR and Brewer O_3 total column with data from two different IASI algorithms and from OMI and GOME-2 satellite instruments, *Atmospheric Measurements Techniques Discussions*, **3**, 5833–5865.

Vigouroux, C., De Mazière, M., Errera, Q., Chabrillat, S., Mahieu, E., Duchatelet, P., Wood, S., Smale, D., Mikuteit, S., Blumenstock, T. *et al.* (2007). Comparisons between ground-based FTIR and MIPAS N_2O and HNO_3 profiles before and after assimilation in BASCOE, *Atmospheric Chemistry and Physics*, **7**, 377–396.

Vigouroux, C., De Mazière, M., Demoulin, P., Servais, C., Hase, F., Blumenstock, T., Kramer, I., Schneider, M., Mellqvist, J., Strandberg, A. *et al.* (2008). Evaluation of tropospheric and stratospheric ozone trends over Western Europe from ground-based FTIR network observations, *Atmospheric Chemistry and Physics*, **8**, 6865–6886.

Vinnikov, K.Y. and Grody, N.C. (2003). Global warming trend of mean tropospheric temperature observed by satellites, *Science*, **302**, 269–272.

Volkovitzky, O.A., Gavrilov, A.A., and Kaidalov, V. (1996). Is man really destroying the planetary ozone layer? *Proceedings of the Russian Academy of Sciences*, **66**, 783–786 [in Russian].

von Clarmann, T. (2006). Validation of remotely sensed profiles of atmospheric state variables: Strategies and terminology, *Atmospheric Chemistry and Physics*, **6**, 4311–4320.

von der Gathen, P., Rex, M., Harris, N.R.P., Lucic, D., Knudsen, B.M., Braathen, G.O., De Backer, H., Fabian, R., Fast, H., Gil, M. *et al.* (1995). Observational evidence for chemical ozone depletion over the Arctic in winter 1991–92, *Nature*, **375**, 131–134.

von der Gathen, P., Rex, M., Harris, N.R.P., Lucic, D., Knudsen, B.M., Braathen, G.O., De Backer, H., Fabian, R., Fast, H., Gil, M. *et al.* (2006). The ASSET intercomparison of ozone analyses: Method and first results, *Atmospheric Chemistry and Physics*, **6**, 5445–5474.

von Savigny, C., Rozanov, A., Bovensmann, H., Eichmann, K.U., Noel, S., Rozanov, V.V., Sinnhuber, B.M., Weber, M., and Burrows, J.P. (2005). The ozone hole breakup in September 2002 as seen by SCIAMACHY on ENVISAT, *Journal of the Atmospheric Sciences*, **62**, 721–734.

von Scheele, F. (1997). *Odin Programme Status and First Test Results* (IAF-97IAA.11.1.08), Swedish Space Corporation, Solna, Sweden.

Vyushin, D.I., Fioletov, V.E., and Shepherd, T.G. (2007). Impact of long-range correlations on trend detection in total ozone, *Journal of Geophysical Research*, **112**, doi: 10.1029/2006JD008168.

Walker, K.A., Randall, C.E., Trepte, C.R., Boone, C.D., and Bernath, P.F. (2005). Initial validation comparisons for the Atmospheric Chemistry Experiment (ACE-FTS), *Geophysical Research Letters*, **32**, L16S04, doi: 10.1029/2005GL022388.

Wallington, T.J., Schneider, W.F., Sehested, J., and Nielsen, O.J. (1995). Hydrofluorocarbons and stratospheric ozone, *Faraday Discussions*, **100**, 55–64.

Walshaw, C.D. (1989). G.M.B. Dobson: The man and his work, *Planetary and Space Science*, **37**, 1485–1507.

Wamsley, P.R., Elkins, J.W., Fahey, D.W., Dutton, G.S., Volk, C.M., Myers, R.C., Montzka, S.A., Butler, J.H., Clarke, A.D., Fraser, P.J. *et al.* (1998). Distribution of halon-1211 in the upper troposphere and lower stratosphere and the 1994 total bromine budget, *Journal of Geophysical Research*, **103**, 1513–1526.

Wang, C., Prinn, R.G., and Sokolov, A. (1998). A global interactive chemistry and climate model: Formulation and testing, *Journal of Geophysical Research*, **103**, 3399–3418.

Wang, D.Y., von Clarmann, T., Fischer, H., Funke, B., Garcia-Comas, M., Gil-Lopez, S., Glatthor, N., Grabowski, U., Hopfner, M., Kellmann, S. *et al.* (2005). Longitudinal variations of temperature and ozone profiles observed by MIPAS during the Antarctic stratosphere sudden warming of 2002, *Journal of Geophysical Research*, **110**, D20101.

Wang, D.Y., Hopfner, H., and Mengistu Tsidu, G. (2007). Validation of nitric acid retrieved by the IMK-IAA processor from MIPAS/ENVISAT measurements, *Atmospheric Chemistry and Physics*, **7**, 1–18.

Wang, H.J., Cunnold, D.M., Thomason, L.W., Zawodny, J.M., and Bodeker, G.E. (2002). Assessment of SAGE version 6.1 ozone data quality, *Journal of Geophysical Research*, **107**, 4691–4709.

Wang, H.S., Cunnold, D.M., and Bao, X. (1996). A critical analysis of Stratospheric Aerosol and Gas Experiment ozone trends, *Journal of Geophysical Research*, **101**, 12495–12514.

Wang, P., Stammes, P., van der A, R., Pinardi, G., and van Roozendael, M. (2008). FRESCO+: An improved O_2 A-band cloud retrieval algorithm for tropospheric trace gas retrievals, *Atmospheric Chemistry and Physics*, **8**, 6565–6576.

Wang, W.-C. and Isaksen, I.S.A. (Eds.) (1995). *Atmospheric Ozone as a Climate Gas: General Circulation Model Simulations*, Springer-Verlag, Berlin, 461 pp.

Wang, W.-C. and Sze, N.D. (1980). Coupled effects of atmospheric N_2O and O_3 on the Earth's climate, *Nature*, **286**, 589–590.

Wang, W.-C., Zhuang, Y.C., and Bojkov, R.D. (1993). Climate implications of observed changes in ozone vertical distributions in middle and high latitudes of the Northern Hemisphere, *Geophysical Research Letters*, **20**, 1567–1570.

Wang, W.-C., Liang, X.-Z, Dudek, M.P., Pollard, D., and Thompson, S.L. (1995). Atmospheric ozone as a climate gas, *Atmospheric Research*, **37**, 247–256.

Wang, W.-C., Mao, H., Isaksen, I.S.A., Fuglestvelt, J.S., and Karlsdottir, S. (1998). Indirect effects of increasing atmospheric methane on the radiative forcing through climate–chemistry interactions, in: R.D. Bojkov and G. Visconti (Eds.), *Atmospheric Ozone: Proceedings of the XVIII Quadrennial Ozone Symposium, L'Aquila, Italy, September 12–21, 1996*, Consorzio Parco Scientifico e Tecnologico d'Abruzzo, L'Aquila, Italy, Vol. 2, pp. 821–826.

Wang, Y.M., Wang, Y.J., Wang, W.H., Zhang, Z.M., Liu, J.G., Fu, L.P., Jiang, F., Chen, J., Wang, J.H., Guan, F.J. *et al.* (2009). FY-3 satellite Ultraviolet Total Ozone Unit, *Chinese Science Bulletin*, 1–6, doi: 10.1007/s11434-009-0335-8.

Wardel, D.I., Kerr, J.B., McElroy, C.T., and Francis, D.R. (Eds.) (1997). *Ozone Science: A Canadian Perspective on the Changing Ozone Layer*, University of Toronto Press, Toronto, Canada, 119 pp.

Waters, J.W., Froidevaux, L., Read, W.G., Manney, G.L., Elson, L.S., Flower, D.A., Jarnot, R.F., and Harwood, R.S. (1993). Stratospheric ClO and ozone from the Microwave Limb Sounder on the Upper Atmosphere Research Satellite, *Nature*, **362**, 597–602.

Waters, J.W., Froidevaux, L., Harwood, R.S., Jarnot, R.F., Pickett, H.M., Read, W.G., Siegel, P.H., Cofield, R.E., Filipiak, M.J., Flower, D.A. *et al.* (2006). The Earth Observing System Microwave Limb Sounder (EOS MLS) on the Aura satellite, *IEEE Transactions on Geoscience and Remote Sensing*, **44**, 1075–1092.

Watson, R.T., Geller, M.A., Stolarski, R.S., and Hampson, R.F. (1986). Present state of knowledge of the upper atmosphere: An assessment report, *Processes that Control Ozone and Other Climatically Important Trace Gases* (NASA RP-1162), NASA GSFC, Greenbelt, MD.

Wauben, W.M.F., Fortuin, J.P.F., and Kelder, H.M. (1998). Sensitivity of radiative forcing due to changes in the global distribution of ozone with application to ozone changes from aviation, in: R.D. Bojkov and G. Visconti (Eds.), *Atmospheric Ozone: Proceedings of the XVIII Quadrennial Ozone Symposium, L'Aquila, Italy, September 12–21, 1996*, Consorzio Parco Scientifico e Tecnologico d'Abruzzo, L'Aquila, Italy, Vol. 2, pp. 827–830.

Waugh, D.W. and Hall, T.M. (2002). Age of stratospheric air: Theory, observations, and models, *Reviews of Geophysics*, **40**, 1010, doi: 10.1029/2000RG000101.

Waugh, D.W., Oman, L., Kawa, S.R., Stolarski, R.S., Pawson, S., Douglass, A.R., Newman, P.A., and Nielsen, J.E. (2009). Impacts of climate change on stratospheric ozone recovery, *Geophysical Research Letters*, **36**, L03805, doi: 10.1029/2008GL036223.

Weatherhead, E.C. and Andersen, S.B. (2006). The search for signs of recovery of the ozone layer. *Nature*, **441**, 39–45.

Weatherhead, E.C., Reinsel, G.C., Tiao, G.C., Jackman, C.H., Bishop, L., Frith, S.M.H., DeLuisi, J., Keller, K., Oltmans, S., Fleming, E. *et al.* (2000). Detecting the recovery of the total column ozone, *Journal of Geophysical Research*, **105**, 22201–22210.

Weber, M., Burrows, J.P., and Cebula, R.P. (1998). GOME solar UV/VIS irradiance measurements between 1995 and 1997: First results on proxy solar activity studies, *Solar Physics*, **177**, 63–77.

Weber, M., Lamsal, L.N., Coldewey-Egbers, M., Bramstedt, K., and Burrows, J.P. (2005). Pole-to-pole validation of GOME WFDOAS total ozone with ground based data, *Atmospheric Chemistry and Physics*, **5**, 1341–1355.

Weber, R.O. and Talkner, P. (2001). Spectra and correlations of climate data from days to decades, *Journal of Geophysical Research*, **106**, 20131–20144.

Webster, E.W. (1923). *Aristotle's "Meteorologica I"*, Oxford University Press, Oxford, U.K., pp. 348b–349a.

Wehr, T. (2002). The European Space Agency's present and future atmospheric composition sounding missions, *Advances in Space Research*, **29**, 1603–1608.

Weinstock, E.M. and Anderson, J.G. (1989). On the continuing disagreement between in situ and satellite stratospheric ozone measurements, in: R.D. Bojkov and P. Fabian (Eds.), *Proceedings of the Quadrennial Ozone Symposium, 1988, Göttingen, Germany*, Deepak Publications, Hampton, VA, pp. 238–242.

Weinstock, E.M., Phillips, M.J., and Anderson, J.G. (1981). In situ observations of ClO in the stratosphere: A review of recent results, *Journal of Geophysical Research*, **86**, 7273–7278.

Weisenstein, D.K., Ko, M.K.W., Dyominov, I.G., Pitari, G., Ricciardulli, L., Visconti, G., and Bekki, S. (1998). The effect of sulfur emissions from HSCT aircraft: A 2-D model intercomparison, *Journal of Geophysical Research*, **103**, 1527–1547.

Wellemeyer, C.G., Taylor, S.L., Jaross, G., Deland, M.T., Seftor, C.J., Labow, G., Swissler, T.J., and Cebula, R.P. (1996). *Final Report on Nimbus-7 TOMS Version 7 Calibration* (NASA Contractor Report 4717), NASA, Washington, D.C.

Werner, J., Rothe, K.W., and Walther, H. (1983). Monitoring of the ozone stratospheric layer by laser radar, *Applied Physics B*, **32**, 113–118.

Werner, R., Stebel, K., Hansen, H.G., Hoppe, U.-P., Gausa, M., Kivi, R., von der Gathen, P., Orsolini, Y., and Kilifarska, N. (2011). Study of the seasonal ozone variations at European high latitudes, *Advances in Space Research*, **47**, 740–747.

Wessel, S., Aoki, S., Winkler, P., Weller, R., Herber, A., Gernandt, H., and Schrems, O. (1998). Tropospheric ozone depletion in polar regions: A comparison of observations in the Arctic and Antarctic, *Tellus B*, **50**, 34–50.

Wetzel, G., Oelhaf, H., Friedl-Vallon, F., Kleinert, A., Lengel, A., Maucher, G., Nordmeyer, H., Ruhnke, R., Nakajima, H., Sasano, Y. *et al.* (2006). Intercomparison and validation of ILAS-II version 1.4 target parameters with MIPAS-B measurements, *Journal of Geophysical Research*, **111**, D11S06, doi: 10.1029/2005JD006287.

Wilcoxon, F. (1945). Individual comparisons by ranking methods, *Biometrics*, **1**, 80–83.

Williamson, D.L. and Rasch, P.J. (1994). Water vapor transport in the NCAR CCM2, *Tellus A*, **46**, 34–51.

Windsor, H.L. and Toumi, R. (2001). Scaling and persistence of UK pollution, *Atmospheric Environment*, **35**, 4545–4556.

Witte, J.C., Schoeberl, M.R., Douglass, A.R., and Thompson, A.M. (2008). The quasi-biennial oscillation and annual variations in tropical ozone from SHADOZ and HALOE, *Atmospheric Chemistry and Physics*, **8**, 3929–3936.

Wittrock, F., Eisinger, M., Ladstatter-Weibenmayer, A., Richter, A., and Burrows, J. P. (1998). Ground based UV–VIS measurements of O_3, NO_2, OClO, and BrO over Ny-Ålesund (79°N, 12°E), in: R.D. Bojkov and G. Visconti (Eds.), *Proceedings of the XVIII Quadrennial Ozone Symposium, September 12–21, 1996, L'Aquila, Italy*, Consorzio Parco Scientifico e Tecnologico d'Abruzzo, L'Aquila, Italy, Vol. 2, pp. 623–626.

WMO-GAW Report No. 29 (1993). *Handbook for Dobson Ozone Data Re-evaluation*, WMO Global Ozone Monitoring Project, WMO/TD-No. 597.

WMO (1957). Definition of the tropopause. *WMO Bulletin*, **6**, 136.

WMO (1975). *Statement on Modification of the Ozone Layer Due to Human Activities and Some Possible Geophysical Consequences* (WM/R/ST2, annex), World Meteorological Organization, Geneva, Switzerland.

WMO (1978). *International Operations Handbook for Measurement of Background Atmospheric Pollution* (WMO-No. 491). World Meteorological Organization, Geneva, Switzerland, 110 pp.

WMO (1981). *The Stratosphere 1981: Theory and MeasurementS* (Report No. 11), World Meteorological Organization, Geneva, Switzerland.

WMO (1986). *Atmospheric Ozone 1985: Assessment of Our Understanding in the Processes Controlling Its Present Distribution and Change* (Global Ozone Research and Monitoring Project, Report No. 16), World Meteorological Organization, Geneva, Switzerland, 1150 pp.

WMO (1989). *Scientific Assessment of Ozone Depletion: 1988* (Global Ozone Research and Monitoring Project, Report No. 20), World Meteorological Organization, Geneva, Switzerland.

WMO (1990). *Trends in Total Column Ozone Measurements* (1990 Report of the International Ozone Trends Panel, Report No. 18), World Meteorological Organization, Geneva, Switzerland.

WMO (1991). *Scientific Assessment of Ozone Depletion: 1991* (Report No. 25), World Meteorological Organization, Geneva, Switzerland.

WMO (1995). *Scientific Assessment of Ozone Depletion: 1994* (Global Ozone Research and Monitoring Project, Rep. 37), Ozone Secretariat, World Meteorological Organization, Geneva, Switzerland, 457 pp.

WMO (1996). *Report of the Fifth WMO Meting of Experts on Quality Assurance/Science Activity Centers (QA/SACs) of the Global Atmosphere Watch, 1996* (WMO/GAW No. 114), World Meteorological Organization, Geneva, Switzerland.

WMO (1999). *Scientific Assessment of Ozone Depletion: 1998* (Global Ozone Research and Monitoring Project, Report No. 44), World Meteorological Organization, Geneva, Switzerland.

WMO (2003). *Scientific Assessment of Ozone Depletion: 2002* (WMO Global Ozone Research and Monitoring Project, Report No. 47), World Meteorological Organization, Geneva, Switzerland, 498 pp.

WMO (2007). *Scientific Assessment of Ozone Depletion: 2006* (Global Ozone Research and Monitoring Project, Report No. 50), World Meteorological Organization, Geneva, Switzerland, 572 pp.

WMO (2008). *Operations Handbook: Ozone Observations with a Dobson Spectrophotometer* (Revised 2008, GAW Rep. 183), World Meteorological Organization, Geneva, Switzerland, 93 pp.

WMO (2010a). *Scientific Assessment of Ozone Depletion: 2010* (Global Ozone Research and Monitoring Project, Report No. 52), World Meteorological Organization, Geneva, Switzerland.

WMO (2010b). Executive summary, *Scientific Assessment of Ozone Depletion: 2010*, Scientific Assessment Panel of the Montreal Protocol on Substances that Deplete the Ozone Layer (*http://www.unep.org/PDF/PressReleases/898_ExecutiveSummary_EMB.pdf*) or (*http://ozone.unep.org/Assessment_Panels/SAP/ExecutiveSummary_SAP_2010.pdf*

Wofsky, S.C. and McElroy, M.B. (1974). HO_x, NO_x and ClO_x: Their role in atmospheric photochemistry, *Canadian Journal of Chemistry*, **52**, 1582–1591.

Wofsy, E., Gottlieb, W., Lait, L.R., Newman, P.A., Schoeberl, M.R., Loewenstein, M., Podolske, J.R., Strahan, S.E., Proffitt, M.H., Webster, C.R. *et al.* (1993). Chemical loss of ozone in the Arctic polar vortex in the winter of 1991–1992, *Science*, **261**, 1146–1149.

Wold, H. (1938). *A Study in the Analysis of Stationary Time Series*, Almqvist & Wiksell, Uppsala, Sweden, 214 pp.

Wuebbles, D.J. (1983). Chlorocarbon emission scenarios: Potential impact on stratospheric ozone, *Journal of Geophysical Research*, **88**, 1433–1443.

Wuebbles, D.J. (1996). Three-dimensional chemistry in the greenhouse: An editorial comment, *Climate Change*, **34**, 397–404.

Wuebbles, D. and Chang, J.S. (1979). A theoretical study of stratospheric trace species variations during a solar eclipse, *Geophysical Research Letters*, **6**, 179–182.

Wuebbles, D.J., Wei, C.-F., and Patten, K.O. (1998). Effects of stratospheric ozone and temperature during the Maunder Minimum, *Geophysical Research Letters*, **25**, 523–526.

Wyser, K. and Ström, J. (1998). A possible change in global radiative forcing due to aircraft exhaust, *Geophysical Research Letters*, **25**, 1673–1676.

Yang, E.-S., Cunnold, D.M., Newchurch, M.J., and Salawitch, R. J. (2005). Change in ozone trends at southern high latitudes, *Geophysical Research Letters*, **32**, L12812, doi: 10.1029/2004GL022296.

Yang, H. and Tung, K.K. (1994). Statistical significance and pattern of extratropical QBO in column ozone, *Geophysical Research Letters*, **21**, 2235–2238.

Yang, H. and Tung, K.K. (1995). On the phase propagation of extratropical ozone quasi-biennial oscillation in observational data, *Journal of Geophysical Research*, **100**, 9091–9100.

Yang, S-K., Long, C.S., Miller, A.J., He, X., Yang, Y., Wuebbles, D.J., and Tiao, G. (2009). Modulation of natural variability on the trend analysis of the updated cohesive SBUV(/2) total ozone, *International Journal of Remote Sensing*, **30**, 3975–3986.

Yates, S.R., Wang, D., Gan, J., Ernst, F.F., and Yury, W.A. (1998). Minimizing methyl bromide emissions from soil fumigation, *Geophysical Research Letters*, **25**, 1633–1636.

Yela, M., Rodriguez, S., Gil, M., and Caceneuve, H. (1998). NO_2 and O_3 total columns at three different latitudes in the Antarctic region from ground-based visible spectroscopy, in: R.D. Bojkov and G. Visconti (Eds.), *Proceedings of the XVIII Quadrennial Ozone Symposium, L'Aquila, Italy, September 12–21, 1996*, Consorzio Parco Scientifico e Tecnologico d'Abruzzo, L'Aquila, Italy, Vol. 2, 237–240.

Yela, M., Parrondo, C., Gil, M., Rodriguez, S., Araujo, J., Ochoa, H., Deferrari, G., and Diaz, S. (2005). The September 2002 Antarctic vortex major warming as observed by visible spectroscopy and ozone soundings, *International Journal of Remote Sensing*, **26**, 3361–3376.

Yoshiki, M. and Sato, K. (2000). A statistical study of gravity waves in the polar regions based on operational radiosonde data, *Journal of Geophysical Research*, **105**, 17995–18011.

Yushkov, V., Dorokhov, V., Khattatov, V., Lukyanov, A., Zaitcev, I., Zvetkova, N., Nakane, H., Akiyoshi, H., and Ogawa, T. (1998). Evidence of ozone depletion over Yakutsk, Eastern Siberia, in 1995, in: R.D. Bojkov and G. Visconti (Eds.), *Proceedings of the XVIII Quadrennial Ozone Symposium, L'Aquila, Italy, September 12–21, 1996*, Consorzio Parco Scientifico e Tecnologico d'Abruzzo, L'Aquila, Italy, Vol. 1, pp. 241–246.

Yushkov, V., Oulanovsky, A., Lechenuk, N., Roudakov, I., Arshinov, K., Tikhonov, F., Stefanutti, L., Ravegnani, F., Bonafe, U., and Georgiadis, T. (1999). A chemiluminescent analyzer for stratospheric measurements of the ozone concentration (FOZAN), *Journal of Atmospheric and Oceanic Technology*, **16**, 1345–1350.

Zahn, A., Barth, V., and Platt, U. (1998). Airborne high-resolution ozone and potential vorticity to study stratospheric–tropospheric exchange, in: R.D. Bojkov and G. Visconti (Eds.), *Atmospheric Ozone: Proceedings of the XVIII Quadrennial Ozone Symposium, L'Aquila, Italy, September 12–21, 1996*, Consorzio Parco Scientifico e Tecnologico d'Abruzzo, L'Aquila, Italy, Vol. 1, pp. 183–185.

Zanis, P., Maillard, E., Staehelin, J., Zerefos, C., Kosmidis, E., Tourpali, K., and Wohltmann, I. (2006). On the turnaround of stratospheric ozone trends deduced from the reevaluated Umkehr record of Arosa, Switzerland, *Journal of Geophysical Research*, **111**, D22307, doi: 10.1029/2005JD006886.

Zeng, G. and Pyle, J.A. (2005). Influence of El Niño Southern Oscillation on stratosphere/troposphere exchange and the global troposheric ozone budget, *Geophysical Research Letters*, **32**, doi: 10.1029/2004GL021353.

Zepp, R.G., Callaghan, T.V., and Erickson, D.J. III (2003). Interactive effects of ozone depletion and climate change on biogeochemical cycles, *Photochemical and Photobiological Sciences*, **2**, 51–61.

Zerefos, C.S. and Bais, A.F. (Eds.) (1997). *Solar Ultraviolet Radiation: Modelling, Measurements and Effect* (NATO ASI Series I: Global Environmental Change), Springer, Berlin, Vol. 52, 332 pp.

Zerefos, C.S. and Ghazi, A. (Eds.) (1985). *Atmospheric Ozone: Proceedings of the Quadrennial Ozone Symposium, September 3–7, 1984, Halkidiki, Greece*, D. Reidel, Dordrecht, The Netherlands, 842 pp.

Zerefos, C.S., Bais, A.F., Ziomas, I.C., and Bojkov, R.D. (1992). On the relative importance of quasi-biennial oscillation and El Niño Southern oscillation in the revised Dobson total ozone records, *Journal of Geophysical Research*, **97**, 10135–10144.

Zerefos, C., Meleti, C., Balis D., Tourpali K., and Bais, A.F. (1998). Quasibiennial and longer-term changes in clear-sky UV-B solar irradiance, *Geophysical Research Letters*, **25**, 4345–4348.

Zerefos, C.S., Balis, D.S., Meleti, C., Bais, A.F., Tourpali, K., Kourtidis, K., Vanicek, K., Cappellani, F., Kaminski, U., Colombo, T. *et al.* (2000). Changes in surface solar UV irradiances and total ozone during the solar eclipse of August 11, 1999, *Journal of Geophysical Research*, **105**, D21, doi: 10.1029/2000JD900412.

Zerefos, C., Balis, D., Tzortziou, M., Bais, A., Tourpali, K., Meleti, C., Bernhard, G., and Herman, J. (2001a). A note on the interannual variations of UV-B erythemal doses and solar irradiance from ground-based and satellite observations, *Annales Geophysicae*, **19**, 115–120.

Zerefos, C.S., Balis, D.S., Zanis, P., Meleti, C., Bais, A.F., Tourpali, K., Melas, D., Ziomas, I., Galani, E., Kourtidis *et al.* (2001b). Changes in surface UV solar irradiance and ozone over the Balkans during the eclipse of August 11, 1999, *Advances in Space Research*, **27**, 1955–1963.

Zerefos, C.S., Gerasopoulos, E., Tsagouri, I., Psiloglou, B., Belehaki, A., Herekakis, T., Bais, A., Kazadzis, S., Eleftheratos, C., Kalivitis, N. *et al.* (2007). Evidence of gravity waves into the atmosphere during the March 2006 total solar eclipse, *Atmospheric Chemistry and Physics Discussions*, **7**, 7603–7624.

Zerefos, C., Contopoulos, G., and Skalkeas, G. (2009). *Twenty Years of Ozone Decline: Proceedings of the Symposium for the 20th Anniversary of the Montreal Protocol*, Springer-Verlag, Berlin.

Zhang, L., Jacob, D.J., Bowman, K.W., Logan, J.A., Turquety, S., Hudman, R.C., Li, Q., Beer, R., Worden, H.M., Worden, J.R. *et al.* (2006). Ozone–CO correlations determined by the TES satellite instrument in continental outflow regions, *Geophysical Research Letters*, **33**, L18804, doi: 10.1029/2006GL026399.

Zhang, L., Jacob, D.J., Liu, X., Logan, J.A., Chance, K., Eldering, A., and Bojkov, B.R. (2010). Intercomparison methods for satellite measurements of atmospheric composition: Application to tropospheric ozone from TES and OMI, *Atmospheric Chemistry and Physics Discussions*, **10**, 1417–1456.

Zhong, W.R., Toumi, R., and Haigh, J.D. (1996). Climate forcing by stratospheric ozone depletion calculated from observed temperature trends, *Geophysical Research Letters*, **23**, 3183–3186.

Zhu, J. and Liu, Z. (2003). Long-range persistence of acid deposition, *Atmospheric Environment*, **37**, 2605–2613.

Ziemke, J.R., Chandra, S., McPeters, R.D., and Newman P.A. (1997). Dynamical proxies of column ozone with applications to global trend models, *Journal of Geophysical Research*, **102**, 6117–6129.

Ziemke, J.R., Herman, J.R., Stanford, J.L., and Bhartia, P.K. (1998). Total ozone/UVB monitoring and forecasting: Impact of clouds and horizontal resolution of satellite retrievals, *Journal of Geophysical Research*, **103**, 3865–3871.

Ziemke, J.R., Chandra, S., Herman, J., and Varotsos, C. (2000). Erythemally weighted UV trends over northern latitudes derived from Nimbus-7 TOMS measurements, *Journal of Geophysical Research*, **105**, 7373–7382.

Ziemke, J.R., Chandra, S., and Bhartia, P.K. (2005). A 25-year data record of atmospheric ozone in the Pacific from Total Ozone Mapping Spectrometer (TOMS) cloud slicing: Implications for ozone trends in the stratosphere and troposphere, *Journal of Geophysical Research*, **110**, D15105, doi: 10.1029/2004JD005687.

Ziereis, H., Schlager, H., Schulte, P., and Klemm, O. (1998). Aircraft observations of the NO_x abundance and variability in the upper troposphere over the North Atlantic, in: R.D. Bojkov and G. Visconti (Eds.), *Atmospheric Ozone: Proceedings of the XVII Quadrennial Ozone Symposium, L'Aquila, Italy, September 12–21, 1996*, Consorzio Parco Scientifico e Tecnologico d'Abruzzo, L'Aquila, Italy, Vol. 2, pp. 647–650.

Zillman, J.W. (1997). Atmospheric science and public policy, *Science*, **276**, 1084–1086.

Ziomas, I.C., Tzoumaka, P., Balis, D., Melas, D., Zerefos, C.S., and Klemm, O. (1998). Ozone episodes in Athens, Greece: A modelling approach using data from the medcaphot-trace, *Atmospheric Environment*, **32**, 2313–2321.

Zommerfeld, W. and Kunzi, K.F. (1989). Operational ground-based ozone sensor (OGOS), in: D. Bojkov and P. Fabian (Eds.), *Proceedings of the Quadrennial Ozone Symposium, 1988, Göttingen, Germany*, R. Deepak Publications, Hampton, VA, pp. 784–787.

Zou, H., Zhou, L.B., and Jian, Y.X. (2002). An observational study on the vertical distribution and synoptic variation of ozone in the Arctic, *Advances in Atmospheric Sciences*, **19**, 855–862.

Zuev, V.V. (2005). Reconstruction and prediction of long-period variations of ozonosphere using ozonometric and dendrochronological data, *International Journal of Remote Sensing*, **26**, 3631–3639.

Zuev, V. and Efstathiou, M. (2006). Is there any more long-memory effect in the total ozone content? *Kocmoca*, **3**, 37–39.

Zuev, V.E. and Kabanov, M.V. (1987). *Optics of Atmospheric Aerosols*, Gidrometeoizdat, Leningrad, 259 pp. [in Russian].

Zwick, A. (1998). *Climate Change Research and Policy* (Update No. 11, EUR 18088 EN), Joint Research Centre, Seville, Spain, 78 pp.

Index

above sea level, 45
absorption, 3, 4, 5, 7, 8, 21, 27, 35, 37, 39, 40, 41, 42, 43, 44, 45, 46, 49, 50, 51, 52, 53, 55, 56, 58, 66, 69, 70, 72, 73, 74, 77, 78, 82, 83, 84, 85, 87, 89, 91, 92, 93, 95, 97, 99, 101, 107, 108, 110, 120, 150, 165, 178, 185, 198, 208, 211, 214, 223, 232, 246,282, 283, 365, 396, 436, 508, 518, 541
Advanced Earth Observing Satellite, 93
Advanced Pollution Instruments, 78
Advanced TOVS, 84
Aerosol Limb Absorption, 119
aerosols, 9, 19, 24, 53, 56, 73, 75, 86, 92, 97, 99, 101, 108, 112, 115, 119, 120, 124, 129, 255, 328, 346, 347, 393, 403, 415, 465, 471, 472, 486, 495, 506, 507, 508, 524, 532, 541, 544, 547, 549, 556
  atmospheric, 55
  stratospheric, 54, 108, 255, 462, 486
air pollution, 70, 321, 322, 324, 544
airborne, 50, 58, 61, 68, 71, 74, 80, 120, 202, 204, 205, 241, 283, 340, 360, 362, 363
Airborne Arctic Stratosphere Expedition, 362
Airborne Submillimeter Receiver, 75
albedo, 87, 91, 261, 501, 508, 517, 519, 527, 528, 529, 537
Alfred Wegener Institut, 470
Alps, 45, 57, 166, 167, 349, 390
Antarctic, 3, 45, 80, 85, 101, 104, 110, 124, 130, 143, 145, 164, 172, 234, 235, 272, 285, 286, 339, 344, 346, 351, 352, 354, 355, 356, 357, 358, 359, 360, 361, 362, 363, 364, 381, 382, 383, 385, 386, 388, 389, 400, 406, 407, 411, 413, 414, 415, 421, 425, 427, 428,429, 430, 431, 432, 433, 434, 435, 437, 438, 439, 440, 441, 442, 447, 449, 450, 451, 453, 454, 457, 467, 468, 469, 470, 471, 491, 493, 501, 517, 525, 526, 527, 528, 531, 533, 542, 554, 555, 557
Antarctic Oscillation, 400
Applications Explorer Mission-2, 107
Arctic, 46, 47, 58, 74, 75, 80, 103, 104, 112, 124, 130, 140, 141, 142, 143, 144, 145, 172, 190, 198, 199, 201, 204, 208, 209, 219, 250, 251, 256, 272, 274, 285, 286, 339, 362, 363, 388, 400, 405, 406, 407, 408, 409, 410, 413, 414, 425, 431, 432, 433, 434, 435, 436, 437, 438, 439, 440, 458, 469, 470, 471, 477, 491, 498, 526, 531, 533, 534, 555, 557
Arctic Oscillation, 256, 400
Arosa, 7, 10, 12, 13, 14, 15, 16, 17, 51, 149, 259, 315, 316, 317, 349

Assessment Report, 465, 534
atmosphere, 1, 2, 3, 4, 6, 7, 8, 9, 10, 12, 14, 16, 18, 20, 21, 22, 24, 25, 26, 28, 30, 32, 34, 35, 36, 37, 38, 40, 42, 44, 45, 46, 48, 50, 51, 52, 54, 55, 56, 58, 60, 62, 63, 64, 66, 68, 70, 72, 73, 74, 76, 78, 79, 80, 82, 84, 85, 87, 94, 95, 97, 98, 99, 101, 106, 107, 108, 109, 110, 115, 119, 120, 121, 122, 124, 125, 126, 127, 128, 129, 141, 142, 149, 151, 157, 161, 188, 208, 229, 235, 254, 256, 257, 272, 300, 301, 306, 314, 317, 322, 323, 340, 341, 342, 343, 346, 351, 358, 378, 379, 381, 382, 383, 385, 387, 395, 399, 406, 419, 429, 431, 442, 451, 454, 457, 458, 462, 470, 477, 478, 479, 487, 490, 492, 498, 503, 505, 506, 507, 515, 517, 525, 526, 528, 531, 532, 534, 535, 536, 538, 539, 544, 545, 546, 547, 548, 551
atmospheric circulation, 5, 261, 280, 283, 485, 506, 531, 546
  pressure, 21, 298
  temperature, 101, 537, 542
Atmospheric Chemistry & Climate, 268
Atmospheric Environment Service, 5, 35
Atmospheric Infrared Sounder, 250
Atmospheric Trace Molecule Spectroscopy, 45, 108
Aura Validation Experiment, 249

backscattered, 56, 73, 82, 87, 89, 92, 94, 95, 96, 97, 118, 156, 246, 258, 399
Backscattered Ultraviolet, 93
Balloon Experiment on Standards for Ozone Sondes, 63
Balloon Intercomparison Campaign, 120
Balloon Microwave Limb Sounder, 68
Beer's law, 47, 69
biogeochemical cycles, 491, 549
biosphere, 407, 491, 547
boundary
  conditions, 459, 466, 467, 542
  layer, 146, 324, 494, 496, 516
Brewer, Alan W, 34
British Antarctic Survey, 352
catalytic reaction, 342, 357
Central Aerological Observatory, 89
CFCs, 3, 151, 340, 341, 343, 344, 345, 346, 347, 348, 349, 351, 355, 357, 358, 361, 363, 366, 367, 368, 369, 372, 373, 376, 378, 381, 385, 386, 413, 425, 455, 459, 462, 479, 480, 481, 489, 492, 498, 502, 504, 542, 545, 556
chaotic, 537, 553
chemical reactions, 1, 126, 130, 321, 363, 415, 420, 421, 488, 490, 545, 546
chemiluminescence, 67, 71, 77
CHEmistry of Ozone in the Polar Stratosphere, 362
chronology, 344, 366
climate
  change, 1, 73, 92, 118, 143, 164, 273, 297, 314, 318, 371, 372, 442, 449, 451, 477, 478, 485, 490, 491, 492, 493, 496, 497, 498, 503, 512, 515, 528, 532, 534, 535, 537, 542, 543, 544, 548, 549, 553
  modeling, 455, 515, 524, 526, 556
  system, 144, 515, 520, 528, 547, 548, 549, 551, 553, 554
Climate Monitoring and Diagnostics Laboratory, 10
clouds, 9, 24, 25, 43, 54, 68, 85, 87, 90, 97, 98, 110, 115, 121, 130, 140, 141, 142, 143, 156, 225, 226, 227, 231, 235, 236, 243, 346, 357, 359, 363, 364, 407, 415, 433, 448, 486, 492, 495, 501, 506, 508, 510, 533, 537, 549, 550, 556
Committee on Earth Observation Satellites, 255
Community Atmosphere Model, 268
correlation coefficient, 52, 72, 110, 152, 153, 156, 157, 158, 160, 188, 216, 249, 250, 257, 260, 273, 293, 306, 307, 309, 312, 314, 317, 335, 356, 418, 419, 422, 425, 428, 437, 474, 475, 476, 538, 547
Cross-track Infrared Sounder, 100
Cryogenic Infrared Radiance Instrument for Shuttle, 124
Cryogenic Limb Array Etalon Spectrometer, 122
cyclones, 306, 408

database, 45, 70, 74, 129, 130, 144, 178, 202, 223, 244, 268, 271, 305, 331, 467
Department of Defense, 99
deterministic, 306, 443, 553
development, 5, 43, 70, 75, 79, 80, 98, 150, 161, 164, 255, 257, 261, 275, 299, 301, 312, 322, 340, 344, 363, 371, 372, 389, 407, 455, 456, 457, 463, 464, 467, 517, 532, 534, 550
Direct Sun, 10, 15
Dobson Units, 13, 52

Earth Observing System, 2, 68, 90, 108, 305
Earth Radiation Budget, 68, 108, 211
Earth Radiation Budget Satellite, 108
Earth Resources Satellite, 96
East, 38, 103, 263, 403, 405, 410, 454, 479, 507, 508, 510, 523, 525
eclipse, 91, 323, 324, 325, 326, 327, 328
electrochemical, 61, 63, 64, 65, 75, 187
Electrochemical Ozone Sensor, 75
Electron Capture Detector, 343
emissions, 1, 118, 122, 128, 271, 272, 273, 293, 321, 346, 351, 372, 382, 389, 390, 451, 462, 465, 478, 479, 480, 481, 482, 489, 490, 491, 492, 493, 495, 502, 503, 510, 511, 512, 515, 516, 523, 524, 530, 536, 537, 539, 541, 544, 545, 554
environment, 99, 143, 367, 491
Environment Canada, 5, 35, 113, 187, 251
Environmental Data Records, 100
ENVIronmental SATellite, 79, 157
Environmental Science Corporation, 254
ENVISAT Stratospheric Aircraft and Balloon Campaigns, 162
Europe, 5, 11, 12, 14, 15, 36, 38, 61, 92, 100, 103, 140, 142, 143, 259, 279, 299, 307, 321, 362, 408, 409, 413, 414, 454, 472, 481, 506, 507, 508, 509, 510, 511, 516, 524, 539
European Arctic Stratospheric Ozone Experiment, 141, 362
European Space Agency, 79, 82, 83, 125, 126, 127, 157, 468
Everest-Pyramid, 297
exchange, 125, 142, 156, 161, 274, 281, 282, 293, 294, 295, 296, 297, 304, 305, 317, 346, 348, 410, 413, 486, 492, 493, 530, 532, 533, 534, 545, 547
experiment, 14, 15, 17, 25, 55, 68, 82, 107, 109, 112, 128, 253, 470, 471, 512, 526, 528, 535, 538

Fast Fourier Transform, 109
Finnish Meteorological Institute, 92
forcings, 151, 256, 328, 332, 336, 458, 462, 538, 539, 545
Fourier Transform Far-Infrared, 193
Fourier Transform Infrared, 252
Fourier Transform Spectrometer, 125, 251
Free University, 470

gas, 22, 40, 64, 67, 69, 73, 78, 82, 83, 90, 107, 110, 112, 123, 216, 223, 317, 343, 357, 364, 382, 387, 421, 437, 449, 452, 458, 459, 487, 503, 505, 508, 510, 511, 533, 535, 539, 547, 549
Gas and Aerosol Monitoring Sensorcraft, 75
Global Atmosphere Composition Mission, 120
Global Atmospheric Watch, 5, 291
Global Modeling Initiative, 275
Global Ozone Monitoring by Occultation of Stars, 98
Global Ozone Monitoring Experiment, 90, 229, 284, 452
Global Ozone Observing System, 5, 14, 287
Goddard Institute for Space Studies, 520
Goddard Space Flight Center, 80, 89, 95, 153, 531
GOME Data Processor, 246, 452
GOME Direct-FITting, 246
Governing Council, 345, 346, 366
Greece, 13, 14, 15, 19, 21, 22, 28, 48, 104, 153, 154, 155, 156, 157, 158, 159, 187, 276, 295, 321, 323, 325, 327, 329, 330, 331, 332, 333, 334, 336, 391, 392, 474, 475, 478
greenhouse gas, 5, 74, 256, 272, 318, 332, 351, 371, 372, 389, 451, 462, 477, 478, 480, 482, 490, 497, 500, 502,

greenhouse gas (*cont.*)
505, 506, 507, 508, 510, 511, 515, 520, 525, 528, 530, 531, 532, 536, 544, 547, 548, 549, 554, 555
greenhouse gases, 5, 74, 256, 318, 332, 351, 371, 451, 462, 477, 478, 497, 500, 502, 505, 506, 507, 508, 510, 511, 515, 520, 525, 547, 548, 549, 554, 555

HALogen Occultation Experiment, 223
High Resolution Dynamics Limb Scanner, 122
High Resolution Infrared Sounder, 84
Hohenpeissenberg, 11, 12, 13, 14, 15, 19, 153, 166, 167, 233, 281, 282, 295, 330, 335, 390, 475, 477, 498, 499, 500
Huggins band, 6, 7, 37, 87

Improved Stratospheric and Mesospheric Sounder, 122
infrared, 3, 44, 45, 57, 58, 59, 68, 69, 72, 84, 85, 87, 93, 108, 109, 110, 112, 115, 116, 117, 118, 119, 121, 122, 123, 124, 125, 127, 128, 129, 150, 178, 192, 201, 211, 223, 250, 421, 463, 513
Infrared, 45, 58, 69, 71, 72, 80, 100, 102, 114, 117, 119, 123, 125, 128, 193, 252, 462
Infrared Atmospheric Sounding Interferometer, 102, 125, 252
Infrared Balloon Experiment, 193
Infrared Grating Spectrometer, 72
instrumentation, 4, 5, 7, 9, 11, 13, 15, 17, 19, 21, 23, 25, 27, 29, 31, 33, 39, 41, 43, 45, 47, 49, 50, 55, 61, 63, 65, 67, 69, 71, 73, 75, 82, 83, 93, 94, 95, 107, 109, 111, 113, 115, 117, 119, 121, 123, 125, 127, 129, 156, 157, 158, 159, 171, 263, 283, 299, 361, 362, 472
Inter Tropical Convergence Zone, 530
intercomparisons, 10, 11, 12, 14, 35, 83, 96, 152, 161, 245, 247, 249, 251, 253, 254, 282, 283
Intercontinental Chemical Transport Experiment, 145, 469
Interface Adapter Model, 87
interferometer, 44, 46, 69, 101, 102, 108, 109, 113, 116, 125, 127, 250
Interferometric Monitor for Greenhouse Gases, 101
Intergovernmental Panel on Climate Change, 465, 526
International Geophysical Year, 1, 5, 164, 355
International Ozone Commission, 10, 290
INTEX Ozonesonde Network Study, 145

Japanese Experiment Module, 128
Jet Propulsion Laboratory, 59, 80

Kalavryta, 14, 15, 19

Laser Heterodyne Spectrometer, 85
lidar, 55, 56, 73, 109, 120, 163, 164, 165, 168, 172, 176, 178, 232, 233, 237, 241, 242, 243, 248, 249, 252, 253, 255, 283, 285, 390, 396
Limb Infrared Monitor of the Stratosphere, 118, 298
Limb Ozone Retrieval Experiment, 116
Limb Radiance Inversion Radiometer, 118
long-term memory, 307, 320, 549
Lower Stratosphere, 115

mapping, 82, 90, 91, 163, 296
Match, 144, 145, 146, 441, 469, 470, 471, 472
measurements
- aircraft, 204, 241, 545, 547
- ground-based, 35, 59, 90, 93, 96, 98, 150, 151, 153, 161, 164, 176, 219, 220, 232, 255, 437, 439, 449, 453, 469
- satellite, 7, 37, 183, 190, 201, 209, 211, 213, 215, 217, 219, 221, 222, 223, 225, 227, 229, 231, 233, 234, 237, 244, 263, 440, 449, 453, 555
- total ozone, 153, 155, 157, 159, 160

Meetings of Parties, 369
Meteorological Observatory of Hohenpeissenberg, 12
Meteorological Research Institute, 56

methyl bromide, 385
Michelson Interferometer for Passive Atmospheric Sounding, 119, 161
Microwave Limb Sounder, 92, 120, 248, 300, 406, 462, 472
Microwave Radiometer, 56, 219
MIcrowave RAdiometry, 58
Microwave Sounding Units, 312
Middle Atmosphere High Resolution Spectrographic Investigation, 124
Middle Atmosphere Nitrogen TRend Assessment, 187
monitoring, 4, 10, 42, 58, 59, 78, 79, 82, 83, 84, 90, 93, 98, 100, 118, 130, 143, 150, 152, 161, 164, 245, 247, 249, 251, 253, 255, 294, 323, 324, 339, 346, 352, 358, 363, 366, 368, 370, 379, 431, 449, 451, 453, 455, 457, 459, 461, 463, 465, 467, 469, 471, 472, 473, 475, 477, 486, 535
Montreal Protocol, 3, 4, 100, 149, 247, 289, 339, 340, 342, 344, 345, 346, 348, 350, 352, 354, 356, 357, 358, 360, 361, 362, 364, 366, 367, 368, 369, 370, 371, 372, 373, 374, 375, 376, 377, 378, 379, 380, 381, 382, 384, 385, 386, 388, 389, 391, 398, 404, 406, 413, 431, 437, 438, 440, 448, 450, 478, 479, 480, 481, 482, 534, 555
Multilateral Fund, 369, 371, 375
Multi-sensor Upper Tropospheric Ozone Product, 121
muon, 417, 418, 419, 420

Nambu dynamics, 550, 551
National Aeronautics and Space Administration, 150
National Air Pollution Monitoring Network, 321
National Institute of Standards and Technology, 62
National Meteorological Center, 105
National Oceanic and Atmospheric Administration, 14, 80
National Ozone Expedition, 359, 361
National Space Development Agency of Japan, 93
Near-Infrared Spectrometer, 117, 119
Network for the Detection of Atmospheric Composition Change, 164
Network for the Detection of Stratospheric Changes, 59, 164
North Pole, 178, 408
Northern Hemisphere, 83, 104, 151, 152, 153, 156, 172, 178, 204, 212, 213, 230, 258, 259, 260, 264, 268, 272, 273, 274, 275, 278, 282, 288, 291, 292, 293, 294, 296, 297, 305, 306, 329, 334, 335, 337, 349, 350, 389, 396, 403, 407, 410, 417, 429, 431, 432, 433, 451, 456, 459, 465, 470, 477, 492, 496, 498, 499, 502, 507, 508, 509, 516, 517, 521, 523, 525, 532, 534, 535, 537, 540, 547, 555
Northern Hemisphere Ground Ozone Stations, 152
NPOESS Preparatory Project, 99

Orbiting Ozone and Aerosol Monitor, 110
ozone
  atmospheric, 2, 4, 10, 52, 62, 79, 80, 81, 82, 84, 85, 86, 87, 88, 90, 92, 94, 96, 98, 100, 102, 104, 106, 108, 110, 112, 114, 116, 118, 120, 122, 124, 126, 128, 130, 132, 134, 136, 138, 140, 142, 144, 146, 149, 150, 151, 152, 154, 156, 158, 160, 162, 164, 166, 168, 170, 172, 174, 176, 178, 180, 182, 184, 186, 188, 190, 192, 194, 196, 198, 200, 202, 204, 206, 208, 210, 212, 214, 216, 218, 220, 222, 224, 226, 228, 230, 232, 234, 236, 238, 240, 242, 244, 245, 246, 248, 250, 252, 254, 255, 256, 258, 260, 262, 264, 266, 268, 270, 272, 274, 276, 278, 280, 282, 284, 286, 288, 290, 292, 294, 296, 298, 300, 302, 304, 306, 308, 310, 312, 314, 316, 318, 320, 322, 324, 326, 328, 330, 332, 334, 336, 339, 340, 342, 343, 350, 351, 355, 367, 378, 379, 380, 382, 384, 386, 388, 390, 392, 394, 396, 398, 400, 402, 404, 406, 408, 410, 412, 414, 416, 418, 420, 422, 424, 426, 428, 430, 432, 434, 436, 438, 440, 442, 444, 446, 448, 450, 452, 454, 456, 458, 460, 462, 463, 464, 465, 466, 468, 469, 470, 472, 474, 476, 477, 478, 480, 482, 495, 499, 536, 547

ozone (*cont.*)
concentration, 2, 4, 6, 8, 10, 12, 14, 16, 18, 20, 22, 24, 26, 28, 30, 32, 34, 36, 38, 40, 42, 44, 46, 48, 50, 51, 52, 54, 56, 58, 60, 61, 62, 64, 66, 68, 70, 71, 72, 74, 76, 77, 78, 84, 87, 95, 116, 117, 119, 121, 127, 128, 130, 141, 149, 151, 161, 172, 201, 211, 226, 228, 232, 247, 248, 273, 279, 280, 293, 295, 305, 320, 343, 350, 352, 361, 379, 385, 388, 390, 393, 396, 401, 405, 406, 412, 448, 454, 455, 457, 458, 459, 462, 463, 470, 493, 517, 518, 524, 527, 531, 533, 534, 535, 536, 541, 543, 544
depletion, 3, 4, 100, 149, 164, 257, 259, 272, 273, 275, 277, 278, 291, 318, 332, 333, 336, 339, 340, 346, 347, 348, 349, 351, 352, 354, 356, 358, 359, 360, 361, 362, 368, 369, 372, 376, 378, 381, 384, 385, 386, 388, 389, 390, 391, 392, 393, 395, 397, 399, 401, 403, 405, 406, 407, 408, 410, 414, 415, 425, 429, 431, 435, 437, 441, 453, 454, 455, 469, 470, 475, 477, 478, 479, 481, 483, 491, 497, 499, 501, 502, 503, 504, 505, 512, 515, 527, 528, 534, 535, 549, 550, 554, 555, 556
hole, 45, 80, 140, 171, 172, 287, 288, 289, 301, 315, 339, 344, 346, 349, 351, 352, 353, 354, 355, 356, 357, 358, 359, 360, 361, 363, 381, 385, 388, 393, 406, 407, 413, 414, 415, 417, 421, 425, 426, 427, 428, 429, 430, 431, 437, 438, 439, 440, 441, 442, 443, 444, 447, 448, 449, 450, 451, 457, 467, 468, 469, 493, 501, 517, 525, 526, 527, 528, 531, 536, 542, 554, 555
loss, 58, 104, 140, 141, 142, 143, 144, 267, 272, 279, 299, 301, 306, 346, 349, 350, 351, 357, 360, 362, 363, 365, 382, 387, 388, 392, 406, 407, 411, 412, 413, 417, 425, 431, 433, 434, 435, 436, 437, 441, 448, 450, 453, 457, 462, 463, 469, 470, 471, 472, 499, 501, 503, 532, 546, 554, 555
low-ozone pockets, 299, 300, 301, 305, 306
surface, 3, 77, 298, 321, 323, 325, 326, 474, 493, 511
total trends, 85, 106, 259, 266, 292, 458
ozone layer, 3, 4, 7, 14, 15, 50, 92, 123, 130, 142, 143, 149, 150, 161, 164, 247, 256, 261, 262, 277, 283, 285, 312, 340, 341, 342, 343, 344, 346, 348, 351, 354, 366, 367, 368, 369, 373, 379, 381, 385, 388, 407, 409, 411, 413, 415, 417, 419, 421, 423, 425, 427, 429,É431, 433, 434, 435, 437, 438, 439, 440, 441, 443, 445, 447, 448, 449, 450, 451, 452, 453, 454, 455, 457, 458, 459, 461, 463, 465, 467, 469, 471, 473, 474, 475, 477, 485, 497, 505, 534, 537, 544, 545, 548, 554
tropospheric, 3, 35, 42, 92, 145, 147, 156, 244, 245, 259, 263, 272, 294, 296, 297, 406, 472, 473, 478, 492, 493, 495, 496, 498, 499, 501, 502, 506, 507, 508, 510, 511, 512, 513, 515, 517, 519, 520, 521, 523, 528, 529, 530, 539, 544, 546
Ozone and Aerosol Fine Structure, 107
Ozone Dynamics Ultraviolet Spectrometer, 101
Ozone Layer Monitoring Experiment, 101
Ozone Mapping and Profiler Suite, 99, 100
Ozone Monitoring Instrument, 90, 244
Ozone Sounding Network, 282
ozone–climate modeling, 550
ozonesonde, 4, 36, 37, 53, 54, 59, 61, 62, 63, 64, 65, 66, 67, 102, 124, 145, 152, 163, 164, 170, 171, 172, 187, 188, 189, 201, 211, 223, 237, 243, 244, 246, 248, 249, 250, 251, 254, 268, 275, 281, 282, 284, 285, 294, 333, 334, 336, 353, 359, 361, 389, 390, 393, 394, 396, 406, 407, 411, 412, 413, 436, 437, 450, 454, 467, 469, 470, 471, 495, 540
OZON-MIR, 86
ozonometer, 5, 37, 38, 39, 43, 48, 67, 69, 247, 435, 454

Pacific Exploratory Mission-West, 282
parameterization, 98, 262, 486
photochemical, 3, 21, 22, 109, 280, 281, 282, 283, 291, 299, 300, 301, 304, 321, 322, 323, 324, 326, 333, 382, 390, 441, 442, 449, 450, 455, 456, 463, 468, 471, 487, 491, 503, 505, 510, 511, 521, 522, 527, 544, 546

Pinatubo, 103, 256, 258, 267, 398, 399, 404, 456, 459, 469, 475, 533, 547, 556
Polar Operational Environmental Satellites, 99
Polar Ozone and Aerosol Measurement, 213, 441
Polar Ozone and Aerosol Measurement III, 213
Polar Stratospheric Clouds, 56
PROFile FIT, 178

Quasi-Biennial Oscillation, 102, 120, 250, 256, 260, 308, 316, 318, 410, 449, 455, 534, 538
Quasi-Triennial Oscillation, 260

radiation
- scattered, 7
- solar, 6, 14, 35, 45, 47, 82, 94, 109, 112, 118, 314, 323, 324, 326, 490, 527, 544
- ultraviolet, 3, 7, 37, 70, 92, 95, 323, 407, 490

radiometer, 47, 57, 58, 59, 60, 64, 68, 75, 86, 93, 94, 100, 102, 110, 115, 118, 120, 122, 124, 128, 129, 207, 223, 327
Radiometer for Atmospheric Measurements, 59
ratification, 357, 368, 372
Rayleigh Scattering Attitude Sensor, 116
remote sensing, 80, 112, 339, 340, 356, 358, 368, 491, 535, 547
Rocket Ozonesonde, 66
Rutherford Appleton Laboratory, 58, 229

satellite, 1, 4, 7, 10, 11, 14, 17, 37, 41, 46, 62, 70, 73, 74, 79, 82, 83, 84, 86, 87, 89, 90, 92, 93, 94, 96, 98, 99, 100, 101, 102, 106, 107, 109, 110, 112, 115, 116, 117, 118, 119, 120, 121, 122, 124, 125, 126, 127, 128, 129, 143, 145, 146, 147, 149, 150, 151, 152, 153, 156, 157, 159, 160, 161, 162, 163, 169, 182, 183, 190, 196, 201, 202, 204, 205, 208, 209, 210, 211, 213, 215, 217, 219, 221, 222, 223, 225, 227, 228, 229, 231, 232, 233, 234, 237, 240, 241, 242, 243, 244, 245, 246, 247, 249, 251, 252, 255, 256, 257, 258, 262, 263, 264, 265, 268, 272, 273, 281, 283, 284, 288, 294, 298, 306, 312, 345, 346, 347, 350, 351, 352, 353, 354, 356, 357, 358, 362, 363, 381, 388, 391, 392, 393, 396, 399, 401, 406, 407, 410, 411, 415, 425, 437, 438, 439, 440, 441, 444, 448, 449, 452, 453, 463, 465, 469, 472, 473, 533, 535, 545, 552, 555, 556
- observations, 17, 83, 151, 245, 255, 268, 284, 288, 298, 306, 381, 406, 407, 410, 465
- remote sensing, 1, 79, 82
- systems, 153, 156, 161, 294, 452, 453

scaling behavior, 322
Scientific Assessment, 264, 272, 378, 379, 380, 381, 383, 388, 393, 450, 466, 479, 481, 532, 545, 554, 555, 557
seasonal variations, 102, 259, 283, 401, 403, 404, 468, 489, 517
Second European Stratospheric Arctic and Mid-latitude Experiment, 144
Shuttle Ozone Limb Sounding Experiment, 116
Shuttle SBUV, 96
Solar Absorption Balloon Experiment, 74
Solar Backscatter Ultraviolet, 94, 95, 100
Solar Maximum Mission, 107, 117
Solar Mesosphere Explorer, 116, 117, 119
Sounding of the Atmosphere using Broadband Emission Radiometry, 128
South Pole, 110, 285, 353, 356, 361, 388, 437, 438, 441, 450, 453, 469, 471, 506, 554
Southern Annular Mode, 542
Southern Oscillation Index, 261
Special Report on Emissions Scenarios, 271
spectrometer, 12, 35, 42, 43, 44, 45, 46, 47, 48, 49, 57, 67, 69, 71, 72, 75, 76, 77, 83, 85, 86, 90, 93, 96, 97, 98, 100, 101, 107, 108, 115, 116, 117, 119, 122, 123, 124, 125, 160, 192, 201, 208, 217, 220, 229, 241, 251, 252, 356, 365, 452, 454
SPectrometer for Atmospheric TRAcers Monitoring, 48, 49

spectrophotometer, 5, 7, 14, 17, 19, 187, 247, 352, 358
Spectrophotometer
  Brewer, 5, 6, 35, 36, 43, 48, 54, 55
  Dobson, 5, 7, 8, 9, 10, 11, 12, 14, 15, 17, 19, 21, 22, 24, 27, 28, 32, 35, 36, 37, 38, 51, 54, 62, 87, 149, 151, 152, 153, 157, 158, 160, 349, 356, 389, 403, 406, 437, 440
spectroscopy, 39, 43, 46, 58, 59, 69, 92, 198, 246
Spectroscopy of the Atmosphere using Far IR Emission, 68
Star Pointing Spectrometer, 46
stochastic, 320, 443, 553
stratosphere
  lower, 57, 59, 61, 74, 85, 98, 102, 104, 114, 120, 121, 123, 125, 130, 144, 170, 212, 219, 223, 228, 234, 246, 249, 251, 252, 275, 276, 277, 278, 280, 281, 282, 294, 295, 296, 297, 305, 306, 315, 317, 327, 333, 335, 336, 342, 359, 361, 388, 390, 392, 396, 403, 405, 406, 407, 410, 411, 412, 413, 427, 429, 431, 441, 450, 451, 454, 457, 469, 475, 477, 478, 486, 495, 497, 498, 499, 502, 506, 518, 519, 524, 526, 527, 532, 533, 534, 538, 540, 541, 542, 546, 556
  upper, 142, 170, 212, 243, 249, 261, 263, 274, 280, 284, 304, 387, 388, 395, 396, 398, 411, 412, 413, 429, 451, 497, 500, 526, 528, 531, 534, 535, 540, 546, 556
Stratosphere and Mesosphere Sounder, 118
Stratospheric Aerosol and Gas Experiment, 107, 211, 296, 396, 499
Stratospheric Aerosol Measurement, 108
stratospheric ozone dynamics, 545
Stratospheric Ozone Experiment, 46
Stratospheric Ozone Monitoring Radiometer, 59
Stratospheric Processes and their Role in Climate, 468
Stratospheric Sounding by Infrared Heterodyne Spectroscopy, 57
Stratospheric Wind Interferometer for Transport Studies, 126
stray light, 7, 10, 24, 25, 26, 27, 28, 29, 30, 31, 32, 33, 35, 43, 47, 55, 114
Submillimeter Atmospheric Sounder, 75
sudden stratospheric warmings, 417
Sun–Photometer Earth Atmosphere Measurement, 111
Superconducting Submillimeter-Wave Limb-Emission Sounder, 128
System for Analysis of Observation at Zenith, 36, 42
System for the Monitoring of Stratospheric Compounds, 46

Technical and Economic Assessment Panel, 370
Television InfraRed Observational Satellite, 83
temperature, 3, 21, 22, 45, 47, 51, 52, 53, 54, 55, 60, 61, 62, 63, 64, 68, 69, 72, 73, 75, 84, 94, 104, 105, 108, 112, 114, 115, 116, 117, 118, 119, 120, 122, 123, 124, 125, 126, 128, 140, 141, 142, 146, 161, 163, 164, 165, 183, 187, 192, 210, 213, 216, 217, 221, 223, 225, 229, 243, 250, 259, 260, 261, 262, 277, 278, 279, 280, 281, 283, 284, 295, 296, 301, 307, 312, 314, 317, 318, 320, 323, 327, 329, 331, 332, 335, 336, 337, 356, 358, 389, 394, 395, 401, 407, 410, 413, 415, 416, 417, 418, 419, 420, 421, 422, 423, 424, 428, 431, 432, 434, 435, 436, 448, 454, 458, 459, 463, 466, 471, 473, 474, 476, 477, 485, 491, 497, 498, 499, 501, 502, 503, 504, 505, 506, 508, 517, 518, 519, 520, 521, 522, 524, 525, 527, 528, 529, 531, 533, 534, 535, 536, 537, 541, 542, 544, 545, 546, 547, 548, 549, 552, 556, 557
  instrument, 12, 84
  internal, 48
  mean, 22, 417, 476, 524, 537
  potential, 196, 200, 300, 304, 407, 441
  radiosonde, 22
  room, 6, 68
  standard, 2
  stratospheric, 42, 68, 104, 105, 359, 385, 408, 415, 431, 436, 448, 456, 497, 499, 512, 517, 518, 521, 534, 540, 555, 556
  surface, 21, 84, 271, 320, 462, 507, 520, 527, 531

Temperature Humidity Infrared Radiometer, 87
terms of reference, 371
Third European Stratospheric Experiment on Ozone, 142, 362
TIROS Operational Vertical Sounder, 83, 152, 352
total ozone content, 150, 154, 156, 187, 314, 352, 398, 417, 439, 492, 501, 502
Total Ozone Monitoring Spectrometer, 82, 86, 352
Total Ozone Portable Spectrometer, 42
Total Ozone Unit, 118
trace gases, 39, 40, 41, 45, 59, 74, 79, 82, 86, 91, 92, 112, 114, 115, 117, 121, 122, 123, 153, 161, 193, 208, 221, 223, 225, 247, 351, 358, 359, 363, 437, 489, 492, 516, 547
Tropical Ozone Transport Experiment, 283
tropopause properties, 327, 329, 330, 331, 333, 335, 336, 337, 474
troposphere
  lower, 297
  middle, 108, 281, 494
  upper, 74, 82, 112, 114, 115, 116, 120, 121, 125, 156, 164, 201, 214, 216, 219, 228, 237, 243, 251, 252, 262, 281, 294, 295, 296, 297, 298, 305, 306, 317, 320, 327, 336, 390, 410, 413, 420, 454, 473, 475, 486, 493, 494, 496, 498, 516, 530, 531, 532, 534, 537
Tropospheric Emission Spectrometer, 120, 244, 472
tropospheric ozone, 3, 35, 42, 92, 145, 147, 156, 244, 245, 259, 263, 272, 294, 296, 297, 406, 472, 473, 478, 492, 493, 495, 496, 498, 499, 501, 502, 506, 507, 508, 510, 511, 512, 513, 515, 517, 519, 520, 521, 523, 528, 529, 530, 539, 544, 546
Tropospheric Ozone Experiment, 282
tropo-stratospheric variability, 545
Tunable Diode LHS, 85

U.S.S.R., 13, 37, 38, 267, 403
Ultraviolet Spectrometer, 116
Umkehr, 10, 32, 36, 50, 51, 52, 53, 54, 109, 140, 249, 252, 253, 255, 258, 279, 284, 393, 395, 397, 399, 411, 412, 413, 414, 429
  Brewer, 54, 55, 284
  Dobson, 51, 54, 258, 428
United Kingdom, 13, 149, 216
Universal Processor for UV/VIS Atmospheric Spectrometers, 247
University of California, Irvine, 294
Upper Atmospheric Research Satellite, 111, 120
Upper Troposphere, 115

validation, 10, 17, 37, 46, 70, 80, 82, 145, 146, 147, 150, 152, 153, 161, 162, 163, 165, 166, 167, 169, 171, 173, 175, 177, 179, 181, 183, 185, 187, 189, 190, 192, 193, 200, 201, 202, 207, 208, 211, 214, 223, 224, 231, 232, 233, 234, 237, 240, 241, 242, 244, 245, 247, 248, 250, 282, 283, 294, 390, 449, 472, 526
vertical profile, 50, 51, 53, 55, 57, 58, 59, 61, 69, 71, 72, 74, 82, 84, 86, 92, 93, 94, 99, 101, 112, 114, 119, 124, 129, 171, 172, 176, 182, 185, 187, 189, 201, 202, 210, 211, 242, 245, 249, 273, 275, 277, 279, 281, 283, 284, 285, 287, 290, 291, 295, 329, 393, 395, 411, 416, 419, 501
Vienna Convention, 346, 348, 367, 370, 372, 380
volcanic eruptions, 408
vortex, 58, 59, 75, 104, 110, 130, 140, 142, 199, 202, 204, 274, 275, 278, 296, 298, 299, 300, 304, 305, 315, 356, 357, 359, 360, 363, 393, 407, 408, 410, 411, 412, 413, 415, 417, 426, 431, 432, 434, 435, 436, 437, 441, 453, 454, 456, 464, 467, 468, 469, 470, 501, 526, 531, 532, 536, 554, 555

warming
  global, 1, 74, 314, 368, 371, 459, 493, 504, 529, 531
  ocean, 262

warming (*cont.*)
  stratospheric, 140, 142, 417
  tropospheric, 262
wavelength
  long, 14, 40
  short, 26, 40, 94
West, 82, 282, 351, 431, 530
Working Group, 346, 366
World Meteorological Organization, 164, 165, 351
World Ozone and Ultraviolet Data Center, 5, 473

zenith angle, 7, 9, 24, 36, 38, 40, 42, 43, 46, 51, 52, 54, 94, 98, 101, 153, 234, 245, 246, 248, 258, 304, 363, 434, 440, 503, 508
zenith sky, 25, 34, 35, 37, 38, 42, 43, 47, 48, 50, 51, 52, 54, 160, 247

MIX
Papier aus verantwortungsvollen Quellen
Paper from responsible sources
FSC® C105338

Printed by Books on Demand, Germany